GENERATION
OF SURFACES

KINEMATIC
GEOMETRY
OF SURFACE
MACHINING

STEPHEN P. RADZEVICH

CRC Press
Taylor & Francis Group
Boca Raton London New York

CRC Press is an imprint of the
Taylor & Francis Group, an **informa** business

CRC Press
Taylor & Francis Group
6000 Broken Sound Parkway NW, Suite 300
Boca Raton, FL 33487-2742

First issued in paperback 2017

© 2014 by Taylor & Francis Group, LLC
CRC Press is an imprint of Taylor & Francis Group, an Informa business

No claim to original U.S. Government works

Version Date: 20130729

ISBN 13: 978-1-4822-2211-1 (hbk)
ISBN 13: 978-1-138-07443-9 (pbk)

Library of Congress Cataloging-in-Publication Data

Radzevich, S. P. (Stepan Pavlovich)
   Generation of surfaces : kinematic geometry of surface machining / Stephen P. Radzevich.
      pages cm
   Includes bibliographical references.
   ISBN 978-1-4822-2211-1 (alk. paper)
   1. Machining--Mathematics. 2. Surfaces (Technology)--Mathematical models. 3. Machine parts--Computer-aided design. 4. Kinematic geometry. I. Title.
TJ1185.R236 2013
671.3'5--dc23
                                    2013028017

Visit the Taylor & Francis Web site at
http://www.taylorandfrancis.com

and the CRC Press Web site at
http://www.crcpress.com

This book is dedicated to my son Andrew.

Contents

Section II   Fundamentals

Section III   Application

Preface

This book, based on intensive research I have conducted since the late 1970s, is my attempt to cover in one monograph the modern theory of part surface generation with a focus on kinematic geometry of part surface machining on a multiaxis (*NC*) machine. Although the orientation of this book is toward *computer-aided design* (*CAD*) and *computer-aided machining* (*CAM*), it is also useful for solving problems that pertain to the generation of part surfaces on machine tools of conventional design (for example, on gear generators and so forth).

Machining of part surfaces can be interpreted as the transformation of a work into the machined part having the desired shape and design parameters. The major characteristics of the machined part surface—its shape and actual design parameters as well as the properties of the subsurface layer of the part material—strongly depend upon the parameters of the part *surface-generating process*. It is the primary purpose of this book, to present the key methods that lie at the heart of the *DG/K*-based approach of part surface generation.

In addition to the part surface-generating process, there are, of course, many other engineering considerations—namely, wear of the cutting tool, stiffness of the machine tool, tool chatter, heat generation, coolant and lubricant supply, and so forth. The analysis in this book is limited to those parameters of the part surface-machining process that can be expressed in terms of part surface geometry and kinematics of relative motion of the cutting tool. The assumption has been made that the reader is familiar with the concept of the *vector* and has a basic knowledge of *coordinate geometry*.

I hope that this book will be useful for anyone embarking on a *CAD/CAM*-oriented program, or starting work on new software packages for sculptured part surface-machining on a multiaxis *NC* machine.

Acknowledgments

I share the credit for any research success with my numerous doctoral students with whom I have tested the proposed ideas and applied them in industry. The contributions of many friends, colleagues, and students in overwhelming numbers cannot be acknowledged individually, and as much as our benefactors have contributed, their kindness and help must go unrecorded.

My thanks also go to those at CRC Press who took over the final stages and who will have to cope with the marketing and sales of the fruit of my efforts.

Author

Stephen P. Radzevich is a professor of mechanical engineering and manufacturing engineering. He earned his MSc in 1976, PhD in 1982, and Dr.(Eng) Sc. in 1991, all in mechanical engineering. Dr. Radzevich has extensive industrial experience in gear design and manufacture. He has developed numerous software packages dealing with *CAD* and *CAM* of precise gear finishing for a variety of industrial sponsors. His main research interest is *kinematic geometry of surface generation*, particularly with focus on a precision gear design, high–power density gear trains, torque share in multiflow gear trains, design of special purpose gear cutting/finishing tools and design and machining (finishing) of precision gears for low-noise/noiseless transmissions of cars, light trucks, etc. Dr. Radzevich has spent about 40 years developing software, hardware, and other processes for gear design and optimization. Besides his work for industry, he trains engineering students at universities and gear engineers in companies. He has authored and coauthored over 30 monographs, handbooks, and textbooks. The monographs entitled *Generation of Surfaces* (2001), *Kinematic Geometry of Surface Machining* (CRC Press, 2008), *CAD/CAM of Sculptured Surfaces on Multiaxis NC Machine: The DG/K-Based Approach* (M&C Publishers, 2008), *Gear Cutting Tools: Fundamentals of Design and Computation* (CRC Press, 2010), *Precision Gear Shaving* (Nova Science Publishers, 2010), *Dudley's Handbook of Practical Gear Design and Manufacture* (CRC Press, 2012), *Theory of Gearing: Kinematics, Geometry, and Synthesis* (CRC Press, 2012), *Geometry of Surfaces: A Practical Guide for Mechanical Engineers* (Wiley, 2012) are among the recently published volumes. He also authored and coauthored over 250 scientific papers and holds more than 220 patents on inventions in the field.

Notations

$\mathscr{A}n\,(P)$	*Andrew's* indicatrix of normal curvature of the part surface P
$\mathscr{A}n_k(P/T)$	Combined *Andrew's* indicatrix of normal curvature of the part surfaces P and T
$\mathscr{A}n_k(T)$	*Andrew's* indicatrix of normal curvature of the generating surface T of the form-cutting tool
$\mathscr{A}n_R(P)$	*Andrew's* indicatrix of radii of normal curvature of the part surface P
$\mathscr{A}n_R(P/T)$	Combined *Andrew's* indicatrix of radii of normal curvature of the part surfaces P and T
$\mathscr{A}n_R(T)$	*Andrew's* indicatrix of radii of normal curvature of the generating surface T of the form-cutting tool
CC point	Cutter contact point (current point of contact between the part surface P and the generating surface T of the form-cutting tool)
$Cnf_k(P/T)$	Indicatrix of conformity of the part surface P and of the generating surface T of the form-cutting tool at a current contact point, K, of the surfaces P and T (this characteristic curve is expressed in terms of normal curvatures of the contacting surfaces P and T)
$Cnf_R(P/T)$	Indicatrix of conformity of the part surface P and of the generating surface T of the form-cutting tool at a current contact point, K, of the surfaces P and T (this characteristic curve is expressed in terms of radii of normal curvature of the contacting surfaces P and T)
$Crv(P)$	Curvature indicatrix of the part surface P
$Crv(T)$	Curvature indicatrix of the generating surface T of the form-cutting tool
$C_{1.P},\,C_{2.P}$	The first and the second principal plane sections respectively of the part surface P
$C_{1.T},\,C_{2.T}$	The first and the second principal plane sections respectively of the generating surface T of the form-cutting tool
$\mathbf{Ds}(P/T)$	Matrix of the resultant displacement of the form-cutting tool with respect to the part surface P
$Dup(P)$	*Dupin's* indicatrix at a point within the part surface P
$Dup(P/T)$ or $Dup\mathscr{R}_{cl}$	*Dupin's* indicatrix at a point of the surface of relative curvature, $\mathscr{R}_{cl}$, constructed at a current point of contact, K, the part surface P and of the generating surface T of the form-cutting tool
$Dup(T)$	*Dupin's* indicatrix at a point within the generating surface T of the form-cutting tool
$\mathscr{E}$	A characteristic line (the limit configuration of the line of intersection between two consecutive positions of a moving surface when the distance between the surfaces is approaching to zero)
$E_P,\,F_P,\,G_P$	Fundamental magnitudes of the first order at a point of the part surface P (or, in other terminology, coefficients of the first fundamental form, $\Phi_{1.P}$, at a point of the part surface P)

E_T, F_T, G_T	Fundamental magnitudes of the first order at a point of the generating surface T of the form-cutting tool (or, in other terminology, coefficients of the first fundamental form, $\Phi_{1.T}$, at a point of the generating surface T of the form-cutting tool)
$\mathbf{Eu}(\psi, \theta, \varphi)$	Operator of the *Eulerian* transformation
$\bar{F}_{fr}$	Feed rate per tooth of the form-cutting tool
$[\bar{F}_{fr}]$	Limit value of the feed rate per tooth of the form-cutting tool
$\mathbf{F}_{fr}$	Vector of the feed-rate motion of the form-cutting tool
$\bar{F}_{ss}$	Magnitude of the side step of the form-cutting tool
$[\bar{F}_{ss}]$	Limit value of magnitude of the side step of the form-cutting tool
$\mathbf{F}_{ss}$	Vector of the side-step motion of the form-cutting tool
$\mathscr{F}_1, \mathscr{F}_2, \mathscr{F}_3$	The degree of conformity functions
$GInd(P)$	*Gauss'* indicatrix of the part surface P
$GInd(T)$	*Gauss'* indicatrix of the generating surface T of the form-cutting tool
$GMap(P)$	*Gauss'* map of the part surface P
$GMap(T)$	*Gauss'* map of the generating surface T of the form-cutting tool
$\mathscr{G}_P, \mathscr{G}_T$	Full (*Gaussian*) curvature at a point of the part surface P, and of the generating surface T of the form-cutting tool
H_P	Discriminant of the first fundamental form, $\Phi_{1.P}$, of the part surface P
H_T	Discriminant of the first fundamental form, $\Phi_{1.T}$, of the generating surface T of the form-cutting tool
$\mathfrak{Ir}_k(P/T)$	A planar curvature characteristic curve (parameters of this characteristic curve are expressed in terms of normal curvatures at a point of contact, K, of the part surface P and of the generating surface T of the form-cutting tool)
$\mathfrak{Ir}_R(P/T)$	A planar radii of curvature characteristic curve (parameters of this characteristic curve are expressed in terms of radii of normal curvatures at a point of contact, K, of the part surface P and of the generating surface T of the form-cutting tool)
K	Point of contact between the part surface P and of the generating surface T of the form-cutting tool (or a point within the line of contact, L_c, between the surfaces P and T)
L_c	Line of contact, L_c, between the part surface P and of the generating surface T of the form-cutting tool (line of contact, L_c, is the limit position of the characteristic line, E)
L_P, M_P, N_P	Fundamental magnitudes of the second order at a point of the part surface P (or, in other terminology, coefficients of the second fundamental form, $\Phi_{2.P}$, at a point of the part surface P)
L_T, M_T, N_T	Fundamental magnitudes of the second order at a point of the generating surface T of the form-cutting tool (or, in other terminology, coefficients of the second fundamental form, $\Phi_{2.T}$, at a point of the generating surface T of the form-cutting tool)
$Mch(P/T)$	Indicatrix of machinability of the part surface P with the form-cutting tool T
$\mathscr{M}_P, \mathscr{M}_T$	Mean curvature at a point of the part surface P, and of the generating surface T of the form-cutting tool
$\mathbf{N}_P$	Perpendicular to the part surface P at a point of interest

$\mathbf{N}_T$	Perpendicular to the generating surface T of the form-cutting tool at a point of interest
P	Smooth regular part surface to be machined
$\mathcal{P}_{mr}$	Chip (material) removal output when machining a sculptured part surface P
$\mathcal{P}_{sg}$	Part surface generation, P, output
$\mathcal{P}\ell_k(P)$	*Plücker*'s indicatrix of normal curvature at a point within the part surface P
$\mathcal{P}\ell_k(T)$	*Plücker*'s indicatrix of normal curvature at a point within the generating surface T of the form-cutting tool
$\mathcal{P}\ell_R(P)$	*Plücker*'s indicatrix of normal radii of curvature at a point within the surface part P
$\mathcal{P}\ell_R(T)$	*Plücker*'s indicatrix of radii of normal radii of curvature at a point within the generating surface T of the form-cutting tool
R_e-surface	Reversibly enveloping surface
$\mathcal{R}_{el}$	Surface of relative curvature constructed at a point of contact, K, between the part surface P and between the generating surface T of the form-cutting tool
$\mathbf{Rf}_x(Y_iZ_i)$	Operator of reflection with respect to Y_iZ_i–coordinate plane
$\mathbf{Rf}_y(Z_iX_i)$	Operator of reflection with respect to Z_iX_i–coordinate plane
$\mathbf{Rf}_z(X_iY_i)$	Operator of reflection with respect to X_iY_i–coordinate plane
$\mathbf{RPY}(\varphi_x, \varphi_y, \varphi_z)$	Operator of *roll/pitch/yaw* transformation
$\mathbf{Rs}(A \mapsto B)$	Operator of the resultant coordinate system transformation, say from the coordinate system A to the coordinate system B
$\mathbf{Rt}(\varphi_x, X)$	Operator of rotation through an angle φ_x about X axis
$\mathbf{Rt}(\varphi_y, Y)$	Operator of rotation through an angle φ_y about Y axis
$\mathbf{Rt}(\varphi_z, Z)$	Operator of rotation through an angle φ_z about Z axis
$\mathbf{Rt}(\varphi_A, \mathbf{A})$	Operator of rotation through an angle φ_A about $\mathbf{A}$ axis not through origin of the coordinate system
$\mathbf{Rt}(\varphi_A, \mathbf{A}_0)$	Operator of rotation through an angle φ_A about $\mathbf{A}_0$ axis through origin of the coordinate system
$\mathbf{Rt}_\omega(A \mapsto B)$	Operator of nonorthogonal coordinate system transformation
$R_{1.P}, R_{2.P}$	The first and the second principal radii of curvature at a point within the part surface P
$R_{1.T}, R_{2.T}$	The first and the second principal radii of curvature at a point the generating surface T of the form-cutting tool
T	Generating surface of the form-cutting tool
T_P	Discriminant of the second fundamental form, $\Phi_{1.P}$, of the part surface P
T_T	Discriminant of the second fundamental form, $\Phi_{1.T}$, of the generating surface T of the form-cutting tool
$\mathbf{Tl}(P/T)$	Resultant matrix of tolerances on relative configuration of the form-cutting tool with respect to the part surface P
$\mathbf{Tr}(a_x, X)$	Operator of translation at a distance a_x along X axis
$\mathbf{Tr}(a_y, Y)$	Operator of translation at a distance a_y along Y axis
$\mathbf{Tr}(a_z, Z)$	Operator of translation at a distance a_z along Z axis

$\mathbf{T}_{1.P}, \mathbf{T}_{2.P}$	Tangent vectors of the principal directions at a point within the part surface P
$\mathbf{T}_{1.T}, \mathbf{T}_{2.T}$	Tangent vectors of the principal directions at a point within the generating surface T of the form-cutting tool
U_P, V_P	Curvilinear (*Gaussian*) coordinates of a point of the part surface P
U_T, V_T	Curvilinear (*Gaussian*) coordinates of a point of the generating surface T of the form-cutting tool
$\mathbf{U}_P, \mathbf{V}_P$	Tangent vectors to the curvilinear coordinate lines on the part surface P
$\mathbf{U}_T, \mathbf{V}_T$	Tangent vectors to the curvilinear coordinate lines on the generating surface T of the form-cutting tool
$\mathbf{V}_\Sigma$	Vector of the resultant motion of the generating surface T of the form-cutting tool with respect to the part surface P
X_{NC}, Y_{NC}, Z_{NC}	*Cartesian* coordinates of a point in the coordinate system associated with the multiaxis *NC* machine
X_P, Y_P, Z_P	*Cartesian* coordinates of a point of the part surface P
X_T, Y_T, Z_T	*Cartesian* coordinates of a point of the generating surface T of the form-cutting tool
$\mathbf{ds}_i[i \mapsto (i \pm 1)]$	Matrix of an elementary ith displacement of the form-cutting tool with respect to the part surface P
$[h]$	Tolerance on accuracy of the machined part surface P
h_{fr}	Height of the part surface, P, waviness
h_{ss}	Height of the part surface, P, cusps in the direction of vector $\mathbf{F}_{ss}$ of the side-step motion
h_Σ	Resultant deviation of the machined part surface, P, from the desired part surface
$k_{1.P}, k_{2.P}$	The first and the second principal curvatures at a point within the part surface P
$k_{1.T}, k_{2.T}$	The first and the second principal curvatures at a point the generating surface T of the form-cutting tool
m	Current point of interest
$\mathbf{n}_P$	Unit normal vector to the part surface P at a current point of interest, m, within the surface P
$\mathbf{n}_T$	Unit normal vector to the generating surface T of the form-cutting tool at a current point of interest, m, within the surface T
r_{cnf}	Position vector of a point of the indicatrix of conformity $Cnf_R(P/T)$ of the part surface P and of the generating surface T of the form-cutting tool at a current contact point, K, of the surfaces P and T
$\mathbf{r}_P$	Position vector of a point of the part surface P
$\mathbf{r}_T$	Position vector of a point of the generating surface T of the form-cutting tool
$\mathbf{tl}_i[i \mapsto (i \pm 1)]$	Matrix of the ith element of the resultant tolerance on configuration of the cutting tool with respect to the part surface P
$\mathbf{t}_{1.P}, \mathbf{t}_{2.P}$	Unit tangent vectors of the principal directions at a point on the part surface P
$\mathbf{t}_{1.T}, \mathbf{t}_{2.T}$	Unit tangent vectors of the principal directions at a point on the generating surface T of the form-cutting tool

$\mathbf{u}_P, \mathbf{v}_P$	Unit tangent vectors to the curvilinear coordinate lines at a point on the part surface P
$\mathbf{u}_T, \mathbf{v}_T$	Unit tangent vectors to the curvilinear coordinate lines at a point on the generating surface T of the form-cutting tool
$x_P y_P z_P$	Associated with the part surface, P, local orthogonal reference system having origin at the point of contact, K, between the part surface P and the generating surface T of the form-cutting tool (sometimes, it is convenient to designate this coordinate system as $x_K y_K z_K$, or just as xyz)
$x_T y_T z_T$	Associated with the generating surface, the T of the form-cutting tool local orthogonal reference system having origin at the point of contact, K, between the part surface P and the generating surface T of the form-cutting tool

Greek Symbols

$\Phi_{1.P}, \Phi_{2.P}$	The first and second fundamental forms of the part surface P
$\Phi_{1.T}, \Phi_{2.T}$	The first and second fundamental forms of the generating surface T of the form-cutting tool
Σ	Crossed-axis angle
α	Clearance (flank) angle of the cutting tool
β	Tool wedge angle
γ	Rake angle of the cutting tool
γ_{cf}	Rake angle of the cutting tool in the chip-flow direction
δ	Cutting angle
ε	Tool tip (nose) angle
φ_e	Major cutting-edge approach angle
φ_{e1}	Minor cutting-edge approach angle
ϕ_n	Normal pressure angle of a gear-cutting tool
λ	Angle of inclination of the cutting edge of the form-cutting tool
μ	Angle of the surfaces P and T local relative orientation
τ_P	Torsion of a curve through the point of interest, m, within the part surface P
τ_T	Torsion of a curve through the point of interest, m, within the generating surface T of the form-cutting tool
ψ, θ, φ	The *Eulerian* angles (namely, the angles in the operator, $\mathbf{Eu}(\psi, \theta, \varphi)$, of the *Eulerian* transformation)
ζ_T	Setting angle of the gear-finishing tool
ω_P	Coordinate angle at a point within the part surface P
ω_T	Coordinate angle at a point within the part generating surface T of the form-cutting tool

Subscripts

$\mathscr{R}$	Surface of relative curvature
cnf	Conformity
max	Maximum
min	Minimum
opt	Optimal
P	Part surface being machining
T	Generating surface of the form-cutting tool

Introduction

"Gaining time is gaining everything."

John Shebbeare, 1709–1788

People have been concerned for centuries with the generation of part surfaces. Any machining operation is aimed at the generation of a part surface that has appropriate shape and parameters. Enormous practical experience has been accumulated so far in this particular area of engineering. Improvements to the part surface-machining operation are based mostly on generalization of the accumulated practical experience. The elements of the theory of part surface generation began to appear in the late 1950s–early 1960s.

Historical Background. For a long time, scientific developments in the field of part surface generation were aimed at solving those problems that are relatively simple in nature. This was mostly because no scientific theory of part surface generation had been developed.

It is well known that *kinematics*, the English version of the word *cinématique* coined by A.M. Ampère* from the Greek κίνημα, or *movement*, is that branch of science that deals with motion on its own in isolation from the forces associated with motion. Displacements, both linear and angular—and their combination—and the successive derivatives with respect to time of such displacements, namely, velocities, accelerations, hyper-accelerations, all combine into *kinematics*.

Kinematic geometry is confined even more narrowly within dynamics and lies within kinematics itself. T. Olivier,[†] T.M. Chasles,[‡] and F. Savary[§] contributed much to kinematic geometry. It deals with only the first and simplest segment of kinematics, *displacements*. Usually time, as a variable, need not be brought into account at all; the displacements or movements in a mechanism can, as far as this book is concerned, be performed as slowly, or as quickly, as we may wish. Yet on occasion, it is convenient, more in the interest of brevity and terminology than anything else, to introduce speed (or velocity, where it is needed as a vector), and, more rarely, acceleration.

In the part surface generation process, *geometry* and *kinematics* are in a type of symbiosis. As *geometry* and *kinematics* are inseparable in the part surface generation process, then the term *kinematic geometry* is logical to be used to specify the *part surface generation process*.

---

* André-Marie Ampère (January 20, 1775–June 10, 1836), a famous French physicist and mathematician.
† Théodore Olivier (January 21, 1793–August 5, 1853), a French mathematician.
‡ Michel Flréal Chasles (November 15, 1793–December 18, 1880), a French mathematician.
§ Felix Savary (October 4, 1797–July 15, 1841), a French astronomer and mathematician.

The term *kinematic geometry* is a bit confusing: why *kinematic geometry* and not *geometrical kinematics*? To avoid this indefiniteness, the term *generation of surfaces* is used in this book instead of the term *kinematic geometry*.

In the late 1970s and early 1980s, the idea of the synthesis of the optimal* (or, in other words, the most favorable) part surface-machining operation was, in a manner of speaking, mentioned by the author for the first time. After a decade of gestation, original articles on the subject began to appear. Now, with the passing of a fourth decade, it is appropriate to attempt a consolidated story of some of the many efforts of European and American researchers.

The Importance of the Subject. The machining of sculptured part surfaces on a multiaxis *NC* machine is a commonly used process in many industries. The automotive, aerospace, and some other industries are the most advanced in this respect.

The ability to quickly introduce new quality products is a decisive factor in capturing market share. For this purpose, the use of multiaxis *NC* machines is vital. Multiaxis *NC* machines of modern design are extremely costly. Because of this, the machining of sculptured part surfaces is costly as well. To decrease the cost of machining a sculptured part surface on a multiaxis *NC* machine, the machining time must be as short as possible. Definitely, this is the case where the phrase "time is money!" applies.

Reduction of the machining time is a critical issue when machining sculptured part surfaces on multiaxis *NC* machines. It is also an important consideration when machining part surfaces on machine tools of conventional design, namely, on lathes, gear generators, and so forth.

The optimization of part surface generation on a multiaxis *NC* machine results in time savings. Remember, *gaining time is gaining everything.*

Certainly, the subject of this book is of great importance for contemporary industry and engineering.

Uniqueness of This Publication. Literature on the theory of part surface generation on a multiaxis *NC* machine is lacking. A limited number of texts on the topic are available for the English-speaking audience. Conventional texts provide an adequate presentation and analysis of a given operation of sculptured part surface machining. The problem of part surface generation

---

* The difference between *optimization* and *synthesis* of a certain object (system or process) should be emphasized here. First, *optimization* relates to an object having a certain *known (definite) structure.* The structure of the object is prespecified and will not undergo any changes. After the optimization procedure is over, a given structure of the object remains the same as it was before the optimization. Optimization means a determination (that is, calculation) of a set of input parameters of the object aiming to achieve the desired value of a specified criterion of the optimization. Second, *synthesis* relates to an object of *unknown (indefinite) structure.* The desired structure of the object is to be determined along with the simultaneous determination of its parameters that ensure the best desired functioning of the object. Third, when saying "synthesis of an *optimal* object," this should be understood in the sense that the object is *synthesized* (and *not optimized*) to meet the best possible value of a given criterion, namely, the criterion under which performance of the synthesized object is the best possible.

is treated in all recently published books on the topic from the standpoint of *analysis*, and not of *synthesis*, of the optimal (the most favorable) process of part surface generation machining.

In the past 30 years, a wealth of new journal papers relating to the synthesis of optimal part surface-generation process have been published both in this country and abroad. The rapid intensification of research in the theory of part surface generation for *CAD* and *CAM* applications and new needs for advanced technology inspired me to complete this work.

The present text is an attempt to present a well-balanced and intelligible account of some of the geometric and algebraic procedures, filling in as necessary, making comparisons, and elaborating on the implications to give a well-rounded picture.

In this book, various procedures for handling particular problems constituting the *synthesis* of optimal (the most favorable) part surface generation processes on a multiaxis *NC* machine are investigated, compared, and applied. To begin, definitions, concepts, and notations are reviewed and established, and familiar methods of sculptured part surface analysis are recapitulated.* The fundamental concepts of sculptured part surface geometry are introduced, and known results in the theory of multiparametric motion of a rigid body in E_3 space are presented.

It is postulated in this text that the part surface to be machined is the primary element of the surface-generation process. Other elements, for example, the generating surface of the form-cutting tool and kinematics of their relative motion, are the secondary elements; thus, their optimal parameters must be determined in terms of design parameters of the part surface to be machined.

To the best of my knowledge, I was the first to formulate the problem of synthesizing the optimal part surface generation process in the late 1970s and early 1980s. At the beginning, the problem was understood mostly intuitively.

In current practice, the following input information is commonly used for the development of a sculptured part surface-generation process, S_{gp}, namely:

- The set of design parameters of a sculptured part surface to be machined, P
- The set of design parameters of generation surface T of the form-cutting tool to be implemented for machining of a given part surface

Reasonable (not optimal) kinematics of the part surface-machining process, K_{in}, is determined as a function of the two sets of design parameters of the surfaces P and T: the function $K_{in} = K_{in}(P, T)$ is expressed in terms of design parameters of the part surface P and of the cutting tool surface

---

* Recall here the old Chinese proverb, *The beginning of wisdom is to call things by their right names.*

T. Ultimately, all the parameters of the part surface-generation process (not optimal) are expressed in the form

$$S_{gp} = S_{gp}(P, T, K_{in}) = S_{gp}[P, T, K_{in}(P, T)] = S_{gp}(P, T) \qquad (I.1)$$

Another approach is disclosed below in this book. The discussed approach is targeting

(a) Minimization of the required input information for the development of a sculptured part surface-generation process
(b) Synthesizing of the *best possible* (the most favorable) sculptured part surface-generation process

The first principal achievements in the field* allowed an expression of the optimal parameters of kinematics of the sculptured part surface machining on a multiaxis *NC* machine in terms of the geometry of the part surface and of the generating surface of the form-cutting tool. This means that the kinematics, K_{in}, can be interpreted as a function of a specified geometry of the part surface *P* to be machined and of a specified geometry of the generating surface *T* of the form-cutting tool to be implemented, namely: $K_{in} = K_{in}(P, T)$.

A bit later, a principal solution to the problem of profiling the form-cutting tool was derived [104]. This solution makes possible the determination of the generating surface of the form-cutting tool as the $\mathbb{R}$-mapping of the sculptured part surface to be machined. Therefore, optimal parameters of the generating surface of the form-cutting tool, *T*, can be expressed in terms of design parameters of the part surface, *P*, to be machined. In other words, the invention yields the determination of a function $T = T(P)$.

These two above-discussed inventions make evident that the problem of synthesizing optimal part surface-generation process is solvable in nature.

Taking into account that the optimal parameters of kinematics of part surface machining, K_{in}, are already specified in terms of design parameters of the part surface *P* and generating surface *T* of the form-cutting tool [namely, in the form of the function $K_{in} = K_{in}(P, T)$], the last solution allows an analytical representation of the entire part surface-generation process in terms of design parameters of the sculptured part surface *P*. This means that the necessary input information for solving the problem of synthesizing the optimal part surface-machining operation is reduced just to the set of design parameters of the sculptured part surface *P* to be machined. This input information is the minimum possible, and it cannot be reduced further.

---

* Radzevich, S.P., *A Method of Sculptured Surface Machining on a Multi-Axis NC Machine*, Patent 1185749, USSR, B23C 3/16, filed October 24, 1983, and Radzevich, S.P., *A Method of Sculptured Surface Machining on a Multi-Axis NC Machine*, Patent 1249787, USSR, B23C 3/16, filed November 27, 1984.

Ultimately, all the parameters of the *optimal* (the most favorable) process of the entire sculptured part surface-generation process, S_{gp}, can be expressed in terms of design parameters only of the part surface to be machined:

$$S_{gp} = S_{gp}(P, T, K_{in}) = S_{gp}\{P, T(P) \ K_{in}[(P, T(P)]\} = S_{gp}(P) \qquad (I.2)$$

This input information in Equation I.2 is the minimum feasible, and it cannot be decreased further.

On the premises of these principal results (Patent No. 1185749, USSR; Patent No. 1249787; and Patent application 4242296/08, USSR), dozens of novel methods and means for part surface machining have been developed, and many of them are successfully used in the current industry (see Chapter 11).

It is important to stress that the decrease in required input information for synthesizing the most favorable part surface-generation process indicates that the theory is getting closer to ideal. This concept, which this book strictly adheres to, is widely known as the principle of *Occam's razor*. The principle of *Occam's* razor is one of the first principles that ever allowed the evaluation of how a theory becomes ideal. Minimal feasible input information indicates the strength of the proposed theory. *Occam's* razor principle states that the explanation of any phenomenon should make as few assumptions as possible, eliminating, or "shaving off," those that make no difference in the observable predictions of the explanatory hypothesis of theory. In short, when given two equally valid explanations for a phenomenon, one should embrace the less-complicated formulation. The principle is often expressed in Latin as the *lex parsimoniae* (law of succinctness):

> *Entia non sunt multiplicanda praeter necessitatem,*

which translates to

> *Entities should not be multiplied beyond necessity.*

This is often paraphrased as "All things being equal, the simplest solution tends to be the best one." In other words, when multiple competing theories are equal in other respects, the principle recommends selecting the theory that introduces the fewest assumptions and postulates the fewest hypothetical entities. It is in this sense that *Occam's razor* is usually understood.

Following the fundamental philosophical principle of *Occam's razor*, one can calculate optimal values of all the major parameters of the sculptured part surface-machining process on a multiaxis *NC* machine. Previous practical experience in the field (if any) is helpful, but not mandatory for solving the problem of synthesizing the optimal part surface-machining operation.

Important new topics help the reader to solve the challenging problems of synthesizing optimal methods and means of the part surface generation process. To employ the disclosed approach, limited input information is required: for this purpose, only analytical representation of the part surface to be generated is necessary. No other known theory of part surface generation is capable of solving the problems of synthesizing optimal methods of part surface generation. Moreover, no other known theory is capable of treating the problem on only the premise of the geometrical information of the part surface being generated alone—lots of additional information is commonly required when an engineer is following a known theory or approach for sculptured part surface generation.

The theory of part surface generation has been substantially complemented in this book by recent discoveries made primarily by myself and my colleagues. I have made a first attempt to summarize the obtained results of the research in the field in 1991. That year, my first monograph in the field of part surface generation (in Russian) was courageously introduced to the engineering community [Radzevich, S.P., *Sculptured Part Surface Machining on a Multiaxis NC Machine*, Kiev, Vishcha Shkola, 1991, 192 pp.]. Ten years later, a much more comprehensive summary was carried out [Radzevich, S.P., *Fundamentals of Surface Generation*, Kiev, Rastan, 2001, 592 pp.]. Both of these monographs are widely used in Europe as well as in the United States.

The first translation of the 2001 book into English was published by CRC Press in December 14, 2007 [Radzevich, S.P., *Kinematic Geometry of Surface Machining*, CRC Press, Boca Raton, Florida, 2007, 508 pp.]. An abridged version of the theory (mostly for tutorial purposes) is also known [Radzevich, S.P., *CAD/CAM of Sculpture Surfaces on Multiaxis NC Machine: The DG/K-Based Approach*, M&C Publishers, San Rafael, California, 2008, 114 pp.].

This second edition of my 2007 monograph (published by CRC Press) is significantly revised and enhanced.

There is a concern that some of today's mechanical engineers, manufacturing engineers, and engineering students may not be learning enough about the *theory of part surface generation*. Although containing some vitally important information, books to date do not provide methodological information on the subject, which can be helpful in making critical decisions in the process design, design and selection of form-cutting tools, and implementation of the proper machine tools. The most important information is dispersed throughout a great number of research and application papers and articles. Commonly, isolated theoretical and practical findings for a particular part surface-generation process are reported instead of the methodology, so the question "What would happen if the input parameters are altered?" remains unanswered. Therefore, a broad-based book on the theory of part surface generation is needed.

The purpose of the book is twofold:

- To summarize the accumulated information on part surface generation with a critical review of previous work, thus helping specialists and practitioners to separate facts from myths. The major problem in the theory of part surface generation is the absence of methods by the use of which the challenging problem of optimal part surface generation can be successfully solved. Other known problems are just consequences of the absence of the said proper methods of part surface generation.

- To present, explain, and exemplify a novel principal concept in the theory of part surface generation, namely, that the part surface is the primary element of the part surface-machining operation. The rest of the elements are secondary elements of the part surface-machining operation; thus, all of them can be expressed in terms of the desired (the most favorable) design parameters of the part surface to be machined.

The distinguished feature of this book is that the practical ways of synthesizing and optimizing the part surface-generation process are considered using just one set of input parameters—the design parameters of the part surface to be machined. The desired design parameters of the part surface to be machined are known in a research laboratory as well as in a shop floor environment. This makes this book not just another book on the subject. For the first time, the theory of part surface generation is presented as a science that really works. It is the right place to recall the words by famous Scottish physicist James C. Maxwell[*]: "There is nothing more practical than a good theory."

This book is written based on my over 40 years of experience in research, practical application, and teaching the theory of part surface generation, applied mathematics and mechanics, fundamentals of *CAD/CAM*, and engineering systems theory. Emphasis is placed on the practical application of the results in the everyday practice of part surface machining and cutting tool design. The application of these recommendations will increase the competitive position of the users through machining economy and productivity. This helps in designing better quality form-cutting tools and processes and in enhancing technical expertise and levels of technical services.

Intended Audience. Many readers will benefit from this book: mechanical and manufacturing engineers involved in continuous process improvement, research workers who are active or intend to become active in the field, and senior undergraduate and graduate university students of mechanical engineering and manufacturing.

This book is intended to be used as a reference book as well as a textbook. Chapters that cover the geometry of sculptured part surfaces and elementary kinematics of part surface generation and some sections that pertain to the design of the form-cutting tools can be used for graduate study. The content of the book arises from a postgraduate course on *CAD/CAM* for sculptured part surface machining on a multiaxis *NC* machine as well as from the experience of working on the development of efficient methods of part surface machining.

I have used this book for graduate study in my lectures at the National Technical University of Ukraine "Kyiv Polytechnic Institute" (Kyiv, Ukraine).

---

[*] James Clerk Maxwell (June 13, 1831–November 5, 1879), a famous Scottish physicist and mathematician.

The design chapters and practical implementation of the proposed theory (Section III) will be of interest for mechanical and manufacturing engineers and for researchers.

The Organization of This Book. The book comprises three sections entitled "Basics," "Fundamentals," and "Application," respectively:

Section I: Basics

This section of the book includes analytical descriptions of part surfaces, basics of differential geometry of sculptured part surfaces, along with principal elements of the theory of multiparametric motion of a rigid body in E_3 space. The applied coordinate systems and linear transformations are briefly considered. The selected material focuses on the solution to the problem of synthesizing optimal machining operation of sculptured part surfaces on a multiaxis *NC* machine.

The chapters and their contents are as follows:

Chapter 1. Part Surfaces: Geometry

In this book only two assumptions are made about the reader's mathematical prowess: first, that you have an appreciation of the *Cartesian* frame of reference, which is used to map three-dimensional space; and, second, that you know the rules for manipulating vectors and matrices. It will be comforting to know that if you have knowledge of the *vector* and the *matrix* there is nothing in this book that you cannot readily appreciate.*

The basics of differential geometry of sculptured part surfaces are explained. The focus here is on the difference between *classical differential geometry* and the *engineering geometry of surfaces*. Numerous examples of the calculation of major part surface elements are provided. A feasibility of classification of part surfaces is discussed, and a scientific classification of local patches of sculptured part surfaces is proposed.

Chapter 2. Principal Kinematics of Part Surface Generation

The generalized analysis of kinematics of sculptured part surface generation is presented. Here, generalized kinematics of instant relative motion of the cutting tool relative to the work is proposed. The principle of *inverse kinematic* of a machining operation is widely used in this section of the book. Here, the principle is presented in a way specific to the generation of part surfaces when machining. For the purpose of the profound investigation of the part surface-generation process, novel kinds of relative

---

* For any readers who wish to brush up on their vector geometry, a general introduction to the topic can be found in Appendix A.

motions of the cutting tool are discovered, including generating motion of the cutting tool, motions of orientation of the cutting tool, and relative motions that cause sliding of a surface over itself. Vector representation of kinematic schemes of part surface generation on conventional machine tools is widely used in this section of the book. The chapter concludes with a discussion on all feasible kinematic schemes of part surface generation. Several particular issues of kinematics of part surface generation are discussed in this chapter as well.

Chapter 3. Applied Coordinate Systems and Linear Transformations

The definitions and determinations of major applied coordinate systems are introduced in this chapter. The matrix approach for the coordinate system transformations is briefly discussed. Here, useful notations and practical equations are provided. Two issues of critical importance are introduced here. The first is chains of consequent linear transformations and a *closed loop* (*closed circuit*) of consequent coordinate system transformations. The impact of the coordinate system transformations on fundamental forms of the surfaces is the second. Both of the above-mentioned issues make possible significant simplifications when developing an analytical description of the entire part surface-generation process as well as any or all of its components.

These tools, rust-covered for many readers (the voice of experience), are resharpened in an effort to make the book a self-sufficient unit suited for self-study.

Section II: Fundamentals

The fundamentals of the theory of part surface generation are the core of the monograph. This part of the monograph includes a novel powerful method of an analytical description of the *geometry of contact* (or in other terminology, the *contact geometry*) of two smooth, regular surfaces in the first order of tangency; a novel type of mapping of one surface onto another surface; a novel analytical method of investigation of the cutting tool geometry; and a set of analytically described conditions of proper part surface generation. A solution to the challenging problem of synthesizing optimal part surface machining begins here. Consideration is based on the analytical results presented in the first part of the book.

The following chapters are included in this section:

Chapter 4. Geometry of Contact between a Sculptured Part Surface and Generating Surface of the Form-Cutting Tool

The local characteristics of contact of two smooth, regular surfaces that make tangency of the first order are considered in this chapter of the monograph. The sculptured part surface is one of the contacting surfaces, and the generating surface of the form-cutting tool is the second. The performed analysis includes an analytical description for local relative orientation of the contacting surfaces and the first- and second-order analyses. The concept of conformity of two smooth, regular surfaces in the first order of tangency is introduced and explained in this chapter. For an analytical description of contact geometry of the surfaces P and T, a characteristic curve of a novel type is introduced. This planar characteristic curve of the fourth order is referred to as *indicatrix of conformity*, $Cnf_R(P/T)$, of the cutting tool surface T to the part surface P at a current point of contact of the surfaces. For the purposes of analyses, the properties of Plücker's conoid are implemented. Ultimately, all feasible kinds of contact of the part surface, P, and of generating surface of the form-cutting tool, T, are classified.

Chapter 5. Profiling of Form-Cutting Tools of Optimal Design for Machining a Given Part Surface

A novel method of profiling the form-cutting tools for sculptured part surface machining on a multiaxis *NC* machine is disclosed in this chapter of the book. The method is based on the analytical representation of the contact geometry of the part surface P and the generating surface T of the form-cutting tool that is disclosed in the previous chapter. Methods of profiling form-cutting tools for machining part surfaces on conventional machine tools are also considered. These methods are based on elements of the theory of enveloping surfaces. The theory of enveloping surfaces is also enhanced in this chapter. Numerous particular issues of profiling form-cutting tools are discussed at the end of the chapter.

Chapter 6. The Geometry of the Active Part of a Cutting Tool

The generating body of the form-cutting tool is bounded by the generating surface of the cutting tool. Methods of transformation of the generating body of the form-cutting tool into a workable cutting tool are discussed in detail. In addition to two known methods, one novel method for this purpose is proposed. Results of the analytical investigation of the geometry of the active part of cutting tools in both the *Tool-in-Hand* reference system as well as in the *Tool-in-Use* reference system are represented. Numerous practical examples of the calculations are also presented in this chapter of the book.

Chapter 7. Conditions of Proper Part Surface Generation

The satisfactory conditions necessary and sufficient for proper part surface generation are proposed and examined. The conditions include the optimal workpiece orientation on the worktable of a multiaxis *NC* machine and the set of six analytically described conditions of proper part surface generation. The chapter concludes with novel methods for the global verification of satisfaction of the conditions of proper part surface generation.

Chapter 8. Accuracy of Surface Generation

The accuracy of part surface generation is an important issue for the manufacture of machined part surfaces. Analytical methods for the analysis and calculation of the deviations of a machined part surface from the desired part surface are discussed here. Two principal kinds of deviations of the machined part surface from the nominal part surface are distinguished. Methods for the calculation of the elementary surface deviations are proposed. The total displacements of the cutting tool with respect to the part surface are analyzed. Effective methods for reduction of elementary surface deviations are proposed. The conditions under which the *principle of superposition* of elementary surface deviations is applicable are established.

Section III: Application

This section illustrates the capabilities of a novel and powerful tool for the development of highly efficient methods of part surface generation. Numerous practical examples of implementation of the proposed theory of part surface generation are disclosed in this chapter of the monograph.

This section of the book is organized as follows:

Chapter 9. Selection of the Criterion of Optimization

To implement in practice the proposed *differential geometry/ kinematics* (*DG/K*)-based method of part surface generation, an appropriate criterion of efficiency of part surface-machining process is necessary. This helps answer the question of what we want to obtain when performing a certain machining operation. It is of critical importance to stress here that regardless of the criterion it is referred to as the criterion of *optimization*; in this book, the criterion is implemented not for the purpose of optimization, instead, it is implemented for the purpose of *synthesis* of the most favorable part surface-generation processes. Various criteria of efficiency of machining operation are considered. The tight connection of the economical criteria of optimization with geometrical analogues (as established in Chapter 4) is illustrated. The part surface generation output is expressed in terms of functions of conformity. The last significantly simplifies the synthesizing of the most favorable operations of part surface machining.

Chapter 10. Synthesis of the Part Surface-Machining Operations

The synthesis of optimal operations of actual part surface machining on both the multiaxis *NC* machine and a conventional machine tool is explained. For this purpose, three steps of the analysis are distinguished. They are local analysis, regional analysis, and global analysis. The possibility of development of the *DG/K*-based *CAD/CAM* system for the optimal sculptured part surface machining is shown.

Chapter 11. Examples of Implementation of the *DG/K*-Based Method of Part Surface Generation

This chapter demonstrates numerous novel methods of part surface machining—those developed on the premise of implementation of the proposed *DG/K*-based method of part surface generation. Addressed are novel methods of machining sculptured part surfaces on a multiaxis *NC* machine, novel methods of machining surfaces of revolution, and a novel method of finishing involute gears. It is shown here that the developed *DG/K*-based method of part surface generation possesses unique capabilities for significant improvements of part surface-machining operations as well as for accuracy of the machined part surfaces.

The proposed theory of part surface generation is oriented on extensive application of multiaxis *NC* machines of modern design. In particular cases, the implementation of the theory can be useful for machining part surfaces on conventional machine tools.

Taken as a whole, the topics covered will enable you to synthesize optimal (the most favorable) part surface-machining operation and optimal means for these purposes.

A book of this size is likely to contain omissions and errors. If you have any constructive suggestions, please communicate them to Dr. S. Radzevich (radzevich@gmail.com).

Section I

Basics

This section of the book includes analytical description of part surfaces, basics on differential geometry of sculptured part surfaces, along with principal elements of the theory of multiparametric motion of a rigid body in the *Euclidian E_3* space. *Geometry* is the foundation on which the theory of part surface generation is built.

The applied coordinate systems and linear transformations are briefly considered.

This section of the book is not a general introduction to the subject of part surface geometry. If you want to acquire background knowledge of part surface geometry or the current state of what is now a huge industry, there are many excellent sources of information available.

The selected material focuses on the solution to the problem of synthesizing the most favorable (optimal) machining operation of sculptured part surfaces on a multiaxis *NC* machine, as essentially, this book is targeted at those that work on applications under the broad heading of sculptured part surface machining on a multiaxis *NC* machine.

I hope to be able to introduce all this material in such a way that it can readily be put into practice.

1

Part Surfaces: Geometry

The generation of part surfaces is one of the major purposes of every machining operation. An enormous variety of parts are manufactured in various industries. Every part to be machined is bounded with two or more surfaces.* Each of the part surfaces is a smooth, regular surface, or it can be composed with a certain number of patches of smooth, regular surfaces that are properly linked to one another. An example of such a composite working surface of a die is illustrated in Figure 1.1. Here, the complex working surface comprises planes, cylinders of revolution, cones of revolution, torus surfaces, and so forth.

This chapter describes the essential mathematical concepts that form the basis for the theory of part surface generation. It also establishes the notation and conventions that will be used through the book.

To be machined on a numerical control (NC) machine and for computer-aided design (CAD) and computer-aided manufacturing (CAM) applications, a formal (analytical) representation of a part surface is the required prerequisite. Analytical representation of a part surface (the surface P) to be machined is based on analytical representation of surfaces in geometry, specifically, in the

a. *Differential geometry* of surfaces

b. *Engineering geometry* of surfaces

The second is based on the first.

For further consideration, it is convenient to briefly consider the principal elements of differential geometry of surfaces. Methods of classical differential geometry of surfaces are widely used in this text.

If experienced in differential geometry of surfaces, the following section may be skipped. Then, proceed directly to Section 1.2.

1.1 Elements of Differential Geometry of Surfaces

Discussion in this book is focused primarily on generation of sculptured part surfaces on a multiaxis NC machine. In many cases, part surfaces to

---

* The ball of a ball bearing is one of just a few examples of a part surface, which is bounded with the only surface that is the sphere.

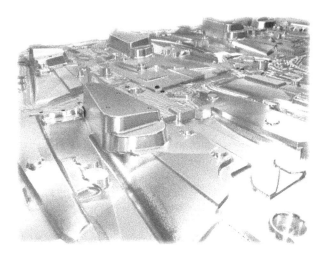

FIGURE 1.1
An example of a complex working surface of a die comprising planes, cylinders of revolution, cones of revolution, torus surface, and so forth.

FIGURE 1.2
Working surface of an impeller blade is a perfect example of a smooth, regular sculptured part surface P.

be machined on a multiaxis *NC* machine are referred to as *sculptured part surfaces*. Sculptured part surfaces are also often referred to as *free-form part surfaces*. As shown in Figure 1.2, the working surface of an impeller blade is a perfect example of a smooth, regular sculptured part surface P.

Surfaces and their motion in space are analytically described in a reference system. An orthogonal *Cartesian** reference system is the major type of

---

* René Descartes (March 31, 1596–February 11, 1650) (Latinized form: Renatus Cartesius), a French mathematician, philosopher, and writer.

reference systems that is commonly used for this purpose. Mutually perpendicular coordinate axes of a *Cartesian* coordinate system are conventionally labeled as X, Y, and Z.

In the *Cartesian* system, the axes can be oriented in either a left- or right-handed sense. A right-handed *Cartesian* reference system is preferred, and all algorithms and formulae used in this book assume a right-handed convention.

A coordinate system provides a numerical frame of reference for the three-dimensional space in which we will develop the theory. Two coordinate systems are particularly useful to us: the ubiquitous *Cartesian* (XYZ) rectilinear system and the spherical polar (r, θ, φ) or angular system. *Cartesian* coordinate systems are the most commonly used, but angular coordinates are often helpful as well.

If you wish, you can skip directly to Chapter 2 where the main topics of the book begin with details of algorithms for the rendering pipeline.

1.1.1 Specification of a Part Surface

A surface could be uniquely determined by two independent variables. Therefore, we give a part surface P (Figure 1.3), in most cases, by expressing

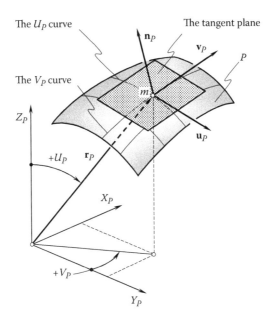

FIGURE 1.3
Principal parameters of local topology of a part surface P.

its rectangular coordinated X_P, Y_P, and Z_P, as functions of two *Gaussian** coordinates U_P and V_P in a certain closed interval:

$$\mathbf{r}_P = \mathbf{r}_P(U_P, V_P) = \begin{bmatrix} X_P(U_P, V_P) \\ Y_P(U_P, V_P) \\ Z_P(U_P, V_P) \\ 1 \end{bmatrix} \tag{1.1}$$

$$U_{1.P} \le U_P \le U_{2.P};\ V_{1.P} \le V_P \le V_{2.P}$$

where $\mathbf{r}_P$ is the position vector of a point of the part surface P, U_P and V_P are the curvilinear (*Gaussian*) coordinates of the point of the surface P, X_P, Y_P, Z_P are the *Cartesian* coordinates of the point of surface P, $U_{1.P}$ and $U_{2.P}$ are the boundary values of the closed interval of the U_P parameter, $V_{1.P}$ and $V_{2.P}$ are the boundary values of the closed interval of the V_P parameter.

The parameters U_P and V_P must enter into Equation 1.1 independently, which means that the matrix

$$\mathbf{M} = \begin{bmatrix} \dfrac{\partial X_P}{\partial U_P} & \dfrac{\partial Y_P}{\partial U_P} & \dfrac{\partial Z_P}{\partial U_P} \\[2ex] \dfrac{\partial X_P}{\partial V_P} & \dfrac{\partial Y_P}{\partial V_P} & \dfrac{\partial Z_P}{\partial V_P} \end{bmatrix} \tag{1.2}$$

has a rank 2.

Positions, where the rank is 1 or 0 are singular points; when the rank at all points is 1, then Equation 1.1 represents a curve.

Other methods of surface specification are known as well. Specification of a part surface by

- An equation in explicit form
- An equation in implicit form
- A set of parametric equations

are among the most frequently used in practice methods of part surface specification.

It is assumed here and below that any given type of part surface specification can be converted either into the vector form, or into the matrix form of specification as it is following from Equation 1.1.

---

* Johan Carl Friedrich Gauss (April 30, 1777–February 23, 1855), a famous German mathematician and physical scientist.

1.1.2 Tangent Vectors and Tangent Plane; Unit Normal Vector

The following notation is proven to be convenient in the consideration below.

The first derivatives of $\mathbf{r}_p$ with respect to *Gaussian* coordinates U_P and V_P are designated as

$$\frac{\partial \mathbf{r}_p}{\partial U_p} = \mathbf{U}_p \tag{1.3}$$

$$\frac{\partial \mathbf{r}_p}{\partial V_p} = \mathbf{V}_p \tag{1.4}$$

and for the unit tangent vectors

$$\mathbf{u}_p = \frac{\mathbf{U}_P}{|\mathbf{U}_P|} \tag{1.5}$$

$$\mathbf{v}_p = \frac{\mathbf{V}_P}{|\mathbf{V}_P|}, \tag{1.6}$$

respectively.*

The direction of the tangent line to the U_p coordinate line through a given point m on the part surface P is specified by the unit tangent vector $\mathbf{u}_p$ (as well as by the tangent vector $\mathbf{U}_p$). Similarly, direction of the tangent line to the V_P coordinate line through that same point m on the part surface P is specified by the unit tangent vector $\mathbf{v}_p$ (as well as by the tangent vector $\mathbf{V}_p$).

Significance of the unit tangent vectors $\mathbf{u}_p$ and $\mathbf{v}_p$ becomes evident from the following considerations.

First, unit tangent vectors $\mathbf{u}_p$ and $\mathbf{v}_p$ yield an equation of the tangent plane to the part surface P at a specified point m:

$$\text{Tangent plane} \quad \Rightarrow \quad \begin{bmatrix} [\mathbf{r}_{t.p} - \mathbf{r}_P^{(m)}] \\ \mathbf{u}_p \\ \mathbf{v}_p \\ 1 \end{bmatrix} = 0 \tag{1.7}$$

where $\mathbf{r}_{t.p}$ is the position vector of a point of the tangent plane to the part surface P at a specified point m and $\mathbf{r}_P^{(m)}$ is position vector of the point m on the part surface P.

---

* It is the right point to underline here that the unit tangent vectors $\mathbf{u}_p$ and $\mathbf{v}_p$ are dimensionless values following from Equations 1.5 and 1.6.

Second, tangent vectors yield an equation of the perpendicular $\mathbf{N}_p$ and of the unit normal vector $\mathbf{n}_p$ to the part surface P at a given point m:

$$\mathbf{N}_p = \mathbf{U}_p \times \mathbf{V}_p \tag{1.8}$$

and

$$\mathbf{n}_p = \frac{\mathbf{N}_p}{|\mathbf{N}_p|} = \frac{\mathbf{U}_p \times \mathbf{V}_p}{|\mathbf{U}_p \times \mathbf{V}_p|} = \mathbf{u}_p \times \mathbf{v}_p \tag{1.9}$$

When the order of the multipliers in Equations 1.8 and 1.9 is chosen properly, then the unit normal vector $\mathbf{n}_p$ (as well as the normal vector $\mathbf{N}_p$) is pointed outward of the bodily side of the surface P.

1.1.3 Local Frame

Two unit tangent vectors $\mathbf{u}_p$ and $\mathbf{v}_p$ along with the unit normal vector $\mathbf{n}_p$ comprise a local frame $\mathbf{u}_p, \mathbf{v}_p, \mathbf{n}_p$ having origin at a current point m on the part surface P. This frame is schematically illustrated in Figure 1.4. Unit tangent vector $\mathbf{u}_p$ is perpendicular to the unit normal vector $\mathbf{n}_p$ (i.e., $\mathbf{u}_p \perp \mathbf{n}_p$), and unit tangent vector $\mathbf{v}_p$ is also perpendicular to the unit normal vector $\mathbf{n}_p$ (i.e., $\mathbf{v}_p \perp \mathbf{n}_p$). Generally, the unit tangent vectors $\mathbf{u}_p$ and $\mathbf{v}_p$ are not perpendicular to one another, they form a certain angle ω_p. To construct an orthogonal local frame, either the unit

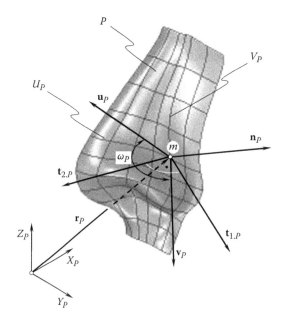

FIGURE 1.4
An example of *Darboux* frame ($\mathbf{t}_{1.P}$, $\mathbf{t}_{2.P}$, $\mathbf{n}_p$) at a point m of the sculptured surface P.

tangent vector $\mathbf{u}_P$ in the local frame $(\mathbf{u}_P, \mathbf{v}_P, \mathbf{n}_P)$ must be substituted with a unit tangent vector $\mathbf{u}_P^*$, or the unit tangent vector $\mathbf{v}_P$ in that same local frame $(\mathbf{u}_P, \mathbf{v}_P, \mathbf{n}_P)$ must be substituted with a unit tangent vector $\mathbf{v}_P^*$ (the vectors $\mathbf{u}_P^*$ and $\mathbf{v}_P^*$ are not shown in Figure 1.4). For the calculation of the newly introduced unit tangent vectors $\mathbf{u}_P^*$ and $\mathbf{v}_P^*$, the following equations can be used:

$$\mathbf{u}_P^* = \mathbf{u}_P \times \mathbf{n}_P \tag{1.10}$$

$$\mathbf{v}_P^* = \mathbf{v}_P \times \mathbf{n}_P \tag{1.11}$$

It is convenient to chose the order of the multipliers in Equations 1.10 and 1.11 that preserves the orientation (the hand) of the original local frame $(\mathbf{u}_P, \mathbf{v}_P, \mathbf{n}_P)$, namely, if the original local frame $(\mathbf{u}_P, \mathbf{v}_P, \mathbf{n}_P)$ is right-hand oriented, then the newly constructed local frame [either the local frame $(\mathbf{u}_P^*, \mathbf{v}_P, \mathbf{n}_P)$, or the local frame $(\mathbf{u}_P, \mathbf{v}_P^*, \mathbf{n}_P)$] should also be a right-hand oriented local frame, and vice versa.

It should be pointed out here that another possibility to construct an orthogonal local frame is also available. The local frame of this type is commonly referred to as *Darboux** frame and is briefly considered below in this section of the book.

Unit tangent vectors $\mathbf{u}_P$ and $\mathbf{v}_P$ to a surface P at a point m are of critical importance when solving practical problems in the field of part surface generation. This statement is proven by numerous examples shown below.

1.1.4 Fundamental Forms of a Surface

Consider two other important issues concerning part surface geometry—both relate to intrinsic geometry in differential vicinity of a current surface point m.

1.1.4.1 First Fundamental Form of a Surface

The first issue is the so-called *first fundamental form* $\Phi_{1.P}$ of a part surface P. The metric properties of the part surface P are described by the first fundamental form, $\Phi_{1.P}$, of the surface. Usually, the first fundamental form, $\Phi_{1.P}$, is represented as the quadratic form

$$\Phi_{1.P} \Rightarrow ds_P^2 = E_P \, dU_P^2 + 2F_P \, dU_P \, dV_P + G_P \, dV_P^2 \tag{1.12}$$

where s_P is the linear element on the part surface P (s_P is equal to the length of a segment of a certain curve on the part surface P) and E_P, F_P, and G_P are the fundamental magnitudes of the first order.

---

* Jean Gaston Darboux (August 14, 1842–February 23, 1917), a French mathematician.

Equation 1.12 for the first fundamental form, $\Phi_{1.P}$, is known from many advanced sources. In the theory of part surface generation, another form of analytical representation of the first fundamental form, $\Phi_{1.P}$, has proven to be useful:

$$\Phi_{1.P} \Rightarrow ds_P^2 = [\ dU_P \quad dV_P \quad 0 \quad 0\ ] \cdot \begin{bmatrix} E_P & F_P & 0 & 0 \\ F_P & G_P & 0 & 0 \\ 0 & 0 & 1 & 0 \\ 0 & 0 & 0 & 1 \end{bmatrix} \cdot \begin{bmatrix} dU_P \\ dV_P \\ 0 \\ 0 \end{bmatrix}$$

(1.13)

This type of analytical representation of the first fundamental form $\Phi_{1.P}$ is proposed by Radzevich [110].

The practical advantage of Equation 1.13 is that it can easily be incorporated into computer programs when multiple coordinate system transformations are used. The last is vital for both for the theory of part surface generation as well as for most of *CAD/CAM* applications.

The fundamental magnitudes of the first order, E_P, F_P, and G_P, can be calculated from the set of the following equations:

$$E_P = \mathbf{U}_P \cdot \mathbf{U}_P \tag{1.14}$$

$$F_P = \mathbf{U}_P \cdot \mathbf{V}_P \tag{1.15}$$

$$G_P = \mathbf{V}_P \cdot \mathbf{V}_P \tag{1.16}$$

Equations 1.14 through 1.16 can be represented in the expanded form:

$$E_P = \frac{\partial \mathbf{r}_P}{\partial U_P} \cdot \frac{\partial \mathbf{r}_P}{\partial U_P} = \frac{\partial X_P}{\partial U_P} \cdot \frac{\partial X_P}{\partial U_P} + \frac{\partial Y_P}{\partial U_P} \cdot \frac{\partial Y_P}{\partial U_P} + \frac{\partial Z_P}{\partial U_P} \cdot \frac{\partial Z_P}{\partial U_P} \tag{1.17}$$

$$F_P = \frac{\partial \mathbf{r}_P}{\partial U_P} \cdot \frac{\partial \mathbf{r}_P}{\partial V_P} = \frac{\partial X_P}{\partial U_P} \cdot \frac{\partial X_P}{\partial V_P} + \frac{\partial Y_P}{\partial U_P} \cdot \frac{\partial Y_P}{\partial V_P} + \frac{\partial Z_P}{\partial U_P} \cdot \frac{\partial Z_P}{\partial V_P} \tag{1.18}$$

$$G_P = \frac{\partial \mathbf{r}_P}{\partial V_P} \cdot \frac{\partial \mathbf{r}_P}{\partial V_P} = \frac{\partial X_P}{\partial V_P} \cdot \frac{\partial X_P}{\partial V_P} + \frac{\partial Y_P}{\partial V_P} \cdot \frac{\partial Y_P}{\partial V_P} + \frac{\partial Z_P}{\partial V_P} \cdot \frac{\partial Z_P}{\partial V_P} \tag{1.19}$$

The fundamental magnitudes of the first order, E_P, F_P, and G_P, are functions of the U_P and V_P coordinates of a point of the part surface P. In general form, these relationships can be represented in the form

$$E_P = E_P(U_P, V_P) \tag{1.20}$$

$$F_P = F_P(U_P, V_P) \tag{1.21}$$

$$G_P = G_P(U_P, V_P) \tag{1.22}$$

It is important to point out here that fundamental magnitudes E_P and G_P are always positive ($E_P > 0$, $G_P > 0$), and the fundamental magnitude F_P can equal zero ($F_P \geq 0$). The result is that the first fundamental form, $\Phi_{1.P}$ of a part surface P, is always positively defined ($\Phi_{1.P} \geq 0$), and it cannot be of a negative value.

Using the first fundamental form, $\Phi_{1.P}$, the following major parameters of geometry of the part surface P can be calculated:

(a) Length of a curve-line segment on the part surface P
(b) Square of the part surface P portion that is bounded by a closed curve on the surface
(c) Angle between any two directions on the part surface P

The length, s_P, of a curve-line segment

$$U_P = U_P(t) \tag{1.23}$$

$$V_P = V_P(t) \tag{1.24}$$

on the part surface P is given by the equation

$$s_P = \int_{t_0}^{t} \sqrt{ E_P \left(\frac{dU_P}{dt} \right)^2 + 2F_P \frac{dU_P}{dt} \frac{dV_P}{dt} + G_P \left(\frac{dV_P}{dt} \right)^2 } \, dt \tag{1.25}$$

$$t_0 \leq t \leq t_1$$

To calculate the square, $\mathcal{S}_P$, of the part surface P patch Σ, which is bounded by a closed curve on the surface P, the following equation can be used:

$$\mathcal{S}_P = \iint_{\Sigma} \sqrt{E_P G_P - F_P^2} \, dU_P \, dV_P \tag{1.26}$$

Ultimately, the value of the angle, ω_P, between two given directions through a certain point m on the part surface P can be calculated from one of these equations:

$$\cos \omega_P = \frac{F_P}{\sqrt{E_P G_P}}$$
(1.27)

$$\sin \omega_P = \frac{H_P}{\sqrt{E_P G_P}}$$
(1.28)

$$\tan \omega_P = \frac{H_P}{F_P}$$
(1.29)

For the calculation of the discriminant, H_P, of the first fundamental form, $\Phi_{1.P}$, the following equation can be used:

$$H_P = \sqrt{E_P G_P - F_P^2}$$
(1.30)

It is assumed here that the discriminant, H_P, is always nonnegative—that is, $H_P = +\sqrt{E_P G_P - F_P^2}$.

The first fundamental form, $\Phi_{1.P}$, represents the length of a curve-line segment, and thus, it is always nonnegative—that is, the inequality $\Phi_{1.P} \geq$ is always valid.

The first fundamental form, $\Phi_{1.P}$, remains the same when the surface is banding. This is another important feature of the first fundamental form $\Phi_{1.P}$. The feature can be employed for designing a three-dimensional cam for finishing a turbine blade with an abrasive strip as a cutting tool.

1.1.4.2 Second Fundamental Form of a Surface

The *second fundamental form* $\Phi_{2.P}$ of a part surface P is another of the two above-mentioned issues. Second fundamental form $\Phi_{2.P}$ describes curvature of a smooth regular surface P.

Consider point K on a smooth regular part surface P (Figure 1.5). The location of point K is specified by two coordinates U_P and V_P. A line through point K is entirely located within the surface P. A nearby point m is located within the line through point K. Location of the point m is specified by the coordinates $U_P + dU_P$ and $V_P + dV_P$ as it is infinitesimally close to point K. The closest distance of approach of the point m to the tangent plane through point K is expressed by the second fundamental form $\Phi_{2.P}$. The torsion of the curve Km is ignored. Therefore, distance a is assumed equal to zero ($a = 0$).

The second fundamental form, $\Phi_{2.P}$, describes the curvature of a smooth, regular part surface P. Usually, it is represented as the quadratic form (Figure 1.5)

$$\Phi_{2.P} \quad \Rightarrow \quad -d\mathbf{r}_P \cdot d\mathbf{n}_P = L_P \, dU_P^2 + 2 M_P \, dU_P \, dV_P + N_P \, dV_P^2$$
(1.31)

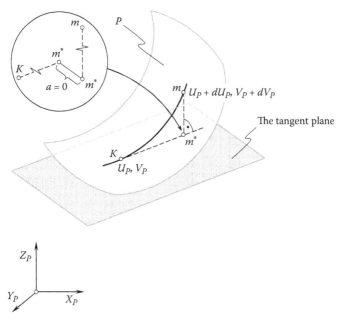

FIGURE 1.5
Definition of second fundamental form, $\Phi_{2.P}$, of a smooth regular part surface P.

Equation 1.31 is known from many advanced sources.

In the theory of part surface generation, another analytical representation of the second fundamental form, $\Phi_{2.P}$, has proven to be useful:

$$\Phi_{2.P} \Rightarrow [\ dU_P \quad dV_P \quad 0 \quad 0\ ] \cdot \begin{bmatrix} L_P & M_P & 0 & 0 \\ M_P & N_P & 0 & 0 \\ 0 & 0 & 1 & 0 \\ 0 & 0 & 0 & 1 \end{bmatrix} \cdot \begin{bmatrix} dU_P \\ dV_P \\ 0 \\ 0 \end{bmatrix}$$

(1.32)

This analytical representation of the second fundamental form, $\Phi_{2.P}$, is proposed by Radzevich [110].

Similar to Equation 1.13, the practical advantage of Equation 1.32 is that it can easily be incorporated into computer programs when multiple coordinate system transformations are used. The last is vital for both for the theory of part surface generation as well as for most *CAD/CAM* applications.

In Equation 1.32, the parameters L_P, M_P, N_P designate fundamental magnitudes of the second order.

By definition (see page 173 in [5]), fundamental magnitudes of the second order are equal:

$$L_P = -\mathbf{U}_P \cdot \frac{\partial \mathbf{n}_P}{\partial U_P} = \mathbf{n}_P \cdot \frac{\partial \mathbf{U}_P}{\partial U_P} \tag{1.33}$$

$$M_P = -\frac{1}{2}\left(\mathbf{U}_P \cdot \frac{\partial \mathbf{n}_P}{\partial V_P} + \mathbf{V}_P \cdot \frac{\partial \mathbf{n}_P}{\partial U_P} \right) = \mathbf{n}_P \cdot \frac{\partial \mathbf{U}_P}{\partial V_P} = \mathbf{n}_P \cdot \frac{\partial \mathbf{V}_P}{\partial U_P} \tag{1.34}$$

$$N_P = -\mathbf{V}_P \cdot \frac{\partial \mathbf{n}_P}{\partial V_P} = \mathbf{n}_P \cdot \frac{\partial \mathbf{V}_P}{\partial V_P} \tag{1.35}$$

For the calculation of the fundamental magnitudes of the second order of a smooth regular part surface P, the following equations can be used:

$$L_P = \frac{\dfrac{\partial \mathbf{U}_P}{\partial U_P} \times \mathbf{U}_P \cdot \mathbf{V}_P}{\sqrt{E_P G_P - F_P^2}} \tag{1.36}$$

$$M_P = \frac{\dfrac{\partial \mathbf{U}_P}{\partial V_P} \times \mathbf{U}_P \cdot \mathbf{V}_P}{\sqrt{E_P G_P - F_P^2}} = \frac{\dfrac{\partial \mathbf{V}_P}{\partial U_P} \times \mathbf{U}_P \cdot \mathbf{V}_P}{\sqrt{E_P G_P - F_P^2}} \tag{1.37}$$

$$N_P = \frac{\dfrac{\partial \mathbf{V}_P}{\partial V_P} \times \mathbf{U}_P \cdot \mathbf{V}_P}{\sqrt{E_P G_P - F_P^2}} \tag{1.38}$$

Equations 1.36 through 1.38 can be represented in the expanded form:

$$L_P = \frac{\begin{vmatrix} \dfrac{\partial^2 X_P}{\partial U_P^2} & \dfrac{\partial^2 Y_P}{\partial U_P^2} & \dfrac{\partial^2 Z_P}{\partial U_P^2} \\[2mm] \dfrac{\partial X_P}{\partial U_P} & \dfrac{\partial Y_P}{\partial U_P} & \dfrac{\partial Z_P}{\partial U_P} \\[2mm] \dfrac{\partial X_P}{\partial V_P} & \dfrac{\partial Y_P}{\partial V_P} & \dfrac{\partial Z_P}{\partial V_P} \end{vmatrix}}{\sqrt{E_P G_P - F_P^2}} \tag{1.39}$$

$$M_P = \dfrac{\begin{vmatrix} \dfrac{\partial^2 X_P}{\partial U_P \partial V_P} & \dfrac{\partial^2 Y_P}{\partial U_P \partial V_P} & \dfrac{\partial^2 Z_P}{\partial U_P \partial V_P} \\[2ex] \dfrac{\partial X_P}{\partial U_P} & \dfrac{\partial Y_P}{\partial U_P} & \dfrac{\partial Z_P}{\partial U_P} \\[2ex] \dfrac{\partial X_P}{\partial V_P} & \dfrac{\partial Y_P}{\partial V_P} & \dfrac{\partial Z_P}{\partial V_P} \end{vmatrix}}{\sqrt{E_P G_P - F_P^2}} \tag{1.40}$$

$$N_P = \dfrac{\begin{vmatrix} \dfrac{\partial^2 X_P}{\partial V_P^2} & \dfrac{\partial^2 Y_P}{\partial V_P^2} & \dfrac{\partial^2 Z_P}{\partial V_P^2} \\[2ex] \dfrac{\partial X_P}{\partial U_P} & \dfrac{\partial Y_P}{\partial U_P} & \dfrac{\partial Z_P}{\partial U_P} \\[2ex] \dfrac{\partial X_P}{\partial V_P} & \dfrac{\partial Y_P}{\partial V_P} & \dfrac{\partial Z_P}{\partial V_P} \end{vmatrix}}{\sqrt{E_P G_P - F_P^2}} \tag{1.41}$$

The fundamental magnitudes of the second order, L_P, M_P, N_P, are also functions of the U_P and V_P coordinates of a point of the part surface P. In general form, these relationships can be represented in the form:

$$L_P = L_P(U_P, V_P) \tag{1.42}$$

$$M_P = M_P(U_P, V_P) \tag{1.43}$$

$$N_P = N_P(U_P, V_P) \tag{1.44}$$

The discriminant, T_P, of the second fundamental form, $\Phi_{2.P}$, can be calculated from the following equation:

$$T_P = \sqrt{L_P N_P - M_P^2} \tag{1.45}$$

We now come to the theorem, which is essential justification for considering the differential geometry of surfaces in connection with the six fundamental magnitudes. It was first proven (1867) by Bonnet* [6], and may be enunciated as follows:

---

* Pierre Ossian Bonnet (December 22, 1819–June 22, 1892), a French mathematician.

Theorem 1.1:

When six fundamental magnitudes E_P, F_P, G_P, and L_P, M_P, N_P are given, and when they satisfy the Gauss characteristic equation, and the two Mainardi*–Codazzi[†] relations, they determine a surface P uniquely as to its position and orientation in space.   ∎

This theorem is commonly referred to as the *main theorem in the theory of surface*, or simply as *Bonnet theorem*. According to the main theorem, two surfaces that have identical first and second fundamental forms must be either congruent or symmetrical to one another.

Using six fundamental magnitudes, all parameters of local geometry of a given part surface can be calculated.

1.1.5 Principal Directions of a Part Surface

The direction of vectors of principal directions, $\mathbf{T}_{1.P}$ and $\mathbf{T}_{2.P}$ at a point on the part surface P, can be specified in terms of the ratio dU_P/dV_P. For the vectors of the first, $\mathbf{T}_{1.P}$, and for the second, $\mathbf{T}_{2.P}$, principal directions at a point m of a smooth, regular part surface P the corresponding values of the ratio dU_P/dV_P are calculated as roots of the quadratic equation:

$$\begin{vmatrix} E_P\,dU_P + F_P\,dV_P & F_P\,dU_P + G_P\,dV_P \\ L_P\,dU_P + M_P\,dV_P & M_P\,dU_P + N_P\,dV_P \end{vmatrix} = 0 \qquad (1.46)$$

The first principal plane section, $C_{1.P}$, is perpendicular to the part surface P at a current surface point m, and passes through the vector of the first principal direction $\mathbf{T}_{1.P}$. The second principal plane section, $C_{2.P}$, is orthogonal to the part surface P at a current surface point m, and passes through the vector of the second principal direction $\mathbf{T}_{2.P}$.

The principal directions $\mathbf{T}_{1.P}$ and $\mathbf{T}_{2.P}$ can be identified at any and all points of a smooth, regular part surface P except of umbilic points, and in flatten points of the surface. At umbilic points of a surface as well as at flatten points, the principal directions cannot be identified.

In the theory of part surface generation, it is often preferred to use not the vectors $\mathbf{T}_{1.P}$ and $\mathbf{T}_{2.P}$ of the principal directions, but, instead, to use the unit vectors $\mathbf{t}_{1.P}$ and $\mathbf{t}_{2.P}$ of the principal directions. The unit tangent vectors $\mathbf{t}_{1.P}$ and $\mathbf{t}_{2.P}$ are calculated from the equations

---

* Gaspare Mainardi (June 27, 1800–March 9, 1879), an Italian mathematician.
[†] Delfino Codazzi (March 7, 1824–July 21, 1873), an Italian mathematician.

$$\mathbf{t}_{1.P} = \frac{\mathbf{T}_{1.P}}{|\mathbf{T}_{1.P}|} \tag{1.47}$$

$$\mathbf{t}_{2.P} = \frac{\mathbf{T}_{2.P}}{|\mathbf{T}_{2.P}|}, \tag{1.48}$$

respectively.

Unit tangent vectors $\mathbf{t}_{1.P}$ and $\mathbf{t}_{2.P}$ of principal directions at a point m on the part surface P along with unit normal vector $\mathbf{n}_P$ at that same point m comprise an orthogonal local frame $(\mathbf{t}_{1.P}, \mathbf{t}_{2.P}, \mathbf{n}_P)$. All three unit vectors $\mathbf{t}_{1.P}, \mathbf{t}_{2.P}$, and $\mathbf{n}_P$ are mutually perpendicular to one another as it is depicted in Figure 1.4. The local frame $(\mathbf{t}_{1.P}, \mathbf{t}_{2.P}, \mathbf{n}_P)$ is commonly referred to as *Darboux* frame.

1.1.6 Curvatures of a Part Surface

The first, $R_{1.P}$, and the second, $R_{2.P}$, principal radii of curvature at a point of the part surface P are measured within the first and in the second principal plane sections, $C_{1.P}$ and $C_{2.P}$, respectively. For the calculation of values of the principal radii of curvature, the following equation is commonly used:

$$R_P^2 - \frac{E_P N_P - 2F_P M_P + G_P L_P}{T_P} R_P + \frac{H_P}{T_P} = 0 \tag{1.49}$$

Remember that the algebraic values of the radii of principal curvature, $R_{1.P}$ and $R_{2.P}$, relate to one another as $R_{2.P} > R_{1.P}$. In particular cases, at umbilic points on the part surface P, no principal curvatures can be identified as all normal curvatures of the surface P at an umbilic point are equal to one another.

Another two important parameters of local topology of a part surface P are mean curvature, $\mathcal{M}_P$, and intrinsic (*Gaussian* or full) curvature, $\mathcal{G}_P$.

For the calculation of the curvatures $\mathcal{M}_P$ and $\mathcal{G}_P$, the following equations are commonly used:

$$\mathcal{M}_P = \frac{k_{1.P} + k_{2.P}}{2} = \frac{E_P N_P - 2F_P M_P + G_P L_P}{2 \cdot (E_P G_P - F_P^2)} \tag{1.50}$$

$$\mathcal{G}_P = k_{1.P} \cdot k_{2.P} = \frac{L_P N_P - M_P^2}{E_P G_P - F_P^2} \tag{1.51}$$

The expressions for the mean curvature $\mathcal{M}_P$ and for the *Gaussian* curvature $\mathcal{G}_P$:

$$\mathcal{M}_P = \frac{k_{1.P} + k_{2.P}}{2} \tag{1.52}$$

$$\mathcal{G}_P = k_{1.P} \cdot k_{2.P} \tag{1.53}$$

considered together yield a quadratic equation with respect to principal curvatures $k_{1.P}$ and $k_{2.P}$:

$$k_P^2 - 2\mathcal{M}_P k_P + \mathcal{G}_P = 0 \tag{1.54}$$

The following formulae

$$k_{1.P} = \mathcal{M}_P + \sqrt{\mathcal{M}_P^2 - \mathcal{G}_P} \tag{1.55}$$

$$k_{2.P} = \mathcal{M}_P - \sqrt{\mathcal{M}_P^2 - \mathcal{G}_P} \tag{1.56}$$

are the solutions to Equation 1.54.

Here, in Equations 1.55 and 1.56, the first principal curvature of the part surface P at a current point m is designated as $k_{1.P}$, and $k_{2.P}$ designates the second principal curvature of the part surface P at that same point m.

The principal curvatures $k_{1.P}$ and $k_{2.P}$ are the reciprocals to the corresponding principal radii of curvature $R_{1.P}$ and $R_{2.P}$:

$$k_{1.P} = \frac{1}{R_{1.P}} \tag{1.57}$$

$$k_{2.P} = \frac{1}{R_{2.P}} \tag{1.58}$$

The first principal curvature, $k_{1.P}$, is always larger than the second principal curvature, $k_{2.P}$, of the part surface P at a current point m—that is the inequality

$$k_{1.P} > k_{2.P} \tag{1.59}$$

is always valid.

This brief consideration of major elements of part surface geometry makes it possible the introduction of two the definitions that are of critical importance for further discussion.

Definition 1.1

Sculptured part surface P is a smooth, regular surface with major parameters of local geometry that differ in differential vicinity of any two infinitely closed points. ∎

It is instructive to point out here that sculptured part surface P does not allow sliding *"over itself"*.

While machining a sculptured part surface, the cutting tool rotates about its axis of rotation and is traveling relative to the sculptured part surface P. While rotating with a certain angular velocity ω_T or while performing relative cutting motion of another type (e.g., straight motion), the cutting edges of the form-cutting tool generate a certain surface. This surface can be interpreted as a loci of successive positions of the cutting edges of the cutting tool. We refer to that surface represented by successive positions of cutting edges as to *generating surface of the cutting tool* [112,117,136]:

Definition 1.2

The generating surface of a form-cutting tool is a surface represented as the set of successive positions of the cutting edges of the cutting tool in their motion relative to the stationary reference system, associated with the cutting tool itself. ∎

In most practical cases, the generating surface T of a form-cutting tool allows for sliding *over itself*. This means that the enveloping surface to consecutive positions of the surface T that is performing such a motion is congruent to the surface T itself. The screw surfaces of constant pitch, surfaces of revolution, and cylinders (not cylinders of revolution only) are perfect examples of surfaces that allow for sliding *over itself*.

When machining a part surface P, the generating surface T of the form-cutting tool is in permanent tangency to the sculptured part surface P.

As already mentioned earlier in this section of the book, it was proven by Bonnet [6] that the specification of the first and the second fundamental forms determines a unique surface if the *Gauss'* characteristic equation and the *Mainardi–Codazzi's* relations of compatibility are satisfied, and those two surfaces that have identical the first and the second fundamental forms are congruent.* Six fundamental magnitudes determine a surface uniquely, except as to position and orientation in space.

---

* Two surfaces with the identical first and second fundamental forms might also by symmetrical. Refer to the literature, J. J. Koenderink [30], on differential geometry of surfaces for details about this specific issue.

The specification of a surface in terms of the first and the second fundamental forms is usually called the *natural type* of surface representation. In general form, this type of part surface representation can be expressed by a set of two equations:

$$\text{The natural form of a} \atop \text{surface } P \text{ representation} \Bigg| \Rightarrow P = P(\Phi_{1.P}, \Phi_{2.P}) \begin{cases} \Phi_{1.P} = \Phi_{1.P}(E_P, F_P, G_P) \\ \Phi_{2.P} = \Phi_{2.P}(E_P, F_P, G_P, L_P, M_P, N_P) \end{cases}$$

(1.60)

Equation 1.60 can be derived from Equation 1.1. A given part surface P can be expressed in both forms, namely, either by Equation 1.19 or by Equation 1.1.

In further consideration, the natural representation of the part surface P plays an important role.

1.1.7 Illustrative Example

Consider an example of how an analytical representation of a surface in a *Cartesian* reference system can be converted into the natural representation of that same surface [117].

A *Cartesian* coordinate system $X_g Y_g Z_g$ is associated with a gear tooth flank $\mathcal{G}$ as it is schematically shown in Figure 1.6.

The position vector of a point, $\mathbf{r}_g$, of the gear tooth flank $\mathcal{G}$ can be represented as summa of three vectors

$$\mathbf{r}_g = \mathbf{A} + \mathbf{B} + \mathbf{C} \tag{1.61}$$

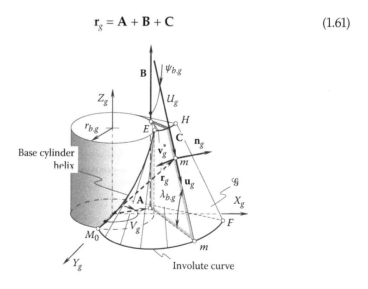

FIGURE 1.6
Derivation of the natural form of representation of a gear tooth flank, $\mathcal{G}$. (From Radzevich, S.P., *ASME J. Mech. Des.*, 124, 772–786, 2002. With permission.)

Each of the vectors **A**, **B**, and **C** can be expressed in terms of projections onto the axes of the reference system $X_g Y_g Z_g$. Then, Equation 1.61 casts into the equation

$$\mathbf{r}_g(U_g, V_g) = \begin{bmatrix} r_{b.g}\cos V_g + U_g \cos \tau_{b.g} \sin V_g \\ r_{b.g}\sin V_g - U_g \sin \tau_{b.g} \sin V_g \\ r_{b.g}\tan \tau_{b.g} - U_g \sin \tau_{b.g} \\ 1 \end{bmatrix} \tag{1.62}$$

This yields the calculation of two tangent vectors $\mathbf{U}_g(U_g, V_g)$ and $\mathbf{V}_g(U_g, V_g)$, which are respectively equal to

$$\mathbf{U}_g(U_g, V_g) = \begin{bmatrix} \cos \tau_{b.g} \sin V_g \\ -\cos \tau_{b.g} \cos V_g \\ -\sin \tau_{b.g} \\ 0 \end{bmatrix} \tag{1.63}$$

$$\mathbf{V}_g(U_g, V_g) = \begin{bmatrix} -r_{b.g}\sin V_g + U_g \cos \tau_{b.g} \cos V_g \\ r_{b.g}\cos V_g + U_g \cos \tau_{b.g} \sin V_g \\ r_{b.g}\tan \tau_{b.g} \\ 0 \end{bmatrix} \tag{1.64}$$

Substituting the derived vectors $\mathbf{U}_g$ and $\mathbf{V}_g$ into Equation 1.14, one can come up with formulae for the calculation of the fundamental magnitudes of the first order:

$$E_g = 1 \tag{1.65}$$

$$F_g = -\frac{r_{b.g}}{\cos \tau_{b.g}} \tag{1.66}$$

$$G_g = \frac{U_g^2 \cos^4 \tau_{b.g} + r_{b.g}^2}{\cos^2 \tau_{b.g}} \tag{1.67}$$

These expressions can be substituted directly to Equation 1.12 for the first fundamental form $\Phi_{1.g}$ of the gear tooth flank, $\mathcal{G}$:

$$\Phi_{1.g} \Rightarrow dU_g^2 - 2\frac{r_{b.g}}{\cos \tau_{b.g}} dU_g dV_g + \frac{U_g^2 \cos^4 \tau_{b.g} + r_{b.g}^2}{\cos^2 \tau_{b.g}} dV_g^2 \qquad (1.68)$$

The derived expressions for the fundamental magnitudes E_g, F_g, and G_g (see Equations 1.6 through 1.67) can also be substituted to Equation 1.13. In this way, a corresponding matrix representation of the first fundamental form $\Phi_{1.g}$ of the gear tooth flank, $\mathcal{G}$, can be calculated. The interested reader may wish to complete these formulae on his or her own.

The discriminant, H_g, of the first fundamental form of the gear tooth flank, $\mathcal{G}$, can be calculated from the expression:

$$H_g = U_g \cos \tau_{b.g} \qquad (1.69)$$

To derive an equation for the second fundamental form, $\Phi_{2.g}$, of the gear tooth flank, $\mathcal{G}$, the second derivatives of the position vector of a point, $\mathbf{r}_g(U_g, V_g)$, with respect to U_g and V_g parameters are necessary. The above derived equations for the tangent vectors $\mathbf{U}_g$ and $\mathbf{V}_g$ (see Equations 1.63 and 1.64) make possible the following expressions for the derivatives under consideration:

$$\frac{\partial \mathbf{U}_g}{\partial U_p} = \begin{bmatrix} 0 \\ 0 \\ 0 \\ 1 \end{bmatrix} \qquad (1.70)$$

$$\frac{\partial \mathbf{U}_g}{\partial V_g} \equiv \frac{\partial \mathbf{V}_g}{\partial U_g} = \begin{bmatrix} \cos \tau_{b.g} \cos V_g \\ \cos \tau_{b.g} \sin V_g \\ 0 \\ 1 \end{bmatrix} \qquad (1.71)$$

$$\frac{\partial \mathbf{V}_g}{\partial V_g} = \begin{bmatrix} -r_{b.g} \cos V_g - U_g \cos \tau_{b.g} \sin V_g \\ -r_{b.g} \sin V_g + U_g \cos \tau_{b.g} \cos V_g \\ 0 \\ 1 \end{bmatrix} \qquad (1.72)$$

Further, substitute these expressions (see Equations 1.70 through 1.72) into Equations 1.36 through 1.38. After the necessary formula transformations are complete, cast Equations 1.36 through 1.38 into the set of formulae for the calculation of the fundamental magnitudes of the second order of the gear tooth flank, $\mathcal{G}$. This set of formulae is as follows:

$$L_g = 0 \tag{1.73}$$

$$M_g = 0 \tag{1.74}$$

$$N_g = -U_g \sin \tau_{b.g} \cos \tau_{b.g} \tag{1.75}$$

Further, after substituting Equations 1.73 through 1.75 into Equation 1.31, an equation for the calculation of the second fundamental form of the gear tooth flank, $\mathcal{G}$, can be represented in the form:

$$\Phi_{2.g} \quad \Rightarrow \quad -d\mathbf{r}_g \cdot d\mathbf{N}_g = -U_g \sin \tau_{b.g} \cos \tau_{b.g} dV_g^2 \tag{1.76}$$

Similar to Equation 1.68, the derived expressions for the fundamental magnitudes L_g, M_g, and N_g of the second order can be substituted into Equation 1.32 for the second fundamental form $\Phi_{2.g}$. In this way, a corresponding matrix representation of the second fundamental form, $\Phi_{2.g}$, of the surface $\mathcal{G}$ can be derived. The interested reader may wish to complete this formulae transformation on his or her own.

For the calculation of the discriminant, T_g, of the second fundamental form, $\Phi_{2.g}$, of the gear tooth flank, $\mathcal{G}$, the following expression can be used:

$$T_g = U_g \sin \tau_{b.g} \cos \tau_{b.g} \tag{1.77}$$

The natural representation of the gear tooth flank, $\mathcal{G}$, can be expressed in terms of the derived set of six equations for the calculation of the fundamental magnitudes of the first E_g, F_g, G_g, and of the second L_g, M_g, and N_g (Table 1.1).

All the major elements of the local geometry of the gear tooth flank, $\mathcal{G}$, can be calculated based on the fundamental magnitudes, E_g, F_g, G_g, of the first, $\Phi_{1.P}$, and L_g, M_g, N_g, of the second $\Phi_{2.g}$, fundamental forms. The location and orientation of the gear tooth flank, $\mathcal{G}$, are the two parameters that remain indefinite.

TABLE 1.1

Fundamental Magnitudes of the First and the Second Order of the Gear Tooth Flank, $\mathcal{G}$

$E_g = 1$	$L_g = 0$
$F_g = -\dfrac{r_{b.g}}{\cos \tau_{b.g}}$	$M_g = 0$
$G_g = \dfrac{U_g^2 \cos^4 \tau_{b.g} + r_{b.g}^2}{\cos^2 \tau_{b.g}}$	$N_g = -U_g \sin \tau_{b.g} \cos \tau_{b.g}$

Once a part surface is represented in natural form—that is, it is expressed in terms of six fundamental magnitudes of the first and of the second order—then further calculation of parameters of the part surface P becomes much easier. To demonstrate significant simplification of the calculation of parameters of the part surface P, several useful equations are presented below as examples.

1.1.8 Few More Useful Equations

Many calculations of parameters of geometry can be significantly simplified by use of the first and of the second fundamental forms of a smooth, regular part surface P.

1. For the calculation of value of radius, R_P, of normal curvature within a normal plane section through a current point m on the part surface P and at a given direction, the following equation can be used:

$$R_P = \frac{\Phi_{1.P}}{\Phi_{2.P}} \qquad (1.78)$$

2. Euler's formula for the calculation of normal curvature, $k_{\theta.P}$, at a point m in a direction that is specified by the angle, θ, can be represented as follows:

$$k_{\theta.P} = k_{1.P} \cos^2\theta + k_{2.P} \sin^2\theta \qquad (1.79)$$

Here, in Equation 1.79, θ is the angle that the normal plane section, C_P, makes with the first principal plane section, $C_{1.P}$. In other words, $\theta = \angle(\mathbf{t}_P, \mathbf{t}_{1.P})$; here $\mathbf{t}_P$ designates the unit tangent vector within the normal plane section C_P.

Equation 1.79 also is a good illustration of significant simplification of the calculations when fundamental magnitudes, E_g, F_g, G_g, of the first and L_g, M_g, N_g of the second order are used.

1.1.9 Shape Index and Curvedness of the Surface at a Point

The shape index and the curvedness of the surface are two other useful properties that are also drawn from the principal curvatures at a surface point m.

The shape index, $S_{h.P}$, is a generalized measure of concavity and convexity. It can be defined [30] by

$$S_{h.P} = -\frac{2}{\pi} \arctan \frac{k_{1.P} + k_{2.P}}{k_{1.P} - k_{2.P}} \qquad (1.80)$$

The shape index varies within the interval from −1 to +1. It describes the local shape at a surface point m independent of the scale of the surface.

A shape index value of +1 corresponds to a concave local portion of the part surface P for which the principal directions are unidentified; thus, normal radii of curvature in all directions are identical.

A shape index of 0 corresponds to a saddle-like local portion of the part surface P with principal curvatures of equal magnitude but opposite sign.

The curvedness, $\mathcal{R}_{cv.P}$, at a surface point, m, is another measure derived from the principal curvatures [30]:

$$\mathcal{R}_{cv.P} = \frac{\sqrt{k_{1.P}^2 + k_{2.P}^2}}{2} \qquad (1.81)$$

The curvedness describes the scale of the part surface P independent of its shape.

The quantities $\mathcal{S}_{h.P}$ and $\mathcal{R}_{cv.P}$ are the primarily differential properties of the part surface P. Note that they are the properties of the surface itself and do not depend upon its parameterization except for a possible change of sign.

To get a profound understanding of differential geometry of surfaces, the interested reader may wish to go to advanced monographs in the field. Systematic discussion of the topic is available from many sources. The author would like to turn the reader's attention to the books by doCarmo [14], Eisenhart [15], Stuik [158], and others.

1.2 On the Difference between Classical Differential Geometry and Engineering Geometry of Surfaces

Classical differential geometry is developed mostly for the purpose of investigation of smooth, regular surfaces. Engineering geometry also deals with the surfaces. What is the difference between these two geometries?

The difference between *classical differential geometry of surfaces* and *engineering geometry of surfaces* is mostly due to how the surfaces are interpreted.

Only *imaginary* (*phantom*) surfaces are studied in classical differential geometry. Surfaces of this type do not exist in reality. They may be imagined as a film of an appropriate shape and with zero thickness. Such a film may be accessed from both sides of the surface. This causes the following indefiniteness.

As an example, consider a surface having positive *Gaussian* curvature, $\mathcal{G}_P$, at a surface point m: $\mathcal{G}_P > 0$. Classical differential geometry returns no answer to the question of whether the surface P at this point m is convex (in this case the surface mean curvature is of a positive value: $\mathcal{M}_P > 0$), or it is concave (in this case the surface mean curvature is of a negative value: $\mathcal{M}_P < 0$).

In classical differential geometry, the answer to this question may be given only by convention.

A similar observation is made when *Gaussian* curvature, $\mathscr{G}_p$, at a certain surface point m is negative: $\mathscr{G}_p < 0$.

Surfaces in classical differential geometry strictly follow the equation they are specified by. No deviation of the surface shape from what is

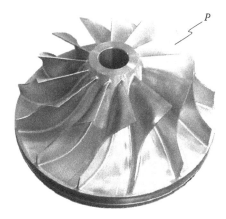

FIGURE 1.7
Example of a surface that mechanical engineers are dealing with. (Reprinted from *Computer-Aided Design*, 37, Radzevich, S.P., A cutting-tool-dependent approach for partitioning of sculptured surface, 767–778, Copyright 2005, with permission from Elsevier.)

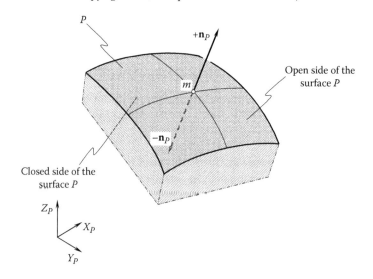

FIGURE 1.8
Open and closed sides of a part surface P. (Reprinted from *Computer-Aided Design*, 34, Radzevich, S.P., Conditions of proper sculptured surface machining, 727–740, Copyright 2002, with permission from Elsevier.)

predetermined by the equation is allowed. More examples can be found in following chapters of the book.

In turn, surfaces that are treated in engineering geometry, bound a part (or a machine element). This part can be called a real object (Figure 1.7). The real object is the bearer of the surface shape.

Surfaces that bound real objects are accessible from only one side as is schematically illustrated in Figure 1.8. We refer to this side of the surface as the *open side of a surface*. The opposite side of the surface P is not accessible. Because of this, we refer to the opposite side of the surface P as the *closed side of a surface*.

The positively directed unit normal vector $+\mathbf{n}_p$ is pointed outward from the part body—that is, from its bodily side to the void side. The negative unit normal vector $-\mathbf{n}_p$ is pointed opposite to $+\mathbf{n}_p$. The existence of open and closed sides of a part surface P eliminates the problem of identifying whether the surface is convex or concave. No convention is required in this concern.

Another principal difference is due to the nature of the real object.

No real object can be machined or manufactured precisely without deviations of its actual shape from the desired shape of the real object. Smaller or larger deviations of shape of the real object from the desired shape of it are inevitable in nature. We do not go into detail here about this concern, as this could be a subject for another book.

Because of the deviations, the actual part surface P^{act} deviates from its nominal surface P as is illustrated in Figure 1.9. However, the deviations do not exceed a reasonable prespecified range. Otherwise, the real object will become useless. In practice, the selection of appropriate tolerances on shape and dimensions of the actual surface P^{act} easily solve this particular problem.

Similar to measuring deviations, the tolerances are measured in the direction of the unit normal vector, $\mathbf{n}_p$, to the nominal part surface P. Positive tolerance, δ^+, is measuring along the positive direction of the unit normal vector

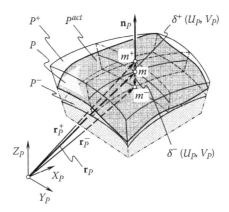

FIGURE 1.9
The analytical description of an actual part surface P^{act} that is located in between the boundary surfaces P^+ and P^-.

$+\mathbf{n}_P$, and negative tolerance, δ^-, is measuring along the negative direction of the unit normal vector, $-\mathbf{n}_P$. In a particular case, one of the tolerances, either the upper tolerance, δ^+, or the lower tolerance, δ^-, can be equal to zero.

Often, values of the upper tolerance δ^+, and of the lower tolerance δ^- are constant within the entire patch of the part surface P. However, in special cases, for example when machining a sculptured part surface on a multiaxis numerical control machine, the actual value of the tolerances δ^+ and δ^- may be set as functions of coordinates of current point m on the part surface P. This results in the tolerances being represented in terms of curvilinear coordinates, U_P and V_P, of a point m on the part surface P, say in the form

$$\delta^+ = \delta^+(U_P, V_P) \tag{1.82}$$

$$\delta^- = \delta^-(U_P, V_P) \tag{1.83}$$

The endpoint of the vector $\delta^+ \cdot \mathbf{n}_P$ at a current part surface point m produces a point m^+. Similarly, the endpoint of the vector $\delta^- \cdot \mathbf{n}_P$ produces the corresponding point m^-.

The bounding surface, P^+, of the upper tolerance is represented by loci of the points m^+ (i.e., by loci of endpoints of the vector $\delta^+ \cdot \mathbf{n}_P$). This makes possible an analytical expression for the position vector of a point $\mathbf{r}_P^+$ of the bounding surface of the upper tolerance P^+ in the form:

$$\mathbf{r}_P^+(U_P, V_P) = \mathbf{r}_P + \delta^+ \cdot \mathbf{n}_P \tag{1.84}$$

Usually, the bounding surface P^+ of the upper tolerance is located above the nominal part surface P, that is, from the void side of the surface P.

Similarly, the bounding surface, P^-, of the lower tolerance is represented by loci of the points m^- (i.e., by loci of endpoints of the vector $\delta^- \cdot \mathbf{n}_P$). This makes possible an analytical expression for the position vector of a point $\mathbf{r}_P^-$ of the bounding surface of the lower tolerance P^- in the form:

$$\mathbf{r}_P^-(U_P, V_P) = \mathbf{r}_P + \delta^- \cdot \mathbf{n}_P \tag{1.85}$$

In common practice, the bounding surface, P^-, of the lower tolerance is located beneath the nominal part surface P, that is, from the bodily side of the surface P.

The actual part surface, P^{act}, cannot be represented analytically.* Moreover, the above-considered parameters of local topology of the part

---

* Actually, the surface P^{act} is unknown—any surface that is located within the gap between the surfaces of upper tolerance, P^+, and lower tolerance, P^-, satisfies the requirements of the part surface blueprint; thus, every such surface may be considered as an actual part surface, P^{act}. An equation of the part surface P^{act} cannot be represented in the form $P^{act} = P^{act}(U_P, V_P)$, because the actual value of the deviation, δ^{act}, at a current surface point is not known. CMM data (coordinate-measuring machine data) yields only an approximation for the deviation δ^{act}, as well as the corresponding approximation for the part surface P^{act}.

surface P cannot be calculated for the surface P^{act}. However, because the tolerances δ^+ and δ^- are small compared with the normal radii of curvature of the nominal part surface P, it is assumed here and below that the actual part surface P^{act} possesses those same geometrical properties as the nominal part surface P does, and that the difference in corresponding geometrical parameters of the surfaces P^{act} and P is negligibly small, and thus, it may be ignored. In further consideration, this makes possible the replacement of the actual part surface P^{act} with the nominal part surface P, which is much more convenient for performing the required calculations.

The consideration above in this section of the book illustrates the second principal difference between classical differential geometry and the engineering geometry of surfaces. Because of the differences, engineering geometry often presents problems that were not envisioned in classical (pure) differential geometry of surfaces.

1.3 On the Classification of Surfaces

The number of different surfaces that bound real objects is infinitely large. A systematic consideration of surfaces for the purposes of part surface generation is of critical importance.

1.3.1 Surfaces That Allow Sliding over Themselves

In industry, a small number of surface types featuring relatively simple geometry are in wide use. Surfaces of this particular type allow for *sliding over themselves*. The property of a surface that allows sliding over itself means that for a certain part surface P, a special type of corresponding motion exists. When performing this motion, the enveloping surface to the consecutive positions of the moving surface P is congruent to the surface P itself. The motions of the type mentioned can be either monoparametric, or biparametric, or, finally, triparametric motions.

A screw surface of constant axial pitch ($p_x = Const$) is the most general type of surface that allow for sliding over itself. When performing the screw motion of that same axial pitch, p_x, the part surface P is sliding over itself, similar to the "bolt-and-nut" pair. In the case under consideration, the screw surface of the bolt is congruent to the screw surface of the nut. When the bolt is performing the screw motion, the screw surfaces of the bolt and of the nut are sliding over one another.

When the axial pitch of a screw surface reduces to zero ($p_x = 0$), then the screw surface degenerates to the corresponding surface of revolution. Every surface of revolution is sliding over itself when rotating.

When axial pitch of a screw surface increases to an infinitely large value, then the screw surface degenerates into a general type of cylinder (not mandatory to a cylinder of revolution). Surfaces of this particular type allow for straight motion along straight generating lines of the surface.

The considered types of part surface motion are listed below:

(a) A screw motion of constant axial pitch ($p_x = Const$)
(b) A rotation about the centerline of the surface of revolution
(c) A straight motion correspondingly

All of these three motions are monoparametric motions.

Surfaces, such as that of a round cylinder, allow for rotation as well as for straight motion along the centerline of the cylinder. In this particular case, the surface motion is an example of biparametric (both rotation and translation of the round cylinder can be performed independently) motion of the surface.

A sphere allows for rotations about three axes independently. A plane surface allows for straight motion in two different directions as well as for rotation about an axis that is perpendicular to the plane. Again, all three of these motions may be performed independently from one another. The surface motions in the last two cases (for a sphere and for a plane) are perfect examples of triparametric motion of a part surface P.

Ultimately, one can summarize that surfaces allowing for sliding over themselves are limited to screw surfaces of constant axial pitch, general types of cylinders, surfaces of revolution, circular cylinders, spheres, and planes. It has been proven [116,117,136,147] that there are no other types of surfaces that allow for sliding over themselves.

Surfaces that allow for sliding over themselves have proven to be very convenient in engineering applications, in manufacturing as well as in other industrial applications. Most of the surfaces being machined in various industries are examples of surfaces of this nature.

1.3.2 Sculptured Surfaces

Many products are designed with aesthetic sculptured surfaces to enhance their aesthetic appeal, an important factor in customer satisfaction, especially for the automotive and consumer-electronics products. In other cases, products have sculptured surfaces to meet functional requirements. Examples of functional surfaces can be easily found in aerodynamic, gas dynamic, and hydrodynamic applications (turbine blades), optical (lamp reflector), and medical (parts of anatomic reproduction) applications, manufacturing surfaces (molding die, die face), and so forth. Functional surfaces interact with the environment or with other surfaces. Due to this, functional surfaces can also be called *dynamic surfaces*.

A functional surface does not possess the property to slide over itself. This causes significant complexity in the machining of sculptured part surfaces.

The application of a multiaxis *NC* machine is the only practical way to efficiently machine-sculpt part surfaces.

In the most general case, at every instant of part surface machining on a multiaxis *NC* machine, the sculptured part surface being machined and the generating surface of the form-cutting tool make point contact with each other. To develop an advanced technology of sculptured part surface generation, a comprehensive understanding of the local topology of a sculptured part surface is highly desired.

1.3.3 Circular Diagrams

For the purpose of precisely describing the local topology of a part surface *P*, circular diagrams* can be implemented. Circular diagrams are a powerful tool for the analysis and in-depth understanding of the topology of a sculptured part surface. A circular diagram reflects the principal properties of a sculptured surface in differential vicinity of a part surface point.

Euler's formula for normal part surface curvature,[†]

$$k_{\theta.P} = k_{1.P} \cos^2\theta + k_{2.P} \sin^2\theta \tag{1.86}$$

together with *Germain's*[‡] equation (or *Bertrand's*[§] equation in other interpretations),

$$\tau_{\theta.P} = (k_{2.P} - k_{1.P})\sin\theta \cos\theta \tag{1.87}$$

are the foundation of circular diagrams at a point of a sculptured part surface *P*.

Here, in Equation 1.87, the torsion of a curve through a surface point *m* in the direction specified by the value of angle *θ*, is designated as $\tau_{\theta.P}$.

An example of graphical interpretation of the function $k_{\theta.P} = k_{\theta.P}(\theta)$ (see Equation 1.86), and of the function $\tau_{\theta.P} = \tau_{\theta.P}(\theta)$ (see Equation 1.87) is illustrated in parts a and b, respectively, of Figure 1.10.

---

* Initially proposed by Christian Otto Mohr (October 8, 1835–October 2, 1918), a German civil engineer, for the purpose of solving problems in the field of strength of materials, circular diagrams later gained wider application. The origin of application of circular diagrams for the purposes of differential geometry of surfaces can be traced back to the publication by Miron [40] and Vaisman [160]. Lowe [36,37] applied circular diagrams in studying surface geometry with special reference to twist, as well as in developing plate theory. A profound analysis of properties of circular diagrams can be found out in publications by Nutbourn [46], and Nutbourn and Martin [47]. The application of circular diagrams in the field of sculptured part surface machining on a multiaxis *NC* machine is known from the monograph by Radzevich [117].

[†] Leonhard Euler (April 15, 1707–September 18, 1783), a famous Swiss mathematician and physicist who spent most of his life in Russia and Germany.

[‡] Marie-Sophie Germain (April 1, 1776–June 27, 1831), a French mathematician, physicist, and philosopher.

[§] Joseph Louis François Bertrand (March 11, 1822–April 5, 1900), a French mathematician.

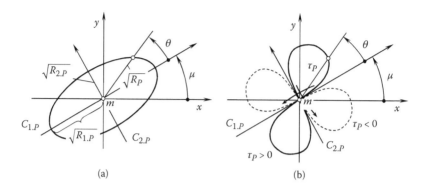

FIGURE 1.10
Radius of normal surface curvature, $k_P(\theta)$ (Equation 1.86) (a), and torsion, $\tau_P(\theta)$ (Equation 1.87) (b), of a surface curve at a point m of a part surface P.

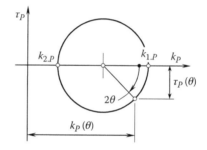

FIGURE 1.11
An example of a circular diagram constructed for a convex elliptic local patch of a smooth, regular sculptured part surface P.

An example of a circular diagram that is constructed for a convex local patch of elliptic type is schematically illustrated in Figure 1.11. It is important to point out here that due to the algebraic value of the first principal curvature, $k_{1.P}$ (that is, the inequality $k_{1.P} > k_{2.P}$ is always observed), the algebraic value of the second principal curvature, $k_{2.P}$, of a sculptured part surface P, is always greater the circular diagram point with coordinates $(0, k_{1.P})$ is always located at the far right relative to the circular diagram point with coordinates $(0, k_{2.P})$.

The application of circular diagrams enables one to easily identify the type of local part surface patch* in the differential vicinity of a current point m on a sculptured part surface.

---

* We refer to a sculptured surface area in differential vicinity of a surface point m as the *local part surface patch*, and not as a *surface point*. The name of local part surface patch corresponds to the name of surface point. However, from the standpoint of part surface machining, usage of the term "local surface patches" instead of "surface points" is preferred.

Circular diagrams of convex ($\mathcal{M}_p > 0$, $\mathcal{G}_p > 0$) and concave ($\mathcal{M}_p < 0$, $\mathcal{G}_p > 0$) local surface patches of elliptic type are depicted in Figure 1.12a. In this particular case, the circular diagrams are represented by two circles, which are remote at a certain distance from the τ_p axis. The radii of the circles are equal to half a summa of the part surface principal curvatures $k_{1.p}$ and $k_{2.p}$ at a specified surface point m.

For local part surface patches of umbilic type, principal directions at a surface point m are not identified. Thus, values of normal curvatures in all directions through the surface point m are identical. This means that for local part surface patches of umbilic type, circular diagrams shrink into points as is shown in Figure 1.12b. The coordinates of the points are

- $(k_p > 0, 0)$ for convex ($\mathcal{M}_p > 0$, $\mathcal{G}_p > 0$) local surface patch
- $(k_p < 0, 0)$ for concave ($\mathcal{M}_p < 0$, $\mathcal{G}_p > 0$) surface patch

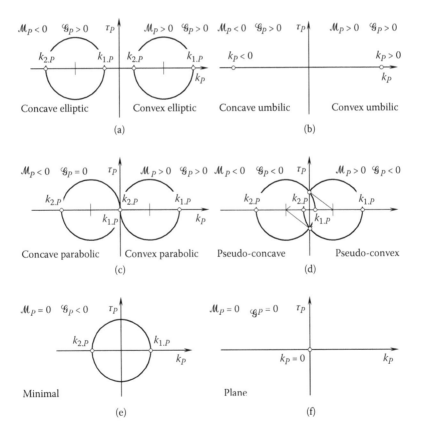

FIGURE 1.12
Circular diagrams of smooth, regular local patches of a sculptured part surface P.

Circular diagrams of convex ($\mathcal{M}_p > 0$, $\mathcal{G}_p = 0$), and of concave ($\mathcal{M}_p < 0$, $\mathcal{G}_p = 0$) local part surface patches of parabolic type are passing through the origin of the coordinate system $k_p\tau_p$ as is illustrated in Figure 1.12c.

It is recommended that quasi-convex ($\mathcal{M}_p > 0$) and quasi-concave ($\mathcal{M}_p < 0$) local part surface patches be distinguished when they are of hyperbolic type—that is, for saddle-like local part surface patches ($\mathcal{G}_p < 0$). Circular diagrams of saddle-like local part surface patches intersect τ_p axis. This is schematically depicted in Figure 1.12d.

A particular case of saddle-like local part surface patches is distinguished for the part surface patches of hyperbolic type ($\mathcal{G}_p < 0$) of zero mean curvature ($\mathcal{M}_p = 0$). Part surface local patches of this particular type are referred to as *minimal* surface local patches. An example of a circular diagram of minimal local part surface patch is shown in Figure 1.12e.

Finally, the circular diagram of a part surface patch of planar type ($\mathcal{M}_p = 0$, $\mathcal{G}_p = 0$) degenerates into the point that coincides with the origin of the coordinate system $k_p\tau_p$. A circular diagram of part surface patch of this particular type is shown in Figure 1.12f. All local patches of a plane may be considered as local part surface patches of parabolic-umbilic type.

As follows from the analysis of Figure 1.12, the circular diagram clearly illustrates the major local properties of a sculptured part surface geometry. Principal curvatures, normal curvatures, and torsion of a curve on the part surface may be easily seen from the corresponding circular diagram.

Moreover, actual values of mean, $\mathcal{M}_p$, and *Gaussian*, $\mathcal{G}_p$, curvatures may also be gained from the circular diagram. Examples of how the mean, $\mathcal{M}_p$, and the *Gaussian*, $\mathcal{G}_p$, curvatures can be constructed are shown in Figure 1.13. These examples are given for convex and concave local part surface patches of elliptic type shown in Figure 1.12a as well as for quasi-convex and quasi-concave saddle-like local part surface patches depicted in Figure 1.12b.

The above performed analysis makes possible a conclusion that the circular diagram is a simple characteristic image that provides the researcher

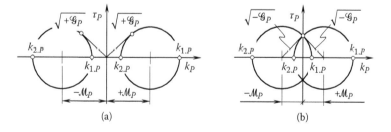

FIGURE 1.13
Geometric interpretation of mean $\mathcal{M}_p$ and of full $\mathcal{G}_p$ (*Gaussian*) curvature at a point of a smooth, regular sculptured part surface P.

with comprehensive information on the local topology of the part surface. This information includes

- Principal curvatures, $k_{1.P}$ and $k_{2.P}$, at a part surface point m
- Normal curvature, k_P, in a specified direction through the part surface point m
- Extremum values of torsion, $\tau_P^{\min}$ and $\tau_P^{\max}$, of a curve through the part surface point m
- Torsion of a curve, τ_P, in a specified direction through the part surface point m
- Mean curvature, $\mathcal{M}_P$, at a part surface point m, and
- Full (*Gaussian*) curvature, $\mathcal{G}_P$, at a part surface point m

No other characteristic image of that simple nature (as the circular diagram is) provides the researcher with such comprehensive information on local topology at a point of a sculptured part surface. Use of the circular diagrams is helpful for solving plurality of problems in the theory of part surface generation.

One such problem pertains to the classification of surfaces. The classification of sculptured part surfaces is necessary for developing efficient technology of part surface machining on a multiaxis *NC* machine. We will take a brief look at surface classification from this point of view.

Sculptured part surfaces are geometrical objects of complex nature. It is shown [30] that no scientific classification of sculptured part surfaces can be developed. It is recommended [30] that classification of local part surface patches be considered instead of classification of sculptured part surface as no classification of sculptured part surface in global is feasible. The recommendation is base on the consideration of point contact that a sculptured part surface makes with the generating surface of the form-cutting tool when machining.

Figure 1.14 is insightful for understanding the relationship between local part surface patches of different types. The shift of a circular diagram along k_P axis in $\mathcal{V}_-^+$, or in $\mathcal{V}_+^-$ direction reflects a corresponding change of shape and type of local part surface patches. On the premises of these changes, a classification of local part surface patches can be worked out.

Depicted in Figure 1.14a, the initial circular diagram represents a convex elliptic local patch with certain values of the part surface principal curvatures. Due to a shift in $\mathcal{V}_+^-$ direction, the shape of the convex elliptic surface patch changes; however, to some extent, it remains of elliptic type. When the circular diagram passes through the origin of the coordinate system $k_P \tau_P$, this leads to a dramatic change in the shape of the local part surface patch. Instead of being of elliptic type, it transforms to a local part surface patch of the parabolic type.

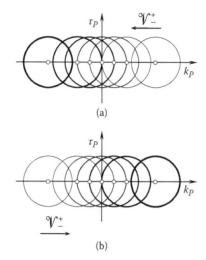

FIGURE 1.14
Various locations of the circular diagram correspond to local part surface patches of different types.

A further shift of the circular diagram in the $\mathcal{V}_+^-$ direction results in the consequent change of shape of the local part surface patch to

(a) A pseudo-convex saddle-like local part surface patch
(b) A minimum saddle-like local part surface patch

and so on up to a concave elliptic local part surface patch of certain values of principal curvatures at a specified point m on the surface.

Similar changes in shape and type of local part surface patch are observed when the circular diagram shifts in the direction of $\mathcal{V}_+^-$, which is opposite to the direction of $\mathcal{V}_+^-$ (Figure 1.14b).

Figure 1.14 can be enhanced to a three-dimensional case. For this purpose, k_P and τ_P axes are complemented with d_k axis as illustrated in Figure 1.15. Here, d_k is a parameter that specifies diameter of the circular diagram at a part surface P current point m ($d_k = k_{1.P} - k_{2.P}$). As an example, the diagram shown in Figure 1.14a is plotted in Figure 1.15. As the circular diagram is centering at a point that is traveling, $\mathcal{V}_+^-$, parallel to k_P axis, corresponding circular diagrams reflect local geometry of part surface patches of the following types:

(a) Convex elliptical
(b) Convex parabolic
(c) Pseudo-convex hyperbolic
(d) Minimal

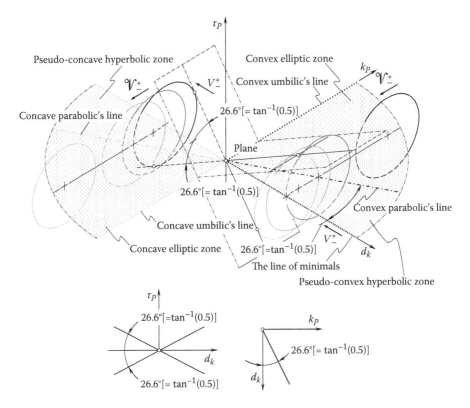

FIGURE 1.15
A diagram that illustrates variation of the geometry of a local patch of a smooth, regular part surface P when the circular diagram is centering at different points within the coordinate plane $k_P d_k$.

(e) Pseudo-concave hyperbolic

(f) Concave parabolic, and

(g) Concave elliptic

It should be pointed out here that circular diagram at a part surface P current point m can be centering at any point within two quadrants shaded in Figure 1.15. So, in addition to traveling in the directions $\mathcal{V}_+^+$ and $\mathcal{V}_-^-$, traveling in the directions $\mathcal{V}_-^+$ and $\mathcal{V}_+^-$ is permissible as well. The directions of traveling $\mathcal{V}_-^+$ and $\mathcal{V}_+^-$ are parallel to d_k axis of the reference system $k_P d_k \tau_P$.

A performed analysis of the schematic shown in Figure 1.15 makes possible the following conclusions.

First, several special boundary lines can be noticed within the coordinate plane $k_P d_k$.

Umbilic's line is one of the said special lines. This line can be interpreted as comprising two portions, namely, of *convex umbilic's line* (for the points having coordinates $\tau_p = 0$, $d_k = 0$, $0 < k_p < +\infty$) and *concave umbilic's line* (for the points having coordinates $\tau_p = 0$, $d_k = 0$, $-\infty < k_p < 0$). All points within umbilic's line correspond to local surface patches of umbilic type having different value of normal curvature k_p.

The line of minimals is the second of the said special lines. This boundary line is aligned with d_k axis of the reference system $k_p d_k \tau_p$. The can be split in two portions. Points having coordinates ($\tau_p = 0$, $0 < d_k < +\infty$, $k_p = 0$) are located within the first portion of the line of umbilic's. Accordingly, points having coordinates ($\tau_p = 0$, $-\infty < d_k < 0$, $k_p = 0$) are located within the second portion of the line of umbilic's as is schematically depicted in Figure 1.15.

Parabolic's line is the third of the said special lines. This line can be also interpreted as comprising two portions, namely, of *convex parabolic's line* [for the points having coordinates $\tau_p = 0$, $d_k = +k_p \cot^{-1}(0.5)$] and *concave parabolic's line* [for the points having coordinates $\tau_p = 0$, $d_k = -k_p \cot^{-1}(0.5)$]. All points within umbilic's line correspond to local surface patches of parabolic type having zero mean curvature ($M_p = 0$).

The said boundary straight lines form several sectors within the coordinate plane $k_p d_k$ within which circular diagrams for part surface of different types are centering. All of these zones, namely,

- Convex elliptic zone
- Pseudo-convex hyperbolic zone
- Concave elliptic zone
- Pseudo-concave hyperbolic zone

are depicted in Figure 1.15. As follows from the name of a zone, circular diagrams for convex local surface patches of elliptic type are centering within the convex elliptic zone. The same is valid with respect to three other zones listed above.

The circular diagram for a plane is centering at the origin of the coordinate system $k_p d_k \tau_p$. In this particular case, the circular diagram is shrunk to a point that is coincident with origin of the coordinate system $k_p d_k \tau_p$.

When the center of a circular diagram is traveling within the coordinate plane $k_p d_k$, and during such travel it is approaching the origin of the coordinate system $k_p d_k \tau_p$, local part surface patch is getting more and more flattened. This is observed for local surface patches of all types: for local part surface patches of umbilic, elliptic, parabolic type, etc. When approaching the origin of the coordinate system $k_p d_k \tau_p$ the corresponding circular diagram reduces in diameter, and ultimately it shrinks to a point that is coincident with origin of the coordinate system $k_p d_k \tau_p$. This means that plane can be interpreted as a degenerate case not only of umbilics but as a degenerate case of all possible types of local part surface patches as well.

1.3.4 Circular Chart for Local Patches of Smooth, Regular Part Surfaces

To understand the relationship between the local part surface patches of different types, it is convenient to also consider the shift of a circular diagram together with change of ratio between principal curvatures, $k_{1.P}/k_{2.P}$, at a part surface point m. Following this, one can come up with the idea of circular distribution* of the circular diagrams.

As circular diagrams properly reflect the geometry of local part surface patches, a circular chart for local patches of smooth regular part surfaces can be constructed on the premise of the circular diagrams. An example of such a circular chart is depicted in Figure 1.16.

Transition from a local surface patch (Figure 1.16) to a nearby local surface patch either in the radial direction of the circular chat, or circumferentially shows how normal curvatures of the local surface patch change.

The implementation of circular diagrams of a smooth regular part surface P makes it easier to tell the difference between local surface patches of different types.

1.3.5 On Classification of Local Patches of Smooth, Regular Part Surfaces

Local patches of sculptured part surfaces can be classified. Figure 1.16 is helpful for understanding the local topology of the sculptured part surface being machined. It also yields a scientific classification of local patches of smooth regular part surface P (Figure 1.17). The classification includes 10 types of local surface patches and is an accomplished one.

Based on the analysis of sculptured part surface geometry as well as on classification of local surface patches (Figure 1.17), a profound scientific classification of all feasible types of local surface patches is developed by Radzevich [110,116]. It has been proven that total number of feasible types of local part surface patches is limited (and is not infinite). Hence, local part surface patches of every type can be investigated separately. No part surface patches would be left out of consideration. This indicates that the problem of synthesis of optimal machining operation of a specified sculptured part surface is solvable. (See reference [110] for details on the developed scientific classification of local part surface patches and its use in the theory of part surface generation.)

---

* Reading of the monograph by J. J. Koenderink [30] inspired the author to implement the circular disposition of the *circular diagrams* of local part surface patches to the needs of the theory of part surface generation. In this way, a circular chart for local part surface patches is created. In his monograph (see Figure 268 on page 321 in [30]), Koenderink proposed to distribute 9 images of local surface patches circumferentially. This was done for illustrating the relationship among local surface patches of different geometries. The proposed circular chart for local part surface patches is significantly enhanced compared with that done by J. J. Koenderink.

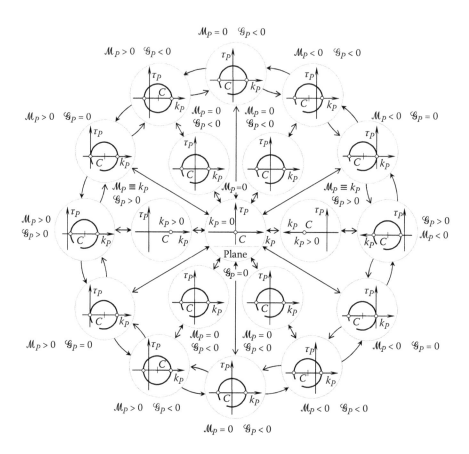

FIGURE 1.16
Circular chart comprising circular diagrams for ten different types of local patches of a real smooth, regular part surface *P*.

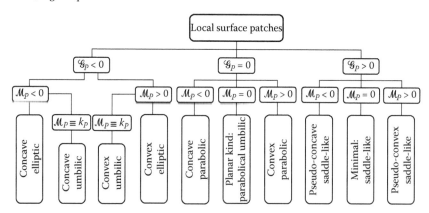

FIGURE 1.17
Ten types of local patches of a real smooth, regular sculptured part surface *P*.

2

Principal Kinematics of Part Surface Generation

A motion of the form-cutting tool relative to the work is necessary for machining a given part surface on the machine tool. Relative motions of the work and of the form-cutting tool are necessary for the generation of a part surface being machined. No machining operation can be performed without a certain relative motion of the work or the cutting tool. The work and the cutting tool relative motion are observed for all methods of part surface machining on a machine tool.

The work and the cutting tool relative motions of different natures are observed. The following relative motions, among others, may be commonly recognized:

- A setup motion of the work and of the cutting tool. Motions of that type are required for proper positioning of the form-cutting tool relative to the work being machined in its initial position prior to start of the machining operation.
- Cutting motions are necessary for the removal of stock out of workpiece.
- Feed-rate motions are required for continuation of the process of stock removal.
- The part surface-generating motions. Motions of this type are required for generating the entire part surface being machined.
- Orientational motions of the form-cutting tool.

In addition to those listed above, relative motions of other types may also be distinguished in particular cases.

In this chapter, the generation motions, feed-rate motions, and orientational motions of the form-cutting tool are discussed.

Cutting tools are used for machining various part surface shapes and geometries. When machining, the cutting wedge of the cutting tool cuts off a stock from the work. The shape and parameters of the machined part surface may be considered as a result of interaction of the work and of the cutting edge of a certain shape, which is performing a certain motion relative to the work. This means that the kinematics of the part surface machining operation directly affects the shape, accuracy, and surface quality of the machined part surface. Kinematic structure of a conventional machine tool is pre-determined by the required kinematics of a machining operation. The exact requirements for an appropriate code for controlling a multiaxis numerical control (NC) machine are also established by the required kinematics of a machining operation.

To synthesize the most efficient machining operation of a specified part surface, it is necessary to determine the optimal parameters of the relative motion of the work and of the cutting tool at every instant of the machining operation. A corresponding criterion for the synthesis is of critical importance when synthesizing the most favorable methods and means for machining a given part surface.

The relative motion of the work and of the cutting tool can be represented as a superposition of a certain number of elementary motions that are performed by the machine tool of a certain design. The mentioned elementary motions are the translations along and the rotations about various axes that are differently oriented relative to one another.

Ultimately, a combination of a certain number of the translations and the rotations can result in a complex resultant relative motion of the work and of the cutting tool. Elementary motions generated by cams, copiers, and computer codes can be incorporated into the resultant motion of the work and of the cutting tool as well. All the elementary motions are timed (synchronized) with one another in a timely proper manner.

Consider a simple example of drilling a hole in a drilling press. To drill the hole, it is necessary to rotate a twist drill, and feed it into the work. A combination of the rotation and of the translation of the twist drill results in the hole being machined in the work.

The machining of that same hole may be performed under another scenario. The twist drill is rotated as in the previous machining operation. However, not the twist drill, but the worktable with the work is feeding into the rotating twist drill. The combination of the rotation of the twist drill and of the translation of the worktable, which is carrying the work, results in the hole being machined in the work.

The drilling of that same hole in the part may be performed on a lathe. Under such a scenario, the work is rotating about its axis, and the twist drill is feeding into the rotating work. This also results in the hole being machined in the work.

The possible combinations of the translation and of the rotation of the work and of the twist drill are not limited to the three scenarios just considered. However, it is important to point out here that in all cases of drilling the hole, the relative motion of the work and of the twist drill remains the same. This motion is represented with a screw relative motion regardless of the absolute motions* performed by the twist drill, and of the absolute motions performed by the work itself.

---

* The absolute motions of the cutting tool are those performed in relation to the machine tool body.

The example reveals that even in cases of machining a relatively simple part surface (the surface of the hole in the particular case under consideration), the number of possible combinations of elementary relative motions in an actual machining operation can be large. In the case of machining a sculptured part surface on a multiaxis *NC* machine, the number of possible combinations of elementary motions increases dramatically. Moreover, parameters of the elementary motions may vary in time.

The *differential geometry/kinematics* (*DG/K*) approach of part surface generation is based on the consideration of the *relative motion* of the work and of the cutting tool regardless of the type of elementary absolute motions that are performing the work and the cutting tool. Use of the concept of *relative motion* makes possible calculation of the optimal parameters of the desired relative motion of the form-cutting tool in relation to the work. After being calculated, the optimal relative motion of the form-cutting tool in relation to the work may be further decomposed into elementary absolute motions that the machine tool of a certain design is capable of executing.

The machining of a sculptured part surface on a multiaxis *NC* machine represents the most general case of kinematics of part surface machining. Use of *NC* makes any desired relative motion of the work and of the cutting tool possible. Therefore, it is reasonable to begin the discussion of the kinematics of part surface generation from consideration of the most general case of part surface machining—that is, from the case of machining a sculptured part surface on a multiaxis *NC* machine.

2.1 Kinematics of Sculptured Part Surface Generation

For machining of sculptured part surfaces on a multiaxis *NC* machine, cutting tools of relatively simple design are commonly used. When machining a sculptured part surface, the cutting tool is performing a complex motion relative to the work. This relative motion may be interpreted as a summa of numerous instant translations and rotations along and about the instant axes. Due to the complex geometry of sculptured part surfaces, the relative translations as well as the relative rotations vary in time. Because of this, not a kinematic scheme of part surface generation, but a principal *instant kinematics* of sculptured part surface generation has to be discussed instead.

The kinematics of sculptured part surface generation features *instant* values of the relative motions of the cutting tool and of the work. Accelerations and decelerations, both linear and angular, are inherent in the instant kinematics of sculptured part surface generation.

To investigate the principal instant kinematics of part surface generation, it is necessary to specify the sculptured part surface *P* to be machined. It is assumed below that the equation of the part surface *P* is represented either in vector form $\mathbf{r}_P = \mathbf{r}_P(U_P, V_P)$, or in matrix form (see Equation 1.1). In cases when the part surface P is specified in some other manner, conversion of actual specification of the part surface into it, parameterization in terms of U_P and V_P parameters is required in addition.

It is also assumed that the generating surface T of a form-cutting tool is specified. This means that the equation of the tool surface T is represented either in vector form $\mathbf{r}_T = \mathbf{r}_T(U_T, V_T)$, or in matrix form.

The analysis of kinematics of the multiparametric sculptured part surface generation is based on the presumption that any desired relative motion of the form-cutting tool in relation to the work can be executed on the *NC* machine.

Without loss of generality, the analysis of kinematics of sculptured part surface generation can be substantially simplified if the *principle of inversion of motions* is implemented. To implement this fundamental principle of mechanics for the investigation of the part surface generation process, an additional motion is applied to both the work and the cutting tool. The speed and direction of the additional motion is equal to the speed of the motion that the work is performing in a certain machining operation. The direction of the additional motion is reversed to the motion that the work is performing in the machining operation. When summarized with the initial motions, then the work becomes stationary. However, the resultant motion of the cutting tool in relation to the part gets more complex. This motion is composed now of the initial motions of the cutting tool and the reversed motions of the work.

Implementation of the principle of inversion of motions has proven to be convenient for the investigation of the multiparametric part surface generation process.

2.1.1 Establishment of a Local Reference System

A reference coordinate system is necessary for the development of an analytical description of the instant motion of the cutting tool relative to the work. A right-hand-oriented *Cartesian* coordinate system is used below as the local reference system. Several approaches can be employed to get the local coordinate system orthogonal.

First, a convenient way to establish an appropriate local coordinate system is to use a coordinate system that is naturally embedded to the part surface P. For this purpose, *Darboux*[*] frame is commonly used (see Figure 1.4).

Darboux frame is a type of frame associated with a point on a part surface P to be machined, and defined by three vectors, namely, by the unit normal vector, $\mathbf{n}_P$, to the part surface, and by two mutually orthogonal principal unit tangent vectors $\mathbf{t}_{1.P}$ and $\mathbf{t}_{2.P}$ to the part surface such that $\mathbf{n}_P = \mathbf{t}_{1.P} \times \mathbf{t}_{2.P}$. Local properties of the part surface can be analytically described in terms of displacement of the *Darboux* frame when its base point moves over the surface P [11].

---

[*] Jean Gaston Darboux (August 14, 1842–February 23, 1917), a French mathematician.

The unit vectors $\mathbf{t}_{1.P}$, $\mathbf{t}_{2.P}$, and $\mathbf{n}_P$ are schematically depicted in Figure 2.1. These three vectors make up a right-hand-oriented frame (otherwise the direction of the parametric curves must be reversed), which is naturally associated with the part surface P.

The use of the unit vectors $\mathbf{t}_{1.P}$, $\mathbf{t}_{2.P}$, and $\mathbf{n}_P$ makes possible construction of the moving orthogonal *Cartesian* coordinate system, $x_P y_P z_P$, having the origin at a point, K, of contact of the sculptured part surface P to be machined, and of the generating surface T of the form-cutting tool. The contact point K is also referred to as the *cutter-contact point* (CC-point).

The axes x_P, y_P, and z_P of the local reference system $x_P y_P z_P$ are directed along the corresponding unit vectors $\mathbf{t}_{1.P}$, $\mathbf{t}_{2.P}$, and $\mathbf{n}_P$, as shown in Figure 2.1. Thus, the moving coordinate system $x_P y_P z_P$ is established at every point of the sculptured part surface P.

Most of the equations get significantly simplified when *Darboux* frame is used. However, it should be kept in mind, that calculation of the unit tangent vectors of the principal directions $\mathbf{t}_{1.P}$, $\mathbf{t}_{2.P}$ could be a computation-consuming procedure.

Second, to construct a right-hand-oriented local *Cartesian* coordinate system $x_P y_P z_P$ having origin at the point of contact, K, of the sculptured part surface P, and of the generating surface T of the form-cutting tool, the set of unit tangent vectors $\mathbf{u}_P$, $\mathbf{v}_P$, and $\mathbf{n}_P$ can be also employed. Generally speaking, pairs of the unit vectors $\mathbf{u}_P$ and $\mathbf{n}_P$ as well as $\mathbf{v}_P$ and $\mathbf{n}_P$ are always perpendicular to

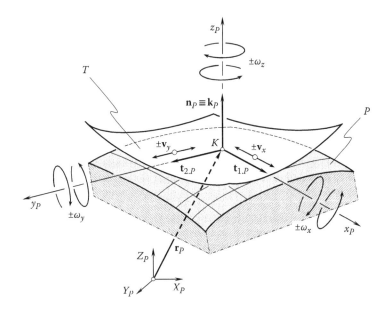

FIGURE 2.1
The principal instant kinematics of sculptured part surface generation on a multiaxis *NC* machine.

each other, while the unit tangent vectors $\mathbf{u}_p$ and $\mathbf{v}_p$ are not always orthogonal to one another. Aiming the construction of the right-hand-oriented local *Cartesian* reference system, the original unit tangent vector $\mathbf{v}_p$ can be substituted with the third unit vector $\mathbf{v}_p^*$, which is equal to $\mathbf{v}_p^* = \mathbf{u}_p \times \mathbf{n}_p$.

The set of three unit vectors $\mathbf{u}_p$, $\mathbf{v}_p^*$, and $\mathbf{n}_p$ make up a right-hand-oriented frame. This frame can be constructed at any point of a sculptured part surface *P*.

Third, the initial parameterization of the sculptured part surface *P* can be altered to an orthogonal parameterization of the surface. Under such a scenario, the unit tangent vectors $\mathbf{u}_p$ and $\mathbf{v}_p$ are always perpendicular to one another. It should be pointed out here that the orthogonal local frame, ($\mathbf{u}_p$, $\mathbf{v}_p$, $\mathbf{n}_p$), constructed in this manner is generally not identical to the *Darboux* frame constructed at that same part surface point *m*.

To alter the initial parameterization of the sculptured part surface *P*, it is necessary to replace the part surface parameters U_P and V_P with a set of new surface parameters

$$U_P^* = U_P^*(U_P, V_P) \tag{2.1}$$

$$V_P^* = V_P^*(U_P, V_P) \tag{2.2}$$

The first derivatives of the position vector of a point, $\mathbf{r}_p$, of the sculptured part surface *P* with respect to the new surface parameters U_P^* and V_P^*

$$\frac{\partial \mathbf{r}_p}{\partial U_P^*} = \frac{\partial \mathbf{r}_p}{\partial U_P} \cdot \frac{\partial U_P}{\partial U_P^*} + \frac{\partial \mathbf{r}_p}{\partial V_P} \cdot \frac{\partial V_P}{\partial U_P^*} \tag{2.3}$$

$$\frac{\partial \mathbf{r}_p}{\partial V_P^*} = \frac{\partial \mathbf{r}_p}{\partial U_P} \cdot \frac{\partial U_P}{\partial V_P^*} + \frac{\partial \mathbf{r}_p}{\partial V_P} \cdot \frac{\partial V_P}{\partial V_P^*} \tag{2.4}$$

can be drawn from Equations 2.3 and 2.4; thus, to meet the equality

$$\mathbf{A}^* = \mathbf{A} \cdot \mathbf{J} \tag{2.5}$$

where

$$\mathbf{A} = \begin{bmatrix} \dfrac{\partial X_P}{\partial U_P} & \dfrac{\partial Y_P}{\partial U_P} & \dfrac{\partial Z_P}{\partial U_P} \\[2ex] \dfrac{\partial X_P}{\partial V_P} & \dfrac{\partial Y_P}{\partial V_P} & \dfrac{\partial Z_P}{\partial V_P} \end{bmatrix}^T = \begin{bmatrix} \dfrac{\partial \mathbf{r}_p}{\partial U_P} & \dfrac{\partial \mathbf{r}_p}{\partial V_P} \end{bmatrix} \tag{2.6}$$

and

$$
\mathbf{J} = \begin{bmatrix} \dfrac{\partial U_P}{\partial U_P^*} & \dfrac{\partial U_P}{\partial V_P^*} \\[3mm] \dfrac{\partial V_P}{\partial U_P^*} & \dfrac{\partial V_P}{\partial V_P^*} \end{bmatrix}
$$

(2.7)

is called the *Jacobian* matrix of the transformation.*

After the parameterization of a sculptured part surface P is altered to an orthogonal parameterization of that same surface, the new fundamental matrix of the part surface P is given by

$$
\begin{vmatrix} E_P^* & F_P^* \\ F_P^* & G_P^* \end{vmatrix} = \mathbf{A}^{*T} \cdot \mathbf{A}^* = \mathbf{J}^T \cdot \mathbf{A}^T \cdot \mathbf{A} \cdot \mathbf{J} = \mathbf{J}^T \cdot \begin{vmatrix} E_P & F_P \\ F_P & G_P \end{vmatrix} \cdot \mathbf{J}
$$

(2.8)

By the properties of determinants, it can be seen from Equation 2.8 that the following equality is valid:

$$
\begin{vmatrix} E_P^* & F_P^* \\ F_P^* & G_P^* \end{vmatrix} = |\mathbf{J}|^2 \cdot \begin{vmatrix} E_P & F_P \\ F_P & G_P \end{vmatrix}
$$

(2.9)

It can be shown based on the premise of Equations 2.3, 2.4, and 2.9 that the unit normal vector, $\mathbf{n}_P$, at a part surface point m is invariant under the transformation of the surface parameters, as could be expected.

The transformation of the second fundamental matrix of the sculptured part surface P can similarly be shown to be given by

$$
\begin{vmatrix} L_P^* & M_P^* \\ M_P^* & N_P^* \end{vmatrix} = |\mathbf{J}|^2 \cdot \begin{vmatrix} L_P & M_P \\ M_P & N_P \end{vmatrix}
$$

(2.10)

by differentiating Equations 2.3 and 2.4, and using the invariance of the unit normal vector $\mathbf{n}_P$. It can be shown from Equations 2.9 and 2.10 that the principal curvatures and the principal directions at a sculptured part surface point m are invariant under the transformation of the surface parameters.

Equations 2.9 and 2.10 yield the natural form of the sculptured part surface P representation with the new U_P^* and V_P^* parameters.

It can be concluded that the unit normal vector, $\mathbf{n}_P$, as well as the principal directions, $\mathbf{t}_{1.P}$ and $\mathbf{t}_{2.P}$, and principal curvatures, $k_{1.P}$ and $k_{2.P}$, at a point m of a smooth, regular part surface P are independent of the U_P and V_P parameters

---

* Carl Gustav Jacob Jacobi (December 10, 1804–February 18, 1851), a well-known German mathematician.

and are therefore geometric properties of the part surface. All these parameters should be continuous if the part surface is to be both a tangent- and a curvature-continuous surface.

At a point of a smooth, regular part surface P, it is required to establish a set of constraints on relationships Equations 2.1 and 2.2, under which an orthogonal parameterization of the sculptured part surface P can be obtained.

To obtain an orthogonally parameterized sculptured part surface P, the relations (see Equations 2.1 and 2.2) have to satisfy the following two conditions, which are the necessary and sufficient conditions for the orthogonally parameterized sculptured part surface P:

$$F_P = 0 \qquad\qquad (2.11)$$

$$M_P = 0 \qquad\qquad (2.12)$$

Once reparameterized, the equation of the part surface P yields easy calculation of the set of orthogonal frame, $(\mathbf{u}_P, \mathbf{v}_P, \mathbf{n}_P)$, at every part surface point.

The presented consideration clearly illustrates the feasibility of composing the right-hand-oriented frame, $(\mathbf{u}_P, \mathbf{v}_P, \mathbf{n}_P)$, using for this purpose one of the methods discussed above in this chapter of the book.

The three unit vectors, $\mathbf{u}_P$, $\mathbf{v}_P$, and $\mathbf{n}_P$, further can be used as the direct vectors for the axes of the local *Cartesian* coordinate system, $x_P y_P z_P$, having its origin at a point of contact, K, of the sculptured part surface P, and of the generating surface T of the form-cutting tool.

2.1.2 Elementary Relative Motions

Instant relative motion of the form-cutting tool can be interpreted as an instant screw motion. Depending on the actual configuration of the local reference system, $x_P y_P z_P$, the instant screw motion of the cutting tool can be decomposed to not more than six elementary motions—that is, to not more than three translations along, and not more than to three rotations about axes of the local coordinate system $x_P y_P z_P$. Not all of the six elementary relative motions are feasible. The translational motion of the cutting tool along z_P axis is not feasible. This elementary relative motion is eliminated from the instant kinematics of sculptured part surface generation because of two reasons.[*]

First, the elementary motions of the cutting tool in the $+\mathbf{n}_P$ direction (outward the part surface P) results in interruption of the generation process of the part surface. No interruption of the part surface generation process is permissible.[†]

---

[*] Further, attention will be focused on special cases of sculptured part surface machining when a motion of the cutting tool along the z_P axis is permissible.

[†] In a case of discrete generation of a part surface P, generation of the part surface is interrupted only in between two consecutive cuts by two neighboring cutting edges of the form-cutting tool.

Second, the motion of the cutting tool in the $-\mathbf{n}_P$ direction (toward the part surface P) results in inevitable interference of the part surface P and of the generating surface T of the cutting tool. No interference of the part surface P and of the generating surface T of the cutting tool is permissible.

Hence, the speed of the translational motion of the cutting tool along the common perpendicular must be equal to zero (see Figure 2.1):

$$V_z \equiv V_n = \frac{\partial z_P}{\partial t} \cdot \mathbf{k}_P = 0 \tag{2.13}$$

where time is denoted by t.

The speed of translational motion of the cutting tool along x_P axis as well as along y_P axis is designated as $\mathbf{V}_x$ and $\mathbf{V}_y$ respectively. Then $\boldsymbol{\omega}_x$, $\boldsymbol{\omega}_y$, and $\boldsymbol{\omega}_z$ designate rotations about the axes of the local coordinate system $x_P y_P z_P$. Figure 2.1 shows that rotations ω_x, ω_y, and ω_z are equal to $\omega_x = |\boldsymbol{\omega}_x|$, $\omega_y = |\boldsymbol{\omega}_y|$, and $\omega_z = |\boldsymbol{\omega}_z|$, respectively.

According to the principal instant kinematics of sculptured part surface generation, schematically shown in Figure 2.1, instant screw motion of the cutting tool relative to the part surface P can be decomposed to not more than five elementary instant motions. This set of five elementary relative motions includes two translations,

$$V_x = \frac{\partial x_P}{\partial t} \cdot \mathbf{u}_P = \frac{\partial^2 \mathbf{r}_P}{\partial U_P \, \partial t}, \tag{2.14}$$

$$V_y = \frac{\partial y_P}{\partial t} \cdot \mathbf{v}_P = \frac{\partial^2 \mathbf{r}_P}{\partial V_P \, \partial t}, \tag{2.15}$$

along the axes of the local reference system $x_P y_P z_P$, and three rotations,

$$\boldsymbol{\omega}_x = \frac{\partial \varphi_x}{\partial t} \cdot \mathbf{u}_P, \tag{2.16}$$

$$\boldsymbol{\omega}_y = \frac{\partial \varphi_y}{\partial t} \cdot \mathbf{v}_P, \tag{2.17}$$

$$\boldsymbol{\omega}_z = \frac{\partial \varphi_z}{\partial t} \cdot \mathbf{n}_P, \tag{2.18}$$

on the axes of the local reference system $x_P y_P z_P$.

Here, in Equations 2.16 through 2.18, the angles of instant rotation of the cutting tool about axes of the local coordinate system $x_P y_P z_P$ are designated as φ_x, φ_y, and φ_z, respectively.

2.2 Generating Motions of the Cutting Tool

After being machined on a multiaxis *NC* machine, the sculptured part surface is represented as a series of tool paths. Generation of a sculptured part surface by successive tool paths is a principal feature of sculptured part surface machining on a multiaxis *NC* machine.

The motion of the cutting tool along a tool path can be considered as a *continuous* following motion. A side-step motion of the cutting tool (in the direction that is perpendicular to the tool path) can be considered as a *discrete* following motion of the cutting tool.

The part surface *P* can be generated as an enveloping surface to successive positions of the moving generating surface *T* of the form-cutting tool, when the surfaces *P* and *T* make either line contact, or they make point contact with one another. When line contact between the surfaces *P* and *T* is observed, then a *one-degree-of-freedom relative motion* of the cutting tool is sufficient for generating of the part surface *P*. Such a relative motion is referred to as *one-parametric motion* of the cutting tool.

When a part surface *P* and the generating surface *T* of the form-cutting tool make point contact, then a *two-degrees-of-freedoms relative motion* of the cutting tool is required for generation the entire part surface *P*. Such a relative motion is referred to as *two-parametric motion* of the cutting tool.

The total number of available degrees of freedom can be greater than two degrees. This results in a *multiparametric motion* of the cutting tool.

The known approaches for the development of part surface machining operations do not return a unique solution to the problem of synthesizing the most efficient (i.e., the optimal) technology of the part surface machining. Use of the known methods returns a variety of solutions to the problem under consideration, of which the efficiency of each is not the highest possible. For the calculation of parameters of a motion of the cutting tool relative to the work, known methods of part surface generation are based solely on the Shishkov's* equation of contact[†]:

$$\mathbf{n}_P \cdot \mathbf{V}_\Sigma = 0 \qquad (2.19)$$

Here in Equation 2.19, the speed of the resultant motion of the cutting tool relative to the part surface *P* is designated as $\mathbf{V}_\Sigma$.

---

* V.A. Shishkov, a soviet scientist mostly known for his accomplishments in the field of gearing and gear machining.

[†] To the best of the author's knowledge, the condition of contact of interacting working surfaces of two components in a mechanism has been known at least since publication of the monograph by R. Willis, Principles of Mechanism, Designed for the Use of Students in the Universities and for Engineering Students Generally, London, John W. Parker, West Stand, Cambridge: J. & J.J. Deighton, 1841, 446 pp. [162]. It is shown by Radzevich [111] that Shishkov was the first (circa 1948) to propose representing the condition of contact of two interacting part surfaces in the form of dot product, $\mathbf{n}_P \cdot \mathbf{V}_\Sigma = 0$ [156].

Restrictions onto only one component of the resultant motion of the cutting tool relative to the work are imposed by the equation of contact, $\mathbf{n}_p \cdot \mathbf{V}_\Sigma = 0$. Namely, projections of speed of the resultant motion of the cutting tool relative to the work onto the direction specified by the unit normal vector, $\mathbf{n}_p$, to the part surface P at a current contact point, K, must be equal to zero,

$$Pr_{\mathbf{n}} \cdot \mathbf{V}_\Sigma \equiv 0 \tag{2.20}$$

No constraints are imposed by the condition of contact (see Equation 2.19) onto other projections of the vector of resultant motion $\mathbf{V}_\Sigma$ of the cutting tool relative to the work. In compliance with the equation of contact, the magnitude and actual direction of the velocity vector $\mathbf{V}_\Sigma$ within the common tangent plane can be arbitrary.

Evidently, the infinite number of the vectors $\mathbf{V}_\Sigma$ of the resultant motion of the cutting tool relative to the work satisfies the equation of contact, $\mathbf{n}_p \cdot \mathbf{V}_\Sigma = 0$. All of these directions are within the tangent plane to the sculptured part surface P at the current contact point K. Indefiniteness of the most favorable direction of the velocity vector $\mathbf{V}_\Sigma$ of the resultant relative motion is the root cause for why the implementation of known methods of part surface generation returns an infinite number of solutions to the problem under consideration. No doubt, performance of a sculptured part surface generation process strongly depends on actual direction of the vector $\mathbf{V}_\Sigma$ of the resultant relative motion of the cutting tool and of the work. For a certain direction of the velocity vector $\mathbf{V}_\Sigma$, the performance of the part surface machining process is better; while for another directions of the velocity vector $\mathbf{V}_\Sigma$ it is poor.

This makes it possible intermediate conclusions that

- The optimal (or the *most favorable*, in other words) direction of the velocity vector $\mathbf{V}_\Sigma$ of the resultant relative motion of the cutting tool and of the work exists.
- The optimal direction of the velocity vector $\mathbf{V}_\Sigma$ satisfies the condition of contact that is specified by the Shishkov's equation of contact, $\mathbf{n}_p \cdot \mathbf{V}_\Sigma = 0$.
- Parameters of the optimal (the *most favorable*) of the velocity vector $\mathbf{V}_\Sigma$ of the resultant relative motion of the cutting tool and of the work can be somehow calculated.

For the calculation of parameters of the velocity vector $\mathbf{V}_\Sigma$ of the optimal instant relative motion of the work and of the cutting tool, an appropriate criterion of optimization is necessary (just to synthesize the part surface generation process at a certain instant of time). The major purpose of the criterion of the optimization is to select the most favorable (optimal in a certain sense) direction of the velocity vector $\mathbf{V}_\Sigma$, which is unique, from the infinite number of feasible directions of this vector that satisfy the condition of contact $\mathbf{n}_P \cdot \mathbf{V}_\Sigma = 0$.

To meet the condition of contact, the velocity vector $\mathbf{V}_\Sigma$, of the resultant relative motion of the cutting tool and of the work must be located within the common tangent plane to the surfaces P and T at a current contact point, K. This is the geometrical interpretation of the equation of contact, $\mathbf{n}_P \cdot \mathbf{V}_\Sigma = 0$.

Consider a local reference system, $x_P y_P z_P$, having origin at a current CC point K as schematically depicted in Figure 2.2. In the coordinate system $x_P y_P z_P$, the vector $\mathbf{V}_\Sigma$ can be analytically described by the expression

$$\mathbf{V}_\Sigma = \frac{\partial x_P}{\partial t} \cdot \mathbf{u}_P + \frac{\partial y_P}{\partial t} \cdot \mathbf{v}_P + \frac{\partial z_P}{\partial t} \cdot \mathbf{n}_P \tag{2.21}$$

In that same local reference system, the equality $\mathbf{n}_P = \mathbf{k}_P$ is observed. Substituting Equation 2.21 and the relationship $\mathbf{n}_P = \mathbf{k}_P$ along with Equation 2.13 into the equation of contact, $\mathbf{n}_P \cdot \mathbf{V}_\Sigma = 0$, one can obtain

$$\mathbf{n}_P \cdot \mathbf{V}_\Sigma = \frac{\partial z_P}{\partial t} = 0 \tag{2.22}$$

To satisfy the condition of contact, projection of the velocity vector $\mathbf{V}_\Sigma$ of the resultant relative motion of surfaces P and T onto direction of the common perpendicular, $\mathbf{n}_P$, must be equal to zero. This is the proof of the statement that the velocity vector $\mathbf{V}_\Sigma$ must be located within the common tangent plane to the surfaces P and T at every CC point K.

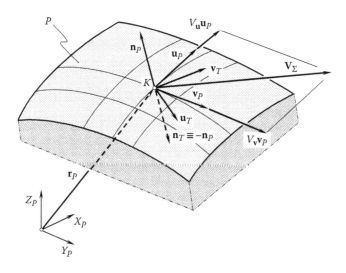

FIGURE 2.2
Feasible elementary relative motions of the cutting tool relative to a part surface P.

It is important to point out here that according to Shishkov [156], the following three cases should be distinguished:

$$\mathbf{n}_P \cdot \mathbf{V}_\Sigma < 0 \qquad (2.23)$$

$$\mathbf{n}_P \cdot \mathbf{V}_\Sigma = 0 \qquad (2.24)$$

$$\mathbf{n}_P \cdot \mathbf{V}_\Sigma > 0 \qquad (2.25)$$

The first condition (see Equation 2.23) can be considered as the condition of rough machining of the part surface P, namely when the chip removal takes place but the part surface P is not generated yet (at this instant of time the surfaces P and T do not make contact with one another). Cutting edges located within the portions of the generating surface T of the cutting tool that performs such a relative motion remove the stock from the work when machining, but they do not generate the part surface P.

The second condition (see Equation 2.24) corresponds to the instants of time when the part surface P is generated by the tool surface T. The cutting edges of the form-cutting tool located within this portion of the cutting tool surface T remove negligibly small amounts of the stock from the work. The main purpose of the cutting edges of this type is to generate the part surface P.

Finally, the third condition (see Equation 2.25) is related to the portions of the generating surface T of the cutting tool that are departing from the already generated part surface P. These cutting edges neither remove the stock from the work nor generate the part surface P.

These conditions are discussed in more detail in Chapter 5.

It should be clear now that instant kinematics of sculptured part surface generation cannot be uniquely specified by the equation of contact, $\mathbf{n}_P \cdot \mathbf{V}_\Sigma = 0$. Therefore, to be able to calculate the most favorable direction of the velocity vector $\mathbf{V}_\Sigma$ of resultant relative motion of surfaces P and T at every CC point K, an additional criterion is necessary to be introduced. The elimination of the indefiniteness in the determination of direction of the velocity vector $\mathbf{V}_\Sigma$ is the main purpose of the criterion just mentioned. This criterion is based on a newly introduced characteristic curve discussed in detail in Chapter 4.

The location and orientation of the unit normal vector, $\mathbf{n}_P$, are both uniquely specified by the geometry of the part surface P to be machined. In most cases, the direction of vector $\mathbf{n}_P$ cannot be altered. However, in particular cases of part surface machining, the orientation of the common perpendicular, $\mathbf{n}_P$, can be altered for the part surface machining purposes. For example, when machining a thin-wall part schematically shown in Figure 2.3, an elastic deformation can be applied to the work. Under the applied load, unit normal vectors $\mathbf{n}_m^a$, $\mathbf{n}_m^b$, and $\mathbf{n}_m^c$ to the part surface P at points a, b, and c became parallel to one another.

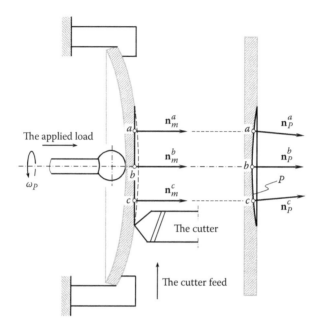

FIGURE 2.3
An example of application of elastic deformation against the work for machining a concave part surface P.

In the deformed stage, the work is machined on a lathe with simple motion of the cutter relative to the work. The elastic deformation means that a plane is machined instead of machining of the concave part surface of complex geometry (for this purpose, the magnitude of the applied load may vary according to the distance traveled by the cutter with a certain feed rate; a distributed load of variable magnitude can be applied as well). After being released, in undeformed stage, the machined work gets its original shape, and the machined plane turns into a concave part surface P having the required complex geometry. The unit normal vectors $\mathbf{n}_P^a$, $\mathbf{n}_P^b$, and $\mathbf{n}_P^c$ at the corresponding points a, b, and c of the machined part surface P are not parallel to each other.

An intentional elastic deformation of the work is proposed to be implemented when shaping thin-wall ring gears used in design of planetary gearboxes. This novel method of gear shaping is schematically illustrated in Figure 2.4. Prior to discussing the proposed method of machining of thin-wall ring gears, the following consideration must be taken into account.

When operating in the gearbox, thin-wall ring gears are deformed by the operating loads. When the deformation is large enough, the operating base pitch $p_{b.op}$ of the ring gear differs from that in the mating planet pinion, which is not allowed. The operating base pitches of mating gears must be equal to one another under any circumstances. To resolve the issue, it is proposed to

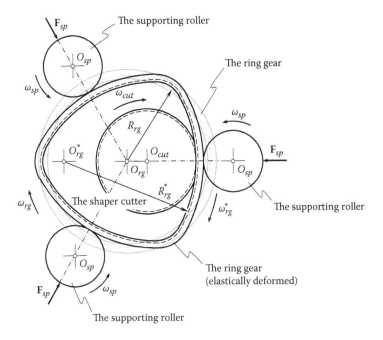

FIGURE 2.4
Schematics of shaping of a thin-wall ring gear in an elastically deformed stage.

cut ring gears in deformed stage. The last could be possible in cases when the ring gear is flexible enough.

According to the proposed method* (Figure 2.4), a ring gear of pitch radius R_{rg} is cut by the shaper cutter. The ring gear is rotated about its axis of rotation O_{rg} at a constant angular velocity ω_{rg}. The shaper cutter is rotated about its axis of rotation, O_{cut}, at a certain angular velocity ω_{cut}. The rotations ω_{rg} and ω_{cut} are timed with one another. When being machined, the ring gear is intentionally deformed to that same stage at which it is deformed when operating in the gearbox. Three rollers can be used for the elastic deformation of the ring gear. The supporting rollers are rotated about their axis of rotation, O_{sp}, at an angular velocity ω_{sp}. The supporting rollers are pressed into the ring gear by radial forces $\mathbf{F}_{sp}$. Because of the applied forces $\mathbf{F}_{sp}$, the ring gear is elastically deformed to a stage when the operating pitch radius is R_{rg}^*. The operating pitch radius R_{rg}^* is always larger than that R_{rg} of nondeformed ring gear (that is, the inequality $R_{rg}^* > R_{rg}$ is always observed). It could be imagined that a fake ring gear of pitch radius R_{rg}^* is rotated about a fake axis of rotation O_{rg}^* with a certain fake angular velocity ω_{rg}^*. The rotations of the ring gear and of the shaper cutter are timed with one another so as to ensure equal linear velocities at the contact point of the circle of radii R_{rg}^* and the pitch circle of the shaper cutter.

---

* This method of shaping of thin-wall (flexible) ring gears is proposed by Radzevich (2012).

In the deformed stage of the ring gear, the shaper cutter generates the gear teeth having the desired value of base pitch equal to that of the planet pinion. Other methods of machining of thin-wall (flexible) ring gears can be proposed based on the disclosed concept.

In cases when a certain intentional elastic deformation of the work is used for manufacturing purposes, the condition of contact must be satisfied in the deformed stage of the part surface being machined.

The intentional elastic deformation of a work for manufacturing purposes is observed when machining flex-spline that is an essential machine element of a harmonic drive as well as in other applications.

The capability to alter the orientation of the unit normal vector to the part surface is limited, as elastic deformation of most of materials to be machined is commonly small. However, such a capability is feasible, and it affects the generation of the part surface P, which is of principal importance and should be taken into account.

Capabilities of alteration of orientation of the unit normal vector, $\mathbf{n}_T$, of a cutting tool are significantly wider, especially when implementing for machining a given sculptured part surface P of cutting tools of special design having variable shape and parameters of the generating surface T [108,112,136]. Form-cutting tools of this particular type are in wide use in aerospace industry, for example.

When machining a sculptured part surface on a multiaxis NC machine, the cutting tool performs a continuous following motion along every tool path [116,136]. Therefore, the generating motion can be considered as a continuous following motion of the cutting tool relative to the work. This motion means that the CC point is traveling along the tool path.

After the machining of a certain tool path is complete, then the cutting tool feeds across the tool path in a new position. The machining of another tool path begins from the new position of the cutting tool. Hence, the feed motion of the cutting tool can be represented as a discrete following motion of the cutting tool relative to the work. This motion means the CC point is traveling across the tool path.

The generating motion of the cutting tool can be described analytically. For this purpose, the elementary motions that make up the principal instant kinematics of sculptured part surface generation are used (see Figure 2.1).

The elementary relative motions of the cutting tool are properly timed (synchronized) with one another to produce the desired instant generation motion of the cutting tool as it is schematically illustrated in Figure 2.5.

The following equations can easily be composed based on the premise of the analysis of the instant kinematics of sculptured part surface generation:

$$|\mathbf{V}_x| = |\mathbf{\omega}_y| \cdot R_{P.x} \tag{2.26}$$

$$|\mathbf{V}_y| = |\mathbf{\omega}_x| \cdot R_{P.y} \tag{2.27}$$

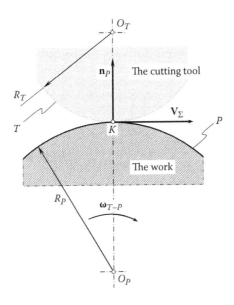

FIGURE 2.5
Definition of the *generating motion* of the cutting tool.

Here, in Equations 2.26 and 2.27, $R_{P.x}$ and $R_{P.y}$ designate the normal radii of curvature of the sculptured part surface P in the section by coordinate planes $x_P z_P$ and $y_P z_P$ of the local reference system $x_P y_P z_P$, respectively.

Equations 2.26 and 2.27 can be generalized and can be represented in the form

$$|\mathbf{V}_\Sigma| = |\boldsymbol{\omega}_{T-P}| \cdot R_{P.\Sigma} \tag{2.28}$$

where $\mathbf{V}_\Sigma$ is the velocity vector of the resultant motion of the CC point along the tool path, $|\boldsymbol{\omega}_{T-P}|$ is the vector of instant rotation of the generating surface T about an axis that is perpendicular to the normal plane through the velocity vector $\mathbf{V}_\Sigma$, and $R_{P.\Sigma}$ is radius of normal curvature of the part surface P in the section by normal plane through the velocity vector $\mathbf{V}_\Sigma$.

After the relative motion of the cutting tool in the direction of the unit normal vector, $\mathbf{n}_P = \mathbf{k}_P$, is eliminated from further analysis, then the velocity vector, $\mathbf{V}_\Sigma$, of the resultant motion of the CC point along the tool path can be expressed in terms of projections on axes of the local coordinate system $x_P y_P z_P$ in the form

$$\mathbf{V}_\Sigma = \frac{\partial x_P}{\partial t} \cdot \mathbf{u}_P + \frac{\partial y_P}{\partial t} \cdot \mathbf{v}_P \tag{2.29}$$

Generally, to meet the conditions, which are analytically expressed by Equations 2.26 and 2.27, both of the additives in Equation 2.29, are necessary.

Otherwise, the relative motion of the cutting tool and of the work cannot be identified as the generating motion of the cutting tool.

The condition (see Equations 2.26 and 2.27) is satisfied if

$$\frac{\partial x_P}{\partial t} \cdot \mathbf{u}_P = \mathbf{\omega}_y \times (R_{P.x} \cdot \mathbf{n}_P) \tag{2.30}$$

$$\frac{\partial y_P}{\partial t} \cdot \mathbf{v}_P = \mathbf{\omega}_x \times (R_{P.y} \cdot \mathbf{n}_P) \tag{2.31}$$

Under this result, Equation 2.29 casts into

$$\mathbf{V}_\Sigma = \mathbf{\omega}_y \times (R_{P.x} \cdot \mathbf{n}_P) + \mathbf{\omega}_x \times (R_{P.y} \cdot \mathbf{n}_P) \tag{2.32}$$

The generating motion of the cutting tool satisfies both the equation of contact, $\mathbf{n}_P \cdot \mathbf{V}_\Sigma = 0$, and Equation 2.32 at every instant of time when machining a sculptured part surface on a multiaxis *NC* machine.

The component, $\mathbf{V}_z$, of the relative motion of the cutting tool (see Figure 2.1) is not completely eliminated from further discussion. Taking into consideration the tolerance for accuracy of the machined part surface *P*, which was schematically illustrated in Figure 1.9, the component $\mathbf{V}_z$ of the relative motion of the cutting tool along the unit normal vector, $\mathbf{n}_P$, is feasible if it is performing within the tolerance bend of $\delta = \delta^+ + \delta^-$. Moreover, because of deviations of actual motion of the cutting tool from the desired cutting tool motion, the component $\mathbf{V}_z$ of the relative motion of the cutting tool is always observed in the actual part surface machining operation. If necessary, the motion $\mathbf{V}_z$ may be incorporated into the principal instant kinematics of the sculptured part surface generation.

This is one more example of the difference between the classical differential geometry of surfaces and the engineering geometry of surfaces.

The principal instant kinematics of part surface generation includes five elementary relative motions. Thus, the part surface *P* can be represented as an enveloping surface to consecutive positions of not more than five-parametric motion of the generating surface *T* of the form-cutting tool.

2.3 Motions of Orientation of the Cutting Tool

As mentioned above, machining of a sculptured part surface on a multiaxis *NC* machine is the most general case of part surface generation. This is due to two surfaces, namely, the sculptured part surface *P* and the generating surface *T* of the form-cutting tool, making point contact at every instant of the part surface machining.

Among various possible types of relative motions of the cutting tool, one more specific motion can be distinguished. When performing relative motion of this particular type, the CC point does not change its position on the sculptured part surface P being machined. Only orientation of the cutting tool relative to the work is changed because of a motion of this type. Motions of this particular type are referred to as *orientational motions* of the cutting tool. Orientational motion of the cutting tool can be interpreted as a component of the resultant motion of the cutting tool relative to the work.

When performing the orientational motion, the CC point may maintain its location on both surfaces, on the part surface P being machined, and on the generating surface T of the form-cutting tool. Orientational motion of this particular type is referred to as the *orientational motion of the first type*.

In another case, when machining a sculptured part surface, the CC point may maintain its location only on the part surface P, while its location on the generating surface T of the cutting tool is altered. Orientational motion of this particular type is referred to as the *orientational motion of the second type*.

The speed of the orientational motion of the cutting tool is a function of variation of normal curvatures of the part surface P at the current CC point and of speed of the generating motion. Orientational motions of the cutting tool do not directly affect the stock removal productivity or intensity of generation of the part surface P. Motions of this type alternate orientation of the cutting tool relative to the work as well as the direction of the generating motion of the cutting tool.

To determine all possible types of the orientational motions of the cutting tool, it is helpful to consider all possible groups of relative motions of the cutting tool. All groups of the relative motions of the cutting tool are represented by the singular relative motions and by the combined relative motions. The singular relative motions comprise one elementary relative motion of the cutting tool. The combined relative motions of the cutting tool comprise two or more elementary relative motions.

There are only five groups of elementary relative motions of the cutting tool. The total number of elementary relative motions at every group of the relative motions is equal to the number of combinations of five elementary motions by i elementary motions (here $i = 1, 2, \ldots, 5$ is an integer number). The total number N of relative motions in the principal instant kinematics of part surface generation can be calculated from the following equation:

$$N = \sum_{i=1}^{5} N_i = \sum_{i=1}^{5} C_5^i = 31 \qquad (2.33)$$

The performed analysis of all 31 possible types of the relative motions of the cutting tool reveals that only a few elementary relative motions and their combinations can be referred to as the orientational motions of the cutting tool. These relative motions of the cutting tool are listed immediately below [116,117,131]

The first group of the motions: $\{\omega_n\}$.

The second group of the motions: $\{\omega_u, V_v\}$, $\{\omega_v, V_u\}$.

The third group of the motions: $\{\omega_u, \omega_n, V_v\}$, $\{\omega_v, \omega_n, V_u\}$.

The fourth group of the motions: $\{\omega_u, V_v, \omega_v, V_u\}$.

The fifth group of the motions: $\{\omega_u, V_v, \omega_n, \omega_v, V_u\}$.

It is assumed here and below in this section that U_P and V_P coordinate lines on the sculptured part surface P are along principal directions at every surface P point. Therefore, the local reference system $x_P y_P z_P$ (see Figure 2.1) is a type of orthogonal *Cartesian* coordinate system. Moreover, the reference system $x_P y_P z_P$ can also be referred to as *Darboux* frame. Because of this, the following designations ω_x, ω_y, ω_z for speeds of the rotations are equivalent to the corresponding designations ω_u, ω_v, and ω_n, respectively. Similarly, the designations V_x, V_y, V_z for speeds of the translations are equivalent to the corresponding designations V_u, V_v, and V_n, respectively.

Ultimately, one can come up with the complete set of the orientational motions of the cutting tool in a part surface machining operation. One of the orientational motions is the *singular* orientational motion of the cutting tool, and six other orientational motions are the *combined* orientational motions of the cutting tool.

The orientational motions of the first type are represented with the only singular orientational motion, $\{\omega_n\}$, of the cutting tool. The rest of the orientational motions of the cutting tool are the orientational motions of the second type.

The orientational motion of the second type, $\{\omega_u, V_v, \omega_n, \omega_v, V_u\}$, is the most general type of orientational motion of the cutting tool under consideration. Other types of the orientational motions of the cutting tool can be interpreted as degenerate types of the orientational motion of this orientational motion of general type.

The elementary motions that comprise a combined orientational motion of the cutting tool are timed (synchronized) with one another. The rotational elementary motion ω_u about x_P axis is timed with the translational motion V_v along y_P axis. Similarly, the rotational elementary motion ω_v about y_P axis is timed with the translational motion V_u along x_P axis. The timing of the elementary motions results in that the generating surface T of the form-cutting tool is rolling and slipping over the sculptured part surface P.

The necessary timing of the elementary motions of the cutting tool can be achieved when the following condition is satisfied.

The orientational motion of the second type can be interpreted as a superposition of the instant translational motion with a certain instant speed V_{T-P}, and of the instant rotation $\boldsymbol{\omega}_{T-P}$ of the cutting tool as schematically illustrated in Figure 2.6. A relative motion of the cutting tool is referred to as orientational motion of the second type if the following equality is satisfied:

$$|\mathbf{V}_{T-P}| = |\boldsymbol{\omega}_{T-P}| \cdot R_T \qquad (2.34)$$

where $\boldsymbol{\omega}_{T-P}$ is the vector of the instant rotation of the cutting tool about the axis O_T that crosses the velocity vector $\mathbf{V}_{T-P}$ at right angle, R_T is the radius of curvature of the generating surface T of the form-cutting tool in the normal plane section through the velocity vector $\mathbf{V}_{T-P}$.

The vector of instant linear velocity $\mathbf{V}_{T-P}$ can be represented in a form similar to that of Equation 2.29:

$$\mathbf{V}_{T-P} = \frac{\partial x_P}{\partial t} \cdot \mathbf{u}_P + \frac{\partial y_P}{\partial t} \cdot \mathbf{v}_P \qquad (2.35)$$

To meet the necessary condition (that is specified by Equation 2.34), both additives in Equation 2.29 must result in a continuous following motion of the CC point over the part surface P being machined. Otherwise, the relative motion of the work and of the cutting tool cannot be referred to as the

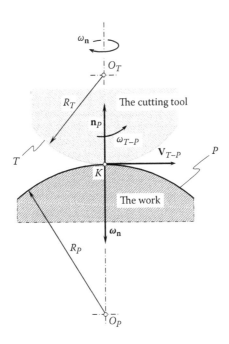

FIGURE 2.6
Motions of orientation of the cutting tool.

orientational motion of the cutting tool. The condition (see Equation 2.34) is satisfied if

$$\frac{\partial x_P}{\partial t} \cdot \mathbf{u}_P = \boldsymbol{\omega}_\mathbf{v} \cdot R_{T.\mathbf{u}} \tag{2.36}$$

$$\frac{\partial y_P}{\partial t} \cdot \mathbf{v}_P = \boldsymbol{\omega}_\mathbf{u} \cdot R_{T.\mathbf{v}} \tag{2.37}$$

where $R_{T.\mathbf{u}}$ and $R_{T.\mathbf{v}}$ are the normal radii of curvature of the generating surface T of the form-cutting tool in the plane sections through the unit tangent vectors $\mathbf{u}_P$ and $\mathbf{v}_P$, respectively, and $\boldsymbol{\omega}_\mathbf{u}$ and $\boldsymbol{\omega}_\mathbf{v}$ are vectors of instant angular velocities (rotations about the axes specified by the directions $\mathbf{u}_P$ and $\mathbf{v}_P$).

Equations 2.36 and 2.37 make possible a representation of the combined orientational motion of the cutting tool, $\mathbf{V}_{orient}$, in the form

$$\{\boldsymbol{\omega}_\mathbf{u}, \mathbf{V}_\mathbf{v}, \boldsymbol{\omega}_\mathbf{v}, \mathbf{V}_\mathbf{u}\} \quad \Rightarrow \quad \begin{cases} \boldsymbol{\omega}_{orient} = \boldsymbol{\omega}_\mathbf{u} + \boldsymbol{\omega}_\mathbf{v} \\ \mathbf{V}_{orient} = \boldsymbol{\omega}_\mathbf{u} \times (R_{T.\mathbf{v}} \cdot \mathbf{n}_P) + \boldsymbol{\omega}_\mathbf{v} \times (R_{T.\mathbf{u}} \cdot \mathbf{n}_P) \end{cases} \tag{2.38}$$

The local mobility of the generating surface T of the form-cutting tool in relation to the sculptured part surface being machined can be expressed in terms of the orientational motions of the cutting tool along with the instant motion of it in the direction of the tool path.

On the premise of the performed analysis of kinematics of the multiparametric motion of the form-cutting tool in relation to the sculptured part surface being machined, a classification of the orientational motions of the cutting tool is developed. The classification is schematically illustrated in Figure 2.7.

When representing a sculptured part surface P as an enveloping surface to consecutive positions of the generating surface T of the form-cutting tool that is performing a five-parametric motion (see Section 2.2), the orientational motions of the cutting tool can be omitted from consideration.

A machining operation of a sculptured part surface on a multiaxis NC machine is getting more agile when orientational motions of the cutting tool are incorporated into the principal kinematics of the part surface machining.

Two principal inventions in the realm of sculptured part surface machining on a multiaxis NC machine are based on the implementation of the orientational motions of the cutting tool. The first invention [73] is based on implementation of the orientational motions of the first type.* The other one [78] is based on implementation of orientational motions of the second type.[†] Both of the inventions were proposed by the author in the late 1970s.

---

* Patent No. 1185749, USSR, *A Method of Sculptured Part Surface Machining on a Multi-Axis NC Machine*. S.P. Radzevich, B23C 3/16, Filed: October 24, 1983.
† Patent No. 1249787, USSR, *A Method of Sculptured Part Surface Machining on a Multi-Axis NC Machine*. S.P. Radzevich, B23C 3/16, Filed: December 27, 1984.

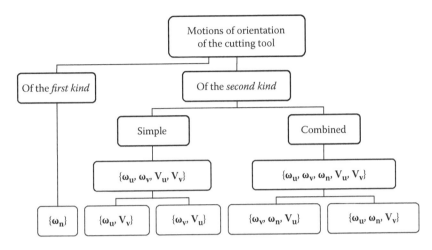

FIGURE 2.7
A classification of possible types of motions of orientation of the cutting tool.

Dozens of inventions in this particular area of mechanical engineering feature the combined orientational motions of the cutting tool.

Keeping control over the orientational motions of the cutting tool makes possible increase in the performance of the part surface machining operation.

Implementation of the discussed approach of part surface generation makes possible calculation of the most favorable (optimal) parameters of all the relative motions of the cutting tool relative to the work. The solution to the problem of synthesis of the optimal kinematics of generation of a sculptured part surface on a multiaxis *NC* machine can be drawn up from the analysis of instant kinematics of multiparametric motion of the cutting tool relative to the work.

2.4 Relative Motions Causing Sliding of a Surface over Itself

Surfaces that allow for sliding over themselves are convenient in many applications in mechanical and manufacturing engineering. For a long while, surfaces of this particular type were under investigation by mechanical engineers and mathematicians as well.

In a classical investigation [12], the problem of determining all possible motions of a smooth regular surface *A* with respect to a smooth regular surface *B* such that the path of every point of *A* is a planar curve was solved by *Darboux*. Trivial examples, for instance, the case in which the planes of the paths are all parallel, are also known.

Surfaces that allow for sliding over themselves can be generated by a corresponding motion of a curve (of a surface profile) of an appropriate geometry. A relative motion of this type is commonly referred to as motion that causes the surface sliding over itself. The required motion of the part surface profile can be easily performed on a machine tool. This is the main reason why surfaces that allow for sliding over themselves found wide application in engineering practice.

Relative motions causing sliding of a surface over itself and part surfaces that feature such a feature, namely, the capability to slide over itself, are tightly connected to one another. Commonly, the relative motions and the surfaces of this type are considered together.

Surfaces that allow for sliding over themselves can be interpreted as the surfaces for which a resultant relative motion of a special type exists. Relative motion of this particular type means that the enveloping surface to consecutive positions of the moving part surface P and the surface P itself are congruent to each other. The same is true with respect to generating surface T of the cutting tool.

As part surfaces that feature the capability to slide over themselves are important for engineering applications, it is of critical importance to find out all possible types of surface of this kind.

It was proven by Radzevich [131] that the most general type of part surfaces that allow for sliding over themselves is represented only by screw surfaces of constant axial pitch. Accordingly, screw motion of constant axial pitch is the most general type of relative motion that allows a part surface P (as well as generating surface T of the cutting tool) to slide over itself.

A screw motion of constant axial pitch can be represented as a superposition of two elementary motions, namely, of a translation and of a rotation. Velocities of these elementary motions are timed (synchronized) with one another. Therefore, regardless of motions of the type under consideration comprise two elementary motions, the last property of motions of this type results in a one-parametric relative motion—that is, a screw motion of constant axial pitch is a type of *one-degree-of-freedom* relative motions (Figure 2.8) and not a type of *two-degrees-of-freedom* relative motion. One degree of freedom in a screw motion of constant axial pitch is also known as *screw-freedom*. Screw freedom is the most general of all freedoms within the lower kinematic pairs. Part surfaces that have freedom of this type are commonly called *general helicoids*.

Other possible types of part surfaces that allow for sliding over themselves can be interpreted as a particular type of the corresponding screw motion of constant axial pitch p_x. Possible reductions are outlined briefly below:

- When the axial pitch of a screw part surface is equal to zero ($p_x =$ 0), then the screw surface reduces to a surface of revolution with a certain axial profile. When a surface of revolution is rotated about its axis of rotation, the enveloping surface to the rotating surface of

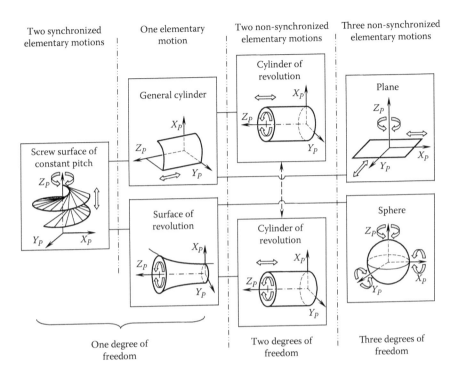

FIGURE 2.8
Part surfaces that are invariant with respect to a group of elementary motions.

revolution is congruent to the surface itself: When a surface of revolution is rotating about its axis of rotation, it is sliding over itself. The rotation of a surface of revolution has to be considered as a one-parametric relative motion similar to that of a screw surface of constant pitch. Part surfaces that have freedom of this type are commonly called *surfaces of revolution*.

- When the axial pitch of a screw part surface is equal to infinity ($p_x \to \infty$), then the screw surface reduces to a cylinder of general type. Translation of the cylinder along its straight generating line causes sliding of the cylinder over itself. The translation of a cylinder has to be considered as a one-parametric relative motion similar to that of a screw surface of constant pitch. Part surfaces having freedom of this type are commonly called *surfaces of translation*.

- When a translation and a rotation are not timed (synchronized) with one another, this combination of elementary motions is feasible only for cylinders of revolution. Both of the motions, namely the translation as well as the rotation of the cylinder of revolution, independently cause sliding of the cylinder of revolution over itself. Because the elementary motions are independent from one another, then the translation and the rotation of a cylinder of revolution is interpreted as a two-parametric motion—that is, as a two-degrees-of-freedom relative motion of a part surface P.

A smooth regular part surface as well as generating surface of the cutting tool can perform three elementary relative motions, which are not timed (synchronized) with one another. Only two types of three-parametric motions are feasible:

- Three-parametric relative motion of the first type comprises three rotations about the axes of a certain *Cartesian* coordinate system. All of these rotations are independent from one another. Only a sphere allows for three independent rotations. A sphere is sliding over itself when it is rotated simultaneously about three axes through the center of the sphere. Three rotations in the case under consideration comprise a type of three-parametric relative motion. Because of all three elementary motions are independent from one another, then three rotations of a sphere is interpreted as a *three-degrees-of-freedom* relative motion of a part surface P.

- Three-parametric relative motion of the second type comprises two translations and of one rotation. All of these elementary relative motions are independent of each other (they are not timed with one another). Only a plane surface allows for two translations and one rotation. A plane surface is sliding over itself when it is simultaneously translating in two different directions (which are parallel to the surface), and is rotating about an axis that is perpendicular to directions of the translations. Two translations and a rotation in the case under consideration comprise another type of three-parametric relative motion. Because all three elementary motions are independent from one another, then two translations and a rotation of a plane surface is interpreted as a *three-degrees-of-freedom* relative motion of a part surface P.

The performed analysis of part surfaces that allow for sliding over themselves (and of corresponding relative motions that cause the surfaces to slide over themselves) makes possible another interpretation of surfaces (motions) that allow for sliding over themselves. The surfaces of that type can also be considered as the surfaces that are invariant with respect to a certain group of elementary motions.

It was proven by Bonnet* that the specification of the first and of the second fundamental forms determines a unique surface and that two surfaces that have identical first and second fundamental forms are congruent to one

---

* Pierre Ossian Bonnet (December 22, 1819–June 22, 1892), a French mathematician.

another (see Section 1.1). Six fundamental magnitudes of the first and of the second order determine a surface uniquely, except as to position and orientation in space if Gauss'* characteristic equation and the Codazzi[†]–Mainardi's[‡] relationships of compatibility are satisfied. This is due to the *main theorem in surface theory.*

It is natural to assume that the property of surfaces that allows for sliding over themselves, can be expressed in terms of six fundamental magnitudes of the first and of the second fundamental forms.

The property of surfaces to allow for sliding over themselves imposes an additional restriction onto relationships among the fundamental magnitudes of the first and of the second order. Thus, a representation of a surface in natural form, that is, in terms of the fundamental magnitudes of the first and of the second order, will possess certain features. A comprehensive analysis and a solution to the problem of analytical representation of surfaces that allow for sliding over themselves can be found out in [116,117,131,136,147]. An analytical criterion for surfaces of the type under consideration is derived. The derived criterion is expressed in terms of the fundamental magnitudes E_P, F_P, G_P of the first, and L_P, M_P, N_P of the second order of a sculptured part surface P. The criterion enable one to identify whether a certain analytically described surface allows for sliding over itself or not, and if it does, what particular type of surface it represents.

The criterion reflects the following properties of a surface:

- For a screw part surface: axial pitch of the surface is constant ($p_x = Const$)
- For a surface of revolution: axial pitch of the surface is constant and is equal to zero ($p_x = 0$)
- For a cylinder of general type: axial pitch of the surface is constant and is equal to infinity ($p_x \to \infty$)

For part surfaces of all the above-listed types, the generating line (the surface profile) is of nonchangeable shape, and its orientation with respect to the directing line of the surface remains the same.

More particular types of part surfaces that allow for sliding over themselves require additional restrictions to be imposed. The interested reader may wish to see references [131,147] for details on analytical descriptions of surfaces that allow for sliding over themselves.

---

* Johann Carl Friedrich Gauss (Latin: Carolus Fridericus Gauss) (April 30, 1777–February 23, 1855), a German mathematician and physical scientist who contributed significantly to many fields, including number theory, statistics, analysis, differential geometry, geodesy, geophysics, electrostatics, astronomy, and optics.
† Delfino Codazzi (March 7, 1824–July 21, 1873), an Italian mathematician.
‡ Gaspare Mainardi (June 27, 1800–March 9, 1879), an Italian mathematician.

Relative motions that cause part surfaces to slide over themselves are widely used in engineering practice. For example, in a method of sculptured part surface machining on a multiaxis *NC* machine,* an additional relative motion of the cutting tool that causes the generating surface of the form-cutting tool to slide over itself makes possible an increase in accuracy of the machined part surface [79].

2.5 Possible Kinematic Schemes of Part Surface Generation

The kinematics of part surface generation is a cornerstone of an infinite variety of various machining operations. Together with the shape and parameters of geometry of the part surface *P* to be machined, the kinematic schemes of part surface generation specify the generating surface of the form-cutting tool as well as the principal part of kinematic structure of a corresponding machine tool.

2.5.1 Essence of the Kinematic Schemes of Part Surface Generation

Nonagile kinematics of part surface generation is usually executed on conventional machine tools. Kinematics of this particular type features relative motion of the cutting tool with constant parameters. It is commonly used for the generation of part surfaces *P* that allow for sliding over themselves. Various possible versions of nonagile types of kinematics of part surface generation can be interpreted as a particular (degenerate) case of agile kinematics of part surface generation, for example, as the kinematics of sculptured part surface machining on a multiaxis *NC* machine shown in Figure 2.1.

It should be pointed out here that no absolute motions, only relative motions of the cutting tool are covered by kinematics of part surface generation.

A combination of certain number of elementary relative motions of the work and of the cutting tool makes up the corresponding nonagile kinematics of part surface generation. Translational and rotational motions of the work and of the cutting tool along and about certain axes are implemented as the elementary relative motions. A combination of translations and of rotations of the cutting tool is commonly referred to as the *kinematic scheme* of part surface generation.

The kinematics of generation of part surfaces that allow for sliding of one or both surfaces *P* and *T* over themselves is usually referred to as *rigid kinematics* of part surface generation. Rigid kinematics of part surface generation

---

* Patent No. 1336366, USSR, *A Method of Sculptured Part Surface Machining on a Multi-Axis NC Machine.* S.P. Radzevich, B23C 3/16, Filed: October 21, 1984.

usually features constant relative translation motions $\mathbf{V}_i$ and rotations $\boldsymbol{\omega}_i$. No acceleration or deceleration is usually occurs. Due to this, the kinematic *schemes* of part surface generation can be composed.

Similar to generation of sculptured part surfaces on a multiaxis *NC* machine, the principle of inversion is implemented for the investigation of the nonagile kinematics of part surface generation. To implement this fundamental principle of mechanics, both the work and the cutting tool are moved with the motions directed opposite to the motions that the work is performing in the machining operation. Ultimately, the implementation of the principle of inversion results in the work becoming motionless, and the cutting tool performing all the required relative motions. Under such a scenario, a *Cartesian* coordinate system $X_P Y_P Z_P$ associated with the work is considered as the stationary coordinate system.

The two principal problems are tightly connected with the kinematics of part surface generation on conventional machine tools. One of them is referred to as the *direct principal problem* of part surface generation, and the other is referred to as the *inverse principal problem* of part surface generation.

The direct principal problem of part surface generation relates to the determination of the shape and parameters of the generating surface of the form-cutting tool for machining a given part surface. Therefore, the set of design parameters of the generation surface T of the form-cutting tool is the solution to the first principal problem of part surface generation. The cutting tool surface T is expressed in terms of the shape and parameters of the part surface P and of parameters of the kinematic scheme of the machining operation.

The inverse principal problem of part surface generation relates to the determination of the shape and parameters of the actually machined part surface P. The set of necessary and sufficient conditions of proper part surface generation* (further, conditions of proper *PSG* for simplicity) are not always satisfied [112,117,136]. Violation of the conditions of proper part surface generation is the root cause of inevitable deviations of the actually machined part surface P_{act} from the desired nominal part surface P_{des}. Therefore, the set of design parameters of the actual generated part surface P_{act} is the solution to the second principal problem of part surface generation. The part surface P_{act} is expressed in terms of parameters of the generating surface T of the form-cutting tool and of parameters of the inverse kinematic scheme of the machining operation.

Based on the performed analysis, it is convenient to distinguish between two different types of kinematic schemes of part surface generation.

First, when a part surface P is considered as the stationary surface, then the combination of elementary motions of the cutting tool relative to the work represents a true kinematic scheme of part surface generation.

---

* The necessary and sufficient conditions of proper part surface generation are considered in Chapter 7.

Second, in another case, the cutting tool (or a reference system to which the cutting tool will be associated after being designed) can be considered as the stationary element. Then, the combination of elementary motions of the work relative to the coordinate system of the cutting tool can be considered as the kinematic scheme of *profiling* of the cutting tool.

The kinematic schemes of profiling of a form-cutting tool are used for solving the direct principal problem of part surface generation. The application of kinematic schemes of this particular type makes it possible the determination of the generating surface T of the form-cutting tool for machining a given part surface P. To solve the inverse principal problem of part surface generation, the true kinematic schemes of part surface generation are used. The application of the true kinematic schemes of part surface generation makes possible the determination of the actually machined part surface P_{act}.

Machining operations of actual part surfaces often include motions that cause sliding of the part surface P, the cutting tool surface T, or both over themselves.

The sliding of a part surface being machined makes possible the generation of the entire part surface P, and so forth. In the production of cutting tools, for example, when releaving the teeth of a gear hob, of a hob for machining splines, and so forth, the generating surface of the cutting tool being machined can perform sliding motion over itself.

The sliding of the generating surface T of the cutting tool (that is, rotation of a form grinding wheel in a part surface finishing operation) makes it possible to obtain the required speed of cutting. Obtaining the required speed of cutting is significantly simplified due to motions of this type.

It is instructive to point out here that the relative motions that cause sliding either of the part surface P or of the tool surface T over them are not a part of the kinematic schemes of part surface generation. They are considered separately from the kinematic scheme of part surface generation.

Various types of kinematic schemes of part surface generation are feasible. Aiming for the development of the easiest possible machining operations, it is recommended that a limited number of elementary motions of simple types (namely, translations and rotations) be used to comprise a kinematic scheme for the generation of a specified part surface. Usually, the total number of elementary relative motions in a kinematic scheme does not exceed five motions. Because of this, the kinematic schemes of part surface generation that are composed of three or fewer translations and of three or fewer rotations are the most widely implemented in practice.

As the total number of elementary relative motions in a kinematic scheme of part surface generation is limited, then the total number of kinematic schemes is not infinite and, moreover, it is relatively small. Kinematic schemes of part surface generation having more than three elementary motions saw limited implementation in the industry. Therefore, the consideration below is limited to the analysis of kinematic schemes of part surface generation having three or fewer elementary motions. Kinematic schemes with more complex structures are considered briefly.

Generally speaking, the instant motion of a rigid body in E_3 space can be represented as a combination of an instant translation and an instant rotation along and about a particular line or axis. The combination of the translation and of the rotation of the rigid body in E_3 space is commonly referred to as an *instant screw motion*. Apparently, this is because the instant screw motion of the rigid body resembles in part the motion that is performed by a bolt or a screw. Consideration of instant motion of a rigid body in E_3 space is an appropriate starting point at which to begin the investigation of the kinematic schemes of part surface generation.

Resultant relative motion of the cutting tool can be reduced to a corresponding kinematic scheme of part surface generation only in cases of non-agile kinematics of part surface machining. Under such a scenario, for the investigation of all possible types of kinematic schemes of part surface generation, rotation vectors that are associated with the work and with the cutting tool can be employed. Implementation of the rotation vectors is useful for the analysis and visualization of mating surfaces in a certain part surface machining operation. Rotation vector of a work, ω_P, is along the axis of rotation of the work, while rotation vector of the cutting tool, ω_T, is along the axis of rotation of the cutting tool.

2.5.2 Representation of the Kinematic Schemes of Part Surface Generation by Means of Rotation Vectors

Rotational motion is a common type of relative motion that is widely used when machining part surfaces on machine tools. For solving practical problems in the field of part surface machining it is often convenient to represent a rotational motion in the form of the corresponding rotation vector, namely, as a straight line segment pointed along the axis of rotation in a proper direction. It is of critical importance to point out from the very beginning that rotation in nature is not a vector. However, rotations can be treated as vectors if proper care is taken. For example, rotation vectors do not obey the commutative law. If this is taken into account, then a rotational motion can be treated as a corresponding vector.

A rotation vector ω_P can be associated with the work, while another rotational vector ω_T can be associated with the cutting tool. Rotations of the work and of the cutting tool about axes that cross each other is the most general case of the relative rotation that is considered in this monograph. An advantage of such a representation of a kinematic scheme of part surface machining is due to the following. At every instance, a pair of rotation vectors ω_P and ω_T is equivalent to a vector of instant screw motion and a vector of relative sliding of the rotating surfaces. Thus, two skew rotation vectors ω_P and ω_T cover the most general type of instant motion of two surfaces relative to one another, namely, an instant screw motion. Moreover, incorporating into consideration of the relative sliding makes such an approach even more general.

More general types of kinematic schemes of part surface generation theoretically are feasible. However, they are too complex and are too far from the practical needs of the current industry. Therefore, the consideration in this book is limited to two rotations ω_P and ω_T about skew axis as well as to reduced cases of such a configuration of two rotation vectors.

Commonly, when a kinematic scheme of part surface machining, which is based on two rotations ω_P and ω_T about skew axis, is employed, a continuously indexing method of part surface machining is observed. Such a kinematics of the relation motion of the work and of the cutting tool is common when machining gears, splines, as well as others regular profiles. Kinematics of part surface generation of this particular type can be interpreted as meshing of two toothed bodies. Under such a scenario, the work gear and the gear-cutting tool perform motions in a properly timed manner. The interaction of the imaginary gear to be machined and of the generating surface of the gear-cutting tool (which also is a phantom surface) can be interpreted as a type of virtual gear mesh. The virtual mesh of the imaginary gear to be machined and of the generating surface of the gear-cutting tool can also be referred to as *gear machining mesh*. Gear machining mesh allows for interpretation in the form of a *work gear-to-generating-surface* mesh or briefly as *P–T* mesh.

The kinematic scheme of part surface generation that is based on two rotations ω_P and ω_T about skew axis plays an important role in investigation of methods of part surface machining, and in designing of cutting tools for this purpose. Because of this, an analysis of all possible types of gear machining meshes having constant tooth ratio is of critical importance for the researchers in the field of mechanical engineering as well as for cutting tool designers. Implementation of the vectors of rotation ω_P and ω_T about skew axis is helpful when investigating kinematic schemes of part surface machining of all types. All possible types of kinematic schemes of part surface machining can be interpreted as a reduced case of this general kinematic scheme.

The relative motions of the work and of the generating surface of the cutting tool, those performed when machining the part, are a type of uniform motion. This means that the rotation of the work ω_P and the rotation of the generating surface of the cutting tool ω_T are timed so that the ratio ω_P/ω_T is constant in time.

A uniform translation and a uniform rotation are perfect examples of uniform motions. Under certain circumstances, a uniform translation and a uniform rotation can be combined so that the resultant motion is a type of a uniform screw motion. Screw motion is the most general type of uniform motion.

A uniform screw motion allows for representation in the form of summa of two motions. Translation is one of the motions. The translation vector is designated as **V**. Rotation is the other of the motion. The rotation vector is designated as ω.

The translation vector **V** and the rotation vector **ω** are sliding vectors. This means that a cutting tool designer is free to choose an appropriate point at which the vectors **V** and **ω** are applied.

Both the velocity vectors **ω** and **V** that comprise the uniform screw motion are collinear vectors. They are acting along the axis O_{sc} of the screw motion. The axis O_{sc} of the screw motion is aligned with the axis O of the rotation vector **ω**.

Actually, the vectors **ω** and **V** are pointed either in the same direction or in opposite directions. Depending on the actual configuration of the velocity vectors **ω** and **V**, the resultant uniform screw motion is recognized either as the *left-hand-oriented* or as the *right-hand-oriented* screw motion. When the vectors **ω** and **V** are pointed in the same direction, then the resultant screw motion is recognized as a left-hand-oriented screw motion. Otherwise (when the vectors **ω** and **V** are of opposite directions), the resultant screw motion is recognized as the right-hand-oriented screw motion.

In a part surface machining process, the work and the cutting tool are rotated about their axis O_P and O_T, respectively. The inequalities $\omega_P \neq 0$ and $\omega_T \neq 0$ are commonly valid. Only in particular cases, that is, when machining a gear with the rack cutter and so forth, does the cutting tool rotation $\omega_T = 0$. In case of "gear-to-rack" mesh, the angular velocity vector ω_T degenerates to zero magnitude, but the direction of this rotation vector remains the same.

The rotation vector ω_P is aligned with the axis O_P of rotation of the work. The vector ω_P is pointed in the way that corresponds to right screw (Figure 2.9). Magnitude ω_P of the rotation vector ω_P of the work ($\omega_P = |\omega_P|$) is proportional to the rotation (*RPM*) of the work. A point at which the rotation vector ω_P is applied is discussed below.

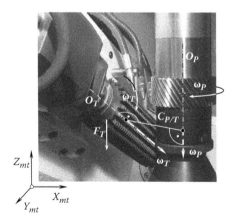

FIGURE 2.9
Configuration of the rotation vectors ω_P and ω_T of the work-gear and of the gear-cutting tool in the gear hobbing operation.

As shown in Figure 2.9, the rotation vector $\boldsymbol{\omega}_T$ is assigned to the hob. The rotation vector $\boldsymbol{\omega}_T$ is similar to the rotation vector $\boldsymbol{\omega}_P$. The vector is $\boldsymbol{\omega}_T$ aligned with the axis of rotation O_T of the hob. The direction of the rotation vector $\boldsymbol{\omega}_T$ corresponds to the hand of screw of the generating surface of the hob. The magnitude ω_P of the rotation vector $\boldsymbol{\omega}_T$ $(\omega_T = |\boldsymbol{\omega}_T|)$ is proportional to the hob rotation. A point at which the vector $\boldsymbol{\omega}_T$ is applied is discussed below.

Magnitudes ω_g and ω_c of the rotations $\boldsymbol{\omega}_P$ and $\boldsymbol{\omega}_T$ relate to one another in compliance with the proportion

$$\omega_P \cdot N_P = \omega_T \cdot N_T \tag{2.39}$$

where N_P designate the number of teeth of the work gear and N_T is the number of teeth/starts of the gear-cutting tool.

Axes O_P and O_T of the rotations $\boldsymbol{\omega}_P$ and $\boldsymbol{\omega}_T$ are apart from one another at a certain distance $C_{P/T}$. The distance $C_{P/T}$ is measured along the *center-line* and is referred to as *center-distance*. The length of the center-distance is equal to the closest distance of approach of the crossing axes O_P and O_T.

The angle between the crossing axes O_P and O_T is an important consideration.

The angle between two skew straight lines L_1 and L_2 is usually measured between one of the lines, say between the line L_1, and some other straight line that intersects the line L_1 and is parallel to the line L_2, as illustrated in Figure 2.10. In this case, either the acute angle Σ, or the obtuse angle Σ^* can be used for the specification of the crossed-angle between the two skew lines L_1 and L_2 (Figure 2.10a). Both cases are equivalent to each other.

When certain vectors $\mathbf{s}_1$ and $\mathbf{s}_2$ are associated with the straight lines L_1 and L_2, the crossed-axis angle between the lines can be specified uniquely. Depending on actual directions of the vectors $\mathbf{s}_1$ and $\mathbf{s}_2$, the crossed-axis angle Σ is either acute as shown in Figure 2.10b or obtuse as illustrated in Figure 2.10c. No ambiguities are observed with specification of the crossed-axis angle Σ if vectors are assigned to the crossing straight lines L_1 and L_2.

The crossing angle between the axes O_P and O_T in the gear machining mesh is designated below as Σ. The crossed-axis angle Σ is the angle that

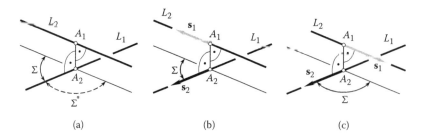

(a)   (b)   (c)

FIGURE 2.10
Crossed-axis angle $\mathbf{s}_1$ between two skew straight lines $\mathbf{s}_1$ and L_2.

the rotation vector of the gear $\boldsymbol{\omega}_P$ and the rotation vector of the hob $\boldsymbol{\omega}_T$ form with each other $[\Sigma = \angle(\boldsymbol{\omega}_P, \boldsymbol{\omega}_T)]$. For an acute crossed-axis angle $\Sigma < 90°$, its actual value is equal to that which is commonly understood under the term *crossed-axis angle* or the term *shaft angle*. In the case of obtuse angle $\Sigma > 90°$, the equality $\bar{\Sigma} = 180° - \Sigma$ is observed. Here, $\bar{\Sigma}$ denotes the crossed-axis angle in common meaning of the term.

Kinematics of most of the methods of gear machining operations is not limited to the motions that are required for machining of the gear. In the example, shown in Figure 2.9, in addition to rotations $\boldsymbol{\omega}_P$ and $\boldsymbol{\omega}_T$, one more motion is performed when machining a gear. This motion is the axial feed F_T of the cutter. The feed motion F_T is performed in the axial direction of the work gear. Depending upon design peculiarities of the hobber, either the hob is traveling in relation to the work gear, or the worktable of the hobber is traveling relative to the rotating hob.

The feed motion F_T of the cutter causes sliding of the gear tooth surface P over itself. Because of this, when investigating the kinematics of a gear machining operation, feed motions of this sort can be omitted from consideration.

The kinematics of the gear-machining process shown in Figure 2.9 is widely used on conventional hobbers for machining cylindrical gears, both spur gears and helical gears as well. It is easy to see that in nature, shown in Figure 2.9, the kinematics of the gear-machining process is an invertible one. The invertibility of the kinematics of the gear-machining process means that the work gear can be replaced with the gear-cutting tool and the gear-cutting tool can be replaced with the work gear as well. For example, in Figure 2.9, the work gear can be replaced with the gear shaper, while the hob can be replaced with blank of a worm. Under such a scenario, the worm can be machined when performing that same rotations $\boldsymbol{\omega}_P$ and $\boldsymbol{\omega}_T$, and moving the cutter in axial direction of the work-gear axis O_P with feed motion F_T. In this case, the feed motion F_T causes sliding of the worm surface over itself. Machining operation of this type is physically possible; however, it is not used in current practice.

The conversion of kinematic scheme of part surface generation into real kinematics of part surface machining similar to that just considered is valid for all types of kinematics of gear-machining processes. When converting a principal kinematics of part surface generation into its practical version, it is recommended to keep in mind the possibility of both principal kinematics of the part surface generation, that is, of the original kinematics and of the converted kinematics of a part surface machining process.

The relative motion of the work and the cutting tool can be specified by the vector $\boldsymbol{\omega}_{pl}$ of instant rotation. For the determination of the rotation vector $\boldsymbol{\omega}_{pl}$, the implementation of the principle of inversion of the motions is convenient. This principle had wide application for solving a variety of problems in kinematics and so forth. The principle of the motion conversion is schematically illustrated in Figure 2.11.

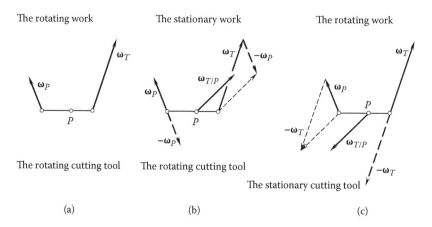

FIGURE 2.11
The principle of inversion of two rotations.

In reality, the work and the cutting tool are rotated about their axes. The rotations are specified by the rotation vectors $\boldsymbol{\omega}_P$ and $\boldsymbol{\omega}_T$ as shown in Figure 2.11a. To get the work stationary, the work must undergo one more rotation $-\boldsymbol{\omega}_P$. The summa of the initial rotation $\boldsymbol{\omega}_P$ and of the additional rotation in the opposite direction $-\boldsymbol{\omega}_P$ is equal to zero $\boldsymbol{\omega}_P + (-\boldsymbol{\omega}_P) \equiv 0$. In this way, the work becomes stationary.

To make no changes to the rest of elements of the part surface machining kinematics, the cutting tool must also perform the additional rotation $-\boldsymbol{\omega}_P$, which is shown in Figure 2.11b. Under such a scenario, the resultant rotation of the cutting tool relative to the motionless work is specified by the vector of instant rotation:

$$\boldsymbol{\omega}_{T/P} = \boldsymbol{\omega}_T + (-\boldsymbol{\omega}_P) \tag{2.40}$$

The velocity vector $\boldsymbol{\omega}_{T/P}$ of the instant rotation of the cutting tool with respect to the work is applied at the pitch point P (Figure 2.11b).

Similar to the manner previously discussed, the vector $\boldsymbol{\omega}_{P/T}$ of instant rotation of the work with respect to the cutting tool can be constructed (Figure 2.11c). This rotation vector is equal:

$$\boldsymbol{\omega}_{P/T} = \boldsymbol{\omega}_P + (-\boldsymbol{\omega}_T) \tag{2.41}$$

Evidently, rotation vectors $\boldsymbol{\omega}_{T/P}$ and $\boldsymbol{\omega}_{P/T}$ are of the same magnitude $[|\boldsymbol{\omega}_{T/P}| = |\boldsymbol{\omega}_{P/T}|]$, and they are pointed opposite to one another.

Assume that the work and the cutting tool together are rotated about the work axis O_P. This rotation is specified by the rotation vector $-\boldsymbol{\omega}_P$. As the rotation $-\boldsymbol{\omega}_P$ is applied, the work becomes stationary:

$$\mathbf{\omega}_P + (-\mathbf{\omega}_P) \equiv 0 \tag{2.42}$$

The resultant rotation of the cutting tool can be specified by the vector $\mathbf{\omega}_T + (-\mathbf{\omega}_P)$, which is equal to the vector $\mathbf{\omega}_{pl}$ of instant rotation of the gear-cutting tool relative to the work:

$$\mathbf{\omega}_{pl} = \mathbf{\omega}_T + (-\mathbf{\omega}_P) \equiv 0 \tag{2.43}$$

Similarly, the rotation vector of the work relative to the cutting tool can be determined. Evidently, this rotation vector is equal to $-\mathbf{\omega}_{pl}$.

The rotation vector $\mathbf{\omega}_{pl}$ is aligned with the instant axis of rotation P_{ln}. The axis of instant rotation is commonly called the *pitch line P_{ln}*. Once the configuration of the rotation vectors $\mathbf{\omega}_P$ and $\mathbf{\omega}_T$ is known, say the rotation vectors $\mathbf{\omega}_P$ and $\mathbf{\omega}_T$ are specified in a certain reference system, then the vector $\mathbf{\omega}_{pl}$ of the instant rotation as well as the pitch line P_{ln} can be determined following common rules known from vector algebra.

As an example, consider two rotation vectors $\mathbf{\omega}_P$ and $\mathbf{\omega}_T$ of a work and of the form-cutting tool, as shown in Figure 2.12. The configuration of the rotation vectors $\mathbf{\omega}_P$ and $\mathbf{\omega}_T$ is initially given in the system of two planes of projections π_1 and π_2. Vectors $\mathbf{\omega}_P$ and $\mathbf{\omega}_T$ are the principal elements of the kinematic scheme of part surface machining. The rest of the important elements of kinematics of

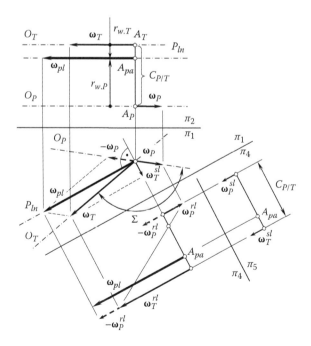

FIGURE 2.12
Vector diagram of the *gear machining mesh.*

the part surface machining process can be drawn up from the *vector diagram*. The vector diagram in Figure 2.12 is constructed on the basis of the rotation vectors ω_P and ω_T of the work and of the form-cutting tool, respectively.

Generally speaking, the rotation vectors ω_P and ω_T can occupy an arbitrary location and orientation in the reference system $\pi_1\pi_2$. Just for convenience of analysis, the vectors ω_P and ω_T are shown parallel to the plane of projections π_1. The crossed-axis angle Σ is projected onto the horizontal plane of projections π_1 with no distortion. Similarly, the center-distance $C_{P/T}$ is projected onto the vertical plane of projections π_2 with no distortion.

In the plane of projections π_1, the vector ω_{pl} of instant rotation is determined in terms of the rotation vectors ω_P and ω_T (see Equation 2.43). The rotation vector ω_{pl} is applied at the apex point A_{pa}, the location of which is not specified yet. It is known only that the apex point A_{pa} is located somewhere within the straight-line segment $C_{P/T}$.

An additional plane of projections π_4 is constructed for determining of the location of the apex point A_{pa}. Within the plane of projections π_1, the axis of projections π_1/π_4 is passing parallel to the vector of instant rotation ω_{pl}. Due to this, the components ω_P^{rl} and ω_T^{rl} of the rotation vectors ω_P and ω_T are projected with no distortions onto the plane of projections π_4. Both the components ω_P^{rl} and ω_T^{rl} are parallel to the vector of instant rotation ω_{pl}:

$$\omega_P^{rl} \left\| \omega_T^{rl} \right\| \omega_{pl} \tag{2.44}$$

The magnitudes ω_P^{rl} and ω_T^{rl} of the components ω_P^{rl} and ω_T^{rl} relate to one another so that the equality

$$\omega_P^{rl} \cdot r_{w.P} = \omega_T^{rl} \cdot r_{w.T} \tag{2.45}$$

is valid. In Equation 2.45, $r_{w.P}$ and $r_{w.T}$ denote the distances of the apex point A_{pa} from the axis of rotation of the part O_P and from the form-cutting tool axis O_T, respectively.

Distances $r_{w.P}$ and $r_{w.T}$ are signed values. Their algebraic summa is equal to the center-distance $C_{P/T}$:

$$r_{w.P} + r_{w.T} = C_{P/T} \tag{2.46}$$

Ultimately, the components ω_P^{rl} and ω_T^{rl} result in pure rolling of the axodes of the work-gear and of the gear-cutting tool over each other.

It is important to point out here that the components ω_P^{rl} and ω_T^{rl} of the vectors ω_P and ω_T represent instant rotations, while the vectors ω_P and ω_T describe continuous rotations.

In the plane of projections π_4, rotation vectors ω_P and ω_T are constructed following the conventional rules developed in descriptive geometry. The location of the apex A_{pa} within the center-distance $C_{P/T}$ is getting evident immediately after the projections of the rotation vectors $\boldsymbol{\omega}_P$ and $\boldsymbol{\omega}_T$ are constructed in the plane of projections π_4. Further, the projection of the point A_{pa} can be constructed in the plane of projections π_2.

Two other components $\boldsymbol{\omega}_P^{sl}$ and $\boldsymbol{\omega}_T^{sl}$ of the rotation vectors $\boldsymbol{\omega}_P$ and $\boldsymbol{\omega}_T$ are perpendicular to the vector of instant rotation $\boldsymbol{\omega}_{pl}$:

$$\boldsymbol{\omega}_P^{sl} \perp \boldsymbol{\omega}_{pl} \tag{2.47}$$

$$\boldsymbol{\omega}_T^{sl} \perp \boldsymbol{\omega}_{pl} \tag{2.48}$$

With no distortions, the components $\boldsymbol{\omega}_P^{sl}$ and $\boldsymbol{\omega}_T^{sl}$ of the rotation vectors $\boldsymbol{\omega}_P$ and $\boldsymbol{\omega}_T$ are projected onto the plane of projections π_5. The axis of projections π_4/π_5 here is erected perpendicular to the axis of projections π_1/π_4. The components $\boldsymbol{\omega}_P^{sl}$ and $\boldsymbol{\omega}_T^{sl}$ rotation vectors $\boldsymbol{\omega}_P$ and $\boldsymbol{\omega}_T$ are equal to each other; therefore, the equality $\boldsymbol{\omega}_P^{sl} = \boldsymbol{\omega}_T^{sl}$ is valid. The components $\boldsymbol{\omega}_P^{sl}$ and $\boldsymbol{\omega}_T^{sl}$ cause pure sliding of the axodes associated with the work and with the form-cutting tool relative to each other.

It is important to point out here that, similar to the vectors $\boldsymbol{\omega}_P^{rl}$ and $\boldsymbol{\omega}_T^{rl}$, the components $\boldsymbol{\omega}_P^{sl}$ and $\boldsymbol{\omega}_T^{sl}$ of the vectors $\boldsymbol{\omega}_P$ and $\boldsymbol{\omega}_T$ also represent instant rotations, while the rotations described by the vectors $\boldsymbol{\omega}_P$ and $\boldsymbol{\omega}_T$ are continuous rotations.

The equality $\boldsymbol{\omega}_P^{rl} = \boldsymbol{\omega}_T^{rl}$ should not be confusing. Besides components $\boldsymbol{\omega}_P^{sl}$ and $\boldsymbol{\omega}_T^{sl}$ are pointed in that same direction, the corresponding linear sliding velocities $\mathbf{V}_P^{sl}$ and $\mathbf{V}_T^{sl}$ possess the following properties:

(a) The vectors $\mathbf{V}_P^{sl}$ and $\mathbf{V}_T^{sl}$ are pointed in opposite directions.
(b) The vectors $\mathbf{V}_P^{sl}$ and $\mathbf{V}_T^{sl}$ are of different magnitudes.

The first is because of the apex A_{pa} is located within the center-distance $C_{P/T}$ and not outside this straight-line segment. The second is because usually the distances $r_{w.P}$ and $r_{w.T}$ are not equal to each other; thus, the inequality $r_{w.P} \neq r_{w.T}$ is commonly observed.

In particular cases, the axodes of the rotating work and of the form-cutting tool are used for the analysis of the part surface machining operation. Axodes are two surfaces of revolution, one of which is associated with the work and another one is associated with the form-cutting tool. When a continuously indexing method of machining is used for machining a work, the axodes roll over each other. It is commonly understood that the axodes are rolling with no sliding between them. It is instructive to note here that the components $\boldsymbol{\omega}_P^{rl}$ and $\boldsymbol{\omega}_T^{rl}$ of the rotations $\boldsymbol{\omega}_P$ and $\boldsymbol{\omega}_T$ result in pure rolling of the axodes (with no slippage between them). However, the components $\boldsymbol{\omega}_P^{sl}$ and $\boldsymbol{\omega}_T^{sl}$ always cause sliding of the axodes when the rotation vectors $\boldsymbol{\omega}_P$ and $\boldsymbol{\omega}_T$ are along two crossing straight lines. This type of sliding is inevitable in nature when two rotations are performed about crossing axes.

Usually, the rotations $\boldsymbol{\omega}_P^{sl}$ and $\boldsymbol{\omega}_T^{sl}$ are strongly undesired when designing, for example, an elementary gear drive. It is common practice for a gear designer to reduce the sliding velocity to the lowest possible range. The components $\boldsymbol{\omega}_P^{sl}$ and $\boldsymbol{\omega}_T^{sl}$ result in the sliding of the axodes of the gear and of the mating pinion. The sliding of the axodes unavoidably entails the corresponding sliding of the tooth flanks of the gear and of the pinion in relation to one another. In this way, sliding components $\boldsymbol{\omega}_P^{sl}$ and $\boldsymbol{\omega}_T^{sl}$ reduce performance of the elementary gear drive.

On the other hand, the components $\boldsymbol{\omega}_P^{sl}$ and $\boldsymbol{\omega}_T^{sl}$ of the rotation vectors $\boldsymbol{\omega}_P$ and $\boldsymbol{\omega}_T$ could be useful when designing a gear-cutting tool. The relative sliding of the axodes can be employed as the primary (cutting) motion. Under such a scenario, it is strongly desired to increase the sliding velocity to the highest possible range; this way, the efficiency of the gear-machining process can be significantly improved.

A conclusion can be drawn from the analysis of Figure 2.12. The conclusion states that pure rolling of the axodes of the work and of the form-cutting tool occurs if and only if the rotation vectors $\boldsymbol{\omega}_P$ and $\boldsymbol{\omega}_T$ are located within a common plane. Otherwise, the sliding of the axodes is inevitable in nature. For kinematic schemes of part surface generation that feature either parallel or intersecting axes of rotation, the axes O_P and O_T are located within a common plane by definition. Therefore, no sliding is observed in these cases. Sliding of the axodes occurs if and only if the work and the form-cutting tool are rotated about skew axis O_P and O_T, or, in another words, only rotations about skew axes are featuring relative sliding of the axodes.

The sliding of the part surface to be machined and of the generating surface of the cutting tool is proportional to the sliding rotation $\omega^{sl} = \omega_P^{sl} \equiv \omega_T^{sl}$. For a given point of interest within the line of contact of the axodes, the sliding between the surfaces P and T is also proportional to the distance of the point of intersect from the pitch line P_{ln}. Keep in mind that the configuration of the pitch line P_{ln} in relation to the axes of rotations O_P and O_T is a function of the center-distance $C_{P/T}$.

The following conclusions can be drawn from the above analysis:

- No sliding of the axodes is observed when axes of rotation of the work and of the form-cutting tool are located within a common plane.
- Only rotations about skew axes are featuring relative sliding of the axodes.
- For the rotations about skew axes, sliding of the axodes in the kinematic scheme of part surface generation becomes more significant when the crossed-axis angle Σ becomes bigger. Thus, the condition $\Sigma \rightarrow 180°$ can be interpreted as the condition for maximal sliding of axodes in the kinematic scheme of part surface generation.

- For the rotations about skew axes, sliding of the axodes in the kinematic scheme of part surface generation becomes smaller when the crossed-axis angle Σ becomes smaller. Thus, the condition $\Sigma \to 0°$ can be interpreted as the condition for minimum sliding of the axodes in the kinematic scheme of part surface generation.

Actually, a kinematic scheme of part surface generation can be specified just by two rotation vectors ω_P and ω_T and by center-distance $C_{P/T}$. The rest of the vectors used for the analysis of the kinematic scheme of part surface generation can be expressed in terms of the vectors ω_P and ω_T and of the center-distance $C_{P/T}$.

2.5.3 Kinematic Relationships for the Kinematic Scheme of Part Surface Generation

Consider two rotations about the skew axes, as schematically illustrated in Figure 2.13. From the geometrical and kinematic standpoints, two rotations about skew axes O_P and O_T can be interpreted as rolling of a hyperboloid of one sheet over another hyperboloid of one sheet. The one-sheet hyperboloids of rotation serve there as axodes in the kinematic scheme of part surface generation. The axes of rotation of the axodes align with the vectors of rotations

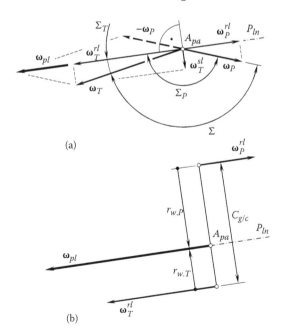

FIGURE 2.13
Derivation of the principal kinematic relationships in a kinematic scheme of part surface generation.

$\boldsymbol{\omega}_P$ and $\boldsymbol{\omega}_T$. The straight generating line of the axodes aligns with the vector of instant rotation $\boldsymbol{\omega}_{pl}$. The vector $\boldsymbol{\omega}_{pl}$ describes the instant rotation of the axodes about the pitch line P_{ln}.

Two rotation vectors $\boldsymbol{\omega}_P$ and $\boldsymbol{\omega}_T$ are crossing at a certain crossed-axis angle Σ. The axes of the rotations $\boldsymbol{\omega}_P$ and $\boldsymbol{\omega}_T$ are apart from each other at a center-distance $C_{P/T}$ as shown in Figure 2.13. For the given configuration of the rotation vectors $\boldsymbol{\omega}_P$, and $\boldsymbol{\omega}_T$ the corresponding vector of instant rotation $\boldsymbol{\omega}_{pl}$ is constructed in Figure 2.13a.

Consider the projections of the rotation vectors $\boldsymbol{\omega}_P$, $\boldsymbol{\omega}_T$, and $\boldsymbol{\omega}_{pl}$ onto a plane that is orthogonal to the straight line segment $C_{P/T}$ (Figure 2.13b). The projections of the rotation vectors $\boldsymbol{\omega}_P$ and $\boldsymbol{\omega}_T$ onto the pitch line P_{ln} are designated as $\boldsymbol{\omega}_P^{rl}$ and $\boldsymbol{\omega}_T^{rl}$, respectively. These components cause pure rotation of the axodes. The projections of the rotation vectors $\boldsymbol{\omega}_P$ and $\boldsymbol{\omega}_T$ onto the direction that is perpendicular to the pitch line P_{ln}, are designated as $\boldsymbol{\omega}_P^{sl}$ and $\boldsymbol{\omega}_T^{sl}$. These components cause sliding of the axodes. It can be proven that the components $\boldsymbol{\omega}_P^{sl}$ and $\boldsymbol{\omega}_T^{sl}$ of the rotation vectors $\boldsymbol{\omega}_P$ and $\boldsymbol{\omega}_T$ are always of the same magnitude and direction. Therefore, the equality $\boldsymbol{\omega}_g^{sl} = \boldsymbol{\omega}_c^{sl}$ is valid.

The components $\boldsymbol{\omega}_P^{rl}$ and $\boldsymbol{\omega}_T^{rl}$ of the rotation vectors are within a common plane through the straight-line segment $C_{P/T}$. The location of the apex point A_{pa} within the straight-line segment $C_{P/T}$ is specified by the ratio

$$\omega_g^{rl} \cdot r_g = \omega_c^{rl} \cdot r_c \tag{2.49}$$

(see Equation 2.43) and is constructed in Figure 2.12.

It is instructive to note here that in particular applications, it is convenient to represent Equation 2.43 in the form

$$\frac{r_{w.g}}{(\omega_g^{rt})^{-1}} = \frac{r_{w.c}}{(\omega_c^{rt})^{-1}} \tag{2.50}$$

where the rotations ω_P^{rl} and ω_T^{rl} themselves are not used, but instead the values $(\omega_P^{rl})^{-1}$ and $(\omega_T^{rl})^{-1}$, which are reciprocal to the rotations. The inverse values $(\omega_P^{rl})^{-1}$ and $(\omega_T^{rl})^{-1}$ are often referred to as *rotation frequencies*.

Due to the equality $r_{w.P} + r_{w.T} = C_{P/T}$ (see Equation 2.46), for the radius $r_{w.T}$, the expression

$$r_{w.T} = C_{P/T} - r_{w.P} \tag{2.51}$$

is valid. After being substituted into the last expression for distance $r_{w.T}$, Equation 2.45 casts into

$$\omega_P^{rl} \cdot r_{w.P} = \omega_T^{rl} \cdot (C_{P/T} - r_{w.P}) \tag{2.52}$$

The elementary formulae transformations yield the expression

$$r_{w.P} = \frac{1 + \omega_T - \omega_P}{1 + \omega_T} \cdot C_{P/T} \tag{2.53}$$

for the calculation of the radius $r_{w.P}$ (Figure 2.13b).

After the actual value of the radius $r_{w.P}$ is calculated, the following formula is used for the calculation of radius $r_{w.T}$:

$$r_{w.T} = C_{P/T} - r_{w.P} \tag{2.54}$$

When solving particular problems of part surface generation, in addition to the vectors $\mathbf{C}_T$ and $\boldsymbol{\omega}_T^{rl}$, it is often convenient to use the corresponding normalized vectors $\bar{\boldsymbol{\omega}}_P^{rl}$ and $\bar{\boldsymbol{\omega}}_T^{rl}$. The vectors $\boldsymbol{\omega}_P^{rl}$ and $\boldsymbol{\omega}_T^{rl}$ are normalized by the value of ω_T^{rl}. In this case, the normalized angular velocity $\bar{\boldsymbol{\omega}}_T^{rl}$ of the form-cutting tool is equal to the unit vector $\bar{\boldsymbol{\omega}}_T^{rl} = \boldsymbol{\omega}_T^{rl} / \omega_T^{rl}$, ($|\bar{\boldsymbol{\omega}}_T^{rl}| \equiv 1$). Here, ω_T^{rl} designates the magnitude of the rotation vector $\boldsymbol{\omega}_T^{rl}$. The normalized angular velocity $\bar{\boldsymbol{\omega}}_P^{rl}$ of the work is equal to a certain value $\bar{\boldsymbol{\omega}}_P^{rl} = \boldsymbol{\omega}_P^{rl} / \omega_T^{rl}$, ($0 \leq \bar{\boldsymbol{\omega}}_P^{rl} < 1$). The normalized value $\bar{\omega}_T^{rl}$ is always greater or equal to the normalized value $\bar{\omega}_P^{rl} = |\bar{\boldsymbol{\omega}}_P^{rl}|$, and the inequality $\bar{\omega}_T^{rl} \geq \bar{\omega}_P^{rl}$ is valid.

The radii $r_{w.P}$ and $r_{w.T}$ of the axodes can also be normalized by $r_{w.T}$. In this case, the normalized radius of the form-cutting tool $\bar{r}_{w.T}$ is equal to unit $\bar{r}_{w.T} = r_{w.T} / r_{w.T} \equiv 1$. Accordingly, the normalized radius of the part surface to be machined $\bar{r}_{w.P}$ is equal to a certain value $\bar{r}_{w.P} = r_{w.P} / r_{w.T} \geq 1$. The normalized radius $\bar{r}_{w.T}$ is always smaller or equal to the normalized radius $\bar{r}_{w.P}$. Thus, the inequality $\bar{r}_{w.T} \leq \bar{r}_{w.P}$ is valid.

A few more formulae listed below are useful for the calculations.

For the calculation of magnitude ω_{pl} of the vector of instant rotation $\boldsymbol{\omega}_{pl}$, the following formula can be used:

$$\omega_{pl} = \sqrt{(\omega_P^{rl})^2 + (\omega_T^{rl})^2 - 2 \cdot \omega_P^{rl} \cdot \omega_T^{rl} \cdot \cos \Sigma} \tag{2.55}$$

The angle Σ_P is the angle that velocity vector $\boldsymbol{\omega}_P$ forms with vector $\boldsymbol{\omega}_{pl}$ of instant rotation [$\Sigma_P = \angle(\boldsymbol{\omega}_P, \boldsymbol{\omega}_{pl})$]. For the calculation of the angle Σ_P, the formula

$$\Sigma_P = \frac{1 + \omega_T - \omega_P}{1 + \omega_T} \cdot \Sigma \tag{2.56}$$

can be used. This formula is similar to Equation 2.53. Equation 2.56 can be derived based on the relationship

$$\frac{\Sigma_P}{\Sigma_T} = \frac{\omega_T}{\omega_P},\tag{2.57}$$

which is evident. As it follows from the analysis of Figure 2.13, for angles Σ_P and Σ_T, the equality

$$\Sigma = \Sigma_P - \Sigma_T\tag{2.58}$$

is valid.

In a particular case, angle Σ_P can be equal to the right angle. The orthogonality $\boldsymbol{\omega}_P^{rl} \perp \boldsymbol{\omega}_{pl}$ occurs when the equality $\omega_{pl} = \sqrt{(\omega_T^{rl})^2 - (\omega_P^{rl})^2}$ is fulfilled.

The condition under which the rotation vectors $\boldsymbol{\omega}_P$ and $\boldsymbol{\omega}_{pl}$ are orthogonal to one another can be represented in the form of dot product of these vectors:

$$\boldsymbol{\omega}_P \cdot \boldsymbol{\omega}_{pl} = 0\tag{2.59}$$

Similar to angle Σ_P, angle Σ_T is the angle that the rotation vector $\boldsymbol{\omega}_T$ makes with the vector $\boldsymbol{\omega}_{pl}$ of instant rotation $[\Sigma_T = \angle(\boldsymbol{\omega}_T, \boldsymbol{\omega}_{pl})]$. Angle Σ_T can be calculated from the formula

$$\Sigma_T = \frac{1 + \omega_P - \omega_T}{1 + \omega_P} \cdot \Sigma\tag{2.60}$$

After angles Σ_P and Σ_T are calculated, it can be shown that the following formulae for the projections $\boldsymbol{\omega}_P^{sl}$ and $\boldsymbol{\omega}_T^{sl}$ are valid:

$$\omega_P^{sl} = \omega_P \cdot \cos(\Sigma_P - 90°) = \omega_P \cdot \sin \Sigma_P\tag{2.61}$$

$$\omega_T^{sl} = \omega_T \cdot \cos(\Sigma_T - 90°) = \omega_T \cdot \sin \Sigma_T \equiv \omega_P^{sl}\tag{2.62}$$

Similarly, for the calculation of projections $\boldsymbol{\omega}_P^{rl}$ and $\boldsymbol{\omega}_T^{rl}$, the following formulae can be used:

$$\omega_P^{rl} = \omega_P \cdot \cos \Sigma_P\tag{2.63}$$

$$\omega_T^{rl} = \omega_T \cdot \cos \Sigma_T\tag{2.64}$$

$$\frac{\omega_P}{\omega_T} = \frac{\omega_P^{rl}}{\omega_T^{rl}} \cdot \frac{\cos \Sigma_T}{\cos \Sigma_P} = \frac{N_T}{N_P}\tag{2.65}$$

Magnitudes $\omega_P^{rl} = \left| \boldsymbol{\omega}_P^{rl} \right|$ and $\omega_T^{rl} = \left| \boldsymbol{\omega}_T^{rl} \right|$ of components $\boldsymbol{\omega}_P^{rl}$ and $\boldsymbol{\omega}_T^{rl}$ are equal, as they follow Figure 2.13a:

$$\omega_P^{rl} = \omega_P \cdot \cos \Sigma_P \tag{2.66}$$

$$\omega_T^{rl} = \omega_T \cdot \cos \Sigma_T \tag{2.67}$$

Magnitudes $\omega_P^{sl} = \left| \boldsymbol{\omega}_P^{sl} \right|$ and $\omega_T^{sl} = \left| \boldsymbol{\omega}_T^{sl} \right|$ of the components $\boldsymbol{\omega}_P^{sl}$ and $\boldsymbol{\omega}_T^{sl}$ are equal to

$$\omega_P^{sl} = \omega_P \cdot \sin \Sigma_P \tag{2.68}$$

$$\omega_T^{sl} = \omega_T \cdot \sin \Sigma_T \tag{2.69}$$

The rotation vector $\boldsymbol{\omega}_{pl}$ is applied at apex point A_{pa}, which is within the center-distance $C_{P/T}$. Location of the apex point A_{pa} meets the condition

$$\frac{r_{w.P}}{\omega_T \cdot \cos \Sigma_T} = \frac{r_{w.T}}{\omega_P \cdot \cos \Sigma_P} = \frac{C_{P/T}}{\omega_{pl}} \tag{2.70}$$

When the principle of inversion is applied to the rotations of the work $\boldsymbol{\omega}_P$ and of the form-cutting tool $\boldsymbol{\omega}_T$, then the work becomes stationary and the form-cutting tool performs the resultant instant rotation $\boldsymbol{\omega}_{pl}$ with respect to the surface being machined. Rotation $\boldsymbol{\omega}_{pl}$ is equal $\boldsymbol{\omega}_{pl} = \boldsymbol{\omega}_T + (-\boldsymbol{\omega}_P)$ (see Equation 2.43). The sliding components ω_g^{sl} and $-\omega_c^{sl}$ of the rotations $\boldsymbol{\omega}_T$ and $-\boldsymbol{\omega}_P$ are of equal magnitude and with opposite directions. They comprise a pair of rotation. This rotation pair is equivalent to a translational motion. The velocity vector $\mathbf{V}^{sl}$ of the translational motion is parallel to the vector $\boldsymbol{\omega}_{pl}$ of the instant rotation. Magnitude V^{sl} of the velocity vector $\mathbf{V}^{sl}$ is shown in the following:

$$V^{sl} = \left| \mathbf{V}^{sl} \right| = C_{P/T} \cdot \omega_P \cdot \sin \Sigma_P = C_{P/T} \cdot \omega_T \cdot \sin \Sigma_T \tag{2.71}$$

The translation vector $\mathbf{V}^{sl}$ is parallel to the rotation vector $\boldsymbol{\omega}_{pl}$. The super-position of two motions, rotation $\boldsymbol{\omega}_{pl}$ and the translation $\mathbf{V}^{sl}$, comprises a screw motion. Parameter p_{pl} of the screw motion is shown in the following:

$$p_{pl} = \frac{V^{sl}}{\omega_{pl}} = \frac{C_{P/T} \cdot \omega_P \cdot \sin \Sigma_P}{\omega_{pl}} = \frac{C_{P/T} \cdot \omega_T \cdot \sin \Sigma_T}{\omega_{pl}} \tag{2.72}$$

The magnitude ω_{pl} of the rotation vector $\boldsymbol{\omega}_{pl}$ is shown in the following:

$$\omega_{pl} = \frac{C_{P/T} \cdot \omega_T \cdot \cos \Sigma_T}{r_{w.P}} = \frac{C_{P/T} \cdot \omega_P \cdot \cos \Sigma_P}{r_{w.T}} \qquad (2.73)$$

Magnitude ω_{pl} is calculated from Equation 2.73 and can be substituted into Equation 2.72, which casts into a final formula

$$p_{pl} = r_{w.P} \cdot \tan \Sigma_T = r_{w.T} \cdot \tan \Sigma_P \qquad (2.74)$$

for the calculation of the parameter p_{pl} of the resultant instant crew motion. Ultimately, an expression for the ratio

$$\frac{r_{w.P}}{r_{w.T}} = \frac{\tan \Sigma_P}{\tan \Sigma_T} \qquad (2.75)$$

can be derived.

The above analysis reveals that the summa of two rotations about skew axes allows for interpretation in the form of an instant screw motion.

2.5.4 Configuration of the Vectors of Relative Motions

The variety of all the possible types of gear machining meshes* strongly depends upon feasible configurations of the rotation vectors $\boldsymbol{\omega}_P$ and $\boldsymbol{\omega}_T$. The rotations $\boldsymbol{\omega}_P$ and $\boldsymbol{\omega}_T$ can be implemented for the development of a scientific classification of all possible types of gear machining meshes.

2.5.4.1 Principal Features of Configuration of the Rotation Vectors

Two arbitrary rotation vectors $\boldsymbol{\omega}_P$ and $\boldsymbol{\omega}_T$ are shown in Figure 2.14. The rotation vectors $\boldsymbol{\omega}_P$ and $\boldsymbol{\omega}_T$ are at a certain center-distance $C_{P/T}$. A *Cartesian* coordinate system XYZ is associated with the rotation vectors $\boldsymbol{\omega}_P$ and $\boldsymbol{\omega}_T$ and with the straight-line segment of the $C_{P/T}$. The origin of the reference system XYZ is located at the apex point A_{pa}. Axis X is aligned with the straight-line segment $C_{P/T}$; Z axis is aligned with the vector of instant rotation $\boldsymbol{\omega}_{pl}$, and axis Y complements axes X and Z to the left-hand-oriented *Cartesian* coordinate system XYZ.

For further analysis, it is convenient to introduce a unit vector c. Vector c is along the X axis, and it is pointed in the positive direction of the X axis.

The newly introduced unit vector c allows for a simple analytical representation of the two vectors, $\mathbf{C}_P$ and $\mathbf{C}_T$, depicted in Figure 2.14. Vector $\mathbf{C}_P$

---

* Here and below, the term *gear machining mesh* should be understood in the sense of all continuously indexing types of part surface generating.

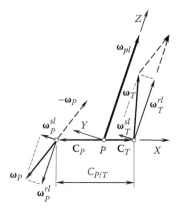

FIGURE 2.14
Specification of the principal parameters of the kinematic scheme of part surface generation.

specifies the center-distance of the axes O_P and P_{ln} (along the rotations ω_g and ω_{pl}). Vector $\mathbf{C}_T$ specifies the center-distance of the axes O_T and P_{ln} (along the rotations ω_T and ω_{pl}).

The vector $\mathbf{C}_T$ is equal to

$$\mathbf{C}_T = r_{w.T} \cdot \mathbf{c} \tag{2.76}$$

This vector is applied at the apex point A_{pa}. The normalized vector $\bar{\mathbf{C}}_T$ can be calculated from the formula

$$\bar{\mathbf{C}}_T = \bar{r}_{w.T} \cdot \mathbf{c} \tag{2.77}$$

Accordingly, the vector $\mathbf{C}_P$ is equal to

$$\mathbf{C}_P = -r_{w.P} \cdot \mathbf{c} \tag{2.78}$$

This vector is also applied at the apex point A_{pa}. The normalized vector $\bar{\mathbf{C}}_P$ can be calculated from

$$\bar{\mathbf{C}}_P = -\bar{r}_{w.P} \cdot \mathbf{c} \tag{2.79}$$

The magnitude of vector $\mathbf{C}_P$ is always greater or equal to the length of the vector $\mathbf{C}_T$, and, thus, the inequality $|\mathbf{C}_P| \geq |\mathbf{C}_T|$ is always valid. The similar is valid with respect to the normalized vectors $\bar{\mathbf{C}}_P$ and $\bar{\mathbf{C}}_T$, that is, the inequality $|\bar{\mathbf{C}}_P| \geq |\bar{\mathbf{C}}_T|$ is always valid as well.

The center-distance $C_{P/T}$ can be expressed in terms of the vectors $\mathbf{C}_P$ and $\mathbf{C}_T$. For this purpose, the following expression can be used:

$$C_{P/T} = |\mathbf{C}| = |\mathbf{C}_T - \mathbf{C}_P| \tag{2.80}$$

The similar equation is valid for the normalized parameters $\bar{\mathbf{C}}_P$, $\bar{\mathbf{C}}_T$, and $\bar{C}_{P/T}$:

$$\bar{C}_{T/P} = |\bar{\mathbf{C}}| = |\bar{\mathbf{C}}_T - \bar{\mathbf{C}}_P| \tag{2.81}$$

By definition, the normalized radius is equal $\bar{r}_{w.T} = 1$. As the inequality $\bar{r}_{w.P} \geq \bar{r}_{w.T}$ is always observed, then the following expression is valid for the normalized center-distance $\bar{C}_{P/T}$:

$$\bar{C}_{P/T} = \bar{r}_{w.P} + \bar{r}_{w.T} \geq 2 \tag{2.82}$$

Equations 2.81 and 2.82 impose strong restrictions on the configuration of axodes of the kinematic scheme of part surface generation.

In general, rotation vectors $\boldsymbol{\omega}_P$ and $\boldsymbol{\omega}_T$ cross each other at a certain crossed-axis angle Σ. For all feasible combinations of the vectors of rotations $\boldsymbol{\omega}_P$ and $\boldsymbol{\omega}_T$, the magnitude of the crossed-axis angle Σ is within the interval $0° \leq \Sigma \leq 180°$. Depending on design parameters of the actual kinematic scheme of part surface generation, three different locations of the apex A_{pa} are recognized:

- First, apex A_{pa} can be located within the straight-line segment $C_{P/T}$.
- Second, apex A_{pa} can be located outside the straight-line segment $C_{P/T}$.
- Third, apex A_{pa} can be located at one of two endpoints of the straight-line segment $C_{P/T}$.

No other locations of the apex A_{pa} are feasible.

The location of the apex A_{pa} within the straight-line segment $C_{P/T}$ indicates that for a given configuration of the rotation vectors $\boldsymbol{\omega}_P$ and $\boldsymbol{\omega}_T$, the corresponding axodes make external tangency with each other. When the apex A_{pa} is located outside the straight-line segment $C_{P/T}$, the axodes are in internal tangency to one another.

The third case, when the apex A_{pa} is located at one of two endpoints of the straight-line segment $C_{P/T}$, requires detailed discussion. Because the configuration of the rotation vectors in the third case relates exactly to what is between the first case (when the axodes are in external tangency) and the second case (when the axodes are in internal tangency), the third case is referred to as that featuring *critical configuration* of the rotation vectors $\boldsymbol{\omega}_P$ and $\boldsymbol{\omega}_T$.

The term "critical configuration" reflects a particular configuration of the rotation vectors $\boldsymbol{\omega}_P$ and $\boldsymbol{\omega}_T$, for which apex A_{pa} is located at one of two ends of the straight-line segment $C_{P/T}$.

For a certain configuration of the rotation vectors $\boldsymbol{\omega}_P$ and $\boldsymbol{\omega}_T$, rotation vector $\boldsymbol{\omega}_P$ is orthogonal to the vector of instant rotation $\boldsymbol{\omega}_{pl}$ as schematically illustrated in Figure 2.15. In this case, apex A_{pa} is at zero-distance $r_{w.T} = 0$ from the axis of rotation O_T of the form-cutting tool. The equality $\boldsymbol{\omega}_P \cdot \boldsymbol{\omega}_{pl} = 0$ is always fulfilled for the orthogonal vectors $\boldsymbol{\omega}_P$ and $\boldsymbol{\omega}_{pl}$.

Setting in Equation 2.60 $\Sigma_T = 90°$, one can come up with the expression

$$\Sigma_{cr}^{(1)} = \frac{1+\omega_P}{1+\omega_P - \omega_T} \cdot \frac{\pi}{2} \tag{2.83}$$

for the calculation of the first critical value $\Sigma_{cr}^{(1)}$ of the crossed-axis angle Σ.

Equation 2.83 pertains to the scenario under which the rotation vector $\boldsymbol{\omega}_T$ of the form-cutting tool is orthogonal to the vector of instant rotation $\boldsymbol{\omega}_{pl}$.

Apex A_{pa} can be located at the opposite end of the straight-line segment $C_{P/T}$. In this second case, by setting $\Sigma_P = 90°$ into Equation 2.56, the expression

$$\Sigma_{cr}^{(2)} = \frac{1+\omega_T}{1+\omega_T - \omega_P} \cdot \frac{\pi}{2} \tag{2.84}$$

can be derived for the second critical value $\Sigma_{cr}^{(2)}$ of the crossed-axis angle Σ.

It should be stressed here that cases with rotation vectors $\boldsymbol{\omega}_P$ and $\boldsymbol{\omega}_T$ either forming the crossed-axis angle $\Sigma_{cr}^{(1)}$ or forming the crossed-axis angle $\Sigma_{cr}^{(2)}$ are poorly investigated in contemporary literature on part surface generation, particularly, in the theory of gearing.

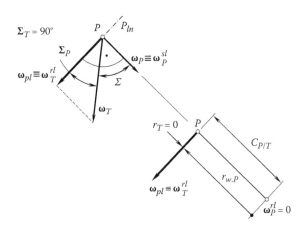

FIGURE 2.15
An example of the *critical* configuration of the rotation vectors $\boldsymbol{\omega}_P$ and $\boldsymbol{\omega}_T$.

2.5.4.2 Classification of Possible Types of Principal Kinematic Schemes of Part Surface Generation

A scientific classification of all possible types of principal kinematic schemes of part surface generation is developed on the premises of the proposed vector representation of kinematics of part surface machining operations. All possible types of the kinematic schemes of part surface generation are classified as shown in Figure 2.16.

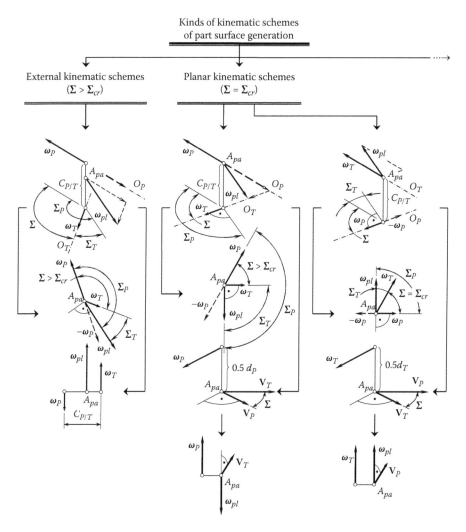

FIGURE 2.16
Classification of types of kinematic schemes of part surface generation.

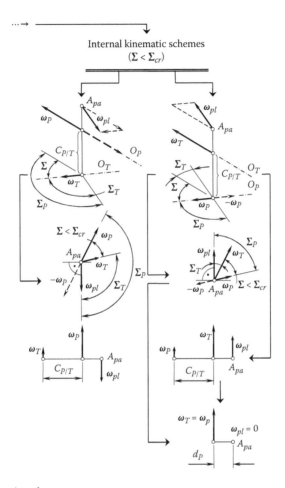

FIGURE 2.16 (Continued)
Classification of types of kinematic schemes of part surface generation.

All possible types of kinematic schemes of part surface generation are sub-divided to three groups:

(a) **External kinematic schemes of part surface generation.** Kinematic schemes of this group feature the crossed-axis angle Σ that is greater that its critical value Σ_{cr}. For the kinematic schemes of part surface generation, the inequality $\Sigma > \Sigma_{cr}$ is valid. The apex A_{pa} is located within the center-distance $C_{P/T}$.

(b) **Planar kinematic schemes of part surface generation.** Kinematic schemes of this group feature the crossed-axis angle Σ that is equal to its critical value Σ_{cr}. For the kinematic schemes of part surface generation, the equality $\Sigma = \Sigma_{cr}$ is valid. The apex A_{pa} is located at one of two ends of the center-distance $C_{P/T}$.

(c) **Internal kinematic schemes of part surface generation.** Kinematic schemes of this group feature the crossed-axis angle Σ that is smaller than its critical value Σ_{cr}. For the kinematic schemes of part surface generation, the inequality $\Sigma < \Sigma_{cr}$ is valid. The apex A_{pa} is located within the center-distance $C_{P/T}$ beyond the end point of the center-distance $C_{P/T}$.

These three groups comprise five different types of kinematic schemes of part surface generation in total.

First, no difference is observed for the configuration of rotation vectors ω_P and ω_T, which represent an external kinematic scheme of part surface generation ($\Sigma_{cr} < \Sigma < 180°$). Any of two rotation vectors can be associated either with the work or with the form-cutting tool and vice versa. Therefore, a vector diagram of only one type can be associated with an external kinematic scheme of part surface generation.

Second, for the planar kinematic scheme of part surface generation ($\Sigma = \Sigma_{cr}$), two different configurations of rotation vectors ω_P and ω_T are distinguished. For one of the configurations of the rotation vectors, the form-cutting tool rotation vector ω_T is perpendicular to the vector of instant rotation ω_{pl}. For another configuration of the rotation vectors, rotation vector ω_P of the work is perpendicular to the vector of instant rotation ω_{pl} instead. Accordingly, different methods of part surface machining can be developed on the premise of the kinematic schemes of part surface generation of these types.

The kinematic schemes of part surface generation that feature the configuration of the rotation vectors, for which the equality $\Sigma = \Sigma_{cr}$ is valid, is of particular importance for designers of form-cutting tools. The values of the design parameters of the form-cutting tool, which entails an increase in components ω_P^{sl} and ω_T^{sl} of the rotation vectors ω_P and ω_T, are often preferred when designing a form-cutting tool.

Third, the internal kinematic schemes of part surface generation ($0° < \Sigma < \Sigma_{cr}$) also feature two different types of configuration of the rotation vectors ω_P and ω_T of the part and of the cutting tool, respectively. For one of the configurations, magnitude ω_P of the rotation vector ω_P of the work is greater than magnitude ω_T of the rotation vector ω_T of the form-cutting tool ($\omega_P > \omega_P$). Kinematic schemes of part surface generation of this type correspond to the machining processes of an external work with the enveloping form-cutting tool. For other types of configuration, magnitude ω_P of the rotation vector ω_P of the work is smaller than magnitude ω_T of the rotation vector ω_T of the form-cutting tool ($\omega_P < \omega_T$). The kinematic schemes of part surface generation of this type correspond to the machining process of an internal work with the external form-cutting tool.

Three design parameters are critical for these five different types of the kinematic schemes of part surface generation. They are as follows:

a. The center-distance $C_{P/T}$
b. The crossed-axis angle Σ
c. Magnitude* of the rotation vectors $\boldsymbol{\omega}_P$ and $\boldsymbol{\omega}_T$

If one of the above listed design parameters has an extremum value, then the corresponding kinematic scheme of part surface generation transforms to one of the particular cases of the kinematic schemes.

Zero value is critical to the center-distance $C_{P/T}$. When center-distance vanishes ($C_{P/T} = 0$), a spatial kinematic scheme of part surface generation reduces to an intersected-axis kinematic scheme of part surface generation of a corresponding type. An external spatial kinematic scheme of part surface generation reduces to an external intersected-axis kinematic scheme of part surface generation. Planar kinematic schemes of part surface generation reduce to two corresponding kinematic schemes, which can be imitated by axodes in the form of a cone of revolution and a plane. An internal kinematic scheme of part surface generation reduces to two corresponding kinematic schemes, which can be imitated by axodes in the form of two cones of revolution. Ultimately, if the center-distance vanishes, five types of spatial kinematic schemes of part surface generation reduce to five corresponding types of planar kinematic schemes.

Meanwhile, the center-distance $C_{P/T}$ between the lines of action of the rotation vectors $\boldsymbol{\omega}_P$ and $\boldsymbol{\omega}_T$ can approach infinity ($C_{P/T} \rightarrow \infty$). In this particular case, either magnitude ω_P of the rotation vector $\boldsymbol{\omega}_P$ or magnitude ω_T of the rotation vector $\boldsymbol{\omega}_T$ is equal to zero (namely, either $\omega_P = 0$ or $\omega_T = 0$). Zero rotations (either $\omega_P = 0$, or $\omega_T = 0$) are a consequence of the infinite center-distance.

Relieving the clearance surfaces of disk type milling cutters of hobs and of face hobs is a good example of practical implementation of the principal kinematic schemes of part surface generation, which feature infinite center-distance. One of the examples is illustrated in Figure 2.17.

When generating the clearance surface P, the hob is rotated ω_P about its axis of rotation O_P. The direction of the rotation is chosen so that the rake surface R_s is faced toward the rotation ω_P. The form cutter is reciprocated toward the axis of rotation O_P of the hob. In Figure 2.17, the reciprocation motion of the cutter is denoted as $\mathbf{V}_{r.T}$. In addition to the reciprocation motion $\mathbf{V}_{r.T}$, the cutter is traveling $\mathbf{V}_{s.T}$ in axial direction of the hob. This motion $\mathbf{V}_{r.T}$ together with the rotation ω_P provides the screw motion of the cutter with

---

* Here and below, zero angular velocities (either $\omega_P = 0$, or $\omega_T = 0$) are considered solely as a consequence of infinite center-distance, $C_{P/T} \rightarrow \infty$. Because of the infinite center-distance magnitude of a rotation vector vanishes, but the direction of the vector remains the same. In this case an angular velocity is substituted by a corresponding linear velocity $\mathbf{V}_P$ or $\mathbf{V}_T$.

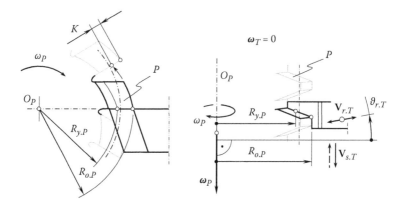

FIGURE 2.17
An example of a principal kinematic scheme of part surface generation that features infinite center-distance, $C_{P/T}$.

respect to the clearance surface P to be cut. This way, external, internal, or face hobs can be relieved.

Each of the three spatial kinematic schemes of part surface generation shown in Figure 2.16 may feature infinite center-distance $C_{P/T} \to \infty$. This gives three principal kinematic schemes of part surface generation in total. Due to lack of space, these kinematic schemes are not shown in Figure 2.16.

Zero value is critical with respect to the crossed-axis angle Σ. When the crossed-axis angle has zero value ($\Sigma = 0°$), then a spatial kinematic scheme of part surface generation reduces to a kinematic scheme with parallel axes of the corresponding type. Three possible types of the kinematic schemes of part surface generation having parallel rotation vectors ω_P and ω_T are depicted in Figure 2.16.

When the magnitude of one of the rotation vectors ω_P and ω_T has zero value (this is observed when the corresponding center-distance, $C_{P/T}$, is of infinite value, $C_{P/T} \to \infty$), then the corresponding spatial kinematic schemes of part surface generation reduce to two kinematic schemes, which are specified either by the pair of the vectors ω_P and ω_T or by the pair of vectors V_P and ω_T. These two types of the kinematic scheme of part surface generation are shown in Figure 2.16. In particular cases, the vectors ω_P and V_Y as well as the vectors V_P and ω_T can be perpendicular to each other ($\omega_P \perp V_T$, and $V_P \perp \omega_T$). These two types of the kinematic scheme of part surface generation are also depicted in Figure 2.16.

In a degenerate case, not just one but two design parameters (namely, $C_{P/T}$ and Σ) can have zero value at that same time ($C_{P/T} = 0$ and $\Sigma = 0°$). An example of the kinematic scheme of part surface generation of this particular type is illustrated in Figure 2.16. For completeness, a few more critical combinations of the design parameters of the kinematic scheme of part surface generation should be mentioned. They are as follows:

a. $C_{P/T} = 0$ and either $\omega_P = 0$ or $\omega_T = 0$
b. $\Sigma = 0°$ and either $\omega_P = 0$ or $\omega_T = 0$
c. $C_{P/T} = 0$, $\Sigma = 0°$ and either $\omega_P = 0$ or $\omega_T = 0$

Ultimately, there is a total of 24 possible types of the principal kinematic schemes of part surface generation. Eighteen types of the kinematic schemes of part surface generation are depicted in Figure 2.16, and six more principal kinematic schemes are specified above by critical values of the design parameters of the kinematic scheme.

The kinematics of all known methods of part surface machining is covered by the proposed classification of principal kinematic schemes of part surface generation. Moreover, the kinematics of novel methods of part surface machining is covered by the proposed classification of principal kinematic schemes as well. Therefore, new methods of part surface machining can be developed based on the proposed classification.*

The classification of all possible types of the kinematic schemes of part surface generation is the key for the development of novel designs of form-cutting tools and new methods of part surface machining.

2.6 Kinematics of Part Surface Machining Processes

The principal kinematics of a part surface generation is the core of the corresponding kinematics of the entire part surface machining process. However, the principal kinematics of part surface generation is not always sufficient for determining the generating surface of the form-cutting tool as well as for machining of the work. Due to that, in addition to the principal kinematics of part surface generation, the kinematics of part surface machining processes also includes more relative motions of the work and of the form-cutting tool.

---

* It is instructive to point out the similarity of the developed classification of principal kinematic schemes of part surface generation with the well-known classification of chemical elements proposed by Prof. Mendeleyev. During the time of Prof. Mendeleyev, all known chemical elements were covered by his proposed *Periodic Table of Elements*. The classification made it possible to discover new elements, as predicted by the *Periodic Table of Elements*. Similarly, as shown in Figure 2.16, the classification of principal kinematic schemes of part surface generation makes possible the development of novel methods of part surface machining, as well as the development of novel designs of form-cutting tools for these purposes. The kinematics of all known methods of part surface machining are covered by the classification (see Figure 2.16), and all the methods of part surface machining to be discovered in the future are covered by the classification (see Figure 2.16) as well.

The additional motions usually are those that allow for sliding of the generating surface over itself. Using motions of this type, a principal kinematic scheme of part surface generation can be complemented in most cases with kinematics of the part surface machining process.

In cases when a generating surface of the form-cutting tool that is created by means of the principal kinematics of profiling of the form-cutting tool does not allow for sliding over itself, a secondary generating surface of the form-cutting tool is created. The secondary generating surfaces of the form-cutting tools can also be implemented just for the convenience of the part surface manufacturer.

Ultimately, designing a form-cutting tool begins with the analysis of the corresponding principal kinematics of part surface generation.

First, the kinematics of a principal kinematic scheme of part surface generation is analyzed from the standpoint of whether a given part surface can be machined using just a certain principal kinematic scheme of part surface generation. If a chosen principal kinematic scheme of part surface generation is capable of generating the surface T of the form-cutting tool as well as for the material removal purpose, then the generated surface T is used for the design of the form-cutting tool.

Second, if the kinematics according to the principal kinematic scheme of part surface generation of no type is capable of machining a given part surface, then the motions of the type that allow for sliding of the surface T over it are added to the principal kinematic scheme of part surface generation. This way, the principal kinematic scheme of part surface generation evolves to a corresponding kinematics of the part surface machining process. Motions that result in the sliding of the part surface P over itself are also often incorporated into the kinematics of the part machining process.

Third, in particular cases, the generating surface of the form-cutting tool does not allow for sliding over itself. Cases of this type are the most inconvenient for both for the designer of the form-cutting tool and for the part surface manufacturer. If a generating surface T of the form-cutting tool does not allow for sliding over itself, then a secondary generating surface T_2 is used for the design of the form-cutting tool.

It is important to stress here that methods of generation of part surfaces cannot be applied for machining for all types of part surfaces but only to particular types of part surfaces. In cases when precise generation of a part surface P is impossible, approximate methods for profiling a form-cutting tool and generation of a particular part surface can be implemented. More detail about this issue is discussed by Radzevich [120]. Here [120] it is shown that three elements are required, namely,

a. A part surface P to be machined

b. The generating surface T of the form-cutting tool

c. A principal kinematic scheme of part surface generation

All of them are tightly connected to one another. If one of three listed above elements is specified, then two others can be derived from that given using, for this purposes, a specified criterion of optimization.

2.7 Axodes and Pitch Surfaces in Generation of Part Surface

The axodes are two surfaces one of which is associated with the part surface to be machined, while the other one is associated with the form-cutting tool to be applied for machining the given part surface. When machining a part surface, the axodes roll over one another with no slip between them. To be more exact, no sliding between the axodes is observed within a plane that is perpendicular to the axis of instant rotation of the axodes. Sliding is often observed in axial plane sections of the part surface being machined, and of the form-cutting tool.

It is instructive to point out here that the concept of axodes is obsolete. Once the rotation vectors ω_P and ω_T are associated with the work and with the form-cutting tool, respectively, then there is no longer a need for the concept of axodes. However, the concept of axodes is widely used by many authors. To make the transition from usage of the concept of axodes to the implementation of the rotation vectors, ω_P and ω_T, it makes sense to consider briefly a possibility of implementation of axodes and of pitch surfaces in the theory of part surface generation.

In many cases of part surface generation, axodes and corresponding pitch surfaces are identical to one another (that is, the pitch surfaces are congruent to the corresponding axodes). Conventional gear shaping operation is a perfect example of part surface generation process at which axodes associated with the work and with the form-cutting tool (which are two cylinders of revolution) and pitch surfaces are congruent to one another.

Unfortunately, there are many cases of part surface generation for which axodes and pitch surfaces are substantially different and the pitch surfaces are not congruent to the corresponding axodes, i.e., in all types of conventional gear shaving operations, in gear hobbing operations, and so forth, when the axodes are two hyperboloids of revolution of one sheet, the pitch surfaces (which are either tow cylinders or two cones of revolution) are not congruent to the corresponding axodes. Two examples are illustrated in Figure 2.18.

The analysis of all possible principal kinematic schemes of part surface generation (the classification is schematically depicted in Figure 2.16) reveals that the axodes of the part surface P to be machined as well as of the form-cutting tool T to be applied can be of simple shape: planes (as is schematically illustrated in Figure 2.18a), cylinders of revolution, cones of revolution (as shown in Figure 2.18b), and so forth. At that same time, in other applications,

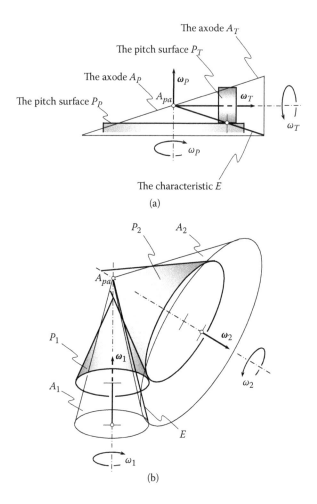

FIGURE 2.18
Examples of implementation of the axodes and of the pitch surfaces for the generation of part surfaces.

axodes of surfaces P and T can be of more complex shape: external and internal noncircular cylinders, external and internal noncircular cones, external and internal hyperboloids of revolution of one sheet, and even surfaces of more complex geometries.

Usually, the implementation of axodes of complex shape is inconvenient. In such cases, it is practical to substitute the axodes of complex shape with corresponding pitch surfaces having simpler geometry. Many methods of part surface machining feature axodes and pitch surfaces that are not congruent to one another. Moreover, these surfaces of two different types might have completely different geometries. To illustrate the difference between the axodes and the pitch surfaces, two examples are considered in Figure 2.18.

First, consider the principal kinematic scheme of part surface generation for the shaping of face gears as is schematically illustrated in Figure 2.18a.

When shaping face gears by means of a standard shaper, the pitch surface P_P of the face gear being machined is rotated about its axis with the angular velocity ω_P. The rotation vector ω_P of the work is pointed outward the apex A_{pa}. The pitch surface P_T of the gear shaper is rotating about its axis with the angular velocity ω_T. The rotation vector of the shaper is designated as ω_T. The rotations ω_P and ω_T of the work and of the cutting tool are timed (synchronized in a timely manner). Because of this, at some point within the line of contact of the gear pitch surface P_P and the shaper cutter pitch surface P_T, no sliding of the pitch surfaces is observed. However, sliding of the pitch surfaces is observed at all other points of the line of contact.

When the configuration of the rotation vectors ω_P and ω_T of the work and of the shaper cutter are specified, then it is possible to construct the corresponding axodes A_P and A_T, associated with the work and with the shaper cutter, respectively. Pure rolling of the axodes A_P and A_T over one another occurs, and no sliding between them is observed.

For the operation of shaping the face gear, Figure 2.18a clearly illustrates the difference between the axodes A_P and A_T and between the corresponding pitch surfaces P_P and P_T.

Second, consider the principal kinematic scheme of part surface generation for the planning of straight bevel gears, as schematically illustrated in Figure 2.18b. In this example, two axodes, A_1 and A_2, share common apex A_{pa}. When planning the straight bevel gear, axodes A_1 and A_2 of the work and of the cutter are rolling over one another with no sliding between them. For manufacturing convenience, axodes A_1 and A_2 are often substituted with the pitch surfaces P_1 and P_2 of the gear and of the cutting tool, respectively. In this case, apexes of the pitch cones are not snapped to the point A_{pa}. The pitch surfaces P_1 and P_2 of the gear and of the cutting tool roll over one another with sliding.

The cylinders of revolution, cones of revolution, and planes are the most widely used types of pitch surfaces when machining part surfaces using the generating method.

2.8 Examples of Implementation of the Principal Kinematic Schemes of Part Surface Generation

The principal kinematics of part surface generation for all known methods of part machining as well as for all those methods of part surface machining to be invented in the future are covered by the classification of principal kinematic schemes of part surface generation shown in Figure 2.16. Using the classification, novel methods of part surface machining can be proposed.

Below, a few methods of part surface machining are discussed to illustrate the capabilities of the classification (Figure 2.16) for the development of novel methods and for designing form-cutting tools for these purposes. Numerous impressive examples in this concern can be found in the industry. These methods of and cutting tools for part surface machining are known to the proficient reader. Just a few examples, those not known by many, from the author's application of the developed classification (Figure 2.16) are presented below.

The principal kinematic schemes of part surface generation that made up the only motion are the simplest possible. This motion results in the sliding of the part surface P to be machined over itself. The principal kinematic schemes of this particular type include no surface-generating motions. Machining operation with the kinematic schemes of this type are used for the broaching of internal and external spur and helical gears, splines, the rotary broaching of bevel gears, and so forth.

More complex principal kinematic schemes of part surface generation are made up of two motions. The principal kinematic schemes of part surface generation of this type are implemented, for example, for gear-shaping operations, both conventional and with modified orientation of the gear shaper axis of rotation, and so forth. The principal kinematic scheme of the conventional gear-shaping operation is made up of two rotations. One of the rotations is performed by the work and the other rotation is performed by the gear shaper. The rotations are timed (synchronized with one another in a timely manner). The synchronization of the rotations results in the axode of the gear rolling without sliding over the axode of the gear shaper. The reciprocation of the gear shaper does not affect the principal kinematic scheme because this motion results in the part surface P sliding over itself.

A novel gear-shaping operation and a novel design of the shaper can be developed based on the analysis of principal kinematic schemes of part surface generation shown in Figure 2.16.

For example, in the gear-shaping operation in Reference [85], the work and the gear shaper rotate about their axes O_{gear} and $O_{sh.c}$, as schematically illustrated in Figure 2.19a, with certain angular velocities ω_{gear} and $\omega_{sh.c}$, respectively. The rotations ω_{gear} and $\omega_{sh.c}$ are synchronized with one another in a timely manner. The shaping cutter reciprocates with a velocity $V_{sh.c}$ along the gear axis of rotation O_{gear}. The shaping cutter axis of rotation, $O_{sh.c}$, is inclined in relation to the axis O_{gear}. In this way, the required value of the clearance angle, $\alpha_{sh.c}$, is created. Thus, the principal kinematic scheme of part surface generation of the method of gear shaping corresponds to the principal kinematic scheme made up by two rotation vectors lines of action that cross in space with one another. The principal kinematic scheme of this particular type is covered by the classification of the principal kinematic schemes of part surface generation shown in Figure 2.16.

The inclination of the axis of rotation of the shaping cutter, $O_{sh.c}$, in relation to the axis of rotation of the gear, O_{gear}, makes possible design and implementation of the shaper cutter having zero clearance angle.

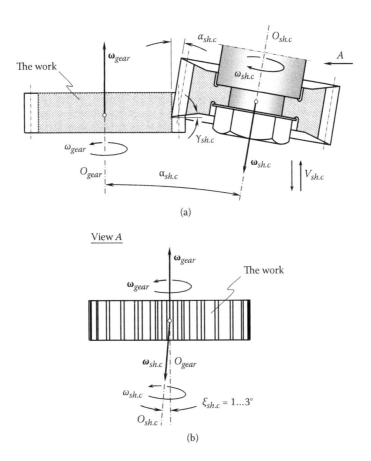

FIGURE 2.19
Schematic of machining of an involute gear by the shaper cutter having a doubly inclined axis of rotation $O_{sh.c}$. (The gear shaper is not shown in view A). (From Radzevich, USSR Patent No. 1.504.903.)

The discussed method of gear shaping operation features two important advantages. First, the implementation of the method of gear shaping makes possible an increase of accuracy of the machined gears. Second, the tool life of the shaper cutter can be easily doubled compared with that when a conventional-design shaper is used.

The additional inclination of the shaper axis of rotation, $O_{sh.c}$, on the angle $\xi_{sh.c} = 1\ldots3°$ (Figure 2.19b) makes an improved kinematics of the gear-machining process possible. Due to the inclination of the shaper axis of rotation, $O_{sh.c}$, on the angle $\xi_{sh.c} = 1\ldots3°$, the clearance angles on the opposite flanks of the shaper tooth are equalized with one another. The additional inclination of the cutting tool axis of rotation, $O_{sh.c}$, additionally increases the tool life of the shaper.

Many examples of novel methods of part surface machining that have various principal kinematic schemes made up on two rotations about skew axes can be found out in many advanced sources [144,148] and others.

Let us consider two methods of finishing gears.

The first method of gear finishing was proposed by Radzevich as early as in 1981 [62,91].

A gear has longitudinally modified teeth, which are schematically depicted in Figure 2.20. In Figure 2.20, the actual value of the tooth flank modification is designated as δ.

A special machine tool is used for finishing the gears. The schematic of the gear finishing machine is illustrated in Figure 2.21.

Gear 1 having modified tooth flank (Figure 2.21a) is finished with a gear-finishing tool 2, which is located inside the bandage 3. A screw 4 is located coaxially with the gear-finishing tool. All the gears being finished are driven by the screw 4. The crossed-axis angle, Σ, between the axis O_T of the gear-finishing tool and of the axis O_g of either spur or helical gear is set equal to the right angle as illustrated in Figure 2.21a.

To maintain the crossed-axis angle Σ equal to $\Sigma = 90°$, a corresponding correlation between the gear helix angle ψ_g, and of the setting angles ζ_T and ζ_w (of the gear-finishing tool and of the driving worm respectively) is required. The necessary correlation between the angles ψ_g, ζ_T, and ζ_w can be derived from the fundamental relationship [131]:

$$d_{b.T} = \frac{mZ_T \cos \alpha_T}{\sqrt{1 - \cos^2 \alpha_T \cos^2 \zeta_T}} \tag{2.85}$$

where $d_{b.T}$ is the base diameter of the gear-finishing tool, m is the modulus, Z_T is the number of starts of the gear-finishing tool, α_T is a normal pressure angle, and ζ_T is the setting angle of the gear-finishing tool.

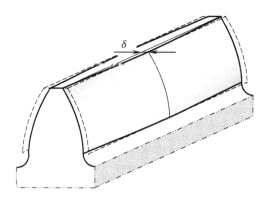

FIGURE 2.20
A gear tooth with longitudinal modification of the tooth flanks.

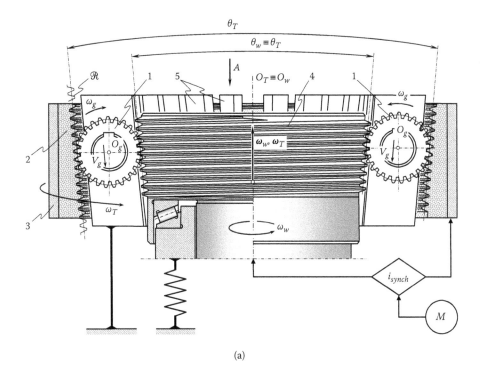

(a)

FIGURE 2.21
The concept of a method and an apparatus for the finishing of a modified gear by means of the tapered gear finishing tool.

The same relationship is valid for the driving worm 4.

A hand of threads of the driving worm is opposite to the hand of threads of the gear-finishing tool. The axis of rotation, O_w, of the driving worm aligns with axis O_T of the gear-finishing tool (namely, $O_T \equiv O_w$ as shown in Figure 2.21a).

The pitch cone angle, θ_T, of the gear-finishing tool 2 (Figure 2.21a) and a similar pitch cone angle, θ_w, of the driving worm 4 are identical to one another ($\theta_T \equiv \theta_w$). The shifting of the conical driving worm 4 up and down along the axis, O_w, of it rotation results in a change of width of the room between the gear-finishing tool 2 and the form 4. In such a way, the gears feed onto the gear-finishing tool.

The supporting screens 5 subdivide the room between the gear-finishing tool 2 and the driving worm 4 into a number of chambers. While finishing, the gear passes through the chamber. Due to the crossed-axis angle $\Sigma = 90°$, the supporting screens have planar working surfaces. They are evenly distributed circumferentially inside the gear-finishing tool 2.

Rotation is transmitted from the electric motor M to the gear-finishing tool 2 and to the driving worm 4. The gear-finishing tool rotates with a certain angular velocity ω_T about the axis O_T. The driving worm 4 rotates with the angular velocity, ω_w. The rotation ω_w of the driving worm 4 is synchronized

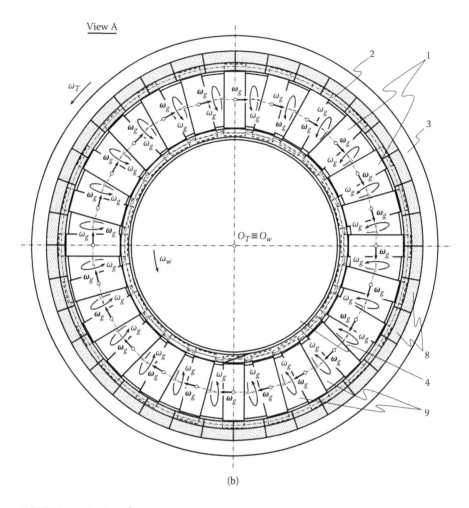

(b)

FIGURE 2.21 (Continued)
The concept of a method and an apparatus for the finishing of a modified gear by means of the tapered gear finishing tool.

with the rotation ω_T of the gear-finishing tool in timely manner. The actual synchronization of the rotations ω_T and ω_w depends upon

- The design parameters of the generating surface T of the gear-finishing tool
- The number of starts of the gear-finishing tool N_T
- The number of starts of the driving worm N_w

While finishing, the gears 1 are traveling through the chambers between the gear-finishing tool 2, the driving worm 4, and the supporting screens 9.

The gear being machined is rotating about its axis of rotation, O_g, with a certain angular velocity, ω_g (here $\omega_g = |\boldsymbol{\omega}_g|$). At the same time, the gear is traveling along the axes $O_T \equiv O_w$ with a certain linear velocity, V_g (here $V_g = |\mathbf{V}_g|$). The rotation of the gear, ω_g, is synchronized with the rotations ω_T and ω_w of the gear-finishing tool and of the driving worm in the manner that allows at least one complete revolution of the gear while traveling through the chamber.

While finishing the gear, the gear-finishing tool performs the required longitudinal modification of the gear tooth flanks. If necessary, profile modification of the gear tooth surfaces can be performed following a conventional approach for modification of the profile of the gear-finishing tool.

Similar to the finishing of gears that have longitudinally modified teeth, the chamfering of the gears, shown in Figure 2.22, can be performed as well [98]. An example of the chamfered gear tooth is shown in Figure 2.23.

To increase the productivity of the gear finishing and chamfering operation, it is possible to apply a multy-start gear-finishing tool and a multi-start driving worm. This way, not one but several (dozens) of gears could be machined in every chamber simultaneously, and an immerse increase in productivity of the gear finishing operation could be observed.

Numerous modifications of the discussed method of gear finishing as well as of gear chamfering are known from References [53,58,59,65,66,68], and others.

It is important to stress here that the accuracy of all methods of part surface machining based on the generation of a part surface P to be machined as an enveloping surface to successive positions of generating surface T of the form-cutting tool to be applied is restricted due to part surfaces P not all geometries can be generated this way. For example, in the case of rotation vectors $\boldsymbol{\omega}_P$ and $\boldsymbol{\omega}_T$ of the work and of the cutting tool are parallel to

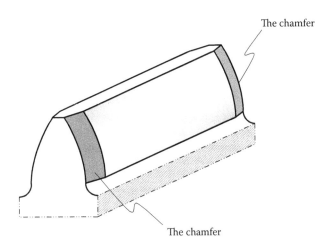

The chamfer

The chamfer

FIGURE 2.22
A chamfered gear tooth.

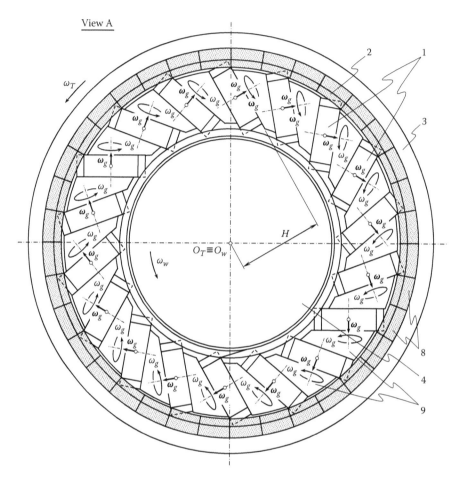

FIGURE 2.23
The concept of a method and an apparatus for the chamfering of gear teeth.

one another, only involute profile can be accurately generated on the work: geometrically accurate profiles of no other geometries can be generated this way (no straight-sided splines, no polygons, as well as no other profiles can be accurately machined using for this purpose generating methods of part surface machining).

The similar constraints occur for all other configurations of the rotation vectors ω_P and ω_T of the work and of the cutting tool. In case of intersected axes of rotation of the work and of the cutting tool, only geometrically accurate spherical involute surfaces can be generated. In the case of crossed axis of rotation of the work and of the cutting tool, part surfaces of geometry similar to that of tooth flanks of R gearing can be generated accurately. (For more in-detail analysis, the interested reader may wish to go to References [120,138]).

The in-detail analysis of the kinematical capabilities of the principal kinematic schemes of part surface generation (see Figure 2.16) might be fruitful when solving complex problems in the realm of part surface machining. Interesting examples of implementation of the classification of the principal kinematic schemes of part surface generation can be found from References [51,52,54–57,61,63,67,71,74,75,77,80,90,91] as well as from many other sources.

The interested reader may wish to apply the analysis of the discussed classification (see Figure 2.16) for the development of novel methods of part surface machining, of novel designs of form-cutting tools, of machine tools of novel designs, and so forth.

3

Applied Coordinate Systems and Linear Transformations

In the theory of part surface generation, part surfaces to be machined are commonly analytically described in a certain reference system associated with the surface. The majority of parts to be machined are bounded by two or more surfaces (see Chapter 1). Each of the part surfaces is usually represented in a reference system associated with the corresponding surface. This is convenient because analytical representation of a surface becomes simpler when the surface is specified in a particular reference system. Ultimately, it is necessary to analytically describe the whole part, namely, a set of part surfaces, in a coordinate system that is common to all of the part surfaces.

This section presents basic results in coordinate transformation. The topics to be discussed are coordinate transformations and frames.

In the theory of part surface generation, preference is given to implementation of the left-hand–oriented orthogonal *Cartesian* coordinate system. Reference systems such as cylindrical coordinates, spherical coordinates, and others are occasionally used as well (see Appendix C).

Transformation of an analytical description of a machining operation and its elements from one coordinate system to another coordinate system is the main purpose of implementation of linear transformations.

Numerous examples of practical implementation of the coordinate system transformation can be found out in the field of

- Part surface generation [116,117,136]
- Machining of part surfaces on a numerically control (*NC*) machine [1,8,9,39]
- Robotics [45,95]
- Computer-aided design (CAD) [16,18,42,43], and others

It must be stressed here that linear transformations are a powerful tool for solving practical problems that pertain to part surface generation.

3.1 Applied Coordinate Systems

Formal analytical representation of a machining operation of the part surface being machined, of the applied cutting tool, and of their relative motions at every instant of time is necessary for machining a given sculptured part surface on a multiaxis *NC* machine.

In the theory of part surface generation, a set of part surfaces being machining is referred to as the *prime element* of the machining operation. The rest of the elements of the machining operation, such as the generating surface of the form-cutting tool, kinematics of the part surface machining operation, and so forth are referred to as the *secondary elements* of the machining operation as they can be derived from the part surface to be machined. $\mathbb{R}$-mapping of a part surface to be machined onto the generating surface of the form-cutting tool (see Chapter 4) is used for this purpose.

With that said, the consideration below begins from reference systems associated with a part being machined.

3.1.1 Coordinate Systems of a Part Being Machined

A part to be machined can be represented as a number of surfaces—planes, cylinders of revolution, cones of revolution, torus surfaces, sculptured surfaces, and so on—that are oriented to one another in a certain manner. To obtain the simplest possible analytical description of a particular part surface, it is necessary to introduce the coordinate system that is properly associated to the surface. Otherwise, analytical representation of the part surface would not be the simplest possible.

Initially, numerous reference systems $X_iY_iZ_i$ are used for the analytical description of a part to be machined. Each of the reference systems, $X_iY_iZ_i$,

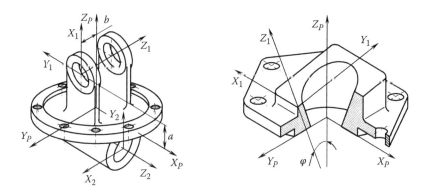

FIGURE 3.1
An example of the coordinate systems associated with the particular part surfaces.

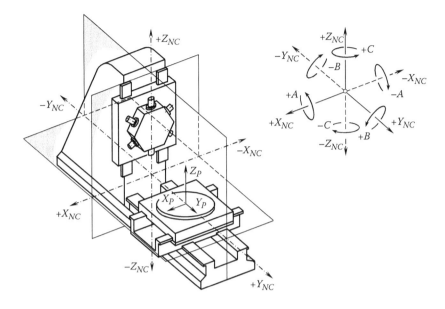

FIGURE 3.2
Coordinate system of a multiaxis numerical control machine (ISO-R 841).

is associated with a corresponding part surface as it is schematically illustrated in Figure 3.1. To machine the part, it is necessary to machine a set of its surfaces. For this purpose, the initial analytical representation of all the part surfaces must be transformed to a common reference system, $X_P Y_P Z_P$, associated with the part. Ultimately, each of the part surfaces is represented not in a particular coordinate system $X_i Y_i Z_i$, but all of them are represented in the common coordinate system $X_P Y_P Z_P$ instead.

3.1.2 Coordinate System of Multiaxis Numerical Control (NC) Machine

For the machining of part surfaces on a multiaxis NC, machine implementation of the right-hand–oriented *Cartesian* coordinate system $X_{NC} Y_{NC} Z_{NC}$ is recommended by ISO-R 841.* The coordinate system $X_{NC} Y_{NC} Z_{NC}$ is illustrated in Figure 3.2. The setup parameters of a part surface to be machined, the cutting tool to be applied, and their relative motion at every instant of time must be analytically described in this coordinate system.

The coordinate systems of at least three types—the coordinate system $X_P Y_P Z_P$ that is associated with the part, the coordinate system $X_T Y_T Z_T$ that

---

* EIA Standard RS-267-B: *Axis and Motion Nomenclature for Numerically Controlled Machines (ANSI/EIA RS-267-B-83)*, Electronic Institute Association, Washington, DC, June 1983. ISO Standard 841-1874: *Axis and Motion Nomenclature for Numerically Controlled Machines*, International Organization for Standardization, Switzerland, 1974.

is associated with the cutting tool, and the coordinate system $X_{NC}Y_{NC}Z_{NC}$ of the multiaxis *NC* machine are used for the analytical description of the procedure of generation of a part surface.

3.2 Coordinate System Transformation

Coordinate system transformation is a powerful tool for solving many geometrical and kinematic problems that pertain to part surface generation. Finite transformation is used to describe the motion of a rigid body point and the motion of the rigid body itself.

Consequent coordinate systems transformations can be easily described analytically with the implementation of matrices. Matrices have a unique algebra that governs operations on them. It was the English mathematician A. Cayley* who developed and refined the algebra of matrices. The use of matrices for the coordinate system transformation† can be traced back to the mid-1940s,‡ when Mozhayev§ described coordinate system transformations using matrices.

Matrices have a natural application to many branches of engineering. The theory of part surface generation can be formulated concisely using matrices, and practical numerical results can be obtained using the theorems of matrix algebra.

Coordinate system transformation is necessary because of two major reasons.

First, as stated above, the implementation of coordinate system transformation is necessary for the representation of all particular part surfaces in a common reference system. Coordinate system transformation makes the transition from the initial coordinate system $X_iY_iZ_i$, which is associated with a particular part surface, to the coordinate system $X_pY_pZ_p$ embedded to the entire part to be machined possible.

Second, coordinate system transformation is necessary for the representation of the cutting tool and its motion relative to the part surface in that above common reference system. At every instant, the configuration (position and

---

* Arthur Cayley (August 16, 1821–January 26, 1895) a Bristish mathematician.
† Matrices provide a compact and flexible notation particularly useful in dealing with linear transformations and present an organized method for the solution of systems of linear differential equations.
‡ Application of matrices for the purposes of analytical representation of coordinate system transformation should be credited to Dr. S.S. Mozhayev (*General Theory of Cutting Tools*, doctoral thesis, Leningrad, Leningrad Polytechnic Institute, 1951, 295 pp.). Dr. S.S. Mozhayev began using matrices for this purpose in the mid of 1940s. Later, the matrix approach for coordinate system transformation was used by Denavit and Hartenberg [13], as well as by many other researchers.
§ S.S. Mozhayev is a soviet scientist mostly known for his accomplishments in the theory of cutting tool design.

orientation) of the cutting tool relative to the work can be described analytically with the help of a homogeneous transformation matrix corresponding
to the displacement of the cutting tool from its current location to its consecutive location.

In this section, coordinate system transformation is briefly discussed from
the standpoint of its implementation in the theory of part surface generation.

3.2.1 Introduction

Homogeneous coordinates utilize a mathematical trick to embed three-
dimensional coordinates and transformations into a four-dimensional matrix
format. As a result, inversions or combinations of linear transformations are
simplified to inversions or multiplication of the corresponding matrices.

3.2.1.1 Homogeneous Coordinate Vectors

Instead of representing each point $\mathbf{r}(x, y, z)$ in a three-dimensional space with
a single three-dimensional vector,

$$\mathbf{r} = \begin{bmatrix} x \\ y \\ z \end{bmatrix} \tag{3.1}$$

homogeneous coordinates allow each point $\mathbf{r}(x, y, z)$ to be represented by any
of an infinite number of four-dimensional vectors:

$$\mathbf{r} = \begin{bmatrix} T \cdot x \\ T \cdot y \\ T \cdot z \\ T \end{bmatrix} \tag{3.2}$$

The three-dimensional vector corresponding to any four-dimensional vector can be calculated by dividing the first three elements by the fourth, and
a four-dimensional vector corresponding to any three-dimensional vector
can be created by simply adding a fourth element and setting it equal to one.

3.2.1.2 Homogeneous Coordinate Transformation
Matrices of the Dimension 4 × 4

Homogeneous coordinate transformation matrices operate on four-dimensional
homogeneous vector representations of traditional three-dimensional coordinate locations. Any three-dimensional linear transformation (translation,
rotation, and so forth) can be represented by a 4 × 4 homogeneous coordinate
transformation matrix. In fact, because of the redundant representation of

three-dimensional space in a homogeneous coordinate system, an infinite number of different 4×4 homogeneous coordinate transformation matrices are available to perform any given linear transformation. This redundancy can be eliminated to provide a unique representation by dividing all elements of a 4×4 homogeneous transformation matrix by the last element (which will become equal to one). This means that a 4×4 homogeneous transformation matrix can incorporate as many as 15 independent parameters. The generic format representation of a homogeneous transformation equation for mapping the three-dimensional coordinate (x_1, y_1, z_1) to the three-dimensional coordinate (x_2, y_2, z_2) is

$$
\begin{bmatrix} T^* \cdot x_2 \\ T^* \cdot y_2 \\ T^* \cdot z_2 \\ T^* \end{bmatrix} = \begin{bmatrix} T^* \cdot a & T^* \cdot b & T^* \cdot c & T^* \cdot d \\ T^* \cdot e & T^* \cdot f & T^* \cdot g & T^* \cdot h \\ T^* \cdot i & T^* \cdot j & T^* \cdot k & T^* \cdot m \\ T^* \cdot n & T^* \cdot p & T^* \cdot q & T^* \end{bmatrix} \cdot \begin{bmatrix} T \cdot x_2 \\ T \cdot y_2 \\ T \cdot z_2 \\ T \end{bmatrix} \tag{3.3}
$$

If any two matrices or vectors of this equation are known, the third matrix (or vector) can be calculated and then the redundant T element in the solution can be eliminated by dividing all elements of the matrix by the last element.

Various transformation models can be used to constraint the form of the matrix to transformations with fewer degrees of freedom.

3.2.2 Translations

Translation of a coordinate system is one of the major linear transformations used in the theory of part surface generation. Translations of the coordinate system $X_2Y_2Z_2$ along axes of the coordinate system $X_1Y_1Z_1$ are depicted in

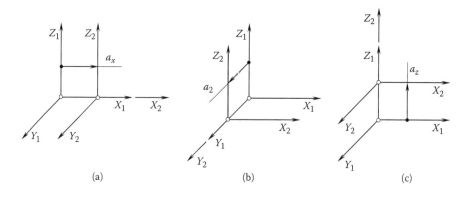

(a)   (b)   (c)

FIGURE 3.3
Analytical description of the operators of translations **Tr**(a_x, X), **Tr**(a_y, Y), **Tr**(a_z, Z) along the coordinate axes of a *Cartesian* reference system XYZ.

Figure 3.3. Translations can be analytically described by the homogeneous transformation matrix of dimension 4×4.

For an analytical description of translation along coordinate axes, the operators of translation $\mathbf{Tr}(a_x, X)$, $\mathbf{Tr}(a_y, Y)$, and $\mathbf{Tr}(a_z, Z)$ are used. These operators yield matrix representation in the form

$$\mathbf{Tr}(a_x, X) = \begin{bmatrix} 1 & 0 & 0 & a_x \\ 0 & 1 & 0 & 0 \\ 0 & 0 & 1 & 0 \\ 0 & 0 & 0 & 1 \end{bmatrix} \tag{3.4}$$

$$\mathbf{Tr}(a_y, Y) = \begin{bmatrix} 1 & 0 & 0 & 0 \\ 0 & 1 & 0 & a_y \\ 0 & 0 & 1 & 0 \\ 0 & 0 & 0 & 1 \end{bmatrix} \tag{3.5}$$

$$\mathbf{Tr}(a_z, Z) = \begin{bmatrix} 1 & 0 & 0 & 0 \\ 0 & 1 & 0 & 0 \\ 0 & 0 & 1 & a_z \\ 0 & 0 & 0 & 1 \end{bmatrix} \tag{3.6}$$

Here, in Equations 3.4 through 3.6, the parameters a_x, a_y, and a_z are assigned values that denote the distance of translation along the corresponding axis.

Consider two coordinate systems $X_1Y_1Z_1$ and $X_2Y_2Z_2$ displaced along X_1 axis at a distance a_x as schematically depicted in Figure 3.3a. A point m in the reference system $X_2Y_2Z_2$ is given by the position vector $\mathbf{r}_2$ (m). In the coordinate system $X_1Y_1Z_1$, that same point m can be specified by the position vector $\mathbf{r}_1$ (m). Then the position vector $\mathbf{r}_1$ (m) can be expressed in terms of the position vector $\mathbf{r}_2$ (m) by the equation

$$\mathbf{r}_1(m) = \mathbf{Tr}(a_x, X) \cdot \mathbf{r}_2(m) \tag{3.7}$$

Equations similar to Equation 3.7 are valid for the operators $\mathbf{Tr}(a_y, Y)$ and $\mathbf{Tr}(a_z, Z)$ of the coordinate system transformation. The latter is schematically illustrated in parts b and c of Figure 3.3.

The use of the operators of translation $\mathbf{Tr}(a_x, X)$, $\mathbf{Tr}(a_y, Y)$, and $\mathbf{Tr}(a_z, Z)$ makes possible an introduction of an operator $\mathbf{Tr}(a, \mathbf{A})$ of a combined transformation. Suppose that a point p on a rigid body goes through a translation describing a straight line from a point p_1 to a point p_2 with a change of coordinates of (a_x, a_y, a_z). This motion of the point p can be analytically described with a resultant translation operator $\mathbf{Tr}(a, \mathbf{A})$:

$$\mathbf{Tr}(a, \mathbf{A}) = \begin{bmatrix} 1 & 0 & 0 & a_x \\ 0 & 1 & 0 & a_y \\ 0 & 0 & 1 & a_z \\ 0 & 0 & 0 & 1 \end{bmatrix} \tag{3.8}$$

The operator $\mathbf{Tr}(a, \mathbf{A})$ of the resultant coordinate system transformation can be interpreted as the operator of translation along an arbitrary axis having the vector A as the direct vector.

An analytical description of translation of the coordinate system $X_1Y_1Z_1$ in the direction of an arbitrary vector $\mathbf{A}$ to the position of $X_2Y_2Z_2$ can be composed from Figure 3.4. The operator of translation $\mathbf{Tr}(a, \mathbf{A})$ of that particular type can be expressed in terms of the operators $\mathbf{Tr}(a_x, X)$, $\mathbf{Tr}(a_y, Y)$, and $\mathbf{Tr}(a_z, Z)$ of elementary translations:

$$\mathbf{Tr}(a, \mathbf{A}) = \mathbf{Tr}(a_z, Z) \cdot \mathbf{Tr}(a_y, Y) \cdot \mathbf{Tr}(a_x, X) \tag{3.9}$$

Evidently, the axis along the vector $\mathbf{A}$ is always the axis through the origins of both the reference systems $X_1Y_1Z_1$ and $X_2Y_2Z_2$.

Any or all coordinate system transformations that do not change the orientation of a geometrical object are referred to as *orientation-preserving*

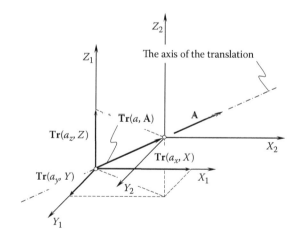

FIGURE 3.4
Analytical description of an operator $\mathbf{Tr}(a, \mathbf{A})$ of translation along an arbitrary axis (vector $\mathbf{A}$ is the direct vector of the axis).

transformation, or *direct transformation.* Therefore, the transformation of translation is an example of a direct transformation.

3.2.3 Rotation about a Coordinate Axis

The rotation of a coordinate system about a coordinate axis is another major linear transformation used in the theory of part surface generation. A rotation is specified by an axis of rotation and the angle of the rotation. It is a simple trigonometric calculation to obtain a transformation matrix for a rotation about one of the coordinate axes.

The possible rotations of the coordinate system $X_2Y_2Z_2$ about the axis of the coordinate system $X_1Y_1Z_1$ are illustrated in Figure 3.5.

For an analytical description of rotation about a coordinate axis, the operators of rotation $\mathbf{Rt}(\varphi_x, X_1)$, $\mathbf{Rt}(\varphi_y, Y_1)$, and $\mathbf{Rt}(\varphi_z, Z_1)$ are used. These operators of linear transformations yield representation in the form of homogeneous matrices:

$$\mathbf{Rt}(\varphi_x, X_1) = \begin{bmatrix} 1 & 0 & 0 & 0 \\ 0 & \cos\varphi_x & \sin\varphi_x & 0 \\ 0 & -\sin\varphi_x & \cos\varphi_x & 0 \\ 0 & 0 & 0 & 1 \end{bmatrix} \tag{3.10}$$

$$\mathbf{Rt}(\varphi_y, Y_1) = \begin{bmatrix} \cos\varphi_y & 0 & \sin\varphi_y & 0 \\ 0 & 1 & 0 & 0 \\ -\sin\varphi_y & 0 & \cos\varphi_y & 0 \\ 0 & 0 & 0 & 1 \end{bmatrix} \tag{3.11}$$

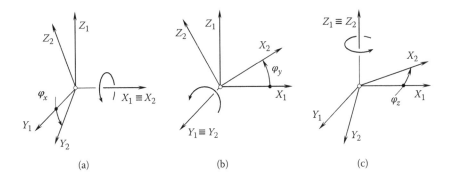

(a)   (b)   (c)

FIGURE 3.5
Analytical description of the operators of rotation $\mathbf{Rt}(\varphi_x, X)$, $\mathbf{Rt}(\varphi_y, Y)$, and $\mathbf{Rt}(\varphi_z, Z)$ about a coordinate axis of the reference system $X_1Y_1Z_1$.

$$\mathbf{Rt}(\varphi_z, Z_1) = \begin{bmatrix} \cos\varphi_z & \sin\varphi_z & 0 & 0 \\ -\sin\varphi_z & \cos\varphi_z & 0 & 0 \\ 0 & 0 & 1 & 0 \\ 0 & 0 & 0 & 1 \end{bmatrix} \qquad (3.12)$$

Here φ_x, φ_y, φ_z are assigned values that denote the corresponding angles of rotations about a corresponding coordinate axis: φ_x is the angle of rotation around the X_1 axis (pitch) of the *Cartesian* coordinate system $X_1Y_1Z_1$; φ_y is the angle of rotation around the Y_1 axis (roll), and φ_z is the angle of rotation around the Z_1 axis (yaw) of that same *Cartesian* reference system $X_1Y_1Z_1$.

The rotation about a coordinate axis is illustrated in Figure 3.5.

Consider two coordinate systems $X_1Y_1Z_1$ and $X_2Y_2Z_2$, which are turned about X_1 axis through an angle φ_x as shown in Figure 3.5a. In the reference system $X_2Y_2Z_2$, a point m is given by a position vector $\mathbf{r}_2(m)$. In the coordinate system $X_1Y_1Z_1$, that same point m can be specified by the position vector $\mathbf{r}_1(m)$. Then, the position vector $\mathbf{r}_1(m)$ can be expressed in terms of the position vector $\mathbf{r}_2(m)$ by the equation

$$\mathbf{r}_1(m) = \mathbf{Rt}(\varphi_x, X)\cdot\mathbf{r}_2(m) \qquad (3.13)$$

Equations that similar to Equation 3.13 are also valid for other operators $\mathbf{Rt}(\varphi_y, Y)$ and $\mathbf{Rt}(\varphi_z, Z)$ of the coordinate system transformation. These elementary coordinate system transformations are schematically illustrated in parts b and c of Figure 3.5.

3.2.3.1 Coupled Linear Transformation

It is right point to note here that a translation, $\mathbf{Tr}(a_x, X)$, along X axis of a *Cartesian* reference system, XYZ, and a rotation, $\mathbf{Rt}(\varphi_x, X)$, about the axis X of that same coordinate system, XYZ, obey the commutative law, that is, these two coordinate system transformations can be performed in different order equally. It makes no difference whether the translation, $\mathbf{Tr}(a_x, X)$, is initially performed, which is followed by the rotation, $\mathbf{Rt}(\varphi_x, X)$, or the rotation, $\mathbf{Rt}(\varphi_x, X)$, is initially performed, which is followed by the translation, $\mathbf{Tr}(a_x, X)$. This is because of the dot products $\mathbf{Rt}(\varphi_x, X) \cdot \mathbf{Tr}(a_x, X)$ and $\mathbf{Tr}(a_x, X) \cdot \mathbf{Rt}(\varphi_x, X)$ are identical to one another:

$$\mathbf{Rt}(\varphi_x, X) \cdot \mathbf{Tr}(a_x, X) \equiv \mathbf{Tr}(a_x, X) \cdot \mathbf{Rt}(\varphi_x, X) \qquad (3.14)$$

This means that the translation from the coordinate system $X_1Y_1Z_1$ to the intermediate coordinate system $X^*Y^*Z^*$) followed by the rotation from the coordinate system $X^*Y^*Z^*$ to the finale coordinate system $X_2Y_2Z_2$ produces that same reference $X_2Y_2Z_2$ as in a case when the rotation from the coordinate system

$X_1Y_1Z_1$ to the intermediate coordinate system $X^*Y^*Z^*$ followed by the translation from the coordinate system $X^*Y^*Z^*$ to the finale coordinate system $X_2Y_2Z_2$.

The validity of Equation 3.14 is illustrated in Figure 3.6. The translation, $\mathbf{Tr}(a_x, X)$, that is followed by the rotation, $\mathbf{Rt}(\varphi_x, X)$, as shown in Figure 3.6a is equivalent to the rotation, $\mathbf{Rt}(\varphi_x, X)$, that is followed by the translation, $\mathbf{Tr}(a_x, X)$ as shown in Figure 3.6b.

Therefore, the two linear transformations, $\mathbf{Tr}(a_x, X)$ and $\mathbf{Rt}(\varphi_x, X)$, can be coupled into a linear transformation:

$$\mathbf{Cp}_x(a_x, \varphi_x) = \mathbf{Rt}(\varphi_x, X) \cdot \mathbf{Tr}(a_x, X) \equiv \mathbf{Tr}(a_x, X) \cdot \mathbf{Rt}(\varphi_x, X) \qquad (3.15)$$

The operator of linear transformation, $\mathbf{Cp}_x(a_x, \varphi_x)$, can be expressed in matrix form:

$$\mathbf{Cp}_x(a_x, \varphi_x) = \begin{bmatrix} 1 & 0 & 0 & a_x \\ 0 & \cos\varphi_x & \sin\varphi_x & 0 \\ 0 & -\sin\varphi_x & \cos\varphi_x & 0 \\ 0 & 0 & 0 & 1 \end{bmatrix} \qquad (3.16)$$

This expression is composed based on Equation 3.4 for the linear transformation $\mathbf{Tr}(a_x, X)$, and on Equation 3.7, which describes the linear transformation $\mathbf{Rt}(\varphi_x, X)$.

Two degenerate cases of operator of the linear transformation, $\mathbf{Cp}_x(a_x, \varphi_x)$, are distinguished.

First, it could happen that in a particular case the component, a_x, of the translation is zero, that is $a_x = 0$. Under such a scenario, the operator of linear transformation, $\mathbf{Cp}_x(a_x, \varphi_x)$, reduces to the operator of rotation, $\mathbf{Rt}(\varphi_x, X)$, and the equality $\mathbf{Cp}_x(a_x, \varphi_x) = \mathbf{Rt}(\varphi_x, X)$ is observed in the case under consideration.

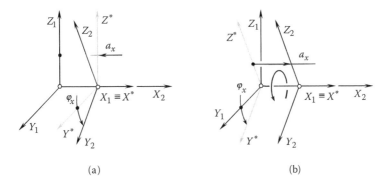

(a)   (b)

FIGURE 3.6
On the equivalency of the linear transformation $\mathbf{Rt}(\varphi_x, X) \cdot \mathbf{Tr}(a_x, X)$ and $\mathbf{Tr}(a_x, X) \cdot \mathbf{Rt}(\varphi_x, X)$, in the operator, $\mathbf{Cp}(a_x, \varphi_x)$, of *coupled linear transformation* of a *Cartesian* reference system XYZ.

Second, in a particular case, the component, φ_x, of the rotation is zero, that is $\varphi_x = 0°$. Under such a scenario, the operator of linear transformation, $\mathbf{Cp}_x(a_x, \varphi_x)$, reduces to the operator of translation, $\mathbf{Tr}(a_x, X)$, and the equality $\mathbf{Cp}_x(a_x, \varphi_x) = \mathbf{Tr}(a_x, X)$ is observed in the case under consideration.

This is valid with respect to the translations and the rotations along and about the axes Y and Z of a *Cartesian* reference system XYZ. The corresponding coupled operators, $\mathbf{Cp}_x(a_x, \varphi_x)$ and $\mathbf{Cp}_z(a_z, \varphi_z)$, for linear transformations of these kinds can also be composed:

$$\mathbf{Cp}_y(a_y, \varphi_y) = \begin{bmatrix} \cos\varphi_y & 0 & \sin\varphi_y & 0 \\ 0 & 1 & 0 & a_y \\ -\sin\varphi_y & 0 & \cos\varphi_y & 0 \\ 0 & 0 & 0 & 1 \end{bmatrix} \tag{3.17}$$

$$\mathbf{Cp}_z(a_z, \varphi_z) = \begin{bmatrix} \cos\varphi_z & \sin\varphi_z & 0 & 0 \\ -\sin\varphi_z & \cos\varphi_z & 0 & 0 \\ 0 & 0 & 1 & a_z \\ 0 & 0 & 0 & 1 \end{bmatrix} \tag{3.18}$$

In the operators of linear transformations, $\mathbf{Cp}_x(a_x, \varphi_x)$, $\mathbf{Cp}_y(a_y, \varphi_y)$, and $\mathbf{Cp}_z(a_z, \varphi_z)$, values of the translations a_x, a_y, and a_z, as well as values of the rotations φ_x, φ_y, and φ_z, are finite values (and not continuous). The linear and angular displacements do not correlate to one another in time, thus, they are not screws. They are just a kind of couple of a translation along, and a rotation about a coordinate axis of a *Cartesian* reference system.

Introduction of the operators of linear transformation, $\mathbf{Cp}_x(a_x, \varphi_x)$, $\mathbf{Cp}_y(a_y, \varphi_y)$, and $\mathbf{Cp}_z(a_z, \varphi_z)$, makes the linear transformations easier as all the operators of the linear transformations become uniform.

The operators of linear transformation $\mathbf{Cp}_x(a_x, \varphi_x)$, $\mathbf{Cp}_y(a_y, \varphi_y)$, and $\mathbf{Cp}_z(a_z, \varphi_z)$ do not obey the commutative law. This is because rotations are not vectors in nature. Therefore, special care should be undertaken when treating rotations as vectors—when implementing coupled operators of linear transformations in particular.

3.2.4 Rotation about an Arbitrary Axis through the Origin

When a rotation is to be performed around an arbitrary vector based at the origin, the transformation matrix must be assembled from a combination of rotations about the *Cartesian* coordinate.

Two different approaches for analytical description of a rotation about an arbitrary axis through the origin are discussed below.

3.2.4.1 Conventional Approach

The analytical description of rotation of the coordinate system $X_1Y_1Z_1$ about an arbitrary axis through the origin to the position of a reference system $X_2Y_2Z_2$ is illustrated in Figure 3.7. It is assumed here that the rotation is performed about the axis having a vector $\mathbf{A_0}$ as the direction vector. The operator $\mathbf{Rt}(\varphi_A, \mathbf{A_0})$ of that type of rotation can be expressed in terms of the operators $\mathbf{Rt}(\varphi_x, X)$, $\mathbf{Rt}(\varphi_y, Y)$, and $\mathbf{Rt}(\varphi_z, Z)$ of elementary rotations:

$$\mathbf{Rt}(\varphi_A, \mathbf{A_0}) = \mathbf{Rt}(\varphi_z, Z) \cdot \mathbf{Rt}(\varphi_y, Y) \cdot \mathbf{Rt}(\varphi_x, X) \qquad (3.19)$$

Evidently, the axis of rotation (a straight line along the vector $\mathbf{A_0}$) is always an axis through the origin.

The operators of translation and rotation also yield other types of linear transformations as well.

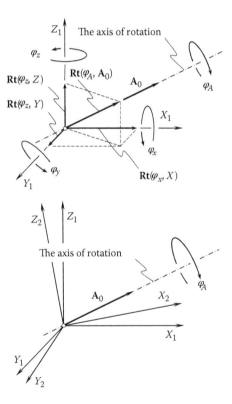

FIGURE 3.7
Analytical description of the operator $\mathbf{Rt}(\varphi_A, \mathbf{A})$ of rotation about an arbitrary axis through the origin of a *Cartesian* coordinate system $X_1Y_1Z_1$ (the vector A is the directing vector of the axis of rotation).

3.2.4.2 Eulerian Transformation

Eulerian transformation is a well-known type of linear transformation used widely in mechanical engineering. This type of linear transformation is analytically described by the operator $\mathbf{Eu}(\psi, \theta, \varphi)$ or *Eulerian** transformation.

The operator $\mathbf{Eu}(\psi, \theta, \varphi)$ is expressed in terms of three *Euler* angles (or *Eulerian* angles) ψ, θ, and φ. The configuration of an orthogonal *Cartesian* coordinate system $X_1Y_1Z_1$ in relation to another orthogonal *Cartesian* coordinate system $X_2Y_2Z_2$ is defined by the *Euler* angles ψ, θ, and φ. These angles are shown in Figure 3.8.

The line of intersection of the coordinate plane X_1Y_1 of the first reference system by the coordinate plane X_2Y_2 of the second reference system is commonly referred to as the *line of nodes*. In Figure 3.8, the line OK is the line of nodes. It is assumed here and below that the line of nodes, OK, and the axes Z_1 and Z_2 form a frame of the same orientation as the reference systems $X_1Y_1Z_1$ and $X_2Y_2Z_2$ do.

The *Euler* angle φ is referred to as the *angle of pure rotation*. This angle is measured between the X_1 axis and the line of nodes, OK. The angle of pure rotation, φ, is measured within the coordinate plane X_1Y_1 in the direction of shortest rotation from the axis X_1 to the axis Y_1.

The *Euler* angle θ is referred to as the *angle of nutation*. The angle of nutation, θ, is measured between the axes Z_1 and Z_2. The actual value of this angle never exceeds 180°.

The *Euler* angle ψ is referred to as *angle of precession*. The angle of precession, ψ, is measured in the coordinate plane X_2Y_2. This is the angle between the line of nodes, OK, and the X_2 axis. The direction of the shortest rotation from the axis X_2 to the axis Y_2 is the direction in which the angle of precession is measured.

In such case when the angle of nutation is equal either $\theta = 0°$ or $\theta = 180°$, the *Euler* angles are not defined.

Operator $\mathbf{Eu}(\psi, \theta, \varphi)$ of *Eulerian* transformation allows for the following matrix representation:

$$\mathbf{Eu}(\psi,\theta,\varphi) = \begin{bmatrix} -\sin\psi\cos\theta\sin\varphi + \cos\psi\cos\varphi & \cos\psi\cos\theta\sin\varphi + \sin\psi\cos\varphi & \sin\theta\sin\varphi & 0 \\ -\sin\psi\cos\theta\cos\varphi - \cos\psi\sin\varphi & \cos\psi\cos\theta\cos\varphi - \sin\psi\cos\psi & \sin\theta\cos\varphi & 0 \\ \sin\theta\sin\varphi & -\cos\psi\cos\theta & \cos\theta & 0 \\ 0 & 0 & 0 & 1 \end{bmatrix} \quad (3.20)$$

It is important to stress here the difference between the operator $\mathbf{Eu}(\psi, \theta, \varphi)$ of *Eulerian* transformation and the operator $\mathbf{Rt}(\psi_A, \mathbf{A}_0)$ of rotation about an arbitrary axis through the origin.

---

* Leonhard Euler (April 15, 1707–September 18, 1783), a famous Swiss mathematician and physicist who spent most of his life in Russia and Germany.

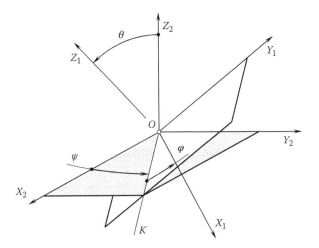

FIGURE 3.8
Euler angles.

The operator $\mathbf{Rt}(\psi_A, \mathbf{A})$ of rotation about an arbitrary axis through the origin can result in the same final orientation of the coordinate system $X_2Y_2Z_2$ in relation to the coordinate system $X_1Y_1Z_1$ as the operator $\mathbf{Eu}(\psi, \theta, \varphi)$ of *Eulerian* transformation does. However, the operators of linear transformations $\mathbf{Rt}(\psi_A, \mathbf{A}_0)$ and $\mathbf{Eu}(\psi, \theta, \varphi)$ are operators of completely different nature. They can result in identical coordinate system transformation, but they are not equal to one another.

3.2.5 Rotation about an Arbitrary Axis Not through the Origin

The transformation corresponding to rotation of an angle φ around an arbitrary vector not through the origin cannot readily be written in a form similar to the rotation matrices about the coordinate axes.

The desired transformation matrix is obtained by combining a sequence of elementary translation and rotation matrices. (Once a single 4×4 matrix has been obtained representing the composite transformations, it can be used in the same way as any other transformation matrix.)

The rotation of the coordinate system $X_1Y_1Z_1$ to a configuration that the coordinate system $X_2Y_2Z_2$ possesses can be performed about a corresponding axis that features an arbitrary configuration in space (Figure 3.9). Vector A is the direction vector of the axis of the rotation. The axis of the rotation is not a line through the origin.

The operator of linear transformation of this particular type of $\mathbf{Rt}(\psi_A, \mathbf{A})$ can be expressed in terms of the operator $\mathbf{Tr}(a, \mathbf{A})$ of translation along and of the operator $\mathbf{Rt}(\psi_A, \mathbf{A}_0)$ of rotation about an arbitrary axis through the origin:

$$\mathbf{Rt}(\varphi_A, \mathbf{A}) = \mathbf{Tr}(-b, \mathbf{B}^*) \cdot \mathbf{Rt}(\varphi_A, \mathbf{A}_0) \cdot \mathbf{Tr}(b, \mathbf{B}) \tag{3.21}$$

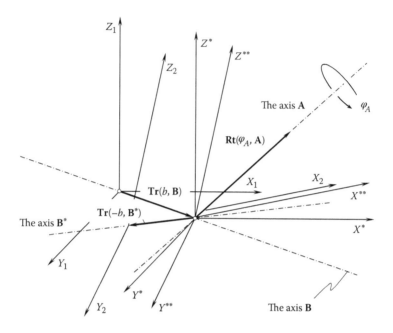

FIGURE 3.9
Analytical description of the operator, $\mathbf{Rt}(\varphi_A, \mathbf{A})$, of rotation about an arbitrary axis not through the origin (vector $\mathbf{A}$ is the direct vector of the axis of the rotation).

where $\mathbf{Tr}(b, \mathbf{B})$ is the operator of translation along the shortest distance of approach of the axis of rotation and origin of the coordinate system and $\mathbf{Tr}(-b, \mathbf{B}^*)$ is the operator of translation in the direction opposite to the translation $\mathbf{Tr}(b, \mathbf{B})$ after the rotation $\mathbf{Rt}(\psi_A, \mathbf{A})$ is completed.

To determine the shortest distance of approach, B, of the axis of rotation (that is, the axis along the directing vector B) and origin of the coordinate system, consider the axis (B) through two given points $\mathbf{r}_{B.1}$ and $\mathbf{r}_{B.2}$.

The shortest distance between a certain point $\mathbf{r}_0$ and the straight line through the points $\mathbf{r}_{B.1}$ and $\mathbf{r}_{B.2}$ can be calculated from the following formula:

$$B = \frac{|(\mathbf{r}_2 - \mathbf{r}_1) \times (\mathbf{r}_1 - \mathbf{r}_0)|}{|\mathbf{r}_2 - \mathbf{r}_1|} \tag{3.22}$$

For the origin of the coordinate system, the equality $\mathbf{r}_0 = 0$ is observed. Then,

$$B = |\mathbf{r}_1| \cdot \sin \angle[\mathbf{r}_1, (\mathbf{r}_2 - \mathbf{r}_1)] \tag{3.23}$$

The matrix representation of the operators of translation $\mathbf{Tr}(a_x, X)$, $\mathbf{Tr}(a_y, Y)$, $\mathbf{Tr}(a_z, Z)$ along the coordinate axes, together with the operators of rotation $\mathbf{Rt}(\varphi_x, X)$, $\mathbf{Rt}(\varphi_y, Y)$, $\mathbf{Rt}(\varphi_z, Z)$ about the coordinate axes is convenient for implementation in the theory of part surface generation. Moreover, the use of the operators is the simplest possible way to analytically describe the linear transformations.

3.2.6 Resultant Coordinate System Transformation

The operators of translation $\mathbf{Tr}(a_x, X)$, $\mathbf{Tr}(a_y, Y)$, and $\mathbf{Tr}(a_z, Z)$ together with the operators of rotation $\mathbf{Rt}(\varphi_x, X)$, $\mathbf{Rt}(\varphi_y, Y)$, and $\mathbf{Rt}(\varphi_z, Z)$ are used for the purpose of composing the operator $\mathbf{Rs}(1 \mapsto 2)$ of the resultant coordinate system transformation. The transition from the initial *Cartesian* reference system $X_1Y_1Z_1$ to other *Cartesian* reference system $X_2Y_2Z_2$ is analytically described by the operator $\mathbf{Rs}(1 \mapsto 2)$ of the resultant coordinate system transformation.

For example, the expression

$$\mathbf{Rs}(1 \mapsto 5) = \mathbf{Tr}(a_x, X) \cdot \mathbf{Rt}(\varphi_z, Z) \cdot \mathbf{Rt}(\varphi_x, X) \cdot \mathbf{Tr}(a_y, Y) \qquad (3.24)$$

indicates that the transition from the coordinate system $X_1Y_1Z_1$ to the coordinate system $X_5Y_5Z_5$ is executed in the following four steps (Figure 3.10):

- Translation $\mathbf{Tr}(a_y, Y)$ followed by rotation $\mathbf{Rt}(\varphi_x, X)$, followed by
- Second rotation $\mathbf{Rt}(\varphi_z, Z)$, and finally followed by the translation $\mathbf{Tr}(a_x, X)$

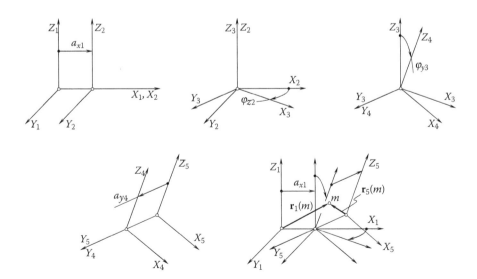

FIGURE 3.10
An example of the resultant coordinate system transformation, analytically expressed by the operator $\mathbf{Rs}(1 \mapsto 5)$.

Ultimately, the equality

$$\mathbf{r}_1(m) = \mathbf{Rs}(1 \mapsto 5) \cdot \mathbf{r}_5(m) \qquad (3.25)$$

is valid.

When the operator $\mathbf{Rs}(1 \mapsto t)$ of the resultant coordinate system transformation is specified, then the transition in the opposite direction can be performed by means of the operator $\mathbf{Rs}\ (t \mapsto 1)$ of the inverse coordinate system transformation. The following equality can be easily proven:

$$\mathbf{Rs}(t \mapsto 1) = \mathbf{Rs}^{-1}(1 \mapsto t) \qquad (3.26)$$

In the example illustrated in Figure 3.10, the operator $\mathbf{Rs}(5 \mapsto 1)$ of the inverse resultant coordinate system transformation can be expressed in terms of the operator $\mathbf{Rs}(1 \mapsto 5)$ of the direct resultant coordinate system transformation. Following Equation 3.26, one can come up with the equation:

$$\mathbf{Rs}(5 \mapsto 1) = \mathbf{Rs}^{-1}(1 \mapsto 5) \qquad (3.27)$$

It is easy to show that the operator $\mathbf{Rs}(1 \mapsto t)$ of the resultant coordinate system transformation allows for representation in the following form:

$$\mathbf{Rs}(1 \mapsto t) = \mathbf{Tr}(a, A) \cdot \mathbf{Eu}(\psi, \theta, \varphi) \qquad (3.28)$$

The linear transformation $\mathbf{Rs}(1 \mapsto t)$ (see Equation 3.28) can also be expressed in terms of rotation about an axis $\mathbf{Rt}(\varphi_A, A)$, not through the origin (see Equation 3.21).

The operators of coupled linear transformations $\mathbf{Cp}_x(a_x,\ \varphi_x)$, $\mathbf{Cp}_y(a_y,\ \varphi_y)$, and $\mathbf{Cp}_z\ (a_z,\ \varphi_z)$ (see Equations 3.16 through 3.18) (Figure 3.11) can be used for the analytical description of a resultant coordinate system transformation. Under such a scenario, the operator, $\mathbf{Rs}(1 \mapsto t)$, of a resultant coordinate system transformation can be expressed in terms of all the operators $\mathbf{Cp}_x\ (a_x, \varphi_x)$, $\mathbf{Cp}_y\ (a_y, \varphi_y)$, and $\mathbf{Cp}_z\ (a_z, \varphi_z)$ by the following expression:

$$\mathbf{Rs}(1 \mapsto t) = \prod_{\substack{i=1 \\ j=x,y,z}}^{t-1} \mathbf{Cp}_j^i(a_j^i, \varphi_j^i) \qquad (3.29)$$

In Equation 3.29, only operators of coupled linear transformations are used.

3.2.7 Linear Transformation Describing a Screw Motion about a Coordinate Axis

Operators for analytical description of screw motions about an axis of the *Cartesian* coordinate system are a particular case of the operators of the resultant coordinate system transformation.

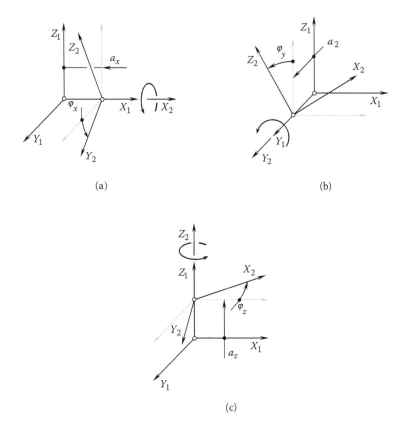

FIGURE 3.11
Analytical description of the operators $\mathbf{Cp}(a_x, \varphi_x)$, $\mathbf{Cp}(a_y, \varphi_y)$, and $\mathbf{Cp}(a_z, \varphi_z)$, of linear transformation of a *Cartesian* reference system XYZ.

By definition (Figure 3.12), the operator $\mathbf{Sc}_x(\varphi_x, p_x)$ of a screw motion about X–axis of the *Cartesian* coordinate system XYZ is equal to

$$\mathbf{Sc}_x(\varphi_x, p_x) = \mathbf{Rt}(\varphi_x, X)\cdot\mathbf{Tr}(a_x, X) \tag{3.30}$$

After substituting the operator of translation $\mathbf{Tr}(a_x, X)$ (see Equation 3.4) through Equation 3.6, and the operator of rotation $\mathbf{Rt}(\varphi_x, X)$ (Equation 3.10), Equation 3.30 casts into the expression

$$\mathbf{Sc}_x(\varphi_x, p_x) = \begin{bmatrix} 1 & 0 & 0 & p_x \cdot \varphi_x \\ 0 & \cos\varphi_x & \sin\varphi_x & 0 \\ 0 & -\sin\varphi_x & \cos\varphi_x & 0 \\ 0 & 0 & 0 & 1 \end{bmatrix} \tag{3.31}$$

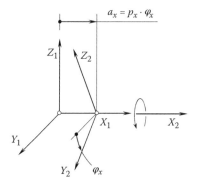

FIGURE 3.12
Analytical description of the operator of screw motion $\mathbf{Sc}_x(\varphi_x, p_x)$.

for the calculation of the operator of the screw motion $\mathbf{Sc}_x(\varphi_x, p_x)$ about X axis.

The operators of screw motions $\mathbf{Sc}_y(\varphi_y, p_y)$ and $\mathbf{Sc}_z(\varphi_z, p_z)$ about the Y and Z axes, respectively, are defined in a way similar to the way the operator of the screw motion $\mathbf{Sc}_x(\varphi_x, p_x)$ is defined:

$$\mathbf{Sc}_y(\varphi_y, p_y) = \mathbf{Rt}(\varphi_y, Y) \cdot \mathbf{Tr}(a_y, Y) \tag{3.32}$$

$$\mathbf{Sc}_z(\varphi_z, p_z) = \mathbf{Rt}(\varphi_z, Z) \cdot \mathbf{Tr}(a_z, Z) \tag{3.33}$$

Using Equations 3.5 and 3.6 together with Equations 3.11 and 3.12, one can come up with the following expressions:

$$\mathbf{Sc}_y(\varphi_y, p_y) = \begin{bmatrix} \cos\varphi_y & 0 & -\sin\varphi_y & 0 \\ 0 & 1 & 0 & p_y \cdot \varphi_y \\ \sin\varphi_y & 0 & \cos\varphi_y & 0 \\ 0 & 0 & 0 & 1 \end{bmatrix} \tag{3.34}$$

$$\mathbf{Sc}_z(\varphi_z, p_z) = \begin{bmatrix} \cos\varphi_z & \sin\varphi_z & 0 & 0 \\ -\sin\varphi_z & \cos\varphi_z & 0 & 0 \\ 0 & 0 & 1 & p_z \cdot \varphi_z \\ 0 & 0 & 0 & 1 \end{bmatrix} \tag{3.35}$$

for the calculation of the operators of the screw motion $\mathbf{Sc}_y(\varphi_y, p_y)$ and $\mathbf{Sc}_z(\varphi_z, p_z)$ about the Y and Z axes.

The screw motions about a coordinate axis, as well as screw surfaces are common in the theory of part surface generation. This makes it practical to use the operators of the screw motion $\mathbf{Sc}_x(\varphi_x, p_x)$, $\mathbf{Sc}_y(\varphi_y, p_y)$, and $\mathbf{Sc}_z(\varphi_z, p_z)$ in the theory of part surface generation.

In case of necessity, an operator of the screw motion about an arbitrary axis whether through the origin of the coordinate system or not through the origin of the coordinate system can be derived following the method similar to that used for the derivation of the operators $\mathbf{Sc}_x(\varphi_x, p_x)$, $\mathbf{Sc}_y(\varphi_y, p_y)$, and $\mathbf{Sc}_z(\varphi_z, p_z)$.

3.2.8 Linear Transformation Describing a Rolling Motion of a Coordinate System

One more practical combination of rotation and translation is often used in the theory of part surface generation.

Consider a *Cartesian* coordinate system $X_1Y_1Z_1$ (Figure 3.13). The coordinate system $X_1Y_1Z_1$ is traveling in the direction of X_1 axis. Velocity of the translation is denoted by V. The coordinate system $X_1Y_1Z_1$ is rotating about its Y_1 axis simultaneously with the translation. The speed of the rotation is denoted as ω. Assume that the ratio V/ω is constant. Under such a scenario, the resultant motion of the reference system $X_1Y_1Z_1$ to its arbitrary position $X_2Y_2Z_2$ allows interpretation in the form of rolling with no sliding of a cylinder of radius R_w over the plane. The plane is parallel to the coordinate X_1Y_1 plane, and it is remote from it at distance R_w. For the calculation of radius of the rolling cylinder, the expression $R_w = V/\omega$ can be used.

Because the rolling of the cylinder of radius R_w over the plane is performed with no sliding, a certain correspondence between the translation and the rotation of the coordinate system is established. When the coordinate system

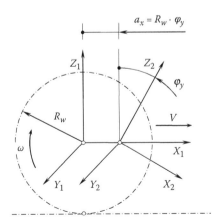

FIGURE 3.13
Illustration of the transformation of rolling $\mathbf{Rl}_x(\varphi_y, Y)$ of a coordinate system.

turns through a certain angle φ_y, then the translation of origin of the coordinate system along X_1 axis is equal to $a_x = \varphi_r \cdot R_w$.

The transition from the coordinate system $X_1Y_1Z_1$ to the coordinate system $X_2Y_2Z_2$ can be analytically described by the operator of the resultant coordinate system transformation **Rs** $(1 \mapsto 2)$. The **Rs** $(1 \mapsto 2)$ is equal to

$$\mathbf{Rs}(1 \mapsto 2) = \mathbf{Rt}(\varphi_y, Y_1) \cdot \mathbf{Tr}(a_x, X_1) \tag{3.36}$$

where $\mathbf{Tr}(a_x, X_1)$ designates the operator of the translation along X_1 axis and $\mathbf{Rt}(\varphi_y, Y_1)$ is the operator of the rotation about the Y_1 axis.

The operator of the resultant coordinate system transformation of the type (see Equation 3.36) is referred to as an *operator of rolling motion over a plane*.

When the translation is performed along the X_1 axis and the rotation is performed about Y_1 axis, the operator of rolling is denoted as $\mathbf{Rl}_x(\varphi_y, Y)$. In this particular case, the equality $\mathbf{Rl}_x(\varphi_y, Y) = \mathbf{Rs} \ (1 \mapsto 2)$ (see Equation 3.36) is valid. Based on this equality, the operator of rolling over a plane $\mathbf{Rl}_x(\varphi_y, Y)$ can be calculated from the following:

$$\mathbf{Rl}_x(\varphi_y, Y) = \begin{bmatrix} \cos\varphi_y & 0 & -\sin\varphi_y & a_x \cdot \cos\varphi_y \\ 0 & 1 & 0 & 0 \\ \sin\varphi_y & 0 & \cos\varphi_y & a_x \cdot \sin\varphi_y \\ 0 & 0 & 0 & 1 \end{bmatrix} \tag{3.37}$$

While rotation remains about the Y_1 axis, the translation can be performed not along the X_1 axis but along the Z_1 axis instead. For rolling of this type, the operator of rolling is equal:

$$\mathbf{Rl}_z(\varphi_y, Y) = \begin{bmatrix} \cos\varphi_y & 0 & -\sin\varphi_y & -a_z \cdot \sin\varphi_y \\ 0 & 1 & 0 & 0 \\ \sin\varphi_y & 0 & \cos\varphi_y & a_z \cdot \cos\varphi_y \\ 0 & 0 & 0 & 1 \end{bmatrix} \tag{3.38}$$

For the cases when the rotation is performed about the X_1 axis, the corresponding operators of rolling are as follow:

$$\mathbf{Rl}_y(\varphi_x, X) = \begin{bmatrix} 1 & 0 & 0 & 0 \\ 0 & \cos\varphi_x & \sin\varphi_x & a_y \cdot \cos\varphi_x \\ 0 & -\sin\varphi_x & \cos\varphi_x & -a_y \cdot \sin\varphi_x \\ 0 & 0 & 0 & 1 \end{bmatrix} \tag{3.39}$$

for the case of rolling along the Y_1 axis, and

$$\mathbf{Rl}_z(\varphi_x, X) = \begin{bmatrix} 1 & 0 & 0 & 0 \\ 0 & \cos\varphi_x & \sin\varphi_x & a_z \cdot \sin\varphi_x \\ 0 & -\sin\varphi_x & \cos\varphi_x & a_z \cdot \cos\varphi_x \\ 0 & 0 & 0 & 1 \end{bmatrix} \tag{3.40}$$

for the case of rolling along the Z_1 axis.

Similar expressions can be derived for the case of rotation about the Z_1 axis:

$$\mathbf{Rl}_x(\varphi_z, Z) = \begin{bmatrix} \cos\varphi_z & \sin\varphi_z & 0 & a_x \cdot \cos\varphi_z \\ -\sin\varphi_z & \cos\varphi_z & 0 & a_x \cdot \sin\varphi_z \\ 0 & 0 & 1 & 0 \\ 0 & 0 & 0 & 1 \end{bmatrix} \tag{3.41}$$

$$\mathbf{Rl}_y(\varphi_z, Z) = \begin{bmatrix} \cos\varphi_z & \sin\varphi_z & 0 & a_y \cdot \sin\varphi_z \\ -\sin\varphi_z & \cos\varphi_z & 0 & a_y \cdot \cos\varphi_z \\ 0 & 0 & 1 & 0 \\ 0 & 0 & 0 & 1 \end{bmatrix} \tag{3.42}$$

The use of the operators of rolling Equations 3.37 through 3.42 significantly simplifies analytical description of the coordinate system transformations.

3.2.9 Linear Transformation Describing Rolling of Two Coordinate Systems

In the theory of part surface generation, combinations of two rotations about parallel axes are of particular interest.

As an example, consider two *Cartesian* coordinate systems $X_1Y_1Z_1$ and $X_2Y_2Z_2$ shown in Figure 3.14. The coordinate systems $X_1Y_1Z_1$ and $X_2Y_2Z_2$ are rotated about their axes Z_1 and Z_2. The axes of the rotations are parallel to each other ($Z_1 \parallel Z_2$). The rotations ω_1 and ω_2 of the coordinate systems can be interpreted so that a circle of a certain radius R_1 that is associated with the coordinates system $X_1Y_1Z_1$, is rolling with no sliding over a circle of the corresponding radius R_2 that is associated with the coordinate system $X_2Y_2Z_2$. When the center-distance C is known, the radii R_1 and R_2 of

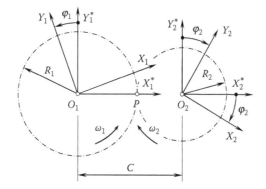

FIGURE 3.14
Derivation of the *operator of rolling* $\mathbf{Rr}_u(\varphi_1, Z_1)$ of two coordinate systems.

the circles (that is, of centrodes) can be expressed in terms of the center-distance C and of the given rotations ω_1 and ω_2. For the calculations, the following formulae

$$R_1 = C \cdot \frac{1}{1+u} \tag{3.43}$$

$$R_2 = C \cdot \frac{u}{1+u} \tag{3.44}$$

can be used. Here, the ratio ω_1/ω_2 is denoted by u.

In the initial configuration, the X_1 and X_2 axes align to each other. The Y_1 and Y_2 axes are parallel to each other. As shown in Figure 3.14, the initial configuration of the coordinate systems $X_1Y_1Z_1$ and $X_2Y_2Z_2$ is labeled as $X_1^*Y_1^*Z_1^*$ and $X_2^*Y_2^*Z_2^*$.

When the coordinate system $X_1Y_1Z_1$ turns through a certain angle φ_1, the coordinate system $X_2Y_2Z_2$ turns through the corresponding angle φ_2. When the angle φ_1 is known, the corresponding angle φ_2 is equal to $\varphi_2 = \varphi_1/u$.

The transition from the coordinate system $X_2Y_2Z_2$ to the coordinate system $X_1Y_1Z_1$ can be analytically described by the operator of the resultant coordinate system transformation $\mathbf{Rs}(1 \mapsto 2)$. In the case under consideration, the operator $\mathbf{Rs}(1 \mapsto 2)$ can be expressed in terms of the operators of the elementary coordinate system transformations

$$\mathbf{Rs}(1 \mapsto 2) = \mathbf{Rt}(\varphi_1, Z_1) \cdot \mathbf{Rt}(\varphi_1/u, Z_1) \cdot \mathbf{Tr}(-C, X_1) \tag{3.45}$$

Other equivalent combinations of the operators of elementary coordinate system transformations can result in that same operator $\mathbf{Rs}(1 \mapsto 2)$ of the resultant coordinate system transformation. The interested reader may wish to consider the derivation of the equivalent expressions for the operator $\mathbf{Rs}(1 \mapsto 2)$.

The operator of the resultant coordinate system transformations of the type (see Equation 3.45) are referred to as *operators of rolling motion over a cylinder*.

When rotations are performed around Z_1 and Z_2 axes, the operator of rolling motion over a cylinder is designated as $\mathbf{Rr}_u(\varphi_1, Z_1)$. In this particular case, the equality $\mathbf{Rr}_u(\varphi_1, Z_1) = \mathbf{Rs}(1 \mapsto 2)$ (see Equation 3.45) is valid. Based on this equality, the operator of rolling $\mathbf{Rr}_u(\varphi_1, Z_1)$ over a cylinder calculated from the equation

$$\mathbf{Rr}_u(\varphi_1, Z_1) = \begin{bmatrix} \cos\left(\varphi_1 \cdot \dfrac{u+1}{u}\right) & \sin\left(\varphi_1 \cdot \dfrac{u+1}{u}\right) & 0 & -C \\ -\sin\left(\varphi_1 \cdot \dfrac{u+1}{u}\right) & \cos\left(\varphi_1 \cdot \dfrac{u+1}{u}\right) & 0 & 0 \\ 0 & 0 & 1 & 0 \\ 0 & 0 & 0 & 1 \end{bmatrix} \quad (3.46)$$

is derived.

For the inverse transformation, the inverse operator of rolling of two coordinate systems $\mathbf{Rr}_u(\varphi_2, Z_2)$ can be used. It is equal to $\mathbf{Rr}_u(\varphi_2, Z_2) = \mathbf{Rr}_u^{-1}(\varphi_1, Z_1)$. In terms of the operators of the elementary coordinate system transformations, the operator $\mathbf{Rr}_u(\varphi_2, Z_2)$ can be expressed as follows:

$$\mathbf{Rr}_u(\varphi_2, Z_2) = \mathbf{Rt}(\varphi_1/u, Z_2) \cdot \mathbf{Rt}(\varphi_1, Z_2) \cdot \mathbf{Tr}(C, X_1) \quad (3.47)$$

Other equivalent combinations of the operators of elementary coordinate system transformations can result in that same operator $\mathbf{Rr}_u(\varphi_2, Z_2)$ of the resultant coordinate system transformation. The interested reader may wish to derive the equivalent expressions for the operator $\mathbf{Rr}_u(\varphi_2, Z_2)$.

For the calculation of the operator of rolling of two coordinate systems $\mathbf{Rr}_u(\varphi_2, Z_2)$, the following equation can be used:

$$\mathbf{Rr}_u(\varphi_2, Z_2) = \begin{bmatrix} \cos\left(\varphi_1 \cdot \dfrac{u+1}{u}\right) & -\sin\left(\varphi_1 \cdot \dfrac{u+1}{u}\right) & 0 & C \\ \sin\left(\varphi_1 \cdot \dfrac{u+1}{u}\right) & \cos\left(\varphi_1 \cdot \dfrac{u+1}{u}\right) & 0 & 0 \\ 0 & 0 & 1 & 0 \\ 0 & 0 & 0 & 1 \end{bmatrix} \quad (3.48)$$

Similar to the expression (see Equation 3.46) derived for the calculation of the operator of rolling $Rr_u(\varphi_1, Z_1)$ around Z_1 and Z_2 axes, corresponding formulae can be derived for the calculation of the operators of rolling $Rr_u(\varphi_1, X_1)$ and $Rr_u(\varphi_1, Y_1)$ about parallel axes X_1 and X_2 as well as about parallel axes Y_1 and Y_2.

Using the operators of rolling about the two axes, $Rr_u(\varphi_1, X_1)$, $Rr_u(\varphi_1, Y_1)$, and $Rr_u(\varphi_1, Z_1)$ substantially simplifies analytical description of the coordinate system transformations.

3.2.10 Example of Nonorthogonal Linear Transformation

Consider a nonorthogonal reference system $X_1Y_1Z_1$ having certain angle ω_1 between the axes X_1 and Y_1. Axis Z_1 is perpendicular to the coordinate plane X_1Y_1. Another reference system, $X_2Y_2Z_2$, is identical to the first coordinate system $X_1Y_1Z_1$ and is turned about the Z_1 axis through a certain angle φ. The transition from the reference system $X_1Y_1Z_1$ to the reference system $X_2Y_2Z_2$ can be analytically described by the operator of linear transformation

$$Rt_\omega(1 \rightarrow 2) = \begin{bmatrix} \dfrac{\sin(\omega_1 + \varphi)}{\sin \omega_1} & \dfrac{\sin \varphi}{\sin \omega_1} & 0 & 0 \\[2mm] -\dfrac{\sin \varphi}{\sin \omega_1} & \dfrac{\sin(\omega_1 - \varphi)}{\sin \omega_1} & 0 & 0 \\[2mm] 0 & 0 & 1 & 0 \\[2mm] 0 & 0 & 0 & 1 \end{bmatrix} \qquad (3.49)$$

To distinguish the operator of rotation in the orthogonal linear transformation $Rt(1 \rightarrow 2)$ from the similar operator of rotation in a nonorthogonal linear transformation $Rt_\omega(1 \rightarrow 2)$, the subscript ω is assigned to the last. When $\omega = 90°$, Equation 3.49 casts into Equation 3.12.

3.2.11 Conversion of a Coordinate System Hand

Application of the matrix method of coordinate system transformation presumes that both of the reference systems i and $(i \pm 1)$ are of the same hand. This means that it assumed from the very beginning that both of them are either right-hand– or left-hand–oriented *Cartesian* coordinate systems. In the event the coordinate systems i and $(i \pm 1)$ are of the opposite hand, say one of them is the right-hand–oriented coordinate system while the other one is the left-hand–oriented coordinate system, one of the coordinate systems must be converted into the oppositely oriented *Cartesian* coordinate system.

For the conversion of a left-hand–oriented *Cartesian* coordinate system into a right-hand–oriented coordinate system, or vice versa, the operators of reflection are commonly used.

To change the direction of the X_i axis of the initial coordinate system i to the opposite direction (in this case, in the new coordinate system $(i \pm 1)$, the equalities $X_{i\pm1} = -X_i$, $Y_{i\pm1} \equiv Y_i$, and $Z_{i\pm1} \equiv Z_i$ are observed), the operator of reflection $\mathbf{Rf}_x(Y_i\,Z_i)$ can be applied. The operator of reflection yields representation in matrix form:

$$\mathbf{Rf}_x(Y_iZ_i) = \begin{bmatrix} -1 & 0 & 0 & 0 \\ 0 & 1 & 0 & 0 \\ 0 & 0 & 1 & 0 \\ 0 & 0 & 0 & 1 \end{bmatrix} \tag{3.50}$$

Similarly, the implementation of the operators of reflections $\mathbf{Rf}_y(X_iZ_i)$ and $\mathbf{Rf}_z(X_iY_i)$ change the directions of Y_i and Z_i axes onto opposite directions. The operators of reflections $\mathbf{Rf}_y(X_iZ_i)$ and $\mathbf{Rf}_z(X_iY_i)$ can be expressed analytically in the form:

$$\mathbf{Rf}_y(X_iZ_i) = \begin{bmatrix} 1 & 0 & 0 & 0 \\ 0 & -1 & 0 & 0 \\ 0 & 0 & 1 & 0 \\ 0 & 0 & 0 & 1 \end{bmatrix} \tag{3.51}$$

$$\mathbf{Rf}_z(X_iY_i) = \begin{bmatrix} 1 & 0 & 0 & 0 \\ 0 & 1 & 0 & 0 \\ 0 & 0 & -1 & 0 \\ 0 & 0 & 0 & 1 \end{bmatrix} \tag{3.52}$$

A linear transformation that reverses direction of the coordinate axis is an *opposite transformation*. Transformation of reflection is an example of *orientation-reversing transformation*.

3.3 Useful Equations

The sequence of successive rotations can vary depending on the intention of the researcher. Several special types of successive rotations are known, including *Eulerian, Cardanian*, two types of *Euler–Krilov* transformations, and so forth. The sequence of successive rotations can be chosen from a total of 12 different combinations. Even though the *Cardanian* transformation is different from the *Eulerian* transformation in terms of the combination of rotations, they both use a similar approach to calculate the orientation angles.

3.3.1 *RPY*-Transformation

A series of rotations can be performed in the order *roll matrix (R)* by *pitch matrix (P)* and by *yaw matrix (Y)*. The linear transformation of this type is commonly referred to as *RPY-transformation*. The resultant transformation of this type can be represented by the homogenous coordinate transformation matrix:

$$RPY(\varphi_x, \varphi_y, \varphi_z) = \begin{bmatrix} \cos\varphi_y\cos\varphi_z + \sin\varphi_x\sin\varphi_y\sin\varphi_z & \cos\varphi_y\sin\varphi_z - \sin\varphi_x\sin\varphi_y\cos\varphi_z & \cos\varphi_x\sin\varphi_y & 0 \\ -\cos\varphi_x\sin\varphi_z & \cos\varphi_x\cos\varphi_z & \sin\varphi_x & 0 \\ \sin\varphi_x\cos\varphi_y\sin\varphi_z - \sin\varphi_y\cos\varphi_z & -\sin\varphi_x\cos\varphi_y\cos\varphi_z - \cos\varphi_y\sin\varphi_z & \cos\varphi_x\cos\varphi_y & 0 \\ 0 & 0 & 0 & 1 \end{bmatrix}$$

$$(3.53)$$

The *RPY*-transformation can be used for solving problems in the field of part surface generation.

3.3.2 Rotation Operator

A spatial rotation operator for the rotational transformation of a point about a unit axis $a_0(\cos\alpha, \cos\beta, \cos\gamma)$ passing through the origin of the coordinate system can be described as follows, with $a_0 = A_0/|A_0|$ designating the unit vector along the axis of rotation A_0.

Suppose the angle of rotation of the point about a_0 is θ, the *rotation operator* is expressed by

$$Rt(\theta, a_0) = \begin{bmatrix} (1-\cos\theta)\cos^2\alpha + \cos\theta & (1-\cos\theta)\cos\alpha\cos\beta - \sin\theta\cos\gamma & (1-\cos\theta)\cos\alpha\cos\gamma + \sin\theta\cos\beta & 0 \\ (1-\cos\theta)\cos\alpha\cos\beta + \sin\theta\cos\gamma & (1-\cos\theta)\cos^2\beta + \cos\theta & (1-\cos\theta)\cos\beta\cos\gamma - \sin\theta\cos\alpha & 0 \\ (1-\cos\theta)\cos\alpha\cos\gamma - \sin\theta\cos\beta & (1-\cos\theta)\cos\beta\cos\gamma + \sin\theta\cos\alpha & (1-\cos\theta)\cos^2\gamma + \cos\theta & 0 \\ 0 & 0 & 0 & 1 \end{bmatrix} \quad (3.54)$$

The solution to a problem in the field of part surface generation can be significantly simplified by implementation of the rotational operator $Rt(\theta, a_0)$ (see Equation 3.54).

3.3.3 A Combined Linear Transformation

Suppose a point p on a rigid body rotates with an angular displacement θ about a unit axis a_0 passing through the origin of the coordinate system at first and then followed by a translation at a distance B in the direction of a unit vector b. The linear transformation of this type can be analytically described by the homogenous matrix

$$\mathbf{Rt}(\theta_{a0}, B_b) = \begin{bmatrix} (1-\cos\theta)\cos^2\alpha + \cos\theta & (1-\cos\theta)\cos\alpha\cos\beta - \sin\theta\cos\gamma & (1-\cos\theta)\cos\alpha\cos\gamma + \sin\theta\cos\beta & B\cos\alpha \\ (1-\cos\theta)\cos\alpha\cos\beta + \sin\theta\cos\gamma & (1-\cos\theta)\cos^2\beta + \cos\theta & (1-\cos\theta)\cos\beta\cos\gamma - \sin\theta\cos\alpha & B\cos\beta \\ (1-\cos\theta)\cos\alpha\cos\gamma - \sin\theta\cos\beta & (1-\cos\theta)\cos\beta\cos\gamma + \sin\theta\cos\alpha & (1-\cos\theta)\cos^2\gamma + \cos\theta & B\cos\gamma \\ 0 & 0 & 0 & 1 \end{bmatrix}$$

$$(3.55)$$

More operators of particular linear transformations can be found in the literature.

3.4 Chains of Consequent Linear Transformations and a Closed Loop of Consequent Coordinate Systems Transformations

Consequent coordinate system transformations form chains (circuits) of linear transformations. The elementary chain of coordinate system transformation is composed of two consequent transformations. Chains of linear transformations play an important role in the theory of part surface generation.

Two different types of chains of consequent coordinate system transformations are distinguished. First, the transition from the coordinate system $X_P Y_P Z_P$ associated with the part surface P to be machined to the local *Cartesian* coordinate system $x_P y_P z_P$ having the origin at a point K of contact of the part surface P and the generating surface T of the form-cutting tool. This linear transformation is also made up of numerous operators of intermediate coordinate system transformations $(X_{in} Y_{in} Z_{in})$. It forms a chain of direct consequent coordinate system transformations illustrated in Figure 3.15a.

The local coordinate system $x_P y_P z_P$ is associated with the part surface P. The operator $\mathbf{Rs}(P \rightarrow K_P)$ of the resultant coordinate system transformation for a direct chain of linear transformations can be composed using for this purpose a certain number of the operators of translations (see Equations 3.4 through 3.6), and a corresponding number of the operators of rotations (see Equations 3.10 through 3.12).

Second, the transition from the coordinate system $X_P Y_P Z_P$ to the local *Cartesian* coordinate system $x_T y_T z_T$, with the origin at a point K of contact of surfaces P and T. The coordinate system $x_T y_T z_T$ is associated to generating surface T of the cutting tool. This linear transformation is also made up of numerous intermediate coordinate system transformations $(X_j Y_j Z_j)$, for example, transitions from the coordinate system $X_{NC} Y_{NC} Z_{NC}$ associated with a multiaxis NC machine, to numerous intermediate coordinate system $X_i Y_i Z_i$. The linear transformation of

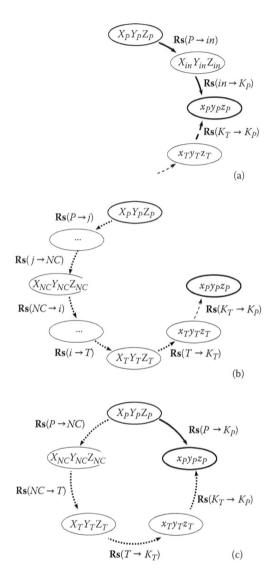

FIGURE 3.15
An example of direct chain (a) of reverse chain (b) and a closed loop (c) of consequent coordinate system transformations.

this type forms a chain of inverse consequent coordinate system transformations shown in Figure 3.15b). The operator $\mathbf{Rs}(P \rightarrow K_T)$ of the resultant coordinate system transformations for the inverse chain of transformations can be composed using for this purpose a certain number of the operators of translations (see Equations 3.4, 3.5, and 3.6) and a corresponding number of the operators of rotations (see Equations 3.10 through 3.12).

Chains of the direct and reverse consequent coordinate system transformations together with the operator of transition from the local coordinate system $x_T y_T z_T$ to the local coordinate system $x_P y_P z_P$ form a closed loop (a closed circuit) of the consequent coordinate system transformations depicted in Figure 3.15c.

If a closed loop of the consequent coordinate system transformations is complete, the implementation of a certain number of the operators of translations (see Equations 3.4, 3.5, and 3.6), and a corresponding number of the operators of rotations (see Equations 3.10 through 3.12) returns a result that is identical to the input data. This means that analytical description of a machining process specified in the original coordinate system remains the same after the implementation of the operator of the resultant coordinate system transformations. This condition is the necessary and sufficient condition for the existence of a closed loop of consequent coordinate system transformations.

Implementation of the chains as well as of the closed loops of consequent coordinate system transformations makes possible consideration of the operation of machining of the part surface P in any or all of reference systems that make up the loop. Therefore, for consideration of a particular problem of part surface generation, the most convenient reference system can be chosen.

To complete the construction of a closed loop of a consequent coordinate system transformation, an operator of transformation from the local coordinate system $x_T y_T z_T$ to the local coordinate system $x_P y_P z_P$ must be composed. Usually, the local reference systems $x_P y_P z_P$ and $x_T y_T z_T$ are types of semi-orthogonal coordinate systems. This means that axis z_P is always orthogonal to the coordinate plane $x_P y_P$, while the axes x_P and y_P can be either orthogonal or not orthogonal to each other. The same is valid with respect the local coordinate system $x_T y_T z_T$.

Two possible ways for performing the required transformation of the local reference systems $x_P y_P z_P$ and $x_T y_T z_T$ are considered below.

In the first way, the operator $\mathbf{Rt}_\omega(t \rightarrow p)$ of the linear transformation of semi-orthogonal coordinate systems (Figure 3.16) must be composed. The operator $\mathbf{Rt}_\omega(t \rightarrow p)$ can be represented in the form of the homogenous matrix:

$$\mathbf{Rt}_\omega(t \rightarrow p) = \begin{bmatrix} \dfrac{\sin(\omega_T + \alpha)}{\sin \omega_T} & -\dfrac{\sin(\omega_P - \omega_T - \alpha)}{\sin \omega_T} & 0 & 0 \\ -\dfrac{\sin \alpha}{\sin \omega_T} & \dfrac{\sin(\omega_P - \alpha)}{\sin \omega_T} & 0 & 0 \\ 0 & 0 & -1 & 0 \\ 0 & 0 & 0 & 1 \end{bmatrix} \quad (3.56)$$

where ω_P is the angle that made by the U_P and V_P coordinate lines on the part surface P (see Equation 1.29), ω_T is the angle made by U_T and V_T coordinate lines on generating surface T of the form-cutting tool (see Equation 1.29), and

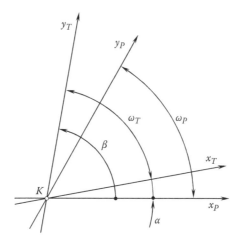

FIGURE 3.16
Local coordinate systems $x_P y_P z_P$ and $x_T y_T z_T$ with the origin at cutter-contact-point K.

α is the angle made by axes x_P and x_T of the local coordinate systems $x_P y_P z_P$ and $x_T y_T z_T$. The auxiliary angle β in Figure 3.16 is equal to $\beta = \omega_T + \alpha$.

The inverse coordinate system transformation, that is, the transformation from the local coordinate system $x_P y_P z_P$ to the local coordinate system $x_T y_T z_T$, can be analytically described by the operator $\mathbf{Rt}_\omega(p \to t)$ of the inverse coordinate system transformation. The operator $\mathbf{Rt}_\omega(p \to t)$ can be represented in the form of the homogenous matrix:

$$\mathbf{Rt}_\omega(p \to t) = \begin{bmatrix} \dfrac{\sin(\omega_P - \alpha)}{\sin \omega_P} & \dfrac{\sin(\omega_P - \omega_T - \alpha)}{\sin \omega_P} & 0 & 0 \\[2mm] \dfrac{\sin \alpha}{\sin \omega_P} & \dfrac{\sin(\omega_T + \alpha)}{\sin \omega_P} & 0 & 0 \\[2mm] 0 & 0 & -1 & 0 \\[1mm] 0 & 0 & 0 & 1 \end{bmatrix} \quad (3.57)$$

Following the second way of transformation of the local coordinate systems, the auxiliary orthogonal local coordinate system must be constructed.

For the purposes of part surface generation, it has proven to be convenient to represent an analytical description of a machining operation of the part surface P in the moving local coordinate system $x_P y_P z_P$. The implementation of the moving coordinate system $x_P y_P z_P$ having orthogonal normalized basis is preferred. The use of the orthogonal local coordinate system $x_P y_P z_P$ results in significant simplification of the linear transformations; it also makes possible the final result of computation, which is not hidden within the bulky equations.

Let us consider an approach in which a closed loop (a closed circuit) of the consequent coordinate system transformations can be composed.

To construct an orthogonal normalized basis of the coordinate system $x_P y_P z_P$, an intermediate coordinate system $x_1 y_1 z_1$ is used. Axis x_1 of the coordinate system $x_1 y_1 z_1$ is pointed along the unit vector $\mathbf{u}_P$, which is tangent to the U_P coordinate curve (Figure 3.17). Axis y_1 is directed along vector $\mathbf{v}_P$, which is tangent to the V_P coordinate line on P. Axis z_1 aligns with unit normal vector $\mathbf{n}_P$ and is pointed outward the surface P body.

For a part surface P having orthogonal parameterization (for which $F_P = 0$, and therefore $\omega_P = \pi/2$), analytical description of coordinate system transformations is significantly simpler. A further simplification of the coordinate system transformation is possible when the coordinate U_P and V_P lines are congruent to the lines of curvature on the part surface P. Under such a scenario, the local coordinate system is represented by *Darboux* frame.

To construct a *Darboux* frame, the principal directions on the part surface P must be calculated. The determination of the unit tangent vectors $\mathbf{t}_{1.P}$ and $\mathbf{t}_{2.P}$ of the principal directions on the part surface P is considered in Chapter 1.

In the common tangent plane, the orientation of the unit vector $\mathbf{t}_{1.P}$ of the first principal direction on the part surface P can be uniquely specified by the included angle $\xi_{1.P}$ that the vector $\mathbf{t}_{1.P}$ forms with the U_P coordinate curve. This angle depends on both on the part surface P geometry and on the part surface P parameterization. Figure 3.18 shows the relationship between the tangent vectors $\mathbf{U}_P$ and $\mathbf{V}_P$ and the included angle $\xi_{1.P}$. From the law of sine,

$$\frac{\sqrt{G_P}}{\sin \xi_{1.P}} = \frac{\sqrt{E_P}}{\sin[\pi - \xi_{1.P} - (\pi - \omega_P)]} = \frac{\sqrt{F_P}}{\sin(\omega_P - \xi_{1.P})} \tag{3.58}$$

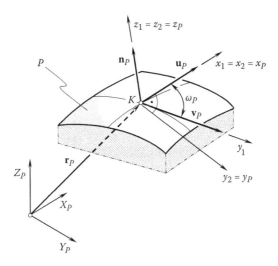

FIGURE 3.17
Local coordinate system $x_P y_P z_P$ associated with the part surface P.

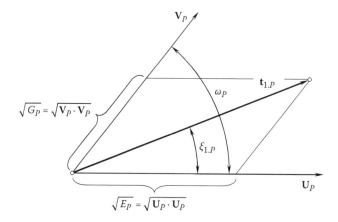

FIGURE 3.18
Differential relationships among the tangent vectors $\mathbf{U}_P$ and $\mathbf{V}_P$, the fundamental magnitudes of the first order, the included angle $\xi_{1.P}$, and the direction of $\mathbf{t}_{1.P}$.

Here, $\omega_P = \cos^{-1}\left(F_P/\sqrt{E_P G_P}\right)$.
Solving the expression above for the included angle $\xi_{1.P}$ results in

$$\xi_{1.P} = \tan^{-1}\frac{\sqrt{E_P G_P - F_P^2}}{E_P + F_P} \tag{3.59}$$

Another possible way of constructing the orthogonal local basis of the local *Cartesian* coordinate system $x_P y_P z_P$ is based on the following consideration.

Consider an arbitrary nonorthogonal and non-normalized basis $\mathbf{x}_1 \mathbf{x}_2 \mathbf{x}_3$ (Figure 3.19a). Let us construct an orthogonal and normalized basis based on the initial given basis $\mathbf{x}_1 \mathbf{x}_2 \mathbf{x}_3$.

The cross-product of any two of three vectors $\mathbf{x}_1$, $\mathbf{x}_2$, $\mathbf{x}_3$, for example, the cross-product of vectors $\mathbf{x}_1 \times \mathbf{x}_2$, determines a new vector $\mathbf{x}_4$ (Figure 3.19b). Evidently, vector $\mathbf{x}_4$ is orthogonal to the coordinate plane $\mathbf{x}_1 \mathbf{x}_2$. Then use the calculated vector $\mathbf{x}_4$ and one of two original vectors $\mathbf{x}_1$ or $\mathbf{x}_2$, for instance, use the vector $\mathbf{x}_2$. This yields calculation of a new vector $\mathbf{x}_5 = \mathbf{x}_4 \times \mathbf{x}_2$ (Figure 3.19c). The calculated basis $\mathbf{x}_1 \mathbf{x}_4 \mathbf{x}_5$ is orthogonal. To convert it into a normalized basis, each of the vectors $\mathbf{x}_1$, $\mathbf{x}_4$, and $\mathbf{x}_5$ must be divided by its magnitude:

$$\mathbf{e}_1 = \frac{\mathbf{x}_1}{|\mathbf{x}_1|} \tag{3.60}$$

$$\mathbf{e}_4 = \frac{\mathbf{x}_4}{|\mathbf{x}_4|} \tag{3.61}$$

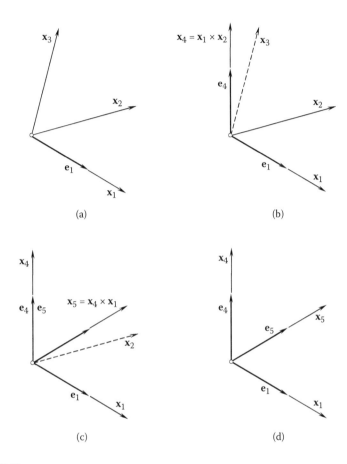

FIGURE 3.19
A normalized and orthogonally parameterized basis $\mathbf{e}_1\mathbf{e}_4\mathbf{e}_5$ that is constructed from an arbitrary basis $\mathbf{x}_1\mathbf{x}_2\mathbf{x}_3$.

$$\mathbf{e}_5 = \frac{\mathbf{x}_5}{|\mathbf{x}_5|} \qquad (3.62)$$

The resultant basis $\mathbf{e}_1\mathbf{e}_4\mathbf{e}_5$ (see Figure 3.19d) is always orthogonal and normalized.

To complete the analytical description of a closed loop of consequent coordinate system transformations, it is necessary to compose the operator $\mathbf{Rs}(K_T \rightarrow K_P)$ of the transformation from the local reference system $x_T y_T z_T$ to the local reference system $x_P y_P z_P$ (Figure 3.15c).

In the case under consideration, axes z_P and z_T align with the common unit normal vector $\mathbf{n}_P$. Axis z_P is pointed from the bodily side to the void side of the part surface P. Axis z_T is opposite, and because of that, the following equality is observed:

$$\mathbf{Rs}(K_T \rightarrow K_P) = \mathbf{Rt}(\varphi_z, z_T) \tag{3.63}$$

The inverse coordinate system transformation can be analytically described by the operator

$$\mathbf{Rs}(K_P \rightarrow K_T) = \mathbf{Rs}^{-1}(K_T \rightarrow K_P) = \mathbf{Rt}(-\varphi_z, z_T) \tag{3.64}$$

Implementation of the discussed results allows for

(a) Representation of the part surface P and of the generating surface T of the form-cutting tool, as well as their relative motion in a common coordinate system.

(b) Consideration of the machining process of the part surface P in any desired coordinate systems that is a component of the chain and/or the closed loop of consequent coordinate system transformations.

The transition from one coordinate system to another coordinate system can be performed in both of two feasible directions, say in direct as well as in inverse directions.

3.5 Impact of the Coordinate Systems Transformations on Fundamental Forms of the Surface

Every coordinate system transformation results in a corresponding change to equations of the part surface P and/or generating surface T of the form-cutting tool. Because of this, it is often necessary to recalculate coefficients of the first $\Phi_{1.P}$ and second $\Phi_{2.P}$ fundamental forms of the part surfaces P (as many times as the coordinate system transformation is performed). This routing and time-consuming operation can be eliminated if the operators of coordinate system transformations are used directly to the fundamental forms $\Phi_{1.P}$ and $\Phi_{2.P}$.

After being calculated in an initial reference system, the fundamental magnitudes E_P, F_P, and G_P of the first $\Phi_{1.P}$ and the fundamental magnitudes L_P, M_P, and N_P of the second $\Phi_{2.P}$ fundamental forms can be determined in any new coordinate system using for this purpose the operators of translation, rotation, and resultant coordinate system transformation [123]. The transformation of such types of fundamental magnitudes $\Phi_{1.P}$ and $\Phi_{2.P}$ becomes possible due to implementation of a formula, which can be derived below.

Let us consider a part surface P that is given by equation $\mathbf{r}_P = \mathbf{r}_P(U_P, V_P)$, where $(U_P, V_P) \in G$.

For the analysis below, it is convenient to use the equation of the first fundamental form $\Phi_{1.P}$ of the part surface P represented in matrix form (see Equation 1.13):

$$[\Phi_{1.P}] = \begin{bmatrix} dU_P & dV_P & 0 & 0 \end{bmatrix} \cdot \begin{bmatrix} E_P & F_P & 0 & 0 \\ F_P & G_P & 0 & 0 \\ 0 & 0 & 1 & 0 \\ 0 & 0 & 0 & 1 \end{bmatrix} \cdot \begin{bmatrix} dU_P \\ dV_P \\ 0 \\ 0 \end{bmatrix} \quad (3.65)$$

Similarly, the equation of the second fundamental form $\Phi_{2.P}$ of the part surface P can be given by Equation 1.38:

$$[\Phi_{2.P}] = \begin{bmatrix} dU_P & dV_P & 0 & 0 \end{bmatrix} \cdot \begin{bmatrix} L_P & M_P & 0 & 0 \\ M_P & N_P & 0 & 0 \\ 0 & 0 & 1 & 0 \\ 0 & 0 & 0 & 1 \end{bmatrix} \cdot \begin{bmatrix} dU_P \\ dV_P \\ 0 \\ 0 \end{bmatrix} \quad (3.66)$$

The coordinate system transformation that is performed by the operator of linear transformation $\mathbf{Rs}(1 \rightarrow 2)$ transfers the equation $\mathbf{r}_P = \mathbf{r}_P(U_P, V_P)$ of the part surface P initially given in $X_1Y_1Z_1$ to the equation $\mathbf{r}_P^* = \mathbf{r}_P^*(U_P^*, V_P^*)$ of that same part surface P in a new coordinate system $X_2Y_2Z_2$. It is clear that $\mathbf{r}_P \neq \mathbf{r}_P^*$.

In the new coordinate system, the part surface P is analytically described by the following expression:

$$\mathbf{r}_P^*(U_P^*, V_P^*) = \mathbf{Rs}(1 \rightarrow 2) \cdot \mathbf{r}_P(U_P, V_P) \quad (3.67)$$

The operator of resultant coordinate system transformation $\mathbf{Rs}(1 \rightarrow 2)$ casts the column matrices of variables in Equations 3.65 and 3.66 to the form

$$\begin{bmatrix} dU_P^* & dV_P^* & 0 & 0 \end{bmatrix}^T = \mathbf{Rs}(1 \rightarrow 2) \cdot \begin{bmatrix} dU_P & dV_P & 0 & 0 \end{bmatrix}^T. \quad (3.68)$$

The substitution of Equation 3.68 into Equations 3.65 and 3.66 makes possible the expressions for $\Phi_{1.P}^*$ and $\Phi_{2.P}^*$ in the new coordinate system:

$$[\Phi_{1.P}^*] = \begin{bmatrix} \mathbf{Rs}(1 \rightarrow 2) \cdot \begin{bmatrix} dU_P & dV_P & 0 & 0 \end{bmatrix}^T \end{bmatrix}^T \cdot [\Phi_{1.P}] \cdot \mathbf{Rs}(1 \rightarrow 2) \cdot \begin{bmatrix} dU_P & dV_P & 0 & 0 \end{bmatrix}^T \quad (3.69)$$

$$[\Phi_{2.P}^*] = \begin{bmatrix} \mathbf{Rs}(1 \rightarrow 2) \cdot \begin{bmatrix} dU_P & dV_P & 0 & 0 \end{bmatrix}^T \end{bmatrix}^T \cdot [\Phi_{2.P}] \cdot \mathbf{Rs}(1 \rightarrow 2) \cdot \begin{bmatrix} dU_P & dV_P & 0 & 0 \end{bmatrix}^T \quad (3.70)$$

The following equation is valid for multiplication:

$$\left[\mathbf{Rs}(1 \rightarrow 2) \cdot \left[dU_P \;\; dV_P \;\; 0 \;\; 0 \right]^T \right]^T = \mathbf{Rs}^T(1 \rightarrow 2) \cdot \left[dU_P \;\; dV_P \;\; 0 \;\; 0 \right] \qquad (3.71)$$

Therefore,

$$[\Phi_{1.P}^*] = \left[dU_P \;\; dV_P \;\; 0 \;\; 0 \right]^T \cdot \left\{ \mathbf{Rs}^T(1 \rightarrow 2) \cdot [\Phi_{1.P}] \cdot \mathbf{Rs}(1 \rightarrow 2) \right\} \cdot \left[dU_P \;\; dV_P \;\; 0 \;\; 0 \right] \qquad (3.72)$$

$$[\Phi_{2.P}^*] = \left[dU_P \;\; dV_P \;\; 0 \;\; 0 \right]^T \cdot \left\{ \mathbf{Rs}^T(1 \rightarrow 2) \cdot [\Phi_{2.P}] \cdot \mathbf{Rs}(1 \rightarrow 2) \right\} \cdot \left[dU_P \;\; dV_P \;\; 0 \;\; 0 \right] \qquad (3.73)$$

It can be easily shown that matrices $[\Phi_{1.P}^*]$ and $[\Phi_{2.P}^*]$ in Equations 3.72 and 3.73 represent quadratic forms with respect to dU_P and dV_P.

The operator of transformation $\mathbf{Rs}(1 \rightarrow 2)$ of the part surface P having the first $\Phi_{1.P}$ and second $\Phi_{2.P}$ fundamental forms from the initial coordinate system $X_1Y_1Z_1$ to the new coordinate system $X_2Y_2Z_2$ results in a new coordinate system, with the corresponding fundamental forms expressed in the form

$$[\Phi_{1.P}^*] = \mathbf{Rs}^T(1 \rightarrow 2) \cdot [\Phi_{1.P}] \cdot \mathbf{Rs}(1 \rightarrow 2) \qquad (3.74)$$

$$[\Phi_{2.P}^*] = \mathbf{Rs}^T(1 \rightarrow 2) \cdot [\Phi_{2.P}] \cdot \mathbf{Rs}(1 \rightarrow 2) \qquad (3.75)$$

Equations 3.74 and 3.75 reveal that after the coordinate system transformation is completed, the first $\Phi_{1.P}^*$ and second $\Phi_{2.P}^*$ fundamental forms of the part surface P in the coordinate system $X_2Y_2Z_2$ are expressed in terms of the first $\Phi_{1.P}$ and second $\Phi_{2.P}$ fundamental forms initially represented in the coordinate system $X_1Y_1Z_1$. To do that, the corresponding fundamental form (either the form $\Phi_{1.P}$ or the form $\Phi_{2.P}$) must be premultiplied by $\mathbf{Rs}(1 \rightarrow 2)$ and postmultiplied by $\mathbf{Rs}^T(1 \rightarrow 2)$.

The implementation of Equations 3.74 and 3.75 significantly simplifies formulae transformations.

Equations similar to Equations 3.74 and 3.75 are valid with respect to the generating surface T of the form-cutting tool.

In case of use of the third $\Phi_{3.P}$ and fourth $\Phi_{4.P}$ fundamental forms, their coefficients can be expressed in terms of the fundamental magnitudes of the first and second orders.

Section II

Fundamentals

Fundamentals of part surface generation on a machine tool are covered in this section of the book. In a certain sense, this is the core of the *DG/K*-based method of part surface generation. The consideration below is focused on

- Contact geometry of a sculptured part surface to be machined and of the generating surface of the form-cutting tool to be applied
- Profiling of the most favorable design of the form cutting tool for machining a given part surface
- Cutting tool geometry, which is vital when converting a body bounded by the generating surface of the form-cutting tool into the workable cutting tool
- Conditions of proper part surface generation on a machine tool

The above listed topics cover the most important aspects of part surface generation that reflect geometric and kinematic aspects of part surface machining on a machine tool.

4

Geometry of Contact between a Sculptured Part Surface and Generating Surface of the Form-Cutting Tool

In the theory of part surface generation, the part surface to be machined is considered as the prime element of the machining process. Other important elements of the part surface generation process, namely, (a) kinematics of the part surface generation and (b) shape and geometry of the generating surface of the cutting tool (as well as numerous others), are considered as the secondary elements of the part surface generation process. This does not mean that the importance of the secondary elements is lower than that of the primary element. No, this is incorrect. This means just that the most favorable parameters of the secondary elements can be expressed in terms of the geometric elements of the prime element. Ultimately, the entire part surface generation process can be synthesized on the premise just of the prime element—that is, on the premise of design parameters of the part surface to be machined. In other words, merely having the geometry of the part surface P to be machined, the implementation of the DG/K-based method of part surface generation makes it possible to solve to the problem of synthesizing of the most favorable machining operation. The only geometry of the part surface P is used for the purposes of synthesizing the best possible machining operation of a given part surface.

The concept that establishes priority of the part surface to be machined over the other elements of the machining process is the cornerstone concept of the DG/K-based method of part surface generation.

To solve the problem of synthesizing the optimal part surface generation process, an appropriate analytical description of contact geometry of the part surface P and the generating surface T of the form-cutting tool is necessary. The problem of analytical description of geometry of contact between two smooth regular surfaces in the first order of tangency is a sophisticated one.

The investigation of contact geometry of curves and surfaces can be traced back to the 18th century. The study of the contact of curves and surfaces in considerable detail has been undertaken by J.L. Lagrange[*] in his *Theorié des Fonctions Analytiques* (1797) [33] and A.L. Cauchy[†] in his *Leçons sur les*

---

[*] Joseph-Louis Lagrange (January 25, 1736–April 10, 1813), a famous Italian-born [born Giuseppe Lodovico (Luigi) Lagrangia] French mathematician and mechanician.
[†] Augustin-Louis Cauchy (August 21, 1789–May 23, 1857), a famous French mathematician.

Applications du Calcul Infinitésimal á la Geometrie (1826) [7]. Later, in the 20th century, an investigation in the realm of contact geometry of curves and surfaces was undertaken by J. Favard* in his *Course de Gèomètrie Diffèrentialle Locale* (1957) [17]. There are a few more names of the researchers in the field that need to be mentioned.

The results of research obtained in the field of contact geometry of two smooth regular surfaces are widely used in the theory of part surface generation. The problem of synthesizing the most favorable technology of part surface machining can be solved on the premise of analysis of topology of the contacting surfaces in the differential vicinity of the point of their contact. In sculptured part surface machining on a multiaxis *NC* machine, the point of contact of surfaces *P* and *T* is often referred to as *cutter-contact point* (*CC* point).

Various methods for analytical description of contact geometry between two smooth regular surfaces have been developed by now. An overview of the methods can be found out in the monograph by Radzevich [126]. The latest achievements in the field are discussed in different articles [116,126,130] and monograph [154].

A detailed analysis of known methods of analytical description of geometry of contact between two smooth regular surfaces uncovered the poor capability of known methods for solving problems in the field of efficient part surface generation. Therefore, an accurate method for analytical description of contact geometry between two smooth regular surfaces in the first order of tangency that fits the needs of the theory of optimal part surface generation is necessary and worked out in this chapter.

It is convenient to begin the discussion by starting with the analytical description of local relative orientation of the part surface *P* to be machined and generating surface *T* of the form-cutting tool to be applied. The proposed analytical description is relevant to the differential vicinity of the point of contact of surfaces *P* and *T*.

4.1 Local Relative Orientation of a Part Surface and of the Cutting Tool

When machining a part on a multiaxis *NC* machine, a part surface *P* being machined and generating surface *T* of the form-cutting tool are in permanent tangency with one another (Figure 4.1). Locally, the contacting surfaces *P* and *T* can be approximated by the corresponding quadrics as schematically illustrated in Figure 4.2. The requirement to be permanently in tangency to each

---

* Jean Favard (August 28, 1902–January 21, 1965), a French mathematician.

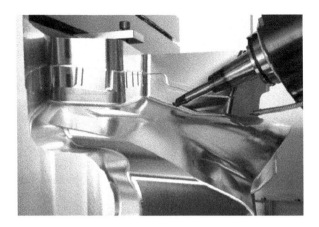

FIGURE 4.1
Example of machining of a sculptured part surface P on a multiaxis NC machine: The part surface P and the generating surface T of the form-cutting tool make point contact at every instant of the surface machining.

other imposes a type of restriction on the relative configuration (location and orientation) of surfaces P and T and on their relative motion.

In the theory of part surface generation, a quantitative measure of relative orientation of the part surface P and generating surface T of the form-cutting tool is established [116].

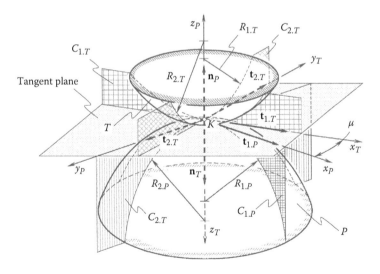

FIGURE 4.2
Local configuration of two quadrics tangent to a part surface P and the generating surface T of the form-cutting tool at a point K of their contact. (Reprinted from *Mathematical and Computer Modeling*, 39, Radzevich, S.P., Mathematical modeling of contact of two surfaces in the first order of tangency, 1083–1112, Copyright 2004, with permission from Elsevier.)

The relative orientation of the part surface P and generating surface T of the cutting tool is specified by the angle μ local* relative orientation of the surfaces. By definition, angle μ is equal to the angle that the unit tangent vector $\mathbf{t}_{1.P}$ of the first principal direction of the part surface P forms with the unit tangent vector $\mathbf{t}_{1.T}$ of the first principal direction of the generating surface T of the cutting tool. That same angle μ can also be determined as the angle formed by the unit tangent vectors $\mathbf{t}_{2.P}$ and $\mathbf{t}_{2.T}$ of the second principal directions of surfaces P and T at point K of their contact. This immediately yields equations for the calculation of the angle μ:

$$\sin \mu = |\mathbf{t}_{1.P} \times \mathbf{t}_{1.T}| = |\mathbf{t}_{2.P} \times \mathbf{t}_{2.T}|, \tag{4.1}$$

$$\cos \mu = \mathbf{t}_{1.P} \cdot \mathbf{t}_{1.T} = \mathbf{t}_{2.P} \cdot \mathbf{t}_{2.T}, \tag{4.2}$$

$$\tan \mu = \frac{|\mathbf{t}_{1.P} \times \mathbf{t}_{1.T}|}{\mathbf{t}_{1.P} \cdot \mathbf{t}_{1.T}} \equiv \frac{|\mathbf{t}_{2.P} \times \mathbf{t}_{2.T}|}{\mathbf{t}_{2.P} \cdot \mathbf{t}_{2.T}} \tag{4.3}$$

where $\mathbf{t}_{1.P}$, $\mathbf{t}_{2.P}$ is the unit vectors of principal directions on the part surface P measured at a contact point K and $\mathbf{t}_{1.T}$, $\mathbf{t}_{2.T}$ are the unit vectors of principal directions on the generating surface T of the cutting tool at that same contact point K of surfaces P and T.

The directions of the unit tangent vectors $\mathbf{t}_{1.P}$ and $\mathbf{t}_{2.P}$ of the principal directions on the part surface P (as well as directions of the unit tangent vectors $\mathbf{t}_{1.T}$ and $\mathbf{t}_{2.T}$ of the principal directions on the generating surface T of the cutting tool) can be specified in terms of the ratio dU_P/dV_P (or in terms of the ratio dU_T/dV_T in case of the cutting tool surface T). The corresponding values of the ratio $dU_{P(T)}/dV_{P(T)}$ are calculated as roots of the quadratic equation

$$\begin{vmatrix} E_{P(T)} \dfrac{dU_{P(T)}}{dV_{P(T)}} + F_{P(T)} & F_{P(T)} \dfrac{dU_{P(T)}}{dV_{P(T)}} + G_{P(T)} \\ L_{P(T)} \dfrac{dU_{P(T)}}{dV_{P(T)}} + M_{P(T)} & M_{P(T)} \dfrac{dU_{P(T)}}{dV_{P(T)}} + N_{P(T)} \end{vmatrix} = 0 \tag{4.4}$$

In the case of point contact of surfaces P and T, the actual value of the angle μ is calculated at the point K of contact of the surfaces. If surfaces P and T are in line contact, then the actual value of the angle μ can be calculated at every

---

* The surface orientation is local in nature because it relates only to the differential vicinity of the point K of contact of the surfaces P and T.

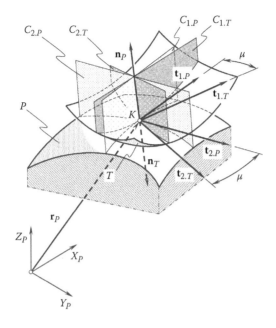

FIGURE 4.3
Specification of the angle μ of local relative orientation of the part surface P and generating surface T of the form-cutting tool.

point within the line of contact.* The line of contact of surfaces P and T is commonly referred to as *characteristic line E* or just as *characteristic E*.

Determination of the angle μ of local relative orientation of surfaces P and T is illustrated in Figure 4.3.

To calculate actual value of the angle μ of local relative orientation of surfaces P and T, unit vectors of the principal directions $\mathbf{t}_{1.P}$ and $\mathbf{t}_{1.T}$ are employed.

Consider two surfaces P and T in point contact, which are represented in a common reference system. The surfaces make contact at a point K. For further analysis, an equation of the common tangent plane to the surfaces P and T at the contact point K is necessary (Figures 4.2 and 4.4).

$$(\mathbf{r}_{tp} - \mathbf{r}_K) \cdot \mathbf{u}_P \cdot \mathbf{v}_P = 0 \tag{4.5}$$

where $\mathbf{r}_{tp}$ is the position vector of a point of the common tangent plane, $\mathbf{r}_K$ is the position vector of the contact point K, and $\mathbf{u}_P$ and $\mathbf{v}_P$ are unit vectors that are tangent to U_P and V_P coordinate lines on the part surface P at the contact point K.

---

* It is worth pointing out that in a case of line contact, the relative orientation of the surfaces P and T is predetermined in a *global* sense. However, the actual value of the angle μ of the surfaces *local* relative orientation at different points of the characteristic E is different.

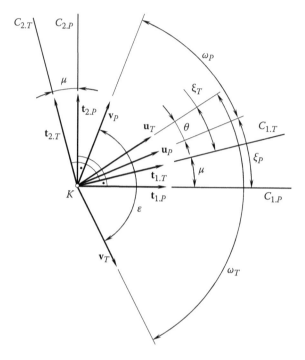

FIGURE 4.4
Local *relative* orientation of the part surface P and generating surface T of the cutting tool considered in a common tangent plane.

The actual value of the angle ω_P can be calculated from one of the following equations (see Equation 1.29):

$$\sin \omega_P = \frac{\sqrt{E_P G_P - F_P^2}}{\sqrt{E_P G_P}} \tag{4.6}$$

$$\cos \omega_P = \frac{F_P}{\sqrt{E_P G_P}} \tag{4.7}$$

$$\tan \omega_P = \frac{\sqrt{E_P G_P - F_P^2}}{F_P} \tag{4.8}$$

Equations similar to Equations 4.6 through 4.8 are also valid for calculation of the angle ω_T on the generating surface T of the form-cutting tool.

Tangent directions $\mathbf{u}_P$ and $\mathbf{v}_P$ to the U_P and V_P coordinate lines on the part surface P, and tangent directions $\mathbf{u}_T$ and $\mathbf{v}_T$ to the U_P and V_P coordinate lines on the generating surface T of the cutting tool are specified in terms of the angles θ and ε. For the calculation of actual values of the angles θ and ε, the following equations can be used:

$$\cos \theta = \mathbf{u}_P \cdot \mathbf{u}_T \tag{4.9}$$

$$\cos \varepsilon = \mathbf{v}_P \cdot \mathbf{v}_T \tag{4.10}$$

The angle ξ_P is the angle that the first principal direction $\mathbf{t}_{1.P}$ on the part surface P forms with the unit tangent vector $\mathbf{u}_P$ (see Figure 4.4). The equation for the calculation of actual value of the angle ξ_P is derived by Radzevich [116,117,136]:

$$\sin \xi_P = \frac{\eta_P}{\sqrt{\eta_P^2 - 2\eta_P \cos \omega_P + 1}} \sin \omega_P \tag{4.11}$$

where η_P designates the ratio $\eta_P = \dfrac{\partial U_P}{\partial V_P}$.

In the event $F_P = 0$, the following equality $\tan \xi_P = \eta_P$ is observed. Here, the ratio η_P is equal to the root of the quadratic equation:

$$(F_P L_P - E_P M_P)\eta_P^2 + (G_P L_P - E_P N_P)\eta_P + (G_P M_P - F_P N_P) = 0 \tag{4.12}$$

which immediately follows from Equation 1.46.

The equation for calculation of the actual value of the angle ξ_P allows for another representation. Following the chain rule, $d\mathbf{r}_P$ can be represented in the form:

$$d\mathbf{r}_P = \mathbf{U}_P dU_P + \mathbf{V}_P dV_P \tag{4.13}$$

By definition, $\tan \xi_P = \dfrac{\sin \xi_P}{\cos \xi_P}$. The functions $\sin \xi_P$ and $\cos \xi_P$ yield representation as

$$\sin \xi_P = \frac{|\mathbf{U}_P \times d\mathbf{r}_P|}{|\mathbf{U}_P| \cdot |d\mathbf{r}_P|} \tag{4.14}$$

$$\cos \xi_P = \frac{\mathbf{U}_P \cdot d\mathbf{r}_P}{|\mathbf{U}_P| \cdot |d\mathbf{r}_P|} \tag{4.15}$$

The last expressions yield

$$\tan \xi_P = \frac{\sin \xi_P}{\cos \xi_P} = \frac{|\mathbf{U}_P \times d\mathbf{r}_P|}{\mathbf{U}_P \cdot d\mathbf{r}_P} = \frac{|\mathbf{U}_P \times d\mathbf{r}_P|}{\mathbf{U}_P \cdot (\mathbf{U}_P dU_P + \mathbf{V}_P dV_P)}$$

$$= \frac{|\mathbf{U}_P \times d\mathbf{r}_P| \cdot dV_P}{\mathbf{U}_P \cdot \mathbf{U}_P dU_P + \mathbf{U}_P \cdot \mathbf{V}_P dV_P} \tag{4.16}$$

By definition (see Equation 1.14),

$$\mathbf{U}_P \cdot \mathbf{U}_P = E_P \tag{4.17}$$

$$\mathbf{U}_P \cdot \mathbf{V}_P = F_P \tag{4.18}$$

$$|\mathbf{U}_P \times \mathbf{V}_P| = \sqrt{E_P G_P - F_P^2} \tag{4.19}$$

Equations 4.13 through 4.19 yield the formula

$$\tan \xi_P = \frac{\sqrt{E_P G_P - F_P^2}}{\eta_P \cdot E_P + F_P} \tag{4.20}$$

for the calculation of the actual value of the angle ξ_P.

Equations similar to Equations 4.11 and 4.20 are also valid for calculation of the actual value of the angle ξ_T that the first principal direction $\mathbf{t}_{1.T}$ on the generating surface T of the cutting tool makes with the unit tangent vector $\mathbf{u}_T$.

The performed analysis makes possible the following equations for the calculation of the unit vectors of principal directions $\mathbf{t}_{1.P}$ and $\mathbf{t}_{2.P}$:

$$\mathbf{t}_{1.P} = \mathbf{Rt}(\xi_P, \mathbf{n}_P) \cdot \mathbf{u}_P \tag{4.21}$$

$$\mathbf{t}_{2.P} = \mathbf{Rt}\left[\left(\xi_P + \frac{\pi}{2}\right), \mathbf{n}_P\right] \cdot \mathbf{u}_P \tag{4.22}$$

for the part surface P, and similar equations for the calculation of the unit vectors of principal directions $\mathbf{t}_{1.T}$ and $\mathbf{t}_{2.T}$

$$\mathbf{t}_{1.T} = \mathbf{Rt}(\xi_T, \mathbf{n}_P) \cdot \mathbf{u}_T \tag{4.23}$$

$$\mathbf{t}_{2.T} = \mathbf{Rt}\left[\left(\xi_T + \frac{\pi}{2}\right), \mathbf{n}_P\right] \cdot \mathbf{u}_T \tag{4.24}$$

for the generating surface T of the form-cutting tool.

Equation 3.14 for the operator $\mathbf{Rt}(\varphi_A, A_0)$ through an angle φ_A about an axis A_0 is employed for the calculation of the operators of rotation in Equations 4.21 through 4.24.

4.2 First-Order Analysis: Common Tangent Plane

Two significantly different approaches for the investigation of contact geometry between two smooth regular surfaces are distinguished. First, contact geometry of surfaces having *various orders of tangency* is investigated in differential geometry of surfaces. Second, conditions of contact of two smooth regular surfaces in the *first order of tangency* are investigated in engineering geometry of surfaces.

Let us begin the discussion from the analysis of the results of the research obtained so far in differential geometry of surfaces.

Various methods can be implemented for the purposes of analytical description of the contact geometry of two smooth regular surfaces in the first order of tangency. The first-order analysis is one of them.

First-order analysis is the simplest method that is used for the analysis of the contact geometry. Implementation of the first-order analysis returns limited information about the contact geometry of the surfaces in the differential vicinity of the point of their contact. Accurate analytical description of the geometry of contact of two smooth regular surfaces is possible only when the first-order analysis is used together with a type of higher-order analysis. Under such a scenario, the first-order analysis is incorporated as a first step into the higher-order analysis.

A common tangent plane can be defined as a plane through the contact point K. The tangent plane is perpendicular to the common unit normal vector $\mathbf{n}_P$ to the part surface P and to the generating surface T of the form-cutting tool at the contact point K. For analytical description of the common tangent plane, the vector equation (see Equation 1.7) can be employed:

$$(\mathbf{r}_{tp} - \mathbf{r}_K) \cdot \mathbf{u}_P \cdot \mathbf{v}_P = 0 \tag{4.25}$$

Any pair of unite vectors chosen from the set of vectors $\mathbf{u}_P$, $\mathbf{v}_P$, $\mathbf{u}_T$, $\mathbf{v}_T$, $\mathbf{t}_{1.P}$, $\mathbf{t}_{2.P}$, $\mathbf{t}_{1.T}$, $\mathbf{t}_{2.T}$ can be used to replace the unit tangent vectors $\mathbf{u}_P$ and $\mathbf{v}_P$ in Equation 4.25.

Unite normal vector $\mathbf{n}_P$ and the tangent plane (see Equation 4.25) is the main output of implementation of the first-order analysis.

The first-order analysis provides very limited information about the contact geometry of two smooth regular surfaces in the first order of tangency. Because of this, the first-order analysis itself has limited application. However,

the importance of the first-order analysis becomes valuable, as it is incorporated into the second-order analysis and/or into the higher-order analysis.

4.3 Second-Order Analysis: Second Fundamental Form

Known methods for analytical description of contact geometry of two
smooth regular surfaces are applicable in degenerate cases of surface contact,
namely, either when the principal directions of one of the contacting surfaces
align with the principal directions of the other of the contacting surfaces, or
when the principal directions of one or of both contacting surfaces are not
identified (that is, in cases of umbilics and/or a plane).

In more general cases of contact of two smooth regular surfaces, namely,
when the angle of local relative orientation of the contacting surfaces is not
zero and it differs from $0.5\pi n$ (here n designates an integer number, $n =$
$1,2,3,...$), known methods for analytical description of contact geometry of
surfaces return only approximate results.

Currently, engineering geometry of surfaces requires in an accurate solution to the problem of analytical description of the contact geometry of two
smooth regular surfaces. The solution must be valid for the most general
cases of surface contact. A possible solution to the problem under consideration is discussed below.

The second fundamental form of a smooth regular surface had been proposed by *Gauss* for analytical description of surface curvature at a point
within a surface patch. Thus, it makes sense to begin the development of
a method for the analytical description of contact geometry of two smooth
regular surfaces starting from an analysis of the second fundamental forms
of both the surfaces at a point of their contact.

The second fundamental form $\Phi_{2.P}$ of a smooth regular part surface P is
analytically represented by an expression (see Equation 1.31) in Section 1.1:

$$\Phi_{2.P} \quad \Rightarrow \quad L_P\, dU_P^2 + 2M_P\, dU_P\, dV_P + N_P\, dV_P^2 \qquad (4.26)$$

where L_P, M_P, N_P are the fundamental magnitudes of the second order of the
part surface P and U_P, V_P are *Gaussian* coordinate lines in the part surface P.

Geometrical interpretation of the second fundamental form $\Phi_{2.P}$ at a point
of a smooth regular part surface P is schematically depicted in Figure 1.5.
Deviation of the part surface P from the tangent plane in a specified direction is analytically described in terms of the fundamental form $\Phi_{2.P}$, namely,
the displacement is equal to $\sqrt{\Phi_{2.P}}$.

The aforementioned with regard to the part surface P is valid, of course,
with regard to the generating surface T of the form-cutting tool. The second

fundamental form $\Phi_{2.T}$ of the smooth regular surface T of the cutting tool is analytically represented by the expression

$$\Phi_{2.T} \quad \Rightarrow \quad L_T\, dU_T^2 + 2M_T\, dU_T\, dV_T + N_T\, dV_T^2 \qquad (4.27)$$

where L_T, M_T, N_T is the fundamental magnitudes of the second order of generating surface T of the cutting tool amd U_T, V_T is the *Gaussian* coordinate lines in the cutting tool surface T.

That said, consider two smooth regular part surfaces P and T, which contact each other at a point K (Figure 4.5).

For the contacting surfaces P and T, the second fundamental forms $\Phi_{2.P}$ and $\Phi_{2.T}$ are specified by Equations 4.26 and 4.27. A common tangent plane is a plane through the contact point K. Consider a spatial line l_P that is entirely located within the part surface P. This line is the line through the contact point K. Direction of the line l_P within the tangent plane is specified by unit tangent vector $\mathbf{t}_P$. A point m_P within the curve l_P deviates from the tangent plane at a distance $m_P m_P^*$. The deviation $m_P m_P^*$ can be calculated from the formula

$$m_P m_P^* = \sqrt{\left|\Phi_{2.P}\right|}\, \operatorname{sgn} \Phi_{2.P} \qquad (4.28)$$

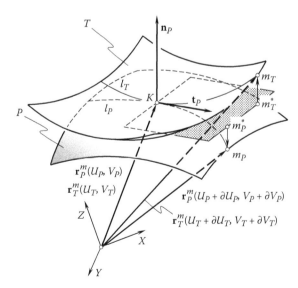

FIGURE 4.5
Definition of degree of conformity of a smooth regular part surface P and generating surface T of the cutting tool at a point K of their contact.

Due to the torsion of the curve l_P, the point m_P^* within the common tangent plane is located aside of the straight line along the unit vector $\mathbf{t}_P$. However, as the point m_P approaches the contact point K, displacement of point m_P^* from the straight line approaches zero [116]. Therefore, in further consideration, torsion of the curve l_P can be ignored.

Similarly, in that same direction that is specified by the unit tangent vector $\mathbf{t}_P$ (actually this direction is specified by unit tangent vector $\mathbf{t}_T \equiv \mathbf{t}_P$), a three-dimensional line l_T within the part surface T is passing. A point m_T within the curve l_T deviates from the tangent plane at a distance $m_T m_T^*$. The deviation $m_T m_T^*$ can be calculated from the formula

$$m_T m_T^* = \sqrt{|\Phi_{2.T}|}\, \mathrm{sgn}\, \Phi_{2.T} \tag{4.29}$$

Due to the torsion of the curve l_T, the point m_T^* within the common tangent plane is located aside of the straight line along the unit vector $\mathbf{t}_P$. However, as point m_T approaches contact point K, the displacement of point m_T^* from the straight line approaches zero [116]. Similar to the case of the curve l_P, torsion of the curve l_T can be ignored as well.

The deviation $m_P m_P^*$ of the part surface P from the tangent plane together with the deviation $m_T m_T^*$ of the cutting tool surface T from the tangent plane can be employed to quantify deviation of the part surface P and the generating surface T of the form-cutting tool from one another in a direction that is specified by the unit tangent vector $\mathbf{t}_P$.

Now refer to Figure 4.6, where a section of the contacting part surface P and generating surface T of the cutting tool by a plane through the unit vector of the common perpendicular $\mathbf{n}_P$ is schematically shown. The torsion of the lines l_P and l_T is ignored here as when points m_P and m_T approach the contact point K, the corresponding projections m_P^* and m_T^* of these points onto the tangent plane approach the straight line along the unit tangent vector $\mathbf{t}_P$.

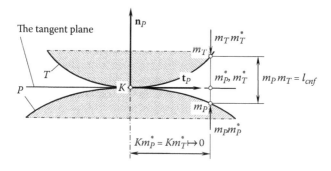

FIGURE 4.6
Specification of the deviation l_{cnf} in terms of the second fundamental forms $\Phi_{2.P}$ and $\Phi_{2.T}$ of the contacting part surface P and the generating surface T of the cutting tool, respectively.

Having calculated the distances $m_P m_P^*$ (see Equation 4.28) and $m_T m_T^*$ (see Equation 4.29), the distance $l_{cnf} = m_P m_T$ between the contacting smooth regular part surface P and generating surface T of the cutting tool can be expressed in terms of the second fundamental forms $\Phi_{2.P}$ and $\Phi_{2.T}$, namely [116]

$$l_{cnf} = \sqrt{|\Phi_{2.P}|} \operatorname{sgn} \Phi_{2.P} + \sqrt{|\Phi_{2.T}|} \operatorname{sgn} \Phi_{2.T} \qquad (4.30)$$

For this purpose, both the fundamental forms, $\Phi_{2.P}$, and $\Phi_{2.T}$, must be represented in a common reference system.

The resultant deviation l_{cnf} can be employed as a quantitative measure of deviation of two smooth regular surfaces P and T from one another in the direction specified by the unit vector $\mathbf{t}_P$. It could be anticipated that the quantitative criterion l_{cnf} be the best possible criterion for evaluation of the degree of conformity of two smooth regular surfaces in the first order of tangency.

To implement the resultant deviation l_{cnf} in practice,

(a) Both fundamental magnitudes $\Phi_{2.P}$, and $\Phi_{2.T}$ must be represented in a common reference system.

(b) Both fundamental magnitudes $\Phi_{2.P}$ and $\Phi_{2.T}$ must be expressed in that same U and V parameters.

For the second fundamental form $\Phi_{2.P}$ specified in the original reference system associated with the part surface P, a corresponding expression $\Phi_{2.P}^*$ in another reference system (which is common to the contacting surfaces P and T) can be derived from the Equation 3.69:

$$[\Phi_{2.P}^*] = \mathbf{Rs}^T (P \to C) \cdot [\Phi_{2.P}] \cdot \mathbf{Rs}(P \to C) \qquad (4.31)$$

The operator of the resultant coordinate system transformation, namely the operator of the transformation from the original reference system associated with the part surface P to the common reference system (the reference system C) in Equation 4.31 is designated as $\mathbf{Rs}(P \to C)$.

An expression similar to Equation 4.31 is valid with respect to the second fundamental form $\Phi_{2.T}$ of generating surface T of the form-cutting tool [116]:

$$[\Phi_{2.T}^*] = \mathbf{Rs}^T (T \to C) \cdot [\Phi_{2.T}] \cdot \mathbf{Rs}(T \to C) \qquad (4.32)$$

Here, in Equation 4.32, the operator of the resultant coordinate system transformation, namely of the transformation from the original reference system associated with the generating surface T of the form-cutting tool to the common reference system (the reference system C) is designated as $\mathbf{Rs}(T \to C)$.

Refer to Appendix B for details on change of the surface parameters to express equations for both of the contacting surfaces P and T in terms of same U and V parameters.

The discussed measure l_{cnf} of degree of conformity of two smooth regular surfaces P and T in the first order of tangency, is legitimate but it is computationally inefficient. For practical applications, another quantitative measure of degree of conformity of two smooth regular part surfaces P and T that make contact at a point K should be developed [116].

4.4 Second-Order Analysis: Planar Characteristic Images

For more accurate analytical description of contact geometry of the part surface P and generating surface T of the cutting tool, the second-order parameters should be considered. The second-order analysis incorporates elements of both of the first order as well as elements of the second-order analysis. For performing the second-order analysis, familiarity with Dupin's indicatrix is highly desired. Dupin's indicatrix is a perfect start point for the second-order analysis.

4.4.1 Preliminary Remarks: Dupin's Indicatrix

At any point of a smooth regular part surface P (as well as at any point of a smooth regular generating surface T of the form-cutting tool), corresponding Dupin's indicatrices can be constructed. Dupin's indicatrices $Dup\ (P)$ at a point of a part surface P and Dupin's indicatrix $Dup\ (T)$ at a point of the machining surface T of the form-cutting tool are planar characteristic curves of the second order. They are used for graphical interpretation of the distribution of normal curvatures of a surface in differential vicinity of a surface point.

Dupin's indicatrices of surfaces P and T are of critical importance in the theory of part surface generation. The generation of this planar characteristic curve is illustrated with a diagram shown in Figure 4.7.

A plane W through the unit normal vector $\mathbf{n}_P$ to the part surface P at a point m is rotating about $\mathbf{n}_P$. While rotating, the plane occupies consecutive positions W_1, W_2, W_3, and others. The radii of normal curvature of the line of intersection of the part surface P by normal planes W_1, W_2, W_3 are equal to $R_{P,1}$, $R_{P,2}$, $R_{P,3}$, and so forth. The part surface P is intersected by a plane Q. The plane Q is orthogonal to the unit normal vector $\mathbf{n}_P$. This plane is at a certain small distance δ from the point m. When distance δ approaches zero ($\delta \rightarrow 0$) and when scale of the line of intersection of the part surface P by the plane Q approaches infinity, the line of intersection of the part surface P by the plane Q approaches the planar characteristic curve that is commonly referred to as Dupin's indicatrix $Dup\ (P)$.

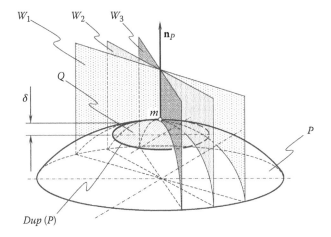

FIGURE 4.7
Dupin's indicatrix at a point of a smooth regular part surface P.

In differential geometry of surfaces, Dupin's indicatrices of the following five different types are distinguished (Figure 4.8):

- Elliptic (Figure 4.8a)
- Umbilic (Figure 4.8b)
- Parabolic (Figure 4.8c)
- Hyperbolic (Figure 4.8d)
- Minimal (Figure 4.8e)

Dupin's indicatrix for a plane local surface patch does not exist. In the case of a plane, all points of the Dupin's indicatrix are remote to infinity.

For local surface patches having negative full curvature ($\mathcal{G}_P < 0$), phantom branches (that is the branches that are not intersected by a plane perpendicular to the unite normal vector to the part surface P at a point m) of the characteristic curve Dup (P) in Figure 4.8d and e are shown in dashed lines.

An easy way to derive an equation of the characteristic curve Dup (P) is discussed immediately below.

Euler's formula

$$k_{1.P} \cos^2 \varphi + k_{2.P} \sin^2 \varphi = k_P \tag{4.33}$$

yields representation in the form:

$$\frac{k_{1.P}}{k_P} \cos^2 \varphi + \frac{k_{2.P}}{k_P} \sin^2 \varphi = 1 \tag{4.34}$$

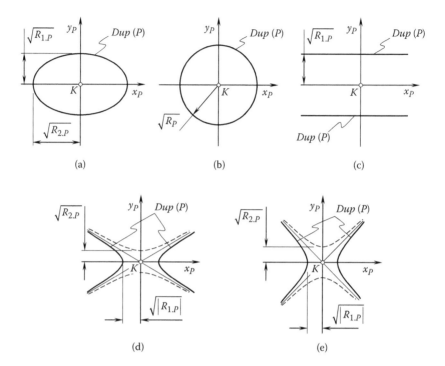

FIGURE 4.8

Five different types of Dupin's indicatrices, *Dup* (P), at a point of a smooth regular surface.

Transition from polar coordinates to *Cartesian* coordinates can be performed using well-known formulae:

$$x_P = \rho \cos \varphi \tag{4.35}$$

$$y_P = \rho \sin \varphi \tag{4.36}$$

These formulae make the following expressions possible for $\cos^2 \varphi = x_P^2/\rho^2$ and $\sin^2 \varphi = y_P^2/\rho^2$. After substituting the last formulae into Equation 4.34, one can come up with the equation

$$\frac{k_{1.P}}{k_P} \cdot \frac{x_P^2}{\rho^2} + \frac{k_{2.P}}{k_P} \cdot \frac{y_P^2}{\rho^2} = 1. \tag{4.37}$$

It is convenient to designate $\rho = \sqrt{k_P^{-1}}$. Principal curvatures $k_{1.P}$ and $k_{2.P}$ are the roots of the quadratic equation:

$$\begin{vmatrix} L_P - E_P k_P & M_P - F_P k_P \\ M_P - F_P k_P & N_P - G_P k_P \end{vmatrix} = 0 \tag{4.38}$$

Substituting the calculated values of the principal curvatures $k_{1.P}$ and $k_{2.P}$ into Equation 4.37, and after performing necessary formulae transformations, an equation* for the Dupin's indicatrix *Dup* (*P*) can be represented in the form

$$k_{1.P}x_P^2 + k_{2.P}y_P^2 = 1 \qquad (4.39)$$

Equation 4.39 describes a particular case of the Dupin's indicatrix, which is represented in Darboux' frame.

The general form of the equation of Dupin's indicatrix at a part surface point *m* is often represented as

$$Dup\,(P) \Rightarrow \frac{L_P}{E_P}x_P^2 + \frac{2M_P}{\sqrt{E_P G_P}}x_P y_P + \frac{N_P}{G_P}y_P^2 = 1 \qquad (4.40)$$

In Equation 4.40, the characteristic curve *Dup* (*P*) is expressed in terms of the fundamental magnitudes E_P, F_P, and G_P of the first, $\Phi_{1.P}$, and in terms of the fundamental magnitudes L_P, M_P, and N_P of the second order, $\Phi_{2.P}$ of the part surface *P*.

4.4.2 Surface of Relative Curvature

The implementation of the surface of relative curvature for the purposes of analytical description of the contact geometry of two smooth regular surfaces in the first order of tangency is a perfect example of the second-order analysis. This idea can be traced back to the publications by H. Hertz.[†]

4.4.2.1 Comments on Analytical Description of the Local Geometry of Two Contacting Surfaces Loaded by a Normal Force

4.4.2.1.1 Hertz Proportional Assumption

It makes sense to begin the discussion on the contact geometry of two smooth regular surfaces from the assumptions made by the founder of contact mechanics of materials, the German physicist Heinrich Hertz. The theory proposed by Hertz is based on implementation of an imaginary surface, which (after Hertz) is referred to as *surface of relative curvature*.

---

* The same equation of the Dupin's indicatrix could be derived in another way. Coxeter [25] considers a pair of conics obtained by expanding an equation in Monge's form $z = z(x, y)$ in a Maclauren series $z = z(0,0) + z_1 x + z_2 y + \frac{1}{2}(z_{11}x_1^2 + 2z_{12}xy + z_{22}y^2) + \ldots = \frac{1}{2}(b_{11}x^2 + 2b_{12}xy + b_{22}y^2)$. This gives the equation $(b_{11}x^2 + 2b_{12}xy + b_{22}y^2) = \pm 1$ of the Dupin's indicatrix.

[†] Heinrich Rudolf Hertz (February 22, 1857–January 1, 1894), a famous German physicist.

The investigation of geometry of interacting surfaces under an applied normal load can be traced back to the fundamental research by Hertz (1881–1896) on contact of solid elastic bodies [23].

In 1886–1889, Hertz published two articles on what was to become known as the field of *contact mechanics* of materials. His work basically summarizes how two axisymmetric objects placed in contact will behave under loading. The developed theory is based on Hertz' observation of elliptical *Newton's rings* formed upon placing a glass sphere upon a lens as the basis of the assumption that the pressure exerted by the sphere follows an elliptical distribution.

The interaction of an elastic sphere and a plane is schematically illustrated in Figure 4.9a. The initial contact of the sphere of a certain radius R_{sphere} and the plane can be assumed at point K. Then, after a normal load, F_n, is applied; the contact point, K, is spread to a round contact patch of contact of radius, r_{pc}, as schematically illustrated in Figure 4.9b.

It is important to stress here the two features of the theory developed by Hertz.

First, the theory developed by Hertz is based on the assumption that the radius of contact patch, r_{pc}, is much smaller compare to the radius of the sphere, R_{sphere}. The theory returns reasonable results of the calculation of contact stress if the radius, r_{pc}, is 10 (or more) times smaller than the radius of

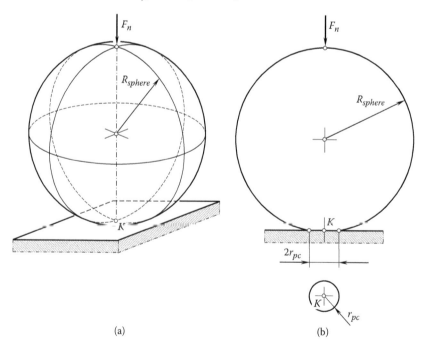

(a)   (b)

FIGURE 4.9
Interaction of a sphere of radius, R_{sphere}, and of a plane under a normal load, F_n.

the sphere, R_{sphere}. If the inequality, $R_{sphere} \gg r_{pc}$, is not fulfilled, then the Hertz theory is not valid.

Assumption 4.1

The dimensions of the contact patch are much smaller in comparison to corresponding radii of curvature of the contacting elastic bodies. ■

A conclusion that can be immediately drawn from that statement is as follows: It is necessary to be very careful when applying the Hertz theory for the calculation of contact stress and so forth in case of contact of elastic bodies bounded by *convex* and *concave* surfaces because in this particular case the inequality $R_{sphere} \gg r_{pc}$ is commonly not fulfilled.

Second, Hertz has considered the interaction of two elastic bodies having simple geometries of the contacting surfaces. A plane, spheres of various radii, and so forth are common in the research undertaken by Hertz. It should be noted that for surfaces with a simple geometry, the principal directions at the point of contact, K, are either not identified (as observed for a sphere and a plane) or they are aligned to one another. For surfaces with a simple geometry, the concept of surface of relative curvature is applicable.

In the simplest case of contact of a plane and a sphere (Figure 4.10a), the actual value of radius of the sphere, R_{sphere}, is sufficient for an analytical description of geometry of contact of the sphere and the plane. No radius

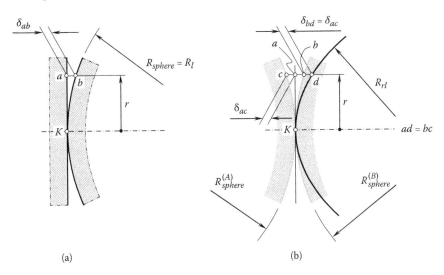

(a)   (b)

FIGURE 4.10
The definition of radius of relative curvature, R_{rl}, of two smooth regular surfaces, A and B, making point contact at K.

of relative curvature, R_{rl}, is necessary in this simplest case of surface contact as these two radii are identical to one another ($R_{sphere} \equiv R_{rl}$). The results of analytical description of the contact geometry of a plane and of a sphere can be enhanced to similar problems when two bodies with more complex geometries make contact, for example, to the case of contact of two spheres of different radii. For this purpose, *radius of relative curvature* is introduced into consideration. In the case of contact depicted in Figure 4.10a, two points, *a* and *b*, are taken at a reasonably short distance, *r*, from a straight line through the contact point, *K*, that is perpendicular to the plane. The points, *a* and *b*, are at a certain distance, δ_{ab}, from one another.

In a case of contact of two spheres, *A* and *B*, of radii, $R_{sphere}^{(A)}$ and $R_{sphere}^{(B)}$, respectively (Figure 4.10b), two points, *c* and *d*, are taken into consideration. These two points are equivalent to the points, *a* and *b*, in the aforementioned case of contact of a sphere and a plane. The distance between points *c* and *d* (δ_{cd}, not shown in Figure 4.10b) significantly exceeds the distance δ_{ab}. A surface of relative curvature of radius, R_{rl}, for spheres *A* and *B* is designed so to ensure equality of the distances, δ_{ad}, to distance δ_{ab} in the case depicted in Figure 4.10a. If equality $\delta_{ad} = \delta_{ab}$ is fulfilled (and the equality $ad = bc$ is met as well), then the problem of contact of two spheres of radii, $R_{sphere}^{(A)}$ and $R_{sphere}^{(B)}$ (Figure 4.10b), can be substituted with the equivalent problem of contact of a plane and a sphere (Figure 4.10a).

To construct a *surface of relative curvature*, the following manipulations with radii of curvatures must be performed.

The geometry of contact of the two surfaces can be expressed in terms of curvature of the sphere. For the sphere of the radius, R_{sphere}, its curvature is expressed by the parameter inverse to the radius of the sphere, $k_{sphere} = R_{sphere}^{-1}$.

To accommodate the obtained results for the case of contact of two elastic bodies bounded by two spheres, *A* and *B*, a concept of the *surface of relative curvature* has been introduced. At any normal section through the point of contact, *K*, of the bodies bounded by two spheres of radii, $R_{sphere}^{(A)}$ and $R_{sphere}^{(B)}$ (and having normal curvatures k_A and k_B, respectively), normal curvature, k_r, of the surface of relative curvature can be calculated from the following formula:

$$k_r = k_A + k_B \qquad (4.41)$$

This formula is composed under the assumption that the deviation, δ_{ad}, of the surface of relative curvature from the plane in the case depicted in Figure 4.10b, is equal to the deviation, δ_{ab}, of the sphere from the plane as illustrated in Figure 4.10a. The equality $\delta_{ad} = \delta_{ab}$ is fulfilled (as well as the equality $ad = bc$) when the deviations δ_{bd} and δ_{ac} are equal ($\delta_{bd} = \delta_{ac}$). Such an equality $\delta_{ad} = \delta_{ab}$ is reasonable if and only if the inequality $R_{sphere} \gg r_{pc}$ is fulfilled. Otherwise, the application of the Hertz theory is invalid.

The contact of two elastic bodies bounded by surfaces having more complex geometry has not been investigated by Hertz.

4.4.2.2 Construction of Surface of Relative Curvature

Let us consider a Taylor expansion of equations of the part surface P and generating surface T of the cutting tool at a point of their contact K. Implementation of Taylor expansion is convenient for the determination of the surface of relative curvature.

In the local coordinate system $x_P y_P z_P$, equation of the part surface P in the differential vicinity of point K can be represented in the form

$$P \Rightarrow z_P = \frac{x_P^2}{2R_{1.P}} + \frac{y_P^2}{2R_{2.P}} + \dots, \tag{4.42}$$

where $R_{1.P}$ and $R_{2.P}$ are the principal radii of curvature of the part surface P at the contact point K. They are positive if the corresponding center of curvature is located within the body bounded by the part surface P, that is, the center of curvature is located on the negative portion of z_P–axis of the reference system $x_P y_P z_P$.

Similarly, in the local coordinate system $x_T y_T z_T$, the equation of generating surface T of the form-cutting tool in the differential vicinity of the contact point K yields representation in the form

$$T \Rightarrow z_T = \frac{x_T^2}{2R_{1.T}} + \frac{y_T^2}{2R_{2.T}} + \dots. \tag{4.43}$$

The quantities $R_{1.T}$ and $R_{2.T}$ have the same meaning for the generating surface T as the quantities $R_{1.P}$ and $R_{2.P}$ for the part surface P.

A surface for which the equality $z_{\mathfrak{R}} = z_P - z_T$ is observed is referred to as *surface of relative curvature*. Further, the surface of relative curvature is designated as $\mathfrak{R}$.

To employ the equation $z_{\mathfrak{R}} = z_P - z_T$, it is necessary

(a) To represent both the surfaces P and T in a common reference system having origin at the contact point K

(b) To align positive direction of $z_{\mathfrak{R}}$ axis with the unit normal vector $\mathbf{n}_P$ to the part surface P at the contact point K

The local coordinate system $x_P y_P z_P$ is suitable for the purposes of analytical description of the surface $\mathfrak{R}$ of relative curvature ($x_P y_P z_P \equiv x_{\mathfrak{R}} y_{\mathfrak{R}} z_{\mathfrak{R}}$). In this reference system, the equation of the surface of relative curvature $\mathfrak{R}$ yields a representation in the form

$$\mathfrak{R} \Rightarrow z_P = \frac{x_P^2}{2R_{1.\mathfrak{R}}} + \frac{y_P^2}{2R_{2.\mathfrak{R}}} + \dots \tag{4.44}$$

The surface of relative curvature, $\Re$, reflects the topology of the contacting surfaces P and T, and geometry of their contact only locally, just in differential vicinity of the contact point K. The terms in Equation 4.42, which are taken into account and are written down explicitly, do not allow for more general interpretation of the geometry of the part surface P and generating surface T of the cutting tool.

In Darboux's frame, this result makes possible the simplified equation for the surface of relative curvature:

$$\Re \Rightarrow 2z_P = k_{1.\Re}x_P^2 + k_{2.\Re}y_P^2 \tag{4.45}$$

The relative curvature is the major analytical tool that is used in contemporary practice for the purpose of analytical description of the geometry of contact in higher kinematic pairs.

Relative normal curvature* $k_\Re$ at a point of contact of a part surface P and generating surface T of the cutting tool is defined as the summa of normal curvatures k_P and k_T of both surfaces P and T. It is taken in a common normal plane section of surfaces P and T and is equal to

$$k_\Re = k_P + k_T \tag{4.46}$$

where $k_\Re$ designates normal curvature of the surface of relative curvature $\Re$, k_P is the normal curvature of the part surface P, and k_T is the normal curvature of the generating surface T of the cutting tool.

All three curvatures, k_P, k_T, and $k_\Re$, are calculated in a common normal plane section of the surfaces P, T, and $\Re$ through the contact point K.

Consider a common normal plane section through the contact point K of the part surface P and the generating surface T of the form-cutting tool. This plane section forms a certain angle ϕ with the unit tangent vector $\mathbf{t}_{1.P}$. That same common normal plane section makes the angle $(\phi + \mu)$ with the unit tangent vector $\mathbf{t}_{1.T}$. Recall that the angle μ of local relative orientation of the surfaces is the angle that form the first $\mathbf{t}_{1.P}$ and $\mathbf{t}_{1.T}$ (or the same, the second $\mathbf{t}_{2.P}$ and $\mathbf{t}_{2.T}$) principal directions of surfaces P and T (Figure 4.3).

Euler's equation yields representation of the normal curvatures k_P and k_T in the form:

$$k_P = k_{1.P} \cos \varphi + k_{2.P} \sin \varphi \tag{4.47}$$

$$k_T = k_{1.T} \cos(\varphi + \mu) + k_{2.T} \sin(\varphi + \mu) \tag{4.48}$$

---

* Similarly, relative normal radius of curvature $R_\Re$ of two contacting surfaces P and T at a given point K could be defined as the difference of the normal radii of curvature R_P and R_T. It is taken in a common normal plane section of the surfaces P and T and is equal to $R_\Re = R_\Re - R_T$.

where $k_{1.P}$ and $k_{2.P}$ are the first and the second principal curvatures at a point of the part surface P, $k_{1.T}$ and $k_{2.T}$ are the first and the second principal curvatures at a point of the generating surface T of the cutting tool, φ is the angular parameter, and μ is the angle of local relative orientation of surfaces P and T.

It is important to point out here that the inequality $k_{1.P} > k_{2.P}$ is always observed.*

Thus, Equation 4.46 makes an equation for the calculation of the normal curvature, $k_{\Re}$, of the surface, $\Re$, of relative curvature possible:

$$k_{\Re} = k_{1.P} \cos^2 \varphi + k_{2.P} \sin^2 \varphi + k_{1.T} \cos^2(\varphi + \mu) + k_{2.T} \sin^2(\varphi + \mu) \qquad (4.49)$$

Equation 4.49 is expressed in terms

(a) Of principal curvatures $k_{1.P(T)}$, $k_{2.P(T)}$ of surfaces P and T contacting with one another at the contact point K

(b) Of angle μ of local relative orientation of surfaces P and T

(c) Of the angular parameter ϕ

Equation 4.49 can then be cast to

$$k_{\Re} = a \cdot \cos^2 \varphi + b \cdot \sin(2\varphi) + c \cdot \sin^2 \varphi \qquad (4.50)$$

Here, in Equation 4.50, parameters a, b, and c can be calculated from the following expressions:

$$a = k_{1.P} + k_{1.T} \cos^2 \mu + k_{2.T} \sin^2 \mu \qquad (4.51)$$

$$b = \frac{(k_{2.T} - k_{1.T})}{2} \cdot \sin(2\mu) \qquad (4.52)$$

$$c = (k_{2.P} + k_{1.T} \sin^2 \mu + \cos^2 \mu) \qquad (4.53)$$

The principal curvatures $k_{1.\Re}$ and $k_{2.\Re}$ of the surface of relative curvature, $\Re$, are the extreme values of function $k_{\Re}(\varphi)$ (see Equation 4.50).

---

* Commonly, this inequality is loosely represented in the form $k_{1.P(T)} \geq k_{2.P(T)}$, which is incorrect. In case of equality, that is, if $k_{1.P(T)} = k_{2.P(T)}$, then all normal curvatures at the point of contact K of the surface P and T are of the same value (and of the same sign). This is observed for umbilics as well as for plane surfaces. Because of this, at an umbilic point, the principal directions are **undefined**. Therefore, principal curvatures are also undefined. This means that the inequality $k_{1.P(T)} \geq k_{2.P(T)}$ properly reflects the correspondence between the principal curvatures $k_{2.P(T)}$ and $k_{2.P(T)}$.

The principal directions of the surface $\mathfrak{R}$ are those directions in the common tangent plane for which the following equation is satisfied:

$$\frac{\partial k_{\mathfrak{R}}(\varphi)}{\partial \varphi} = 0 \tag{4.54}$$

Equation 4.54, together with the Equation 4.50, leads to

$$\tan(2\varphi) = \frac{2b}{c - a} \tag{4.55}$$

Two solutions for the angle φ, φ_1 and $\varphi_2 = \varphi_1 + 90°$, are determined by Equation 4.55. This means that there are two perpendicular directions for the principal directions of the surface of relative curvature. The principal curvatures of the surface of relative curvature can be calculated from

$$k_{1.2,\mathfrak{R}} = \frac{(a+c) \pm \sqrt{(a+c)^2 + 4b^2}}{2} \tag{4.56}$$

The surface of relative curvature $\mathfrak{R}$ is important in many engineering applications.

Note again that all three normal curvatures $k_{\mathfrak{R}}$, k_P, and k_T in Equation 4.46 are taken in a common normal plane section through the point* of contact K of the surfaces P, T, and $\mathfrak{R}$.

Based on the calculated values of principal curvatures $k_{1.\mathfrak{R}}$ and $k_{2.\mathfrak{R}}$, the implicit equation of the surface of relative curvature yields the following representation:

$$2Z_{\mathfrak{R}} = k_{1.\mathfrak{R}} X_{\mathfrak{R}}^2 + k_{2.\mathfrak{R}} Y_{\mathfrak{R}}^2 \tag{4.57}$$

A characteristic surface similar to the surface $\mathfrak{R}$ of relative curvature can be defined as the surface for which the equality $R_{\mathfrak{R}} = R_P - R_T$ is observed. Evidently, this equality is similar in nature to Equation 4.46.

4.4.3 Dupin's Indicatrix of the Surface of Relative Curvature

Dupin's indicatrix can be constructed at any point of a smooth regular part surface P and/or at a point of a smooth regular generating surface T of the

---

* In case of line contact of a part surface P and generating surface T of the cutting tool, point K is the point within the line of the surface contact at which the normal curvatures $k_{\mathfrak{R}}$, k_P, and k_T are required to be calculated.

cutting tool. Similar to that, Dupin's indicatrix can be constructed at a point of a surface of relative curvature, $\mathfrak{R}$.

To construct Dupin's indicatrix at a point of a surface of relative curvature, $\mathfrak{R}$, consider intersection of the surface of relative curvature by a plane, parallel to the tangent plane at the contact point K, but only at a small distance away from it. Then, project the intersection on the tangent plane. In the coordinate plane $x_P y_P$, principal part of the intersection will be given by the equation of Dupin's indicatrix* [14]:

The distribution of normal relative curvature within differential vicinity of the contact point K is described by equation of Dupin's indicatrix $Dup\,(\mathfrak{R})$ [116,117,136]:

$$Dup\,(\mathfrak{R}) \equiv Dup\,(P/T) \quad \Rightarrow \quad \frac{L_\mathfrak{R}}{E_\mathfrak{R}} x_\mathfrak{R}^2 + \frac{2M_\mathfrak{R}}{\sqrt{E_\mathfrak{R} G_\mathfrak{R}}} x_\mathfrak{R} y_\mathfrak{R} + \frac{N_\mathfrak{R}}{G_\mathfrak{R}} y_\mathfrak{R}^2 = \pm 1 \quad (4.58)$$

Here, $E_\mathfrak{R}$, $F_\mathfrak{R}$, and $G_\mathfrak{R}$ designate fundamental magnitudes of the first order, and $L_\mathfrak{R}$, $M_\mathfrak{R}$, and $N_\mathfrak{R}$ designate fundamental magnitudes of the second order of the surface of relative curvature $\mathfrak{R}$ at the point of contact K of the part surface P and the generating surface T of the form-cutting tool.

If axes $x_\mathfrak{R}$ and $y_\mathfrak{R}$ of the local coordinate system $x_\mathfrak{R} y_\mathfrak{R}$ align with the principal directions $\mathbf{t}_{1.\mathfrak{R}}$ and $\mathbf{t}_{2.\mathfrak{R}}$ of the surface of relative curvature $\mathfrak{R}$, then Equation 4.58 is reduced to

$$Dup\,(P/T) \quad \Rightarrow \quad k_{1.\mathfrak{R}} x_\mathfrak{R}^2 + k_{2.\mathfrak{R}} y_\mathfrak{R}^2 = \pm 1 \qquad (4.59)$$

An important intermediate conclusion immediately follows from Equation 4.58:

Conclusion 4.1

The direction $\mathbf{t}_{1.\mathfrak{R}}$ for the maximum $k_{1.\mathfrak{R}}$ and the direction $\mathbf{t}_{2.\mathfrak{R}}$ for the minimum $k_{2.\mathfrak{R}}$ values of normal curvature of the surface of relative curvature $\mathfrak{R}$ are always orthogonal to one another, and, therefore, the condition $\mathbf{t}_{1.\mathfrak{R}} \perp \mathbf{t}_{2.\mathfrak{R}}$ is always observed.   ∎

---

* To be more exact, Dupin's indicatrix $Dup\,(\mathfrak{R}) \equiv Dup\,(P/T)$ does not reflect the distribution of normal relative curvature $k_\mathfrak{R}$ itself, but the distribution of normal relative radii of curvature $R_\mathfrak{R}$. Thus, it could be designated as $Dup_R\,(\mathfrak{R})$. However, equation of the indicatrix $Dup_R\,(\mathfrak{R})$ of a surface of normal curvature could also be composed. Similar to that, the corresponding equations for the normalized $Dup_R\,(\overline{\mathfrak{R}})$ indicatrix of relative normal radius of curvature and indicatrix of normal curvature $Dup_k\,(\mathfrak{R})$ could also be easily derived.

The major axes of the Dupin's indicatrix Dup $(\Re)$ make the angles φ_{min} and φ_{max} with the principal directions $\mathbf{t}_{1.P}$ and $\mathbf{t}_{2.P}$.

4.4.4 Curvature Indicatrix

It is convenient to begin a discussion on curvature indicatrix at a point of the surface of relative curvature, $\Re$, with a brief introduction of curvature indicatrix that is constructed at a point of a smooth regular part surface P.

Five different types of the characteristic curve Dup (P) are distinguished in differential geometry of surfaces (Figure 4.8). They are

- Elliptic, for which Gauss' curvature is always positive, $(\mathcal{G}_P > 0)$
- Umbilic $(\mathcal{G}_P > 0)$
- Parabolic $(\mathcal{G}_P = 0)$
- Hyperbolic $(\mathcal{G}_P < 0)$
- Minimal hyperbolic $(\mathcal{G}_P < 0, |R_{1.P}| = R_{2.P})$

For planar local patch of a part surface P the characteristic curve Dup (P) does not exist. All points of this characteristic curve for planar local patch of a part surface P are remote to infinity.

As shown in Chapter 1, the surfaces considered in engineering geometry differ from the surfaces considered in differential geometry of surface and an illustration of that is discussed below.

In differential geometry of surfaces, Dupin's indicatrix is implemented for the purpose of illustration of the distribution of the surface normal curvature. There are only five different types of Dupin's indicatrices* of a smooth regular surface. All of them are depicted in Figure 4.8.

The bodily and void side of a part surface P must be distinguished when the DG/K-based method of part surface generation is used [110,116,117,136]. Unfortunately, Dupin's indicatrix for a convex part surface P can be identical to the Dupin's indicatrix for a concave part surface P. This is observed in particular when both the concave and convex surfaces are represented by a *mathematical* surface that is described by that same equation. Therefore, by means of Dupin's indicatrix no difference can be found between the convex and the concave surfaces. The above can be summarized by the following conclusion:

---

* Dupin's indicatrix Dup (P) is completely equivalent to the second fundamental form $\Phi_{2.P}$ of the surface P. The second fundamental form $\Phi_{2.P}$ is also known as an operator of the surface shape. Koenderink [30] recommends that the characteristic curve Dup (P) be considered as a rotation of operator of the surface shape $\Phi_{2.P}$.

Conclusion 4.2

Dupin indicatrix *Dup* (*P*) at a point of a part surface *P* possesses no capability to distinguish whether the part surface *P* is convex at this point or the surface *P* is concave. ∎

For the purpose of distinguishing whether a part surface *P* is convex or the surface *P* is concave, a characteristic image of a novel type can be used. This newly introduced characteristic image is referred to as the *curvature indicatrix Crv(P)* at a point of the part surface *P*.

The curvature indicatrix at a point of the part surface *P* can be described analytically by the inequality

$$Crv(P) \Rightarrow \frac{L_P}{E_P} x_P^2 + \frac{2M_P}{\sqrt{E_P G_P}} x_P y_P + \frac{N_P}{G_P} y_P^2 \geq 1 \qquad (4.60)$$

when mean curvature of the part surface at this point is non-negative ($\mathcal{M}_P \geq 0$) and by the inequality:

$$Crv(P) \Rightarrow \frac{L_P}{E_P} x_P^2 + \frac{2M_P}{\sqrt{E_P G_P}} x_P y_P + \frac{N_P}{G_P} y_P^2 \leq 1 \qquad (4.61)$$

when mean curvature of the part surface at this point is non-positive ($\mathcal{M}_P \leq 0$).

The inequalities 4.60 and 4.61 are composed based on the Dupin's indicatrix *Dup* (*P*) constructed at a point of the part surface *P*.

The curvature indicatrix can be used for the purpose of analysis of conditions of contact of a part surface *P* and generating surface *T* of the cutting tool.

Expressions, similar to the inequalities (see Equations 4.60 and 4.61) are valid for generating surface *T* of the cutting tool.

Although the analytical description of the curvature indicatrix *Crv(P)* (see Equations 4.60 and Equation 4.61) resembles analytical description of the Dupin's indicatrix *Dup* (*P*) (see Equation 4.40), these two characteristic images are different in nature. Dupin's indicatrix is *a planar curve* of the second order. The curvature indicatrix is *a portion of plane* that is bounded by Dupin's indicatrix. This portion of plane is located either inside the characteristic curve *Dup* (*P*) [if $\mathcal{M}_{P(T)} \geq 0$] or outside the corresponding Dupin's indicatrix [if $\mathcal{M}_{P(T)} \leq 0$].

When plotting the curvature indicatrix of a part surface *P*, the use of mean curvature, $\mathcal{M}_{P(T)}$, of the surface along with it *Gaussian* curvature, $\mathcal{G}_P$, is helpful.

Curvature indicatrix can be used for the purposes of determining whether or not two contacting surfaces interfere into one another. When machining a part surface on a multiaxis *NC* machine, interference of a smooth regular part surface *P* and generating surface *T* of the cutting tool is not permissible under any circumstances. Depending on the geometry and parameters of curvature indicatrix of the surface of relative curvature, $\mathfrak{R}$, a conclusion can be made on whether or not the contacting surfaces *P* and *T* interfere into one another.

Curvature indicatrices, *Crv* ($\mathfrak{R}$), at a point of the surface of relative curvature, $\mathfrak{R}$, which are constructed at a point of physically feasible contact of a smooth regular part surface *P* and generating surface *T* of the cutting tool are illustrated in Figure 4.11. [For the reader's convenience, all the possible types of the curvature indicatrices *Crv* ($\mathfrak{R}$) at a point of the surface of relative curvature, $\mathfrak{R}$, are listed below together with corresponding sign of the mean $\mathcal{M}_{\mathfrak{R}}$ and of the *Gaussian* $\mathcal{G}_{\mathfrak{R}}$ curvature (see Figure 4.11)]:

- Convex elliptic ($\mathcal{M}_{\mathfrak{R}} > 0$, $\mathcal{G}_{\mathfrak{R}} > 0$) in Figure 4.11a
- Convex umbilic ($\mathcal{M}_{\mathfrak{R}} > 0$, $\mathcal{G}_{\mathfrak{R}} > 0$) in Figure 4.11b
- Convex parabolic ($\mathcal{M}_{\mathfrak{R}} > 0$, $\mathcal{G}_{\mathfrak{R}} = 0$) in Figure 4.11c

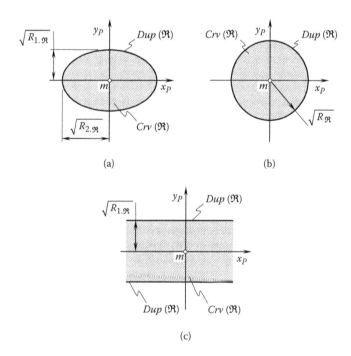

FIGURE 4.11
Three types of curvature indicatrices *Crv* ($\mathfrak{R}$) at a point *m* of a surface $\mathfrak{R}$ of relative curvature for the cases when physical contact of a part surface *P* and generating surface *T* of the cutting tool is feasible.

The contact between a smooth regular part surface P and the generating surface T of the cutting tool is physically feasible in all cases of positive mean curvature, $\mathcal{M}_{\Re} > 0$, at a point of the surface of relative curvature, $\Re$. Surfaces P and T can contact one another physically if the curvature indicatrix $Crv\,(\Re)$ is convex elliptic (Figure 4.11a), convex umbilic (Figure 4.11b), or convex parabolic (Figure 4.11c). Under such circumstances, the curvature indicatrix $Crv\,(\Re)$ at a point of the surface $\Re$ of relative curvature can be described analytically by the inequality

$$Crv\,(\Re) \quad \Rightarrow \quad \frac{L_{\Re}}{E_{\Re}}x_P^2 + \frac{2M_{\Re}}{\sqrt{E_{\Re}G_{\Re}}}x_P y_P + \frac{N_{\Re}}{G_{\Re}}y_P^2 > 1 \tag{4.62}$$

In all of these cases, the mean curvature, $\mathcal{M}_{\Re}$, at a point of the surface $\Re$ is of positive value, and full curvature, $\mathcal{G}_{\Re}$, of the surface of relative curvature is either positive ($\mathcal{G}_{\Re} > 0$) or zero ($\mathcal{G}_{\Re} = 0$).

In a particular case, the surface of relative curvature reduces either to a point or to a line. These degenerate cases of the curvature indicatrices of the surface $\Re$ are schematically illustrated in panels a and b, respectively, of Figure 4.12. The following two types of surface of relative curvature $\Re$ feature curvature indicatrices of these geometries, namely

- Parabolic ($\mathcal{M}_{\Re} \to +\infty$, $\mathcal{G}_{\Re} = 0$) in Figure 4.12a
- Umbilic ($\mathcal{M}_{\Re} = 0$, $\mathcal{G}_{\Re} > 0$) in Figure 4.12b

In the cases shown in Figure 4.12, it is impossible to make a decision whether or not contact of the part surface P and generating surface T of the cutting tool is physically feasible. The curvature indicatrix provides

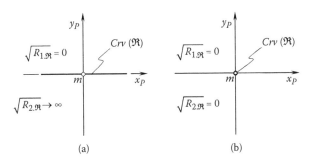

FIGURE 4.12
Two types of curvature indicatrices $Crv\,(\Re)$ at a point m of a surface $\Re$ of relative curvature for the cases when physical contact of part surface P and generating surface T of the cutting tool is unidentified.

insufficient information in this concern. Additional required information for making a decision can be obtained if higher order members in Equation 4.44 are taken into account.

Curvature indicatrices, Crv ($\mathfrak{R}$), constructed at a point of the surface of relative curvature, $\mathfrak{R}$, corresponding to the cases when contact of a smooth regular part surface P and generating surface T of the cutting tool is physically infeasible, are illustrated in Figure 4.13:

- Concave elliptic ($\mathcal{M}_{\mathfrak{R}} < 0$, $\mathcal{G}_{\mathfrak{R}} > 0$) in Figure 4.13a
- Concave umbilic ($\mathcal{M}_{\mathfrak{R}} < 0$, $\mathcal{G}_{\mathfrak{R}} > 0$) in Figure 4.13b
- Concave parabolic ($\mathcal{M}_{\mathfrak{R}} < 0$, $\mathcal{G}_{\mathfrak{R}} = 0$) in Figure 4.13c
- Quasi-convex hyperbolic ($\mathcal{M}_{\mathfrak{R}} > 0$, $\mathcal{G}_{\mathfrak{R}} < 0$) in Figure 4.13d
- Quasi-concave hyperbolic ($\mathcal{M}_{\mathfrak{R}} < 0$, $\mathcal{G}_{\mathfrak{R}} < 0$) in Figure 4.13e
- Minimal hyperbolic ($\mathcal{M}_{\mathfrak{R}} = 0$, $\mathcal{G}_{\mathfrak{R}} < 0$) in Figure 4.13f

Phantom branches of the characteristic curves in Figure 4.13d through f are shown in dashed lines.

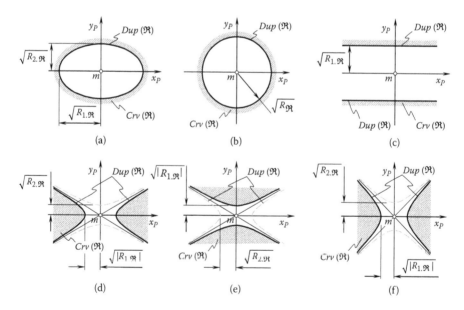

FIGURE 4.13
Types of curvature indicatrices Crv ($\mathfrak{R}$) at a point m of a surface $\mathfrak{R}$ of relative curvature for the cases when physical contact of a part surface P and generating surface T of the cutting tool is not feasible.

Curvature indicatrices Crv ($\mathfrak{R}$) at a point of a surface of relative curvature $\mathfrak{R}$ in Figure 4.13a through f fulfill the following inequality:

$$Crv\,(\mathfrak{R}) \;\Rightarrow\; \frac{L_{\mathfrak{R}}}{E_{\mathfrak{R}}}x_P^2 + \frac{2M_{\mathfrak{R}}}{\sqrt{E_{\mathfrak{R}}G_{\mathfrak{R}}}}x_P y_P + \frac{N_{\mathfrak{R}}}{G_{\mathfrak{R}}}y_P^2 < 1 \tag{4.63}$$

Ultimately, two degenerate types of curvature indicatrices Crv ($\mathfrak{R}$) at a point of a surface of relative curvature $\mathfrak{R}$ are represented in Figure 4.14:

- Parabolic ($\mathcal{M}_{\mathfrak{R}} \to -\infty$, $\mathcal{G}_{\mathfrak{R}} = 0$) in Figure 4.14a
- Umbilic ($\mathcal{M}_{\mathfrak{R}} = 0$, $\mathcal{G}_{\mathfrak{R}} > 0$) in Figure 4.14b

In the two last cases shown in Figure 4.14, the contact of the part surface P and generating surface T of the cutting tool is physically infeasible.

Ultimately, as shown in Figures 4.13 and 4.14, the curvature indicatrices Crv ($\mathfrak{R}$) indicate that surfaces P and T cannot contact one another physically.

The performed analysis reveals that a smooth regular part surface P and generating surface T of the cutting tool can physically contact each other only in the cases for which curvature indicatrices Crv ($\mathfrak{R}$) of the surface of relative curvature $\mathfrak{R}$ corresponds to one of three possible types illustrated in Figure 4.11.

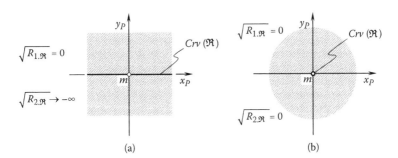

FIGURE 4.14
Three types of curvature indicatrices (b) at a point (a) of a surface $\mathfrak{R}$ of relative curvature for the cases when physical contact of a part surface P and generating surface T of the cutting tool is infeasible.

4.4.5 Matrix Representation of Equation of the Dupin's Indicatrix of the Surface of Relative Curvature

Like any other quadratic form, Equation 4.58 of the Dupin's indicatrix of the surface of relative curvature $\mathfrak{R}$ can be represented in matrix form:

$$Dup\,(P/T) \Rightarrow [x_P \;\; y_P \;\; 0 \;\; 0] \cdot \begin{bmatrix} \dfrac{L_{\mathfrak{R}}}{E_{\mathfrak{R}}} & \dfrac{2M_{\mathfrak{R}}}{\sqrt{E_{\mathfrak{R}}G_{\mathfrak{R}}}} & 0 & 0 \\[2ex] \dfrac{2M_{\mathfrak{R}}}{\sqrt{E_{\mathfrak{R}}G_{\mathfrak{R}}}} & \dfrac{N_{\mathfrak{R}}}{G_{\mathfrak{R}}} & 0 & 0 \\[2ex] 0 & 0 & \mp 1 & 0 \\[1ex] 0 & 0 & 0 & 1 \end{bmatrix}$$

$$\cdot \begin{bmatrix} x_P \\ y_P \\ 0 \\ 0 \end{bmatrix} = \pm 1 \tag{4.64}$$

In the Darboux frame, this equation reduces to

$$Dup\,(P/T) \Rightarrow [x_P \;\; y_P \;\; 0 \;\; 0] \cdot \begin{bmatrix} L_{\mathfrak{R}} & M_{\mathfrak{R}} & 0 & 0 \\ M_{\mathfrak{R}} & N_{\mathfrak{R}} & 0 & 0 \\ 0 & 0 & \mp 1 & 0 \\ 0 & 0 & 0 & 1 \end{bmatrix}$$

$$\cdot \begin{bmatrix} x_P \\ y_P \\ 0 \\ 0 \end{bmatrix} = \pm 1 \tag{4.65}$$

It is convenient to implement matrix representation of the equation of the Dupin's indicatrix (see above), for instance, when developing software for machining a sculptured surface on a multiaxis *NC* machine when multiple coordinate system transformations are required.

The equation of Dupin's indicatrix can be represented in the form

$$r_{Dup}(\varphi) = \sqrt{|R_P(\varphi)|} \cdot \mathrm{sgn}\,\Phi_{2.P}^{-1} \tag{4.66}$$

The last equation reveals that the position vector of a point of the indicatrix *Dup* (P) in any direction is equal to the square root of the radius of curvature in that same direction.*

4.4.6 Surface of Relative Normal Radii of Curvature

The normal curvatures $k_{\mathfrak{R}}$, k_P, and k_T can be represented in the form: $k_{\mathfrak{R}} = R_{\mathfrak{R}}^{-1}$, $k_P = R_P^{-1}$, and $k_T = R_T^{-1}$, respectively, where $R_{\mathfrak{R}}$, R_P, and R_T are the corresponding radii of normal curvature of the surfaces P, T, and $\mathfrak{R}$. They are also taken in a common normal plane section through the point of contact K of the surfaces P, T, and $\mathfrak{R}$.

The radius of relative normal curvature is another widely known tool that is used in contemporary practice for analytical description of the contact geometry of surfaces when performing the second-order analysis. The radius $R_{\mathfrak{R}}$ of relative normal curvature can be defined by the following expression:

$$R_{\mathfrak{R}} = R_P - R_T \tag{4.67}$$

In many applications, Equation 4.67 for the radius of relative normal of curvature $R_{\mathfrak{R}}$ is equivalent to Equation 4.46 for the relative normal curvature $k_{\mathfrak{R}}$.

4.4.7 Normalized Relative Curvature

When performing the second-order analysis, usually it is preferred to operate with unitless values rather than with values having units. To eliminate unit values, it is recommended to use normalized relative normal curvature $\bar{k}_{\mathfrak{R}}$ of the part surface P and generating surface T of the cutting tool. The normalized relative curvature $\bar{k}_{\mathfrak{R}}$ of surfaces P and T is referred to as the value determined by

$$\bar{k}_{\mathfrak{R}} = \frac{k_P + k_T}{|k_{1.P}|} \tag{4.68}$$

Similarly, the normalized radius of relative normal curvature $\bar{R}_{\mathfrak{R}}$ of the part surface P and generating surface T of the cutting too can be introduced here

---

* Similar to Dupin's indicatrix *Dup* (P), a planar characteristic curve of another type can be introduced. An equation of this characteristic curve can be postulated in the form $r_{Dup.k}(\varphi) = \sqrt{|k_P(\varphi)|} \cdot \text{sgn } \Phi_{2.P}^{-1}$. The application of curvature indicatrix in the form $r_{Dup.k}(\varphi)$ makes it possible to avoid uncertainty in cases of plane surface. For the plane surface, the characteristic curve *Dup* (P) does not exist while $r_{Dup.k}(\varphi)$ exists. It shrinks to the contact point K.

based on Equation 4.67. The normalized relative radius of normal curvature $\bar{R}_{\Re}$ of surfaces P and T is referred to as the value determined by

$$\bar{R}_{\Re} = \frac{R_P - R_T}{|R_{1.P}|} \tag{4.69}$$

The implementation of the unitless parameters $\bar{k}_{\Re}$, $\bar{R}_{\Re}$, and others permits avoiding of operation with unit values. Equations made up of unitless parameters are often more convenient in application.

Dupin's indicatrix can be constructed for all of the above-considered characteristic surfaces:

(a) For surface of normal relative radii of curvature $Dup_R\,(\Re)$

(b) For normalized surface of relative curvature $Dup\,(\tilde{\Re})$

(c) For normalized surface of normal radii of relative curvature $Dup_R\,(\tilde{\Re})$

4.4.8 $\Im r_k (P/T)$ Characteristic Curve

For the purpose of analytical description of the distribution of normal curvature in differential vicinity of a point on a smooth regular surface, Böhm recommends [5] to employ the following characteristic curve.

Setting $\eta = dV_P / dU_P$ at a given point of a sculptured part surface P, one can rewrite the equation

$$k_P = \frac{\Phi_{2.P}}{\Phi_{1.P}} = \frac{L_P \, dU_P^2 + 2\,M_P \, dU_P \, dV_P + N_P \, dV_P^2}{E_P \, dU_P^2 + 2\,F_P \, dU_P \, dV_P + G_P \, dV_P^2} \tag{4.70}$$

for normal curvature in the form of

$$k_P = \frac{L_P + 2M_P\eta + N_P\eta^2}{E_P + 2F_P\eta + G_P\eta^2} \tag{4.71}$$

In the particular case when $L_P : M_P : N_P = E_P : F_P : G_P$, the normal curvature k_P is independent of η. Surface points with this property are known as *umbilic* points and *flatten* points.

In general cases when k_P changes as η changes, the function $k_P = k_P(\eta)$ is a rational quadratic form, as illustrated in Figure 4.15. The extreme values $k_{1.P}$ and $k_{2.P}$ of the function $k_P = k_P(\eta)$ occur at the roots η_1 and η_2 of

$$\begin{vmatrix} \eta^2 & -\eta & 1 \\ E_P & F_P & G_P \\ L_P & M_P & N_P \end{vmatrix} = 0 \tag{4.72}$$

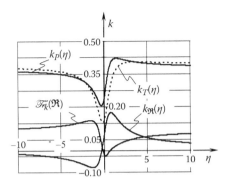

FIGURE 4.15
Derivation of an equation of the characteristic curve $\mathscr{T}_{k}(\mathfrak{R})$.

It can be shown that η_1 and η_2 are always real.

The quantities η_1 and η_2 define directions that align with the principal directions on the part surface P.

The distribution of normal curvature of the surface P (Figure 4.15) at a surface point m is specified by the characteristic curve $k_P = k_P(\eta)$. The distribution of normal curvature of generating surface T of the cutting tool that makes contact with P at m is specified by another characteristic curve $k_T = k_T(\eta)$. For surfaces P and T, the surface of relative curvature $\mathfrak{R}$ can be constructed. The distribution of normal curvature of the surface $\mathfrak{R}$ at K is described by the characteristic curve $k_{\mathfrak{R}} = k_{\mathfrak{R}}(\eta)$.

The characteristic curve $\mathscr{T}_{t_k}(P/T)$ of this type is defined here as

$$\mathscr{T}_{t_k}(P/T) \quad \Rightarrow \quad k_{\mathscr{T}_t} = k_P(\eta) + k_T(\eta + \mu) \tag{4.73}$$

Similarly, a characteristic curve $\mathscr{T}_{t_R}(P/T)$ of another sort is defined as

$$\mathscr{T}_{t_R}(P/T) \quad \Rightarrow \quad R_{\mathscr{T}_t} = R_P(\eta) - R_T(\eta + \mu) \tag{4.74}$$

The developed methods for analytical description of contact geometry of two smooth regular surfaces in the first order of tangency are not limited to the methods disclosed above [125,126,130].

4.5 Degree of Conformity of Two Smooth Regular Surfaces in the First Order of Tangency

The accuracy of the discussed methods of the second-order analysis is not sufficient for accurate analytical description of the contact geometry of two smooth regular surfaces in the first order of tangency. To increase the accuracy, higher-order analysis must be done.

The methods of higher-order analysis discussed below target the development of an analytical description of degree of conformity of generating surface T of the cutting tool to the part surface P at a current point K of their contact. The higher degree of conformity of surfaces P and T, the closer these surfaces to each other in the differential vicinity of point K. This qualitative (*intuitive*) definition of degree of conformity of two smooth regular surfaces needs a corresponding quantitative measure.

4.5.1 Preliminary Remarks

Implementation of the resultant deviation l_{cnf} of two smooth regular surfaces in contact (see Section 4.3 for details in this concern) for the analytical description of contact geometry of two surfaces in contact is a type of straightforward solution to the problem under consideration. This approach is proven to be computationally ineffective. However, the approach gives an insight into how an effective method for solving the problem under consideration can be developed.

As seen in Figure 4.16, three geometrical parameters are interrelated when deviation of a surface from the tangent plane is considered in the differential vicinity of a surface point. They are

(a) The measure of the deviation, $m_p m_p^*$, of a part surface P from the tangent plane, l_{cnf}
(b) The distance, $K m_p^*$, of a current point m_p from the contact point K
(c) Radius of normal curvature R_P of the part surface P at the contact point K

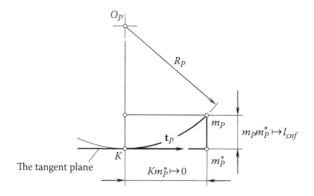

FIGURE 4.16
Transition from the resultant deviation, l_{cnf} of two surfaces to indicatrix of conformity $Cnf(P/T)$ at a contact point K between a smooth regular part surface P and the generating surface T of the form-cutting tool.

As a consequence of the relationship among parameters $m_P m_P^*$, Km_P^*, and R_P, any one of them can be used for the quantitative evaluation of degree of conformity of the contacting part surface P and generating surface T of the cutting tool. Following Figure 4.16,

$$m_P m_P^* = R_P - \sqrt{R_P^2 - (Km_P^*)^2}\ \Big|_{m_P \to K} \quad \longmapsto \quad l_{cnf} \qquad (4.75)$$

Inversely, for the radius of normal curvature R_P at a point of the part surface P, the following expression is valid:

$$R_P = \frac{(m_P m_P^*)^2 + (Km_P^*)^2}{2 \cdot m_P m_P^*} \qquad (4.76)$$

Ultimately, one may conclude that any legitimate analytical function of normal radii of curvature R_P and R_T at a point of contact of the part surface P and generating surface T of the cutting tool can be used for this particular purpose.

Consider two smooth regular surfaces P and T in the first order of tangency that make contact at a point K. The degree of conformity of surfaces P and T can be interpreted as a function of radii of normal curvature R_P and R_T of the contacting surfaces. The radii of normal curvature R_P and R_T of surfaces P and T are taken in a common normal plane section through the point K. For a specified radius of normal curvature R_P of the part surface P, the degree of conformity of the surfaces depends upon the corresponding value of radius of normal curvature R_T of the surface generating surface T of the cutting tool.

In most cases of part surface generation, the degree of conformity of surfaces P and T is not constant, and it is changing as coordinates of the contact point change. The degree of surface conformity to one another depends on orientation of the normal plane section through the contact point K, and changes as the normal plane section turns about the common perpendicular $\mathbf{n}_P$. This statement immediately follows from the above conclusion that degree of conformity of surfaces P and T yields interpretation in terms of radii of normal curvature R_P and R_T.

The change of degree of conformity of a part surface P and generating surface T of the cutting tool due to turning of the normal plane section about the common perpendicular $\mathbf{n}_P$ is illustrated in Figure 4.17. Here, in Figure 4.17, only two-dimensional examples for which that same normal plane section of the part surface P makes contact with different plane sections $T^{(i)}$ of the generating surface T of the form-cutting tool are shown.

In the example shown in Figure 4.17a, the radius of normal curvature $R_T^{(1)}$ of the convex plane section $T^{(1)}$ of the cutting tool surface T is positive $[R_T^{(1)} > 0]$. The convex normal plane section of the surface T makes contact with the convex

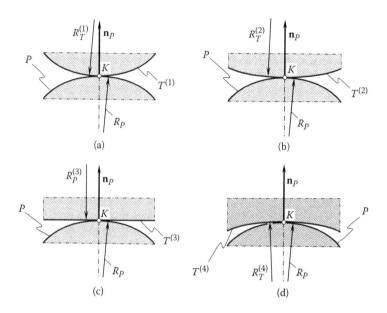

FIGURE 4.17
Sections of two smooth regular surfaces $T^{(4)}$ and T by a plane through common perpendicular $\mathbf{n}_P$. (Reprinted from *Mathematical and Computer Modeling*, 39, Radzevich, S.P., Mathematical modeling of contact of two surfaces in the first order of tangency, 1083–1112, Copyright 2004, with permission from Elsevier.)

normal plane section ($R_P > 0$) of a part surface P. The degree of conformity of the generating surface T of the cutting tool to the part surface P in Figure 4.17a is relatively low as both the contacting surfaces are convex.

Another example is shown in Figure 4.17b. Radius of normal curvature $R_T^{(2)}$ of the convex plane section $T^{(2)}$ of the cutting tool surface T also is positive [$R_T^{(2)} > 0$]. However, its value exceeds the value $R_T^{(1)}$ of radius of normal curvature in the first example [$R_T^{(2)} > R_T^{(1)}$]. This means the degree of conformity of the cutting tool surface T to the part surface P (Figure 4.17b) is higher compare to what is shown in Figure 4.17a.

In the next example depicted in Figure 4.17c, the normal plane section $T^{(3)}$ of the cutting tool surface T is represented with a locally flattened section. The radius of normal curvature $R_T^{(3)}$ of the flattened plane section $T^{(3)}$ approaches infinity [$R_T^{(3)} \to \infty$]. Thus, the inequality $R_T^{(3)} > R_T^{(2)} > R_T^{(1)}$ is valid. Therefore, the degree of conformity of the cutting tool surface T to the part surface P in Figure 4.17c is also getting higher.

Finally, for a concave normal plane section $T^{(4)}$ of the cutting tool surface T that is illustrated in Figure 4.17d, radius of normal curvature $R_T^{(4)}$ is of negative value [$R_T^{(4)} < 0$]. In this case, the degree of conformity of the cutting tool surface T to the part surface P is the highest of four examples considered in Figure 4.17.

The examples shown in Figure 4.17 qualitatively illustrate what is known intuitively regarding the different degree of conformity of two smooth

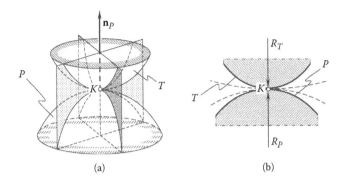

(a)   (b)

FIGURE 4.18
Analytical description of the contact geometry of two smooth regular surfaces T and R_P. (Reprinted from *Mathematical and Computer Modeling*, 39, Radzevich, S.P., Mathematical modeling of contact of two surfaces in the first order of tangency, 1083–1112, Copyright 2004, with permission from Elsevier.)

regular surfaces in the first order of tangency. Intuitively, one can realize that in the examples shown in Figure 4.17a through d, the degree of conformity of two surfaces P and T is gradually increased.

A similar observation is made for a given pair of surfaces P and T when different sections of the surfaces by a plane surface through the common perpendicular $\mathbf{n}_P$ are considered (Figure 4.18a). When rotating the plane section about the common perpendicular, $\mathbf{n}_P$, it can be observe that, degree of conformity of surfaces P and T is different in different configurations of the cross-sectional plane (Figure 4.18b).

The above examples provide an intuitive understanding of what degree of conformity at a point of contact of two smooth regular surfaces P and T means. The examples cannot be employed directly to evaluate in quantities the degree of conformity at a point of contact of two smooth regular surfaces P and T. The next necessary step is to introduce an appropriate quantitative evaluation of degree of conformity of two surfaces in the first order of tangency. In other words, how can a certain degree of conformity of two smooth regular surfaces be described analytically?

4.5.2 Indicatrix of Conformity at a Point of Contact of Part Surface *P* and Generating Surface *T* of the Cutting Tool

This section introduces a quantitative measure of degree of conformity at a point of contact between two smooth regular surfaces. The degree of conformity of two surfaces P and T indicates how the cutting tool surface T is close to the surface P in the differential vicinity of a point K of their contact, say, how much surface T is *congruent* to the surface P in the differential vicinity of point K. This particular type of congruency between the contacting surfaces P and T can also be understood as the *local congruency* of the contacting surfaces.

Quantitatively, the degree of conformity at a point of contact of a surface T to another surface P can be expressed in terms of difference between the corresponding radii of normal curvature of the contacting surfaces. To develop a quantitative measure of degree of conformity of surfaces P and T, it is convenient to implement Dupin's indicatrices Dup (P) and Dup (T), constructed at a point of contact of the part surface P and generating surface T of the cutting tool respectively.

It is natural to assume that the higher degree of conformity of surfaces P and T is due to the smaller difference between the normal curvatures of surfaces P and T in a common cross-section by a plane through the common normal vector $\mathbf{n}_P$.

Dupin's indicatrix Dup (P) indicates the distribution of radii of normal curvature at a point of the part surface P as had been shown, for example, for a concave elliptic patch of the part surface P (Figure 4.19). For part surface P, the equation of this characteristic curve (see Equation 4.58) in polar coordinates can be represented in the form

$$Dup\ (P) \Rightarrow r_P(\varphi_P) = \sqrt{|R_P(\varphi_P)|} \tag{4.77}$$

where r_P is the position vector of a point of the Dupin's indicatrix Dup (P) at a point of the part surface P and φ_P is the polar angle of the indicatrix Dup (P).

The similar is true with respect to the Dupin's indicatrix Dup (T) at a point of generating surface T of the cutting tool as it had been shown, for instance, for a

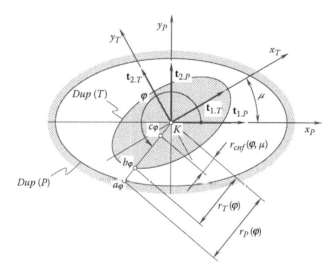

FIGURE 4.19
Derivation of equation of indicatrix of conformity $Cnf_R(P/T)$ at a point of contact of a smooth regular part surface P and generating surface T of the form-cutting tool, which are in the first order of tangency.

convex elliptical patch of the cutting tool surface T (Figure 4.19). The equation of this characteristic curve in polar coordinates can be represented in the form

$$Dup\,(T) \quad \Rightarrow \quad r_T(\varphi_T) = \sqrt{|R_T(\varphi_T)|} \qquad (4.78)$$

where r_T the is position vector of a point of the Dupin's indicatrix $Dup\,(T)$ at a point of generating surface T of the cutting tool and φ_T is the polar angle of the indicatrix $Dup\,(T)$.

In the coordinate plane $x_P y_P$ of the local reference system $x_P y_P z_P$, the equalities $\varphi_P = \varphi$ and $\varphi_T = \varphi + \mu$ are valid. Therefore, in the coordinate plane $x_P y_P$, Equations 4.77 and 4.78 cast into

$$Dup\,(P) \quad \Rightarrow \quad r_P(\varphi) = \sqrt{|R_P(\varphi)|} \qquad (4.79)$$

$$Dup\,(T) \quad \Rightarrow \quad r_T(\varphi, \mu) = \sqrt{|R_T(\varphi, \mu)|} \qquad (4.80)$$

When the degree of conformity at a point of contact of the cutting tool surface T to the part surface P is higher, then the difference between the functions $r_P(\varphi)$ and $r_T(\varphi, \mu)$ becomes smaller and vice versa. The last validates the following conclusion:

Conclusion 4.3

The distance between the corresponding* points of the Dupin's indicatrices $Dup\,(P)$ and $Dup\,(T)$ constructed at a point of contact of a part surfaces P and generating surface T of the cutting tool can be employed for the purpose of indication of degree of conformity at a point of contact of the part surface P and the generating surface T of the form-cutting tool at the contact point K. ■

The equation of the indicatrix of conformity $Cnf_R(P/T)$ at a point of contact of a part surface P and generating surface T of the cutting tool is postulated for the following structure:

$$Cnf_R(P/T) \quad \Rightarrow \quad r_{cnf}(\varphi, \mu) = \sqrt{|R_P(\varphi)|}\,\operatorname{sgn} R_P(\varphi) + \sqrt{|R_T(\varphi, \mu)|}\,\operatorname{sgn} R_T(\varphi, \mu)$$

$$= r_P(\varphi)\operatorname{sgn} R_P(\varphi) + r_T(\varphi, \mu)\operatorname{sgn} R_T(\varphi, \mu) \qquad (4.81)$$

---

* Corresponding points of the Dupin's indicatrices $Dup\,(P)$ and $Dup\,(T)$ share the same straight line through the contact point K of the surfaces P and T and are located at the same side of the point K.

Because the location of a point a_φ of the Dupin's indicatrix Dup (P) at a point of the part surface P is defined by the position vector $r_P(\varphi)$ and the location of a point b_φ of the Dupin's indicatrix Dup (T) at a point of generating surface T of the cutting tool is defined by the position vector $r_T(\varphi, \mu)$, then the location of a point c_φ (see Figure 4.19) of the indicatrix of conformity $Cnf_R(P/T)$ at a point of contact K of surfaces P and T is defined by the position vector $r_{cnf}(\varphi, \mu)$. Therefore, the equality $r_{cnf}(\varphi, \mu) = Kc_\varphi$ is observed, and the length of the straight line segment Kc_φ is equal to the distance $a_\varphi b_\varphi$.

In Equation 4.81, $r_P = \sqrt{|R_P|}$ is the position vector of a point of Dupin's indicatrix of the surface P at a point K of contact with generating surface T of the form-cutting tool and $r_T = \sqrt{|R_T|}$ is the position vector of a corresponding point of the Dupin's indicatrix of generating surface T of the cutting tool. Also, the multipliers sgn $R_P(\varphi)$ and sgn $R_T(\varphi, \mu)$ are assigned to each of the functions $r_P(\varphi) = \sqrt{|R_P(\varphi)|}$ and $r_T(\varphi, \mu) = \sqrt{|R_T(\varphi, \mu)|}$, respectively, for the sole purpose of retaining corresponding sign of the functions, that is, to retain that same sign that the radii of normal curvature $R_P(\varphi)$ and $R_T(\varphi, \mu)$ have.

Ultimately, one can conclude that position vector r_{cnf} of a point of the indicatrix of conformity $Cnf_R(P/T)$ can be expressed in terms of position vectors r_P and r_T of the Dupin's indicatrices Dup (P) and Dup (T).

For the calculation of the current value of the radius of normal curvature $R_P(\varphi)$ at a point of the part surface P, the following equality can be used:

$$R_P(\varphi) = \frac{\Phi_{1.P}}{\Phi_{2.P}} \tag{4.82}$$

Similarly, for the calculation of the current value of the radius of normal curvature $R_T(\varphi, \mu)$ at a point of generating surface T of the cutting tool, the following equality can be employed:

$$R_T(\varphi, \mu) = \frac{\Phi_{1.T}}{\Phi_{2.T}} \tag{4.83}$$

The use of angle μ of local relative orientation of surfaces P and T indicates that the radii of normal curvature $R_P(\varphi)$ and $R_T(\varphi, \mu)$ are taken in a common normal plane section through the contact point K.

Further, it is well known that inequalities $\Phi_{1.P} \geq 0$ and $\Phi_{1.T} \geq 0$ are always valid. Therefore, Equation 4.81 can be rewritten in the following form:

$$r_{cnf} = r_P(\varphi)\,\mathrm{sgn}\,\Phi_{2.P}^{-1} + r_T(\varphi, \mu)\,\mathrm{sgn}\,\Phi_{2.T}^{-1} \tag{4.84}$$

For the derivation of an equation of the indicatrix of conformity $Cnf_R(P/T)$, it is convenient to use the Euler's equation for normal radius of curvature $R_P(\varphi)$ at a point of the part surface P (see Equation 1.79):

$$R_P(\varphi) = \frac{R_{1.P} \cdot R_{2.P}}{R_{1.P} \cdot \sin^2 \varphi + R_{2.P} \cdot \cos^2 \varphi} \tag{4.85}$$

Here, the radii of principal curvature $R_{1.P}$ and $R_{2.P}$ are the roots of the quadratic equation:

$$\begin{vmatrix} L_P \cdot R_P - E_P & M_P \cdot R_P - F_P \\ M_P \cdot R_P - F_P & N_P \cdot R_P - G_P \end{vmatrix} = 0 \tag{4.86}$$

Recall that the inequality $R_{1.P} < R_{2.P}$ is always observed.

Equations 4.85 and 4.86 allow for the expression of the radius of normal curvature $R_P(\varphi)$ of the part surface P in terms of the fundamental magnitudes of the first-order E_P, F_P, and G_P, and of the fundamental magnitudes of the second-order L_P, M_P, and N_P.

A similar consideration is applicable for the generating surface T of the form-cutting tool. Omitting the routing analysis, one can conclude that the radius of normal curvature $R_T(\varphi, \mu)$ of the generating surface T of the cutting tool can be expressed in terms of the fundamental magnitudes of the first order E_T, F_T, and G_T and of the fundamental magnitudes of the second-order L_T, M_T, and N_T.

Finally, on the premise of the above-performed analysis, the following equation for the indicatrix of conformity $Cnf_R(P/T)$ at a point of contact of surfaces P and T can be derived:

$$
\begin{aligned}
r_{cnf}(\varphi, \mu) = &\sqrt{\left| \frac{E_P G_P}{L_P G_P \cos^2 \varphi - M_P \sqrt{E_P G_P} \sin 2\varphi + N_P E_P \sin^2 \varphi} \right|} \, \text{sgn} \, \Phi_{2.P}^{-1} \\
&+ \sqrt{\left| \frac{E_T G_T}{L_T G_T \cos^2(\varphi+\mu) - M_T \sqrt{E_T G_T} \sin 2(\varphi+\mu) + N_T E_T \sin^2(\varphi+\mu)} \right|} \, \text{sgn} \, \Phi_{2.T}^{-1}
\end{aligned}
$$

$$\tag{4.87}$$

Equation 4.87 of the characteristic curve* $Cnf_R(P/T)$ was published in Reference [78] and, in a hidden form, in Reference [73].

---

* Equation of this characteristic curve is known from (a) Pat. No. 1249787, USSR, *A Method of Sculptured Part Surface Machining on a Multi-Axis NC Machine*, S.P. Radzevich, B23C 3/16, filed December 27, 1984 [78] and (in hidden form) from (b) Patent No. 1185749, USSR, *A Method of Sculptured Part Surface Machining on a Multi-Axis NC Machine*, S.P. Radzevich, B23C 3/16, filed October 24, 1983 [73].

The analysis of Equation 4.87 reveals that the indicatrix of conformity $Cnf_R(P/T)$ at a point of contact of a part surface P and generating surface T of the cutting tool is represented by a planar centrosymmetrical curve of the fourth order. In particular cases, this characteristic curve also possesses a property of mirror symmetry. Mirror symmetry of the indicatrix of conformity is observed, for example, when the angle μ of local relative orientation of surfaces P and T is equal $\mu = \pm\pi \cdot n/2$, where n designates an integer number.

It is important to note here that even for the most general case of part surface generation, the position vector of a point $r_{cnf}(\varphi, \mu)$ of the indicatrix of conformity $Cnf_R(P/T)$ is not dependent on the fundamental magnitudes F_P and F_T. The independence of the characteristic curve $Cnf_R(P/T)$ of the fundamental magnitudes F_P and F_T is because of the following.

The coordinate angle ω_P on the part surface P can be calculated by the formula

$$\omega_P = \arccos \frac{F_P}{\sqrt{E_P G_P}} \tag{4.88}$$

It is natural that the position vector $r_{cnf}(\varphi, \mu)$ of a point of the indicatrix of conformity $Cnf_R(P/T)$ is not a function of the coordinate angle ω_P. Although the position vector $r_{cnf}(\varphi, \mu)$ depends on the fundamental magnitudes E_P, G_P and E_T, G_T, the above analysis makes it clear why the position vector $r_{cnf}(\varphi, \mu)$ does not depend upon the fundamental magnitudes F_P and F_T.

Two illustrative examples of the indicatrix of conformity $Cnf_R(P/T)$ at a point of contact of a part surface P and generating surface T of the cutting tool are shown in Figure 4.20. The first example (Figure 4.20a) relates to the cases of contact of a saddle-like local patch of the part surface P and of a convex elliptic-like local patch of generating surface T of the cutting tool. The second one (Figure 4.20b) is for the case of contact of a convex parabolic-like local patch of the part surface P and of a convex elliptic-like local patch of generating surface T of the cutting tool. For both cases (see Figure 4.20), the corresponding curvature indicatrices $Crv(P)$ and $Crv(T)$ at the point of contact of surfaces P and T are depicted in Figure 4.20 as well. The imaginary (phantom) branches of the Dupin's indicatrix $Dup(P)$ for the saddle-like local patch of the part surface P are shown in dashed line (see Figure 4.20a).

Part surface P and generating surface T of the form-cutting tool can make contact geometrically, but the physical conditions of their contact could be violated. Violation of the physical condition of contact results in bodies bounded by the contacting surfaces P and T to interfere with one another. The implementation of the indicatrix of conformity $Cnf_R(P/T)$ immediately uncovers the surfaces interference if there is any. Three illustrative examples of the violation of physical condition of contact are illustrated in Figure 4.21. When correspondence between the radii of normal curvature of the contacting surface P and T is inappropriate, then the indicatrix of conformity

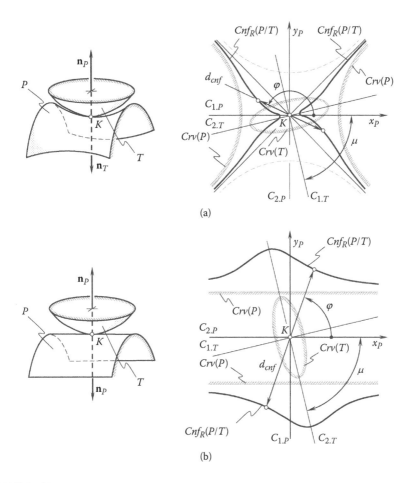

FIGURE 4.20

Examples of indicatrix of conformity y_P at a point of contact of a smooth regular part surface $Cnf_R(P/T)$ and generating surface T of the cutting tool in the first order of tangency. (Reprinted from *Mathematical and Computer Modeling*, 39, Radzevich, S.P., Mathematical modeling of contact of two surfaces in the first order of tangency, 1083–1112, Copyright 2004, with permission from Elsevier.)

$Cnf_R(P/T)$ either intersects itself (Figure 4.21a) or all of it diameters become negative (Figure 4.21b, c).

Another interpretation of satisfaction and of violation of physical condition of contact is illustrated in Figure 4.22. The condition of physical contact is fulfilled when all diameters of the indicatrix of conformity $Cnf_R(P/T)$ are positive. In this case, the part surface P and generating surface T of the cutting tool may contact one another like two rigid bodies do. An example of indicatrix of conformity $Cnf_R(P/T)$ for such a case is depicted in Figure 4.22a. In cases when this planar characteristic curve has negative diameters, as schematically shown in Figure 4.22b, physical contact between the surfaces P and T is not feasible.

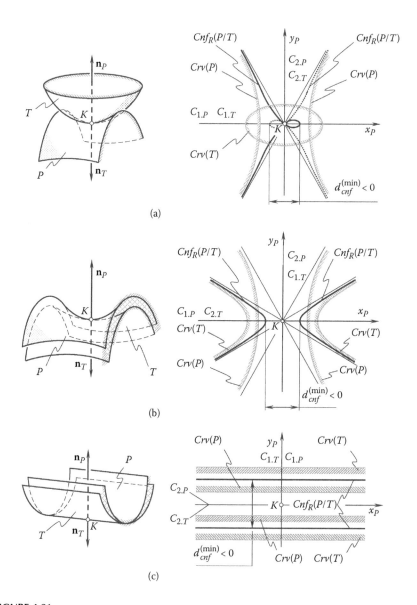

FIGURE 4.21
Examples of violation of physical condition of contact of a smooth regular part surface $Crv(P)$ and generating surface T of the form-cutting tool. (Reprinted from *Mathematical and Computer Modeling*, 39, Radzevich, S.P., Mathematical modeling of contact of two surfaces in the first order of tangency, 1083–1112, Copyright 2004, with permission from Elsevier.)

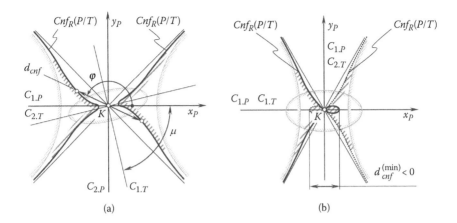

FIGURE 4.22
Another interpretation of satisfaction (a) and of violation (b) of condition of physical contact of a smooth regular part surface P and generating surface T of the cutting tool.

The value of the current diameter* d_{cnf} of the indicatrix of conformity $Cnf_R(P/T)$ indicates degree of conformity of the part surface P and generating surface T of the cutting tool in a corresponding cross section of the surfaces by normal plane through the common perpendicular. The orientation of the normal plane section with respect to the surfaces P and T is defined by the corresponding central angle ϕ.

For the orthogonally parameterized part surface P and generating surface T of the cutting tool, the equation of Dupin's indicatrices $Dup\,(P)$ and $Dup\,(T)$ simplifies to

$$L_P x_P^2 + 2M_P x_P y_P + N_P y_P^2 = \pm 1 \tag{4.89}$$

$$L_T x_T^2 + 2M_T x_T y_T + N_T y_T^2 = \pm 1 \tag{4.90}$$

After being represented in a common reference system, Equations 4.89 and 4.90 make possible a simplified equation of the indicatrix of conformity $Cnf_R(P/T)$ at the point of contact of surfaces P and T:

$$r_{cnf}(\varphi, \mu) = (L_P \cos^2 \varphi - M_P \sin 2\varphi + N_P \sin^2 \varphi)^{-\frac{1}{2}} \operatorname{sgn} \Phi_{2.P}^{-1}$$

$$+ [L_T \cos^2(\varphi + \mu) - M_T \sin 2(\varphi + \mu) + N_T \sin^2(\varphi + \mu)]^{-\frac{1}{2}} \operatorname{sgn} \Phi_{2.T}^{-1} \tag{4.91}$$

---

* Diameter of a centrosymmetrical curve can be defined as a distance between two points of the curve, measured along the corresponding straight line through the center of symmetry of the curve.

Equation 4.87 of the indicatrix of conformity $Cnf_R(P/T)$ at a point of contact of a part surface P and generating surface T of the cutting tool yields an equation of one more characteristic curve. This characteristic curve is referred to as the *curve of minimal values of the position vector* r_{cnf}, which is expressed in terms of the angular parameter φ. Generally, the equation of the curve of minimal values of the position vector r_{cnf} can be represented in the form $r_{cnf}^{min} = r_{cnf}^{min}(\mu)$.

For the derivation of the equation of the characteristic curve $r_{cnf}^{min} = r_{cnf}^{min}(\mu)$, the following method can be employed.

A given relative orientation of a part surface P and generating surface T of the cutting tool is specified by value of the angle μ of local relative orientation of surfaces P and T. The minimal value of the position vector r_{cnf}^{min} is observed when the angular parameter φ is equal to the root φ_1 of equation

$$\frac{\partial}{\partial \varphi} r_{cnf}(\varphi, \mu) = 0 \qquad (4.92)$$

The following additional condition must be fulfilled as well:

$$\frac{\partial^2}{\partial \varphi^2} r_{cnf}(\varphi, \mu) > 0 \qquad (4.93)$$

To find the necessary value of the angle φ_1, it is required to solve the equation $\frac{\partial}{\partial \varphi} r_{cnf}(\varphi, \mu) = 0$ with respect to μ. After substituting the obtained solution μ^{min} to Equation 4.72 of the indicatrix of conformity $Cnf_R(P/T)$, the equation $r_{cnf}^{min} = r_{cnf}^{min}(\varphi)$ of the curve of minimal diameters of the characteristic curve $Cnf_R(P/T)$ can be derived.

In a similar way, one more characteristic curve, say the characteristic curve $r_{cnf}^{max} = r_{cnf}^{max}(\varphi)$ can be derived as well. The last characteristic curve reflects the distribution of the maximal values of the position vector r_{cnf} of the indicatrix of conformity $Cnf_R(P/T)$ in terms of φ.

4.5.3 Directions of the Extremum Degree of Conformity of a Part Surface *P* and Generating Surface *T* of the Cutting Tool

The directions along which degree of conformity at a point of contact of a part surface P and generating surface T of the cutting tool is extremum, i.e., degree of conformity reaches either maximal of its value or minimal of its value, are of prime importance for engineering applications. This issue is especially important when designing blend surfaces, for computation of parameters of optimal tool-paths for machining of sculptured surfaces on a multiaxis NC machine, for improving the accuracy of solution to the problem of two elastic bodies in contact, and for many other applications in applied science and engineering.

The directions of extremal degree of conformity of the contacting smooth regular surfaces P and T, i.e., the directions pointed along the extremal diameters d_{cnf}^{min} and d_{cnf}^{max} of the indicatrix of conformity $Cnf_R(P/T)$, can be found from the equation of the indicatrix of conformity $Cnf_R(P/T)$. For the reader's convenience, Equation 4.72 of this characteristic curve is transformed and is represented in the form

$$r_{cnf}(\varphi, \mu) = \sqrt{\left| r_{1.P} \cos^2 \varphi + r_{2.P} \sin^2 \varphi \right|} \operatorname{sgn} \Phi_{2.P}^{-1}$$

$$+ \sqrt{\infty \left| r_{1.T} \cos^2 (\varphi + \mu) + r_{2.T} \sin^2 (\varphi + \mu) \right|} \operatorname{sgn} \Phi_{2.T}^{-1} \sqrt{a^2 + b^2} \qquad (4.94)$$

Two angles, φ_{min} and φ_{max}, specify two directions within the common tangent plane, along which degree of conformity of the cutting tool surface T to the part surface P reaches its extremal values. These angles are the roots of equation

$$\frac{\partial}{\partial \varphi} r_{cnf}(\varphi, \mu) = 0. \qquad (4.95)$$

It can be easily proven that in general case of contact of two sculptured surfaces, the difference between the angles φ_{min} and φ_{max} is not equal to 0.5π. This means that the equality

$$\varphi_{min} - \varphi_{max} = \pm 0.5\pi n \qquad (4.96)$$

is not always observed, and in most cases, the relationship

$$\varphi_{min} - \varphi_{max} \neq \pm 0.5\pi n \qquad (4.97)$$

is valid (here n is an integer number). The condition (see Equation 4.96) $\varphi_{min} = \varphi_{max} \pm 0.5\pi n$ is fulfilled only in cases when the angle μ of local relative orientation of the contacting surfaces P and T is equal to $\mu = \pm 0.5\pi n$, and thus the principal directions $\mathbf{t}_{1.P}$ and $\mathbf{t}_{2.P}$ of the part surface P, and the principal directions $\mathbf{t}_{1.T}$ and $\mathbf{t}_{2.T}$ of the cutting tool surface T either aligned to each other or they are directed oppositely.

This enables the following statement to be made:

Statement 4.1

In the general case of contact of two sculptured surfaces, the directions along which degree of conformity of two smooth regular surfaces P and T is extremal are not orthogonal to one another. ∎

This statement is important for engineering applications.

The solution to Equation 4.49 returns two extremal angles φ_{min} and $\varphi_{max} = \varphi_{min} + 90°$. Equation 4.95 allows for two solutions φ^*_{min} and φ^*_{max}. Therefore, the extremal difference $\Delta\varphi_{min} = \varphi_{min} - \varphi^*_{min}$ as well as the extremal difference $\Delta\varphi_{max} = \varphi_{max} - \varphi^*_{max}$ can be easily calculated.

Generally speaking, neither the extremal difference $\Delta\varphi_{min}$ nor the extremal difference $\Delta\varphi_{max}$ is equal to zero. They are equal to zero only in particular cases, say when the angle μ of local relative orientation of surfaces P and T fulfills the relationship $\mu = \pm 0.5\,\pi n$.

Let us consider an example that illustrates Statement 4.1.

Example 1. As an illustrative example, let us describe analytically contact geometry of two convex parabolic patches of the contacting surfaces P and T (Figure 4.23). The example pertains to finishing a helical involute gear by a disk-type shaving cutter. In the example under consideration, the design parameters of the gear and of the shaving cutter, along with the specified the gear and the shaving cutter configuration yield the following numerical data for the calculation. At the point K of surface contact, the principal curvatures of the surface P are equal $k_{1.P} = 4\ \text{mm}^{-1}$ and $k_{2.P} = 0$. Principal curvatures of the surface T are equal $k_{1.T} = 1\ \text{mm}^{-1}$ and $k_{2.T} = 0$. The angle μ of local relative orientation of surfaces P and T is equal $\mu = 45°$.

Two approaches can be implemented for the analytical description of the contact geometry of surfaces P and T. The first one is based on implementation

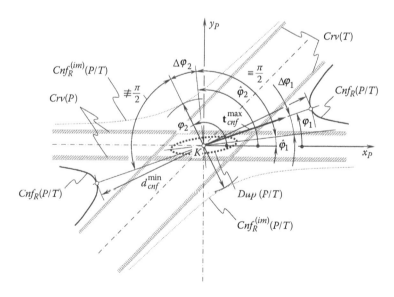

FIGURE 4.23
Example 4.1: Determination of the optimum instant kinematics for a gear shaving operation. (Reprinted from *Mathematical and Computer Modeling*, 39, Radzevich, S.P., Mathematical modeling of contact of two surfaces in the first order of tangency, 1083–1112, Copyright 2004, with permission from Elsevier.)

of Dupin's indicatrix of the surface of relative curvature. The second one is based on application of the indicatrix of conformity $Cnf_R(P/T)$ constructed at a contact point K of surfaces P and T.

The first approach. In the case under consideration Equation 4.49 reduces to

$$k_{\Re} = k_{1.P} \cos^2 \varphi - k_{1.T} \cos^2(\varphi + \mu) \qquad (4.98)$$

Therefore, the following equality is valid:

$$\frac{\partial k_{\Re}}{\partial \varphi} = -2k_{1.P} \sin \varphi \cos \varphi + 2k_{1.T} \sin(\varphi + \mu) \cos(\varphi + \mu) = 0 \qquad (4.99)$$

For the directions of the extremal degree of conformity at the point of contact of the part surfaces P and generating surface T of the cutting tool, Equation 4.99 yields the calculation of the extremal values $\varphi_{min} = 7°$ and $\varphi_{max} = \varphi_{min} + 90° = 97°$ of the angles φ_{min} and φ_{max}.

The direction that is specified by the angle $\varphi_{min} = 7°$ indicates the direction of the minimal diameter of the Dupin's indicatrix of the surface of relative curvature, $\Re$. That same direction corresponds to the maximal degree of conformity at the point of contact of surfaces P and T. Another direction, which is specified by the angle $\varphi_{max} = 97°$, indicates the direction of the minimum degree of conformity of the contacting surfaces P and T at that same contact point.

The second approach. For the case under consideration, use of Equation 4.87 of the indicatrix of conformity $Cnf_R(P/T)$ at a point of contact of the part surface P and generating surface T of the cutting tool makes possible calculation of the extremal angles $\varphi^*_{min} = 19°$ and $\varphi^*_{max} = 118°$.

Imaginary branches of the indicatrix of conformity $Cnf_R(P/T)$ at the point of contact of surfaces P and T in Figure 4.23 are depicted in dashed lines.

It is important to direct the readers' attention here onto two issues.

First, the extremal angles φ_{min} and φ_{max} that are calculated using the first approach are not equal to the corresponding extremal angles φ^*_{min} and φ^*_{max} that are calculated using the second approach. The relationships $\varphi_{min} \neq \varphi^*_{min}$ and $\varphi_{max} \neq \varphi^*_{max}$ are generally observed.

Second, the difference $\Delta\varphi^*$ between the extremal angles φ^*_{min} and φ^*_{max} is not equal to half of π. Therefore, the relationship $\varphi^*_{max} - \varphi^*_{min} \neq 90°$ between the extremal angles φ^*_{min} and φ^*_{max} is observed. In the general case of contact of two sculptured surfaces, the directions of the extremal degree of conformity of the part surface P and generating surface T of the cutting tool are not orthogonal to one another.

The discussed example reveals that in general cases of contact of two smooth regular sculptured surfaces, the indicatrix of conformity $Cnf_R(P/T)$ can be implemented for the purpose of accurate analytical description of the contact geometry of the surfaces. The Dupin's indicatrix of the surface of relative curvature can be implemented for this purpose only in particular cases of surfaces

P and T relative orientation. The application of the Dupin's indicatrix of the surface of relative curvature enables only approximate analytical description of the geometry of contact of the surfaces. The Dupin's indicatrix of the surface of relative curvature could be equivalent to the indicatrix of conformity only in degenerate cases of contact of the surfaces. Advantages of the indicatrix of conformity over the Dupin's indicatrix of the surface of relative curvature are due to that this characteristic curve, $Cnf_R(P/T)$, is a curve of the fourth order.

4.5.4 Asymptotes of the Indicatrix of Conformity $Cnf_R(P/T)$

In the theory of part surface generation, asymptotes of the indicatrix of conformity $Cnf_R(P/T)$ play an important role. The indicatrix of conformity could have asymptotes when a certain combination of parameters of shape of the part surface P and generating surface T of the cutting tool is observed. These special cases are of particular importance for the theory of part surface generation.

Straight lines that possess the property of becoming and staying infinitely close to the curve as the distance from the origin increases to infinity are referred to as the *asymptotes*. This definition of the asymptotes is helpful for derivation of the equation of asymptotes of the indicatrix of conformity, $Cnf_R(P/T)$, at a point of contact of the part surface P and generating surface T of the cutting tool.

In polar coordinates, indicatrix of conformity $Cnf_R(P/T)$ is analytically described by Equation 4.87. For the readers' convenience, the equation of this characteristic curve is represented below in the form $r_{cnf} = r_{cnf}(\varphi, \mu)$.

The derivation of the equation of the asymptote(s) of the characteristic curve $r_{cnf} = r_{cnf}(\varphi, \mu)$ can be accomplished in just a few steps:

(a) For a given indicatrix of conformity $r_{cnf} = r_{cnf}(\varphi, \mu)$ constructed at a point of contact of a part surface P and generating surface T of the cutting tool, compose a function $r^*_{cnf}(\varphi, \mu)$ that is equal

$$r^*_{cnf}(\varphi, \mu) = \frac{1}{r_{cnf}(\varphi, \mu)} \tag{4.100}$$

(b) Solve the equation $r^*_{cnf}(\varphi, \mu) = 0$ with respect to φ. Solution φ_0 to this equation specifies the direction of the asymptote.

(c) Calculate the value of the parameter m_0. The value of the parameter m_0 is equal $m_0 = \left(\dfrac{\partial g(\varphi, \mu)}{\partial \varphi} \right)^{-1}$ under the condition $\varphi = \varphi_0$.

(d) The asymptote(s) is the line through the point $(m_0, \varphi_0 + 0.5\pi)$ and with the direction φ_0. Its equation is

$$r(\varphi) = \frac{m_0}{\sin(\varphi - \varphi_0)} \tag{4.101}$$

In particular cases, the asymptotes of the indicatrix of conformity $Cnf_R(P/T)$ at a point of contact of surfaces P and T can coincide either with the asymptotes of the Dupin's indicatrix Dup (P) of the part surface P, of the Dupin's indicatrix Dup (T) of the generating surface T of the cutting tool, or finally with Dupin's indicatrix Dup (P/T) of the surface of relative curvature R constructed at a point of contact of surfaces P and T.

4.5.5 Comparison of Capabilities of the Indicatrix of Conformity $Cnf_R(P/T)$ and of the *Dupin's* Indicatrix *Dup* (*R*) of the Surface of Relative Curvature

Both characteristic curves, namely the indicatrix of conformity $Cnf_R(P/T)$ at a point of contact K of a part surface P and generating surface T of the cutting tool and the Dupin's indicatrix Dup (P/T) of the surface of relative curvature constructed at that same point K can be used with the same goal for analytical description of the contact geometry of surfaces P and T in the first order of tangency. Therefore, it is important to compare the capabilities of these characteristic curves with one another in order to identify appropriate areas of their correct implementation.

An in detail analysis of the capabilities of the characteristic curves, namely,

- Of the indicatrix of conformity $Cnf_R(P/T)$ at a point of contact of surfaces P and T (this characteristic curve is given by Equation 4.87)
- Of the Dupin's indicatrix of the surface of relative curvature Dup (P/T) (this characteristic curve is given by Equation 4.58)

has been performed. This analysis allows for making the following conclusions.

From the perspective of completeness and effectiveness of analytical description of the contact geometry of two smooth regular surfaces in the first order of tangency, the indicatrix of conformity $Cnf_R(P/T)$ is a more informative characteristic curve compared to the Dupin's indicatrix Dup (P/T) of the surface of relative curvature. Important features of the geometry of contact of surfaces P and T in the differential vicinity of the contact point K are reflected by the indicatrix of conformity $Cnf_R(P/T)$ more accurately. Thus, implementation of the indicatrix of conformity $Cnf_R(P/T)$ for scientific and for engineering purposes has advantages over the Dupin's indicatrix of the surface of relative curvature Dup (P/T). This conclusion directly follows from

(a) Directions of the extremal degree of conformity at a point of contact of a part surface P and generating surface T of the cutting tool that are specified by the Dupin's indicatrix Dup (P/T) are always orthogonal to one another. In reality, in the general case of contact of two sculptured surfaces these directions are not orthogonal to each other. They could be orthogonal only in particular cases of surface contact. The indicatrix of conformity $Cnf_R(P/T)$ at a point of contact of surfaces P and T properly specifies the actual directions of the extremal degree of conformity of surfaces P and T. This is particularly (but not only) because the characteristic curve $Cnf_R(P/T)$ is a curve of the fourth order, while the Dupin's indicatrix Dup (P/T) of the surface of relative curvature is a curve of the second order.

(b) Accounting for members of higher order in the equation of the Dupin's indicatrix Dup (P/T) of the surface of relative curvature does not enhance the capabilities of this characteristic curve and is useless from a practical standpoint. Accounting for members of higher order in Taylor's expansion of the equation of the Dupin's indicatrix gives nothing more for proper analytical description of the contact geometry of two smooth regular surfaces in the first order of tangency. The principal features of the equation of this characteristic curve cause a principal disadvantage to the Dupin's indicatrix Dup (P/T). The disadvantage above is inherited by the Dupin's indicatrix and it cannot be eliminated.

In addition to these two principal reasons, a few more particular reasons can be outlined as well.

4.5.6 Important Properties of the Indicatrix of Conformity $Cnf_R(P/T)$ at a Point of Contact of Surfaces P and T

The performed analysis of Equation 4.87 of the indicatrix of conformity $Cnf_R(P/T)$ at a point of contact of a sculptured part surface and generating surface of the cutting tool reveals that this characteristic curve possesses the following important properties.

1. The indicatrix of conformity $Cnf_R(P/T)$ at a point of contact of surfaces P and T is a planar characteristic curve of the fourth order. It possesses the property of central symmetry, and in particular cases, it also possesses the property of mirror symmetry.

2. The indicatrix of conformity $Cnf_R(P/T)$ is closely related to the surfaces P and T second fundamental forms $\Phi_{2.P}$ and $\Phi_{2.T}$. This characteristic curve is invariant with respect to the type of parameterization of surfaces P and T, but its equation is not. A change in surfaces P and T parameterization leads to changes in the equation of the indicatrix of conformity $Cnf_R(P/T)$, while the shape and parameters of this characteristic curve remained unchanged.

3. The characteristic curve $Cnf_R(P/T)$ is independent of the actual value of the coordinate angle ω_P that forms the coordinate lines U_P and V_P on the part surface P. It is also independent of actual value of the

coordinate angle ω_T that forms the coordinate lines U_T and V_T on the generating surface T of the cutting tool. However, the parameters of the indicatrix of conformity $Cnf_R(P/T)$ depend upon the angle μ of local relative orientation of surfaces P and T. Therefore, for the given pair of surfaces P and T, the degree of conformity of the surface varies correspondingly to variation of the angle μ while the surface T of the cutting tool is spinning around the unit vector of the common perpendicular.

More properties of the indicatrix of conformity $Cnf_R(P/T)$ at a point of contact of a part surface P and generating surface T of the cutting tool can be outlined.

4.5.7 Converse Indicatrix of Conformity at a Point of Contact of Surfaces *P* and *T* in the First Order of Tangency

For the Dupin's indicatrix $Dup\,(P/T)$ [or $Dup\,(\mathfrak{R})$ in other designation] at a point of the surface of relative curvature, $\mathfrak{R}$, there exists a corresponding inverse Dupin's indicatrix $Dup_k\,(P/T)$. Similarly, for the indicatrix of conformity $Cnf_R(P/T)$ at a point of contact of surfaces P and T, there exists a corresponding converse indicatrix of conformity $Cnf_k\,(P/T)$. This characteristic curve can be expressed directly in terms of surfaces P and T normal curvatures k_P and k_T:

$$Cnf_k\,(P/T) \quad \Rightarrow \quad r_{cnf}^{cnv}(\varphi,\mu) = \sqrt{|k_P(\varphi)|} \cdot \mathrm{sgn}\,\Phi_{2.P}^{-1} - \sqrt{|k_T(\varphi,\mu)|} \cdot \mathrm{sgn}\,\Phi_{2.T}^{-1}$$

(4.102)

For a derivation of the equation of the converse indicatrix of conformity $Cnf_k\,(P/T)$, the Euler's formula for a surface normal curvature is used in the following representation:

$$k_P(\varphi) = k_{1.P}\cos^2\varphi + k_{2.P}\sin^2\varphi$$

(4.103)

$$k_T(\varphi,\mu) = k_{1.T}\cos^2(\varphi+\mu) + k_{2.T}\sin^2(\varphi+\mu)$$

(4.104)

In Equations 4.103 and 4.104, the principal curvatures of the part surface P are designated as $k_{1.P}$ and $k_{2.P}$, whereas $k_{1.T}$ and $k_{2.T}$ designate the principal curvatures of the generating surface T of the cutting tool.

After substituting of Equations 4.103 and 4.104 into Equation 4.102, one can come up with the equation

$$r_{cnf}^{cnv}(\varphi,\mu) = \sqrt{|k_{1.P}\cos^2\varphi + k_{2.P}\sin^2\varphi|}\,\mathrm{sgn}\,\Phi_{2.P}^{-1}$$

$$-\sqrt{|k_{1.T}\cos^2(\varphi+\mu) + k_{2.T}\sin^2(\varphi+\mu)|}\,\mathrm{sgn}\,\Phi_{2.T}^{-1} \qquad (4.105)$$

for the converse indicatrix of conformity Cnf_k (P/T) at a point of contact of surfaces P and T in the first order of tangency.

Here, in Equation 4.105, the principal curvatures $k_{1.P}$, $k_{2.P}$, and $k_{1.T}$, $k_{2.T}$ can be expressed in terms of the corresponding fundamental magnitudes E_P, F_P, G_P of the first and L_P, M_P, N_P of the second order of the part surface P, and in terms of the corresponding fundamental magnitudes E_T, F_T, G_T of the first and L_T, M_T, N_T of the second order of the generating surface T of the cutting tool. Following this way, Equation 4.105 of the inverse indicatrix of conformity Cnf_k (P/T) can be cast to the form similar to Equation 4.87 of the ordinary indicatrix of conformity Cnf_R (P/T) at a point of contact of surfaces P and T.

It can be shown that similar to the indicatrix of conformity $Cnf_R(P/T)$, the characteristic curve Cnf_k (P/T) also possesses the property of central symmetry. In particular cases of surface contact, it also possesses the property of mirror symmetry. The directions of the extremal degree of conformity of the part surface P and the generating surface T of the cutting tool are orthogonal to one another only in degenerate cases of surface contact.

Equation 4.105 of the converse indicatrix of conformity Cnf_k (P/T) is convenient for implementation when (a) the part surface P, (b) the cutting tool surface T, or (c) both of them feature point(s) or line(s) of inflection. In the point(s) or line(s) of inflection, the radii of normal curvature $R_{P(T)}$ of the surface $P(T)$ are equal to infinity. Points/lines of inflection cause indefiniteness when calculating the position vector $r_{cnf}(\varphi, \mu)$ of a point of the characteristic curve $Cnf_R(P/T)$. Equation 4.105 of the converse indicatrix of conformity Cnf_k (P/T) is free of the disadvantages of such type, and therefore, it is recommended for practical applications.

4.6 *Plücker* Conoid: More Characteristic Curves

More characteristic curves for the purposes of analytical description of the contact geometry of two smooth regular surfaces in the first order of tangency can be derived on the premise of a Plücker conoid* [96].

4.6.1 *Plücker* Conoid

Several definitions for Plücker conoid are known.

First, Plücker conoid is a smooth regular ruled surface. A ruled surface sometimes is also called the *cylindroid*, which is the inversion of the cross-cap.

---

* Plücker's conoid is a ruled surface, which bears the name of a famous German mathematician and physicist Julius Plücker (1802–1868) known for his research in the field of a new geometry of space [96].

Second, a Plücker conoid can also be considered as an example of right conoid. A ruled surface is called a right conoid if it can be generated by moving a straight line intersecting a fixed straight line such that the lines are always perpendicular.

As with the *cathenoid*, another ruled surface, a Plücker conoid must be reparameterized to see the rulings. The illustrative examples of various Plücker conoids are considered in Reference [106].

4.6.1.1 Basics

A ruled surface can be swept out by moving a line in space and therefore has a parameterization of the form

$$\mathbf{x}(u, v) = \mathbf{b}(u) + v\boldsymbol{\delta}(u), \tag{4.106}$$

where $\mathbf{b}$ is called the directrix (also referred to as the *base curve* or as the *directing curve*) and $\boldsymbol{\delta}$ is the director curve. The straight lines themselves are called rulings. The rulings of a ruled surface are asymptotic curves. Furthermore, the *Gaussian* curvature on a ruled regular surface is everywhere nonpositive. The surface is known for the presence of two or more folds formed by the application of a cylindrical equation to the line during this rotation. This equation defines the path of the line along the axis of rotation.

4.6.1.2 Analytical Representation

For a Plücker conoid, von Seggern [161] gives the general functional form as

$$ax^2 + by^2 - zx^2 - zy^2 = 0 \tag{4.107}$$

whereas Fischer [19] and Gray [22] give

$$z = \frac{2xy}{x^2 + y^2} \tag{4.108}$$

Another form of *Cartesian* equation $z = a\dfrac{x^2 - y^2}{x^2 + y^2}$ for a twofold Plücker conoid is known as well [24].

The last equation yields the following matrix representation of nonpolar parameterization of a Plücker conoid:

$$\mathbf{r}_{pc}(u, v) = \begin{bmatrix} u & v & \dfrac{2uv}{u^2 + v^2} & 0 \end{bmatrix}^T \tag{4.109}$$

A Plücker conoid could be represented by the polar parameterization:

$$\mathbf{r}_{pc}(r, \theta) = [r \cos \theta \quad r \sin \theta \quad 2 \cos \theta \sin \theta \quad 0]^T \tag{4.110}$$

A more general form of Plücker conoid is parameterized below, with "n" folds instead of just two. A generalization of a Plücker conoid to n folds is given by [22]

$$\mathbf{r}_{pc}(r, \theta) = [r \cos \theta \quad r \sin \theta \quad \sin(n\theta) \quad 0]^T \tag{4.111}$$

The difference between these two forms is the function in the z axis. The polar form is a specialized function that outputs only one type of curvature with two undulations while the generalized form is more flexible with the number of undulations of the outputted curvature being determined by the value of n.

Cartesian parameterization of equation of the multifold Plücker conoid (see Equation 4.111) therefore gives [24]:

$$z\left(\sqrt{x^2 + y^2}\right)^n = \sum_{0 \leq k \leq \frac{n}{2}} (-1)^k C_n^{2k} x^{n-2k} y^{2k}. \tag{4.112}$$

The surface appearance depends upon the actual number of folds [106].

To represent a Plücker conoid as a ruled surface, it is sufficient to represent the above Equation 4.111 in the form of Equation 4.112:

$$\mathbf{r}_{pc}(r, \theta) = \begin{bmatrix} r \cos \theta \\ r \sin \theta \\ \sin(n\theta) \\ 0 \end{bmatrix} = \begin{bmatrix} r \cos \theta \\ r \sin \theta \\ 2 \cos \theta \sin \theta \\ 0 \end{bmatrix} = \begin{bmatrix} 0 \\ 0 \\ 2 \cos \theta \sin \theta \\ 0 \end{bmatrix} + r \begin{bmatrix} \cos \theta \\ \sin \theta \\ 0 \\ 0 \end{bmatrix} \tag{4.113}$$

Taking the perpendicular plane as the xy plane and take the line to be the x axis gives the parametric equation [22]:

$$\mathbf{r}_{pc} = \begin{bmatrix} v \cdot \cos v(u) \\ v \cdot \sin v(u) \\ h(u) \\ 0 \end{bmatrix} \tag{4.114}$$

Equation in cylindrical coordinates [24]:

$$z = a \cos(n\theta), \tag{4.115}$$

which simplifies to

$$z = a \cos 2\theta \tag{4.116}$$

if $n = 2$.

4.6.1.3 Local Properties

Following Bonnet's theorem (see Chapter 1), the local properties of a Plücker conoid could be analytically expressed in terms of the first and second fundamental forms of the surface. For a practical application, some useful auxiliary formulae are also necessary.

The first and the second fundamental forms [24] of a Plücker conoid can be represented as

$$\Phi_1 \Rightarrow ds^2 = d\rho^2 + (\rho^2 + n^2a^2 \sin^2(n\theta))d\theta^2 \tag{4.117}$$

$$\Phi_2 \Rightarrow \frac{na}{H}[\sin(n\theta)d\rho - n\rho\cos(n\theta)d\theta]d\theta \tag{4.118}$$

Asymptotes are given by the equation

$$\rho'' = ka^n \sin(n\theta). \tag{4.119}$$

They strictly correlate to *Bernoulli's* lemniscates [24].

For the simplified case of a Plücker conoid, when $n = 2$, the first and the second fundamental forms reduce to [24]

$$\Phi_1 \Rightarrow ds^2 = d\rho^2 + (\rho^2 + 4a^2)\cos^2 2\theta d\theta^2 \tag{4.120}$$

$$E = 1, \tag{4.121}$$

$$F = 0, \tag{4.122}$$

$$G = \rho^2 + 4a^2 \cos^2 2\theta, \tag{4.123}$$

$$H = \sqrt{G} \tag{4.124}$$

$$\Phi_2 \Rightarrow -\frac{4a}{H}[\sin 2\theta d\rho - n\rho\cos 2\theta d\theta]d\theta \tag{4.125}$$

$$L = 0, \tag{4.126}$$

$$M = -\frac{2a\cos 2\theta}{H},$$ (4.127)

$$N = -\frac{4a\rho\sin 2\theta}{H}$$ (4.128)

$$T = \sqrt{-M^2}.$$ (4.129)

Because the consideration below is limited only to the case of $n = 2$, auxiliary formulae for references would be helpful.

4.6.1.4 Auxiliary Formulae

At $u = u_0$, $v = v_0$, the tangent to the surfaces is parameterized by

$$\mathbf{r}_{pc}(u,v) = \begin{bmatrix} u + u_0 \\ v + v_0 \\ \dfrac{2(-u\,u_0^2 v_0 + u\,v_0^3 + u_0 v_0^2(-v + v_0) + u_0^3(v + v_0))}{u_0^2 + v_0^2} \\ 0 \end{bmatrix}$$ (4.130)

The surface normal is its double line [106,158].
The infinitesimal area of a patch on the surface is given by

$$\Phi_1 \;\Rightarrow\; ds = \sqrt{1 + \frac{4(u-v)^2(u+v^2)}{(u^2+v^2)^3}}\,du\,dv$$ (4.131)

Gaussian curvature at a point of a Plücker conoid could be calculated from

$$G(u,v) = -\frac{4(u^4 - v^4)^2}{[u^6 + v^4(4+v^2)+u^2v^2(-8+3v^2)+u^4(4+3v^2)]^2}$$ (4.132)

Mean curvature at a point of a Plücker conoid is equal to

$$\mathcal{M}(u,v) = -\frac{4uv}{(u^2+v^2)^2\left(1 + \dfrac{4(u-v)^2(u+v)^2}{(u^2+v^2)^3}\right)^{\frac{3}{2}}}$$ (4.133)

The above listed equations are proven to be useful when performing an analysis of local properties of a surface of a Plücker conoid.

4.6.2 Analytical Description of Local Topology at a Point of a Smooth Regular Part Surface *P*

For the consideration below, the following characteristics at a point of a smooth regular part surface *P* are of prime importance, namely

(a) Tangent plane to the part surface *P*

(b) Unit normal $\mathbf{n}_P$ at a surface point *m*

(c) Principal curvatures $k_{1.P}$ and $k_{2.P}$ at a surface point *m* as well as normal curvature k_P at the prespecified direction on the part surface the part surface *P*

A Plücker conoid can be used for the visualization of the distribution of normal curvature at a given point *m* within the part surface *P*. The corresponding Plücker conoid can be determined at *every point* of a smooth regular part surface *P*. The surface unit normal vector $\mathbf{n}_P$ can be employed as the axis of a Plücker conoid.

The rulings are the straight lines that intersect *z* axis at right angle. The generating straight-line segments of a Plücker conoid are always parallel to the tangent plane to the part surface *P* at the point, at which a Plücker conoid is erected. In the consideration below, other applications of the tangent plane to the part surface *P* are of importance as well.

Consequently, the above makes possible the natural method for connecting Plücker conoid to the part surface *P*.

4.6.2.1 Preliminary Remarks

An example of implementation of a Plücker conoid is given by Struik [158]. He considers a cylindroid, which is represented by the locus of the curvature vectors at a point *m* of a surface *P* belonging to all curves passing through *m*:

$$z(x^2 + y^2) = k_{1.P}x^2 + k_{2.P}y^2 \qquad (4.134)$$

Here in Equation 4.134, $k_{1.P}$ and $k_{2.P}$ designate principal curvatures at a point of the part surface *P* (the inequality $k_{1.P} > k_{2.P}$ is always observed).

The curvature vector is defined in the following way. According to [158], a proportionality factor k_P such that

$$\mathbf{k}_P = d\mathbf{t}_P/dS = k_P\mathbf{n}_P \qquad (4.135)$$

can be introduced.

The vector $\mathbf{k}_P = d\mathbf{t}_P/dS$ expresses the rate of change of the tangent when we proceed along the curve. It is called the *curvature vector*. The factor k_P is called the *curvature*; $|\mathbf{k}_P|$ is the length (magnitude) of the curvature vector. Although the sense of $\mathbf{n}_P$ may be arbitrary chosen, that of $d\mathbf{t}_P/dS$ is perfectly determined by the curve, independent of its orientation; when S changes sign, $\mathbf{t}_P$ also changes sign. When $\mathbf{n}_P$ (as it often done) is taken in the sense of S, then κ_P is always positive, but we shall note adhere to this convention.

4.6.2.2 Plücker Conoid

To develop an appropriate graphical interpretation of a Plücker conoid $\mathbf{Pl}_R(P)$ at a point of a part surface P, let us consider a smooth regular part surface P that is given by vector equation in the form $\mathbf{r}_P = \mathbf{r}_P(U_P, V_P)$.

From the perspective of natural connection of a Plücker conoid to the part surface P itself, the axis of a Plücker conoid $\mathbf{Pl}_R(P)$ is aligned to the unit normal vector $\mathbf{n}_P$ to the part surface P at a surface point m.

For further consideration, it is necessary to calculate normal radii of curvature $R_P = k_P^{-1}$ of the part surface P at the point m. To simplify the calculations, the equation

$$R_P = \frac{\Phi_{1.P}}{\Phi_{2.P}} \tag{4.136}$$

for the radius R_P can be reduced to the Euler formula for normal radii of curvature:

$$R_P(\varphi) = (R_{1.P}^{-1} \cos^2 \varphi + R_{2.P}^{-1} \sin^2 \varphi)^{-1} \tag{4.137}$$

where $R_{1.P}$ and $R_{2.P}$ are the principal radii of curvature at a point m of the part surface P and φ is the angle that the normal plane section for $R_P(\varphi)$ forms with the first principal direction $\mathbf{t}_{1.P}$.

Point C_1 is coincident with the curvature center of the part surface P in the first principal plane section of P at m. It is located on the axis of a Plücker conoid, $\mathbf{Pl}_R(P)$. The straight-line segment of length $R_{1.P}$ extends from the point C_1 in the direction of unit tangent vector, $\mathbf{t}_{1.P}$, of the first principal direction at m. The unit tangent vector $\mathbf{t}_{1.P}$ makes right angle with the axis of the surface $\mathbf{Pl}_R(P)$. The straight-line segment of the same length $R_{1.P}$ also extends from C_1 in the opposite direction of $-\mathbf{t}_{1.P}$.

Point C_2 is coincident with the curvature center of the surface P in the second principal plane section of P at m. It is remote from C_1 at a distance $(R_{1.P} - R_{2.P})$. (Remember, that the normal radius of curvature R_P and the principal radii of curvature $R_{1.P}$ and $R_{2.P}$ are the algebraic values in nature.) The straight-line segment of the length $R_{2.P}$ extends from C_2 in the direction of

$t_{2.P}$. The unit tangent vector $\mathbf{t}_{2.P}$ indicates the second principal direction of the part surface P at m. It also makes right angle with the axis of the characteristic surface $\mathbf{Pl}_R(P)$. The straight-line segment of the same length $R_{2.P}$ also extends from the point C_2 in the direction of $-\mathbf{t}_{2.P}$.

A certain point C is coincident with the center of curvature of the part surface P in the normal plane section of P at m in an arbitrary direction that is specified by the corresponding value of central angle φ. The point C is located on the axis of the characteristic surface $\mathbf{Pl}_R(P)$. The normal radius of curvature $R_P(\varphi)$ corresponds to the principal radii of curvature $R_{1.P}$ and $R_{2.P}$ in the manner $R_{1.P} < R_P(\varphi) < R_{2.P}$. Again, the equality $R_{1.P} = R_{2.P}$ is observed under no circumstances (umbilics, flatten point of a part surface) as in this case principal directions at a point of the part surface P cannot be identified.

The straight-line segment of length $R_P = R_P(\varphi)$ is spinning about and is traveling up and down the axis of the Plücker conoid $\mathbf{Pl}_R(P)$. In such a way, a Plücker conoid could be represented as a locus of consecutive positions of the straight-line segment $R_P = R_P(\varphi)$.

Figure 4.24 reveals* that the topology of the surface P in the differential vicinity of point m is perfectly reflected by a Plücker conoid. Therefore, the surface $\mathbf{Pl}_R(P)$ can be implemented as a tool for the visualization of the change of its local parameters.

To plot a Plücker conoid $R_r = R_P + R_T$ together with the part surface P itself (Figure 4.24), it is necessary to represent equations of both the surfaces in a common reference system, for example, in the coordinate system $X_S Y_S Z_S$. For this purpose, it is necessary to compose the operator of the resultant coordinate system transformation $\mathbf{Rs}(S \rightarrow P)$ (see Equation 3.19).

After been constructed at a point m of a smooth regular part surface P, the characteristic surface $\mathbf{Pl}_R(P)$ clearly indicates

- Actual values of principal radii of curvature $R_{1.P}$ and $R_{2.P}$
- Location of the curvature centers $O_{1.P}$ and $O_{2.P}$
- Orientation of the principal plane sections $C_{1.P}$ and $C_{2.P}$ (that is, the directions of the unit tangent vectors $\mathbf{t}_{1.P}$ and $\mathbf{t}_{2.P}$ of the principal directions)
- Current value of normal radii of curvature $R(\varphi)$
- Location of curvature center O_P for any given section of the part surface P by a normal plane C_P through the given direction $\mathbf{t}_P(\varphi)$

Therefore, a Plücker conoid can be considered as an example of a *characteristic surface* that potentially could be used in the theory of part surface generation

---

* It is important to point out here for the reader's convenience that the Plücker's conoid in Figure 4.24 is scaled along the axes of the local coordinate system (with the sole goal of better visualization of local geometrical properties at a point of the part surface P).

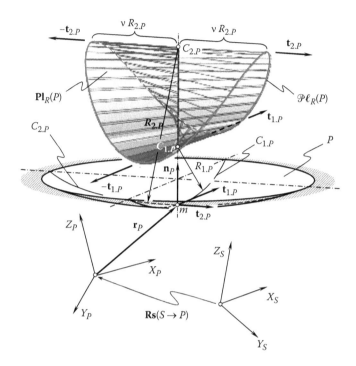

FIGURE 4.24
Plücker conoid, $\mathbf{Pl}_R(P)$, and Plücker curvature indicatrix, $\mathcal{Pl}_R(P)$, naturally associated with the concave patch of a smooth regular part surface P. (Reprinted from *Mathematical and Computer Modeling*, 42, Radzevich, S.P., A possibility of application of Plücker's conoid for mathematical modeling of contact of two smooth regular surfaces in the first order of tangency, 999–1022, Copyright 2005, with permission from Elsevier.)

for the purposes of synthesizing the most favorable operations of sculptured part surface generation.

In addition to Plücker conoid $\mathbf{Pl}_R(P)$ of the type described above (Figure 4.24), a characteristic surface $\mathbf{Pl}_k(P)$ of inverse type can be introduced as well.

When constructing a Plücker conoid $\mathbf{Pl}_k(P)$, the straight-line segment of the length $k_p(\varphi)$ is used instead of the straight-line segment of the length $R_p(\varphi)$ shown in Figure 4.24. [Here $k_p(\varphi) = R_P^{-1}(\varphi)$ is normal curvature at a point m of the part surface P in the given normal plane section through m]. This makes possible construction of the characteristic surface $\mathbf{Pl}_k(P)$ of the inverse type.

The characteristic surfaces $\mathbf{Pl}_R(P)$ and $\mathbf{Pl}_k(P)$ resemble one another from many aspects. They also appear similar, except in the cases, when $R_p(\varphi)$ and/or $k_p(\varphi)$ is equal either to zero (0) or to infinity (∞).

The characteristic surface $\mathbf{Pl}_R(P)$ is referred to as a Plücker conoid *of the first type* and the characteristic surface $\mathbf{Pl}_k(P)$ is referred to as a Plücker conoid *of the second type*.

The conoids $\mathbf{Pl}_R(P)$ and $\mathbf{Pl}_k(P)$ are *inverse* to one another [$\mathbf{Pl}_R(P) = \mathbf{Pl}_k^{inv}(P)$ and vice versa].

Plücker conoids—that is, $\mathbf{Pl}_R(P)$ and $\mathbf{Pl}_k(P)$—clearly indicate the change of parameters of local topology in the differential vicinity of a point of smooth regular part surface P.

4.6.2.3 Plücker's Curvature Indicatrix

The boundary curve of a Plücker conoid contains all the necessary information on the distribution of normal curvatures in the differential vicinity of a point m of the part surface P, while the surface $\mathbf{Pl}_R(P)$ itself represents additional information about local topology of the part surface P. This additional information is of importance from the standpoint of its implementation in the theory of part surface generation. We are again reminded of the principle of *Ockham's razor*.*

Thus, without loss of generality, a Plücker conoid itself can be replaced with the boundary curve of the characteristic surface $\mathbf{Pl}_R(P)$. The boundary curve $\mathcal{P}\ell_R(P)$ of the characteristic surface $\mathbf{Pl}_R(P)$ is referred to as Plücker's *curvature indicatrix of the first type* at a point m of the part surface P.

Plücker's curvature indicatrix is represented therefore by end-points of the position vector of the length of $R_P(\varphi)$ when the vector is rotating about and travels up and down the axis of the characteristic surface $\mathbf{Pl}_R(P)$. This immediately leads to an equation of this characteristic curve:

$$\mathcal{P}\ell_R(P) \quad \Rightarrow \quad \mathbf{r}_R(\varphi) = \begin{bmatrix} R_P(\varphi)\cos\varphi \\ R_P(\varphi)\sin\varphi \\ R_P(\varphi) \\ 1 \end{bmatrix} \tag{4.138}$$

where $R_P(\varphi)$ is given by the Euler's formula

$$R_P(\varphi) = (R_{1.P}^{-1}\cos^2\varphi + R_{2.P}^{-1}\sin^2\varphi)^{-1}. \tag{4.139}$$

The performed analysis [106] reveals that for most types of smooth regular part surface P Plücker's curvature indicatrix $\mathcal{P}\ell_R(P)$ of the first type is a *closed* regular 3D curve. For the surface local patches of parabolic and saddle-like type, Plücker's indicatrix $R_r = R_P + R_T$ separates onto two and four branches, respectively. In particular cases, it could be reduced even to a planar curve—to a circle, for example, for umbilic local patches of the part surface P.

---

* *William of Ockham*, also spelled *Occam* (b.c. 1285, Ockham, Surrey(?), England, d. 1347/49, Munich, Bavaria [now in Germany]). He is remembered well because of the tools of logic he developed. He insisted that we should always look for the simplest explanation that fits all the facts, instead of inventing complicated theories. The rule, which said "plurality should not be assumed without necessity," is called "Ockham's razor".

The equation, similar to the above Equation 4.138 is valid for Plücker's curvature indicatrix $\mathcal{P}\ell_k(P)$ of the second type:

$$\mathcal{P}\ell_k(P) \quad \Rightarrow \quad \mathbf{r}_k(\varphi) = \begin{bmatrix} k_P(\varphi)\cos\varphi \\ k_P(\varphi)\sin\varphi \\ k_P(\varphi) \\ 1 \end{bmatrix} \tag{4.140}$$

where

$$k_P(\varphi) = k_{1.P}\cos^2\varphi + k_{2.P}\sin^2\varphi. \tag{4.141}$$

Usually, Plücker's curvature indicatrix $\mathcal{P}\ell_k(P)$ is represented by a closed curve.

A further possible simplification of analytical description of local topology at a point of a smooth regular part surface P could be based on the following consideration.

4.6.2.4 $\mathcal{A}n_R(P)$-Indicatrix of the Surface P

Aiming further simplification of the analytical description of local topology at a point of contact of two smooth regular surfaces in the first order of tangency, Plücker's curvature indicatrix can be replaced with a planar characteristic curve of a novel type.

As follows from Equation 4.138, the first two elements $R_P(\varphi)\cos\varphi$ and $R_P(\varphi)\sin\varphi$ at the right-hand side portion of the above equation contain all the necessary information on the distribution of normal radii of curvature at a point m of a part surface P. Hence, a planar characteristic curve $\mathcal{A}n_R(P)$ of simpler structure can be implemented instead of Plücker's curvature indicatrix $\mathcal{P}\ell_k(P)$ (see Equation 4.138) for the purpose of analytical description of the geometry of contact of two smooth regular surfaces. The equation of this characteristic curve yields representation in the form

$$\mathcal{A}n_R(P) \quad \Rightarrow \quad \mathbf{r}_{iR}(\varphi) = \begin{bmatrix} R_P(\varphi)\cos\varphi \\ R_P(\varphi)\sin\varphi \\ 0 \\ 1 \end{bmatrix} \tag{4.142}$$

This planar characteristic curve is referred to as $\mathcal{A}n_R(P)$ *indicatrix of the first type* at a point of a smooth, regular part surface *P*.

The distribution of normal curvature at a point of the part surface P also can be given by another planar characteristic curve:

$$\mathscr{A}_{n_k}(P) \quad \Rightarrow \quad \mathbf{r}_{ik}(\varphi) = \begin{bmatrix} k_P(\varphi)\cos\varphi \\ k_P(\varphi)\sin\varphi \\ 0 \\ 1 \end{bmatrix} \tag{4.143}$$

This planar characteristic curve (see Equation 4.143) is referred to as to $\mathscr{A}_{n_k}(P)$-indicatrix of the second type at a point of a smooth, regular part surface P.

An example of $\mathscr{A}_{n_k}(P)$-indicatrix is shown in Figure 4.25. The parameters of the characteristic curve $\mathscr{A}_{n_R}(P)$ are calculated at a point of the part surface P having principal radii of curvature equal to $R_{1.P} = 3$ mm and $R_{2.P} = 15$ mm. It is

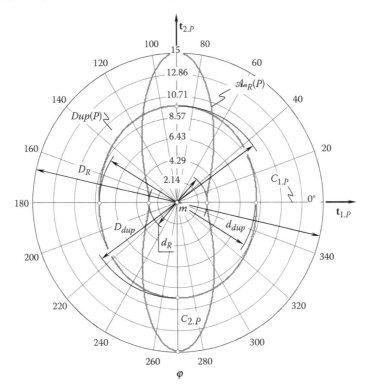

FIGURE 4.25

$\mathscr{A}_{n_R}(P)$-indicatrix at a point m of a part surface P ($R_{1.P} = 3$ mm, $R_{2.P} = 15$ mm) plotted together with the corresponding Dupin indicatrix Dup (P) that is magnified 10 times. (Reprinted from *Mathematical and Computer Modeling*, 42, Radzevich, S.P., A possibility of application of Plücker's conoid for mathematical modeling of contact of two smooth regular surfaces in the first order of tangency, 999–1022, Copyright 2005, with permission from Elsevier.)

import to point out here that the direction of minimal diameter d_R of the characteristic curve $\mathscr{An}_R(P)$ is aligned with the first principal direction $\mathbf{t}_{1.P}$ of the part surface P and the direction of maximal diameter D_R of the characteristic curve $\mathscr{An}_R(P)$ is aligned with the second principal direction $\mathbf{t}_{2.P}$ on the part surface P at that same point. Therefore, the directions for the extremum diameters d_R and D_R of the characteristic curve $\mathscr{An}_k(P)$ are always orthogonal to one another.

It is of interest to compare the $\mathscr{An}_k(P)$-indicatrix with the corresponding Dupin's indicatrix $Dup\,(P)$. To make the comparison, the characteristic curve $Dup\,(P)$ is calculated at that same point m of the part surface P at which principal radii of curvature are equal to $R_{1.P} = 3$ mm and $R_{2.P} = 15$ mm, respectively.

The characteristic curve $Dup\,(P)$ is also plotted in Figure 4.25. For the reader's convenience, the characteristic curve $Dup\,(P)$ is magnified 10 times with respect to its original (calculated) parameters. The direction of the minimal diameter d_{Dup} of the Dupin indicatrix $Dup\,(P)$ is aligned with the first principal direction $\mathbf{t}_{1.P}$, and the direction of maximal diameter D_{Dup} is aligned with the second principal direction $\mathbf{t}_{2.P}$ on the part surface P at the point m.

The performed analysis of Figure 4.25 makes it clear that both characteristic curves—that is, the $\mathscr{An}_R(P)$-indicatrix and Dupin indicatrix $Dup\,(P)$ indicate the same directions for the first $R_{1.P}$ as well as for the second $R_{2.P}$ principal radii of curvature at the point m of the part surface P. However, the difference in shape between the characteristic curves $\mathscr{An}_R(P)$ and $Dup\,(P)$ is observed. Dupin's indicatrix is a planar smooth regular curve of the second order. In the case under consideration, it is always convex. This characteristic curve features uniform change of curvature. The $\mathscr{An}_R(P)$-indicatrix is also a planar smooth regular curve. However, the points of inflection are inherited to this characteristic curve in nature. This is because the $\mathscr{An}_R(P)$-indicatrix is a curve of the fourth order. The higher order of the curve enhances capabilities of the characteristic curve $\mathscr{An}_R(P)$ for the purposes of analytical description of local topology at a point of a sculptured part surface P.

Because the order of equation for the characteristic curve $\mathscr{An}_R(P)$ is higher, the distribution of normal radii of curvature is indicated by the $\mathscr{An}_R(P)$-indicatrix, whereas the distribution of square root of normal radii of curvature of a sculptured part surface P at a point m is reflected by Dupin's indicatrix $Dup\,(P)$. To make the difference clear, it is sufficient to represent the equation of the Dupin's indicatrix in the form that is similar to the $\mathscr{An}_R(P)$-indicatrix (see Equation 4.142):

$$Dup\,(P) \Rightarrow \mathbf{r}_{Dup}(\varphi) = \begin{bmatrix} \sqrt{|R_P(\varphi)|}\cos\varphi\,\mathrm{sgn}\,R_P(\varphi) \\ \sqrt{|R_P(\varphi)|}\sin\varphi\,\mathrm{sgn}\,R_P(\varphi) \\ 0 \\ 1 \end{bmatrix} \qquad (4.144)$$

Evidently, Equations 4.142 and 4.144 are similar to one another.

4.6.3 Relative Characteristic Curves

The discussed properties of a Plücker conoid can be employed for derivation of the equation of a planar characteristic curve for analytical description of the contact geometry of two smooth regular surfaces for the needs of the theory of part surface generation.

4.6.3.1 Possibility of Implementation of Two Plücker Conoids

At first glimpse, the implementation of two Plücker conoids sounds promising for solving the problem of analytical description of the contact geometry between two smooth regular surfaces.

To develop an appropriate solution to the problem under consideration, the characteristic surface $\mathbf{Pl}_R(P/T)$ that reflects the *summa* of the corresponding normal radii of curvature of the part surface P and generating surface T of the form-cutting tool can be introduced. The following matrix representation of the equation of the characteristic surface $\mathbf{Pl}_R(P/T)$ immediately follows from the above consideration:

$$\mathbf{Pl}_R(P/T) \quad \Rightarrow \quad \mathbf{R}_R^*(\varphi) = \begin{bmatrix} (R_P + R_T)\cos\varphi \\ (R_P + R_T)\sin\varphi \\ 2\sin\varphi\cos\varphi \\ 1 \end{bmatrix} \qquad (4.145)$$

Below, the characteristic surface $\mathbf{Pl}_R(P/T)$ is referred to as Plücker's *relative conoid*.

Because the centers of the principal curvature $c_{1.P}$ and $c_{2.P}$ at a point of the part surface P as well as centers of principal curvatures $c_{1.T}$ and $c_{2.T}$ at a point of generating surface T of the cutting tool in general case do not coincide with one another, actual reciprocation of the straight-line segment of the length $(R_P - R_T)$ could be restricted by different pairs of the limiting points $c_{1.P}, c_{2.P}, c_{1.T},$ and $c_{2.T}$. Various locations of the limiting points along the axis of rotation of the straight-line segment $(R_P - R_T)$ result in the *deformation* of the characteristic surface $\mathbf{Pl}_R(P/T)$ in its axial direction. The deformation of such a type does not affect the surface appearance in the direction of $(R_P + R_T)$, which is of critical importance for the theory of part surface generation.

The characteristic surface $\mathbf{Pl}_R(P/T)$ is analytically described by Equation 4.100. This indicates that degree of conformity at a point of contact K between the part surface P and generating surface T of the cutting tool is properly indicated by Plücker's relative conoid. However, the characteristic surface $\mathbf{Pl}_R(P/T)$ itself is inconvenient for implementation in engineering geometry of surfaces. To fix this particular unfavorable problem, one may decide to follow the same method as disclosed above, and introduce the Plücker's relative indicatrix $\mathbf{Pl}_R(P/T)$. The equation of this three-dimensional characteristic curve immediately follows from Equation 4.145:

$$\mathcal{Pl}_R(P/T) \quad \Rightarrow \quad \mathbf{R}_R(\varphi) = \begin{bmatrix} (R_P + R_T)\cos\varphi \\ (R_P + R_T)\sin\varphi \\ (R_P + R_T) \\ 1 \end{bmatrix} \qquad (4.146)$$

Further, the characteristic curve $\mathcal{Pl}_R(P/T)$ could be reduced to a corresponding planar characteristic curve. To shorten the discussion below, the intermediate considerations are omitted, and one can wish to go directly to the $\mathcal{An}_R(P/T)$-*relative indicatrix* at a point of contact K between the part surface P and generating surface T of the cutting tool.

4.6.3.2 $\mathcal{An}_R$(P/T)-*Relative Indicatrix of Surfaces* P *and* k$_r$ = k$_P$ − k$_T$

Aiming for further simplification of analytical description of contact geometry between the part surface P and generating surface $k_r = k_P - k_T$ of the cutting tool, Plücker's relative indicatrix $k_r = k_P - k_T$ can be replaced with the planar characteristic curve of a simpler structure. The equation of the two-dimensional $\mathcal{An}_R(P/T)$-relative indicatrix at a point of contact K between surfaces P and T can be derived on the premises of Equation 4.146:

$$\mathcal{An}_R(P/T) \quad \Rightarrow \quad \mathbf{R}_{iR}(\varphi) = \begin{bmatrix} (R_P + R_T)\cos\varphi \\ (R_P + R_T)\sin\varphi \\ 0 \\ 1 \end{bmatrix} \qquad (4.147)$$

This planar characteristic curve is referred to as the $\mathcal{An}_R(P/T)$-*relative indicatrix of the first type*. The distribution of summa of normal radii of curvature at a point of contact K between the part surface P and generating surface T of the cutting tool is analytically described by the $\mathcal{An}_R(P/T)$-relative indicatrix of the first type.

An example of the $\mathcal{An}_R(P/T)$-relative indicatrix at a point of contact between the surfaces P and T is illustrated in Figure 4.26. The parameters of the characteristic curve $\mathcal{An}_R(P/T)$ are calculated for the case of contact of the convex elliptic local patch of the part surface P ($R_{1.P} = 3$ mm and $R_{2.P} = 15$ mm) with the concave elliptic local patch of the surface T ($R_{1.T} = -2$ mm and $R_{2.T} = -5$ mm). The contacting surfaces P and T are turned through the angle $\mu = 45°$ relative to one another around the common perpendicular $\mathbf{n}_P$. The corresponding $\mathcal{An}_R(P)$-indicatrix as well as the $\mathcal{An}_R(T)$-indicatrix are also

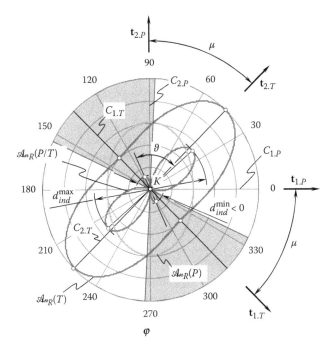

FIGURE 4.26
An example of the $\mathscr{A}n_R(P/T)$-relative indicatrix at a point of contact K between a part surface P and generating surface T of the cutting tool having principal radii of curvature $R_{1.P} = 3$ mm, $R_{2.P} = 15$ mm and $R_{1.T} = -2$ mm, $R_{2.T} = -5$ mm, respectively, and turned in relation to one another through the angle of local relative orientation $\mu = 45°$. (Reprinted from *Mathematical and Computer Modeling*, 42, Radzevich, S.P., A possibility of application of Plücker's conoid for mathematical modeling of contact of two smooth regular surfaces in the first order of tangency, 999–1022, Copyright 2005, with permission from Elsevier.)

plotted in Figure 4.26. It is important to point out here that the direction of the minimal diameter d_{ind}^{min} and the direction of the maximal diameter d_{ind}^{max} of the characteristic curve $\mathscr{A}n_R(P/T)$ do not align neither with the principal directions $\mathbf{t}_{1.P}$ and $\mathbf{t}_{2.P}$ on the part surface P nor with the principal directions $\mathbf{t}_{1.T}$ and $\mathbf{t}_{2.T}$ on the generating surface T of the cutting tool. The extremal directions of the $\mathscr{A}n_R(P/T)$-relative indicatrix are not orthogonal to each other. In a general case of surface contact, they create an angle, $\vartheta \neq 90°$.

The following conclusion can be drawn from the consideration above:

Conclusion 4.4

In the general case of contact of two smooth, regular surfaces, directions of the extremal (that is, of the maximal and of the minimal) degree of conformity at a point K of contact of the part surface P and generating surface T of the cutting tool are not orthogonal to one another. The directions of the extremal degree of conformity of surfaces P and T could be orthogonal to one another only in particular (degenerate) cases of surface contact. ∎

The shape and parameters of the $\mathscr{A}_{nR}(P/T)$-relative indicatrix depend upon algebraic values of the principal radii of curvature $R_{1.P}$, $R_{2.P}$, and $R_{1.T}$, $R_{2.T}$ at the point of contact of the surfaces $k_r = k_P - k_T$ and T as well as on the actual value of the angle μ of surfaces P and T local relative orientation.

Dupin's indicatrix $Dup (P/T)$ at a point of the surface of relative curvature indicates orthogonality of the directions of the extremal degree of conformity at the point of contact K of the part surface P and generating surface T of the cutting tool. The above discussion reveals that in the general case of contact of two smooth, regular surfaces this is not correct, and results in calculation errors.

The characteristic curve $\mathscr{A}_{nR}(P/T)$ is of simpler structure than that of Plücker's relative indicatrix $\mathscr{Pl}_R(P/T)$. The $\mathscr{A}_{nR}(P/T)$-relative indicatrix is always a planar curve, and the Plücker's relative indicatrix $\mathscr{Pl}_R(P/T)$ is a three-dimensional curve. This makes the use of the characteristic curve $\mathscr{A}_{nR}(P/T)$ more preferred rather than the Plücker's relative indicatrix $\mathscr{Pl}_R(P/T)$.

The distributions of the differences between normal curvatures at a point of contact K of the part surface P and generating surface T can be analytically described by a planar characteristic curve of another type:

$$
\mathscr{A}_{n_k}(P/T) \quad \Rightarrow \quad \mathbf{R}_{ik}(\varphi) =
\begin{bmatrix}
(k_P - k_T)\cos\varphi \\
(k_P - k_T)\sin\varphi \\
0 \\
1
\end{bmatrix}
\tag{4.148}
$$

The characteristic curve (see Equation 4.148) is referred to as the $\mathscr{A}_{n_k}(P/T)$-*relative indicatrix of the second type*.

The difference between the $\mathscr{A}_{nR}(P/T)$-relative indicatrix and the Dupin's indicatrix of the surface of relative curvature $Dup_R (P/T)$ is clearly illustrated in Figure 4.27. The principal directions $\mathbf{t}_{1.r}$ and $\mathbf{t}_{2.r}$ those defined by the Dupin's indicatrix of the surface of relative curvature $Dup_R (P/T)$ are always orthogonal to one another. The directions $\mathbf{t}_{\min}$ and $\mathbf{t}_{\max}$ of minimum and maximum degree of conformity at a point of contact K of the part surface P and generating surface T of the cutting tool are not orthogonal to each other in a general case of surface contact. They may be orthogonal in a degenerate case of surface contact only.

The planar characteristic curves $\mathscr{A}_{nR}(P)$ and $\mathscr{A}_{n_k}(P)$ as well as the characteristic curves $\mathscr{A}_{nR}(P/T)$ and $\mathscr{A}_{n_k}(P/T)$ originate from a Plücker conoid. Equations 4.142, 4.143, 4.147, and 4.148 of the corresponding indicatrices $\mathscr{A}_{nR}(P)$, $\mathscr{A}_{nR}(P/T)$, and $\mathscr{A}_{n_k}(P)$, $\mathscr{A}_{n_k}(P/T)$ are derived on the premise of Equation 4.113 of the surface of a Plücker conoid (see Reference [136] and Section 4.9, pp. 257–260, of Reference [117]).

It is proven analytically that both planar characteristic curves, say the characteristic curve $\mathscr{A}_{nR}(P/T)$ as well as the indicatrix of conformity $Cnf_R(P/T)$ at

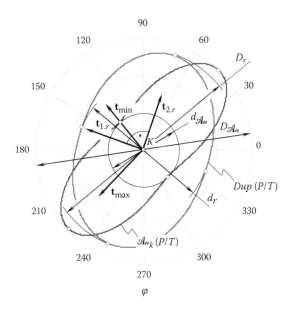

FIGURE 4.27
An example of the $\mathscr{A}n_k(P/T)$-relative indicatrix at the point of contact, K, of a part surface P (with $R_{1.P} = 2$ mm and $R_{2.P} = 3$ mm), and generating surface T of the cutting tool (with $R_{1.T} = -2$ mm and $R_{2.T} = -5$ mm). Surfaces P and T are turned in relation to one another at an angle of local relative orientation $\mu = 45°$. The $\mathscr{A}n_k(P/T)$-relative indicatrix is plotted together with the corresponding Dup (P/T)-indicatrix. (Reprinted from *Mathematical and Computer Modeling*, 42, Radzevich, S.P., A possibility of application of Plücker's conoid for mathematical modeling of contact of two smooth regular surfaces in the first order of tangency, 999–1022, Copyright 2005, with permission from Elsevier.)

a point of contact of a smooth regular part surface P and generating surface T of the cutting tool specify the same direction $\mathbf{t}_{cnf}^{max}$ at which the degree of conformity of the contacting surfaces P and T is maximal. Both characteristic curves, $Cnf_R(P/T)$ and $\mathscr{A}n_R(P/T)$ are powerful tools in the theory of part surface generation. They are widely used for the analysis of contact geometry of two smooth regular surfaces P and T.

4.7 Feasible Types of Contact of a Part Surface P and Generating Surface T of the Cutting Tool

The analysis and classification of all feasible types of contact of a part surface P and the generating surface T of the cutting tool are critical issues in the theory of part surface generation. Development of a scientific classification of all possible types of contact of the surfaces P and T can be considered as the ultimate point in the analysis of contact geometry of a part surface P and generating surface T of the cutting tool.

Before discussing these important issues, a few more particular issues must be discussed. The first is related to the possibility of implementation of the indicatrix of conformity for identification of the actual type of contact of the smooth, regular surfaces P and T. The second issue is related to the impact of accuracy of the calculations on the desired parameters of the indicatrix of conformity $Cnf_R(P/T)$ at a point of contact of two smooth regular surfaces P and T in the first order of tangency.

4.7.1 Possibility of Implementing the Indicatrix of Conformity $Cnf_R(P/T)$ for the Identification of a Type of Contact of Surfaces P and T

A smooth regular part surface P and generating surface T of the cutting tool can make contact

- At a point
- Along a line (that is, along the characteristic curve E)
- Over a surface patch

Features in the shape and in the parameters of the indicatrix of conformity $Cnf_R(P/T)$ at a point of contact of a part surface P and generating surface T of the cutting tool can be interpreted as the result of the actual type of contact of surfaces P and T.

It follows from the analysis of Equation 4.87 that special features in the shape and in the parameters of the indicatrix of conformity $Cnf_R(P/T)$ are given in every type of contact of surfaces P and T. For example, when the surfaces P and T make contact:

(a) At point K (Figure 4.28a), the minimal diameter $d_{cnf}^{\min}$ of the indicatrix of conformity $Cnf_R(P/T)$ at a point of contact of surfaces P and T (as well as all other diameters of this characteristic curve) is always of positive value ($d_{cnf}^{\min} > 0$).

(b) Along characteristic curve E (Figure 4.28b), the minimal diameter $d_{cnf}^{\min}$ of the indicatrix of conformity $Cnf_R(P/T)$ at all points of contact of surfaces P and T is always identical to zero ($d_{cnf}^{\min} \equiv 0$) while all other diameters of this characteristic curve are of positive values ($d_{cnf} > 0$).

(c) Over a surface patch (Figure 4.28c), then the indicatrix of conformity of surfaces P and T shrinks to a point, which coincides with the point K within the patch of contact of surfaces P and T.

The examples depicted in Figure 4.28 are worked out for the cases of surfaces P and T contact, when in the local vicinity of point K both the surfaces P and T are smooth regular surfaces of saddled type. The same is observed

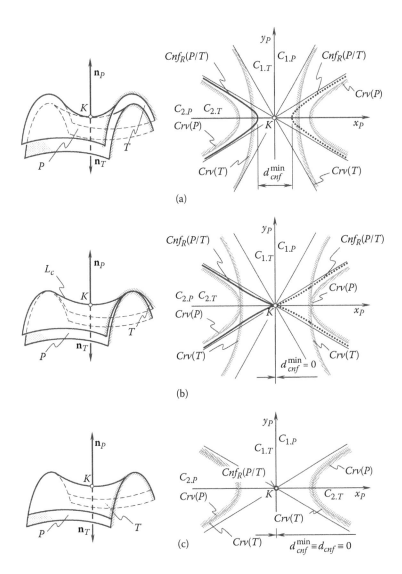

FIGURE 4.28
A correspondence between the shape of the indicatrix of conformity $Cnf_R(P/T)$ at a point of contact of a smooth regular part surface P and generating surface T of the cutting tool and a type of contact between the surfaces P and T: (a) point contact, (b) line contact, and (c) surface-to-surface contact. (Reprinted from *Mathematical and Computer Modeling*, 39, Radzevich, S.P., Mathematical modeling of contact of two surfaces in the first order of tangency, 1083–1112, Copyright 2004, with permission from Elsevier.)

for all other types of local patches of a part surface P and generating surface T of the cutting tool.

In the cases when the surfaces P and T intersect one another—that is, they interfere into each other as illustrated in Figure 4.21, the minimal diameter of the indicatrix of conformity $Cnf_R(P/T)$ is always of negative value ($d_{cnf}^{min} < 0$).

It is important to make a distinction between partial and full interference of a part surface P and generating surface T of the cutting tool in the differential vicinity of the contact point K.

For instance, in the differential vicinity of contact point K, a convex elliptic local patch of the cutting tool surface T can partially intersect a hyperbolic local patch of the part surface P as illustrated in Figure 4.21a. In this case, the minimal diameter d_{cnf}^{min} of the indicatrix of conformity $Cnf_R(P/T)$ is of negative value ($d_{cnf}^{min} < 0$). Because of this, the indicatrix of conformity $Cnf_R(P/T)$ at a point of contact of surfaces P and T not only intersects itself, but the intersection of each of its branches also occurs. Varying the angular parameter φ within the interval $0 \le \varphi \le \pi$, positive ($d_{cnf} > 0$) as well as negative ($d_{cnf} < 0$) values of current diameter d_{cnf} of the characteristic curve $Cnf_R(P/T)$ are observed.

In another example, the differential vicinity of contact point K, the local patch of generating surface T of the cutting tool interferes into a local patch of the part surface P totally as shown for two hyperbolic local patches (Figure 4.21b) and for two parabolic (Figure 4.21c) local patches of surfaces P and T. For cases of total local interference of surfaces P and T, the minimal diameter d_{cnf}^{min} of the indicatrix of conformity $Cnf_R(P/T)$ is always negative ($d_{cnf}^{min} < 0$); all other diameters d_{cnf} of this characteristic curve are also of negative value regardless of the actual value of the angular parameter φ.

The examples illustrated in Figure 4.21 reveal that every type of contact of a part surface P and generating surface T of the form-cutting tool feature important peculiarities of the shape and parameters of the indicatrix of conformity $Cnf_R(P/T)$ at a point of contact of surfaces P and T. The shape and the parameters of the characteristic curve $Cnf_R(P/T)$ uniquely follow the actual type of contact of surfaces P and T and are completely determined by the actual type of contact of the surfaces. In this concern, it is natural to assume that the inverse statement could be also true. The question can be formulated as follows:

> Are the features of shape and parameters of the indicatrix of conformity $Cnf_R(P/T)$ necessary and sufficient for making a conclusion regarding the type of surfaces P and T contact: contact at a point, along a characteristic curve E, or over a surface patch?

In another words, could the value and sign of the minimal diameter d_{cnf}^{min} of the indicatrix of conformity and features of its shape be implemented as a criterion for the determination uniquely the actual type of contact of surfaces P and T?

The investigation in detail of this particular subproblem a distinction the following conclusion:

Conclusion 4.5

The actual value and sign of minimal diameter d_{cnf}^{min} of the indicatrix of conformity at a point of contact of a part surface P and generating surface T of the cutting tool as well as features of it shape cannot be implemented as a sufficient criterion for determining uniquely the actual type of contact of smooth, regular surfaces P and T. ∎

The positive value of the minimal diameter d_{cnf}^{min} of the indicatrix of conformity $Cnf_R(P/T)$ at a point of contact of surfaces P and T (i.e., $d_{cnf}^{min} > 0$) is a sufficient but not necessary requirement for contact of surfaces P and T at a point.

The indicatrix of conformity $Cnf_R(P/T)$, that is shrunk to the point K, is not sufficient to identify the type of contact of surfaces P and T over a surface patch. However, if the surfaces P and T are congruent to each other within a certain surface patch, then the indicatrix of conformity shrinks to a point that coincides with the contact point K. The inverse statement is not correct. In the event the indicatrix of conformity $Cnf_R(P/T)$ shrinks to a point, then the surfaces P and T can be congruent to one another only locally. Thus, if the indicatrix of conformity $Cnf_R(P/T)$ shrinks to a point K, this indicates only necessary but not sufficient condition of surfaces P and T contact over a certain surface patch. In the case under consideration, surfaces P and T can make contact along a characteristic curve E and at a point K as well.

If the minimal diameter d_{cnf}^{min} of the indicatrix of conformity $Cnf_R(P/T)$ at a point of contact of surfaces P and T is equal to zero, then this does not necessarily mean that the surfaces P and T make contact along the characteristic curve E. This requirement is only necessary but not sufficient for the line contact of surfaces. In the event $d_{cnf}^{min} = 0$, the surfaces P and T can make contact at a point.

As follows from the discussion above, if part surface P and generating surface T of the cutting tool make contact along a characteristic curve E, then the direction along the minimal diameter d_{cnf}^{min} aligns with the tangent line to the characteristic curve E at the contact point K. This issue is important for the theory of part surface generation because it follows directly from the statement made above, according to which in differential vicinity of the contact point K the direction, along which the minimal diameter d_{cnf}^{min} can be measured, is aligned with the direction, along which degree of conformity of surfaces P and T gets the maximal value. Therefore, at the contact point K, the direction along the minimal diameter d_{cnf}^{min} and the direction that is tangent to the characteristic curve E align to one another. For this reason, the contact point K is a point of tangency:

(a) Of the straight line that aligns with the direction in which the minimal diameter d_{cnf}^{min} is measuring

(b) Of characteristic curve E

The statements above are also true for the inverse indicatrix of conformity $Cnf_k(P/T)$ at a point of contact of the part surface P and generating surface T of the cutting tool.

It is of critical importance to stress here that some of the conclusions, similar to those listed above, could be made based on the known approaches. This indirectly proves that the proposed novel approach for analytical description of the contact geometry of two smooth, regular surfaces in the first order of tangency is in perfect agreement with the earlier developed approaches in the field. In cases, the contacting surfaces are of simple geometry, then the implementation of the discussed approach as well as implementation of known approaches return identical results. (It is assumed here that the principal directions at a point of contact of the surfaces align to one another, respectively.) If the contacting surfaces are either of complex geometry or the principal directions at a point of contact of the surfaces do not align to one another, then the implementation only of the indicatrix of conformity (of both types) or of the relative indicatrix (of both types), and not of any other characteristic curve, ensures correct results of calculation of the contact geometry of two smooth, regular surfaces P and T.

4.7.2 Impact of Accuracy of the Calculations on the Desired Parameters of the Indicatrices of Conformity $Cnf_R(P/T)$

All the computations performed by the NC system of a numerically controlled machine are performed with certain errors of the computations. No computations by NC system are performed with zero error. Errors of the calculations are inevitable for many reasons.

The accuracy of the computations affects the desired parameters of the indicatrix of conformity $Cnf_R(P/T)$ at a point of contact of the part surface P and generating surface T of the cutting tool. For a predetermined error of the computations, the favorable parameters of the characteristic curve $Cnf_R(P/T)$ can be calculated. As is known, the characteristic curve $Cnf_R(P/T)$ is a function of parameters of two surfaces P and T as well as of their relative orientation. The part surface P to be machined is always given and it cannot be changed, except of the scenario illustrated in Figure 2.3. However, the parameters of generating surface T of the cutting tool and configuration of the surface T in relation to P are under control of the designer. It is possible to change both the geometry of the surface T as well as it orientation relative to the part surface P. All these alterations affect the parameters of the characteristic curve $Cnf_R(P/T)$. It is possible to compute such parameters of the indicatrix of conformity (say, it is possible to compute such design parameters of

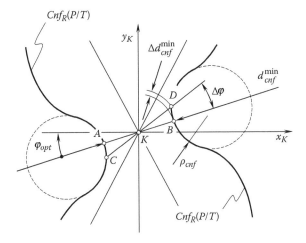

FIGURE 4.29
A local vicinity of the indicatrix of conformity $Cnf_R(P/T)$ at a point of contact K between the part surface P and generating surface T of the cutting tool.

generating surface T of the cutting tool and its configuration) for which the indicatrix of conformity is less vulnerable to the errors of the computations.

Figure 4.29 illustrates a portion of an indicatrix of conformity $Cnf_R(P/T)$ for a certain type of contact of the part surface P and generating surface T of the cutting tool. The actual arc of the indicatrix of conformity within the direction through points A and B along which the minimal diameter d_{cnf}^{min} of the indicatrix of conformity is measured is approximated by a circular arc. The radius of the circular arc $\rho_{r.cnf}$ is equal to the radius of curvature of the characteristic curve $Cnf_R(P/T)$ at the point B. The radius can be calculated from the equation

$$\rho_{r.cnf}(\varphi, \mu) = \frac{\left[r_{cnf}^2 + \left(\dfrac{\partial r_{cnf}}{\partial \varphi} \right)^2 \right]^{\frac{3}{2}}}{r_{cnf}^2 + 2\left(\dfrac{\partial r_{cnf}}{\partial \varphi} \right)^2 - r_{cnf} \cdot \dfrac{\partial r_{cnf}^2}{\partial^2 \varphi}} \tag{4.149}$$

As follows from the analysis of Figure 4.29, the error Δd_{cnf}^{min} of the computation of the minimal diameter d_{cnf}^{min} causes the deviation $\Delta\varphi$ of the direction along which the minimal diameter d_{cnf}^{min} of the indicatrix of conformity $Cnf_R(P/T)$ is measured. The deviation $\Delta\varphi$ can be computed from the equation

$$\Delta\varphi = \cos^{-1}\left(\frac{0.25 \cdot (d_{cnf}^{min} + \Delta d_{cnf}^{min})^2 + (0.5 \cdot d_{cnf}^{min} + \rho_{r.cnf})^2}{2 \cdot (0.5 \cdot d_{cnf}^{min} + \rho_{r.cnf}) \cdot (d_{cnf}^{min} + \Delta d_{cnf}^{min})} \right) \tag{4.150}$$

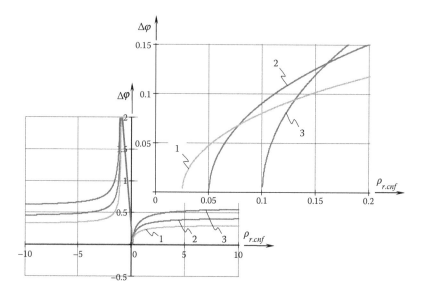

FIGURE 4.30
Illustration of the impact of the errors of the computations on the direction of the maximum degree of conformity of generating surface T of the cutting tool to smooth regular part surface P at a contact point K.

An example of the function $\Delta\varphi = \Delta\varphi(\rho_{r.cnf})$ is plotted in Figure 4.30. It is important to point out here that

- First, the optimal value of the radius $\rho_{r.cnf}$ is not equal to zero.
- Second, the optimal value of the radius $\rho_{r.cnf}$ depends on the geometry of the generating surface T of the cutting tool, and on the tool configuration in relation to the part surface P to be machined.

This means that for the purposes of proper design of the form-cutting tool and to properly configure the cutting tool in relation to the part surface P, minimization of the impact of the errors of the computations on the process of the optimal generation of the part surface P could be helpful.

Similar computations can be performed with respect to favorable parameters of the characteristic curves of other types.

The consideration above can be employed for the enhancement of the classification of types of contact of a part surface P and generating surface T of the cutting tool.

4.7.3 Classification of Possible Types of Contact of a Part Surface P and Generating Surface T of the Cutting Tool

Classification of possible types of contact of two smooth, regular surfaces in the first order of tangency is important for implementation of the methods developed in the theory of part surface generation. The above-performed analysis makes possible the development of a scientific classification of all possible types of contact of a smooth regular part surface P and generating surface T of the form-cutting tool.

The following statement is postulated when developing the classification:

> If a surface P to be machined and machining surface of the form-cutting tool T make contact with one another, then there is at least one point of their contact.

A sculptured part surface P and generating surface T of the form-cutting tool can make contact

(a) At a point K (or at a certain number of distinct points K_i)
(b) Along a characteristic E (or along several characteristics E_i)
(c) Within a certain surface patch

No other types of surface contact are feasible. This is mostly due to fundamental properties of the three-dimensional space we live in.

The following three type of surface contact are evident. They are

(a) Point contact
(b) Line contact
(c) Surface-to-surface contact of two smooth, regular surfaces P and T

These three types of surface contact are not only evident, but they are trivial. It is the right point to turn the readers' attention here to the following:

1. Consider the type of point contact of a part surface P and generating surface T of the cutting tool. When surfaces P and T make contact at a point, three types of point contact can be distinguished.

   1.1. In the first type of point contact of two surfaces, there are no normal plane sections through the contact point K of surfaces P and T, at which normal curvatures k_P and k_T are of the same magnitude and opposite sign. The equality $k_P = -k_T$ is observed in no plane section of surfaces P and T through the common unit normal vector $\mathbf{n}_P$. This type of surface contact is referred to as *true point contact* of the surfaces. If two surfaces make true point contact, then the expression $k_P(\varphi) \neq -k_T(\varphi, \mu)$ is valid for any value of the angle φ.

   1.2. In the second type of point contact of two surfaces, there is only one normal plane section through the contact point K of surfaces P and T, at which normal curvatures k_P and k_T are of the same magnitude and opposite sign. Thus, the equality $k_P = -k_T$ is observed in a single plane section of surfaces P and T through the common unit normal vector $\mathbf{n}_P$. In this plane section, surfaces P and T make contact along an infinitely short arc. Torsion of surfaces P and T along the infinitely short arc of contact is identical to one another—that is, geodesic (relative) torsions of the curves are equal $\tau_{g.P} \equiv \tau_{g.T}$. This type of surface contact is referred to as *local line contact* of the surfaces. When two surfaces are in local line contact, then the expression $k_P(\varphi) = -k_T(\varphi, \mu)$ is valid for a certain value of φ.

As long as the second (and not higher) derivatives are considered, then the local line contact of the surfaces is identical to the true line contact of the surfaces.

1.3. In the second type of point contact of two surfaces, normal curvatures k_P and k_T are of the same magnitude and opposite sign in all normal plane sections through the contact point K of surfaces P and T. Thus, the identity $k_P \equiv -k_T$ is observed in all plane sections of surfaces P and T through the common unit normal vector $\mathbf{n}_P$. In the case under consideration, the surfaces P and T make contact within the infinitely small area. This type of surface contact is referred to as *local surface-to-surface contact* of the first type.

As long as the second (and not higher) derivatives are considered, the local surface-to-surface contact of the first type is identical to true surface-to-surface contact of the surfaces.

2. Consider the type of line of contact of a part surface P and generating surface T of the form-cutting tool. When a sculptured part surface P and generating surface T of the form-cutting tool make contact along a characteristic, then two types of line contact can be distinguished.

2.1. In the first type of line contact of two surfaces, there is the only normal plane section through the contact point K of surfaces P and T, at which normal curvatures k_P and k_T are of the same magnitude and opposite sign. This normal plane section is congruent to the osculate plane to the characteristic E at K. Thus, the equality $k_P = -k_T$ is observed in a single plane section through the common unit normal vector $\mathbf{n}_P$ to the surfaces P and T. The torsion of surfaces P and T along the arc of contact is identical to one another—that is, geodesic (relative) torsions of the curves are equal $\tau_{g.P} \equiv \tau_{g.T}$. This type of surface contact is referred to as *true line contact* of the surfaces. When two surfaces are in true line contact, the expression $k_P(\varphi) = -k_T(\varphi, \mu)$ is valid for a certain value of φ.

The degree of rotation of the tangent plane to the surface P about the curve E is determined by geodesic (relative) torsion $\tau_{g.P}$ of the characteristic curve E. It is assumed that the characteristic curve E and the part surface P are regular, and the degree of rotation of the tangent plane is a function of length s of the characteristic curve E. Relative torsion can be defined by a point on the characteristic curve E and by a direction on the surface P. It is equal to the torsion of the geodesic curve in that same direction

$$\tau_{g.P} = \left[\frac{d\mathbf{r}_E}{ds} \times \mathbf{n}_P \cdot \frac{d\mathbf{n}_P}{ds} \right] = \tau_E + \frac{d\phi}{ds} = (k_{1.P} - k_{2.P}) \sin \kappa \cos \kappa \qquad (4.151)$$

where $\mathbf{r}_E$ is the position vector of a point of the characteristic curve E, $\mathbf{n}_P$ is the unit normal vector to the part surface P, τ_E is the regular torsion of the characteristic curve E, ϕ is the angle that make the osculating plane to E and the tangent plane to the part surface P, $k_{1.P}$ and $k_{2.P}$ are the principal curvatures at the point K of the part surface P, and κ is the angle that make the tangent to E at K and the first principal direction $\mathbf{t}_{1.P}$ of the part surface P

2.2. In the second type of line contact of two surfaces, normal curvatures k_P and k_T are of the same magnitude and opposite sign in all normal plane sections through the contact point K of surfaces P and T. Thus, the identity $k_P \equiv -k_T$ is observed in all plane sections through the common unit normal vector $\mathbf{n}_P$ to the surfaces P and T. In the case under consideration, the surfaces P and T make contact within infinitely small area. This type of surface contact is referred to as *local surface-to-surface contact* of surfaces of the second type.

As long as the second (and not higher) derivatives are considered, the local surface-to-surface contact of the second type is identical to the true surface-to-surface contact of the surfaces. In the differential vicinity of a contact point K, the sculptured part surface P and generating surface T of the form-cutting tool are locally congruent to one another.

3. Consider the surface-to-surface type of contact of a part surface P and generating surface T of the cutting tool. When surfaces P and T make contact within a surface patch, only one type of contact can be recognized.

3.1. In case of surface-to-surface type of contact of two surfaces, normal curvatures k_P and k_T are of the same magnitude and opposite sign in all normal plane sections through the contact point K of surfaces P and T. Thus, the identity $k_P \equiv -k_T$ is observed in all plane sections through the common unit normal vector $\mathbf{n}_P$ to surfaces P and T. In the case under consideration, surfaces P and T make contact within a surface patch. This type of surface contact is referred to as the *true surface-to-surface contact* of surfaces.

Without going into details, just to mention here that for the purposes of efficient part surface generation in a machining operation, it is desired to maintain that type of contact of the part surface P and generating surface T of the form-cutting tool, which features the highest possible degree of conformity of the generating surface T of the cutting tool to the part surface P.

Actually, when machining a part surface, deviations in the cutting tool relative location and orientation with respect to the part surface P are always observed. The deviations in the configuration of the cutting tool are inevitable in nature. Because of the deviations, the desired locally extremal type of contact* of surfaces P and T is substituted by another type of contact of surfaces P and T. Such substitution can be achieved with introduction of the precalculated deviations of the cutting tool principal radii of curvature $R_{1.T}$ and $R_{2.T}$. If the precalculated deviations are *small*, then instead of the desired locally extremal types of contact of the surfaces, the so-called quasi type of contact of a part surface P and generating surface T of the cutting tool occurs. Several types of quasi-contact of two surfaces in the first order of tangency are distinguished below. They are

(a) quasi-line contact of surfaces P and T
(b) quasi-surface-to-surface of the first type
(c) quasi-surface-to-surface of the second type contact of surfaces P and T

The required precalculated values of *small* deviations of the actual normal curvatures from their initially specified values can be determined on the premise of the following consideration. When the maximal deviations in the actual cutting-tool configuration (location and orientation of the cutting tool relative to the part surface being machined) occur, the degree of conformity of the generating surface T with respect to the part surface P must not exceed the degree of their conformity in one of locally extremal type of surface contact. When the actual deviations of the cutting tool configuration do not exceed the corresponding tolerances, then one of the feasible types of quasi-contact of a part surface P and generating surface T of the cutting tool is observed.

Evidently, the larger the deviations in the cutting tool configuration the larger the precalculated corrections in normal curvature of the generating surface of the cutting tool and vice versa.

In the ideal case, when there are no deviations in the cutting tool configuration, it is recommended to assign normal curvatures of the values that enable one of the locally extremal types of contact[†] of the part surface P and

---

* Local extremal type of contact of two smooth regular surfaces stands for one of the following types of contact, namely, (a) local line contact or (b) local surface-to-surface contact (of the first type), and finally, (c) local surface-to-surface contact (of the second type) of a part surface P and generating surface T of the cutting tool.

† When a part surface P and generating surface T of the cutting tool are in a locally extremal type of contact [when they make (a) local line contact, (b) local surface-to-surface of the first type contact, or (c) local surface-to-surface of the second type of contact], then the equality to zero of the minimum diameter d_{cnf}^{min} of the indicatrix of conformity, Cnf_R (P/T), at a point of contact of the surfaces P and T does not indicate whether or not the interacting surfaces P and T interfere with each other. In this particular case of surface contact, a conclusion can be made based (a) on the comparison of intensity of change of curvatures of the surfaces P and T and (b) on the comparison of torsions of the interacting of the surfaces P and T.

generating surface T of the cutting tool. Local surface-to-surface contact of the second type is the preferred type of contact of surfaces P and T. The local surface-to-surface contact of the second type yields the minimal value of the radius $r_{cnf}^{min} = 0$ of the indicatrix of conformity $Cnf_R(P/T)$ at the point of contact of the part surface P and generating surface T of the cutting tool.

When machining an actual part surface, deviations in the cutting tool configuration are inevitable in nature. The pure surface-to-surface type of contact of the surfaces (when the equality $r_{cnf}^{min} = 0$ is observed) is not feasible at all. Due to the deviations in the cutting tool configuration, maintaining of the pure surface-to-surface contact of surfaces P and T would inevitably result in interference of the cutting tool beneath the part surface P. Therefore, it is recommended to maintain a type of quasi-surface-to-surface contact of the second type instead of maintaining a pure surface-to-surface contact. A quasi-surface-to-surface contact of surfaces P and T makes possible avoiding interference of generating surface T of the cutting tool within the interior of the part surface P to be machined. Moreover, the minimal radius r_{cnf}^{min} of the characteristic curve $Cnf_R(P/T)$ could be as close to zero as possible ($r_{cnf}^{min} > 0$, $r_{cnf}^{min} \to 0$, and $r_{cnf}^{min} \neq 0$).

A quasi-contact of a part surface P and generating surface T of the cutting tool is observed only when deviations of the cutting tool configuration are incorporated into consideration.*

Three different types of quasi-contact between a part surface P to be machined and generating surface T of the form-cutting tool are recognized. Each of them is associated with the corresponding locally extremal type of contact of surfaces P and T.

A definition of *quasi-line type of contact* of two smooth, regular surfaces P and T in the first order of tangency can be drawn up based on the similarity between *quasi-line type of contact* and between *local line type of contact* of the surfaces:

Definition 4.1

Quasi-line type of contact of a part surface P and generating surface T of the cutting tool is a type of slightly "corrupted" local line type of contact, for which the actual degree of conformity d_{cnf}^{min} of the interacting surfaces is either equal to or exceeds the limiting degree of conformity $[d_{cnf}^{min}]$ of the contacting surfaces P and T. ∎

A definition of *quasi-surface-to-surface (of the first type) contact* of a part surface P and generating surface T of the cutting tool can be drawn up based on the similarity between quasi-surface-to-surface (of the first type) contact

---

* It is instructive to point out here that the higher the degree of conformity of generating surface T of the cutting tool to a part surface P, the lower the permissible displacement in relation to each other. This entails tighter manufacturing tolerances for part surface P along with severe constraint on the displacements of the part and of the cutting in relation to each other under the load.

and between local surface (of the first type) contact of two smooth regular surfaces:

Definition 4.2

Quasi-surface-to-surface (of the first type) contact of a part surface P and generating surface T of the cutting tool is a type of slightly "corrupted" local surface-to-surface (of the first type) contact, for which the actual degree of conformity d_{cnf}^{min} of the interacting surfaces is either equal to or exceeds the limiting degree of conformity $[d_{cnf}^{min}]$ of the contacting surfaces P and T.   ■

Ultimately, a definition of *quasi-surface-to-surface (of the second type) contact* of a part surface P and generating surface T of the cutting tool can be drawn up based on the similarity between quasi-surface-to-surface (of the second type) contact and local surface-to-surface (of the second type) contact of two smooth regular surfaces:

Definition 4.3

Quasi-surface-to-surface (of the second type) contact of a part surface P and generating surface T of the cutting tool is a type of slightly "corrupted" local surface-to-surface (of the second type) contact, for which the actual degree of conformity d_{cnf}^{min} of the interacting surfaces is either equal to or exceeds the limiting degree of conformity $[d_{cnf}^{min}]$ of the contacting surfaces P and T.   ■

The difference between various type of quasi-contact of a part surface P and generating surface T of the form-cutting tool as well as the difference between the corresponding types of locally extremal contact of the surfaces can be recognized only under the limit values of the allowed deviations in the cutting tool configuration relative to the part surface P. In the event that actual deviations are below the tolerances, then various possible types of quasi-contact of the surfaces cannot be distinguished from other non-quasi types of contact. The only difference is in actual location of point K of contact of the surfaces. Due to the deviations, the contact point K is shifted from the original position to a certain other location.

There are only nine different types of contact of a part surface P and generating surface T of the cutting too. In addition to

   (i)   True point contact

  (ii)   True line contact

 (iii) True surface-to-surface contact

the following three locally extremal types of surface contact are distinguished as well:

(a) local line contact
(b) local surface-to-surface contact of the first type
(c) local surface-to-surface contact of the second type

Three types of quasi-contact of the surfaces are also possible. They are
listed immediately below:

(a) quasi-line contact
(b) quasi-surface-to-surface contact (of the first type)
(c) quasi-surface-to-surface contact (of the second type) of surfaces P
and T

Taking into consideration, that there are only ten different types of local
patches of smooth, regular part surfaces P and generating surfaces T of the
form-cutting tool (see Chapter 1, Figure 1.16), each of the nine types of surface contact can be represented in more detail. For this purpose, a square
morphological matrix of size $10 \times 10 = 100$ can be composed. All possible
combinations of surface contact are covered by the morphological matrix.
One axis of the morphological matrix is represented by 10 types of local
patches of the part surface P, whereas its other axis is represented by 10 types
of local patches of generating surface T of the cutting tool. The morphological matrix contains 100 different combinations of the local patches of a part
surface P and generating surface T of the cutting tool. Only

$$\sum_{m=1}^{9} C_9^m = \frac{9!}{m!(9-m)!} = \frac{100-10}{2} + 10 = 55 \qquad (4.152)$$

of them are necessary to be investigated.

The performed analysis reveals that the following types of contact of part
surface P and generating surface T of the cutting tool are feasible:

(a) 29 types of the true point contact*
(b) 23 types of the true line contact
(c) 6 types of the true surface-to-surface contact
(d) 20 types of the local line contact
(e) 7 types of the local surface-to-surface (of the first type) contact

---

* Results of a more detailed investigation of all possible types of true point contact of two
smooth, regular surfaces in the first order of tangency can be found in Reference [117].

(f) 8 types of the local surface-to-surface (of the second type) contact

(g) 20 types of the quasi-line contact

(h) 7 types of the quasi-surface-to-surface (of the first type) contact

(i) 8 types of the quasi-surface-to-surface (of the second type) contact

This means that only $29 + 23 + 6 + 20 + 7 + 8 + 20 + 7 + 8 = 128$ types of contact of two smooth, regular surfaces P and T are possible in nature. For some types of surface contact, no constraints are imposed on the actual value of the angle μ of local relative orientation of surfaces P and T. For the other types of surface contact a corresponding interval of the allowed value of the angle μ: $[\mu_{min}] \leq \mu \leq [\mu_{max}]$ can be specified. For particular cases of surface contact, the only feasible value $\mu = [\mu]$ is allowed.

On the premise of the above-performed analysis, a scientific classification of all possible types of contact of a part surface P and generating surface T of the cutting tool is developed (Figure 4.31).

As shown in Figure 4.31, the classification is a potentially complete one. It can be further developed and enhanced. The classification can be used for the analysis and qualitative evaluation of the degree of effectiveness of a machining operation. The classification indicates perfect correlation with the earlier developed classification [117].

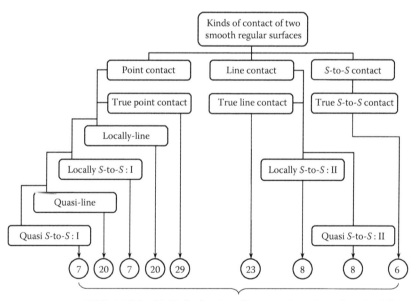

128 (in total) feasible kinds of contact of two smooth regular surfaces P and T

FIGURE 4.31
There are as many as 128 (in total) different types of contact of two smooth regular surfaces P and T (S-to-S stands for "surface-to-surface" type of contact of surfaces P and T).

The replacement of the true point contact of the part surface P and generating surface T of the form-cutting tool with their local line contact and further with the local surface-to-surface of the first and/or of the second types, and finally with the true surface-to-surface contact results in significant alterations of the part surface P generation. Only local extremal types of contact of surfaces P and T are considered here. If deviations in the cutting tool configuration are considering, then the above mentioned local extremal types of surface contact require being replaced with the corresponding quasi-types of contact of surfaces P and T.

To achieve the highest possible productivity of machining of a part surface P, the true surface-to-surface contact of the part surface P and generating surface T of the cutting tool should be maintained. Under such a scenario, all portions of the part surface P are machined in one instant. However, in those cases, large-scale surfaces P cannot be machined. Meanwhile, the machining of the part surface P when the true point contact of the surfaces is maintained is the least efficient one. In this case, the generation of every strip on the part surface P occurs in time.

Depending on the type of contact of the part surface P and generation surface T of the cutting tool that is maintained when machining the part surface P, all possible types of contact of the surface can be arranged in the following order (from the least efficient to the most efficient):

(a) True point contact

(b) Local line and/or quasi-line contact

(c) Local surface-to-surface contact (of the first type) and/or quasi-surface-surface-to-surface contact (of the first type)

(d) True-line contact

(e) Local surface-to-surface contact (of the second type) and/or quasi-surface-to-surface contact (of the second type)

(f) True-surface-to-surface contact of surfaces P and T

The productivity of machining of the part surface P is increased from item (a) to item (f).

However, the productivity of part surface machining is not the only criterion of effectiveness of a machining operation. The agility of the machining operation is another criterion of interest. The agility of the machining operation can be evaluated by the versatility of part surfaces P that can be machined with a given cutting tool. From this standpoint, the most effective is a machining operation under which the true point contact of surfaces P and T is continuously maintained. The least effective is a machining operation under which the true surface-to-surface contact of surfaces P and T is maintained. The methods of part surface machining under which the surfaces P and T contact with other types is maintained can be arranged in the order that is converse to the above—that is, from item (f) to item (a).

Maintaining the true point contact of the surface P and T results not only in the highest possible agility of a machining operation. The true point contact can also be regarded as the most general type of surface contact. When true point contact of the surfaces is observed, the cutting tool can perform five-parametric motion in relation to the part surface P. While under the true surface-to-surface contact, the cutting tool is capable of performing no motion relative the work. A relative motion of surfaces P and T is feasible only as an exclusion, say, when the surfaces P and T yield sliding over themselves. In those cases, an enveloping surface P to consecutive positions of the generating surface T of the cutting tool that is traveling relative to the work is congruent to P. Generally speaking, under such a scenario, surfaces P and T are capable of performing a single-parametric motion and not higher than three-parametric motion (see Chapter 2, Section 2.4).

5

Profiling of Form-Cutting Tools of Optimal Design for Machining a Given Part Surface

Practical machining of a part surface P is performed by means of cutting tools of an appropriate design. The stock removal and generation of the part surface P are the two major functions of the cutting tool to be applied. A high-performance cutting tool is capable of efficient performance of both of these functions. This section of the book is primarily focused on the second function of the cutting tool, namely on generating of a given part surface P.

Determination of the generating surface T of the form-cutting tool is the key point when designing a cutting tool for machining a given part surface P. As shown below, the performance of cutting process is significantly affected by the shape and parameters of cutting tool surface T. The cutting edges of a precision cutting tool are located within the generating surface of the tool. This makes it clear that prior to developing a practical design of the cutting tool, the optimal parameters of the generating surface of the cutting tool must be determined.

In this chapter, profiling of the form-cutting tools for all possible methods of part surface machining is considered. The consideration begins from the theory of profiling of the cutting tools for machining sculptured part surfaces on a multiaxis numerical control (NC) machine. This subject represents the most complex case in the theory of profiling of cutting tools.

5.1 Profiling of the Form-Cutting Tools for Sculptured Surface Machining

The problem of profiling of the form-cutting tool for machining of a sculptured part surface on a multiaxis NC machine has not yet been investigated. Selecting a certain cutting tool among several available designs is often recommended instead of profiling the form-cutting tool of optimal design. The selection of the cutting tool is usually based on minimizing machining time, reducing scallop height, and so forth. This yields a conclusion that a robust mathematical method for designing the optimal form-cutting tool for the maximally productive machining of a given sculptured part surface on a multiaxis NC machine is needed.

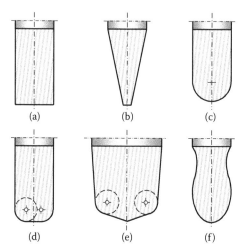

FIGURE 5.1
Examples of milling cutters of conventional design for the machining of sculptured part surfaces on a multiaxis numerical control machine: cylindrical (a), conical (b), ball-end (c), filleted end (d), and form-shaped (e, f).

5.1.1 Preliminary Remarks

Many advanced sources are devoted to the investigation of sculptured part surface generation on a multiaxis *NC* machines. Without going into a detailed review of the previous publications in the field, monographs by Amirouch [1], Chang and Melkanoff [8], Choi and Jerard [9], Marciniak [39], and some others should be mentioned. Unfortunately, the problem of profiling the form-cutting tools for sculptured surface machining has not yet been investigated. Most often, the generation of sculptured surfaces with the milling cutters of conventional designs is considered. The commonly used designs of milling cutters are shown in Figure 5.1.

The following terms (some of which are not new) are introduced below to avoid ambiguities in later discussions*:

Definition 5.1

*Sculptured part surface **P** is a smooth, regular surface, the major parameters of local topology at a point of which are not identical to the corresponding parameters of local topology of any other infinitesimally close point of the surface.*   ∎

It is instructive to point out here that the sculptured part surface *P* does not allow for sliding "over itself."

---

* In fact, our terminology draws inspiration mostly from the *Theory for Mechanisms and Machines* and from the *Theory of Conjugate Surfaces*.

When machining a sculptured part surface, the cutting tool is rotated about its axis of rotation and is translated relative to the sculptured part surface P. A certain surface is generated by cutting edges of the cutting tool when the cutting tool is rotated or when it is performing relative motion of another type. The surface that is represented by consecutive positions of moving cutting edges of the cutting tool is referred to as the *generating surface of the cutting tool* [116,117,136]:

Definition 5.2

Generating surface T of the cutting tool is a surface that can be in permanent tangency to the part surface P being machined.   ■

For a specified sculptured part surface P, there is an infinite number of surfaces T that satisfy Definition 5.2. The use of all of the generating surfaces of the cutting tool mandatorily satisfies Shishkov's equation of contact $\mathbf{n}_p \cdot \mathbf{V}_\Sigma = 0$ (see Chapter 2).

The unit normal vector $\mathbf{n}_p$ to the sculptured part surface P is uniquely determined at a given surface point by shape of the surface. The number of feasible vectors $\mathbf{V}_\Sigma$ of relative motion of the generating surface T relative to the part is equal to infinity: Any and all velocity vectors $\mathbf{V}_\Sigma$ within the common tangent plane satisfy the equation of contact $\mathbf{n}_p \cdot \mathbf{V}_\Sigma = 0$. It is natural to assume that not all of the velocity vectors $\mathbf{V}_\Sigma$ are equivalent to one another apart from the perspective of efficient part surface generation and that some of them could be more favorable. Moreover, it is natural to assume that there exists an optimal direction for the velocity vector $\mathbf{V}_\Sigma$ within the common tangent plane for which efficiency of the part surface machining reaches it maximum value. Because of this, the problem of profiling of a form-cutting tool for machining a specified sculptured part surface on a multiaxis *NC* machine is indefinite in much. However, the indefiniteness is successfully overcome below. The uniquely determined generating surface T of the form-cutting tool is used in further steps of designing of an optimal form-cutting tool for machining a given part surface.*

---

* The procedure of designing a form-cutting tool usually begins from the determination of the generating surface of the cutting tool. This is a common practice. However, if a geometric structure of the part surface to be machined is inconsistent, another procedure is used as well. The operation of relieving of a hob clearance surface, cutting of bevel gears with spiral teeth, machining of non-involute gears of the first and second types are perfect examples of part surface machining when the geometric structure of the part surface to be machined is inconsistent. Under such a scenario, the generating surface of the cutting tool of an appropriate form is selected. Further, the actual shape and parameters of the machined part surface can be determined.

  It is important to keep in mind that the part surface to be machined is the primary element of the machining operation, on the premise that the determination of the optimal machining operation is possible. This includes both profiling of the optimal cutting tool and the optimal kinematics of machining operation. Otherwise, only an approximate solution to the problem of optimal surface generation is possible in this particular case.

In most cases of part surface generation, the generating surface T of the cutting tool does not exist physically. Usually, it is represented as the set of consecutive positions of the cutting edges in their motion relative to a stationary coordinate system, which is associated with the cutting tool itself.

In most practical cases, the generating surface T allows for sliding "over itself." The enveloping surface to consecutive positions of the surface T that is performing such a motion is congruent to the surface T itself.

When machining a part surface P, the generating surface T of the cutting tool is in permanent tangency to the sculptured part surface P.

For the simplification of programming of a machining operation, the *APT* cutting tool is proposed (Figure 5.2). Physically, the *APT* cutting tool does not exist. The generating surface T of the *APT* cutting tool consists of

a. A conical portion that has cone angle α
b. A conical portion that has cone angle β
c. A portion of the surface of a torus

The last is specified by the radius r of the generating circle and by the diameter d of the directing circle. Axial location of the torus surface with respect to the conical surfaces is specified by the parameter that is designated as f. For a certain combination of the parameters α, β, r, d, and f, the generating surface of the virtual *APT* cutting tool transforms to the generating surface T of the actual cutting tool. For example, assuming $\alpha = 0°$, $\beta = 0°$, and $r = 0$, one can come up with a generating surface T of the cylindrical milling cutter (Figure 5.1a). If $r = d/2$ and $\beta = 0°$, the generating surface of the virtual *APT* cutting

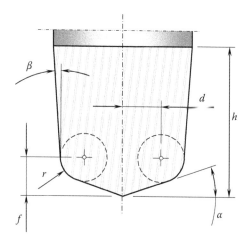

FIGURE 5.2
Major design parameters of the *APT* cutting tool. (Reprinted from *Computer-Aided Design*, 34, Radzevich, S.P., Conditions of proper sculptured surface machining, 727–740, Copyright 2002, with permission from Elsevier.)

tool reduces to generating surface *T* of the ball-end milling cutter (Figure 5.1c). Evidently, this series of elementary transformations of the *APT* cutting tool to the generating surface *T* of the actual cutting tool can be extended.

The major advantage of the *APT* cutting tool is as follows. If a computer program for controlling a multiaxis *NC* machine is developed for implementation of the *APT* cutting tool, this computer program can be easily adapted for implementation of a cutting tool of any other design.

Actually, the major developments in profiling of form-cutting tools for machining a given sculptured part surface on a multiaxis *NC* machine are limited to

a. The selection of an appropriate cutting tool among its available designs

b. The implementation of the *APT* cutting tool for the development of a computer program for controlling a multiaxis *NC* machine.

The above discussion makes clear the necessity of the development of a method for profiling of the most favorable form-cutting tool for machining a given sculptured part surface on a multiaxis *NC* machine.

5.1.2 On the Concept of Profiling of the Optimal Form-Cutting Tool

It is of critical importance to clarify from the very beginning what the term *"cutting tool of optimal design"* stands for. In the consideration below, the term *"cutting tool of optimal design"* means that the design parameters of a certain cutting tool are those whose implementation of the form-cutting tool ensures the required extremum (minimum/minimum) of the prespecified criterion of optimization. Maximal productivity of the part surface machining and minimal deviations of the actual part surface from the desired part surface are perfect examples of candidates for a criterion of optimization.

In the theory of part surface generation, only those criteria of optimization are applicable that could be expressed in terms of

- Geometry of the sculptured part surface *P*
- Geometry of generating surface *T* of the cutting tool
- Kinematics of the machining operation

The following example illustrates the actual meaning of the criterion of optimization in the sense of profiling the most favorable form-cutting tool for machining of a sculptured part surface.

Consider a trivial machining operation—a turning operation of an arbor on a lathe (Figure 5.3). When machining, the work is rotated about its axis of rotation with an angular velocity ω_P. The cutting tool is traveling along the work axis of rotation with a feed rate *S*. The feed rate *S* is of constant magnitude in the examples considered below. A stock *t* is removed in the turning operation.

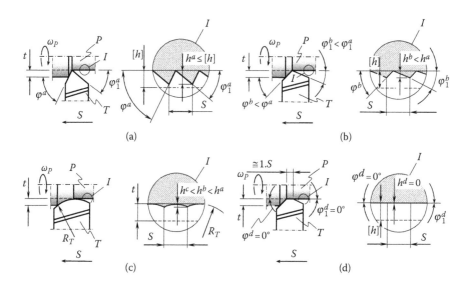

FIGURE 5.3
The machined surface of an arbor after machining on a lathe: on the concept of profiling of a form-cutting tool of optimal (that is, the most favorable) design for machining a given sculptured part surface.

The cusps on the machined part surface P are observed after machining the part by the cutter having the tool cutting edge angle φ^a and the tool minor (end) cutting edge angle φ_1^a (Figure 5.3a). The actual height h^a of the cusps must be less than the tolerance band $[h]$ for the accuracy of the machined part surface P. To satisfy the inequality $h^a \le [h]$, a corresponding relationship among parameters φ^a, φ_1^a, and S must be observed. Otherwise, the part cannot be machined to the part blueprint.

That same arbor can be machined with the cutter having the tool cutting edge angle $\varphi^b < \varphi^a$, and the tool minor cutting edge angle $\varphi_1^b < \varphi_1^a$ as schematically illustrated in Figure 5.3b. Cusps on the machined part surface P are inevitably observed. Elementary calculations of the actual cusp height h^b in this case reveal that the inequality $h^b < h^a$ is valid.

Further, that same part surface P can be machined by the cutter having the cutting edge that is shaped in the form of a circular arc of radius R (Figure 5.3c). The use of the cutter with the curvilinear cutting edge also results in cusps on the machined part surface P. However, if the radius R is chosen properly, the actual cusp height h^c can be smaller than h^b in the case illustrated in Figure 5.3b. In other words, the inequality $h^c < h^b$ could be observed.

Ultimately, let us consider the turning operation of the part surface P by means of the cutter that has an auxiliary cutting edge as shown in Figure 5.3d. The auxiliary cutting edge is parallel to the direction of the feed rate S. The length of the auxiliary cutting edge exceeds the distance that the cutter travels per one revolution of the work. Geometrical parameters of the auxiliary cutting edge can be specified by the tool cutting edge angle $\varphi^d = 0°$ and the tool minor (end) cutting edge angle $\varphi_1^d = 0°$. Under such a scenario, no cusps are observed on the machined part surface P.

The above consideration leads to the following conclusion:

Conclusion 5.1

An appropriate alteration to shape of the cutting edge of the cutter can make possible a reduction in deviations of the actual (machined) part surface with respect to the desired part surface.   ∎

This conclusion is of critical importance for further analysis.

It is the right point now to consider a more general example of part surface generation that supports the aforementioned conclusion.

Consider the generation of a sculptured part surface P by means of the form-cutting tool having arbitrarily shaped generation surface T. The intersection of the part surface P and the tool surface T by a plane through the unit normal vector $\mathbf{n}_P$ is shown in Figure 5.4. This plane section is perpendicular to the tool path on the part surface P at the current contact point K. In Figure 5.4, the width of the tool path is designated as S_T. In all the examples considered in Figure 5.4, the width S_T of the tool path remains identical to each other. Radius of normal curvature R_P at the contact point K on the part surface P also remains the same value. The scallop's height on the machined part surface P is designated as h_P.

The part surface P can be generated by the generating surface T^a of the cutting tool as schematically depicted in Figure 5.4a. The radius of curvature of the generating surface T^a is of a certain positive value $R_T^a > 0$. Because surfaces P and T^a make point contact, the desired part surface P is not generated, but an approximate surface is generated instead. Scallops are inevitably observed on the actually generated surface. For the prespecified width S_T of the tool path, the scallop height is equal to a certain value h_P^a. The scallop height must not exceed the tolerance $[h]$ for the accuracy of the machined part surface P, namely, the inequality $h_P^a \leq [h]$ must be satisfied.

That same part surface P can be generated by means of machining surface T^b of the cutting tool (Figure 5.4b). The radius of curvature of the generating surface T^b in this case exceeds the value of radius of curvature of the surface T^a in the previous case ($R_T^b > R_T^a$). Because surfaces P and T^b are in point contact, then the scallops on the generated part surface are always observed. If width S_T of the tool path is predetermined, the scallop height is of a certain value h_P^b. Under such a scenario, a certain reduction of the scallop height is observed ($h_P^b < h_P^a$). The scallop height reduction is because, in the differential vicinity of the contact point K, the generating surface T^b gets closer to the part surface P than surface T^a does. Locally, the generating surface T^b is more congruent to the part surface P than surface T^a, i.e., the degree of conformity

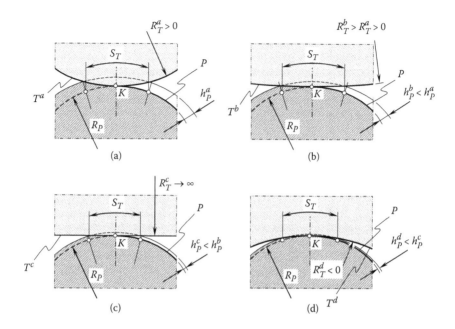

FIGURE 5.4
Examples of contact of a surface P and generating surface T of the form-cutting tool having different degree of conformity in a plane section through the unit vector $\mathbf{n}_p$ of the common perpendicular. (Reprinted from *Mathematical and Computer Modeling*, 36, Radzevich, S.P., $\mathbb{R}$-mapping based method for designing of form cutting tool for sculptured surface machining, 921–938, Copyright 2002, with permission from Elsevier.)

of the generating surface T^b to the part surface P is greater than that of the generating surface T^a to that same part surface P.

Further, that same part surface P can be generated by means of the generating surface T^c of the cutting tool (Figure 5.4c). At contact point K, surface T^c is flattened, and therefore, the radius of curvature R_T^c is equal to infinity ($R_T^c \to \infty$). That value of the radius of curvature R_T^c exceeds the value of the radius of curvature R_T^b. Again, surfaces P and T^c make contact at a point, and therefore scallops on the generated part surface are observed. Because the inequality $R_T^c > R_T^b$ is valid, then the scallop height h_p^c is smaller than the scallop height h_p^b. Reducton of the scallop height in this case is due to an increase of degree of conformity of the generating surface T^c of the tool to the part surface P compared to that that observed with respect to surfaces P and T^b.

Ultimately, let us consider generation of the part surface P by means of concave generating surface T^d of the cutting tool as illustrated Figure 5.4c. Radius of curvature R_T^d of the cutting tool surface T^d is of negative value ($R_T^d < 0$). In this case, the degree of conformity of the generating surface T^d of the cutting tool to the part surface P is the largest of all considered cases (Figure 5.4). Thus, the scallops of the smallest height h_p^d are observed on the generated part surface P.

Summarizing the analysis of Figure 5.4, the following conclusion can be formulated:

Conclusion 5.2

An increase of degree of conformity of generating surface T of the cutting tool to the sculptured part surface P causes a corresponding reduction of height of residual scallop on the machined part surface P. ∎

This conclusion is of critical importance for the development of methods of profiling of form-cutting tools as well as for the entire theory of part surface generation.

Degree of conformity of generating surface T of the form-cutting tool to the part surface P can be used as a mathematical criterion of efficiency of a part surface machining operation. This issue is of prime importance to bypass all the major bottlenecks in design of the optimal form-cutting tools imposed by initial indefiniteness of the problem.

5.1.3 ℝ-Mapping of the Part Surface *P* on Generating Surface *T* of the Form-Cutting Tool

The theory of part surface generation offers a method for profiling of the form-cutting tools of the most favorable (or optimal) design for machining a given sculptured part surface on a multiaxis *NC* machine. This means that it is possible to design the best possible cutting tool for machining of a given sculptured part surface.

The degree of conformity at a point of contact of generating surface T of the cutting tool to the sculptured part surface P significantly affects the efficiency of the part surface machining operation. The higher degree of conformity at a point of contact of the surfaces results in

- Higher productivity of the part surface machining operation
- Smaller residual scallops on the machined part surface
- Shorter machining time and so forth

(See [116,117,136,142] for more details in this concern.)

The generating surface of the optimal (or the most favorable) form-cutting tool is maximally conformal to the part surface to be machined at every point of contact of the surfaces. Because of this, the generating surface T the form-cutting tool for machining a specified sculptured part surface P can be generated as a type of mapping of the part surface P to be machined. The necessary type of mapping of the surface P onto the generating surface T must ensure the desired degree of conformity of surfaces P and T at every point of their contact. This type of mapping of the part surface P onto generating surface T had been proposed initially by Radzevich* [102,133,134,136] in the 1980s. This method of surface mapping is referred to as the $\mathbb{R}$-mapping of a sculptured part surface P onto the generating surface T of the form-cutting tool [104].

Consider a generating surface T of a form-cutting tool that makes contact with a sculptured part surface P at point K. The unit normal vector to the surface P at K is designated as $\mathbf{n}_P$. A pencil of planes can be constructed using the vector $\mathbf{n}_P$ as the directing vector of the axis of the pencil of planes. The maximal degree of conformity of generating surface T of the tool to the sculptured part surface P is observed when, for every plane of the pencil of planes, the equality is valid (in some cases, not the equality $R_T = -R_P$, but the equivalent equality $k_T = -k_P$ can be used instead):

$$R_T = -R_P \tag{5.1}$$

When the equality (see Equation 5.1) is satisfied, surfaces P and T make either surface-to-surface type of contact or local surface-to-surface contact (either local surface-to-surface contact of the first type or local surface-to-surface contact of the second type; see Chapter 4).

Actually, deviations in the configuration of the form-cutting tool with respect to the sculptured part surface P are inevitable. Because of this, in reality, Equation 5.1 cannot be satisfied. This forces the replacement of Equation 5.1 with an equality of the sort:

$$R_T = R_T(R_P) \tag{5.2}$$

The function $R_T(R_P)$ can be expressed in terms of deviations (or tolerances for accuracy) of the actual configuration of the cutting tool with respect to the work. It can be determined using either analytical or experimental methods.

Ultimately, the problem of profiling a form-cutting tool of the optimal design for machining a given sculptured part surface on a multiaxis NC machine is reduced to determination of the generating surface T that is maximally conformal to the given part surface P at every point of surface contact. Equation 5.2 is always satisfied if generating surface T of the form-cutting tool is determined in this way.

Switching from use of the function $R_T = -R_P$ to use of a function $R_T = R_T(R_P)$ results in the ideal locally extremal type of contact of surfaces P and T being replaced with their quasi-type of contact (Chapter 4). Recall that the quasi-types of contact of the part surface P and of generating surface T of

---

* Patent No. 4242296/08 (USSR), *A Method for Designing of the Optimal Form Cutting Tool for Machining a Given Sculptured Part Surface on a Multi-Axis NC Machine*, S. P. Radzevich, filed March 31, 1987.

the form-cutting tool yield that same range of agilability of the part surface machining operation as that of the point contact of the surfaces. Moreover, the productivity of part surface generation when maintaining the quasi-type of contact of surfaces P and T is practically identical to the productivity of part surface generation when the surface-to-surface type of contact of the surfaces is maintained.

Further, consider a sculptured part surface P that is analytically represented by an equation in vector form (see Equation 1.1):

$$\mathbf{r}_P = \mathbf{r}_P(U_P, V_P) \tag{5.3}$$

An equation of the sculptured part surface yields calculation of the fundamental magnitudes E_P, F_P, and G_P of the first order (see Equation 1.14) and of the fundamental magnitudes L_P, M_P, and N_P of the second order (see Equation 1.36) of the part surface P.

Using $\mathbb{R}$-mapping of the surfaces, it is convenient to derive an equation of generating surface T of the form-cutting tool initially in its natural parameterization (see Equation 1.60). For this purpose, it necessary to express the first $\Phi_{1.T}$ and the second $\Phi_{2.T}$ fundamental forms of the generating surface T of the form-cutting tool in terms of the fundamental magnitudes E_P, F_P, G_P and L_P, M_P, N_P of the part surface P.

The $\mathbb{R}$-mapping of surfaces is capable of establishing the desired correspondence between points of the part surface P and generating surface T of the cutting tool. This means that for every point on the part surface P, at which the precalculated principal radii of curvature are equal to $R_{1.P}$ and $R_{2.P}$, a corresponding point on the generating surface T of the cutting tool, with the desired principal radii of curvature $R_{1.T}$ and $R_{2.T}$ can be calculated. (The inverse statement is not valid mandatorily. There could be one or more points on the surface P that correspond to that same point of the generating surface T of the form-cutting tool.)

When two surfaces P and T are given, one can easily calculate degree of conformity of the surfaces at a given CC point (see Chapter 4). In the case under consideration, a problem of another sort arises. This problem can be interpreted as an inverse problem to the problem of calculation of the actual degree of conformity at a point of contact of the generating surface T of the form-cutting tool to the part surface P in a prescribed direction on surface P.

A functional relationship between principal radii of curvature of the part surface P and the generating surface T of the cutting tool in the differential vicinity of the CC point K is established by the $\mathbb{R}$-mapping of surfaces. Use of the $\mathbb{R}$-mapping of surfaces makes it possible to compose two equations for the determination of six unknown fundamental magnitudes E_T, F_T, G_T of the first $\Phi_{1.T}$ and L_T, M_T, N_T of the second $\Phi_{2.T}$ fundamental forms of the generating surface T of the cutting tool to be determined.

The equation $R_T = R_T(R_P)$ can be split into the set of two equations:

$$\begin{cases} \mathcal{M}_T = \mathcal{M}_T(\mathcal{M}_P; \mathcal{G}_P) \\ \mathcal{G}_T = \mathcal{G}_T(\mathcal{M}_P; \mathcal{G}_P) \end{cases}$$

(5.4)

where the mean curvatures of the part surface P and generating surface T at a current CC point K are designated as $\mathcal{M}_P$ and $\mathcal{M}_T$, respectively, and $\mathcal{G}_P$ and $\mathcal{G}_T$ denote *Gaussian* curvatures of surfaces P and T at that same CC point K, respectively.

The expression $R_T = R_T(R_P)$ gives an insight into the significance of a correlation between the radii of normal curvature R_P and R_T. To construct the desired function $R_T = R_T(R_P)$, degree of conformity functions $\mathcal{F}_1$, $\mathcal{F}_2$, and $\mathcal{F}_3$ are implemented. The degree of conformity functions $\mathcal{F}_1$, $\mathcal{F}_2$, and $\mathcal{F}_3$ are of principal importance for the determination of the function $R_T = R_T(R_P)$. The main purpose of the functions $\mathcal{F}_1$, $\mathcal{F}_2$, and $\mathcal{F}_3$ is to ensure the substitution of an initially given type of contact of surfaces P and T by a desired type of quasi-contact of the surfaces: *quasi-line*, *quasi-surface-to-surface* of the first type, *quasi-surface-to-surface* of the second type of contact of a part surface P and of generating surface T of the cutting tool (see Chapter 4 for more details on types of contact of a part surface P and generating surface T of the cutting tool).

As a desired degree of conformity of generating surface T of the cutting tool to the sculptured part surface P at every CC point is specified by the functions $\mathcal{F}_1$, $\mathcal{F}_2$, and $\mathcal{F}_3$, the functions of this type are referred to as *degree of conformity functions*.

By means of the degree of conformity functions $\mathcal{F}_1$, $\mathcal{F}_2$, and $\mathcal{F}_3$, an optimal form-cutting tool of reasonable size can be designed for machining sculptured part surfaces even of large scale.

To satisfy the set of two equations in Equation 5.4, it is necessary to satisfy the following equalities:

$$L_T N_T - M_T^2 = \mathcal{F}_1(L_P N_P - M_P^2)$$

(5.5)

$$E_T N_T - 2F_T M_T + G_T L_T = \mathcal{F}_2(E_P N_P - 2F_P M_P + G_P L_P)$$

(5.6)

$$E_T G_T - F_T^2 = \mathcal{F}_3(E_P G_P - F_P^2)$$

(5.7)

In a particular case, Equations 5.5 through 5.7 can be reduced to the form

$$L_T N_T - M_T^2 = L_P N_P - M_P^2$$

(5.8)

$$E_T N_T - 2F_T M_T + G_T L_T = E_P N_P - 2F_P M_P + G_P L_P$$

(5.9)

$$E_T G_T - F_T^2 = E_P G_P - F_P^2$$

(5.10)

However, this does not require the necessity of the equality $R_T = -R_P$.

These expressions (see Equations 5.5 through 5.7) analytically describe the vital link between the optimal design parameters of the form-cutting tool and the parameters of an actual process of machining of a given sculptured part surface. Equations 5.5 through 5.7 allow for incorporation into the design of the actual form-cutting tool all of the important features of the machining operation: cutting tool performance, cutting tool wear, rigidity of the cutting tool, and so forth.

The degree of conformity functions $\mathcal{F}_1$, $\mathcal{F}_2$, and $\mathcal{F}_3$ are of prime importance for designing the form-cutting tool of optimal design for machining a given sculptured part surface. They can be determined, for instance, using the proposed [83] experimental method of simulating the machining operation of a sculptured part surface. Other approaches for determining the degree of conformity functions $\mathcal{F}_1$, $\mathcal{F}_2$, and $\mathcal{F}_3$ can be used as well. There is much room for research in this concern.

Equations 5.5 through 5.7 are necessary but not sufficient for the determination of six unknown fundamental magnitudes E_T, F_T, G_T of the first $\Phi_{1.T}$ and L_T, M_T, N_T of the second $\Phi_{2.T}$ fundamental forms of generating surface T of the form-cutting tool. The equations of compatibility could be incorporated into the analyses to transform Equations 5.5 through 5.7 to a set of six equations of six unknowns.

Every smooth, regular generating surface T of the form-cutting tool mandatorily satisfies the *Gauss'* equation of compatibility that follows from his famous *theorema egregium* [14,116,117,150,158]:

$$\tilde{G}_T(E_T G_T - F_T^2) = \left[\frac{\partial^2 F_T}{\partial U_T \partial V_T} - \frac{1}{2}\left(\frac{\partial^2 E_T}{\partial V_T^2} + \frac{\partial^2 G_T}{\partial U_T^2}\right)\right] \cdot (E_T G_T - F_T^2)$$

$$+ \begin{vmatrix} 0 & \dfrac{\partial F_T}{\partial V_T} - \dfrac{1}{2}\dfrac{\partial G_T}{\partial U_T} & \dfrac{1}{2}\dfrac{\partial G_T}{\partial V_T} \\[3mm] \dfrac{1}{2}\dfrac{\partial E_T}{\partial U_T} & E_T & F_T \\[3mm] \dfrac{\partial F_T}{\partial U_T} - \dfrac{1}{2}\dfrac{\partial E_T}{\partial V_T} & F_T & G_T \end{vmatrix} - \begin{vmatrix} 0 & \dfrac{1}{2}\dfrac{\partial E_T}{\partial V_T} & \dfrac{1}{2}\dfrac{\partial G_T}{\partial U_T} \\[3mm] \dfrac{1}{2}\dfrac{\partial E_T}{\partial V_T} & E_T & F_T \\[3mm] \dfrac{1}{2}\dfrac{\partial G_T}{\partial U_T} & F_T & G_T \end{vmatrix}$$

$$\tag{5.11}$$

Two other equations of compatibility were independently derived by Mainardi* and by Codacci† [14,116,117,150,158]:

---

* Gaspare Mainardi (June 27, 1800–March 9, 1879), an Italian mathematician.
† Delfino Codazzi (March 7, 1824–July 21, 1873), an Italian mathematician.

$$\frac{\partial L_T}{\partial V_T} - \frac{\partial M_T}{\partial U_T} = L_T \Gamma_{12}^1 + M_T \cdot (\Gamma_{12}^2 - \Gamma_{11}^1) - N_T \Gamma_{11}^2 \qquad (5.12)$$

$$\frac{\partial M_T}{\partial V_T} - \frac{\partial N_T}{\partial U_T} = L_T \Gamma_{22}^1 + M_T \cdot (\Gamma_{22}^2 - \Gamma_{12}^1) - N_T \Gamma_{12}^2 \qquad (5.13)$$

where the Christoffel* symbols of the second type are used. The Christoffel symbols can be calculated from the formulae [14,158]

$$\Gamma_{11}^1 = \frac{G_u \dfrac{\partial E_u}{\partial U_u} - 2F_u \dfrac{\partial F_u}{\partial U_u} + F_u \dfrac{\partial E_u}{\partial V_u}}{2(E_u G_u - F_u^2)} \qquad (5.14)$$

$$\Gamma_{11}^2 = \frac{2E_u \dfrac{\partial F_u}{\partial U_u} - E_u \dfrac{\partial E_u}{\partial V_u} + F_u \dfrac{\partial E_u}{\partial U_u}}{2(E_u G_u - F_u^2)} \qquad (5.15)$$

$$\Gamma_{12}^1 = \frac{G_u \dfrac{\partial E_u}{\partial V_u} - F_u \dfrac{\partial G_u}{\partial U_u}}{2(E_u G_u - F_u^2)} = \Gamma_{21}^1 \qquad (5.16)$$

$$\Gamma_{12}^2 = \frac{E_u \dfrac{\partial G_u}{\partial U_u} - F_u \dfrac{\partial E_u}{\partial V_u}}{2(E_u G_u - F_u^2)} = \Gamma_{21}^2 \qquad (5.17)$$

$$\Gamma_{22}^1 = \frac{2G_u \dfrac{\partial F_u}{\partial V_u} - G_u \dfrac{\partial G_u}{\partial U_u} + F_u \dfrac{\partial G_u}{\partial V_u}}{2(E_u G_u - F_u^2)} \qquad (5.18)$$

$$\Gamma_{22}^2 = \frac{E_u \dfrac{\partial G_u}{\partial V_u} - 2F_u \dfrac{\partial F_u}{\partial V_u} + F_u \dfrac{\partial G_u}{\partial U_u}}{2(E_u G_u - F_u^2)} \qquad (5.19)$$

Equations of compatibility are necessary in order to transform the set of three equations (Equations 5.5 through 5.7) to a set of six equations of six unknowns.

---

* Elwin Bruno Christoffel (November 10, 1829–March 15, 1900), a German mathematician and physicist.

The set of three equations, Equations 5.5 through 5.7, together with three equations of compatibility (Equations 5.11 through 5.13) completely describe the $\mathbb{R}$-mapping of two smooth regular surfaces (that is, they describe the $\mathbb{R}$-mapping of the sculptured part surface P onto generating surface T of the form-cutting tool of the most favorable design).

Thus, the fundamental magnitudes of the first $\Phi_{1.T}$ and the second $\Phi_{2.T}$ fundamental forms of the generating surface T of the form-cutting tool can be determined using the $\mathbb{R}$-mapping of the sculptured part surface P onto the generating surface T of the form-cutting tool. A routing procedure can be used to solve the set of six equations of six unknowns, say, of Equations 5.5 through 5.7 and Equations 5.11 through 5.13 with six unknowns E_T, F_T, G_T and L_T, M_T, N_T. Ultimately, the determined generating surface T of the form-cutting tool is represented in natural parameterization similar to Equation 1.60.

Below, the fundamental magnitudes E_T, F_T, G_T and L_T, M_T, N_T of generating surface T of the form-cutting tool are considered as the known functions.

5.1.4 Reconstruction of the Generating Surface *T* of the Form-Cutting Tool from the Precalculated Natural Parameterization

Analytical representation of generating surface T in the form of Equation 1.60 is inconvenient for application in engineering practice of designing form-cutting tools. However, natural parameterization of the generating surface T can be converted to a parameterization of the surface T in a convenient form, say, to represent in a *Cartesian* reference system.

To transform the natural parameterization of generating surface T of a form-cutting tool to its representation in a *Cartesian* reference system it is necessary to solve the set of two *Gauss-Weingarten's* equations in tensor notation:

$$
\begin{array}{c}
\text{Generating surface } T \\
\text{of the form-cutting tool}
\end{array}
\Leftarrow
\begin{cases}
\mathbf{r}_{ij} = \Gamma_{ij}^{k}\mathbf{r}_P + b_{ij}\mathbf{n}_P \\
\\
\mathbf{n}_i = -b_{ik}g^{kj}\mathbf{r}_j
\end{cases}
\tag{5.20}
$$

The solution to the set of two Equations 5.20 returns a matrix equation of generating surface T of the form-cutting tool of the most favorable design for machining a given sculptured part surface P on a multiaxis *NC* machine. The initial conditions of integration of the set of Equations 5.20 must be selected properly.

In Equations 5.20, $\mathbf{r}_i = \dfrac{\partial \mathbf{r}_T}{\partial U_T}$, $\mathbf{r}_{ij} = \dfrac{\partial^2 \mathbf{r}_T}{\partial U_T \partial V_T}$, $\mathbf{n}_i = \dfrac{\partial \mathbf{n}_T}{\partial U_T}$, $b_{ij} = \mathbf{r}_{ij} \cdot \mathbf{n}_T = -\mathbf{n}_i \mathbf{r}_j - \mathbf{n}_j \cdot \mathbf{r}_i$, g_{ij} is a metric tensor of generating surface T of the form-cutting tool of optimal design, and g^{ij} is a contravariant tensor of the generating surface T of the cutting tool.

Known methods [28] are used for solving the set of Equations 5.20.

The initial conditions for the integration of the set of Equations 5.20 must be established. These conditions, for example, might include coordinates of two points on generating surface T of the cutting tool and direct cosines of the unit normal vector $\mathbf{n}_T$ to the cutting tool surface T at one of these points.

The set of two differential equations in tensor notation (see Equations 5.20) can be converted either to a set of five differential equations in vector notation or to a corresponding set of fifteen differential equations in parametric notation. Conventional mathematical methods can be implemented for solving such a set of five differential equations or a set of fifteen differential equations with a corresponding number of unknowns. This is a type of trivial mathematical problem that follows from the proposed theory of part surface generation.

5.1.5 Method to Determine the Degree of Conformity Functions $\mathscr{F}_1$, $\mathscr{F}_2$, and $\mathscr{F}_3$

To determine the degree of conformity functions, various approaches can be employed. It is possible to implement a method of experimental simulation* of machining of a sculptured part surface on a multiaxis NC machine to determine the degree of conformity functions $\mathscr{F}_1$, $\mathscr{F}_2$, and $\mathscr{F}_3$. The method of simulation is proposed by Radzevich [83], and is illustrated in Figure 5.5.

As an example of implementation of the method of simulation [83], consider machining of a sculptured part surface P on a multiaxis NC machine schematically illustrated in Figure 5.5a. The part surface P is machined by the form milling cutter having generating surface T of certain geometry.

The method of simulation of machining of sculptured surfaces is carried out by the equivalent models of the part surface P and generating surface T of the cutting tool as shown in Figure 5.5b. The working surfaces of the equivalent models of the part surface P and generating surface T of the cutting tool used for the simulation are designated as $P^{(s)}$ and $T^{(s)}$, respectively.

The local topology of the surfaces $P^{(s)}$ and $T^{(s)}$ can be uniquely specified by two parameters—that is, by the mean curvature $\mathcal{M}_{P(T)}$ and the Gaussian curvature $\mathcal{G}_{P(T)}$. As only two parameters of local topology are sufficient to specify local topology of the modeling surfaces, the types of the surfaces $P^{(s)}$ and $T^{(s)}$ are limited only to 10 similar to that illustrated earlier in Figure 1.16.

The modeling surface $P^{(s)}$ as well as the surface $T^{(s)}$ is a type of quadric surface. Both the modeling surfaces $P^{(s)}$ and $T^{(s)}$ make tangency at a point K. The local geometry of tangency of the modeling surfaces $P^{(s)}$ and $Tv^{(s)}$ is identical to that of the sculptured part surface P and generating surface T of the cutting tool. Because of this, unit tangent vectors $\mathbf{t}_{1.P}^{(s)}$ and $\mathbf{t}_{2.P}^{(s)}$ of the principal directions on the quadric surface $P^{(s)}$ align with the corresponding unit tangent

---

* Patent No. 1449246 (USSR), *A Method of Experimental Simulation of Machining of a Sculptured Part Surface on a Multi-Axis NC Machine*, S. P. Radzevich, filed February 17, 1987, Int. Cl. B 23 C, 3/16.

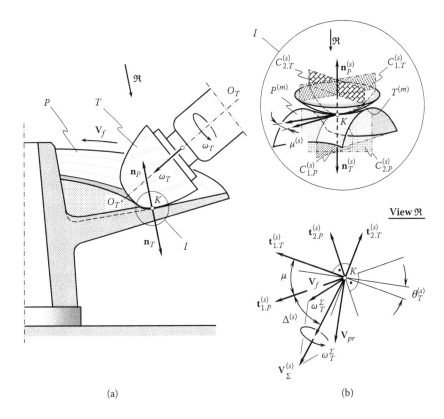

(a)   (b)

FIGURE 5.5
Schematic of an experimental method to determine the desired degree of conformity functions $\mathscr{F}_1$, $\mathscr{F}_2$, and $\mathscr{F}_3$. (After S. P. Radzevich, USSR Patent No. 1449246, 1987.) (Reprinted from *Mathematical and Computer Modeling*, 36, Radzevich, S.P., $\mathbb{R}$-mapping based method for designing of form cutting tool for sculptured surface machining, 921–938, Copyright 2002, with permission from Elsevier.)

vectors $\mathbf{t}_{1.P}$ and $\mathbf{t}_{2.P}$ of the sculptured part surface P. Moreover, the principal radii of curvature $R_{1.P}^{(s)}$ and $R_{2.P}^{(s)}$ of the quadric surface $P^{(s)}$ at every CC point K are equal to the corresponding principal radii of curvature $R_{1.P}$ and $R_{2.P}$ of the sculptured part surface P (that is, the identities $R_{1.P}^{(s)} \equiv R_{1.P}$ and $R_{2.P}^{(s)} \equiv R_{2.P}$ are observed). Because of this, Euler's formula yields the conclusion that in the differential vicinity of the CC point K, the modeling surface $P^{(s)}$ is locally congruent to the sculptured part surface P up to the members of the second order.

The same conclusion is valid for modeling quadric surface $T^{(s)}$ that is used for local simulation of the generating surface T of the form-cutting tool. At the CC point K, unit tangent vectors $\mathbf{t}_{1.T}^{(s)}$ and $\mathbf{t}_{2.T}^{(s)}$ of the principal directions on the quadric surface $T^{(s)}$ align with the corresponding unit tangent vectors $\mathbf{t}_{1.T}$ and $\mathbf{t}_{2.T}$ of the generating surface T of the cutting tool. The principal radii of curvature $R_{1.T}^{(s)}$ and $R_{2.T}^{(s)}$ of the quadric surface $T^{(s)}$ are also equal to the corresponding principal radii of curvature $R_{1.T}$ and $R_{2.T}$ of the tool surface T (that is, the identities $R_{1.T}^{(s)} \equiv R_{1.T}$ and $R_{2.T}^{(s)} \equiv R_{2.T}$ are observed). Therefore, in differential vicinity of the CC point K, the modeling surfaces $T^{(s)}$ is locally congruent to generating surface T of the cutting tool up to the members of the second order.

The local orientation of the cutting edges on the quadric surface $T^{(s)}$ remains the same with respect to the principal directions $\mathbf{t}_{1.T}^{(s)}$ and $\mathbf{t}_{2.T}^{(s)}$ as that for the actual form-cutting tool T [$\mathbf{t}_{1.T}^{(s)} \equiv \mathbf{t}_{1.T}$ and $\mathbf{t}_{2.T}^{(s)} \equiv \mathbf{t}_{2.T}$]. For this purpose, it is necessary at first to determine the orientation of the principal plane sections $C_{1.T}$ и $C_{2.T}$ of the generating surface T of the form-cutting tool with respect to the coordinate U_T and V_T lines. The orientation of the plane section of a surface by a normal plane surface can be determined by the ratio $\dfrac{\partial U_T}{\partial V_T}$ (see Chapter 1).

For the orthogonally (U_T; V_T) parameterized generating surface T of the cutting tool, the ratio $\dfrac{\partial U_T}{\partial V_T}$ determines value of $\tan \xi_T$. Here, angle ξ_T designates the angle of inclination of the principal plane sections $C_{1.T}$ and $C_{2.T}$ relative to the coordinate U_T and V_T lines on the generating surface T. Usually, the parameterization of the cutting tool surface T is not orthogonal. In such a case, the angle ξ_T (is not shown on Figure 5.5b) can be calculated from the formula [19]:

$$\sin \xi_T = \frac{\partial V_T}{\partial U_T} \left(\left(\frac{\partial V_T}{\partial U_T} \right)^2 - 2 \frac{\partial V_T}{\partial U_T} \cos \omega_T + 1 \right)^{-\frac{1}{2}} \tag{5.21}$$

At CC point K, the cutting edge forms a certain angle ζ_T with U_T coordinate line. The angle ζ_T can be calculated from the equation [116,117].

The cutting edge makes a certain angle θ_T with the first principal plane section $C_{1.T}$ of the generating surface T of the cutting tool. The angle θ_T is equal to algebraic sum of the angles ξ_T and ζ_T, that is, $\theta_T = \xi_T + \varsigma_T$. Therefore, the orientation of the cutting edge [angle $\theta_T^{(m)} \equiv \theta_T = \xi_T + \varsigma_T$] relative to the first principal plane section $C_{1.T}^{(s)}$ of the quadric surface $T^{(s)}$ makes local orientation of the cutting edge on the quadric surface $T^{(s)}$ identical to that on the generating surface T of the cutting tool. The accuracy of the simulation is up to the members of the second order or even higher.

The modeling quadric surfaces $P^{(s)}$ and $T^{(s)}$ are turned about common unit normal vector $\mathbf{n}_P^{(s)}$ relative to each other through angle $\mu^{(s)}$. The angle $\mu^{(s)}$ is the angle of local relative orientation of surfaces $P^{(s)}$ and $T^{(s)}$. The angle $\mu^{(s)}$ is identical to the angle μ of local relative orientation of the actual surfaces P and T [$\mu^{(s)} \equiv \mu$]. Angle μ is the angle that form the first $\mathbf{t}_{1.P}$ and $\mathbf{t}_{1.T}$ (or, the same, the second $\mathbf{t}_{2.P}$ and $\mathbf{t}_{2.T}$) principal directions of the surfaces at the CC point [116,117]:

$$\mu^{(s)} \equiv \mu = \tan^{-1} \frac{|\mathbf{t}_{1.P} \times \mathbf{t}_{1.T}|}{\mathbf{t}_{1.P} \cdot \mathbf{t}_{1.T}} \equiv \tan^{-1} \frac{|\mathbf{t}_{2.P} \times \mathbf{t}_{2.T}|}{\mathbf{t}_{2.P} \cdot \mathbf{t}_{2.T}} \tag{5.22}$$

During simulation, the local relative orientation of the quadric modeling surfaces $P^{(s)}$ and $T^{(s)}$ in differential vicinity of the CC point K up to the members of the second order is identical to local relative orientation of the actual sculptured part surface P and generating surface T of the cutting tool when machining the part surface P on a multiaxis NC machine.

The trajectory of a cutting edge point relative to the sculptured part surface P can be represented as a vector sum of the motions that surfaces P and T perform on a multiaxis NC machine.

When simulating a machining operation of the part surface P, the modeling quadric surfaces $P^{(s)}$ and $T^{(s)}$ perform a relative motion with respect to one another. Resultant speed $\mathbf{V}_{\Sigma}^{(s)}$ of the relative motion can be represented as a vector sum of the speed of cutting $\mathbf{V}_{c}^{(s)}$ and of all other partial motion $\mathbf{V}_{i}^{(s)}$, namely

$$\mathbf{V}_{\Sigma}^{(s)} = \mathbf{V}_{c}^{(s)} + \sum_{i=1}^{n-1} \mathbf{V}_{i}^{(s)} \tag{5.23}$$

where n designates total number of all of the partial motions $\mathbf{V}_{i}^{(s)}$. The feed rate motion $\mathbf{V}_{f}^{(s)}$ is an example of the partial motions $\mathbf{V}_{i}^{(s)}$.

When machining a sculptured part surface, instant relative motion of surfaces P and T can be represented as an instant screw motion. Therefore, when simulating the machining operation, in addition to the resultant translation with a linear velocity $\mathbf{V}_{\Sigma}^{(s)}$, the modeling quadric surfaces $P^{(s)}$ and $T^{(s)}$ perform a rotation with a resultant angular velocity $\boldsymbol{\omega}_{\Sigma}^{(s)}$. The resultant screw motion is denoted by $\mathbf{S}_{cr}^{(s)} = \mathbf{V}_{\Sigma}^{(s)} \cup \boldsymbol{\omega}_{\Sigma}^{(s)}$.

During simulation, the resultant relative motion $\mathbf{S}_{cr}^{(s)}$ of the modeling surfaces $P^{(s)}$ and $T^{(s)}$ is identical to the instant resultant relative screw motion $\mathbf{S}_{cr} = \mathbf{V}_{\Sigma} \cup \boldsymbol{\omega}_{\Sigma}$ of the actual sculptured part surface P and the generating surface T of the cutting tool $[\mathbf{S}_{cr}^{(s)} \equiv \mathbf{S}_{cr}]$. For this purpose, the angle $\Delta^{(s)}$ that the vector $\mathbf{V}_{\Sigma}^{(m)}$ makes with the first principal plane section $C_{1.P}^{(s)}$ of the modeling quadric surface $P^{(s)}$ is identical to the similar angle Δ that the vector $\mathbf{V}_{\Sigma}$ makes with the first principal plane section $C_{1.P}$ of the sculptured surface P [i.e., $\Delta^{(s)} \equiv \Delta$].

The instant relative screw motion of the quadric surfaces $P^{(s)}$ and $T^{(s)}$ is identical to the instant relative screw motion of the sculptured surface P and of the generating surface T of the cutting tool $[\mathbf{S}_{cr}^{(s)} \equiv \mathbf{S}_{cr}]$.

When the method of experimental simulation of machining of a sculptured part surface on a multiaxis NC (Figure 5.5) is implemented, the identity to each other of all local geometrical and of all instant kinematical parameters of the machining operation to be simulated to that in the simulating procedure is ensured, namely, at every CC point K [83]:

- The quadric surface $P^{(s)}$ is locally identical to the actual sculptured part surface P
- The quadric surface $T^{(s)}$ is locally identical to the actual generating surface T of the form-cutting tool

- The location of the cutting edge on the quadric surface $T^{(s)}$ is identical to that on the generating surface T of the actual form-cutting tool
- The relative local orientation of the modeling quadric surfaces $P^{(s)}$ and $T^{(s)}$ is identical to the relative local orientation of the actual sculptured part surface P and the generating surface T of the actual form-cutting tool
- The instant motion of modeling quadric surfaces $P^{(s)}$ and $T^{(s)}$ in relation to one another when the simulation is identical to the instant relative motion of sculptured surface P and generating surface T of the cutting tool in actual machining operation (that is, instant kinematics of the relative motion in both cases remains the same).

Ultimately, this results in high efficiency of the method of simulation of machining of a sculptured part surface on a multiaxis *NC* (Figure 5.5) [83], and in high accuracy of the determined degree of conformity functions $\mathscr{F}_1$, $\mathscr{F}_2$, and $\mathscr{F}_3$.

In reality, reproduction of a motion that is identical to an instant relative motion of two surfaces is often inconvenient. Therefore, when simulating a machining operation, it is preferred to perform a continuous relative motion of the modeling quadric surfaces $P^{(s)}$ and $T^{(s)}$, rather than the instant relative motion. The implementation of continuous relative motions leads to significant simplification of the modeling procedure.

To perform the desired continuous relative motion of the modeling quadrics $P^{(s)}$ and $T^{(s)}$, the use of a surface that allows sliding "over itself" is helpful. A screw surface of constant axial pitch, $p = Const$, is the most general type of surfaces $P^{(s)}$ and $T^{(s)}$ that allow for sliding "over itself" (see Chapter 1).

When a screw surface is traveling along and is rotated about its axis with that same parameter of the screw motion as the instant screw parameter of the screw surface, the enveloping surface to successive positions of the screw surface is congruent to the screw surface itself. Particular cases of surfaces that allow for sliding "over itself" [surfaces of revolution (for which $p = 0$), surfaces of translation, or, in other words, general cylinders, and not just cylinders of revolution (for which $p = \infty$)] are considered in Chapter 1. Cylinders of revolution, spherical surfaces, and planes represent examples of the simplest and completely degenerate surfaces that allow for sliding "over itself."

For simulation of a machining operation of a sculptured part surface on a multiaxis *NC* machine, it is convenient to use a screw with external surface $P^{(s)}$ and having either convex or concave thread profile as depicted in Figure 5.6. The application of such a screw surface $P^{(s)}$ allows the simulation of both convex and saddle-like local patches of a given sculptured part surface P.

For the simulation of concave and saddle-like local patches of a given sculptured part surface P, a screw having an internal surface $P^{(s)}$ and with either convex or concave thread profile can be used (Figure 5.7).

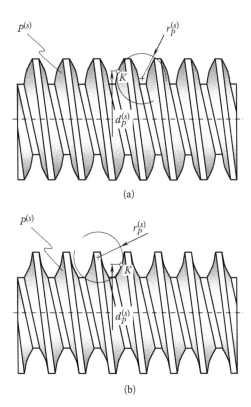

FIGURE 5.6
External screws having convex (a) and saddle-like (b) local patches of the modeling quadric
surface $P^{(s)}$ that are used for the purpose of experimental simulation of machining of a sculptured part surface on a multiaxis numerical control machine.

In both cases shown in Figures 5.6 and 5.7, the corresponding screw might
be single- or multi-threaded as well as a single- or multi-started.

To ensure

a. The necessary parameters of topology of the modeling surface $P^{(s)}$—
   that is, to ensure the identities $R_{1.P}^{(s)} \equiv R_{1.P}$ and $R_{2.P}^{(s)} \equiv R_{2.P}$

b. The necessary radii of principal curvature of the modeling surface
   $T^{(s)}$—that is, to ensure the identities $R_{1.T}^{(s)} \equiv R_{1.T}$ and $R_{2.T}^{(s)} \equiv R_{2.T}$

c. Their local relative orientation—that is, to ensure the identity of the
   angles of local relative orientation $\mu^{(s)} \equiv \mu$ of the surfaces $P^{(s)}$ and $T^{(s)}$
   at a point of their contact,

the design parameters $d_P^{(s)}$, and $r_P^{(s)}$ of the modeling screw must be calculated
in a proper way. For this purpose, Mensnier and Euler formulae are commonly used.

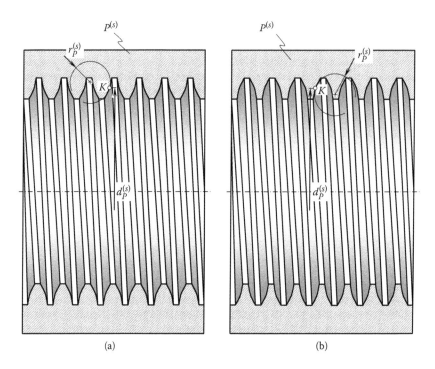

FIGURE 5.7
Internal screws having saddle-like (a) and concave (b) local patches of the modeling quadric surface $P^{(s)}$ that are used for the purpose of experimental simulation of machining of a sculptured part surface on a multiaxis numerical control machine $P^{(s)}$.

Mensnier formula establishes the correspondence between a radius of normal curvature, R_P, of a sculptured part surface P (or a corresponding generating surface T of the form-cutting tool) in a normal plane through a certain direction $\mathbf{t}_P$ on the surface, and between the radius of curvature $R_P^{(\vartheta)}$ of the part surface P (or the cutting tool surface T) through the same direction $\mathbf{t}_P$ on the surface, which is inclined to the normal plane section at a specified angle ϑ_P. Usually, Mensnier formula is represented in the form:

$$R_P^{(\vartheta)} = R_P \cdot \cos \vartheta_P \tag{5.24}$$

Euler formula allows for representation in the form:

$$R_P^{(\varphi)} = [R_{1.P}^{-1} \cos \varphi + R_{2.P}^{-2} \sin \varphi]^{-1} \tag{5.25}$$

where $R_P^{(\varphi)}$ is the normal radius of curvature at a specified point of a sculptured part surface P in a plane section that forms certain angle φ with the first principal direction on the surface, $R_{1.P}$ is the first principal radius of curvature of the

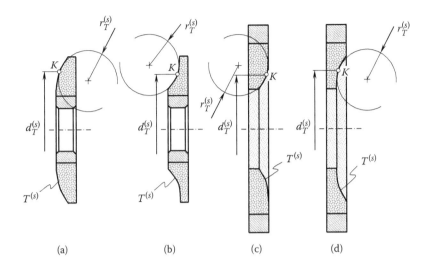

FIGURE 5.8
External (a, b), and internal (c, d) surfaces of revolution featuring convex (a), saddle-like (b, c) and concave (d) local patches of the modeling quadric surface $T^{(s)}$, which are used for the purpose of experimental simulation of machining of a sculptured part surface on a multiaxis numerical control machine.

sculptured part surface P at the specified point, and $R_{2.P}$ is the second principal radius of curvature of the sculptured part surface P at the specified point.

For the simulation, it is much more convenient to model the surface $T^{(s)}$ by an external or by an internal surface of revolution that has either convex or concave axial profile as illustrated in Figure 5.8. The same Mensnier and Euler formulae can be used for calculation of the design parameters $d_T^{(s)}$ and $r_T^{(s)}$ of the cutting tool in order to ensure the identities $R_{1.T}^{(s)} \equiv R_{1.T}$ and $R_{2.T}^{(s)} \equiv R_{2.T}$.

The implementation of the screw modeling surfaces $P^{(s)}$ as illustred in Figures 5.6 and 5.7 as well as the surfaces of revolution depicted in Figure 5.8 makes it possible to achieve the desired type of local topology (Figure 1.16) of the sculptured part surface P to be modeled and of the generating surface T of the form-cutting tool. The identity of the topologies of the modeling surfaces $P^{(s)}$ and $T^{(s)}$ to the topologies of the actual surfaces P and T at a point of their contact is the prerequisite of high efficiency of the simulation.

An example of implementation of the disclosed method* of simulation [83] is schematically illustrated in Figure 5.9. In a particular case, the machining of a convex local patch of the part surface P by the saddle-like local patch of the generating surface T is simulated by the external worm having convex profile of the threads that is machined by the grinding wheel having concave axial profile (Figure 5.9). The design parameters of the worm as well as the design

---

* Patent No. 1449246 (USSR), *A Method of Experimental Simulation of Machining of Sculptured Part Surface on a Multi-Axis NC Machine*, S. P. Radzevich, filed February 17, 1987.

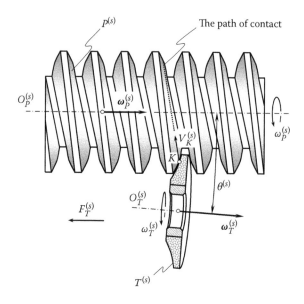

FIGURE 5.9
Schematic of simulation of machining a convex local patch of a sculptured part surface $P^{(s)}$ by a saddle-like local patch of generating surface $T^{(s)}$ of the form-cutting tool on a multiaxis NC machine.

parameters of the grinding wheel are precalculated in tight correlation with the corresponding design parameters of the actual part surface P and the actual generating surface T of the form-cutting tool. The rotation of the worm and the translation of the grinding wheel are timed with one another to get the resultant relative motion of the modeling surfaces $P^{(s)}$ and $T^{(s)}$ identical to that of the actual surfaces P and T in the machining operation being simulated.

In a case when one or both modeling quadric surfaces $P^{(s)}$ and $T^{(s)}$ allow for sliding "over itself," the manufacturing of the specimens for the simulation is simplified. At the same time, this allows two instant relative motions of the modeling surfaces $P^{(s)}$ and $T^{(s)}$ to be substituted with their continuous relative motion. The last is much more convenient for the simulation and makes it possible to obtain more accurate and more reliable results of the experiments. Ultimately, this allows for an accurate determination of the degree of conformity functions $\mathcal{F}_1$, $\mathcal{F}_2$, and $\mathcal{F}_3$. It is important to stress that much experimental data that is necessary to determine the degree of conformity functions $\mathcal{F}_1$, $\mathcal{F}_2$, and $\mathcal{F}_3$, can be collected from already published scientific papers in the field. For this purpose, it is necessary to analyze the published results of the research on efficiency of part surface machining on a machine tool from the standpoint of implementation of $\mathbb{R}$-mapping of the sculptured part surface P onto the generating surface T of the form-cutting tool as well as of other aspects of the theory of part surface generation.

It is critical to notice here that the degree of conformity functions $\mathcal{F}_1$, $\mathcal{F}_2$, and $\mathcal{F}_3$ play an additional important role in the theory of part surface generation. They serve as a *bridge* between the pure geometrical and kinematical theory and real processes of a part surface machining including the physical phenomena that occur.

5.1.6 Algorithm for the Computation of the Design Parameters of the Form-Cutting Tool

The majority of the computations are required to determine the major design parameters of generating surface T of the form-cutting tool for machining of a given sculptured surface on a multiaxis NC machine. Generally, computations of this type can be performed with the help of computers.

An algorithm for the computation of the design parameters of generating surface T of the form-cutting tool is illustrated with the flowchart shown in Figure 5.10. The algorithm works in the following way:

1. Compose an equation of the smooth regular sculptured part surface P. When the part surface P comprises two or more portions, a set of equations of all n surface patches $P_i|_{i=1}^{n}$ should be composed.
2. Compute the first derivatives of equation(s) of the sculptured part surface P.

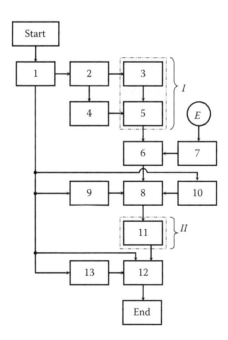

FIGURE 5.10
Flowchart of the algorithm for the computation of the major design parameters of the form-cutting tool of optimal design for machining a given sculptured part surface on a multiaxis numerical control machine.

3. Compute the fundamental magnitudes E_P, F_P, and G_P of the first order of the part surface P.

4. Compute the second derivatives of equation(s) of the part surface P.

5. Compute the fundamental magnitudes L_P, M_P, and N_P of the second order of the part surface P.

Results (3) and (5) can be interpreted as the natural form of parameterization of the sculptured part surface P.

6. Compose a set of three equations that describe the desired degree of conformity at a point of contact of the generating surface T of the form-cutting tool to the sculptured part surface P to be machined.

7. Determine the degree of conformity functions $\mathscr{F}_1$, $\mathscr{F}_2$, and $\mathscr{F}_3$.

8. The use of the $\mathbb{R}$-mapping of the sculptured part surface P onto the generating surface T of the form-cutting tool returns a set of three equations (Equations 5.5 through 5.7) of six unknowns (E_T, F_T, G_T, and L_T, M_T, N_T) for the computation of the fundamental magnitudes of the generating surface T of the form-cutting tool.

9. Incorporate into consideration the *Gauss'* equation of compatibility (see Equation 5.11).

10. Incorporate into consideration two Mainardi–Codazzi's equations of compatibility (see Equations 5.12 and 5.13)—ultimately, this yields a set of six equations (see Equations 5.5 through 5.7 and Equations 5.11 through 5.13) of six unknowns E_T, F_T, G_T and L_T, M_T, N_T.

11. Compute the fundamental magnitudes E_T, F_T, and G_T of the first, and L_T, M_T, and N_T of the second order of the generating surface T of the form-cutting tool.

The results of the computation (11) can be interpreted as the natural form of parameterization of the generating surface T of the form-cutting tool.

12. Solve the set of two Gauss–Weingarten's equations (see Equations 5.20) in tensor notation. The output is a matrix equation of the generating surface T of the form-cutting tool in Cartesian coordinates.

13. Incorporate into consideration the initial conditions for integration of the set of two Gauss–Weingarten's equations.

In the general case, principal curvatures $k_{1.P} = R_{1.P}^{-1}$ and $k_{2.P} = R_{2.P}^{-1}$ at a point of a sculptured surface P depend on surface definition (that is, they depend on the surface topology) and are not usually related by any function. Because of this, principal curvatures $k_{1.T}$ and $k_{2.T}$ for the designed generating surface T of the cutting tool will be not related by a certain function. In cases of milling cutter, grinding wheel, and so forth, the principal curvatures $k_{1.T}$ and $k_{2.T}$ are related by a function—in this case, the surface T is represented by a surface of revolution. For machining of a sculptured part surface P of any geometry, the discussed above generalized solution (see Equation 5.20) allows for an approximation of the most favorable generating surface T (that is featuring optimal topology) by a surface of revolution and so finding a form-cutting tool having all necessary combinations of the principal curvatures $k_{1.T}$ and $k_{2.T}$.

5.1.7 Illustrative Examples of the Computation of the Design Parameters of Generating Surface *T* of the Form-Cutting Tool

Two illustrative examples of the computation of the design parameters of generating surface T of the form-cutting tool, which do not require extensive application of computing are presented below.

Without loss of generality of the developed approach, let us consider a case of the computation of the design parameters of the generating surface T of the form-cutting tool when the six fundamental magnitudes E_T, F_T, G_T of the first and L_T, M_T, N_T of the second fundamental forms of the cutting tool surface T are already found from Equations 5.5 through 5.7 and from equations of compatibility Equations 5.11 through 5.13, as the determination of the fundamental magnitudes E_T, F_T, G_T of the first and L_T, M_T, N_T of the second fundamental forms of the surface T is a trivial task.

Example 5.1

Given two differential forms

$$\Phi_{1.T} \Rightarrow dU_T^2 + \cos^2 U_T dV_T^2 \tag{5.26}$$

$$\Phi_{2.T} \Rightarrow dU_T^2 + \cos^2 U_T dV_T^2 \tag{5.27}$$

Find the generating surface T of the form-cutting tool, for which $\Phi_{1.T}$ and $\Phi_{2.T}$ are the first and second fundamental forms.

Since

$$E_T = 1 \tag{5.28}$$

$$F_T = 0 \tag{5.29}$$

$$G_T = \cos^2 U_T \tag{5.30}$$

and

$$L_T = 1 \tag{5.31}$$

$$M_T = 0 \tag{5.32}$$

$$N_T = \cos^2 U_T \tag{5.33}$$

then Christoffel's symbols are equal to

$$\Gamma^1_{11} = \Gamma^2_{22} = \Gamma^1_{12} = \Gamma^2_{22} = 0 \tag{5.34}$$

$$\Gamma^2_{12} = -\tan U_T \tag{5.35}$$

$$\Gamma^1_{22} = \sin U_T \cos U_T, \tag{5.36}$$

which satisfy the Gauss–Codazzi's equations of compatibility, as direct substitution shows.

The set of Gauss–Weingarten's equations (see Equation 5.20) returns the solution

$$\tilde{\mathbf{r}}_T(U_T, V_T) = \mathbf{r}_{0T} + \mathbf{r}_T(U_T, V_T) = \mathbf{r}_{0T} + \begin{bmatrix} \cos V_T \cos U_T \\ \sin V_T \cos U_T \\ \sin U_T \\ 1 \end{bmatrix} \tag{5.37}$$

which is the equation of the sphere.

The detailed derivation of Equation 5.37 is not covered here. However, it is covered in all details in our earlier publication [105].

Here, $\mathbf{r}_{0T}$ designates the vector that specifies location of the generating surface T of the cutting tool. By the choice of $\mathbf{r}_{0T}$, one can place the cutting tool surface T in any position of space, selecting any orthogonal system of meridians and parallels for U_T and V_T curvilinear coordinates of an arbitrary point m on the tool surface T.

The sphere can be used as generating surface T of the cutting tool, for example, of the grinding wheel (Figure 5.11) for finishing a sculptured part surface P on a multiaxis NC machine.

Example 5.2

In much the same way as above, for the fundamental form

$$\Phi_{1,T} \quad \Rightarrow \quad (r_T \cos\theta_T + R_T)^2 dU_T^2 + (-r_T \sin\theta_T + R_T)^2 + r_T^2 \cos\theta_T^2 dV_T^2 \tag{5.38}$$

and the corresponding fundamental form

$$\Phi_{2,T} \Rightarrow L_T dU_T^2 + 2M_T dU_T dV_T + N_T dV_T^2, \tag{5.39}$$

one can derive an equation of the generating surface T of the form-cutting tool. Without going into details, on solving Equation 5.20, the following expression for the cutting tool surface T is derived:

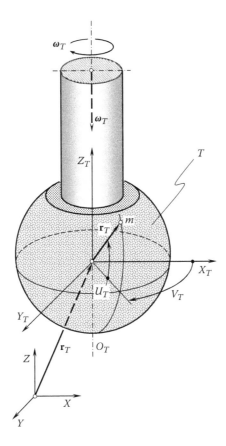

FIGURE 5.11
Example 5.1. Spherical generating surface *T* of a grinding wheel.

$$
\mathbf{r}_T =
\begin{bmatrix}
(r_T \cos\theta_T + R_T)\cos\varphi_T \\
(r_T \cos\theta_T + R_T)\sin\varphi_T \\
r_T \sin\theta_T \\
1
\end{bmatrix}
\tag{5.40}
$$

and the generating surface *T* of the form-cutting tool is a torus surface as illustrated in Figure 5.12.

The presented solution to the problem of the calculation of design parameters of the generating surface of the form-cutting tool is based on implementation of a novel type of surface mapping—on the ℝ-mapping of a sculptured part surface *P* onto the generating surface *T* of the form-cutting tool developed by the author [116,117,136]. The disclosed method is tightly connected to the method of simulation of interaction of the form-cutting tool and the work. The last method is vital to determine the degree of conformity functions, which are of critical importance

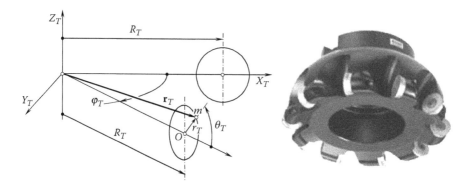

FIGURE 5.12
Example 5.2. Toroidal generating surface T of the milling cutter. (Reprinted from *Mathematical and Computer Modeling*, 36, Radzevich, S.P., ℝ-mapping based method for designing of form cutting tool for sculptured surface machining, 921–938, Copyright 2002, with permission from Elsevier.)

for the use of the ℝ-mapping based method of the form-cutting tool design. The idea and the general concept of implementation of the ℝ-mapping of surfaces for the design of optimal form-cutting tools for machining a sculptured part surface on a multiaxis *NC* machine has been proposed* [102] by the author [134,136] in the 1980s.

5.2 Generation of Enveloping Surfaces

When machining a part surface, surfaces P and T are the conjugate surfaces. At every instant of the machining operation, surfaces P and T are tangent to each other. Commonly, they make contact at either a point or along a characteristic line (cases of contact of surfaces P and T at two or more points, along two and more lines, or at a point and along a line simultaneously are of the theoretical interest only). The tangency of the surfaces is a strong constraint on the parameters of their relative motion. Neither interference of the cutting tool surface T into the part surface P nor interruption of their contact is allowed when the machining is in progress.

When machining a part surface, the interaction of the part surface P and generating surface T of the cutting tool can be interpreted as a type of a virtual mechanism† that is composed of two elements P and T. However, a

---

* Patent No. 4242296/08 (USSR), *A Method for Designing of the Optimal Form Cutting Tool for Machining of a Given Sculptured Part Surface on a Multi-Axis NC Machine*, S. P. Radzevich, filed March 31, 1987.

† If we are to consider parallels between the conjugate action of surface in the theory of part surface generation and between the conjugate action of surfaces in a gear drive, then it is of critical importance to point out here that in part surface generation the surface of action is always congruent to the part surface P to be machined.

principal difference between the conjugate action of the surfaces in machining operations and the conjugate action of the surfaces in a corresponding mechanism certainly occur. The difference is due to the following reason: *The input motion of the mechanism is specified and the desired output motion is known. The interacting surfaces of a mechanism must be determined to ensure the required parameters of the output motion.*

In the theory of part surface generation, two different types of problem are recognized.

When solving problems of the first type, it is assumed that the part surface *P* is machined and the kinematics of the machining operation are known. In this case, it is necessary to determine the generating surface *T* of the form-cutting tool for machining a given part surface. Problems of the first type are commonly referred to as the *direct problems of the theory of part surface generation.*

Problems of the second type are inverse to the direct problems of part surface generation. When solving problems of this type, the generating surface *T* of the cutting tool, and kinematics of the machining operation are assumed to be known. It is necessary to determine the actual parameters of shape of the machined part surface *P*. Commonly, problems of the second type are referred to as the *inverse problems of the theory of part surface generation.*

The total number of types of problem to be solved in the theory of part surface generation is not limited to just the two above-mentioned types of problem. Problems of another nature are covered by the theory as well.

Part surfaces that allow for sliding "over itself" usually can be machined on a conventional machine tool. Part surfaces that do not possess this important property are inconvenient for machining on a conventional machine tool. The conjugate action of the part surface to be machined and of the generating surface of the cutting tool is insightful from the standpoint of implementation of elements of the theory of enveloping curves and enveloping surfaces for the purpose of profiling form-cutting tools. The use of elements of the theory of envelopes could significantly simplify the solution to the problem of profiling form-cutting tools for machining part surfaces that allow for sliding "over itself."

5.2.1 Elements of the Theory of Envelopes

The theory of enveloping curves and the theory of enveloping surfaces both are widely used for profiling form-cutting tools. For convenience, brief presentations from differential geometry of curves and surfaces are made below.

5.2.1.1 Envelope to a Planar Curve

Consider a planar curve that is traveling within the plane of its location. If certain conditions are satisfied, an enveloping curve to consecutive positions of the moving curve could exist [38]. The enveloping curve is commonly called the *envelope*.

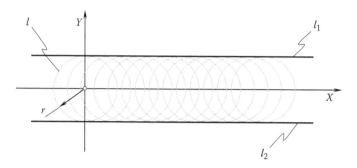

FIGURE 5.13
Generation of the enveloping lines l_1 and l_2 to consecutive positions of a circle l of radius r that is traveling along the X axis.

As an example of a planar curve, a circle l of radius r is shown in Figure 5.13. All points of the circle l are located within the coordinate plane XY. The circle l is traveling along the X axis. When traveling, the circle l occupies consecutive positions. This way, a family of circles is generated. Two straight lines l_1 and l_2 are the enveloping lines to the family of circles of radius r. These straight lines are parallel to the X-axis and are at a distance $2r$ apart from one another.

The enveloping curve to a family of lines is tangent at each of its points to one of the curves of the family of curves. Each circle shown in Figure 5.13 has a common point with the enveloping line l_1 and with another enveloping line l_2.

Consider a family $\mathbf{r}^{fm}(u,\omega)$ of planar curves $\mathbf{r}(u)$

$$\mathbf{r}^{fm}(u,\omega) = \begin{bmatrix} X(u,\omega) \\ Y(u,\omega) \\ 1 \end{bmatrix} \tag{5.41}$$

where $\mathbf{r}(u)$ is the position vector of a point of the initially given planar curve that is traveling within a plane, u is the parameter of the planar curve $\mathbf{r}(u)$, $\mathbf{r}^{fm}(u,\omega)$ is the position vector of a point of the family of the planar curves $\mathbf{r}(u)$, ω is the parameter of motion of the planar curve $\mathbf{r}(u)$, and X, Y are the *Cartesian* coordinates of a point of the planar curve $\mathbf{r}(u)$.

The parameter of motion of the planar curve $\mathbf{r}(u)$ can also be interpreted as the parameter of the family of curves $\mathbf{r}^{fm}(u,\omega)$.

Parallelism of the tangent vectors $\partial \mathbf{r}/\partial u$ and $\partial \mathbf{r}^{fm}/\partial \omega$ is the necessary condition for the existence of the enveloping curve. This condition yields analytical representation in the following form:

$$\frac{\partial \mathbf{r}}{\partial u} \times \frac{\partial \mathbf{r}^{fm}}{\partial \omega} = 0 \tag{5.42}$$

The inequality

$$\frac{\partial \mathbf{r}}{\partial u} \cdot \frac{\partial \mathbf{r}'^{fm}}{\partial \omega} - \frac{\partial \mathbf{r}}{\partial \omega} \cdot \frac{\partial \mathbf{r}'^{fm}}{\partial u} \neq 0 \qquad (5.43)$$

represents the sufficient condition of existence of the enveloping curve.

Example 5.3

Consider a family of planar curves T that is given by the following equation:

$$\mathbf{r}_T(\alpha, R) = \begin{bmatrix} (2 \cdot R + R \cdot \cos\alpha) \\ R \cdot \sin\alpha \\ 1 \end{bmatrix} \qquad (5.44)$$

where $\mathbf{r}_T(\alpha, R)$ is the position vector of a point of the family of the planar curves, R is the radius of a circle of the family of the planar curves, and α is the enveloping parameter of the family of the planar curves.

Partial derivatives of the position vector of a point $\mathbf{r}_T(\alpha, R)$ of the traveling planar curve with respect to α and R parameters are equal:

$$\frac{\partial \mathbf{r}_T}{\partial \alpha}(\alpha, R) = \begin{bmatrix} -R \cdot \sin\alpha \\ R \cdot \cos\alpha \\ 1 \end{bmatrix} \qquad (5.45)$$

and

$$\frac{\partial \mathbf{r}_T}{\partial R}(\alpha, R) = \begin{bmatrix} 2 + \cos\alpha \\ \sin\alpha \\ 1 \end{bmatrix} \qquad (5.46)$$

Equations 5.45 and 5.46 make the following equality possible:

$$\begin{vmatrix} -R \cdot \sin\alpha & R \cdot \cos\alpha \\ 2 + \cos\alpha & \sin\alpha \end{vmatrix} = 0 \qquad (5.47)$$

From the determinant (see Equation 5.47), it is easy to come up with the expression

$$-R \cdot (1 + 2 \cdot \cos\alpha) = 0 \qquad (5.48)$$

Simple formula transformations yield $\cos\alpha = -0.5$ and $\sin\alpha = \sqrt{3}/2$.

After substituting the last equalities into Equation 5.44 of a family of curves and after excluding the parameter R, the equation of the enveloping curve in *Cartesian* coordinates can be represented in the form

$$Y(X) = \frac{1}{\sqrt{3}} \cdot X \qquad\qquad (5.49)$$

Therefore, the enveloping curve is a straight line at the angle ±30° to the X-axis. The family of the curves was a family of circles with centers on the X-axis (Figure 5.14). The family of curves can be generated by a circle having translation along the X-axis, the radius of which increases according to the distance from the origin of the coordinate system XY to the center of a movable circle.

The considered example is of practical importance for machining a sheet metal workpiece with a milling cutter having a conical generating surface T (Figure 5.15). The axis of rotation of the milling cutter is traveling along the X-axis with the feed rate V_{fr}. Simultaneously, the milling cutter is performing a motion V_{ax} in its axial direction along its axis of rotation (Z axis, not shown in Figure 5.15). The actual timing of the motions V_{fr} and V_{ax} depends upon the shape of the part surface P. The functional relation between the motions V_{fr} and V_{ax} can be either linear or nonlinear.

5.2.1.2 Envelope to a One-Parametric Family of Surfaces

Consider a one-parametric family of surfaces. The family of surfaces is dependent on a parameter of motion that is designated as ω. The enveloping surface becomes tangent with each surface of the family of surfaces [38].

For example, the centers of all spheres of a family of spheres of radius r are located within the X-axis of the *Cartesian* coordinate system XYZ (Figure 5.16). The round cylinder of radius r having the X-axis as the axis of its rotation represents the enveloping surface to the family of spheres of radius r.

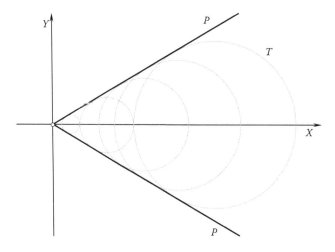

FIGURE 5.14
Enveloping curve to a family of planar curves.

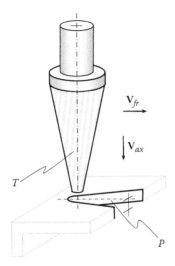

FIGURE 5.15
An example of practical implementation of the solution to the problem of determining an envelope to a family of planar curves.

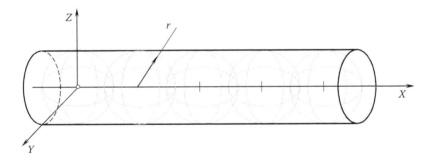

FIGURE 5.16
Generation of the enveloping surface to consecutive positions of a sphere of radius r that is
traveling along the X axis.

Consider a family $\mathbf{r}^{fm}(U, V, \omega)$ of the surfaces $\mathbf{r}(U, V)$ for which the inequality $\dfrac{\partial \mathbf{r}}{\partial U} \times \dfrac{\partial \mathbf{r}}{\partial V} \neq 0$ is valid:

$$
\mathbf{r}^{fm}(U, V, \omega) = \begin{bmatrix} X(U, V, \omega) \\ Y(U, V, \omega) \\ Z(U, V, \omega) \\ 1 \end{bmatrix}
\tag{5.50}
$$

where $\mathbf{r}(U, V)$ is the position vector of a point of a sphere of the family of
spheres, U, V are the *Gaussian* (curvilinear) coordinates on s sphere $\mathbf{r}(U, V)$,

$r^{fm}(U, V, \omega)$ is the position vector of a point of the family of the spheres, and ω is the parameter of motion of the spheres (the enveloping parameter).

The necessary conditions of existence of an enveloping surface are as follows:

$$\mathbf{r}^{fm} = \mathbf{r}^{fm}(U, V, \omega) \tag{5.51}$$

$$\left(\frac{\partial \mathbf{r}}{\partial U} \frac{\partial \mathbf{r}}{\partial V} \frac{\partial \mathbf{r}}{\partial \omega} \right) = 0 \tag{5.52}$$

The line of tangency of a surface $\mathbf{r}(U, V)$ [from a family of surfaces $\mathbf{r}^{fm}(U, V, \omega)$] with the enveloping surface is referred to as *characteristic line E*. The characteristic line E satisfies both of the equations (see Equations 5.48 and 5.49) simultaneously. The enveloping surface yields representation in the form of a family of the characteristic lines.*

If the enveloping surface allows for sliding over itself, the profile of the enveloping surface is described by Equation 5.49.

The satisfaction of the condition $\mathbf{r} \in \omega^2$ of the relationships (see Equations 5.51 and 5.52) together with the conditions

$$\begin{vmatrix} \dfrac{\partial \psi}{\partial U} & \dfrac{\partial \psi}{\partial V} & \dfrac{\partial \psi}{\partial \omega} \\[2mm] \left(\dfrac{\partial \mathbf{r}}{\partial U} \right)^2 & \dfrac{\partial \mathbf{r}}{\partial U} \cdot \dfrac{\partial \mathbf{r}}{\partial V} & \dfrac{\partial \mathbf{r}}{\partial U} \cdot \dfrac{\partial \mathbf{r}}{\partial \omega} \\[2mm] \dfrac{\partial \mathbf{r}}{\partial U} \cdot \dfrac{\partial \mathbf{r}}{\partial V} & \left(\dfrac{\partial \mathbf{r}}{\partial V} \right)^2 & \dfrac{\partial \mathbf{r}}{\partial V} \cdot \dfrac{\partial \mathbf{r}}{\partial \omega} \end{vmatrix} \neq 0, \quad \left| \dfrac{\partial \psi}{\partial U} \right| + \left| \dfrac{\partial \psi}{\partial V} \right| \neq 0 \tag{5.53}$$

is the sufficient condition for the existence of the profile of the enveloping surface.

Violation of the first condition of Equation 5.53 is commonly because an edge of inversion is observed.

The characteristic lines of the part surface P and the generating surface T of the cutting tool satisfy the conditions

---

* In a coordinate system associated with the cutting tool, the family of the characteristic lines, E, determines the generating surface T of the cutting tool. For this purpose, Equation 5.52 must be considered together with the operator that describes the motion of the characteristic line E in the coordinate system associated with the cutting tool. In the event that the inverse problem, not the direct problem of part surface generation is considered, the family of the characteristics E in a coordinate system associated with the work determines the actually machined part surface P.

$$\mathbf{r}_P^{fm} = \mathbf{r}_P^{fm}(U_P, V_P, \omega), \ f[U_P(\omega), V_P(\omega), \omega] = 0, \ \omega = Const \qquad (5.54)$$

$$\mathbf{r}_T^{fm} = \mathbf{r}_T^{fm}(U_T, V_T, \omega), \ f[U_T(\omega), V_T(\omega), \omega] = 0, \ \omega = Const \qquad (5.55)$$

In a stationary coordinate system, for example in a coordinate system that is associated with the machine tool, the family of the characteristic lines can be represented by the set of equations:

$$\frac{\partial \mathbf{r}_2}{\partial f} = \frac{\partial \mathbf{r}_2}{\partial f}(U_2, V_2, \omega), \qquad (5.56)$$

$$f(U_2, V_2, \omega) = 0 \qquad (5.57)$$

where the equality

$$\frac{\partial \mathbf{r}_2}{\partial f}(U_2, V_2, \omega) = \mathbf{Rs}(1 \rightarrow 2) \cdot \mathbf{r}_1(U_1, V_1) \qquad (5.58)$$

is observed.

The operator $\mathbf{Rs}(1 \rightarrow 2)$ of the resultant coordinate system transformation is a function of the parameter of motion ω. The theory of part surface generation deal with surfaces whose enveloping surface is congruent to the moving surface itself.

5.2.1.3 Envelope to a Two-Parametric Family of Surfaces

The two-parametric enveloping surface can be expressed in terms of two parameters, say, ω_1 and ω_2. At every point, the enveloping surface becomes tangent with one of the surfaces of the family of surfaces that is specified by the parameters $\omega_1(U, V)$ and $\omega_2(U, V)$. The parameters ω_1 and ω_2 are of the same value at every point of every surface of the family of surfaces. However, they differ at different points of the enveloping surface.

If the condition $\partial \mathbf{r}/\partial U \times \partial \mathbf{r}/\partial V \neq 0 \pi$ is satisfied, the necessary condition for the existence of the enveloping surface to a family of surfaces $\mathbf{r}(U, V, \omega_1, \omega_2)$ can be represented in the following form [38]:

$$\mathbf{\psi}_1 = \left(\frac{\partial \mathbf{r}}{\partial U} \frac{\partial \mathbf{r}}{\partial V} \frac{\partial \mathbf{r}}{\partial \omega_1} \right) = 0 \qquad (5.59)$$

$$\mathbf{\psi}_2 = \left(\frac{\partial \mathbf{r}}{\partial U} \frac{\partial \mathbf{r}}{\partial V} \frac{\partial \mathbf{r}}{\partial \omega_2} \right) = 0 \qquad (5.60)$$

To obtain a sufficient set of conditions for the existence of the enveloping surface, the above conditions (see Equations 5.59 and 5.60) must be considered together with the following conditions:

$$
\begin{vmatrix}
\dfrac{\partial \boldsymbol{\psi}_1}{\partial u} & \dfrac{\partial \boldsymbol{\psi}_1}{\partial v} & \dfrac{\partial \boldsymbol{\psi}_1}{\partial A} & \dfrac{\partial \boldsymbol{\psi}_1}{\partial B} \\[2mm]
\dfrac{\partial \boldsymbol{\psi}_2}{\partial u} & \dfrac{\partial \boldsymbol{\psi}_2}{\partial v} & \dfrac{\partial \boldsymbol{\psi}_2}{\partial A} & \dfrac{\partial \boldsymbol{\psi}_2}{\partial B} \\[2mm]
\left(\dfrac{\partial \mathbf{r}}{\partial U}\right)^2 & \dfrac{\partial \mathbf{r}}{\partial U}\cdot\dfrac{\partial \mathbf{r}}{\partial V} & \dfrac{\partial \mathbf{r}}{\partial U}\cdot\dfrac{\partial \mathbf{r}}{\partial \omega_1} & \dfrac{\partial \mathbf{r}}{\partial U}\cdot\dfrac{\partial \mathbf{r}}{\partial \omega_2} \\[2mm]
\dfrac{\partial \mathbf{r}}{\partial V}\cdot\dfrac{\partial \mathbf{r}}{\partial U} & \left(\dfrac{\partial \mathbf{r}}{\partial V}\right)^2 & \dfrac{\partial \mathbf{r}}{\partial V}\cdot\dfrac{\partial \mathbf{r}}{\partial \omega_1} & \dfrac{\partial \mathbf{r}}{\partial V}\cdot\dfrac{\partial \mathbf{r}}{\partial \omega_2}
\end{vmatrix} \neq 0, \qquad (5.61)
$$

$$
\frac{D(\boldsymbol{\psi}_1, \boldsymbol{\psi}_2)}{D(\omega_1, \omega_2)} \neq 0 \qquad (5.62)
$$

If a surface $\mathbf{r}(U, V, \omega_1, \omega_2)$

a. Is performing a two-parametric motion
b. Both the motions are independent from each other
c. The characteristics E_1 and E_2 occur for each of the motions

then the point of intersection of the characteristic lines E_1 and E_2 is a point of the enveloping surface. This point is referred to as *characteristic point*. At the characteristic point, the conditions $\mathbf{n}\cdot\mathbf{V}_{1-2}^{(\omega_1)} = 0$ and $\mathbf{n}\cdot\mathbf{V}_{1-2}^{(\omega_2)} = 0$ are always satisfied. Here, n designates a unit normal vector to the enveloping surface.

The conditions $\mathbf{n}\cdot\mathbf{V}_{1-2}^{(\omega_1)} = 0$ and $\mathbf{n}\cdot\mathbf{V}_{1-2}^{(\omega_2)} = 0$ are derived for the cases when the resultant relative motion of the surfaces $\mathbf{V}_\Sigma$ is decomposed onto two components $\mathbf{V}_{1-2}^{(\omega_1)}$ and $\mathbf{V}_{1-2}^{(\omega_2)}$, and both of the components are within the common tangent plane. Definitely, these conditions are sufficient, but they are not necessary. It is possible to decompose the resultant relative motion $\mathbf{V}_\Sigma$ of the surfaces (which is within the common tangent plane) on two particular motions $\mathbf{V}_{1-2}^{(\omega_1)}$ and $\mathbf{V}_{1-2}^{(\omega_2)}$ that are not within the common tangent plane. Consider the following chain of the formulae:

$$
\mathbf{n}\cdot\mathbf{V}_\Sigma = 0 \rightarrow \mathbf{n}\cdot\mathbf{V}_{1-2}^{(\omega_1)} + \mathbf{n}\cdot\mathbf{V}_{1-2}^{(\omega_2)} = 0 \rightarrow \mathbf{n}\cdot[\mathbf{V}_{1-2}^{(\omega_1)} + \mathbf{V}_{1-2}^{(\omega_2)}] = 0 \rightarrow \mathrm{Pr}_{\mathbf{n}}\ \mathbf{V}_{1-2}^{(\omega_1)} = -\,\mathrm{Pr}_{\mathbf{n}}\ \mathbf{V}_{1-2}^{(\omega_2)}]
$$
$$(5.63)$$

However, the location of the velocity vector $\mathbf{V}_\Sigma$ of the resultant relative motion of two surfaces within the common tangent plane is required.

The discussed approach can be employed to determine the envelope (if any) of an arbitrary surface that has motion of any desired type. The interested reader may wish to go to Reference [10] for details of the solution to the problem of computation of an envelope of a sphere that has screw motion. Many practical examples of this approach can be found in other sources.

Elements of the theory of enveloping surfaces can be employed for profiling form-cutting tools (that is, for solving the direct problem of the theory of part surface generation) as well as for determining the actually machined part surface (that is, for solving the inverse problem of the theory of part surface generation). The desired and the actual part surface may differ because of violation of the necessary condition of proper part surface generation (see Chapter 7).

For an illustrative example of a problem that can be solved using elements of the theory of enveloping surfaces, see Figure 5.17.

The generating surface of the first milling cutter is composed of two portions, say, of the cylindrical portion T_{11} and of the spherical portion T_{12}. When performing circular motion, the cylindrical portion T_{11} of the generating surface of the cutting tool is generating the part surface $P_{11}^{(1)}$. The spherical portion T_{12} of the generating surface of the milling cutter generates the surface $P_{12}^{(1)}$, which is a torus. When the circular feed rate motion is stopped, the spherical portion $P_{12}^{(2)}$ is machined on the part.

Similarly, the generating surface of the second milling cutter is composed of two portions, say, of the cylindrical portion T_{21}, and of the flat portion T_{22}. The cylindrical portion T_{21} of the generating surface of the milling cutter generates the plane $P_{21}^{(1)}$, and the circular cylinder $P_{21}^{(2)}$. The plane portion T_{22}

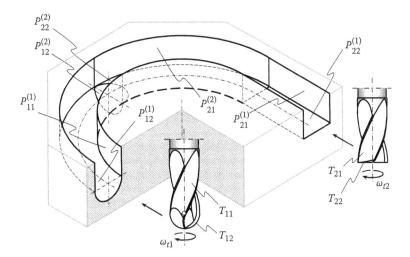

FIGURE 5.17
An example of problems in the theory of part surface generation that can be solved using elements of the theory of enveloping surfaces.

of the generating surface of the cutting tool generates the plane $P_{22}^{(1)}$ and the plane $P_{22}^{(2)}$.

The above-discussed formulae can be used for derivation of equations of all of the machined part surfaces when dimensions of the milling cutters and the parameters of their motion relative to the work are known. Similarly, the above formulae work when necessary to determine the generating surface of a cutting tool for machining of a given part surface.

5.2.2 Kinematic Method to Determine the Enveloping Surfaces

For engineering applications, one more method for the determination of enveloping surfaces is helpful. This method is referred to as the *kinematic method* for the determination of enveloping surfaces. Initially, this method had been proposed by Shishkov as early as in late 1940s [155,156].

The kinematic method is based on the particular location of the velocity vector V_{1-2} of the resultant relative motion of the moving surface and of the enveloping surface. In this particular configuration of the velocity vector V_{1-2}, the Shishkov equation of contact* $n \cdot V_{1-2} = 0$ is fulfilled (here unit normal vector of the common perpendicular is denoted by n).

The velocity vector V_{1-2} is located within the common tangent plane to the surfaces. This condition immediately follows from the following consideration. The motions of only two types are feasible for the moving surface and the enveloping surface. The surfaces can roll and slide over each other. The component of the resultant relative motion V_{1-2} in the direction of common perpendicular to the surfaces is always equal to zero as illustrated in Figure 5.18.

The cutting tool performs a certain motion relative to the work. The part surface P is generated as an enveloping surface to consecutive positions of the generating surface T of the cutting tool. The points of the three different types can be distinguished on the moving tool surface T.

Consider the points of the first type, for example the point A as shown in Figure 5.18. The vector of resultant relative motion of the cutting tool with respect to the work at point A is designated as $V_\Sigma^{(A)}$. The projection $\text{Pr}_n V_\Sigma^{(A)}$ of the velocity vector $V_\Sigma^{(A)}$ onto the unit normal vector $n_T^{(A)}$ to the generating surface T of the cutting tool is pointed to the interior of the work body $[\text{Pr}_n V_\Sigma^{(A)} > 0]$. Therefore, in the vicinity of point A the cutting tool penetrates

---

* Prof. V. A. Shishkov was the first (in 1948 or even earlier) to represent the condition of contact of two smooth regular surfaces in the form of dot product $n \cdot V = 0$. The equation of contact in the form $n \cdot V = 0$ has been published in References [155,156] (see Reference [111] for details). Later, the *Shishkov* equation, $n \cdot V = 0$, saw wide application in science and engineering. For more detailed analysis on the history of the *Shishkov* equation of contact, the interested reader may with to go to the paper, "Concisely on Kinematic Method and on the History of the Equation of Contact in the Form $n \cdot V = 0$" by S. P. Radzevich [*Theory for Machines and Mechanisms*, 2010, No. 1 (15), Vol. 8, pp. 42–51, http://tmm.spbstu.ru].

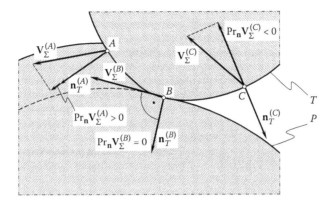

FIGURE 5.18
Concept of the *kinematic method* to determine an enveloping part surface.

the work body. In this way, rough portions of the tool cutting edges cut out the stock.

Further, consider points of the second type, for example point B (Figure 5.18). The velocity vector of resultant relative motion of the cutting tool with respect to the work at point B is designated as $\mathbf{V}_\Sigma^{(B)}$. The projection $\mathrm{Pr_n}\,\mathbf{V}_\Sigma^{(B)}$ of the velocity vector $\mathbf{V}_\Sigma^{(B)}$ onto the unit normal vector $\mathbf{n}_T^{(B)}$ to the generating surface T of the cutting tool is perpendicular to this unit normal vector $[\mathrm{Pr_n}\,\mathbf{V}_\Sigma^{(B)} = 0]$—that is, it is tangential to the part surface P. Therefore, near point B, the cutting tool does not penetrate the part body. The tool cutting edges do not cut out stock. The generating surface T of the cutting tool generates the part surface P near point B.

Ultimately, consider the points of the third type, for example, the point C schematically depicted in Figure 5.18. The velocity vector of the resultant motion of the cutting tool with respect to the work at the point C is designated as $\mathbf{V}_\Sigma^{(C)}$. The projection $\mathrm{Pr_n}\,\mathbf{V}_\Sigma^{(C)}$ of the velocity vector $\mathbf{V}_\Sigma^{(C)}$ onto the unit normal vector $\mathbf{n}_T^{(C)}$ to the generating surface T is pointed outside the part body $[\mathrm{Pr_n}\,\mathbf{V}_\Sigma^{(C)} < 0]$. Therefore, near point C, the cutting tool departs from the machined part surface P. Near the points of the third type, the tool cutting edges do not cut out stock and the generating surface T of the cutting tool does not generate the part surface P.

The considered example unveils the nature of the kinematic method to determine the enveloping surface. Apparently, this method can be employed for solving problems of both types, profiling form-cutting tools for machining a given part surface, and solving the inverse problem of the theory of part surface generation.

As an example, the generation of the plane P with the cylindrical grinding wheel having the generating surface T in the form of a cylinder of revolution is considered in Figure 5.19.

When finishing a plane P, the grinding wheel rotates about its axis of rotation with a certain angular velocity ω_T. Simultaneously, the grinding wheel

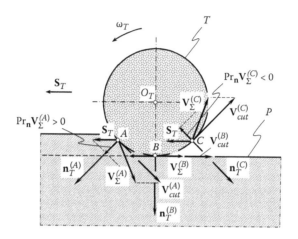

FIGURE 5.19
Analysis of grinding of a plane surface P from the standpoint of the kinematic method for determining the enveloping surface.

travels with a feed rate $\mathbf{S}_T$. At each of the points A, B, and C on the generating surface T of the grinding wheel, the speed of the resultant relative motion of the cutter with respect to the work is designated as $\mathbf{V}_\Sigma^{(A)}$, $\mathbf{V}_\Sigma^{(B)}$, and $\mathbf{V}_\Sigma^{(C)}$, respectively. The speed of the resultant relative motion at every point A, B, and C is equal to the vector sum of the feed rate $\mathbf{S}_T$ and the speed of cutting $\mathbf{V}_{cut}^{(A)}$, $\mathbf{V}_{cut}^{(B)}$, and $\mathbf{V}_{cut}^{(C)}$. The feed rate $\mathbf{S}_T$ is of the same value for all of the points A, B, and C. The velocities $\mathbf{V}_{cut}^{(A)}$, $\mathbf{V}_{cut}^{(B)}$, and $\mathbf{V}_{cut}^{(C)}$ are equal to the linear speed of rotation that is necessary to perform cutting.

It is easy to see that the velocity vectors $\mathbf{V}_{cut}^{(A)}$, $\mathbf{V}_{cut}^{(B)}$, and $\mathbf{V}_{cut}^{(C)}$ are of the same magnitude $\left[\left|\mathbf{V}_{cut}^{(A)}\right| = \left|\mathbf{V}_{cut}^{(B)}\right| = \left|\mathbf{V}_{cut}^{(C)}\right|\right]$. They differ from each other only by directions. However, the velocity vectors $\mathbf{V}_\Sigma^{(A)}$, $\mathbf{V}_\Sigma^{(B)}$, and $\mathbf{V}_\Sigma^{(C)}$ are of different magnitude $\left[\left|\mathbf{V}_\Sigma^{(A)}\right| \neq \left|\mathbf{V}_\Sigma^{(B)}\right| \neq \left|\mathbf{V}_\Sigma^{(C)}\right|\right]$, and they have different directions, $\left[\mathbf{V}_\Sigma^{(A)} \neq \mathbf{V}_\Sigma^{(B)} \neq \mathbf{V}_\Sigma^{(C)}\right]$.

The projections of the velocity vectors $\mathbf{V}_\Sigma^{(A)}$, $\mathbf{V}_\Sigma^{(B)}$, and $\mathbf{V}_\Sigma^{(C)}$ onto the unit normal vectors $\mathbf{n}_T^{(A)}$, $\mathbf{n}_T^{(B)}$, and $\mathbf{n}_T^{(C)}$ are as follows: $\mathrm{Pr_n}\,\mathbf{V}_\Sigma^{(A)} > 0$, $\mathrm{Pr_n}\,\mathbf{V}_\Sigma^{(B)} = 0$, and $\mathrm{Pr_n}\,\mathbf{V}_\Sigma^{(C)} < 0$, respectively. Therefore,

- In the vicinity of point A, the grinding wheel cuts the stock off.
- In the vicinity of point B, the part surface P is generated.
- In the vicinity of point C, the cutting tool departs from the machined plane P.

Similarly, generation of the plane surface P can be performed by the cylindrical milling cutter.

An analysis similar to this example of grinding of a plane P can be performed for any type of machining of a surface on a machine tool.

For a family of surfaces specified by the vector equation of the kind $\mathbf{r} = \mathbf{r}(U, V, \omega)$, equation of the characteristic curve E yields representation in the form

$$\mathbf{r} = \mathbf{r}(U, V, \omega), \tag{5.64}$$

$$\mathbf{n} \cdot \mathbf{V}_{1-2} = 0 \tag{5.65}$$

A perpendicular $\mathbf{N}$ to the surface $\mathbf{r}_1$ can be expressed analytically as

$$\mathbf{N} = \frac{\partial \mathbf{r}}{\partial U_1} \times \frac{\partial \mathbf{r}}{\partial V_1}. \tag{5.66}$$

This means that dot product $\mathbf{N} \cdot \mathbf{V}_{1-2}$ allows for representation in the form of a triple product:

$$\frac{\partial \mathbf{r}_1}{\partial U_1} \times \frac{\partial \mathbf{r}_1}{\partial V_1} \cdot \mathbf{V}_{1-2} \tag{5.67}$$

of three vectors where the velocity vector $\mathbf{V}_{1-2}$ of the resultant relative motion with a certain parameter ω is equal to

$$\mathbf{V}_{1-2}^{(\omega)} = \frac{\partial \mathbf{r}_1}{\partial \omega} = \begin{bmatrix} \dfrac{\partial X_1}{\partial \omega} \\[2mm] \dfrac{\partial Y_1}{\partial \omega} \\[2mm] \dfrac{\partial Z_1}{\partial \omega} \\[2mm] 1 \end{bmatrix} \tag{5.68}$$

Ultimately, the equation of the characteristic curve E

$$\frac{\partial X_1}{\partial U_1}\left(\frac{\partial Y_1}{\partial V_1} \cdot \frac{\partial Z_1}{\partial \omega} - \frac{\partial Y_1}{\partial \omega} \cdot \frac{\partial Z_1}{\partial V_1} \right) - \frac{\partial Y_1}{\partial U_1}\left(\frac{\partial X_1}{\partial V_1} \frac{\partial Z_1}{\partial \omega} - \frac{\partial X_1}{\partial \omega} \frac{\partial Z_1}{\partial V_1} \right)$$

$$+ \frac{\partial Z_1}{\partial U_1}\left(\frac{\partial X_1}{\partial V_1} \frac{\partial Y_1}{\partial \omega} - \frac{\partial X_1}{\partial \omega} \frac{\partial Y_1}{\partial V_1} \right) = 0 \tag{5.69}$$

can be derived for the equation of contact

$$\mathbf{N} \cdot \mathbf{V}_{1-2}^{(\omega)} = \frac{\partial \mathbf{r}_1}{\partial U_1} \times \frac{\partial \mathbf{r}_1}{\partial V_1} \cdot \mathbf{V}_{1-2}^{(\omega)} = 0. \tag{5.70}$$

When using the kinematic method, the sufficient condition for the existence of the enveloping surface can be obtained in the following way.

Consider a smooth, regular surface $\mathbf{r}_1$ that is given in a *Cartesian* coordinate system $X_1Y_1Z_1$. The equation for the position vector of a point $\mathbf{r}_1$ of the surface is represented in the form

$$\mathbf{r}_1 = \mathbf{r}_1(U_1, V_1) \in C^2. \tag{5.71}$$

The family $\mathbf{r}_1^\omega$ of these surfaces in a *Cartesian* coordinate system $X_2Y_2Z_2$ is given by an equation in the form

$$\mathbf{r}_1^\omega = \mathbf{r}_1^\omega(U_1, V_1, \omega), \tag{5.72}$$

where the inequality $\omega^{\min} \leq \omega \leq \omega^{\max}$ is observed.

Then, if either the conditions

$$\left(\frac{\partial \mathbf{r}_1(\omega)}{\partial U_1(\omega)} \times \frac{\partial \mathbf{r}_1(\omega)}{\partial V_1(\omega)} \right) \cdot \frac{\partial \mathbf{r}_1(\omega)}{\partial \omega} = f[U_1(\omega), V_1(\omega), \omega] = 0, \quad f \in C^1 \tag{5.73}$$

or the conditions

$$\left(\frac{\partial \mathbf{r}_1(\omega)}{\partial U_1(\omega)} \times \frac{\partial \mathbf{r}_1(\omega)}{\partial V_1(\omega)} \right) \cdot \mathbf{V}_{1-2} = f[U_1(\omega), V_1(\omega), \omega] = 0, \tag{5.74}$$

$$\left(\frac{\partial f}{\partial U_1} \right)^2 + \left(\frac{\partial f}{\partial V_1} \right)^2 \neq 0 \tag{5.75}$$

$$g_1[U_1(\omega), V_1(\omega), \omega] = \begin{vmatrix} \dfrac{\partial f}{\partial U_1} & \dfrac{\partial f}{\partial V_1} & \dfrac{\partial f}{\partial \omega} \\[2ex] \left(\dfrac{\partial \mathbf{r}_1}{\partial U_1} \right)^2 & \left(\dfrac{\partial \mathbf{r}_1}{\partial U_1} \right) \cdot \left(\dfrac{\partial \mathbf{r}_1}{\partial V_1} \right) & \left(\dfrac{\partial \mathbf{r}_1}{\partial U_1} \right) \cdot \mathbf{V}_{1-2} \\[2ex] \left(\dfrac{\partial \mathbf{r}_1}{\partial V_1} \right) \cdot \left(\dfrac{\partial \mathbf{r}_1}{\partial U_1} \right) & \left(\dfrac{\partial \mathbf{r}_1}{\partial V_1} \right)^2 & \left(\dfrac{\partial \mathbf{r}_1}{\partial V_1} \right) \cdot \mathbf{V}_{1-2} \end{vmatrix} \neq 0$$

$$\tag{5.76}$$

are satisfied at a certain point, then the enveloping surface exists and can be represented by two vector equations in the form

$$\mathbf{r}_1 = \mathbf{r}_1(U_1, V_1, \omega), \tag{5.77}$$

$$\frac{\partial \mathbf{r}_1}{\partial \omega} = 0. \tag{5.78}$$

Determining enveloping surfaces, which are based on implementation of the methods developed in differential geometry of surfaces, makes it possible to determine points of local tangency of the moving surface with the enveloping surface under fixed values of the enveloping parameter ω. However, for a certain value of the enveloping parameter $\omega = Const$, global interference of the surfaces could occur.

Differential methods to determine the enveloping surfaces can be employed only when the equation of the moving surface is differentiable. Because surfaces in engineering applications are not infinite and could be represented by patches and so forth, the part surface P can also be generated by special points on the surfaces.

In the general theory of enveloping surfaces, the family of surfaces that change their shape are considered as well. Results of the research in this area can be used in the theory of part surface generation, particularly for generation of surfaces by the cutting tools that have a changeable generating surface T [108,127,128,152].

Example 5.4

Consider a plane T that is performing a screw motion. The plane T forms a certain angle τ_b with the X_0 axis of the *Cartesian* coordinate system $X_0Y_0Z_0$. The reduced pitch p of the screw motion is given (The reduced pitch can be calculated from the formula $p = |\mathbf{V}|/|\omega|$. Here, $\mathbf{V}$ and ω are the velocity vectors of the translation and the rotation motions of the plane, respectively.)

The axis X_0 is the axis of the screw motion.

The auxiliary coordinate system X_1Y_1 is rigidly associated to the plane T (Figure 5.20).

The equation of the plane T can be represented in the form

$$Y_1 = X_1 \cdot \tan \tau_b \tag{5.79}$$

The auxiliary coordinate system $X_1Y_1Z_1$ performs the screw motion together with the plane T in relation to the motionless coordinate system $X_0Y_0Z_0$. In the coordinate system $X_1Y_1Z_1$, the unit normal vector $\mathbf{n}_T$ to the plane T can be represented as

$$\mathbf{n}_T = \begin{bmatrix} 1 \\ -\tan \tau_b \\ 0 \\ 1 \end{bmatrix} \tag{5.80}$$

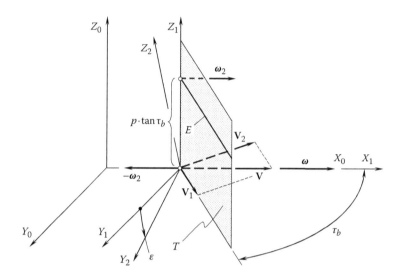

FIGURE 5.20
Generation of a screw involute part surface P as an enveloping surface to consecutive positions of a plane T that is performing a screw motion.

The position vector $\mathbf{r}_T$ of an arbitrary point m of plane T is as follows:

$$\mathbf{r}_T = \begin{bmatrix} X_T \\ Y_T \\ Z_T \\ 1 \end{bmatrix} \tag{5.81}$$

The unit velocity vector, $\mathbf{v}_m$, of an arbitrary point m in its screw motion is

$$\mathbf{v}_m - \mathbf{v} + [\boldsymbol{\omega} \times \mathbf{R}] \tag{5.82}$$

where $\mathbf{v}$ is the unit velocity vector of the translational motion and $\boldsymbol{\omega}$ is the velocity vector of the rotational motion.

To determine the characteristic line E, the direction of the velocity vector $\mathbf{v}_m$ is important, but its magnitude is not of interest. Because of that, it can be assumed that $|\boldsymbol{\omega}| = 1$.

Therefore,

$$\boldsymbol{\omega} = \mathbf{i}, \tag{5.83}$$

$$\mathbf{v} = \mathbf{i} \cdot p \tag{5.84}$$

This makes possible the following expression for the velocity vector $\mathbf{v}_m$:

$$\mathbf{v}_m = \mathbf{i} \cdot p + \begin{vmatrix} \mathbf{i} & \mathbf{j} & \mathbf{k} \\ 1 & 0 & 0 \\ X_1 & Y_1 & Z_1 \end{vmatrix} \tag{5.85}$$

and

$$\mathbf{v}_m = \mathbf{i} \cdot p - \mathbf{j} \cdot Y_1 + \mathbf{k} \cdot Z_1 \tag{5.86}$$

The dot product of the unit normal vector $\mathbf{n}_T$ and of the velocity vector $\mathbf{v}_m$ equals

$$\mathbf{n}_T \cdot \mathbf{v}_m = p \cdot \tan \tau_b - Z_1 = 0 \tag{5.87}$$

Thus, the Shishkov's equation of contact can be represented in the form

$$Z_1 = p \cdot \tan \tau_b \tag{5.88}$$

The above equation of contact together with the equation of the plane T represents the characteristic E.

$$\mathbf{r}_E(t) = \begin{bmatrix} y \\ t \cdot \tan \tau_b \\ p \cdot \tan \tau_b \\ 1 \end{bmatrix} \tag{5.89}$$

where t designates parameter of the characteristic line E.

In the case under consideration, the characteristic line E is the straight line of intersection of the two planes: It is parallel to the coordinate plane $X_1 Z_1$ and distance at $p \cdot \tan \tau_b$.

For a given screw motion, the characteristic line E remains at its location within the plane T in the initial coordinate system $X_0 Y_0 Z_0$.

The angle of rotation of the coordinate system $X_1 Y_1 Z_1$ about the X_0 axis is designated as ε. The translation of the coordinate system $X_1 Y_1 Z_1$ with respect to the reference system $X_0 Y_0 Z_0$ that corresponds to rotation through the angle ε is equal to $p \cdot \varepsilon$. This yields the expression for the operator $\mathbf{Rs}(1 \to 0)$ of the resultant coordinate system transformation

$$\mathbf{Rs}(1 \to 0) = \begin{bmatrix} 1 & 0 & 0 & p \cdot \varepsilon \\ 0 & \cos \varepsilon & \sin \varepsilon & 0 \\ 0 & -\sin \varepsilon & \cos \varepsilon & 0 \\ 0 & 0 & 0 & 1 \end{bmatrix} \tag{5.90}$$

To represent analytically the enveloping surface P, it is necessary to consider the equation $\mathbf{r}_E(t)$ for the position vector of a point of the characteristic line E together with the operator $\mathbf{Rs}(1 \rightarrow 0)$ of coordinate system transformation:

$$\mathbf{r}_P(X_1, \varepsilon) = \begin{bmatrix} X_1 + p \cdot \varepsilon \\ X_1 \cdot \tan \tau_b \cdot \cos \varepsilon + p \cdot \tan \tau_b \cdot \sin \varepsilon \\ -X_1 \cdot \tan \tau_b \cdot \sin \varepsilon + p \cdot \tan \tau_b \cdot \cos \varepsilon \\ 1 \end{bmatrix} \tag{5.91}$$

Consider intersection of the enveloping surface P by the plane

$$X_0 = X_1 + p \cdot \varepsilon = 0. \tag{5.92}$$

The last equation yields $X_1 = -p \cdot \varepsilon$. Therefore,

$$\mathbf{r}_{X_0}(\varepsilon) = \begin{bmatrix} 0 \\ p \cdot \tan \tau_b \cdot (\sin \varepsilon - p \cdot \varepsilon \cdot \cos \varepsilon) \\ p \cdot \tan \tau_b \cdot (\cos \varepsilon + p \cdot \varepsilon \cdot \sin \varepsilon) \\ 1 \end{bmatrix} \tag{5.93}$$

The involute of a circle is analytically described by Equation 5.93. The radius of the base circle of the involute curve is

$$r_b = p \cdot \tan \tau_b \tag{5.94}$$

Therefore, the enveloping surface to consecutive positions of a plane T having a screw motion, is a screw involute surface. The reduced pitch of the screw involute surface equals p and radius of the base cylinder equals $r_b = p \cdot \tan \tau_b$. The screw involute surface intersects the base cylinder. The line of intersection is a helix. The tangent to the helix forms the angle ω_b with the axis of screw motion:

$$\tan \omega_b = \frac{r_b}{p} \tag{5.95}$$

From this, the following two expressions $\tan \omega_b = \tan \tau_b$, and $\omega_b = \tau_b$ are valid. The straight line characteristic E is tangent to the helix of intersection of the enveloping surface P with the base cylinder. For a plane A that is tangent to the base cylinder, this means that

- A straight line E within a plane A makes the base helix angle τ_b with the axis of the screw motion of the plane
- The plane A rolls over the base cylinder without sliding, and the enveloping surface P can be represented as a locus of consecutive positions of the straight line E that rolls without sliding over the base cylinder together with the plane A. The enveloping surface is a screw involute surface.

The obtained screw involute surface (Figure 5.20) is as that shown in Figure 1.6, and as that earlier described analytically by Equation 1.62.

Another solution to the problem of determining the envelope of a plane having screw motion is given by Cormac [10].

The implementation of the kinematic method to determine the enveloping part surface to successive positions of a given smooth, regular part surface that is moving in space in relation to a motionless reference system is especially beneficial in cases when

a. The geometry of the moving part surface is simple (that is, a plane, a cylinder of revolution, a cone of revolution, sphere, and so forth)
b. The motion of the given surface in relation to the motionless reference system is simple (that is, straight motion, rotation, or a combination of these elementary motions).

Due to the simplicity of the surface geometry, unit normal vector **n** to the moving surface can be determined directly with no necessity of differentiating of the equation of the part surface with respect to the surface parameters. Similarly, the simplicity of the relative motion makes it possible to determine the expression for the velocity vector $\mathbf{V}_\Sigma$ of the resultant relative motion of the part surface in relation to the stationary reference system without of operation of differentiation.

Consider an example.

Example 5.5

Consider a screw involute surface $\mathcal{G}$ shown in Figure 1.6. Position vector of a point $\mathbf{r}_g$ of the part surface $\mathcal{G}$ can be expressed in matrix form (see Equation 1.62):

$$\mathbf{r}_g(U_g, V_g) = \begin{bmatrix} 0.5d_{b.g}\cos V_g + U_g\sin\psi_{b.g}\sin V_g \\ 0.5d_{b.g}\sin V_g - U_g\cos\psi_{b.g}\sin V_g \\ 0.5d_{b.g}\tan\psi_{b.g} - U_g\cos\psi_{b.g} \\ 1 \end{bmatrix} \tag{5.96}$$

where U_g, V_g are the *Gaussian* (curvilinear) coordinate lines on the screw involute part surface $\mathcal{G}$, $d_{b.g}$ is the diameter of the base cylinder of the screw involute part surface $\mathcal{G}$, and $\psi_{b.g}$ is the base helix angle of the screw involute surface $\mathcal{G}$.

Once, the equation of the part surface $\mathcal{G}$ is known, the tangent vectors $\mathbf{U}_g$ and $\mathbf{V}_g$ can be derived (see Equations 1.63 and 1.64):

$$\mathbf{U}_g(U_g, V_g) = \frac{\partial \mathbf{r}_g}{\partial U_g}(U_g, V_g) = \begin{bmatrix} \sin\psi_{b.g}\sin V_g \\ -\cos\psi_{b.g}\sin V_g \\ -\cos\psi_{b.g} \\ 0 \end{bmatrix} \tag{5.97}$$

$$\mathbf{V}_g(U_g,V_g) = \frac{\partial \mathbf{r}_g}{\partial V_g}(U_g,V_g) = \begin{bmatrix} -0.5d_{b.g}\sin V_g + U_g\sin\psi_{b.g}\cos V_g \\ 0.5d_{b.g}\cos V_g - U_g\cos\psi_{b.g}\cos V_g \\ 0 \\ 0 \end{bmatrix} \tag{5.98}$$

The derived expressions for the tangent vectors $\mathbf{U}_g$ and $\mathbf{V}_g$ (see Equations 5.97 and 5.98) make possible derivation of an expression for the unit normal vector $\mathbf{n}_g$ to the moving part surface $\mathcal{G}$ (see Equation 1.9):

$$\mathbf{n}_g = \frac{\mathbf{U}_g \times \mathbf{V}_g}{|\mathbf{U}_g \times \mathbf{V}_g|} \tag{5.99}$$

After substituting Equations 5.97 and 5.98 into Equation 5.99 and after making necessary formulae transformations, one can come up with an expression for the unit normal vector $\mathbf{n}_g$:

$$\mathbf{n}_g(U_g,V_g) = \begin{bmatrix} -\cos\psi_{b.g}\cos V_g \\ \cos\psi_{b.g}\sin V_g \\ \sin\psi_{b.g} \\ 0 \end{bmatrix} \tag{5.100}$$

The problem of derivation of an equation for the unit normal vector to a part surface is simple in nature; however, in particular cases, it can be technically inconvenient.

Let us consider how an expression for that same unit normal vector $\mathbf{n}_g$ (see Equation 5.100) can be obtained from the analysis of features of geometry of the part surface $\mathcal{G}$, namely,

- On what the kinematic method of part surface generation is focused
- Based on what implementation of the method can be beneficial

Refer to Figure 5.21 for the more in detail analysis. A straight line along the unit normal vector $\mathbf{n}_g$ is tangent to the base cylinder of the diameter $d_{b.g}$. Moreover, this straight line crosses the axis of the screw involute surface $\mathcal{G}$ at an angle that is equal to $(90° - \omega_{b.g})$. This immediately yields an expression for the unit normal vector $\mathbf{n}_g$:

$$\mathbf{n}_g(U_g,V_g) = \begin{bmatrix} -\cos\psi_{b.g}\cos V_g \\ \cos\psi_{b.g}\sin V_g \\ \sin\psi_{b.g} \\ 0 \end{bmatrix} \tag{5.101}$$

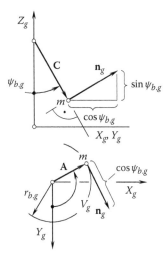

FIGURE 5.21
Determination of the unit normal vector $\mathbf{n}_g$ to a screw involute part surface $\mathcal{G}$, based on analysis of the features of the part surface geometry (in a case when the *kinematic method* for the determination of enveloping part surfaces is used).

It is of critical importance to stress here that Equations 5.100 and 5.101 are identical to one another. However, no derivatives are used for the derivation of Equation 5.101. This advantage of the second approach is utilized when the *kinematic method* for the determination of enveloping part surface is used.

For the determination of the velocity vector $\mathbf{V}_\Sigma$ of the resultant relative motion of the screw involute part surface $\mathcal{G}$ and of the part surface to be determined, let us consider the case when both the surfaces are rotating about parallel axes of rotation.

The given screw involute part surface $\mathcal{G}$ is rotated about the axis O_g as schematically shown in Figure 5.22. A reference system $X_p Y_p Z_p$ of an enveloping part surface to be determined (the surface $\mathcal{P}$, not shown in Figure 5.22) is rotated about the axis O_p. The axes O_g and O_p are parallel to one another. Angle of rotation φ_p of the reference system $X_p Y_p Z_p$ corresponds to an angle of rotation φ_g of the reference system $X_g Y_g Z_g$ associated with the part surface $\mathcal{G}$. The current value of the angle φ_p can be calculated from the expression

$$\varphi_p = \varphi_g \frac{r_g}{r_p} \tag{5.102}$$

where the radii of pitch circles of the part surfaces $\mathcal{G}$ and $\mathcal{P}$ are designated as r_g and r_p, respectively.

The origin of a stationary reference system XYZ is placed at the point of tangency of the pitch circles of radii r_g and r_p, namely, at the pitch point P.

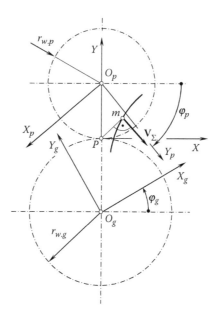

FIGURE 5.22
The determination of velocity vector $\mathbf{V}_\Sigma$ of the resultant motion of the part surface G in relation to a reference system of the enveloping part surface P.

An arbitrary point m is located within the part surface $\mathcal{G}$. In a particular case, this point can be located within the coordinate plane $X_g = 0$. Coordinates of the point m are denoted by X_m and Y_m (remember that in the example under consideration, it is assumed that $Z_g = 0$).

The velocity vector $\mathbf{V}_\Sigma$ of the resultant relative motion of the screw involute part surface $\mathcal{G}$ and of the part surface to be determined can be analytically expressed by equation

$$\mathbf{V} = \mathbf{i} \cdot Y_m + \mathbf{j} \cdot X_m \qquad (5.103)$$

It is important to stress here that no operation of differentiation is used when composing Equation 5.103.

The determined vectors $\mathbf{n}_g$ (see Equation 5.101) and $\mathbf{V}_\Sigma$ (see Equation 5.103) are represented so far in different reference systems. The unit normal vector $\mathbf{n}_g$ is represented in the coordinate system $X_g Y_g Z_g$ associated with the screw involute part surface $\mathcal{G}$, while the velocity vector $\mathbf{V}_\Sigma$ is analytically expressed in the motionless coordinate system XYZ. As the vectors $\mathbf{n}_g$ and $\mathbf{V}_\Sigma$ should be considered together, it is necessary to represent both of them in a common reference system. Then, it will be possible to substitute the derived equations for the vectors $\mathbf{n}_g$ and $\mathbf{V}_\Sigma$ into the Shishkov's equation of contact $\mathbf{n}_g \cdot \mathbf{V}_\Sigma = 0$. Further analysis is trivial and is therefore out of the scope of this monograph. The interested reader may want to consider the formulae transformations required in this particular example.

The discussed example reveals the possibility of the determination of the unit normal vector $\mathbf{n}_g$ as well as of the velocity vector $\mathbf{V}_\Sigma$ of the resultant relative motion using no operation of differentiation for particular surfaces, which are performing certain relatively simple motions in relation to one another. Under such a scenario, namely when no operations of differentiation are used, implementation of the *kinematic method* to determine the enveloping surfaces is becoming reasonable. In this case, implementation of the equation of contact in the form

$$\mathbf{n} \cdot \mathbf{V}_\Sigma = 0 \qquad\qquad (5.104)$$

makes sense.

When expressions either for the unit normal vector $\mathbf{n}$, or for the velocity vector $\mathbf{V}_\Sigma$ of the resultant relative motion, or for both are derived based on derivatives of the moving part surface (see Equations 5.97 and 5.98), then the *kinematic method* is less beneficial. In cases when expressions for the unit normal vector $\mathbf{n}$ and/or for the vector $\mathbf{V}_\Sigma$ of the resultant relative motion are derived based on derivatives, the *kinematic method* reduces to a conventional method that is commonly used in differential geometry of surfaces. To illustrate the last statement, the Shishkov equation of contact (see Equation 5.104) is rewritten in the form

$$\left[\underbrace{\frac{\partial \mathbf{r}}{\partial U} \quad \frac{\partial \mathbf{r}}{\partial V}}_{\mathbf{n}} \quad \underbrace{\frac{\partial \mathbf{r}}{\partial \omega}}_{\mathbf{V}} \right]^{T} = 0 \qquad\qquad (5.105)$$

As in the equation of contact (see Equation 5.104), the dot product of the vectors $\mathbf{n}$ and $\mathbf{V}_\Sigma$ is equalized to zero, the multiplier $\left| \frac{\partial \mathbf{r}}{\partial U} \times \frac{\partial \mathbf{r}}{\partial V} \right|^{-1}$ is omitted in the expression for the unit normal vector $\mathbf{n}$.

5.3 Profiling of the Form-Cutting Tools for Machining Parts on Conventional Machine Tools

Cutting tools for machining parts on conventional machine tools feature a property that allows the generating surface of the cutting tool to slide "over itself." Those surfaces that allow for sliding *"over itself"* are currently the most widely used surfaces in industry. This feature is of importance for the theory of part surface generation. It can be used for the simplification of the solution to the problem of profiling the form-cutting tool. Certain simplification is feasible due to in cases when surfaces allow for sliding *"over themselves,"* it is not necessary to determine the entire generating surface of the cutting tool. It is sufficient to determine either the profile of the generating surface of the cutting tool or the characteristic line along which the generating surface of the cutting tool makes contact with the machined part surface.

5.3.1 On the Looseness of Two Principles by *Theodore Olivier*

When generating a given part surface P, the cutting tool performs certain motions with respect to the work. The part surface P is given, and the generating surface T of the cutting tool is not yet known. Therefore, at the beginning, the corresponding motions of the part surface P and of a certain coordinate system $X_T Y_T Z_T$ are analyzed. After been determined, the generating surface T of the form-cutting tool will be described analytically in the coordinate system $X_T Y_T Z_T$.

The above $\mathbb{R}$-mapping-based method for profiling form-cutting tools (see Section 5.1) is the most general method. This method is a powerful tool for solving the most general problems of profiling of form-cutting tools. However, in particular cases of part surface generation, simpler methods for profiling form-cutting tools for machining a part surface on a conventional machine tool have proven to be convenient. In common practice, when designing form-cutting tools, two Olivier principles* are often used. These principles were proposed by Olivier [48] as early as 1842.

The Olivier principles can be formulated as follows:

First Olivier principle: Two part surfaces that are conjugate to one another can be generated by an auxiliary generating surface. The generating surface in this case differs from both the conjugate surfaces.

Second Olivier principle: A motionless part surface can be generated by another part surface that is traveling in relation to the motionless part surface. In this case, the auxiliary generating surface is congruent to the traveling part surfaces.

It is shown below that both Olivier principles are not valid in the general case of enveloping surfaces, and they are only valid in simple degenerate cases where they are useless.

Prior to solving the problem of profiling a certain form-cutting tool, the geometry of the part surface P (see Chapter 1) and the kinematics of the part surface P generation (see Chapter 2) must be predefined. The operators of the coordinate system transformation (see Chapter 3) are used for the representation of all elements of the part surface generation process in a common

---

* The Olivier principles are often referred to as *fundamental principles* in the theory of part surface generation, which is not correct.

reference system, use of which is preferred for a particular consideration. If we are not only developing a workable cutting tool but also the design of the optimal cutting tool, then the methods of analytical description of the geometry of contact of the part surface P and the generating surface T of the cutting tool (see Chapter 4) are also employed.

A form-cutting tool can be designed on the premise of its generation surface T. Derivation of the generating surface T is the starting point when designing the form-cutting tool.

When profiling a form-cutting tool for machining a given part on a conventional machine tool, it is assumed that the tool surface T is an envelope to successive positions of the part surface P in it motion in relation to the reference system $X_T Y_T Z_T$. Then, when machining the part surface P by the form-cutting tool just designed, the generating surface T reproduces the part surface that is identical to that specified by the part blueprint. In other words, it is assumed here that the part surface P and the generating surface T of the form-cutting tool are a type of the so-called *reversibly-enveloping surfaces* (further, R_e–surfaces) [120,138]. Such an assumption is based on the validity of the Olivier principles. Unfortunately, the assumption is valid only in a few degenerate cases, and it is not valid for all geometries of the part surface P or for all kinematics of the part surface P generation processes.

Generally speaking, the assumption is valid only in the case of the so-called *reversibly-enveloping surfaces* (further, R_e–surfaces) [120,138]. In the later sections of this book, cases of part surface generation when the Olivier principles are either valid or invalid are discussed more in detail.

5.3.1.1 Profiling Form-Cutting Tools for Single-Parametric Kinematic Schemes of Part Surface Generation

Kinematic schemes of part surface generation that feature only one relative motion of the part surface P and the generating surface T of the cutting tool are referred to as *single-parametric kinematic schemes of part surface generation*. The second *Olivier* principle of generation of enveloping surfaces is used when profiling form-cutting tools for single-parametric kinematic schemes of part surface generation.

To demonstrate the capabilities of the Olivier principles in the case under consideration, it is convenient to begin the discussion from the procedure of profiling a form-cutting tool for the generation of a part surface P in the form of a round cylinder.

When machining a round cylinder of radius R_P (Figure 5.23a) [151,150], the work is rotated about its axis O_P with a certain angular velocity ω_P. A coordinate system $X_T Y_T Z_T$ is rotated with a certain angular velocity ω_T. The axis of this rotation O_T crosses at a right angle the axis of rotation O_P of the part surface P. Simultaneously, the coordinate system $X_T Y_T Z_T$ is traveling along the axis O_P with a feed rate S_T. The generating surface T of the cutting tool in this case can be represented as an enveloping surface to consecutive

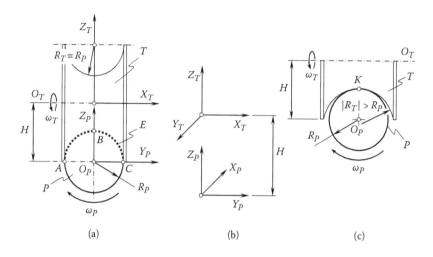

FIGURE 5.23
Example of implementation of a single-parametric kinematic scheme of part surface genera-tion for derivation of the generating surface, *T*, of the cutting tool.

positions of the surface *P* in the coordinate system $X_T Y_T Z_T$. Remember, that the coordinate system $X_T Y_T Z_T$ is the reference system at which the generat-ing surface *T* of the cutting tool could be determined.

After being determined, the generating surface *T* of the cutting tool and the part surface *P* become tangent along the characteristic line *E*. In the case under consideration, the characteristic line *E* is represented with a circular arc $\cup ABC$ of the radius R_P.

The use of a single-parametric kinematic scheme of part surface genera-tion allows a simplification. The generating surface *T* of the cutting tool can be generated as a family of the characteristic lines *E* that rotate about the axis O_T, and not only as enveloping surface to consecutive positions of the part surface *P* in the coordinate system $X_T Y_T Z_T$.

In the example, the generating surface *T* of the cutting tool is shaped in the form of a torus surface. The radius R_T of the generating circle of the torus surface *T* is equal to the radius R_P of the part surface *P* ($R_T = R_P$). The radius of the directing circle of the torus surface *T* is equal to the closest distance of approach *H* of the axes O_P and O_T.

The determined torus surface *T* (Figure 5.23a) can be used for the design of various cutting tools for the machining of the part surface *P*: milling cutters, grinding wheels, and so forth.

The considered example of implementation of the single-parametric kine-matic scheme of part surface generation (see Figure 5.23a) returns a qualita-tive (not quantitative) solution to the problem of profiling of a form-cutting tool. No optimal parameters of the kinematic scheme of part surface genera-tion are determined at this point. The closest distance of approach *H* between the axes O_P and O_T of the rotations and the optimal value of the cross-axis angle χ are those parameters of interest (Figure 5.23b). The optimal values of the parameters H and χ can be calculated on the premise of optimization of the contact geometry between the part surface P and the generating surface T of the cutting tool. The optimal values of the parameters H and χ can be drawn up from the desired degree of conformity of the surface T to the surface P at every point of the characteristic line E.

Actually, in any machining operation, deviations from the actual configuration of the cutting tool with respect to the desired configuration are inevitable. Because of the deviations, it is practical to introduce appropriate alterations to the degree of conformity of the generating surface T of the cutting tool to the part surface P at every point of their contact.

Figure 5.23c illustrates an example when the degree of conformity of the cutting tool surface T to the part surface P is reduced. The reduction of the degree of conformity causes the line contact in the ideal case of part surface generation (Figure 5.23a) being substituted with point contact of surfaces P and T. The depicted schematic (Figure 5.23c) allows for analysis of the impact of the degree of conformity at a point of contact of the surfaces T and P onto the accuracy and quality of part surface generation. Varying values of the parameters H and χ (including the case when $\chi \neq 90°$), one can come up with the solution under which the vulnerability of the machining process to the resultant deviations of configuration of the cutting tool with respect to the part surface P is the smallest possible.

The use of single-parametric kinematic schemes of part surface generation (Figure 5.23) is a perfect example of implementation of the second Olivier principle for profiling form-cutting tools for machining a given part surface on a conventional machine tool.

It should be noticed here that in the example (Figure 5.23) there is no correlation between the rotation of the work, ω_P, and the rotation, ω_T, of the generating surface T of the cutting tool. The derived cutting tool surface T is a type of surface that allows for sliding "over themselves." Under such a scenario, the use of the second Olivier principle for profiling form-cutting tools for machining a given part surface on a conventional machine tool is valid. The use of the second Olivier principle makes possible geometrically accurate generation of the generating surface T of the cutting tool. When solving the inverse problem of part surface generation, the just derived surface T of the cutting tool reproduces the part surface P geometrically accurately.

It should be noticed here that in the trivial case of profiling of the form-cutting tool illustrated in Figure 5.23a, the generating surface T of the cutting tool allows for sliding over itself. Because of this, the use of the second Olivier principle for profiling form-cutting tools is valid.

Consider a more general example of implementation of the second Olivier principle for profiling form-cutting tools for machining a given part surface on a conventional machine tool, namely, for generation of the tool surface T

for a case when relative motions of the part and of the cutting tool are dependent on one another.

The interaction of non-involute profiles is illustrated by an example considered below. This example is adapted from a widely used practice of designing a hob for cutting straight-sided splines. This example clearly illustrates how poorly the industry understands the generation of enveloping surfaces.

A straight profile of the spline is associated with a pitch circle, while the corresponding profile of the conjugate rack is associated with the straight pitch line of the rack.

The hob design is based on an auxiliary rack, the teeth of which are engaged in mesh with splines of the spline-shaft. The tooth profile of the rack is commonly generated as an envelope to successive positions of the spline profile when the pitch circle of the spline-shaft is rolling with no sliding over the pitch line associated with the rack.

The determination of coordinates of points of tooth profile of a rack conjugate to a spline-shaft is executed in practice implementing for this purpose the *method of common perpendiculars*. The tooth profile of a spline hob is generated as an envelope to successive positions of the spline profile when the pitch line of the hob is rolling with no slip over the pitch circle of the spline shaft. An example of solving a problem of this type is illustrated in Figure 5.24.

The profile of the spline is associated with a pitch circle of certain radius $r_{w.sp}$. The pitch line of the conjugate rack to be determined, P_{ln}, is in tangency with the pitch circle of the spline-shaft. The point of tangency of the pitch

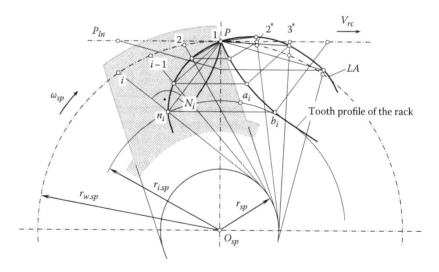

FIGURE 5.24
Interaction between a straight profile associated to a pitch circle with a non-involute profile associated with the pitch line (generation of a rack tooth profile conjugate to lateral profile of a spline of a spline-shaft).

line, P_{ln}, and of the pitch circle, $r_{w.sp}$, is the pitch point in rolling motion of the spline-shaft and of the rack. The pitch point is designated as P.

The spline-shaft is rotated about its axis of rotation, O_{sp}. Angular velocity of the rotation is designated as ω_{sp}. The rack is associated with the pitch line, P_{ln}. The rack travels straight forward together with the pitch line. The linear velocity of the rack is designated as V_{rc}.

Let us assume that at the initial configuration of the pitch circle and the pitch line, the profile of the spline passes through the pitch point, P. This profile is at a distance r_{sp} from the axis of rotation, O_{sp}, of the spline-shaft. Practically, the straight line profile is a tangent to a circle of certain radius r_{sp}. The radius r_{sp} is equal to half the spline thickness of the spline-shaft.

When the spline-shaft rotates, the lateral spline profile is rotated together with the spline-shaft. The spline-shaft lateral profile consequently passes through the points 1, 2, ..., $(i - 1)$, i. The point 1 is coincident with the pitch point, P. The pitch line travels straight forward. In this motion, the pitch point consequently occupies positions 1*, 2*, 3*, ... The distances 1* – 2*, 2* – 3* between consequent locations of the pitch point are equal to lengths of the arcs ⌣1–2, ⌣2–3, ... of the pitch circle of the spline-shaft. This is due to the pitch line of the rack rolling with no slip over the pitch circle of the spline-shaft.

At every chosen location of the lateral profile of the spline, perpendiculars to the profile are constructed so that all of them are through the pitch point, P. For example, a perpendicular N_i is normal to the spline profile in its i-th location (Figure 5.24). The point n_i is the point of tangency of lateral profile of the spline and the rack tooth profile.

The plurality of points constructed in this way for various configurations of the lateral spline profile are within the line of action, LA, in rolling motion of the given spline-shaft and of the rack to be determined. The line of action is determined in a certain stationary reference system.

In a reference system associated with the spline-shaft, all the points are located within lateral spline profile of the spline-shaft.

When the spline-shaft is rotating, the points of the lateral profile consequently pass through the line of action, LA. In these instances of time, these points coincide with the corresponding points of the rack tooth profile. If an arbitrary point n_i within the line of action, that corresponds to the point of contact in i-th location of the lateral profile of the spline, is returned back to the initial position of the spline by means of rotation through the angle of the arc $\overset{\frown}{Pi}$, then this point will occupy the position of the point a_i.

Similarly, in a reference system associated with the rack, contact points are located within the tooth profile of the rack to be determined.

Let us assume that an arbitrary point n_i within the line of action, LA, is associated with the pitch line, P_{ln}. To determine location of this point in the initial instant of time, the pitch line together with the point n_i travels through the distance that is equal to the arc length $\overset{\frown}{Pi}$ in the direction that is opposite to the direction of straight motion of the rack tooth in its rolling motion. After this transition is complete, the point n_i occupies the position of the point b_i. The point b_i is located on the tooth profile of the rack. All points of the rack tooth profile are constructed in the same way that point b_i is constructed. Connecting the constructed points by a smooth curve, the rack tooth profile can be determined.

The above approach for determination of the tooth profile of a rack conjugate to a spline-shaft profile is commonly adopted. However, this method is inaccurate in nature. Deviations of the machined spline profile are inherent in this method. The smaller the number of the splines in the spline-shaft, the larger the deviations, and vice versa.

When the method is used, it is assumed that the generated tooth profile of the rack can generate the spline profile of the spline-shaft when an inverse problem is solved. This is not correct. As the conjugate profiles are not involutes, no straight spline profile can be obtained using inverse rolling of the rack in relation to the spline-shafts. In practice, a certain curved profile of the splines is obtained (Figure 5.25) instead of the desired straight spline profile.

When the obtained hob tooth profile rolls in relation to the spline shaft, some other spline profile is generated instead of the desired original straight-sided spline profile. Then roll the "new" spline profile in relation to the hob, then the hob in relation to the spline, and so on. Under such a scenario, diverging and not converging of the spline profile to the ideal profile is observed. The scenario mentioned above is the observation in the general

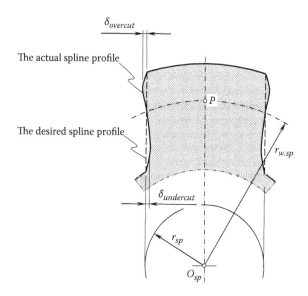

FIGURE 5.25
Deviation of a desired straight profile of the spline (generated as shown in Figure 10.2) from its actual profile.

case of relative motion of two smooth regular part surfaces. This inconsistency is the root cause for troubles when grinding turbine blades using the generating method of grinding, when hobbing non-involute profiles (polygon part profiles as well as the more general case of non-involute part profiles) and so forth.

As the generation of enveloping profiles and surfaces is of critical importance for profiling form-cutting tools and for the generation of part surfaces, the nature of enveloping under a constraint onto the relative motion must be properly understood.

For simplicity, but without loss of generality, consider the generation of a planar profile for the case when both the generating profile and profile to be generated are rotated about parallel axes of rotation.

From elementary gear theory, it can be readily understood that the tooth profiles of all gears operating on parallel axes must obey the well-known conjugate action law: *The common normal at all points of contact must pass through a stationary point on the line of centers.* This motionless point is commonly referred to as the pitch point. This is a key kinematic requirement if one profile is to drive the other at a constant angular speed ratio. It can also be readily understood that a pair of gear profiles contact each other at different positions as the gear rotates. The locus of all possible contact points for a given pair of profiles is called the *path of contact* (or *the line of action*, in other terminology). The *path of action* is a line segment, terminated by the extremities of the gear teeth. Three curves involved in the most fundamental part of the consideration below are

- The profile of the driving gear
- The profile of the driven gear
- The line of action (the path of contact)

A basic geometric statement of great significance is that given a prespecified center-distance and speed ratio, any one of these curves completely determines the other two. For example, if the profile of the driving gear is given, the profile of the driven gear and the line of action are both uniquely determined. Likewise, if the path of contact is shown as some given curve, the profiles of both gears are uniquely determined. Thus it is possible to find out any and all design parameters of the tooth profiles from the given properties of the line of contact.

With that said, consider the generation of a planar profile T of the cutting tool for machining a specified part profile P for the case when both the generating profile P as well as a profile to be generated T are rotated about parallel axes of rotation with uniform angular velocities. Interaction of the planar profiles P and T is schematically depicted in Figure 5.26a.

Axes of the rotations O_P and O_T are at a center distance C apart from one another. The generating part profile P is rotated about the axis O_P with a

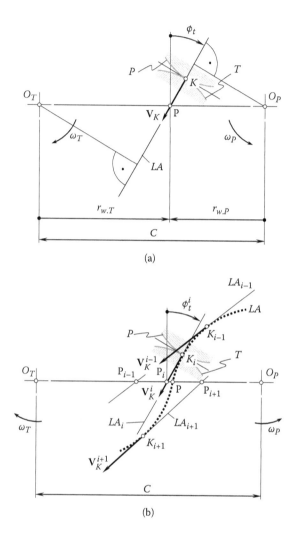

FIGURE 5.26

Schematic of the interaction between conjugate profiles rotating about parallel axes of rotation.

certain angular velocity ω_P. A reference system, which the cutting tool profile will be associated with, is rotated with an angular velocity ω_T. The case of constant velocity ratio u (here $u = \omega_T/\omega_P$) is under consideration. At a current contact point K between the profiles P and T, the part point K_P is traveling in a direction perpendicular to the part profile P with a certain linear velocity $\mathbf{V}_K$. At the instant of time when a corresponding cutting tool profile point K_T is generated, the point K_T of the tool profile T is traveling with that same linear velocity $\mathbf{V}_K$ (as no friction between the profiles P and T is considered in the theory of part surface generation). The velocity vector $\mathbf{V}_K$ is pointed along the line of action LA and intersects the center-line C at the pitch point P. The line of action, LA, forms a certain angle ϕ_t with a perpendicular to the center-line C through the pitch point P. For the cases of constant angular velocity ratio u (u = Const), the line of action, LA, is *a straight line* through the pitch point P, which makes a transverse pressure angle ϕ_t with a perpendicular to the center distance C at P.

The center distance C is divided by the pitch point P into two straight line segments in proportion:

$$r_{w.P}/r_{w.T} = \omega_T/\omega_P \qquad (5.106)$$

where the radii of the pitch circles of the part profile P and of the tool profile T are denoted by $r_{w.P}$ and $r_{w.T}$, respectively ($C = r_{w.P} + r_{w.T}$).

The above conclusion, *"any one of these curves completely determines the other two,"* validates the following statement: *No arbitrary part profile P is capable of generating a cutting tool profile T so that the cutting tool profile T will be capable of generating that same part profile P.* Once the geometry of the line of action is restricted to a straight line segment LA, then only the involute part profile, P, is capable of generating a cutting tool profile T, which is further capable of generating that same part profile P. Moreover, the cutting tool profile T is also not of arbitrary geometry, but it also must be an involute of a circle. Once the velocity ratio u is constant, for a part profile P and for a cutting tool profile T, those interacting on parallel axes, there is not much freedom in selection of the part profile P and in generation of the tool profile T. Only involute profiles P and T meet the requirement of constant velocity ratio u for profiles operating on parallel axes of rotation.

Consider a more general case of generation of a cutting tool profile T by a prespecified part profile P featuring line of action, LA, in the form of a planar curve segment as shown in Figure 5.26b. The line of action is a planar curve through the pitch point P. The part profile P is rotated about the axis O_P. A reference system that the cutting tool profile T will be associated with is rotated about the axis O_T. Let us assume the velocity ratio u is constant (u = Const) as in the previously considered case (Figure 5.26a). As the profiles P and T rotate, the contact point K is traveling along the line of action, LA. The current configuration of the contact point K is designated as K_i. A certain previous location of the contact point is designated as K_{i-1}. When rotation of the profiles P and T is progressing, the contact point occupies the position K_{i+1}, which occurs after the profiles P and T interact with one another at the contact point K.

When the cutting tool profile T is generated in the vicinity of the contact point K_{i-1}, certain portions of the part profile P and of the tool profile T in differential vicinity of the point K_{i-1} are traveling together with a linear velocity $\mathbf{V}_K^{i-1}$. The velocity vector $\mathbf{V}_K^{i-1}$ aligns with the corresponding instant line of action, LA_{i-1}, which is a straight line tangent at K_{i-1} to the line of action, LA. The instant line of action, LA_{i-1}, intersects the center-line C at the instant pitch point, P_{i-1}.

The same analysis is valid with respect to the cutting tool profile T when it is generated in the differential vicinity of the contact points K_i and K_{i+1}. In these two cases, the corresponding instant lines of action, LA_i and LA_{i+1}, are aligned with the instant velocity vectors $\mathbf{V}_K^i$ and $\mathbf{V}_K^{i+1}$, respectively. The instant lines of action, LA_i and LA_{i+1}, intersect the center-line C at the corresponding instant pitch points, P_i and P_{i+1}, respectively. It is clear from this discussion that when the part profile P and the reference system of the cutting tool are rotated at constant angular velocities ω_P and ω_T, the pitch point is not stationary. Instead, the pitch point travels along the center-line C. In such a reciprocating motion, the pitch point occupies locations denoted by P_{i-1}, P_i and P_{i+1}, respectively, which are referred to as *instant pitch points*. Once the pitch point is not motionless, then at a constant velocity ratio ($u = $ Const) the generated cutting tool profile T is not capable of generating the initially specified part profile P. Ultimately, the prespecified part profile P and P_{real} generated by the cutting tool profile T are not identical to each other. The example of generation of a straight-sided spline (which definitely is not of involute geometry) depicted in Figure 5.25 is a perfect illustration of such a type of violation of conditions of profiling of a cutting tool, T, for generating of a given part profile P.

In the considered case of two profiles, interacting with one another on parallel axes of rotation, the profiles P and T are referred to as *reversibly enveloping profiles* (or R_e–*profiles*) only in cases where both of them are conjugate involute profiles.

A possibility of migration of the pitch point within the center-line is illustrated with an example of the required timing of the rotations when determining geometry of centrodes in noncircular gear.

In this example, schematically depicted in Figure 5.27, two noncircular gears to be designed are rotated about the axes of rotation O_g and O_p with angular velocities ω_g and ω_p, respectively. The axes O_g and O_p of the rotations are parallel to one another and are remote from one another at a certain center-distance C.

Consider a case when the driving pinion is rotated at a constant angular velocity, ω_p. Then, angular velocity of the driven gear, ω_g, in the noncircular gearing to be designed is a function of the rotation ω_p, namely, $\omega_g = \omega_g(\omega_p)$. As the current value of angle of rotation of the pinion, $\varphi_{p.i}$ can be expressed in terms of time t and of the pinion angular velocity ω_p, namely, $\varphi_{p.i} = \omega_p t$, then, the instant ratio $r_{p.i}/r_{g.i}$ of pitch radii of the pinion and of the gear is reciprocal to the ratio $\omega_{g.i}/\omega_p$, where $\omega_{g.i}$ is the instant rotation of the driven gear.

Consider a point of contact between axodes of the gear and the pinion. This point is designated as K. The contact point is within a straight line through the axis O_p. The angular location of the straight line is specified by a central angle $\varphi_{p.i}$. Point K is remote from the axis O_p at a certain distance $r_{p.K}$.

Vector $\mathbf{V}_{p.i}$ of linear velocity of the contact point in it rotation with the pinion is perpendicular to the straight line through the axis O_p and the contact point K. This linear velocity $V_{p.i} = |\mathbf{V}_{p.i}|$ is equal to $V_{p.i} = \omega_p r_{p.K}$. Vector $\mathbf{V}_{p.i}$ of the required instant linear velocity of the pitch point P_i can be determined from a prespecified angular velocity ratio:

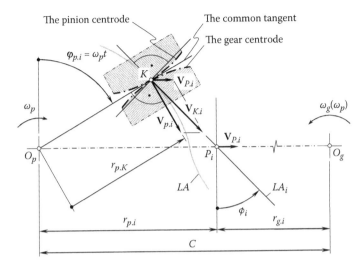

FIGURE 5.27
Construction of a pair of reversibly enveloping centrodes of a driving pinion and of the driven gear.

$$u = \frac{\omega_p}{\omega_g(\omega_p)} = \frac{\varphi_{p.i}}{t} \cdot \frac{t}{\varphi_{g.i}} = u(\varphi_{p.i}) \qquad (5.107)$$

Ultimately, vector $\mathbf{V}_{K.i}$ of the resultant linear velocity of the contact point K is a superposition of the linear velocities $\mathbf{V}_{p.i}$ and $\mathbf{V}_{P.i}$:

$$\mathbf{V}_{K.i} = \mathbf{V}_{p.i} + \mathbf{V}_{P.i} \qquad (5.108)$$

The vector $\mathbf{V}_{K.i}$ of the resultant linear velocity is along the line that goes through the instant pitch point, P_i. Therefore, the instant pitch point in noncircular gearing can be determined as the point of intersection between the center-line and the straight line along the velocity vector $\mathbf{V}_{K.i}$. This straight line is referred to as the *instant line of action* and is designated as LA_i. The common tangent for contacting centrodes is a straight line through the contact point, K, that is perpendicular to the instant line of action.

The line of action, LA, is a planar curve that can be interpreted as a family of instant contact points considered in a motionless reference system.* At every instant of time, the line of action, LA (see Figure 5.27)

---

* It should be pointed out here that instant line of action, la_i, between tooth flanks of the gear and of the pinion should also pass through that same instant pitch point, P_i, through which the instant line of action, LA_i, between the centrodes goes.

- Passes through a current contact point K
- Is tangent to the instant line of action, LA_i
- Is perpendicular to the common tangent through K

Based on the schematic depicted in Figure 5.27, centrodes for a gear pair having prespecified angular velocity ratio $u(\varphi_{p,i})$ can be determined. The centrodes in this case are a type of *reversibly enveloping curves* (R_e–*curves*). The desired geometries of the both centrodes, namely, of the centrode of the gear and of the centrode of the pinion can be determined following common practice:

- The centrode of the gear is a loci of an instant contact point represented in a reference system associated with the gear.
- The centrode of the pinion is a loci of an instant contact point represented in a reference system associated with the pinion.

A pair of centrodes determined in this way is the only possible type of centrode that meets the prespecified angular velocity ratio $u(\varphi_{p,i})$. Centrodes of no other type are capable of fulfilling this requirement.

Another example also pertains to the field of gearing.

Consider centrodes of a pair of noncircular gears as schematically illustrated in Figure 5.28.

In the example shown in Figure 5.28, the pinion centrode is shaped in the form of an ellipse. The axis of rotation of the pinion O_p is the straight line

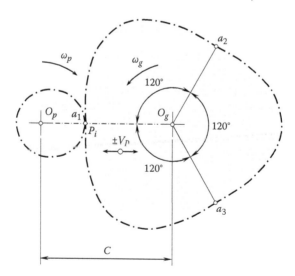

FIGURE 5.28
An example of centrodes for a noncircular gearing: the centrode for the one-lobe pinion and the centrode for the three-lobe gear.

through one of two focuses of the ellipse. The pinion rotates about the axis O_p with certain angular velocity ω_p.

The axis of rotation of the mating gear O_g is parallel to the axis O_p and is remote at a certain center-distance C from the pinion axis of rotation O_p. The rotation of the gear is denoted by ω_g. In the particular case under consideration, the one-lobe pinion is engaged in mesh with a three-lobe gear. This means that the arc length a_1a_2, a_2a_3, and a_1a_3 of the gear centrode are equal to one another ($a_1a_2 = a_2a_3 = a_1a_3$), and all of them are equal to perimeter of the pinion centrode. Each of the arc lengths a_1a_2, a_2a_3, and a_1a_3 span over a central angle of 120°. The perimeter of the gear centrode is triple that of the pinion centrode.

The instant value of the angular velocity of the driven gear ω_g depends on current value of angle of rotation, φ_g, of the driving pinion (here $\varphi_g = t\,\omega_g$, and time is designated as t). Pitch point P in the gear pair travels along the center-line up and down with a certain linear velocity $\pm V_P$. The instant location of the pitch point P within the center-line is designated as P_i.

When designing noncircular gears, it is commonly assumed that one of the centrodes is specified (in the case under consideration it is assumed that the pinion centrode is shaped in the form of an ellipse). The corresponding centrode of the mating member is generated as an envelope to successive positions of the given centrode (in the above-considered example illustrated in Figure 5.28, it is the gear centrode). Unfortunately, in most cases, this is not correct as the centrodes are not reversibly enveloping curves. Therefore, when a driving member having a specified centrode is rotating with a specified constant angular velocity, the centrode of the driven member is generated as an envelope to successive positions of the centrode of the driving member. However, if the driven member turns out to be a driving member, and correspondingly, the driving member turns to be a driven member, the transmitting rotation is not the inverse one to the original rotation.

It should be clearly understood here that the concept of reversibly enveloping surfaces can be easily enhanced to the cases of interaction of two smooth regular part surfaces rotating about intersected axes of rotation as well as those rotating about crossing axes of rotation. Moreover, the rotation to be transmitted is not necessarily steady: it should follow a prescribed time+-dependent function.

5.3.1.2 Profiling of the Form-Cutting Tools for Two-Parametric Kinematic Schemes of Part Surface Generation

The kinematic schemes of part surface generation, those featuring two relative motions of the part surface P and the generating surface T of the cutting tool are referred to as *two-parametric kinematic schemes of part surface generation*.

The resultant motion of the cutting tool with respect to the part surface P can be of complex nature. For simplification, it can be decomposed to two elementary motions. The elementary motions are usually represented with a rotational motion, and with a translational motion. None of these motions caused sliding of the surface "*over itself.*" Under such a scenario, for profiling the form-cutting tool, it is convenient to implement a two-parametric scheme of part surface generation.

Similar to the analysis of implementation of a single-parametric kinematic scheme (Figure 5.23), consider the implementation of two-parametric

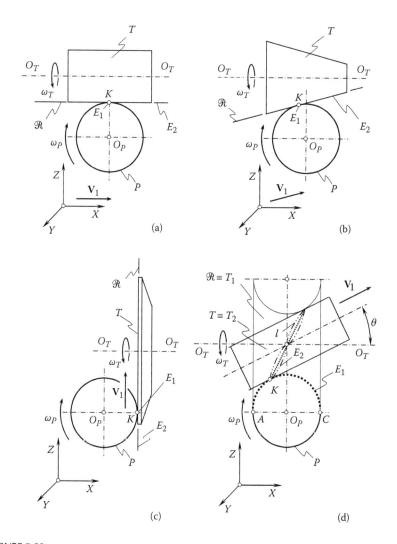

FIGURE 5.29
Examples of two-parametric kinematic schemes of part surface generation for derivation of the generating surface, T, of the cutting tool. (From Rodin, P.R., *Fundamentals of the Theory of Cutting Tool Design*, Mashgiz, Kiev, 1960; and P.R., Rodin, *Issues of the Theory of Cutting Tool Design*, DrSci dissertation, Odessa Polytechnic Institute, Odessa, Ukraine, 1961.)

kinematic schemes of part surface generation for the case of machining a surface of a round cylinder (Figure 5.29).

The work is rotated about the axis O_P (Figure 5.29a) with a certain angular velocity ω_P. The cutting tool coordinate system $X_TY_TZ_T$ (this is the reference system that the cutting tool, after been designed, will be associated with) is rotated with a certain angular velocity ω_T about the axis O_T. The axes of rotations O_P and O_T are at right angles.

The velocity vector $\mathbf{V}_\Sigma$ of the resultant generating motion is decomposed onto two elementary motions $\mathbf{V}_1$ and $\mathbf{V}_2$ (here the equality $\mathbf{V}_\Sigma = \mathbf{V}_1 + \mathbf{V}_2$ is observed).

At the beginning, an auxiliary generating surface $\mathcal{R}$ must be determined. The auxiliary generating surface $\mathcal{R}$ is an enveloping surface to consecutive positions of the part surface P in its motion with the velocity $\mathbf{V}_1$ of the first elementary motion. Further, the generating surface T of the cutting tool is represented in the coordinate system $X_TY_TZ_T$ with the enveloping surface to consecutive positions of the auxiliary generating surface $\mathcal{R}$ in its motion with the velocity $\mathbf{V}_2$ of the second elementary motion. Here, vector $\mathbf{V}_2$ designates linear velocity of rotational motion of the auxiliary surface $\mathcal{R}$ about the axis O_P with the angular velocity ω_T.

The generating surface T of the cutting tool, which is determined following the two-parametric approach, usually makes point contact with the part surface P. This is observed because of the following reason.

The line E_1 is the characteristic line for the part surface P and the auxiliary generating surface $\mathcal{R}$. The auxiliary generating surface $\mathcal{R}$ and the generating surface T of the cutting tool make line contact and the line E_2 is the characteristic to the surfaces $\mathcal{R}$ and T. The characteristic line E_1 and the characteristic line E_2 are within the auxiliary generating surface $\mathcal{R}$. Generally speaking, the characteristics E_1 and E_2 intersect at a point. At least one common point of the characteristics E_1 and E_2 is observed—otherwise, surfaces P and T cannot make contact. Point K is the point of intersection of the characteristics E_1 and E_2. In a particular case, the characteristic lines E_1 and E_2 can be congruent to each other. This results in the part surface P and the generating surface T of the cutting tool making line contact instead of point contact.

When surfaces P and T move relative to each other with the velocity $\mathbf{V}_1$, the characteristic line E_1 occupies a stationary location on the part surface P, but it travels over the generating surface T of the cutting tool. When the plane $\mathcal{R}$ rotates about the axis O_T, the characteristic curve E_2 on the auxiliary surface $\mathcal{R}$ is stationary, but it is traveling over the generating surface T of the cutting tool. Therefore, in the general case of surface generation (Figure 5.29), the point K of the intersection of the characteristic lines E_1 and E_2 is moveable.

The generating surface T of the cutting tool that is determined for a two-parametric kinematic scheme of part surface generation makes point contact with the part surface P. This means that no ideal generation of the part surface P is feasible. The cusps on the machined part surface P are inevitable under such circumstances. Discrete generation of the part surface P is commonly observed when a two-parametric kinematic scheme of part surface generation is implemented.

In all cases of discrete generation, it is necessary to keep the resultant cusps height h_Σ less than the tolerance $[h]$ for the accuracy of generation of the part surface P. The inequality

$$h_\Sigma \leq [h] \qquad\qquad (5.109)$$

must be satisfied.

The location of the point K and the shape of the generating surface T of the cutting tool depend upon the direction of the velocity vector $\mathbf{V}_2$. Depending on the direction of the velocity vector $\mathbf{V}_2$, the cutting tool surface T can be shaped in the form of a round cylinder (Figure 5.29a), a round cone (Figure 5.29b), a plane (Figure 5.29c), and so forth.

In a particular case, the part surface P and the generating surface T of the cutting tool make line contact. Line contact between surfaces P and T is observed when the characteristic lines E_1 and E_2 are congruent. No cusps are observed on the machined part surface P when surfaces P and T are in line contact.

The auxiliary generating surface $\mathfrak{R}$ is a tangent to the part surface P. Therefore, the surface $\mathfrak{R}$ can be employed not only as an auxiliary surface, but also as the generating surface T of the cutting tool. In the last case, it is necessary to consider the surface $\mathfrak{R}$ as the generating surface of the cutting tool that is derived using the single-parametric kinematic scheme of part surface generation. To generate the generating surface T of the cutting tool (Figure 5.29c), the generating surface T_1 can be determined as the enveloping surface to consecutive positions of the part surface P in its motion relative the coordinate system $X_T Y_T Z_T$. By doing this, the approach shown in Figure 5.29a can then be followed. Another motion $\mathbf{V}_2$ of the determined generating surface T_1 is then considered [151,150]. Ultimately, the generating surface T_2 of the cutting tool can be determined as the enveloping surface to consecutive positions of the surface T_1 that is performing the motion $\mathbf{V}_2$.

In the example, schematically illustrated in Figure 5.29c, the generating surface T_2 of the cutting tool is shaped in the form of noncircular cylinder (that is a type of a surface of translation). The generating surface T_2 and the part surface P make point contact. Point of contact K of surfaces P and T_2 is traveling over both surfaces: it is traveling over the part surface P and over the generating surface T_2 of the cutting tool. The closed three-dimensional curve l represents the tool path on the part surface P.

The solution to the problem of profiling the form-cutting tool using two-parametric kinematic schemes of part surface generation yields qualitative (not quantitative) results. No optimal parameters of the kinematic scheme of part surface generation or of the optimal cutting tool can be derived from the considered approach.

An analytical solution to the problem of determining the most favorable (optimal) parameters of the kinematic scheme of part surface generation and of optimal parameters of the geometry of the generating surface of the form-cutting tool can be obtained on the premise of the comprehensive analysis of the geometry of contact of the part surface P and the generating surface T of the cutting tool (see Chapter 4).

The considered example (Figure 5.29) is insightful. It gives impetus to the investigation of the impact of the degree of conformity at a point of contact of the part surface P and generating surface T of the cutting tool on the accuracy and quality of the machined part surface. It is also practical for the analysis of vulnerability of the part surface generation process to the deviations of configuration of the cutting tool with respect to the work.

The use of two-parametric kinematic schemes of part surface generation (Figure 5.29) is a perfect example of implementation of the first Olivier principle for profiling form-cutting tools for machining a given part surface on a conventional machine tool.

As in the previous case of single-parametric kinematic schemes of part surface generation, the implementation of the first Olivier principle in cases of two-parametric schemes of part surface generation is not always valid. For the validity of the first Olivier principle of generation of enveloping surfaces, both pairs of surfaces, namely, the specified moving part surface P and the auxiliary generating surface $\mathcal{R}$ as well as the auxiliary generating surface $\mathcal{R}$ and the generated tool surface T must be pairs of R_e–surfaces. Otherwise, only approximate profiling of the form-cutting tool for machining a given part surface on a conventional machine tool is feasible.

Generation of reversibly enveloping surfaces is illustrated below. As an example, a kinematic scheme of profiling of a form-cutting tool that is based on two rotation vectors $\boldsymbol{\omega}_P$ and $\boldsymbol{\omega}_T$ is considered.[*] The rotation vector $\boldsymbol{\omega}_P$ is associated with the part surface P geometry of which is given. The rotation vector $\boldsymbol{\omega}_T$ is associated with the cutting tool generating surface T which is to be determined.

The rotation vectors, $\boldsymbol{\omega}_P$ and $\boldsymbol{\omega}_T$, are along straight lines, which cross one another. The closest distance of approach between the lines of action of the rotation vectors, $\boldsymbol{\omega}_P$ and $\boldsymbol{\omega}_T$, is denoted by C. This distance is commonly referred to as *center-distance C*.

Consider a vector diagram constructed for an arbitrary configuration of the rotation vectors $\boldsymbol{\omega}_P$ and $\boldsymbol{\omega}_T$. The vector diagram is depicted in Figure 5.30. The diagram corresponds to external interaction between surfaces P and T. The vector diagram shown in Figure 5.30 features an obtuse part surface angle, Σ_P, between the rotation vector, $\boldsymbol{\omega}_P$, of the part and the vector of instant rotation, $\boldsymbol{\omega}_{pl}$.

It should be noticed here that in the case of crossing rotation vectors $\boldsymbol{\omega}_P$ and $\boldsymbol{\omega}_T$, there is no freedom in choosing a configuration of the axis of instant

---

[*] It is important to stress here that angular velocity is considered in this monograph as a vector directed along the axis of rotation in a direction defined by the right-hand screw rule. It is understood here and below that rotations are not vectors in nature. Therefore, special care is required when treating rotations as vectors.

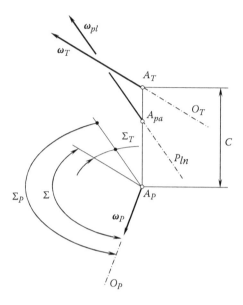

FIGURE 5.30
An example of vector diagram for a kinematic scheme of profiling of a form-cutting tool that is composed of two rotations ω_P and ω_T about crossing axes O_P and O_T, respectively.

rotation, P_{ln}, in relation to the rotation vectors ω_P and ω_T. Once the rotation vectors, ω_P and ω_T, and their relative location and orientation are specified, the configuration of the axis of instant rotation, P_{ln}, can be expressed in terms of the rotations ω_P and ω_T and of the center-distance, C.

A base cone is associated with the part surface P, and another base cone is associated with the generating surface T of the form-cutting tool. This concept is schematically illustrated in Figure 5.31. The axis of rotation of the part surface, O_P, and the axis of rotation of the cutting tool, O_T, cross each other at a shaft angle, Σ. The closest distance of approach of the axes of the rotations, O_P and O_T, is denoted by C.

The vector of instant rotation, ω_{pl}, is the vector through a point, A_{pa}. The point A_{pa} is located within the center-distance, C. Endpoints of the straight line segment, C, are labeled as A_P and A_T. The first, A_P, is the point of intersection of the center-line (that is, the straight line along the closest distance of approach), C, and the part axis of rotation, O_P. The second one, A_T, is the point of intersection of the center-line along the closest distance of approach, C, and the cutting tool axis of rotation, O_T.

The plane of action, PA, is a plane through the axis of instant rotation, P_{ln}. The plane of action, PA, is in tangency with both base cones, namely, with the base cone of the part to be machined as well as with the base cone associated with the cutting tool. Because of this, the plane of action, PA, makes a certain

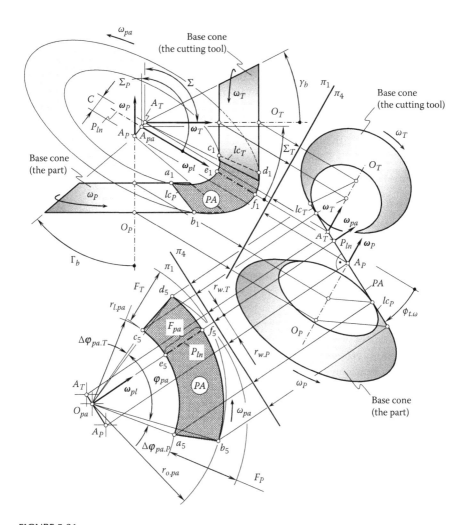

FIGURE 5.31

Base cones and the plane of action, *PA*, in an orthogonal crossed-axis kinematic scheme of profiling of a form-cutting tool.

transverse pressure angle, $\phi_{t.\omega}$, in relation to a perpendicular to a plane associated with the axis of instant rotation, P_{ln}.

The plane of action can be interpreted as a flexible zero thickness film. The film is free to wrap/unwrap from and onto the base cones associated with the part and with the cutting tool. The plane of action, *PA*, is not allowed for any bending about an axis perpendicular to the plane *PA* itself. Under uniform rotation of surfaces *P* and *T*, the plane of action, *PA*, is rotating about the axis, O_{pa}. The rotation vector, $\boldsymbol{\omega}_{pa}$, is along the axis, O_{pa}. The rotation vector, $\boldsymbol{\omega}_{pa}$, is perpendicular to the plane of action, *PA*.

As the axis of instant rotation, P_{ln}, and the axes of rotations of the part, O_P, and of the cutting tool, O_T, cross one another, pure rolling of the base cones of the part and of the cutting over the pitch plane, PA, is not observed, but rolling together with sliding of PA over the base cones is observed instead. This is due to the axes of rotations O_P and O_T crossing one another.

Reversibly enveloping surfaces P and T are generated by a line of contact between these surfaces. The line of contact, LC, is a planar curve that is entirely located within the plane of action, PA.

As surfaces P and T rotate, the line of contact is traveling together with the plane of action, PA, with respect to

a. The part surface P
b. The cutting tool surface T
c. A frame associated with the machine tool

The part surface, P, can be interpreted as loci of consecutive positions of the line of contact, LC, when the line of contact is traveling in relation to a reference system associated with the part. Similarly, generating surface T of the form-cutting tool can be represented as loci of successive positions of that same line of contact, LC, when the line of contact is traveling in relation to a reference system associated with the cutting tool. Ultimately, the loci of consecutive positions of that same line of contact, LC, when the line of contact is traveling in relation to a stationary reference system associated with the machine tool, represents the surface/plane of action. Therefore, once the line of contact is known, the kinematics of part surface machining (Figure 5.31) can be employed for the derivation of an analytical representation of the part surface, P, and of generating surface T of the form-cutting tool in a case when both of surfaces P and T are a type of R_e–surfaces. For convenience, numerous reference systems are used for this purpose. These reference systems are associated with the part, the cutting tool, the housing, and so forth. The transition from one reference system to another reference system is analytically described by a corresponding operator of the coordinate system transformation.

Generally speaking, the position vector of a point, $\mathbf{r}_{lc}$, of the line of contact, LC, can be analytically described by an expression in matrix form:

$$\mathbf{r}_{lc}(v) = \begin{bmatrix} X_{lc}(v) \\ Y_{lc}(v) \\ 0 \\ 1 \end{bmatrix}$$

(5.110)

To represent Equation 5.110 for the position vector of a point, $\mathbf{r}_{lc}$, of the line of contact, LC, in the reference system, $X_P Y_P Z_P$, the operator of the resultant coordinate system transformation, $\mathbf{Rs}(PA \mapsto P)$, can be employed. This makes possible an expression:

$$\mathbf{r}_P(v,\theta_{pa}) = \mathbf{r}_{lc}^P(v,\theta_{pa}) = \mathbf{Rs}(PA \mapsto P) \cdot \mathbf{r}_{lc}(v) \qquad (5.111)$$

for the position vector of a point $\mathbf{r}_P$ of the part surface P.

An expression for the generating surface T of the form-cutting tool can be derived in a similar way to how Equation 5.111 is derived.

The lines of contact of various geometries can be used for the purpose of generation of the part surface P and generating surface T of the form-cutting tool.

The above-considered approach for the determination of the geometry of the part surface, P, and of the generating surface T of the form-cutting tool, is based on generation of the surfaces in the form of a family of consecutive positions of the line of contact, LC, that travels together with the plane of action, PA. This approach does not require a specification of surfaces P and T in the form of enveloping surfaces to successive positions of the generating basic rack. This means that the proposed method for the generation of surfaces P and T does not require implementation of the elements of the theory of enveloping surfaces. This is a significant advantage of the disclosed method for generation of R_e–surfaces, P and T over the methods based on elements of the theory of enveloping surfaces [138].

Generation of reversibly enveloping surfaces by means of a line that is associated with the plane of action, shows that this is the only way to generate R_e–surfaces. No other approach for generation of enveloping surfaces is capable of generating R_e–surfaces.

The discussed method for generation of R_e–surfaces has been used for the development of a novel type of gearing, namely, it has been used for the development of the gearing that is referred to as *R-gearing* (patent pending) [138].

The concept of R_e–surfaces can be enhanced to more general cases of kinematics of profiling of form-cutting tools for machining a given part surface.

5.3.1.3 Concluding Remarks

It should be clear now that there is not much freedom in selection of enveloping surfaces if configuration of the rotation vectors $\boldsymbol{\omega}_P$ and $\boldsymbol{\omega}_T$ associated with the part, P, and with the cutting tool, T, is specified. Once a configuration of the rotation vectors $\boldsymbol{\omega}_P$ and $\boldsymbol{\omega}_T$ is specified, then a limited number of R_e–surfaces can be generated. They differ from one another only by value of transverse pressure angle in relative motion of the surfaces. Thus, in the general case, an arbitrary auxiliary surface cannot be used for generation of a pair of R_e–surfaces. Further, if one of the moving surfaces is given and a configuration of the rotation vectors is specified, then in the general case, an enveloping surface does not exist at all.

Thus, the following conclusion can be derived:

Conclusion 5.3

In the general case of part surface machining, it is a mistake to generate reversibly enveloping surfaces as envelopes either to consecutive positions of an auxiliary generating surface, or of another traveling surface. Both Olivier principles are useless for solving this particular problem of the theory of part surface generation, as reverse enveloping is feasible only in particular cases when the interacting surfaces are a type of R_e–surfaces. ∎

When a pair of R_e–surfaces is generated, then both *Olivier principles* are valid for a pair of already existing smooth, regular reversibly enveloping part surfaces. However, under such a scenario, these two principles become useless.

Ultimately, Olivier principles for generation of enveloping surfaces in general cases are either incorrect or useless when they are valid.

Olivier principles are valid for the generation of only approximate enveloping part surfaces as these principles are geometrically inaccurate.

The application of Olivier principles is valid in degenerate cases, when one of the surfaces allows for sliding "over itself" in the direction of the enveloping motion.

An enveloping surface is specified by the equation of contact ($\mathbf{n} \cdot \mathbf{V} = 0$) only in cases when the envelope exists. Under such a scenario, the equation of contact $\mathbf{n} \cdot \mathbf{V} = 0$ is helpful for derivation of an equation of the enveloping surface. Generally speaking, this equation is fulfilled only in the differential vicinity of a point within the characteristic line, but either before or after the point the properly generated portion of the enveloping surface can be cut off. Once a portion of the generated envelope is cut off, the envelope (in a global sense) does not exist. Prior to implementing the equation of contact, it must be verified whether or not the envelope exists. Such a verification is of critical importance. The conditions of existence of envelopes in general cases of part surface machining have not been investigated thoroughly yet.

The following theorem can be proven:

Theorem 5.1

For a specified geometry of the traveling surface an enveloping surface, if any, can be generated when the given surface is performing not an arbitrary motion, but it is performing a certain unique (in a certain sense) motion; this motion is prespecified by the shape and parameters of the traveling surface. ∎

From the standpoint of this theorem, in the general case of enveloping, there is no freedom for choosing either kinematics of the relative motion, or of geometry of the traveling surface. Namely,

a. For a specified kinematics of the relative motion, there could be a unique family of pairs of R_e–surfaces having a different transverse profile angle, $\phi_{t.\omega}$ and having the line of contact, LC, of various geometries.

b. If the geometry of the traveling part surface is specified, there is no freedom in choosing enveloping motion as well as the geometry of the generated part surface.

The correctness of the theorem is clearly illustrated by the kinematics and geometry of tooth flanks in R-gearing [36]. From the standpoint of R-gearing, the generation of reversibly enveloping surfaces is feasible for a limited number of cases. All types of surfaces for R-gearing are prespecified:

- By a given configuration of the rotation vectors, ω_g and ω_p
- By transverse pressure angle, $\phi_{t.\omega}$

The other types of part surfaces cannot be referred to as R_e–surfaces. Therefore, as the total number of reversibly enveloping surfaces is limited, what is the purpose for implementation in the theory of part surface generation of two principles proposed by Olivier? As there is no freedom to pick an arbitrary type of an auxiliary generating surface, then both the Olivier principles become useless. The principles can be implemented only in the cases of approximate generation of enveloping surfaces.

Conclusion 5.4

In general case of generation of enveloping surfaces, the first and the second Olivier principles are not valid. These two principles are valid only in cases when traveling surfaces allow for sliding in the direction of the enveloping motion. In the theory of part surface generation, both Olivier principles can be implemented for the determination of approximate envelopes only. ∎

The discussion in this section can be enhanced to more general cases of arbitrary relative motion of the given surface in relation to a reference system with which the enveloping surface will be associated.

Further development in the field of generation of enveloping surfaces results in generation of surfaces applied in spatial gearing (CA-gearing) featuring variably disposed axes of rotation of the moving part surface and the enveloping part surface. Reversibly enveloping part surfaces of this particular type are used as tooth flanks in geometrically accurate S_{pr}–gearing [138]. They feature point contact of the interacting smooth regular part surfaces $\mathcal{G}$ and $\mathcal{P}$ and are insensitive to axis misalignment.

When profiling a form-cutting tool for machining a part surface on a conventional machine tool, the discussed inconsistence of the *Olivier* principles should be kept in mind, as in most cases, only approximate profiling of the form-cutting tool can be performed (and geometrically accurate profiling is usually not feasible at all).

5.3.2 Peculiarities of Profiling of the Form-Cutting Tools for Multiparametric Kinematic Schemes of Surface Generation

Kinematic schemes of part surface generation, those featuring more than two relative motions of the part surface P and generating surface T of the cutting tool, are referred to as *multiparametric kinematic schemes of part surface generation*.

The resultant motion of the cutting tool with respect to the part surface P can be of a complex nature. It is not common practice to implement kinematic schemes of part surface generation having more than two elementary relative motions. However, as an example, it is right to notice here that methods of hob relieving comprising up to six elementary motions* are known.

The velocity vector $\mathbf{V}_\Sigma$ of the resultant relative motion of the part surface P and generating surface T of the form-cutting tool can be decomposed on a finite number of elementary motions $\mathbf{V}_i$ (here, i is an integer number and $i > 2$ is considered):

$$\mathbf{V}_\Sigma = \sum_{i=1}^{n} \mathbf{V}_i \tag{5.112}$$

The total number of the elementary motions is designated as n. There are no principal restrictions on number n of the elementary relative motions $\mathbf{V}_i$. In practice, the number n is restricted only by the complexity of the design of a machine tool being capable of reproducing a type of multiparametric kinematic scheme of part surface generation.

The Shishkov equation of contact $\mathbf{n} \cdot \mathbf{V}_\Sigma = 0$ is still valid when the velocity vector $\mathbf{V}_\Sigma$ of the resultant motion (see Equation 5.112) is considered (Figure 5.32):

$$\mathbf{n} \cdot \sum_{i=1}^{n} \mathbf{V}_i = 0 \tag{5.113}$$

---

* Patent No. 965.728, USSR, *A Method of Hob Relieving*, S. P. Radzevich, filed January 21, 1980.

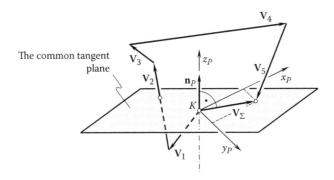

FIGURE 5.32
The velocity vector $\mathbf{V}_\Sigma$ of the resultant motion of the part surfaces P in relation to a reference system, which the form-cutting tool, T, will be associated with, when a case of multiparametric kinematic scheme of profiling of a cutting tool is implemented.

In the event a kinematic scheme of part surface generation comprises n elementary relative motions, it is possible to generate $(n-1)$ auxiliary generating surfaces $\mathcal{R}_1$, $\mathcal{R}_2$, ..., $\mathcal{R}_{n-1}$. The last auxiliary generating surface $\mathcal{R}_n$ is congruent to generating surface T of the cutting tool ($\mathcal{R}_n \equiv T$). Extension of the Olivier principles toward implementation of two auxiliary generating surfaces is supported by the illustrated feasibility of the approximate generation of conjugate surfaces by means of two auxiliary generating surfaces [116,117]. This achievement can be recognized as the *third principle of generation of conjugate surfaces*.

If the multiparametric kinematic scheme of part surface generation is used, the equation of the generating surface T of the form-cutting tool can be derived by solving the set of two equations:

$$\begin{cases} \mathbf{r}_T = \mathbf{r}_T(U_P, V_P, \omega_1, \omega_2, \ldots, \omega_i, \ldots, \omega_n) \\ \mathbf{n}_P \cdot \mathbf{V}_\Sigma = 0 \end{cases} \tag{5.114}$$

In this equation, the velocity vector $\mathbf{V}_\Sigma$ of the resultant motion is represented as summa:

$$\mathbf{V}_\Sigma = \mathbf{V}_1 + \mathbf{V}_2 + \ldots + \mathbf{V}_i + \ldots + \mathbf{V}_n \tag{5.115}$$

However, this does not mean that the particular equalities $\mathbf{n}_P \cdot \mathbf{V}_1 = 0$, $\mathbf{n}_P \cdot \mathbf{V}_2 = 0$, ..., $\mathbf{n}_P \cdot \mathbf{V}_i = 0$, and others must be satisfied. It is wrong to demand that any and all elementary motions $\mathbf{V}_i$ fulfill this requirement, namely, the equality $\mathbf{n} \cdot \mathbf{V}_i = 0$. Projection $Pr_n\mathbf{V}_i$ of a velocity vector $\mathbf{V}_i$ of an elementary relative motion onto the direction of the common perpendicular, $\mathbf{n}$, can differ from zero ($Pr_n\mathbf{V}_i \neq 0$). Satisfaction of the particular equalities is the sufficient, but not the necessary condition. The elementary relative motions $\mathbf{V}_i$ are not mandatory within the common tangent plane. The location of the velocity vector $\mathbf{V}_\Sigma$ of the resultant motion within the common tangent plane is the only mandatory requirement in this concern. To meet this requirement, the summa of all the projections $Pr_n\mathbf{V}_i$ should be equal to zero:

$$\sum_{i=1}^{n} Pr_n\mathbf{V}_i = 0 \qquad\qquad (5.116)$$

In the case of a multiparametric kinematic scheme of profiling of a form-cutting tool (as well as in the case of a multiparametric kinematic scheme of part surface generation), namely, when the total number n of elementary relative motions $\mathbf{V}_i$ is greater than two ($n \geq 2$):

- The characteristic line E_1 for the part surface P and for the first auxiliary generating surface $\mathcal{R}_1$,
- The characteristic line E_2 for the first auxiliary $\mathcal{R}_1$ and for the second generating surface $\mathcal{R}_2$ and so forth

must intersect one another at a common point, which is the contact point K between the part surface P, and the generating surface T of the form-cutting tool.

Conclusion 5.5

In cases of multiparametric kinematic schemes of part surface generation, all the particular characteristic lines $\mathcal{R}_1, \mathcal{R}_2, ..., \mathcal{R}_{n-1}$ intersect one another at a common point, which is the contact point K between the part surface P and between the generating surface T of the form-cutting tool. ∎

This makes the multiparametric kinematic schemes of part surface generation impractical.

An increase in the number of elementary motions $\mathbf{V}_i$ causes kinematic schemes of part surface generation to become more complex. This is the major reason that the generating surface T of the form-cutting tool is generated in most cases using either single- or two-parametric kinematic schemes of part surface generation.

The problem of derivation of an equation of the generating surface T of the form-cutting tool when an implemented kinematic scheme of part surface generation is multiparametric is not complex in principle. However, it often causes technical problems when performing routing computations.

5.4 Characteristic Line E of the Surface P and Generating Surface T of the Cutting Tool

In this section, an advantage of implementation of kinematic schemes of part surface generation is provided. Because sliding of the generating surface of the cutting tool over itself is allowed, it is necessary to derive the equation of the profile of the generating surface of the cutting tool, and not the equation of the entire surface T. Such a substitution results in significant simplification of the problem of profiling of the form-cutting tool.

When the part surface P is given by vector equation in the form $\mathbf{r}_P = \mathbf{r}_P(U_P, V_P)$, the equation of a family of the surfaces P can be represented in the form $\mathbf{r}_P^{fm} = \mathbf{r}_P^{fm}(U_P, V_P, \omega_1, \omega_2, \ldots, \omega_i, \ldots, \omega_n)$. Then, the equation of the characteristic line E can be derived from the set of Equation 5.114. This is a direct approach for the derivation of equation of the characteristic line E for surfaces P and T.

Another way to derive an equation of the characteristic line is based on consideration of instant screw motion of the part surface P relative to the coordinate system $X_T Y_T Z_T$, which the generating surface T will be associated with. The part surface P performs a screw motion about the part axis of the instant screw motion (Figure 5.33). This axis of the screw motion yields analytical representation in the form:

$$\mathbf{P}(p) = \mathbf{P}_0 + \mathbf{p} \cdot p \tag{5.117}$$

where $\mathbf{P}_0$ designates the position vector of a point P_0 on the part axis of the instant screw motion, unit vector $\mathbf{p}$ determines the direction of the part axis, and p designates the *parameter* (the *reduced pitch*) of the part axis.

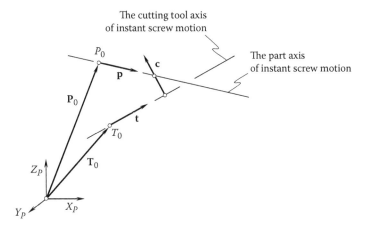

FIGURE 5.33
Interpretation of the instant screw motion of the part surface P and of a reference system $X_T Y_T Z_T$, with which the form-cutting tool (after being designed) will be associated.

Similarly, the coordinate system $X_T Y_T Z_T$ performs the screw motion about the tool axis of the instant screw motion. This axis of the screw motion can be represented by the vector equation:

$$\mathbf{T}(t) = \mathbf{T}_0 + \mathbf{t} \cdot t \qquad (5.118)$$

where $\mathbf{T}_0$ designates the position vector of a point T_0 on the cutting tool axis instant screw motion, unit vector $\mathbf{t}$ determines the direction of the of the tool axis, and t designates parameter (the reduced pitch) of the tool axis.

One restriction for the candidate generating surface T is that the common perpendicular between surfaces P and T in contact and the instant screw axis must satisfy the Ball reciprocity relation [2,3]:

$$a = h \cdot \tan \varepsilon \qquad (5.119)$$

where a the shortest distance of approach between the axis of instant screw motion and the perpendicular to the part surface P at a point of the characteristic line E, h designates the *parameter* (the *reduced pitch*) of the instant screw motion, and ε is the angle between the instant screw axis and the line of action.

If axes of the instant screw motions are not parallel to each other, then the straight line segment C joining the closest points of the part axis and the cutting tool axis is uniquely perpendicular to both axes at the same time. No other straight line segment between the axes possesses this property. That is, the unit direction vector $\mathbf{c}$ is uniquely perpendicular to the line direction vectors $\mathbf{p}$ and $\mathbf{t}$. This is equivalent to the vector $\mathbf{c}$ satisfying the two equations $\mathbf{p} \cdot \mathbf{c} = 0$ and $\mathbf{t} \cdot \mathbf{c} = 0$. To satisfy these equations, the unit vector $\mathbf{c}$ must be equal to $\mathbf{c} = \mathbf{p} \times \mathbf{t}$.

Ultimately, for the calculation of the closest distance of approach of the part axis and of the cutting tool axis, Equation 5.119 can be employed. The direction of the straight line segment through the closest distance of approach is specified by the unit vector $\mathbf{c}$. This makes the calculation of coordinates of points of the characteristic line E possible.

The characteristic line E can be interpreted as the projection of the axis of the instant relative screw motion onto the part surface P. At every point of the characteristic line E, the unit normal vector $\mathbf{n}_P$ to the part surface P makes a certain angle ε with the axis of the screw motion (see Equation 5.119). Following the above consideration, the equation of the characteristic line E can be derived.

Another approach to the determination of the characteristics of a surface having a certain motion can be found in the monograph by Cormac [10].

The parameters of the kinematic scheme of part surface generation can be time-dependent. This means that the parameters of the characteristic line E

could also be time-dependent. The methods of part surface machining with a cutting tool when the characteristic line E changes its shape in time are known from practice [26].

5.5 Selection of the Form-Cutting Tools of Rational Design

For machining a sculptured part surface, cutting tools of standard design are widely used. Form-cutting tools of the most widely used designs for sculptured part surface machining are shown in Figure 5.1. When selecting a certain cutting tool for machining a given sculptured part surface, the user commonly follows conventional rules. This means that the user will want to select the cutting tool design that meets two requirements: First, any and all regions of the part surface must be reachable by the cutting tool without mutual interference of the part surface P and the generating surface T of the cutting tool. To achieve the highest possible productivity of part surface machining, in compliance with the second requirement, the diameter of the cutting tool must be the most feasible.

Versatility of known designs of form-cutting tools for sculptured part surface machining on a multiaxis NC machine is not limited to the designs schematically shown in Figure 5.1. Other designs of form-cutting tools can also be implemented.

For machining sculptured part surfaces, form-milling cutters are commonly used. The machining surface T of the milling cutter is shaped in the form of a surface of revolution. For a better performance of the milling cutter, it is highly desirable to have the principal radii of curvature of the generating surface of the milling cutter equal to (or almost equal to) the extremal values of principal radii of curvature of the part surface taken with the opposite sign (see Chapter 4). For this purpose, the interval of variation of the radius of curvature of the generating curve of the milling cutter surface T and configuration of the generating curve with respect to the cutting tool axis of rotation must be determined in compliance with the interval of variation of principal radii of curvature of the part surface P. It is practical to assign a constant gradient of alteration of radius of curvature of the generating curve.

For the gradient of constant value of the alteration, the linear relationship between the radius of curvature ρ_T of the generating curve and the length of its arc L_T is observed. Here, L_T designates the length of the arc of the generating curve of the cutting tool surface T, which is measured from a certain point on the generating curve to current its point. Therefore, in the case under consideration, the following equality is valid:

$$\rho_T = c \cdot L_T \tag{5.120}$$

where the constant parameter c specifies the intensity of alteration of the radius of curvature ρ_T of the generating curve* of the cutting tool surface T.

In polar coordinates, the equation of the generating curve can be represented in the following way.

Consider the equation of the generating curve in the form $r_T = r_T(\psi)$. Here, r_T designates the position vector of a point of the generating curve of the cutting tool surface T, and ψ designates the parameter of the generating curve. The radius of curvature ρ_T at the current point of the planar curve $r_T = r_T(\psi)$ can be calculated by

$$\rho_T = \frac{\left[r_T^2 + \left(\dfrac{dr_T}{d\psi} \right)^2 \right]^{\frac{3}{2}}}{r_T^2 + 2 \left(\dfrac{dr_T}{d\psi} \right)^2 - r_T \cdot \left(\dfrac{d^2 r_T}{d\psi^2} \right)} \tag{5.121}$$

The length L_T of the arc of the generating curve between two points that are specified by actual values ψ_1 and ψ_2 of the polar angle ψ is equal to

$$L_T = \int_{\psi_1}^{\psi_2} \sqrt{dr_T^2 + r_T^2 d\psi^2} \tag{5.122}$$

After substituting of Equations 5.121 and 5.64 into Equation 5.120, the following equation can be derived:

$$R_T = R_{T.0} \exp(c \cdot \psi) \tag{5.123}$$

where the position vector of a certain zero point (Figure 5.34) is denoted by $R_{T.0}$.

The generating curve (see Equation 5.123) is an isogonal curve to a bunch of straight lines through a point. This point is referred to as the *pole of the curve*. To verify the last statement, it is sufficient to substitute the equation $\varphi \cdot (Y - k \cdot X) - 0$ of a bunch of straight lines with a differential equation:

$$\left(\frac{\partial \varphi}{\partial X} \cos \theta - \frac{\partial \varphi}{\partial Y} \sin \theta \right) dX + \left(\frac{\partial \varphi}{\partial X} \sin \theta + \frac{\partial \varphi}{\partial Y} \cos \theta \right) dY = 0 \tag{5.124}$$

for isogonal trajectories. Here, θ designates the angle at which the curve (see Equation 5.123) intersects all straight lines of the bunch of straight lines. It is important to notice here, that the following equality $\theta = \tan^{-1}(c)$ is observed.

---

* Equation 5.120 is a perfect example of natural parameterization of a planar curve.

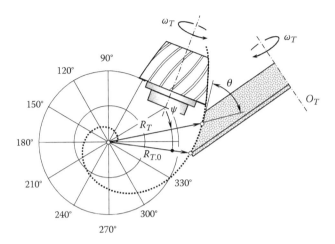

FIGURE 5.34
The generating surface *T* of a form-cutting tool for sculptured part surface machining on a multiaxis *NC* machine, which is shaped in the form of a surface of revolution of a logarithmic spiral curve.

When rotating the generating curve (see Equation 5.123) about the cutting tool axis O_T, the generating surface *T* of the form-cutting tool is generated (Figure 5.34). The cutting tool surface *T* can be further used for designing a form-milling cutter, a form-grinding wheel, or a form-cutting tool of another design.

The equation of the generating surface *T* of the form-cutting tool can be analytically described by a column matrix in the form

$$
\mathbf{r}_T(\psi,\delta) =
\begin{bmatrix}
(r_t + R_o e^{c\cdot\psi} \cos\psi)\cdot\sin\delta \\
(r_t + R_o e^{c\cdot\psi} \cos\psi)\cdot\cos\delta \\
r_t \tan\varphi + R_o e^{c\cdot\psi} \sin\psi \\
1
\end{bmatrix}
\tag{5.125}
$$

The disclosed approach makes it possible to generate the generating surface *T* of the form-cutting tool* that has either a convex- (Figure 5.35a) or a concave-generating curve (Figure 5.35b) as well as the generating curve with a point of inflection (Figure 5.35c), say, the point *m* of tangency of two logarithmic spiral curves given by Equation 5.125. In addition, it makes it possible to create the generating surface of the cutting tool having internal tangency with the part surface to be machined. In the last case, the work is located inside the cutting tool [133].

---

* Patent No. 1.271.680 (USSR), *A Form Cutting Tool for Machining a Sculptured Part Surface on a Multi-Axis NC Machine*, S. P. Radzevich, Int. Cl. B 23 C 5/10, filed August 9, 1984.

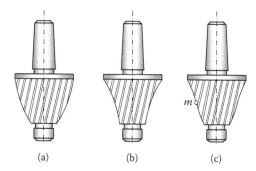

FIGURE 5.35
Examples of form milling cutters for machining a sculptured part surface on a multiaxis numerical control machine. (From USSR Patent 1.271.680.)

The shape of the generating curve of the cutting tool surface T enables us to fit to any desired value of radius of normal curvature at a point of the part surface P. Any desired degree of conformity of the cutting tool surface T to the part surface P can be reached with the cutting tool of the design shown in Figure 5.35. Ultimately, the use of the milling cutters (Figure 5.35) significantly reduces of the height of the cusps, thus improving the quality of the machined part surface.

The fundamental magnitudes of the first order of the generating surface of the form-cutting tool T (see Equation 5.125) are equal:

$$E_T = R_0^2 \cdot e^{2 \cdot c \cdot \psi} \cdot (1 + c^2) \tag{5.126}$$

$$F_T = 0 \tag{5.127}$$

$$G_T = (r_t + R_0 \cdot e^{c \cdot \psi} \cdot \cos \psi)^2 \tag{5.128}$$

The fundamental magnitudes of the second order of the generating surface of the form-cutting tool T (see Equation 5.80) are equal to

$$L_T = -R_0 \cdot e^{c \cdot \psi} \cdot \sqrt{1 + c^2} \tag{5.129}$$

$$M_T = 0 \tag{5.130}$$

$$N_T = -(r_t + R_0 \cdot e^{c \cdot \psi} \cdot \cos \psi) \cdot \frac{c \cdot \sin \psi + \cos \psi}{\sqrt{1 + c^2}} \tag{5.131}$$

The fundamental magnitudes of the first order (see Equations 5.126 through 5.128) and of the second order (see Equations 5.129 through 5.131) make an equation for the Dupin's indicatrix at a point of the generating surface T of the form-cutting tool possible (see Equation 5.125):

$$\frac{1}{R_0 e^{c \cdot \psi} \cdot \sqrt{1+c^2}} \cdot x^2 + \frac{c \cdot \sin \psi + \cos \psi}{r_t + R_0 \cdot e^{c \cdot \psi} \cdot \sqrt{1+c^2} \cdot \cos \psi} \cdot y^2 = \pm 1 \qquad (5.132)$$

Equation 1.49 yields the calculation of the principal radii of curvature at a point of the generating surface T of the form-cutting tool (see Equation 5.125). They can be calculated from the following formulae:

$$R_{1.T} = -\frac{2 \cdot \sqrt{1+c^2} \cdot (r_t + R_0 \cdot e^{c \cdot \theta} \cdot \cos \theta)}{c \cdot \sin \theta + \cos \theta} \qquad (5.133)$$

$$R_{2.T} = -2R_0 \cdot e^{c \cdot \theta} \cdot \sqrt{1+c^2} \qquad (5.134)$$

The calculated values of principal radii of curvature of the generating surface T of the form-cutting tool (see Equations 5.133 and 5.134) were used for the development of novel designs of the form milling cutters* schematically depicted in Figure 5.36.

As shown in Figure 5.36, the generating surface T of the form milling cutter is represented with a surface of revolution (see Equation 5.125). The principal radii of curvature $R_{1.T}$ and $R_{2.T}$ of the generating curve of the cutting tool surface T gradually change from one point of the curve to another point. At least two points are observed on a given generating surface T at which the first principal radius of curvature $R_{1.T}$ reaches it maximal $R_{1.T}^{max}$ and its minimal $R_{1.T}^{min}$ values. There also exist at least two points on the generating surface T of the form-cutting tool at which the second principal radius of curvature $R_{2.T}$ reaches it maximal $R_{2.T}^{max}$ and its minimal $R_{2.T}^{min}$ values. In particular, points (say two pairs of points) at which the principal radii of curvature reach their extremal values could coincide. Two types of coincidence are distinguished:

- In the first case, the principal radii of curvature reach their extremal values $R_{1.T}^{min}$ and $R_{2.T}^{min}$ at one point of the cutting tool surface T, while the extremal values $R_{1.T}^{max}$ and $R_{2.T}^{max}$ are observed in another surface point.

- In the second case, at a certain point of the surface T their values are $R_{1.T}^{min}$ and $R_{2.T}^{max}$, while at another surface T point they are equal to $R_{1.T}^{max}$ and $R_{2.T}^{min}$.

---

* Patent No. 1.355.378 (USSR), *A Form Cutting Tool for Machining a Sculptured Part Surface on a Multi-Axis NC Machine*, S. P. Radzevich, Int. Cl. B 23 C 5/10, filed April 14, 1986.

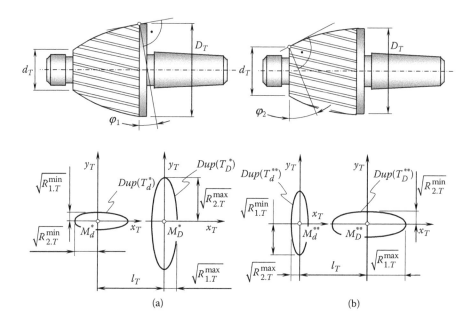

FIGURE 5.36
Form milling cutters for machining a sculptured part surface on a multiaxis numerical control machine. (From USSR Patent 1.355.378.)

The configuration of the generating line of the cutting tool surface T depends upon the diameters d_T and D_T at the ends of the milling cutter, and upon the distance l_T between the faces of the milling cutter. Two options are feasible for the correlation of the design parameters d_T, D_T, and l_T of the milling cutter, and the parameters of the generating curve of the cutting tool surface T.

Following the first option illustrated in Figure 5.36a, the configuration of the generating curve with respect to the milling cutter axis of rotation yields the principal radii of curvature $R_{1.T}^{min}$ and $R_{2.T}^{min}$ at the milling cutter face of smaller diameter d_T, and the principal radii of curvature $R_{1.T}^{max}$ and $R_{2.T}^{max}$ were observed at the milling cutter face of bigger diameter D_T. Dupin's indicatrices $Dup(T_d^*)$ and $Dup(T_D^*)$ show the distribution of radii of normal curvature at a point of the generating surface T of the milling cutter at the end-points of the generating curve.

In compliance with the second option depicted in Figure 5.36b, the configuration of the generating curve with respect to the form milling cutter axis of rotation yields the principal radii of curvature $R_{1.T}^{min}$ and $R_{2.T}^{max}$ at the form milling cutter face of smaller diameter d_T and the principal radii of curvature $R_{1.T}^{max}$ and $R_{2.T}^{min}$ were observed at the form milling cutter face of bigger diameter D_T. Dupin's indicatrices $Dup(T_d^{**})$ and $Dup(T_D^{**})$ show the distribution of radii of normal curvature at a point of the generating surface T of the form milling cutter at the end-points of the generating curve.

Other combinations of the configuration of the generating curve of the surface T relative to the form milling cutter axis of rotation are evident.

The smaller diameter d_T of the form milling cutter is equal to

$$d_T = 2R_{2.T}^{\min} \cos \varphi_2 \tag{5.135}$$

and the bigger diameter D_T is equal to

$$D_T = 2R_{2.T}^{\max} \cos \varphi_1 \tag{5.136}$$

where φ_1 and φ_2 designate the angles between the axis of rotation of the form milling cutter and the tangents to the generating curve at its end-points. It is convenient to show the angles φ_1 and φ_2 between the corresponding perpendiculars to the mentioned lines.

The use of the form milling cutter of the discussed design allows an increase of productivity of part surface machining as well as an enhanced quality of the machined part surface.

That same approach is applicable to the design of finishing tools* for burnishing of sculptured part surface using the method of part surface plastic deformation.

The maximal radius of curvature $R_T^{\max}$ of the generating curve of the generating surface of the form-burnishing tool is observed at the point 3 (Figure 5.37). The minimal radius of curvature $R_T^{\min}$ of the generating curve of the generating surface of the form-burnishing tool is observed at point 4. Two options for the design of the burnishing tool are considered:

- The first option is related to the design of the form-burnishing tool that slides over the part surface. A form-burnishing tool of this design is schematically depicted in Figure 5.37a.
- Another option is related to the design of the form-burnishing tool that rolls over the part surface. A form-burnishing tool of this design is illustrated in Figure 5.37b.

When burnishing a part surface, the form tool (Figure 5.37) performs a following motion along a tool path on the part surface P. Due to the following motion, at every instant of burnishing of the part surface, the desired degree of conformity at a point of contact of the generating surface T of the form-burnishing tool and of the part surface P can be ensured.

---

* Patent No. 1428563 (USSR), *A Tool for Burnishing a Sculptured Part Surface on a Multi-Axis NC Machine*, S. P. Radzevich, Int. Cl. B 24 B 39/00, filed February 11, 1986.

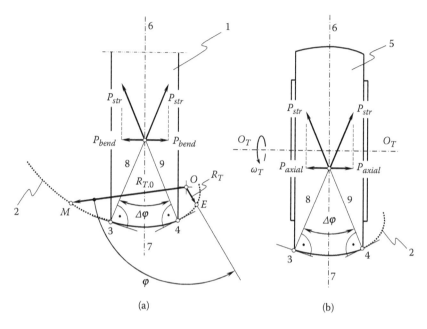

FIGURE 5.37
A tool for burnishing a sculptured part surface on a multiaxis numerical control machine.
(From USSR Patent 1.428.563.)

Use of the burnishing tool of the discussed design shown in Figure 5.37 allows for an increase in productivity of part surface finishing as well as enhanced quality of the machined part surface.

Similarly, the problem of profiling of a form-cutting tool can be solved for more a general case, say, when the correspondence between the parameters ρ_T and L_T is not linear. In such a case, the correspondence between parameters ρ_T and L_T must be given by a certain expression in the form $\rho_T = \rho_T(L_T)$. By means of implementation of a corresponding function $\rho_T = \rho_T(L_T)$, the design parameters of a form tool can be optimized.

5.6 Form-Cutting Tools Having Continuously Changeable Generating Surface *T*

The generating surface *T* of the most known designs of the form-cutting tools is a type of *rigid* surface. No changes to the shape and parameters of the tool surface *T* are feasible. However, for machining a sculptured part surface on a multiaxis *NC* machine, form-cutting tools of special designs are used as well. Cutting tools of the design under consideration feature continuously movable blades [108,127,128,152] and others. The continuous motion of the blades of the cutting tool is under numerical control. Wooden parts as well as parts made of plastics, light alloys, and so forth can be machined with a form-cutting tool of the design.

An equation of the generating surface T of the form-cutting tools of the design under consideration always contains at least one parameter that is under numerical control.

A continuously changeable generating surface T of the form-cutting tool cannot be formed as an envelope to consecutive positions of the part surface P that is performing a motion relative to a coordinate system $X_T Y_T Z_T$ of the form-cutting tool. Unknown kinematics of the part surface generation is the major reason for this: That same sculptured part surface can be machined with that same cutting tool under different kinematics of part surface generation.

For the development of design of the form-cutting tool having a continuously changeable generating surface T as well as for determining the optimal kinematics of part surface generation, the implementation of the $\mathbb{R}$-mapping of surfaces is vital.

5.7 Incorrect Problems in Profiling the Form-Cutting Tools

The use of the methods for derivation of an analytical description of the generating surface of the form-cutting tool based both on $\mathbb{R}$-mapping of the surfaces and on elements of theory of enveloping surfaces returns an accurate solution to the problem of profiling of the form-cutting tool. In the practice of part surface machining, no surface is generated precisely. Deviations of the actual (machined) part surface from the desired part surface are inevitable. To make a decision whether a part surface is machined properly or not, tolerances on the accuracy of the machined part surface are helpful for resolving this issue. If deviations of the machined part surface are within the tolerance, then the part surface P is generated properly.

There are no principal restrictions on what portion of the resultant tolerance on the accuracy should be allowed for deviations of the actual generating surface from the desired generating surface of the form-cutting tool. Moreover, in particular cases, the precise generating surface of the form-cutting tool cannot be generated, and only an approximation to the desired generating surface T exists.*

---

* In a particular case, the approximation of the actual generating surface of the cutting tool to its desired generating surface can even be asymptotically accurate.

The approximation to the desired generating surface T of the form-cutting tool is referred to as the *approximate generating surface of the form-cutting tool,* and it is designated as T_a.

The introduction of the approximate generating surface T_a of the form-cutting tool is necessary because of the following:

- Use of approximate methods of profiling of form-cutting tools for which deviations of the desired generating surface T from the actual generating surface T_a are inevitable.

- Implementation of approximate working surfaces of the cutting tool by surfaces, which are easy in generation when manufacturing the cutting tool.

- Impossibility of generating of geometrically accurate sculptured working surfaces* of the form-cutting tool.

The deviation of profiling of a form-cutting tool is understood to be the distance between the generating surface T of the form-cutting tool and between the approximate generating surface T_a. This distance is measured along the unit normal vector $\mathbf{n}_T$ to the cutting tool surface T in the corresponding surface point.

The following geometrical consideration is valid.

Evidently, for a given part surface P, a particular surface can be found that can be used as approximate generating surface T_a. The cutting tool that is designed on the premise of the approximate surface T_a can be capable of generating the part surface P within any given tolerance for the accuracy of the machined part surface.

Usually, it is convenient to have a generating surface of the cutting tool that is easy to implement. For example, it is commonly desired that the generating surface possesses the property of sliding "over itself." Certainly, this is not mandatory.

The approach under current consideration is not capable of generating precision generating surfaces T of the cutting tool. However, it is capable of generating the approximate generating surface T_a that is asymptotically accurate.

The following two illustrative examples are helpful for understanding the nature of the problem under consideration.

In design of low noise/noiseless transmissions for cars and light trucks, helical gears with topologically modified tooth flanks are implemented. In high-volume production, for finishing the topologically modified gears, plunge shaving cutters of special design are used. The desired tooth flank of a topologically modified shaving cutter is a type of sculptured surface (Figure 5.38a). As an example, consider the generation of the shaving cutter tooth flank in either a grinding or a regrinding operation.

In a grinding operation, the desired tooth flank of the shaving cutter is generated with a surface of revolution of the grinding wheel (Figure 5.38b).

---

* Clearance surfaces of hob teeth are a perfect example of the surfaces of this type.

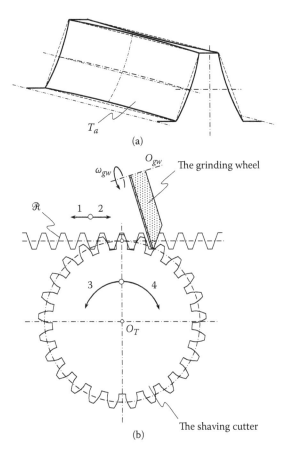

FIGURE 5.38
Generation of the approximate generating surface T_a of a shaving cutter for plunge shaving of topologically modified involute gears.

For precision generation of the tooth flank surface, multiaxis relative motion of the grinding wheel is necessary. However, because the actual number of the *NC*-axis is limited, the shaving cutter tooth flank cannot be generated to blueprint. Deviations of the surface generation in this case are inevitable. Therefore, the generating surface T of the plunge shaving cutter cannot be generated in a shaving cutter grinding operation, but the approximate generating surface T_a is generated instead.

Consider the generation of the surface of an internal circular cone, say, the generation of the rake surface of a broach for machining a hole as schematically illustrated in Figure 5.39. For this purpose, the approximate generating surface T_a of the grinding wheel can be used. Usually, grinding wheels having the generating surface T that is shaped in the form of external round cone are used. Design parameters of the grinding wheel as well as its setup

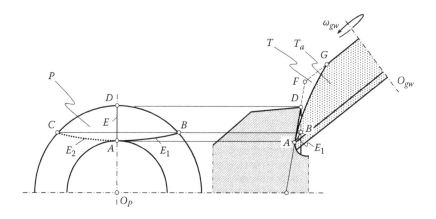

FIGURE 5.39
A particular case of grinding of a broach by a form-grinding wheel (an example of part surface generation when the surface characteristic splits on two portions).

parameters are determined in accordance with accurate generation of the rake surface of the broach to be ground.

The generating straight line AF of the grinding wheel aligns with the generating straight-line AD of the rake face of the broach. In this case, the straight-line segment AD is the surface characteristic line. The necessity of satisfaction of the conditions of proper part surface generation (see Chapter 7 for details) imposes strict constraints onto the maximal diameter of the grinding wheel.

Broaches can be reground with the form-grinding wheels of an appropriate design. The form-grinding wheel rotates about it axis O_T with a certain angular velocity ω_T. The curve AG is used as the generating curve of the approximate generating surface T_a of the grinding wheel (Figure 5.39). The grinding wheel surface T_a, and the rake surface of the broach become tangent along the characteristic curve BAC. The portion BA of this curve is the surface characteristic E_1, and the portion AC serves as the surface characteristic E_2. The approximate generating surface T_a of the form-grinding wheel is represented by a surface of revolution that is tangent to the rake surface of the broach along two symmetrically located surface characteristic lines E_1 and E_2. The parameters of the curvilinear generating curve AG satisfy the conditions under which the deviations of the actually generated part surface P from the desired shape are within the tolerance for accuracy of the part surface P.

Definitely, the number of examples can be extended. Problems of profiling gear cutting tools having zero pressure angle should be noted here,* including rack cutters, hobs [144], shaper cutters† [72], and so forth. The profiling

---

* In particular cases, the approximate generating surface of the cutting tool can be degenerated to a curve, say, into a generating curve of the cutting tool. In cases like the last one, the surface generation is referred to as the *edge-type surface generation*.
† Patent No. 1174187 (USSR), *A Gear Shaper Cutter*, S. P. Radzevich, Int. Cl. B23f 21/10, filed: April 23, 1981.

of skiving gear-cutting tools, multistart disk-type hobs, and others are also dovetailed to the incorrect problems of profiling from cutting tools.

One more example of the problem under consideration is in the field of profiling of form-cutting tools featuring partially completed motion relative to the part. In this case, two portions of the generating surface T are distinguished:

a. The first portion is the ordinary enveloping surface (that is obtained using generating method)
b. The second portion is a *"replica"* of the part surface P—namely, the surface T_P.

Two portions of the generating surface T of the cutting tool are somehow blended to each other. The problem can be formulated as follows: To determine the generating surface T that is *"reduced"* compared to the conventional enveloping surface.

The above-mentioned problems should be investigated in more detail.

5.8 Principal Types of Problems in Profiling Form-Cutting Tools

Profiling of form-cutting tools is a complex engineering problem. The solution to the problem of profiling a form-cutting tool aims to determine the design parameters of the working surfaces of the cutting wedge of the cutting tool that are necessary and sufficient for production of the form-cutting tool.

To solve the problem of profiling a form-cutting tool of relatively simple design, it is commonly sufficient to determine the profile of a certain (usually normal) cross section of the clearance surface of the cutting tool. It is assumed here that the geometry of the rake surface of the cutting tool is prespecified. In other cases, the profile of a certain (usually normal) cross section of the rake surface of the cutting tool is determined. It is assumed here that geometry of the clearance surface of the cutting tool is prespecified. For example, the parameters of the profile of normal cross section of the clearance surface are determined when solving the problem of profiling of a prismatic form tool. Similarly, the parameters of axial cross section of the clearance surface are determined when solving the problem of profiling of a round form tool. The axial cross section of the clearance surface of a round form tool is equivalent to its normal cross section. A determined set of the design parameters of a normal cross section of the clearance surface (or of the rake surface) must be sufficient to produce the form-cutting tool.

The problem of profiling becomes more complex when a form-cutting tool of a complex design has to be designed. Form-cutting tools operating on the generating principle of part surface machining (hobs, shaper cutters, gear cutting heads, and so forth) are good examples of form-cutting tools of complex design.

Prior to solving the problem of profiling a form-cutting tool, the problem must be formulated correctly. When the problem of profiling is formulated correctly, the obtained solution to the problem makes possible proper manufacture of the working surfaces of the cutting edge of the form-cutting tool. Geometrically, the cutting edge, C_e, can be interpreted as the line of intersection of the rake surface, R_s, of the cutting tool by its clearance surface C_s. When the problem of profiling of a form-cutting tool is solved correctly, the cutting edge, C_e, is entirely located within the generating surface T of the form-cutting tool. Generally speaking, all three surfaces, namely, the clearance surface C_s, the rake surface R_s, and the generating surface T must be the surfaces through the common line—that is, through the cutting edge C_e. Again, the latter must be entirely located within the generating surface T of the form-cutting tool. This is the most general formulation of the problem of profiling of a form-cutting tool for machining either a part surface on a conventional machine tool, or for machining a sculptured part surface on a numerical control machine.

Determination of a generating surface T is the first step when solving the problem of profiling a form-cutting tool. The possible types of the problem are discussed below in this section of the book in more detail. A *generating body of a form-cutting tool* is a solid made of an instrumental (cutting) material and is bounded by the generating surface of the cutting tool. A form-cutting tool is designed on the second and the later steps of solving the problem of profiling of a form-cutting tool. For this purpose, rake surfaces, R_s, and clearance surfaces, C_s, are machined in the generating body of the form-cutting tool. This topic is under discussion in Chapter 6 of the book.

The $\mathbb{R}$-mapping–based method of part surface generation allows the determination of the generating surface of the form-cutting tool in terms of the design parameters of the sculptured part surface to be machined. Therefore, the shape and parameters of the generating surface T can be expressed in terms of the shape and parameters of the part surface P being machined. The aforementioned statement can be expressed analytically in the form of a function:

$$T = T(P) \tag{5.137}$$

This means that the $\mathbb{R}$-mapping–based method of part surface generation allows for an expression of the design parameters of the generating surface T of the form-cutting tool in terms of the design parameters of the sculptured surface P to be machined.

The kinematics of part surface generation (namely, the set of motions of the generating surface T of the form-cutting tool relative to the part surface P) is designated as K_{rm}.

Generally speaking, the kinematics of part surface generation on a multiaxis *NC* machine can be analytically expressed in the form of a function:

$$K_{rm} = K_{rm}(P, T) \tag{5.138}$$

Because the $\mathbb{R}$-mapping–based method of part surface generation is valid, the design parameters of the tool surface T can be expressed in terms of the design parameters of the part surface to be machined, namely, by an expression $T = T(P)$. That said, the original function for the kinematics of multiaxis part surface generation

$$K_{rm} = K_{rm}(P, T) \tag{5.139}$$

can be transformed to

$$K_{rm} = K_{rm}[P, T(P)]. \tag{5.140}$$

In the latter expression, the kinematics K_{rm} is a function of P and of T. Taking into account that the expression $T = T(P)$ for T is valid, then the formula $K_{rm} = K_{rm}[P, T(P)]$ for the K_{rm} can be reduced to

$$K_{rm} = K_{rm}(P). \tag{5.141}$$

Ultimately, the entire process of part surface generation can be expressed only in terms of the design parameters of the part surface P to be machined. This is the minimum input information based on the most favorable (optimal) machining process of a part surface that can be developed.

After the generating surface T of the form-cutting tool is determined as well as the kinematics of the part surface generation K_{rm}, it is necessary to solve problems of other types. In nature, problems of these types are inverse to the corresponding problems of part surface generation. To briefly formulate problems of this sort, let us assume that the actually generated part surface is designated as P_a. Then the problems to be solved can be formulated as follows:

a. To determine the function $P_a = P_a[T, K_{rm}^{(a)}]$, taking into account the deviations of the kinematics of the part surface generation, when the actual kinematics of part surface generation $K_{rm}^{(a)}$ differs from the desired kinematics of the part surface generation K_{rm}, say, when the inequality $K_{rm}^{(a)} \neq K_{rm}$ is observed.

b. To determine the function $P_a = P_a[T_a, K_{rm}]$, taking into account the deviations of profiling of the form-cutting tool, say, when the inequality $T_a \neq T$ is observed;

c. To determine the function $P_a = P_a[T_a, K_{rm}^{(a)}]$, taking into account the deviations of the actual kinematics of part surface generation, and the deviations of profiling of the form-cutting tool, say, when the inequalities $K_{rm}^{(a)} \neq K_{rm}$ and $T_a \neq T$ are observed.

Another group of problems to be solved relate to derivation of the approximate generating surface T_a of the form-cutting tool. These problems can be formulated in the following way:

a. The parameters of the nominal part surface P are not given, but the parameters of the approximated part surface P_a are known instead, say, when it is necessary to determine the function $T_a = T_a(P_a, K_{rm})$.

b. The actual kinematics of part surface generation $K_{rm}^{(a)}$ differs from the nominal kinematics of the part surface generation K_{rm}, say, when the function $T_a = T_a(P, K_{rm}^{(a)})$ must be determined.

c. The nominal part surface P and the nominal kinematics of part surface generation K_{rm} are not known, but the actual the part surface P_a, and the actual kinematics $K_{rm}^{(a)}$ of part surface generation are given instead. In this case, it is necessary to determine the function $T_a = T_a(P_a, K_{rm}^{(a)})$.

The problems listed above can be associated with the inverse problem of part surface generation. However, the nature of these problems is completely different [117].

The following 11 principal types of the problem of profiling a form-cutting tool (and of generating a part surface) are distinguished.

The problem of profiling of the first type. Consider a case when the fundamental magnitudes of the first order E_P, F_P, and G_P (of the first fundamental form $\varPhi_{1.P}$) and of the second order L_P, M_P, and N_P (of the second fundamental form $\varPhi_{2.P}$) for a sculptured part surface P are specified.

Then, a desired contact geometry at a point of contact between the known part surface P and a generating surface T to be determined is synthesized. For this purpose, the indicatrix of conformity $Cnf_R(P/T)$ at a point of contact of surfaces P and T is implemented (Figure 5.40).

$\mathbb{R}$-mapping is used for the purpose of designing a form-cutting tool for machining a given part surface. The generating surface T of the form-cutting tool is determined as $\mathbb{R}$-mapping of the part surface P to be machined. The use of $\mathbb{R}$-mapping makes the determination of three fundamental magnitudes E_T, F_T, and G_T of the first order and three fundamental magnitudes L_T, M_T, and N_T of the second order of the generating surface T of the form-cutting tool possible. The fundamental magnitudes E_T, F_T, G_T of the cutting tool surface T are expressed in terms of the fundamental magnitudes E_P, F_P, G_P. The fundamental magnitudes L_T, M_T, N_T are expressed in terms of the fundamental magnitudes E_P, F_P, G_P and L_P, M_P, N_P.

The determined generating surface $T_{\mathbb{R}}^{(1)}$ of the form-cutting tool is initially expressed by an equation in the natural form, namely, it is expressed in terms of the fundamental magnitudes E_P, F_P, G_P and L_P, M_P, N_P of the part surface P. On later steps of profiling the form-cutting tool, the natural representation of the generating surface of the form-cutting tool is converted to another form of surface representation. For this purpose, a set of Gauss–Weingarten equations must be solved.

Further synthesis of the favorable kinematics $K_{rm}^{(1)}$ of the part surface generation is based on (Figure 5.40)

- Geometry of the part surface P (1)
- Favorable contact geometry at current point of contact of the prespecified part surface P and of generating surface of the cutting tool (3)
- Geometry of the generating surface of the cutting tool (4)

The synthesized *fragile kinematics* $K_{rm}^{(1)}$ is the most general type of kinematics of part surface generation. The multiaxis numerical control machines are perfectly suited for the reproduction of the fragile kinematics of part surface generation.

A specified sculptured part surface P can be machined to blueprint by the cutting tool having the determined generating surface T in cases when actual kinematics of part surface generation is different from the optimal one. Such a solution to the problem of generating of a part surface can be referred to as a *positive solution** (and *not* the *optimal solution*) to the problem of part surface generation (see below for the problem of profiling of the fourth type).

The inverse problem of part surface generation then can be solved based on the shape and design parameters of the derived generating surface T of the form-cutting tool (5) and on the synthesized kinematics $K_{rm}^{(1)}$ of part surface generation (6). At this step, the design parameters of the actually machined part surface $P_g^{(1)}$ are determined. A decision on whether or not the part surface is generated properly (namely, to the part blueprint) can be made based on the comparison (7) of the design parameters of the actually machined part surface $P_g^{(1)}$ and that for the nominal part surface P. Such a comparison is also informative regarding the accuracy of the determined generating surface of the form-cutting tool.

The problem of profiling of the second type. It could happen that the determined geometrically accurate generating surface of the form-cutting tool is inconvenient in the production of the cutting tool. Under such a scenario, the geometrically accurate generating surface of the cutting tool can be approximated by a portion of a smooth regular surface, which is convenient in the production of the cutting tool.

---

* The solution to a problem is referred to as a *positive solution* if the obtained solution is not the optimal one (in certain sense).

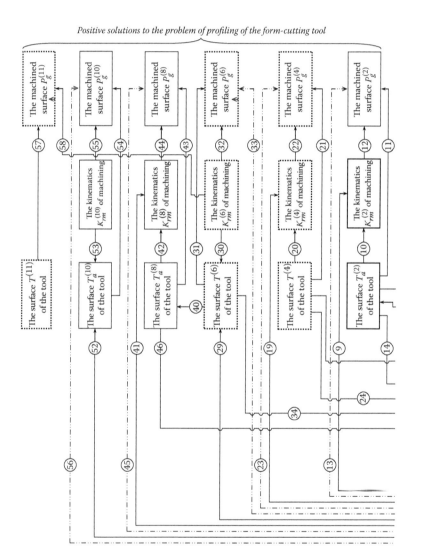

FIGURE 5.40
Principal types of problems in profiling form-cutting tools. (Reprinted from *Mathematical and Computer Modeling*, 36, Radzevich, S.P., ℝ-mapping based method for designing of form cutting tool for sculptured surface machining, 921–938, Copyright 2002, with permission from Elsevier.)

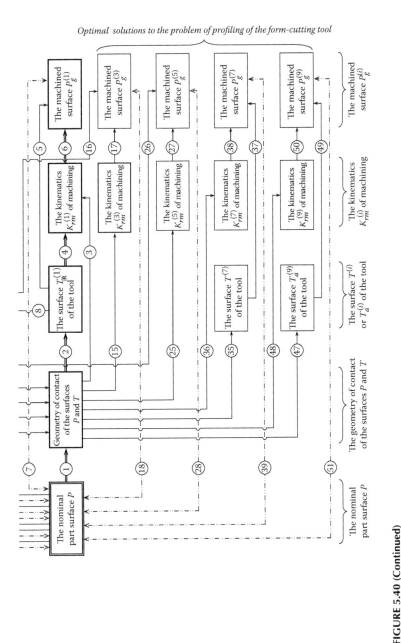

FIGURE 5.40 (Continued)
Principal types of problems in profiling form-cutting tools. (Reprinted from *Mathematical and Computer Modeling*, 36, Radzevich, S.P., $\mathbb{R}$-mapping based method for designing of form cutting tool for sculptured surface machining, 921–938, Copyright 2002, with permission from Elsevier.)

Due to the approximation, the part surface is generated by an approximate generating surface $T_a^{(2)}$, not by geometrically accurate generating surface T of the form-cutting tool. The cutting tool surface T is approximated by surface $T_a^{(2)}$; thus, the deviations of the surface $T_a^{(2)}$ from the surface T must be within a prespecified tolerance for accuracy of the approximation (Figure 5.40).

The parameters of the desired kinematics $K_{rm}^{(2)}$ of part surface generation in this case are calculated on the premise of the design parameters of the desired part surface P and of the design parameters of the approximate generating surface $T_a^{(2)}$ of the form-cutting tool.

Later on, the design parameters of the actually machined part surface $P_g^{(2)}$ are determined based on the shape and parameters of the generating surface $T_a^{(2)}$ of the form-cutting tool (11), and on the parameters of the kinematics $K_{rm}^{(2)}$ of part surface generation (12). Then, the determined parameters of the part surface $P_g^{(2)}$ are compared (13) with the corresponding parameters of the nominal (desired) part surface P. The comparison makes it possible to conclude whether or not the generating surface of the form-cutting tool is approximated properly. Such a solution to the problem of generating of a part surface is a type of positive solution (and *not* the *optimal solution*) to the problem of part surface generation. The most favorable (the optimal) solution to the problem under consideration can be determined by means of a synthesized optimal geometry of contact of the part surface P and the approximate generating surface $T_a^{(2)}$ of the form-cutting tool. For this purpose, consider the problem of profiling of the third type.

The problem of profiling of the third type. The most favorable (the optimal in the certain sense) kinematics of part surface generation $K_{rm}^{(3)}$ is synthesized (15) based on the shape and parameters of the part surface P (1) and of the approximate generating surface $T_a^{(2)}$ of the form-cutting tool (14). This is illustrated in Figure 5.40. Remember that the parameters of the cutting tool surface $T_a^{(2)}$ are correlated with the kinematics of part surface generation. After that, the actually generated part surface $P_g^{(3)}$ can be determined on the premise of the shape and parameters of the approximate generating surface $T_a^{(3)} \equiv T_a^{(2)}$ of the cutting tool and on the parameters of kinematics $K_{rm}^{(3)}$ of part surface generation.

The shape and parameters of the actually generated part surface $P_g^{(3)}$ are compared (18) with that for the nominal part surface P. The comparison makes it possible to conclude whether or not the cutting tool surface $T_a^{(2)}$ is constructed properly and whether or not the parameters of the kinematics of part surface generation $K_{rm}^{(3)}$ are synthesized properly.

The problem of profiling of the fourth type. Machining of a sculptured part surface can be executed by means of a cutting tool selected from several available designs of the cutting tool. In this particular case the design parameters of generating surface $T_a^{(4)}$ do not correlate to the corresponding design parameters of the generating surface that is determined as $\mathbb{R}$-mapping of the part surface P (Figure 5.40).

The parameters of kinematics $K_{rm}^{(4)}$ of part surface machining in this particular case can be determined on the premise of the design parameters of the nominal part surface P (19) and on design parameters of the generating surface $T_a^{(4)}$ of the form-cutting tool (20). The solution to the problem of profiling of the fourth type returns a positive solution to the problem of the part surface generation.

When selecting a cutting tool among available designs, the generating surface of the selected cutting tool $T_a^{(4)}$ can be identical to that of the form-cutting tool determined by means of $\mathbb{R}$-mapping of the part surface P, namely, it could happen that the identity $T_a^{(4)} \equiv T_\mathbb{R}^{(1)}$ is observed (see the problem of profiling of the fourth type). In this particular case, the kinematics of part surface machining can be determined identically to that it is determined in a case when a cutting tool is selected.

The shape and design parameters of the actually generated part surface $P_g^{(4)}$ can be determined (23) based on the set of design parameters of the determined generating surface $T_a^{(4)}$ (21) and on the kinematics of the part surface generation (22). Further, the design parameters of the part surface $P_g^{(4)}$ are compared with that for nominal part surface P. This comparison makes it possible to answer the question whether or not a form-cutting tool is chosen properly and whether or not the designed process of the part surface machining meets the anticipated requirements.

The problem of profiling of the fifth type. For a given part surface P, the most favorable kinematics of part surface machining by a chosen form-cutting tool can be determined (Figure 5.40). For this purpose, the most favorable (the optimal) contact geometry between surfaces P and $T_a^{(5)} \equiv T_a^{(4)}$ must be developed. This can be done (1) on the premise of the shape and design parameters of the nominal part surface P and of the design parameters of the generating surface $T_a^{(5)} \equiv T_a^{(4)}$ (24) of the selected form-cutting tool. Later on, the parameters of the favorable kinematics $K_{rm}^{(5)}$ of the part surface generation are determined (25) on the premise of the calculated parameters of the favorable contact geometry between surfaces P and $T_a^{(5)}$.

In a particular case, when the problem of the fifth type is solved for the generating surface $T_a^{(5)} \equiv T_\mathbb{R}^{(1)}$ of the form-cutting tool, the generated part surface $P_g^{(5)}$ is identical to the $P_g^{(1)}$ already discussed when solving the problem of profiling of the first type (see the problem of profiling of the fourth type).

The generated part surface $P_g^{(5)}$ can be determined using (26) for this purpose the set of design parameters of the determined generating surface $T_a^{(5)} \equiv T_a^{(4)}$ of the form-cutting tool, and (27) parameters of the synthesized kinematics of the part surface machining $K_{rm}^{(5)}$.

The comparison of the design parameters of the part surfaces $P_g^{(5)}$ and P makes it possible to conclude (28) on the correctness of the selection of the form-cutting tool, and on kinematics of the part surface machining.

The considered solution to the problem of profiling of the fifth type is widely used in practice of sculptured part surface machining on a multiaxis numerical control machine.

The problem of profiling of the sixth type. In a certain case, strong constraints can be imposed on the synthesized *agile* kinematics of a sculptured part surface generation. In particular, the synthesized *agile* kinematics can be substituted with a *rigid* kinematics of a part surface generation having the parameters not varying in time.

The total number of types of rigid kinematic schemes of part surface generation is limited to a certain number of types of the principal kinematic schemes (see Figure 2.16). The kinematic scheme of the part surface generation $K_{rm}^{(6)}$ is a reduced case of the *agile* principal kinematics $K_{rm}^{(1)}$ of sculptured part surface generation.

The solution to the problem of profiling of a form-cutting tool for machining a given sculptured part surface can be derived in the following way (Figure 5.40).

The generating surface $T^{(6)}$ is determined (30) based on the shape and design parameters of the nominal part surface P (29) and on parameters of the selected (from Figure 2.16) particular *rigid* kinematic scheme of part surface generation. The cutting tool surface $T^{(6)}$ is an envelope to successive positions of the part surface P when this surface is performing a single-, two-, or multi-parametric motion in relation to a reference system, with which the designed cutting tool will be associated.

The determined generating surface $T^{(6)}$ of the form-cutting tool performs in relation to the part that same single-, two-, or multi-parametric motion, but in the opposite direction. Because of this, the shape and parameters of the actually machined part surface $P_g^{(6)}$ can be determined (31) on the premise of the design parameters of the generating surface $T^{(6)}$ and (32) on the parameters of the principal kinematic scheme of the part surface generation $K_{rm}^{(6)}$. The part surface $P_g^{(6)}$ is an envelope to successive positions of the tool surface $T^{(6)}$ when the cutting tool is traveling in relation to a reference system associated with the part.

The parameters of the generated part surface $P_g^{(6)}$ are compared (33) with the corresponding design parameters of the nominal part surface P. A conclusion on whether the problems of the part surface generation as well as the problem of profiling of the form-cutting tool are solved properly or not can be made based on the obtained results of the comparison.

The considered solution to the problem of profiling of the sixth type is used in the practice of sculptured part surface machining on a machine tool.

The derived solution to the problem of profiling of the sixth type is a *positive* one. The optimal contact geometry between the part surface and the generating surface of the cutting tool must be insured at every instance to come up with the optimal solution to that same problem of the sixth type.

The problem of profiling of the seventh type. To derive the optimal solution to the problem under consideration, the optimal contact geometry between the part surface to be machined and the generating surface of the cutting tool must be synthesized (Figure 5.40). The optimal contact geometry is synthesized using as the input the set of design parameters of the nominal part surface P (1) and the set of design parameters of the generating surface $T^{(6)}$ of the cutting tool (remember that design parameters of the cutting tool surface $T^{(6)}$ are expressed in terms of parameters of the principal kinematic scheme of the part surface generation (34) of the selected type). This makes the determination (35) of the optimal parameters of the principal kinematic scheme $K_{rm}^{(7)}$ of the selected type possible as well as the determination (36) of the corresponding generating surface $T^{(7)}$ that features the optimal design parameters. Later on, the actually machined part surface $P_g^{(7)}$ can be determined (38) based on the set of design parameters of the generating surface $T^{(7)}$ (37) and on the design parameters of the principal kinematic scheme $K_{rm}^{(7)}$ of the part surface generation.

The determined parameters of the generated part surface $P_g^{(7)}$ can be compared (39) with that for the nominal part surface P. This makes possible a conclusion on how successfully the problem of profiling of the form-cutting tool and the problem of part surface generation are solved.

The problem of profiling of the eighth type. It could happen that the determined generating surface $T^{(6)}$ of the form-cutting tool feature complex shapes, and therefore, it is inconvenient for production of the cutting tool. When this is observed, the cutting tool surface $T^{(6)}$ can be approximated (40) by a surface patch $T_a^{(8)}$ that is convenient for production of the cutting tool (Figure 5.40). The approximation of the generating surface $T^{(6)}$ of the form-cutting tool must be executed so as to keep the maximal deviations within the tolerance on accuracy of the cutting tool surface.

Further, the optimal parameters of the kinematic scheme of the part surface generation $K_{rm}^{(8)}$ are calculated (42) using the design parameters of the nominal part surface P (41) and the earlier determined design parameters of the generating surface $T_a^{(8)}$ of the form-cutting tool.

The geometry of the cutting tool surface $T_a^{(8)}$ (43), and the parameters of the kinematic scheme of the part surface generation (44) are used as the input to determine the design parameters of the actually machined part surface $P_g^{(8)}$. The calculated design parameters of the part surface $P_g^{(8)}$ are compared (45) with that for the nominal part surface P. The comparison makes it possible an answer to the question of whether or not the problem under consideration is solved properly.

The problem of profiling of the ninth type. The earlier derived *positive* solution to the problem of profiling of the form-cutting tool and to the problem of part surface generation (see the problem of profiling of the eighth type) can be employed for the purpose of derivation of the *optimal* solution to that same problem. For this purpose, the optimal contact geometry for the part surface to be machined P and for the generating surface $P_g^{(8)}$ of the form-cutting tool must be synthesized. The synthesis can be executed on the premise of the design parameters of the nominal part surface P, and of the design parameters of the generating surface $T_a^{(9)}$ of the form-cutting tool (Figure 5.40). Remember that the design parameters of the generating surface $T_a^{(9)}$ are expressed in terms of the parameters of the principal kinematic scheme of the part surface generation of the selected type (46). Based on that, the optimal parameters of the generating surface $T_a^{(9)}$ of the form-cutting tool can be calculated (47). The optimal parameters of the principal kinematic scheme of the part surface generation $K_{rm}^{(9)}$ are calculated (48) as well. The shape and design parameters of the actually machined part surface $P_g^{(9)}$ can be determined based on the design parameters of the favorable generating surface $T_a^{(9)}$ (49) and on the parameters of the principal kinematic scheme $K_{rm}^{(9)}$ of the part surface generation (50). Further, the corresponding design parameters of the surfaces $P_g^{(9)}$ and P are compared (51) with each other. This makes a decision on correctness of the obtained solution to the problem of profiling of the form-cutting tool and, more generally, to the problem of the part surface generation possible.

The problem of profiling of the tenth type. An approximate generating surface T_a of the form-cutting tool is introduced into consideration not only because of an approximation of the ideal (the desired) generating surface T by a surface T_a featuring more preferred design parameters. When rigid principal kinematic schemes of part surface generation are used in a particular case, the desired generating surface T does not exist at all. For example, the geometrically accurate generating surface does not exist in cases of relieving of gear hobs, the hobs for machining splines, and so forth. More examples in this concern can be found out in References [120,138] and others.

To determine the approximate generating surface $T_a^{(10)}$ of the form-cutting tool in cases such as these, the design parameters of the nominal part surface P (52) and the parameters of the selected principal kinematic scheme of the part surface generation are used (Figure 5.40). The patch of a certain smooth regular surface, that is convenient from the manufacturing standpoint, must somehow be chosen. This surface patch is used further as the approximate generating surface $T_a^{(10)}$ of the form-cutting tool.

The actually machined part surface $P_g^{(10)}$ can be determined based on the design parameters of the generating surface $T_a^{(10)}$ of the form-cutting tool (54) and on the parameters of the principal kinematic scheme of the part surface generation $K_{rm}^{(9)}$ (55).

The design parameters of the part surface $P_g^{(10)}$ generated in this way are further compared (56) with the corresponding design parameters of the nominal part surface P. The obtained results of the comparison are used to make a decision whether or not the problem of profiling of the form-cutting tool is solved accurately. This is also useful to make a more general conclusion whether or not the problem of the part surface generation is solved properly.

The problem of profiling of the eleventh type. Finally, a given part surface can be machined by the cutting tool selected from those available for the user (Figure 5.40). This approach is widely used when

- Generating gears for non-involute gear pairs of the first, and of the second type
- Generating-milling spiral bevel gears by standard gear cutter heads
- Face hobbing spiral bevel gears, and so forth.

In cases like these, a certain design of the cutting tool is selected from cutting tools currently available for the gear manufacturer. Further, a part surface $P_g^{(11)}$ that is machined by the selected cutting tool is investigated under various kinematic schemes of the part surface generation. For this purpose, the part surface $P_g^{(11)}$ is determined as the envelope to successive positions of the generating surface $T_a^{(11)}$ of the form-cutting tool (57) that is traveling in accordance with the principal kinematic scheme $K_{rm}^{(11)} \equiv K_{rm}^{(6)}$ of the part surface generation (58).

A comparison of the surfaces $P_g^{(11)}$ and P makes possible a conclusion on how accurately the cutting form tool is profiled and how accurately the obtained solution to the problem of the part surface generations.

Only a positive solution to the problem of profiling of the form-cutting tool, and to the problem of generation of a given part surface can be derived using the discussed approach. The optimal solution to the problem under consideration cannot be derived, as the exact information of the nominal part surface is not available in this particular case. Because of this, the design parameters of the actually generated part surface $P_g^{(11)}$ cannot be compared with that for the nominal part surface P.

The most favorable part surface cannot be machined at all in this case. However, the generated part surface $P_g^{(11)}$ can deviate from the desired (not known) part surface at any small prespecified value, which is sufficient for engineering applications.

In the general case, the problems of profiling of a form-cutting tool and the problems of part surface generation can be solved in the following five steps:

- Analytical description of the nominal part surface P to be machined
- The synthesis of the most favorable contact geometry of the given part surface P and (not yet known) generating surface T of the form-cutting tool
- The determination of either the geometrically accurate generating surface $T^{(i)}$ or an approximate generating surface $T_a^{(i)}$ of the form-cutting tool
- The determination of the principal kinematics $K_{rm}^{(i)}$ of the part surface generation and of the actually generated part surface $P_g^{(i)}$
- A comparison of the design parameters of the actually generated part surface $P_g^{(i)}$ with that for the nominal part surface P to be machined specified in the part blueprint.

Not all cases of profiling of a form-cutting tool and generating a specified part surface include all of the five of the steps above. In particular cases, certain steps solving the problems under consideration are omitted.

The chart schematically depicted in Figure 5.40 illustrates the correlation among the principal problems of profiling of a form-cutting tool and of part surface generation. The 11 principal problems considered are not isolated from one another, and they are logically connected to each other. The problem of profiling of the first type is the most general one: in this particular case, the solution to the problems under consideration is based on implementation of $\mathbb{R}$-mapping of a part surface P to be machined onto the generating surface T of the form-cutting tool to be applied for the machining of the given part surface. In Figure 5.40, the solutions to the principal problems of all the considered types, those depicted above the first approach, can only be *positive*. The *optimal* solutions to the problems under consideration are shown below the first approach (see Figure 5.40).

In reality, both the generating surfaces of the form-cutting tool and the kinematics of part surface generation (and the kinematics of profiling of the cutting tool) are not accurately reproduced, but they are reproduced with certain deviations from their desired values. The steps illustrated in Figure 5.40 can be significantly evolved and enhanced if the deviations are taken into account. However, the chart in this case becomes too bulky and is not discussed here.

6

Geometry of the Active Part of a Cutting Tool

The active part of a cutting tool appears as a cutting wedge. In a machining operation, the wedge of the cutting tool is properly oriented in relation to the surface of cut as well as in relation to the direction of the velocity vector of the resultant motion of the cutting wedge relative to the surface of cut.

The performance of the cutting tool strongly depends upon the geometry of the cutting wedge. The geometry of the cutting wedge of the cutting tool is critically important from the standpoint of achieving the required surface finish, the productivity of the machining operation, and so forth. The geometry of the active part of cutting tools has remained under careful investigation by researchers since 1870 [159] or even earlier. However, many questions in this concern have no appropriate answers yet.

For the specification of the geometry of the active part of cutting tools, the following geometric parameters are commonly used:

- Rake angle γ
- Clearance angle α
- Angle of inclination of the cutting edge λ
- Major cutting edge approach angle φ_1
- Minor cutting tool approach angle φ_2
- Cutting edge roundness ρ
- Radius of curvature of the cutting edge r_T

In particular cases, some other geometric parameters of the active part of cutting tools are used as well. Prior to designing a high-performance cutting tool, it is necessary to know the optimal (or the desired) values of the geometric parameters of the cutting edge. These geometric parameters of the cutting edge of the active part of a cutting tool are specified in corresponding reference systems.

In the production of cutting tools, for inspection and other purposes, stationary reference systems are widely implemented. Reference systems of this sort are often referred to as the *static reference systems*. They are associated with the reference surface of the cutting tool.

The *tool-in-hand* reference system is a good example of a static reference system. The *tool-in-hand* reference system is recommended for application by the International Standard ISO 3002.

The *tool-in-machine system* (or *setting system*) is another perfect example of a static reference system [32]. The latter reference system incorporates the deviation of the actual configuration of the cutting tool in the machine tool with respect to its desired configuration.

Geometric parameters of the active part of cutting tools that are measured in a static reference system specify the *static geometry of the active part of the cutting tool.*

For the analysis and optimization of the material removal process, moving reference systems are used. Reference systems of this sort are commonly referred to as the *kinematic reference systems.* They are associated with the surface of cut and with the direction of the relative motion of the cutting edge with respect to the surface of cut.

The *tool-in-use* reference system is a good example of a kinematic reference system. This reference system is also recommended for application by the International Standard ISO 3002. Geometric parameters of the active part of a cutting tool, which are measured in a kinematic reference system, specify the *kinematic geometry of the active part of the cutting tool.*

Actual values of the corresponding geometric parameters of the active part of a cutting tool in static and kinematic reference systems usually differ from each other. However, once the geometry of the cutting edge is specified in a certain reference system, it can be converted to another reference system and vice versa. Analytical methods for the conversion are considered below.

6.1 Transformation of the Body Bounded by the Generating Surface *T* into the Cutting Tool

The generating surface *T* of the cutting tool that was determined earlier (see Chapter 5) bounds the body made of a cutting tool material, e.g., high-speed steel. This body is referred to as the *generating body of the cutting tool.* The cutting tool designer uses this body to design a workable cutting tool.

To proceed with further analysis, it is necessary to distinguish two major functions of the cutting tool.

The first major function of the cutting tool is to remove the stock from the work (*roughing* of the part surface). This function can be performed either by roughing cutting edges, or by finishing cutting edges of the cutting tool (in certain cases finishing edges of the cutting tool rough the part surface).

The major purpose of roughing cutting edges is to remove the stock. Roughing cutting edges of the cutting tool are located either within the interior of the generating body of the cutting tool or within the generating surface of the cutting tool. Roughing cutting edges do not generate the part surface *P* even in cases when they are located within the generating surface of the cutting tool. Therefore, the roughing cutting edges can be constructed as a line of intersection of only two surfaces, for example the rake surface and the clearance surface of the cutting tool. Both of these surfaces can be of a shape that is convenient for manufacturing the form-cutting tool.

The second major function of the cutting tool is to generate the part surface (*generating* or *finishing* of the part surface). This function is performed only by finishing cutting edges of the cutting tool. Finishing cutting edges of the cutting tool are located within the generated surface *T* of the cutting tool, and they cannot be located within the interior of the generating body of the cutting tool. Therefore, the finishing cutting edges can be constructed as a line of intersection of three surfaces, for example,

- The generating surface *T* to the cutting tool
- The rake surface
- The clearance surface of the cutting tool

The major purpose of finishing (cleanup) cutting edges is to generate (and to finish) the part surface *P*.

It is important to stress here that different cutting edges of the cutting tool (or their different segments) in different instants of time can perform different functions. An involute gear hob is a perfect example in this concern. Three different types of teeth can be distinguished in the design of an involute hob:

1. Only one function is performed by the cutting teeth at the entering end of the hob—they just remove the stock, which is a roughing operation. These cutting teeth do not generate the gear teeth surface of the gear being machined.

2. The hob teeth that are close enough to the center-distance "the hob–the work" (that is, to the pitch point in the gear machining mesh) remove the stock and generate the gear teeth surface. Gear teeth of this type perform two functions: finishing the gear teeth and generating involute profile.

3. Ultimately, the hob teeth that are beyond the center-distance (that is, beyond the pitch point in the gear machining mesh) remove almost no stock. These teeth mostly generate the gear teeth surface.

More detailed analysis reveals that different segments of that same cutting edge can simultaneously serve as the roughing cutting edges as well as the finishing cutting edges.

The consideration below is limited to the analysis of only two major functions of the cutting tool, namely, roughing and part surface generation. Other important functions of the cutting tool, for example,

- Chip evacuation from the area of cutting
- Chip curling
- Chip braking and its transportation from the area of cutting
- The coolant supply and others are not considered

To design a workable cutting tool on the premise of the generating body of the cutting tool, the latter must be capable of removing the stock from the work.

Three methods for transforming the generating body of the cutting tool into the workable edge-cutting tool are considering below. All of the methods are targeting three surfaces passing through a common line:

- The generating surface of the form-cutting tool
- The cutting tool rake surface
- The cutting tool clearance surface

The line of intersections of three surfaces is referred to as the *cutting edge of the form-cutting tool.*

6.1.1 First Method for the Transformation of the Generating Body of the Form-Cutting Tool into the Workable Edge-Cutting Tool

Consider a scenario under which the generating surface of the cutting tool is already determined (see Chapter 5). The cutting tool rake surface is a type of surface that is convenient for surface machining when manufacturing form-cutting tools, for inspection purposes, and so forth. In most cases, the rake surface R_s is a type of surface that allows for sliding over itself (see Section 2.4).

The rake surface R_s is properly oriented with respect to the generating surface T of the form-cutting tool. It makes an optimal rake angle, γ, with respect to the perpendicular to the surface T at a given point.

Following the first method for the transformation of the generating body of the cutting tool into the workable edge-cutting tool, the cutting edges of the form-cutting tool, C_e, are defined as the line of intersection of the generating surface T of the cutting tool by the rake surface R_s.

Once the cutting edge is constructed, the clearance surface C_s can be constructed in compliance with the following routine:

First, the clearance surface, C_s, is selected among the surfaces that allow for sliding over themselves (see Section 2.4). This requirement is highly desirable, but it is not mandatory. Actually, any surface having reasonable geometry can serve as the clearance surface of the form-cutting tool.

Second, the parameters of the chosen clearance surface C_s must be calculated so as to meet the requirement under which the cutting tool surface C_s is a surface through the earlier calculated cutting edge, C_e, of the form-cutting tool. This requirement imposes strong constraints onto geometry of the rake surface of the form-cutting tool, especially in cases when high accuracy of the form-cutting tool is required. Any deviation of the clearance surface from the line of intersection, C_e, of the generating surface T of the form-cutting tool by the rake surface, R_s, entails a corresponding deviation of the actually machined part surface from the nominal part surface P specified by the part blueprint.

Third, the configuration of the determined clearance surface C_s of the form-cutting tool must be of the sort for which surface C_s makes the optimal clearance angle α with the generating surface T of the form-cutting tool at a given point of the cutting edge.

It is easy to see that clearance surfaces of cutting tools cannot always be shaped in the form that is convenient for manufacturing the cutting tool.

The cutting edge of a precision form-cutting tool can be interpreted as a line of intersection of the three surfaces, for example,

- The generating surface T of the cutting tool
- The rake surface R_s
- The clearance surface C_s

The requirement, three surfaces T, R_s, and C_s being the surfaces through the common line, for example, through the cutting edge of the cutting tool, C_e, can impose strong constraints on the actual shape of the clearance surface of the cutting tool. Under such restrictions, the clearance surface C_s usually cannot allow for sliding over itself. However, the desired surface C_s can be approximated by a surface that allows for sliding over itself; thus, the approximation can be more convenient for the design and manufacture of the form-cutting tool. This means that in certain cases of implementation of the first method, approximation of the desired clearance surface C_s by a surface that features another geometry can be inevitable. The approximation of the desired clearance surface C_s ultimately results in the part surface P being generated with an approximated surface T_a of the cutting tool and not with the precise surface T. The approximated generating surface T_a deviates from the desired surface T. The deviation δ_T is measured along the unit normal vector $\mathbf{n}_T$ to the desired generating surface T of the cutting tool at a corresponding surface point. The application of the form-cutting tool that has approximated the generated surface, T_a, is allowed if and only if the resultant deviation δ_T is within the corresponding tolerance $[\delta_T]$ for the accuracy, that is, when the inequality

$$\delta_T \le [\delta_T] \tag{6.1}$$

is valid.

Summarizing, one can come up with the following generalized procedure for designing the form-cutting tool in compliance with the first method:

1. Determination of the generating surface T of the form-cutting tool (see Chapter 5).
2. Determination of the rake surface R_s: The rake surface is selected among surfaces that are convenient from the manufacturing perspective (the surface R_s is a type of reasonably practical surface). The configuration of the rake surface is specified by the rake angle of the desired value.
3. Determination of the cutting edge, C_e: The cutting edge is represented with the line of intersection of the generating surface T of the form-cutting tool by the rake surface R_s.
4. Construction of the clearance surface C_s that passes through the cutting edge, C_e, and makes the clearance angle, α, of the desired value with the surface of the cut. The clearance surface is selected among surfaces that are convenient from the manufacturing perspective (a type of reasonably practical surface).

For practicality, a normal cross section of the clearance surface required has been determined as well.

Figure 6.1 illustrates an example of implementation of the first method for the transformation of the generating body of the form-cutting tool into the workable edge-cutting tool. For illustration purposes, the round form cutter for external turning of the part is chosen.

In the case under consideration, the part surface P is represented with two separate portions P_1 and P_2. The axial profile of the part is specified by the composite line through the points a_p, b_p, and c_p. For the particular case shown in Figure 6.1, the generating surface T of the form tool is congruent with the part surface P being machined. This statement easily follows from the consideration that is based on the analysis of kinematics of the machining operation (see Chapter 2). Thus, the identity $T \equiv P$ is observed in the case of turning of a part by the round form tool (to be more exact, two identities $T_1 \equiv P_1$ and $T_2 \equiv P_2$ are valid). Ultimately, the generating surface T of the form tool is also represented with two portions T_1 and T_2.

Because the identity $T \equiv P$ is observed, the axial profile of the generating surface T of the form tool is composed of two segments through the points

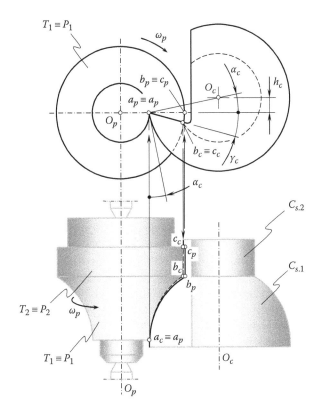

FIGURE 6.1
The concept of the *first method* for the transformation of the generating body of the form-cutting tool into a workable edge-cutting tool.

a_T, b_T, and c_T (not labeled in Figure 6.1), and the identities $a_T \equiv a_p$, $b_T \equiv b_p$, and $c_T \equiv c_p$ are observed.

Then, a plane is chosen as the rake surface R_s of the round form tool. Definitely, the plane allows for sliding over itself. It is convenient to machine the plane when manufacturing the cutting tool. The plane is parallel to the axis of rotation $O_T \equiv O_p$ of the generating surface T of the form tool. The rake surface R_s makes the rake angle γ_c with the perpendicular to the surface T at the base point $a_T \equiv a_c$.

The piecewise line of intersection $a_c b_c c_c$ of the generating surface T of the round form tool by the rake surface R_s serves as the cutting edge C_e of the form tool.

The clearance surface C_s of the form tool is shaped in the form of a surface of revolution. All surfaces of revolution allow for sliding over themselves. The surface of revolution C_s is represented with two separate portions $C_{s.1}$ and $C_{s.2}$. The clearance surface of the round form tool can be generated as a series of consecutive positions of the cutting edge $a_c b_c c_c$ when rotating the cutting edge $a_c b_c c_c$ about the surface C_s axis of rotation O_c. For practical needs, the axial profile of the clearance surface C_s must be determined.

The considered example (Figure 6.1) illustrates the implementation of the *first method* for the transformation of the generating body of the cutting tool into the workable edge-cutting tool. This method has been known for many decades.

The *first method* for the transformation of the generating body of the cutting tool into the workable edge-cutting tool is widely used in many industries. Form edge-cutting tools of most designs can be designed in compliance the first method. Using the first method makes the manufacturing and application of the form-cutting tools convenient.

The major disadvantages of the first method are twofold.

First, in most applications of the first method, no optimal values of the geometrical parameters of the form-cutting tool at every point of the cutting edge can be ensured. The optimization of the geometrical parameters at every point of the cutting edge is a challenging problem. Unfortunately, the solution to the problem of optimization of the geometrical parameters of the cutting edge (if any) is often far from practical needs: It can be feasible, but it is often not practical.

Second, inevitable deviations of the actual approximated generated surface of the form-cutting tool, T_a, from its desired shape, T, often cannot be eliminated when the first method for the transformation of the generating body of the form-cutting tool into the workable edge-cutting tool is employed when designing a form-cutting tool of a particular design.

Worn form-cutting tools designed in compliance with the first method are reground over the rake surface, R_s. This is the main reason why surface R_s is designed to slide over itself.

6.1.2 Second Method for the Transformation of the Generating Body of the Form-Cutting Tool into the Workable Edge-Cutting Tool

Consider another scenario under which the generating surface of the form-cutting tool is also determined (see Chapter 5). The cutting tool clearance surface is chosen among the surfaces that are convenient for machining of the surface when manufacturing the form-cutting tool, for inspection purposes, and so forth. In most cases, the clearance surface C_s is a surface that allows for sliding over itself (see Section 2.4).

The clearance surface C_s is properly oriented in relation to the generating surface T of the form-cutting tool. The surface C_s makes the optimal clearance angle with the cutting tool surface T at a given point.

Following the second method for the transformation of the generating body of the form-cutting tool into the workable edge-cutting tool, the cutting edges of the form-cutting tool are determined as the line of intersection of the generating surface T of the cutting tool by the clearance surface C_s.

Once the cutting edge, C_e, is constructed, the rake surface R_s can be constructed in compliance with the following routing.

First, the rake surface R_s is selected among the surfaces that allow for sliding over themselves (see Section 2.4). This requirement is highly desirable but not mandatory. Actually, any surface having reasonable geometry can serve as the rake surface of the form-cutting tool.

Second, the parameters of the chosen surface R_s must be calculated so to meet the requirement under which the surface R_s passes through the calculated cutting edge, C_e, of the form-cutting tool.

Third, the configuration of the determined rake surface R_s of the form-cutting tool must be the type wherein surface R_s makes the optimal rake angle γ with respect to the perpendicular generating surface T of the form-cutting tool at a specified point of the cutting edge.

It is easy to realize that rake surfaces of a form-cutting tool cannot always be shaped in the form that is convenient for manufacturing the cutting tool. The cutting edge of a precision form-cutting tool can be considered as a line of intersection of three surfaces:

- The generating surface T of the cutting tool
- The rake surface R_s
- The clearance surface C_s

The requirement that three surfaces T, R_s, and C_s be the surfaces through the common line, for example, through the cutting edge of the cutting tool, C_e, can impose strong constraints on the actual shape of the clearance surface of the cutting tool. Under such restrictions, the rake surface R_s usually cannot allow for sliding over itself. However, the desired surface R_s can be approximated by a surface that allows for sliding over itself; thus, the approximation can be more convenient for design and manufacture of the form-cutting tool. This means that in certain cases of implementation of the *second method*, an approximation of the desired rake surface R_s with a surface featuring another geometry, can be inevitable. The approximation of the desired rake surface R_s ultimately means that the part surface P is generated with an approximated surface T_g and not with the precise surface T of the cutting tool. The approximated cutting tool surface T_g deviates from the desired surface T. Deviation δ_T is measured along the unit normal vector $\mathbf{n}_T$ to the cutting tool surface T at a corresponding surface point. The application of the form-cutting tool having approximated the generating surface of the cutting tool is permissible if and only if the resultant deviation δ_T is within the corresponding tolerance $[\delta_T]$ for the accuracy of the part surface P—that is, when the following inequality is observed:

$$\delta_T \leq [\delta_T] \tag{6.2}$$

Summarizing, one can come up with the following generalized procedure for designing of the form-cutting tool in compliance with the second method:

1. Determination of the generating surface T of the form-cutting tool (see Chapter 5).
2. Determination of the clearance surface C_s: The clearance surface is selected among surfaces that are convenient from the manufacturing perspective (the clearance surface C_s is a type of a reasonably practical surface). The configuration of the clearance surface, C_s, is specified by the clearance angle, α, of the desired value.
3. Determination of the cutting edge, C_e: The cutting edge is represented with the line of intersection of the generating surface T of the form-cutting tool by the clearance surface C_s.
4. Construction of the rake surface R_s that passes through the cutting edge, C_e, and makes the rake angle, γ, of the desired value with the perpendicular to the surface of the cut. The rake surface is selected among surfaces that are convenient from the manufacturing perspective (the rake surface R_s is a type of reasonably practical surface).

For practicality, the normal cross section of the clearance surface, C_s, must be determined as well.

Figure 6.2 illustrates an example of implementation of the *second method* for the transformation of the generating body of the form-cutting tool into

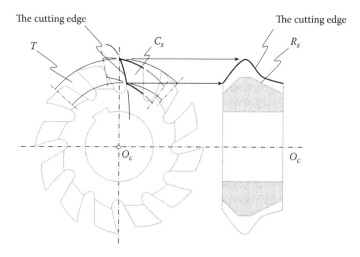

FIGURE 6.2
The concept of the *second method* for the transformation of the generating body of the form-cutting tool into a workable edge-cutting tool.

the workable edge-cutting tool. For illustration purposes, the form milling cutter for machining helical groves is chosen.

Consider that the generating surface T of the cutting tool is already determined. The geometry of the chosen clearance surface C_s is predetermined by kinematics of the operation of relieving of the milling cutter teeth. The cutting edge C_e is represented as the line of intersection of the generating surface T of the milling cutter by the clearance surface C_s.

Further, the constructed cutting edge C_e is used for the generation of the rake face R_s. For this purpose, either the cutting edge, or its projection onto the transverse plane moves along the milling cutter axis of rotation O_c. In the case under consideration, the rake surface R_s is represented as the locus of consecutive positions of the cutting edge C_e in its motion along the axis O_c. The rake surface R_s is shaped in the form a surface of translation (of a general cylinder).

The considered example (Figure 6.2) illustrates implementation of the *second method* for the transformation of the generating body of the form-cutting tool into the workable edge-cutting tool. This method is not as widely used in industry as the first method.

Form edge-cutting tools of most designs can be designed in compliance with the second method. When the second method is employed, the manufacturing and application of form-cutting tools become convenient.

The major disadvantages of the second method are twofold:

First, in most cases of implementation of the second method, no optimal values of all of the geometrical parameters of the form-cutting tool at every point of the cutting edge can be ensured. Optimization of the geometrical parameters at every point of the cutting edge is a challenging problem. The solution (if any) to the problem of optimization of the geometrical parameters of the cutting edge is often far from practical: A solution to the problem can be feasible, but it is often not practical.

Second, inevitable deviations of the actual approximated generated surface T_a of the cutting tool from its desired shape often cannot be eliminated when the second method is employed to design the form-cutting tool.

It is important to stress here that both considered methods for the transformation of the generating body of the form-cutting tool into the workable edge-cutting tool feature a common disadvantage. This disadvantage results in incapability of designing a form-cutting tool that has an optimal value of the angle of inclination λ. The actual value of the angle λ at a current point within the cutting edge is a function of the shape, parameters, and location of the rake R_s surfaces (or the clearance C_s surface) of the form-cutting tool in relation to the generating surface T of the form-cutting tool. Because of this, the optimal values λ_{opt} of angle of inclination of the cutting tool edge become impractical. This is mostly due to significant difficulties in manufacturing of the form-cutting tool.

Worn form-cutting tools designed in compliance with the second method are reground over the clearance surface, C_s. This is the main reason to design the surface C_s to be capable of sliding over itself.

6.1.3 Third Method for the Transformation of the Generating Body of the Form-Cutting Tool into the Workable Edge-Cutting Tool

Ultimately, consider the third scenario under which the generating surface of the form-cutting tool is also determined (see Chapter 5). However, in this case, neither the rake surface R_s nor the clearance surface C_s of a desired geometry is selected at the beginning, instead, the cutting edge of favorable geometry is selected. Such an approach allows for optimization of the angle of inclination λ_{opt} at every point of the cutting edge C_e of the form-cutting tool. The method considered below was proposed by Radzevich [116,117,136].

For the implementation of the third method, it is necessary to construct a special family of lines within the generating surface of the form-cutting tool. Below, this family of lines within the generating surface T (within the surface of cut, S_c) of the form-cutting tool is referred to as the *primary family of lines*. Lines of this family of lines represent the assumed trajectories of points of the cutting edge, C_e, in their motion over the surface of the cut when the work is machined (Figure 6.3). Actually, a family of the lines on the surface of the cut S_c is substituted with a corresponding family of lines on the generating surface T of the form-cutting tool. The analysis of a particular machining operation allows for an analytical representation of the family of lines within the cutting tool surface T. For example, in a case of form milling cutters, the

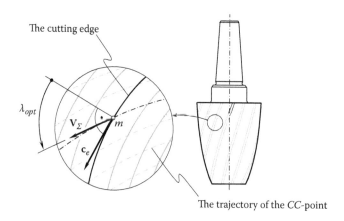

FIGURE 6.3
The concept of the *third method* for the transformation of the generating body of the form-cutting tool into a workable edge-cutting tool. The trajectories of the *CC*-point on the generating surface T of the form-cutting tool represent the first family of lines, and the cutting edges represent the second family of line on the cutting tool surface T.

first family of lines is shaped (approximately, when only rotation of the milling cutter is taken into account) in the form of circles traced by the points of the cutting edge when the milling cutter rotates.

Further, after the primary family of lines is determined, it is necessary to construct a secondary family of lines. The secondary family of lines is also within the generating surface T (within the surface of cut, S_c), and it is isogonal to the primary family of lines. At every point of intersection of the lines of the primary and secondary families, the angle between the lines is set to be equal to $(90° - \lambda_{opt})$. Here, λ_{opt} designates the optimal value of the angle of inclination at a point of the cutting edge, C_e. Therefore, the angle of inclination λ_{opt} is optimal at every point of the cutting edge, C_e. This is due to the primary family of lines within the generating surface T of the form-cutting tool being isogonal to the secondary family of lines at every point of the cutting edge.

Different segments of cutting edge of a form-cutting tool are at different distances from the axis of rotation of the cutting tool. Because of this, they work with different cutting speeds. This means that the optimal value of the angle of inclination can be different for different portions of the cutting edge C_e of the form-cutting tool. Under such a scenario, the actual value of the angle of inclination can be either of constant value within the cutting edge (and thus equal to its average value) or a desired variation of the angle of inclination within the cutting edge C_e can be ensured. In the last case, the problem of design of a form-cutting tool becomes more complex.

An appropriate number of lines from the second family of lines can be selected to serve as the cutting edges C_e of the form-cutting tool to be designed. These lines are uniformly distributed within the generating surface T of the cutting tool and are at a certain distance t from one another. The distance t is equal to the tooth pitch of the form-cutting tool.

The rake surface R_s is a surface through the cutting edge, C_e, of the form-cutting tool. The surface R_s makes the rake angle γ with the perpendicular n_c to the surface of cut. Actually, the perpendicular to the surface of cut deviates from the perpendicular n_T to the generating surface T of the form-cutting tool. Fortunately, this deviation is of negligibly small value, and, thus, it can be neglected. For practical needs of design of the form-cutting tool, the perpendicular n_T to the generating surface T is used instead of the perpendicular to the surface of cut n_c.

The clearance surface C_s is also a surface through the cutting edge C_e of the form-cutting tool. The surface C_s makes the clearance angle α with the generating surface T.

It is possible to formulate the problem of design of the form-cutting tool in such a way that the rake angle γ and clearance angle α can be of optimal value at every point m of the cutting edge C_e of the form-cutting tool. To meet this requirement, both the rake surface R_s and the clearance surface C_s must be of special geometry. The problem of determination of geometry of the rake surfaces R_s and the clearance surface C_s can be solved analytically. Ultimately, appropriate analytical expressions for the cutting tool surfaces R_s and C_s can be derived.

When deriving equations of the rake surface R_s and for the clearance surface C_s, it is necessary to ensure (for the parameters γ, α, and λ) optimal values of the rake angle γ_{opt}, of the clearance angle α_{opt}, and of the angle of inclination λ_{opt} for the new form-cutting tool as well as for the cutting tool after being reground. The optimal values γ_{opt}, α_{opt}, and λ_{opt} for the new form-cutting tool and the reground cutting tool are not necessarily the same.

Summarizing, one can come up with the following generalized procedure for designing of the form-cutting tool in compliance with the third method:

1. Determination of the generating surface T of the form-cutting tool (see Chapter 5).

2. Determination of the cutting edge, C_e. The cutting edge, C_e, is a spatial curve within the generating surface T of the form-cutting tool, which forms the angle of inclination of an optimal value with respect to the direction of speed of the resultant motion of a point m of the cutting edge C_e relative to the surface of cut.

3. Construction of the rake surface R_s, and the clearance surface C_s simultaneously: The rake surface passes through the cutting edge C_e and makes the rake angle of the desired value in relation to the perpendicular to the surface of cut S_c. The clearance surface also passes through the cutting edge, and makes the clearance angle of the desired value with the generating surface of the form-cutting tool. Both the rake surface R_s and the clearance surface C_s are selected among the surfaces based on their convenience for manufacturing purposes and their capability of sliding over themselves (surfaces R_s and C_s are types of reasonably practical surface; see Section 2.4 for more details in this concern).

The *third method* for the transformation of the generating body of the form-cutting tool into the workable edge-cutting tool is a completely novel method [116,117,136]. This method has not been comprehensively investigated yet. Therefore, a more detailed explanation of the method is important.

Consider the generating surface T of a form-cutting tool that is shaped in the form of a surface of revolution. This assumption is practical because, for example, the milling cutters of all designs feature the generating surface T in the form of a surface of revolution.

Using the third method, it is easy to come up with an understanding that the cutting edge of milling cutters of all designs must be shaped in the form of *loxodroma*. By definition, a *loxodrome* is a line that makes equal angles with a given family of lines on a surface. Actually, aloxodrome can be easily defined with respect to coordinate lines on the surface [30].

In the case under consideration, a loxodrome having special parameters of its shape is of particular interest. The loxodrome that makes the angle $(90° - \lambda_{opt})$ with the primary family of lines on the generating surface T of the form-cutting tool can be employed as the cutting edge C_e of the form-cutting tool.

In a particular case, when parameterization of the generating surface T of the form-cutting tool yields the expression

$$\Phi_{1.T} \quad \Rightarrow \quad dS_T^2 = dU_T^2 + G_T(U_T)dV_T^2 \tag{6.3}$$

for the first fundamental form $\Phi_{1.T}$, then the cutting edge, C_e, having an optimal value of the angle of inclination λ_{opt} at every point m, which can be described by the following equation:

$$V_T \cot \lambda_{opt} = \pm \int_{U_{0.T}}^{U_T} \frac{dU_T}{\sqrt{G_T(U_T)}} \tag{6.4}$$

Equation 6.4 of the cutting edge, C_e, is expressed in terms of U_T and V_T parameters of the generating surface T of the form-cutting tool. Using conventional mathematical methods, Equation 6.4 can be converted to *Cartesian* coordinates.

Example 6.1

Consider a ball-nose milling cutter of radius r_T, which is schematically depicted in Figure 6.4. The milling cutter is used for machining a

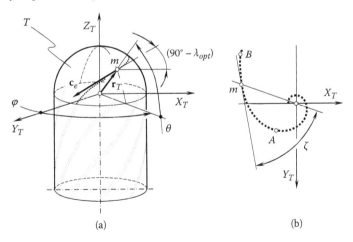

(a)   (b)

FIGURE 6.4
A ball-nose milling cutter having the optimal value of the angle of inclination λ_{opt} at every point m of the cutting edge C_e.

sculptured part surface P on a multiaxis numerical control (NC) machine. It is necessary to derive an equation of the cutting edge, C_e, with the angle of inclination λ equal to its optimal value λ_{opt} at every point m of the cutting edge C_e.

In a *Cartesian* coordinate system $X_T Y_T Z_T$ associated with the milling cutter (Figure 6.4), position vector of a point $\mathbf{r}_T$ of the generating surface T of the milling cutter can be represented in matrix form

$$\mathbf{r}_T(\varphi, \theta) = \begin{bmatrix} r_T \sin \varphi \cos \theta \\ r_T \sin \varphi \sin \theta \\ r_T \cos \varphi \\ 1 \end{bmatrix} \tag{6.5}$$

Under such a parameterization, the cutting edge, C_e, can be interpreted as a line on the generating surface T of the milling cutter, which intersects the meridians at the angle $(90° - \lambda_{opt})$.

A parametric equation of the form $\theta = \theta(\varphi)$ describes a line on the surface. If $\theta = 0°$, the lines $\theta = \theta(\varphi)$ represent meridians on the cutting tool surface T (see Equation 6.5). From another viewpoint, if the equality $\theta = -90°$ is observed, the lines $\theta = \theta(\varphi)$ represent parallels.

The unit tangent vector $\mathbf{c}_e$ to the curve $\theta = \theta(\varphi)$ on the cutting tool surface T makes the angles α_x, β_y, and γ_z with axes of the coordinate system $X_T Y_T Z_T$. The equation for the unit tangent vector $\mathbf{c}_e(\varphi, \theta)$ at a point of the cutting tool surface T can be derived from Equation 6.5:

$$\mathbf{c}_e(\varphi, \theta) = \frac{d\mathbf{r}_T}{dS_T} = \frac{1}{\sqrt{d\varphi^2 + \sin^2 \varphi \, d\theta^2}} \cdot \begin{bmatrix} \cos \varphi \cos \theta \, d\varphi - \sin \varphi \sin \theta \, d\theta \\ \cos \varphi \sin \theta \, d\varphi + \sin \varphi \cos \theta \, d\theta \\ -\sin \varphi \, d\varphi \\ 1 \end{bmatrix} \tag{6.6}$$

where dS_T denotes the differential of the arc segment of the cutting edge C_e. Particularly, when $\theta = \theta_c$ − Const, then Equation 6.6 for the unit tangent vector $\mathbf{c}_e$ reduces to

$$\mathbf{c}_e(\varphi, \theta_c) = \begin{bmatrix} \cos \varphi \cos \theta_c \\ \cos \varphi \sin \theta_c \\ -\sin \varphi \\ 1 \end{bmatrix} \tag{6.7}$$

Under the imposed constraint $\theta = \theta_c$, the following equality is valid:

$$\frac{d\varphi}{\sqrt{d\varphi^2 + \sin^2\varphi\, d\theta^2}} = \cos\wp \tag{6.8}$$

Equation 6.8 immediately yields

$$\frac{d\varphi}{\sin\varphi} = \pm\cot\wp\, d\theta \tag{6.9}$$

where $\wp$ designates a certain angle.

After Equation 6.9 is integrated, one can come up with the solution:

$$\tan\frac{\varphi}{2} = e^{q(\theta+C)} \tag{6.10}$$

where $q = \pm cot\, \wp$ and C are arbitrary constant values.

The implementation of the trivial trigonometric formulae yielded an intermediate result:

$$\sin\varphi = \frac{2\tan\dfrac{\varphi}{2}}{1+\tan^2\dfrac{\varphi}{2}}; \tag{6.11}$$

$$\cos\varphi = \frac{1-\tan^2\dfrac{\varphi}{2}}{1+\tan^2\dfrac{\varphi}{2}} \tag{6.12}$$

$$\sin\varphi = \frac{1}{ch\, q\,(\theta+C)}; \tag{6.13}$$

$$\cos\varphi = th\, q(\theta+C) \tag{6.14}$$

Ultimately, under the assumption $\theta_c = \lambda_{opt}$, the above analysis makes possible an equation for the position vector $\mathbf{r}_{ce}$ of a point of the cutting edge, C_e, of the ball-nose milling cutter (Figure 6.4a)

$$\mathbf{r}_{ce} = \begin{bmatrix} \dfrac{r_T \cos \lambda_{opt}}{\operatorname{ch} q(\lambda_{opt} + C)} \\[2mm] \dfrac{r_T \sin \lambda_{opt}}{\operatorname{ch} q(\lambda_{opt} + C)} \\[2mm] r_T \operatorname{th} q(\lambda_{opt} + C) \end{bmatrix}$$ (6.15)

The angle of inclination for the milling cutter having a cutting edge that is shaped in compliance with Equation 6.15 is constant. At every point m of the cutting edge, C_e, it is equal to its optimal value λ_{opt}.

The entire loxodroma (see Equation 6.15) is not used for design of the cutting edge of the ball-nose milling cutter. Only the arc segment AB is used for this purpose (Figure 6.4b).

Other methods for derivation of Equation 6.15 can be implemented as well [116,117,136].

Consider another approach for the derivation of an analytical description of the cutting edge, C_e, of the ball-nose milling cutter. In a particular case being considered, when parameterization of the equation of the generating surface T of the milling cutter yields the following expression for the first fundamental form $\Phi_{1.T}$,

$$\Phi_{1.T} \quad \Rightarrow \quad dS_T^2 = r_T^2 \left(dU_T^2 + \cos^2 U_T dV_T^2 \right)$$ (6.16)

the cutting edge having optimal value of the angle of inclination λ_{opt} at every point m can be described by the equation

$$V_T \cot \lambda_{opt} = R_T \ln \tan \left(\frac{\pi}{4} + \frac{U_T}{2r_T} \right)$$ (6.17)

It is important to focus on the shape of the loxodrome. A loxodrome makes an infinite number of revolutions about its pole.* It approaches the pole infinitely close: This curve approaches the pole similar to an asymptotic point. The last can cause some inconveniences when manufacturing form-cutting tools. However, several methods have been developed for avoiding such inconvenience [117].

Feasibility of the optimization of the angle of inclination λ is not limited to ball-nose milling cutters. Form-cutting tools having the generating surface T of any feasible shape can be designed with the optimal value of the angle of inclination, λ_{opt} at every point m of the cutting edge C_e. The last statement

---

* The loxodrome's pole is located at the point of intersection of the generating surface T of the milling cutter by the axis of rotation of the cutting tool.

encompasses composite generating surfaces T of the form-cutting tools as well.

As an example, consider the optimization of the angle of inclination of a filleted-end milling cutter shown in Figure 6.5. The generating surface of the filleted-end milling cutter is composed of three portions: the cylindrical portion T_1, the flat end T_2, and the torus surface T_3.

For the cylindrical portion T_1 of the generating surface of the filleted-end milling cutter, the cutting edge AB having the optimal angle of inclination $\lambda_{opt} = Const$ reduces to a helix 1 of constant pitch. For the flat end portion T_2 of the generating surface, the cutting edge is represented in the form of a logarithmic spiral curve 2. Ultimately, the equation of the cutting edge segment BC within the portion T_3 of the generating surface of the milling cutter can be derived on the premise of Equation 6.2. This segment of the cutting edge, C_e, is represented by the arc segment 3 of the loxodrome. The loxodrome is entirely located within the torus surface T_3.

The generalized Equation 6.2 of the cutting edge, C_e, having optimal value of the angle of inclination at every point m of the cutting edge C_e is valid for edge-cutting tools of any possible design. However, in particular cases of the filleted-end milling cutter, significant simplifications are possible.

For example, for the flat-end portion T_2 of the filleted-end milling cutter (Figure 6.5), an equation of the cutting edge, C_e, can be derived following one of two possible ways.

In compliance with the first, the flat-end surface is considered as a surface of revolution that is degenerated into the plane. Further, the equation of the cutting edge, C_e, can be derived as for the surface of revolution.

Following the second way, using the differential equation for isogonal trajectories is preferred. If a planar curve intersects all the curves of the initially given single-parametric family of planar curves $\varphi(x, y, \theta) = 0$ at a

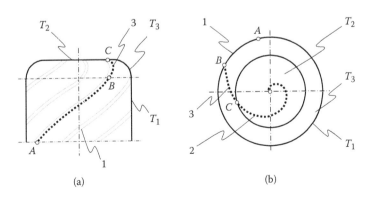

(a) (b)

FIGURE 6.5
The filleted-end milling cutter having an optimized value of the angle of inclination of the cutting edges λ_{opt}.

given constant angle of intersection ς, then the line satisfies the differential equation:

$$\left(\frac{\partial\varphi}{\partial x}\cos\varsigma - \frac{\partial\varphi}{\partial y}\sin\varsigma\right)dx + \left(\frac{\partial\varphi}{\partial x}\sin\varsigma + \frac{\partial\varphi}{\partial y}\cos\varsigma\right)dy = 0 \qquad (6.18)$$

For the flat-end portion T_2 of the generating surface of the filleted-end milling cutter, the initially given single-parametric family of planar curves $\varphi(x, y, \theta) = 0$ is represented by the family of straight lines through the axis of rotation of the milling cutter. Here θ designates the angular parameter of the family of straight lines $\varphi(x, y, \theta) = 0$.

The equation of the family of straight lines can be represented in the form

$$\varphi(x, y, \theta) = y - x\tan\theta = 0 \qquad (6.19)$$

Further, assume that $\varsigma = 90° - \lambda_{opt}$. After substituting Equation 6.19 into Equation 6.18, the latter casts into an equation of the cutting edge, C_e, of the filleted-end milling cutter. This equation describes a logarithmic spiral curve. This means that in the particular case under consideration, the logarithmic spiral curve can be interpreted as the loxodrome for the family of straight lines within the plane.

It is convenient to represent the equation of the cutting edge, C_e, in polar coordinates:

$$\rho = \rho_0\, e^{\varphi\tan\lambda_{opt}}; \quad (\rho_0 > 0, -\infty < \varphi < +\infty) \qquad (6.20)$$

where ρ is the position vector of a point m of the cutting edge, C_e, and ρ_0 is the position vector of a given point m_0 of the cutting edge from which the angle φ is measured.

The cutting edge C_e (see Equation 6.20) intersects all straight lines through point O at that same angle $\varsigma = 90° - \lambda_{opt}$.

The pole of the logarithmic spiral curve coincides with the axis of rotation of the milling cutter. It represents an asymptotic point of this planar curve. Because of this, the cutting edge, C_e, of the flat-end portion T_2 of the generating surface cannot pass through the axis of the cutting tool rotation. It is possible to design the cutting edge in the shape of the logarithmic spiral curve (see Equation 6.20) only within a certain portion, similar to the arc AB (Figure 6.4b). The cutting edge cannot be shaped in the form of the logarithmic spiral curve between a certain point A and the axis of rotation of the filleted-end milling cutter.

In the case under consideration, representation of the equation of the cutting edge, C_e, in the following form has proven to be useful:

$$\rho = \rho_A e^{\varphi \tan \lambda_{opt}} ; \quad (\rho_A > 0, \quad \varphi_A = 0° < \varphi < \varphi_B) \tag{6.21}$$

where ρ_A is the position vector of the point A of the cutting edge AB.

A few other design parameters of the cutting edge C_e of the form-cutting tool, which can be drawn up from the geometrical analysis, are as follows.

Length, S_{AB}, of the cutting edge AB can be calculated from the equation

$$S_{AB} = \frac{(\rho_B - \rho_A)\sqrt{1 + \tan^2 \lambda_{opt}}}{\tan \lambda_{opt}} = \frac{(\rho_B - \rho_A)}{\sin \lambda_{opt}} \tag{6.22}$$

where ρ_B is the position vector of the point B of the cutting edge AB.

The radius of curvature R_T^* at a current point m of the cutting edge AB can be calculated from the equation

$$R_T^*(\rho) = \rho\sqrt{1 + \tan^2 \lambda_{opt}} = \rho_A e^{\varphi \tan \lambda_{opt}} \sqrt{1 + \tan^2 \lambda_{opt}} = \frac{\rho_A e^{\varphi \tan \lambda_{opt}}}{\cos \lambda_{opt}} \tag{6.23}$$

The length S_{AB}, and the radius of curvature R_T^* of the cutting edge, C_e, are often necessary for optimization of performance of the filleted-end milling cutter.

The third method for the transformation of the generating body of the form-cutting tool into the workable edge-cutting tool can be implemented to design cutting tools for machining both sculptured part surfaces on a multi-axis *NC* machine as well as for machining parts on conventional machine tools.

In addition to loxodroma possessing useful properties for a cutting tool designer, this curve can be evolved into two possible areas.

First, a curve similar to loxodroma can be constructed on the generating surface T of a form-cutting tool that is shaped not only in the form of a surface of revolution but also for the cutting tool surface T of another geometry, including surfaces T that allow for sliding over themselves.

Second, the loxodroma can be evolved to a more general area, when the optimal value of the angle of inclination λ_{opt} varies within the cutting edge, C_e. Under such a scenario, the desired current value of the angle λ_{opt} can be expressed in terms of curvilinear coordinates U_T and V_T, for example, by an equation $\lambda_{opt} = \lambda_{opt}(U_T, V_T)$. The desired function $\lambda_{opt} = \lambda_{opt}(U_T, V_T)$ of variation of the optimal value of the angle of inclination λ_{opt} can be determined experimentally.

A form-cutting tool of any type can be designed using any of three methods considered in this section of the book. The cutting tool design engineer makes his or her own decision of which method is preferred to use to design a particular form-cutting tool.

Worn form-cutting tools designed in compliance with the third method are reground either over the rake surface, R_s, or over the clearance surface, C_s, depending on which surface better suits this purpose.

6.1.4 Form-Cutting Tools for Machining Sculptured Part Surfaces on a Multiaxis NC Machine

Having the generating surface T of the form-cutting tool designed and having designed the rake surface, R_s, and the clearance surface, C_s, the cutting tool designer can proceed with design of a form-cutting tool for machining a sculptured part surface, P, on a multiaxis NC machine.

Examples of the form-cutting tools for machining a sculptured part surface on a multiaxis NC machine are illustrated in Figure 6.6. This cutting tools are

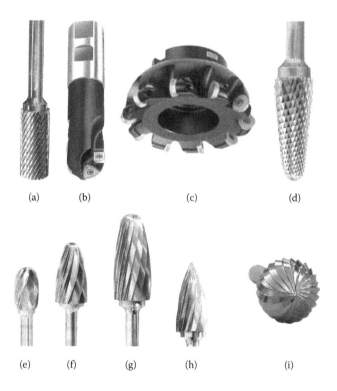

(a)   (b)   (c)   (d)

(e)   (f)   (g)   (h)   (i)

FIGURE 6.6
Milling cutters for machining sculptured part surfaces on a multiaxis NC machine: a cylindrical milling cutter (a), a ball-end milling cutter, having spherical, T_{sph}, and cylindrical, T_{cyl}, portions of the generating surface T (b), a face milling cutter featuring the generating surface T of torus geometry (c), a conical milling cutter, having conical, T_{con}, and spherical, T_{sph} portions of the generating surface T (d), form milling cutters of different design (e–h), ball-nose milling cutter having special configuration of the cutting edges, C_e (i).

designed in compliance with the above discussed approaches. Cutting edges of the milling cutters, C_e, are within the generating surface T of the cutting tool. The rake surfaces, R_s, and the clearance surfaces, C_s, are the surfaces through the cutting edge, C_e.

6.2 Geometry of the Active Part of Cutting Tools in the *Tool-in-Hand* System

The active part of the cutting tool is composed of two surfaces intersecting each other to form the cutting edge. The surface over which the chip flows is known as the *rake surface*, R_s, or more simply as *the face*. That surface, which is faced to the machined part surface, is known as the *clearance surface*, C_s, or *the flank*. In the simplest yet common case, both the surfaces R_s and C_s are planes. The cutting edge, C_e, is represented as the line of intersection of the rake surface R_s and of the clearance surface C_s.

The cutting edges of two types can be distinguished in design of a cutting tool, namely, roughing cutting edges and finishing (cleanup) cutting edges. Roughing cutting edges do not generate the part surface P being machined, but finishing cutting edges do. Finishing cutting edges are always located within the generating surface T of the form-cutting tool. Roughing cutting edges are beneath the surface T and within the generating body of the cutting tool.

The generating surface T of a form-cutting tool can make a point contact with the part surface P. Under such a scenario, roughing portions of the cutting edges may be within the surface T as well.

The major and minor cutting edges of the cutting tool are distinguished. A whole cutting edge or its portion that is faced toward the direction of the feed rate is referred to as the *major cutting edge* of the cutting tool. Another cutting edge or the rest of the whole cutting edge is referred to as the *minor cutting edge* of the cutting tool. The major cutting edge of a cutting tool contacts the chip being cut off. The minor cutting edge of a cutting tool contacts with the uncut portion of the stock. The regular (and not stochastic) residual roughness on the part surface P is caused by both the major and the minor cutting edges of the form-cutting tool.

For a form-cutting tool having curved cutting edges (for example, for a milling cutter), an elementary cutting edge of infinitesimal length $d\,l$ is considered below. Depending upon the actual problem under consideration, the infinitesimal cutting edge $d\,l$ is considered either as a straight-line segment or as a circular-arc segment of the corresponding radius of curvature.

6.2.1 *Tool-in-Hand* Reference System

A reference system associated with the cutting tool is referred to as the *tool-in-hand reference system*. This reference system is often used when designing, manufacturing, regrinding, and inspecting the cutting tool. To accomplish design of a high-performance cutting tool, the geometry of the active part of the cutting tool in various cross sections of the cutting wedge must be known.

The tool-in-hand reference system is made up of planes that are tangents to the generating surface T of the form-cutting tool, to the rake surface R_s, and to the clearance surface C_s. In particular cases, the surfaces T, R_s, and C_s (all or some of them) degenerate to corresponding planes.

The actual values of geometric parameters of the active part of a cutting tool are determined in a coordinate system associated with the cutting tool itself. This coordinate system is referred to as the *static coordinate system*. Various configurations of the static coordinate system with respect to the cutting tool are feasible.

The right-hand–oriented static coordinate system $X_T Y_T Z_T$ is recommended for the application by the International Standard ISO 3002* (Figure 6.7). The axes of this reference system are as follows:

- Axis Z_T aligns with the assumed direction of primary motion $\mathbf{V}_p$
- Axis X_T aligns with the assumed direction of the cutting feed rate motion $\mathbf{V}_f$
- Axis Y_T complements axes X_T and Z_T to a right-hand–oriented *Cartesian* coordinate system $X_T Y_T Z_T$

No principal constraints are imposed on the actual configuration of the static coordinate system. The convenience of performing of the calculations is usually the only recommendation to follow when selecting the coordinate system $X_T Y_T Z_T$.

By definition, the main reference plane P_r is perpendicular to the velocity vector $\mathbf{V}_p$ of the assumed direction of primary motion in the tool-in-hand coordinate system $X_T Y_T Z_T$ (Figure 6.7). In this figure, the vector $\mathbf{V}_f$ designates the assumed direction of the feed rate motion.

For computer-aided design and computer-aided machining (*CAD/CAM*) applications, the analytical description of the reference planes is of critical importance. To describe analytically the reference planes, it is necessary to represent the surfaces T, R_s, and C_s in the tool-in-hand reference system.

---

* ISO 3002. Basic Quantities in Cutting and Grinding—Part 1: Geometry of the Active Part of the Form-Cutting Tool—General Terms, Reference Systems, Tool, and Working Angles, Chip Breakers, 1982, 52 pp. ISO 3002-1/AMD1. Amendment 1 to ISO 3002-1, 1982. 1992, 3 pp. ISO 3002. Basic Quantities in Cutting and Grinding—Part 2: Geometry of the Active Part of Cutting Tools—General Conversion Formulae to Relate Tool and Working Angles, 1982, 35 pp.

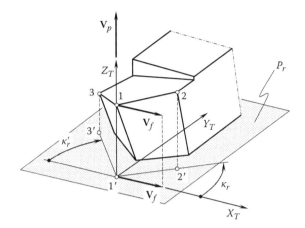

FIGURE 6.7
Definition of the main reference plane, P_r (according to ISO 3002).

In the coordinate systems associated with each of the surfaces T, R_s, and C_s, the surfaces make possible analytical representation in the following form:

$$\mathbf{r}_T = \mathbf{r}_T(U_T, V_T) \tag{6.24}$$

$$\mathbf{r}_{rs} = \mathbf{r}_{rs}(U_{rs}, V_{rs}) \tag{6.25}$$

$$\mathbf{r}_{ce} = \mathbf{r}_{ce}(U_{ce}, V_{ce}) \tag{6.26}$$

For the representation of these equations in the common coordinate system $X_T Y_T Z_T$, the corresponding operators of the resultant coordinate system transformations are used (see Chapter 3).

The equations of the surfaces T, R_s, C_s make possible calculation of unit tangent vectors $\mathbf{u}_T$ and $\mathbf{v}_T$, $\mathbf{u}_{rs}$ and $\mathbf{v}_{rs}$, $\mathbf{u}_{cs}$ and $\mathbf{v}_{cs}$. Ultimately, the following equations can be derived:

- For the tangent plane $\mathbf{r}_{Tt}$ to the generating surface T of the form-cutting tool:

$$[\mathbf{r}_{Tt} - \mathbf{r}^{(m)}] \times \mathbf{u}_T \cdot \mathbf{v}_T = 0 \tag{6.27}$$

- For the tangent plane $\mathbf{r}_{rt}$ to the rake surface R_s of the form-cutting tool:

$$[\mathbf{r}_{rt} - \mathbf{r}^{(m)}] \times \mathbf{u}_{rs} \cdot \mathbf{v}_{rs} = 0 \tag{6.28}$$

- For the tangent plane r_{ct} to the clearance surface C_s of the form-cutting tool:

$$[\mathbf{r}_{ct} - \mathbf{r}^{(m)}] \times \mathbf{u}_{cs} \cdot \mathbf{v}_{cs} = 0 \qquad (6.29)$$

where $\mathbf{r}^{(m)}$ designates the position vector of a point of interest m within the cutting edge, C_e.

The same unit tangent vectors make possible calculation of unit normal vectors $\mathbf{n}_T$, $\mathbf{n}_{rs}$, $\mathbf{n}_{cs}$ at a point of the surfaces T, R_s, C_s at m. They are equal:

$$\mathbf{n}_T = \mathbf{u}_T \times \mathbf{v}_T \qquad (6.30)$$

$$\mathbf{n}_{rs} = \mathbf{u}_{rs} \times \mathbf{v}_{rs} \qquad (6.31)$$

$$\mathbf{n}_{cs} = \mathbf{u}_{cs} \times \mathbf{v}_{cs} \qquad (6.32)$$

accordingly.

All the unit normal vectors $\mathbf{n}_T$, $\mathbf{n}_{rs}$, $\mathbf{n}_{cs}$ are pointed out from the bodily side to the void side of the form-cutting tool (see Chapter 1).

Equations for the tangent planes $\mathbf{r}_{Tt}$, $\mathbf{r}_{rt}$, $\mathbf{r}_{ct}$ together with equations for the unit normal vectors $\mathbf{n}_T$, $\mathbf{n}_{rs}$, $\mathbf{n}_{cs}$ make possible derivation of equations of major reference planes at a current point of interest, m, of the cutting edge, C_e, of a form-cutting tool.

6.2.2 Major Reference Planes: Geometry of the Active Part of a Cutting Tool Defined in a Series of Reference Planes

Figure 6.6 shows a system of the reference planes in the tool-in-hand system defined by the International Standard ISO 3002. The system is made up of five basic planes, each of which is defined relative to the main reference plane P_r:

- By definition, the main reference plane P_r is perpendicular to the velocity vector $\mathbf{V}_p$ of the assumed direction of primary motion (Figure 6.8). (In Figure 6.7, this reference plane is congruent to the $X_T Y_T$ coordinate plane.)
- Perpendicular to the main reference plane P_r and containing the velocity vector $\mathbf{V}_f$ of the assumed direction of feed rate motion is the assumed working plane P_f as shown in Figure 6.8.
- The tool-cutting edge plane P_s is perpendicular to the main reference plane P_r and contains the side (main) cutting edge 1–2. For finishing cutting edges, the plane P_s is a tangent to the generating surface T of the form-cutting tool.

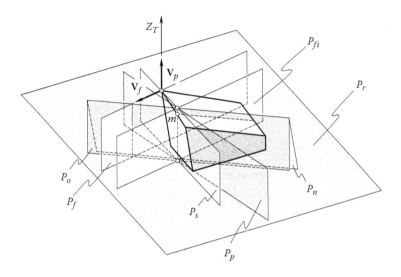

FIGURE 6.8
ISO 3002 system of reference planes in the *tool-in-hand* reference system for a major cutting edge.

- The tool back plane P_p is perpendicular to the main reference P_r and the assumed working P_f planes (Figure 6.8).
- Perpendicular to the projection of the cutting edge C_e onto the main reference plane P_r is the orthogonal plane P_o.
- The cutting edge normal plane P_n is perpendicular to the cutting edge, C_e. This reference plane can also be defined as the plane that is perpendicular to the rake R_s and to the clearance C_s surfaces at a point of interest, m (Figures 6.9 and 6.10).

As an example, the ISO 3002 system of the reference planes for a lathe cutter is depicted in Figure 6.10.

The orthogonal reference plane P_n passes through the unit normal vector $\mathbf{n}_{pc}$ to the cutting edge, C_e, located within the cutting edge reference plane P_s. Configuration of the reference plane P_n can be defined by a pair of unit vectors through a point of interest m within the cutting edge C_e: either by the vectors $\mathbf{n}_{rs}$ and $\mathbf{n}_{cs}$, or by the vectors $\mathbf{n}_T$ and $\mathbf{n}_{pc}$. Other combinations of the above unit vectors and a unit vector $\mathbf{c}_e$ along the cutting edge, C_e (or tangent to the cutting edge, C_e) can be employed for definition of the reference plane P_n as well.

From another perspective, the unit tangent vector $\mathbf{c}_e$ is orthogonal to the unit normal vectors $\mathbf{n}_{rs}$ and $\mathbf{n}_{cs}$ to the rake surface R_s and to the clearance surface C_s, respectively. This yields an equation for $\mathbf{c}_e$:

$$\mathbf{c}_e = \mathbf{n}_{ce} \times \mathbf{n}_{cs} \tag{6.33}$$

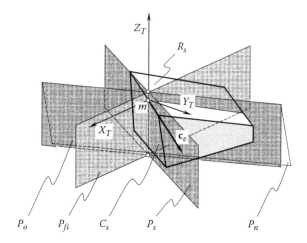

FIGURE 6.9
The derivation of equation of the *normal plane section*, P_n.

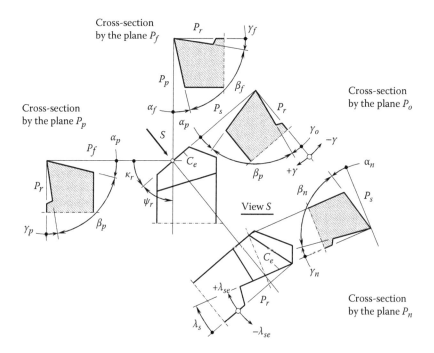

FIGURE 6.10
Example of the cutting tool geometry in the *tool-in-hand* system (ISO 3002).

The corresponding unit normal vectors $\mathbf{n}_{ce}$ and $\mathbf{n}_{cs}$ are equal to

$$\mathbf{n}_{ce} = \mathbf{u}_{ce} \times \mathbf{v}_{ce} \qquad (6.34)$$

$$\mathbf{n}_{cs} = \mathbf{u}_{cs} \times \mathbf{v}_{cs} \qquad (6.35)$$

In the *tool-in-hand* coordinate system $X_T Y_T Z_T$, the equation of the orthogonal reference plane P_n through the point of interest m yields representation in the following form:

$$[\mathbf{r}_n - \mathbf{r}^{(m)}] \times \mathbf{c}_e = 0 \qquad (6.36)$$

where $\mathbf{r}_n$ is the position vector of a point of the orthogonal reference plane P_n and $\mathbf{r}^{(m)}$ is the position vector of a point of interest, m, within the cutting edge, C_e.

Similarly, a corresponding system of reference planes can be attributed to the minor cutting edge. The system of reference planes for the minor cutting edge is composed of five planes P'_f, P'_s, P'_p, P'_o, and P'_n. These reference planes strictly correlate to the reference planes P_f, P_s, P_p, P_o, and P_n of the major cutting edge of a cutting tool.

The ISO 3002 system of reference planes is not consistent for solving a variety of problems that pertain to cutting tool geometry. This is the reason why other reference planes are used in addition to the reference planes P_r, P_f, P_s, P_p, P_o, and P_n for the calculation of the geometry of the active part of cutting tools.

6.2.3 Major Geometric Parameters of the Cutting Edge of a Cutting Tool

The geometry of the active part of a cutting tool is specified by angles and some other geometric parameters. The geometric parameters of the active part of a cutting tool are measured in corresponding reference planes.

6.2.3.1 Main Reference Plane

In the *main reference plane*, P_r, the following geometric parameters of the active part of a cutting tool are measured:

- *The tool cutting edge angle κ_r* (or φ, in other designations) is the acute angle that the reference plane P_s makes with the reference plane P_f. In another words, the angle κ_r is the acute angle between the projection of the major cutting edge, C_e, onto the reference plane P_r and the P_f reference plane as illustrated in Figure 6.7. The tool cutting edge angle, κ_r, is measured in counterclockwise direction from the P_f reference plane. The angle κ_r is always positive.

- *The tool minor (end) cutting edge angle* κ_{r1} (or φ_1 in another designation) is the acute angle that the reference plane P'_s makes with the reference plane P'_f. In another words, the angle κ_{r1} is the acute angle between the projection of the minor (end) cutting edge onto the reference plane P_r and the P'_f reference plane, as depicted in Figure 6.7. The tool minor cutting edge angle κ_{r1} is measured in a clockwise direction from the P'_f reference plane. The angle κ_{r1} is always positive, including zero (Figure 6.7).
- *The tool approach angle* ψ_r is the acute angle that the reference plane P_s makes with the reference plane P_p as shown in Figure 6.10. The angle ψ_r can be calculated from the expression

$$\psi_r = 90° - \kappa_r \tag{6.37}$$

In addition to the angles specified by ISO 3002 and are listed above, the following geometric parameters of the active part of cutting tool are measured in the main reference plane P_r:

- *The tool tip angle* ε is measured within the main reference plane, P_r. This is the angle that makes the projections of the major and the minor cutting edges into the main reference plane, P_r. The tool tip angle ε can be calculated from the equation

$$\varepsilon = 180° - (\kappa_r + \kappa_{r1}) = 90° + \psi_r - \kappa_{r1} \tag{6.38}$$

or

$$\varepsilon = 180° - (\varphi + \varphi_1) \tag{6.39}$$

- *The radius of normal curvature* R_T *of the generating surface of a form-cutting tool T.* The actual radius of curvature of the cutting edge, R_{ce}, can be expressed in terms of the radius R_T.

6.2.3.2 Assumed Reference Plane

In the *assumed reference plane* P_f, the following geometric parameters of the active part of a cutting tool are measured:

- *The tool rake angle* γ_f is the angle between the reference plane P_r (the trace of which appears as the normal to the velocity vector $\mathbf{V}_p$ of the direction of primary motion), and the intersection line formed by the assumed working plane P_f and the tool rake plane R_s. The rake angle is defined as being acute and positive when looking across the rake face from the selected point and along the line of intersection of the face and the assumed working plane P_f. The viewed line of intersection lies on the opposite side of the tool reference plane from the direction of primary motion. The sign of the rake angle is well defined (Figure 6.10).

- *The tool clearance (flank) angle α_f is defined* in a way similar to the tool rake angle γ_f, but here, if the viewed line of intersection lies on the opposite side of the cutting edge plane P_s from the direction of feed rate motion (assumed or actual as the case may be), the clearance angle is positive. In another words, the clearance angle is the angle between the tool cutting edge plane P_s and the tool flank plane C_s (Figure 6.10).
- *The tool wedge angle β_f is* the angle between the two intersection lines formed as the reference plane P_f intersects with the rake R_s and flank C_s planes.

The sum of algebraic values of the rake angle γ_f, the cutting wedge angle β_f, and clearance angle α_f is always equal to 90°:

$$\gamma_f + \beta_f + \alpha_f = 90° \qquad (6.40)$$

Equations similar to Equation 6.40 are valid for other reference planes that cross the cutting edge.

For the minor (side) cutting edge, geometric parameters are specified in a way similar to that considered.

6.2.3.3 Tool Cutting Edge Plane

The orientation and inclination of the cutting edge, C_e, are specified in the *tool cutting edge reference plane*, P_s. In this reference plane, the cutting edge inclination angle λ_{se} is measured. The angle of inclination is measured between the assumed direction of the velocity vector $\mathbf{V}_p$ of primary motion and the unit normal vector $\mathbf{n}_{ce}$ to the cutting edge. The angle λ_{se} is defined as always being acute. It is positive if the cutting edge is turned in counterclockwise direction with respect to the vector $\mathbf{V}_p$. This angle can be defined at any point of the cutting edge. The sign of the inclination angle is well defined in Figure 6.10.

6.2.3.4 Tool Back Plane

In the *tool back plane* P_p, rake angle γ_p, clearance angle α_p, and tool wedge angle β_p are measured. These angles are measured in a way similar to that with which the corresponding angles in the assumed working plane P_f are measured (Figure 6.10). Like in the reference plane P_f, the sum of algebraic values of the rake angle γ_p, the cutting wedge angle β_p, and the clearance angle α_p is always equal to 90°:

$$\gamma_p + \beta_p + \alpha_p = 90° \qquad (6.41)$$

6.2.3.5 Orthogonal Plane

The rake angle γ_o, clearance angle α_o, and wedge angle β_o are measured in the *orthogonal plane* P_o. These angles are measured between lines of intersection of the rake surface R_s, the clearance surface C_s, and the corresponding reference planes by the orthogonal plane P_o. The equality for algebraic values of the angles α_o, β_o, and γ_o to 90° is always observed in the reference plane P_o:

$$\gamma_o + \beta_o + \alpha_o = 90° \tag{6.42}$$

6.2.3.6 Cutting Edge Normal Plane

In the *cutting edge normal plane* P_n, the rake angle γ_n, the clearance angle α_n, and the cutting wedge angle β_n are measured. In addition, the cutting wedge roundness ρ_n is measured in the cutting edge normal plane P_n as well. The sum of algebraic values of the angles γ_n, α_n, and β_n is always equal to 90°:

$$\gamma_n + \beta_n + \alpha_n = 90° \tag{6.43}$$

The angles γ_n, α_n, and β_n are measured between the lines of intersection of the rake surface R_s, the clearance surface C_s, and the corresponding reference planes by the normal plane P_n.

6.2.4 Analytical Representation of the Geometric Parameters of the Cutting Edge of a Cutting Tool

It is the right point now to introduce equations for calculation of actual values of the angles γ_n, α_n, and β_n measured in the normal plane P_n.

The inclination of the rake surface R_s in relation to the tool cutting edge plane P_s is specified by the normal rake angle γ_n. The rake angle γ_n is the angle that forms the rake surface R_s and the unit normal vector $\mathbf{n}_{ps}$ to the reference plane P_s. Its value is accounted from the unit vector $\mathbf{n}_{ps}$ to the line of intersection of the rake surface R_s by the normal plane P_n. The angle γ_n is positive if the unit normal vector $\mathbf{n}_{ps}$ does not pass through the cutting wedge as schematically illustrated in Figure 6.10. Otherwise, the angle γ_n is negative.

The unit normal vector to the rake surface R_s is designated as $\mathbf{n}_{rs}$. It is convenient to define the normal rake angle γ_n as the angle that complements to 90° the angle between the unit normal vectors $\mathbf{n}_{ps}$ and $\mathbf{n}_{rs}$ as shown in Figure 6.10. The above equations $\mathbf{n}_T = \mathbf{u}_T \times \mathbf{v}_T$ (see Equation 6.30) and $\mathbf{n}_{rs} = \mathbf{u}_{rs} \times \mathbf{v}_{rs}$ (see Equation 6.31) for the unit normal vectors $\mathbf{n}_{ps}$ and $\mathbf{n}_{rs}$ (remember that the following approximate equality $\mathbf{n}_{ps} \cong \mathbf{n}_T$ is usually valid) make expressions for the normal rake angle γ_n possible:

$$\gamma_n = 90° - \arccos\,(\mathbf{n}_{rs} \cdot \mathbf{n}_{ps}) = \arcsin\,(\mathbf{n}_{rs} \cdot \mathbf{n}_{ps}) \tag{6.44}$$

$$\gamma_n = 90° - \angle(\mathbf{n}_{rs}, \mathbf{n}_{ps}) = \arctan \frac{\mathbf{n}_{rs} \cdot \mathbf{n}_T}{|\mathbf{n}_{rs} \times \mathbf{n}_T|} \tag{6.45}$$

The cutting tool clearance surface C_s forms a certain normal clearance angle α_n with the tool cutting edge plane P_s. The clearance angle α_n is the angle between the clearance surface C_s and the unit normal vector $\mathbf{n}_{ps}$ to the plane P_s. Its value is accounted from the reference plane P_s toward the clearance surface C_s. The normal clearance angle α_n is always of positive value ($\alpha_n > 0°$). In a particular case, within a narrow strip (land) on the clearance surface along the cutting edge C_e, the clearance angle α_n can be equal to zero, or it can even be of negative value (up to the value of $\alpha_n = -20° - -25°$).

The unit normal vector at a point of the clearance surface C_s is designated as $\mathbf{n}_{cs}$. It is convenient to define the normal clearance angle α_n as the angle that complements to 180° the angle between the unit normal vectors $\mathbf{n}_{ps}$ and $\mathbf{n}_{cs}$ as schematically shown in Figure 6.11. The above equation $\mathbf{n}_T = \mathbf{u}_T \times \mathbf{v}_T$ (see Equation 6.30) and the equation $\mathbf{n}_{cs} = \mathbf{u}_{cs} \times \mathbf{v}_{cs}$ for the unit normal vectors $\mathbf{n}_{ps}$ and $\mathbf{n}_{cs}$ (again, the following approximate equality $\mathbf{n}_{ps} \cong \mathbf{n}_T$ is commonly valid) yield the following formulae for the calculation of the normal clearance angle α_n:

$$\alpha_n = 180° - \angle(\mathbf{n}_{cs}, \mathbf{n}_{ps}) \tag{6.46}$$

$$\alpha_n = -\arctan \frac{|\mathbf{n}_{ps} \times \mathbf{n}_{cs}|}{\mathbf{n}_{ps} \cdot \mathbf{n}_{cs}} \tag{6.47}$$

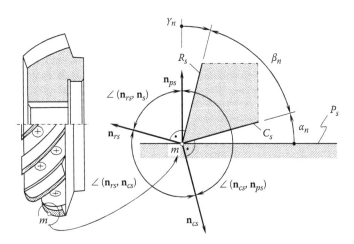

FIGURE 6.11
Cross section of the cutting wedge by a *normal reference plane, P_n*.

The *cutting wedge angle* β_n can be defined as the angle that complements to 180° the angle between the unit normal vectors $\mathbf{n}_{rs}$ and $\mathbf{n}_{cs}$. This immediately leads to the following expressions for the calculation of the angle β_n:

$$\beta_n = 180° - \angle (\mathbf{n}_{rs}, \mathbf{n}_{cs}) \tag{6.48}$$

$$\beta_n = -\arctan \frac{|\mathbf{n}_{rs} \times \mathbf{n}_{cs}|}{\mathbf{n}_{rs} \cdot \mathbf{n}_{cs}} \tag{6.49}$$

Remember that the following expression is the simplest possible for the calculation of the cutting wedge angle β_n:

$$\beta_n = 90° - (\alpha_n + \gamma_n) \tag{6.50}$$

6.2.5 Correspondence between Geometric Parameters of the Active Part of Cutting Tools That Are Measured in Different Reference Planes

Certain geometric parameters of the active part of a cutting tool can be measured directly on the cutting tool of a given design. The following geometric parameters α_n, β_n, γ_n, λ_s, ρ_n, and ε' are among those that can be measured directly.[*] The actual values of other geometric parameters of the active part of a cutting tool can be calculated. For a derivation of equations for calculation of geometry of the active part of cutting tools, the implementation of elements of vector calculus is beneficial. To the best of the author's knowledge, S. S. Mozhayev[†] [44] was the first to implement (in 1948 or even earlier) elements of vector calculus for solving problems that pertain to the geometry of the active part of a cutting tool. Since then (after S.S. Mozhayev proposed the method), the vector approach for the calculation of the cutting tool geometry has become common in investigations undertaken by many researchers in the field.

For illustration of the capabilities of the proposed method, consider an elementary cutting edge of the infinitesimally short length $d\,l$. The geometric parameters of the cutting edge in the normal reference plane P_n and the value of the cutting edge inclination angle λ_{se} are given. It is necessary to determine the rake angle γ_{ce} that is measured in the reference plane P_{ce}. The reference plane P_{ce} is the plane through the velocity vector $\mathbf{V}_p$ of the assumed direction of primary motion perpendicular to the cutting edge plane P_s at a point of interest, m, as illustrated in Figure 6.12.

The *Cartesian* coordinate system $X_T Y_T Z_T$ is associated with the cutting wedge as shown in Figure 6.12.

---

[*] The angle ε' is measured within the rake surface R_s. The projection of the angle ε' onto the main reference plane P_r is equal to the tool tip angle ε.

[†] S. S. Mozhayev, a Soviet scientist and engineer known for his accomplishments in the realm of the theory of cutting tool design.

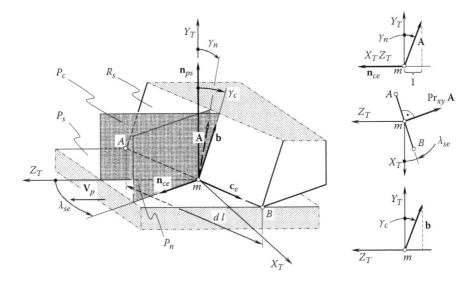

FIGURE 6.12
Geometry of the active part of an infinitesimally short cutting edge, C_e, in cross section by various reference planes.

First, construct a vector A (Figure 6.12). The vector A is a tangent to the line of intersection of the rake surface R_s by the normal reference plane P_n. The vector A is of the length, for which its projection onto the $X_T Y_T$ coordinate plane is equal to $\mathrm{Pr}_{xy}\, A = 1$. An analytical expression for vector A can be represented in matrix form:

$$
A = \begin{bmatrix} -\sin \lambda_{se} \\ \cot \gamma_n \\ -\cos \lambda_{se} \end{bmatrix} \tag{6.51}
$$

Second, construct a unit vector b. The unit vector b is a tangent to the line of intersection of the rake surface R_s by the reference plane P_{ce}. In the coordinate system $X_T Y_T Z_T$, the unit vector b yields representation in the form

$$
b = \begin{bmatrix} 0 \\ -\cos \gamma_{ce} \\ \sin \gamma_{ce} \end{bmatrix} \tag{6.52}
$$

Third, construct a unit vector c_e. The unit vector c_e is directed along the cutting edge of the length $d\, l$. The vector c_e is equal to

$$c_e = \begin{bmatrix} \cos \lambda_{se} \\ -\sin \lambda_{se} \\ 0 \end{bmatrix} \tag{6.53}$$

By construction, all three vectors $\mathbf{A}$, $\mathbf{b}$, and c_e represent a set of coplanar vectors. All three are located within the plane that is a tangent to the cutting edge AB at the point of interest, m. Because of this, triple scalar product $[\mathbf{A}, \mathbf{b}, c_e]$ of the vectors $\mathbf{A}$, $\mathbf{b}$, and c_e is identical to zero. Therefore, the following expression is valid:

$$\mathbf{A} \times \mathbf{b} \cdot c_e = \begin{vmatrix} -\sin \lambda_{se} & \cot \gamma_n & -\cos \lambda_{se} \\ 0 & -\cos \gamma_{ce} & \sin \gamma_{ce} \\ \cos \lambda_{se} & 0 & -\sin \lambda_{se} \end{vmatrix} \equiv 0 \tag{6.54}$$

After the necessary formulae transformations are performed, one can come up with the equation for the calculation of the rake angle γ_{ce}:

$$\gamma_{ce} = \cot^{-1}[\cos \lambda_{se} \cdot \cot \gamma_n] \tag{6.55}$$

Equation 6.55 is derived under the assumption that the cutting edge is infinitesimally short. This means that Equation 6.55 is applicable for the calculation of the rake angle γ_{ce} at any point within the cutting edge C_e of a form-cutting tool of any design.

Considerations similar to those above are valid for other geometrical parameters of the cutting edge of a cutting tool measured in other reference planes.

The angle between the major and the minor cutting edges of a cutting tool is designated as ε'. This angle is measured in the rake surface R_s. The tool tip angle ε can be expressed in terms of the angle ε'. The tool tip angle ε can be thought of as the projection of the angle ε' onto to the *main reference plane* P_r. Conversely, the angle ε' is the projection of the tool tip angle ε onto the rake surface R_s.

The International Standard ISO 3002 recommends the following equations for the calculation of geometry of the active part of a cutting tool:

$$\tan \lambda_{se} = \sin \kappa_r \tan \gamma_p - \cos \kappa_r \tan \gamma_f \tag{6.56}$$

$$\tan \gamma_n = \cos \lambda_{se} \tan \gamma_o \tag{6.57}$$

$$\tan \gamma_o = \cos \kappa_r \tan \gamma_p + \sin \kappa_r \tan \gamma_f \tag{6.58}$$

$$\cot \alpha_n = \cos \lambda_{se} \cot \alpha_o \tag{6.59}$$

Equations 6.56 through 6.59 are derived under the assumptions that the tool side rake angle γ_f, the tool back rake angle γ_p, and the tool cutting edge angle κ_r are the basic angles for the tool face, and the tool side clearance angle α_f, the tool back clearance angle α_p, and the tool cutting edge angle κ_r are the basic angles for the tool flank.

Other equations commonly used for the calculation of the geometry of the active part of a cutting tool are as follows (ISO 3002):

$$\cot \alpha_o = \cos \kappa_r \cot \alpha_p + \sin \kappa_r \cot \alpha_f \tag{6.60}$$

$$\tan \lambda_{se} = -\sin \kappa_r \tan \gamma_p - \cos \kappa_r \tan \gamma_f \tag{6.61}$$

$$\tan \gamma_n = \cos \lambda_{se} \tan \gamma_o \tag{6.62}$$

$$\tan \gamma_o = -\cos \kappa_r \tan \gamma_p + \sin \kappa_r \tan \gamma_f \tag{6.63}$$

$$\cot \alpha_n = \cos \lambda_{se} \cot \alpha_o \tag{6.64}$$

$$\cot \alpha_o = -\cos \kappa_r \cot \alpha_p + \sin \kappa_r \cot \alpha_f \tag{6.65}$$

The implementation of the method of calculation of the geometry of an active part of a cutting tool is illustrated with the following examples [121].

Example 6.2

Consider a twist drill. The core diameter d of the twist drill is 0.2 of the nominal drill diameter D, that is, $d = 0.2D$. The tool cutting edge angle is equal to $2 \cdot \kappa_r = 120°$. It is necessary to determine the cutting edge inclination angle λ_s (Figure 6.13).

Based on definition, the cutting edge inclination angle λ_{se} is equal to the angle that forms the unit velocity vector $\mathbf{v}_p$ of the assumed direction of primary motion and the unit vector $\mathbf{n}_{ce}$ to the cutting edge, C_e. Both vectors $\mathbf{v}_p$ and $\mathbf{n}_{ce}$ are located within the tool cutting edge plane, and the unit vector $\mathbf{n}_{ce}$ is perpendicular to the cutting edge. Therefore, the following equality is valid:

$$\lambda_{se} = \angle\,(\mathbf{n}_{ce},\, \mathbf{v}_p) = [90° - \angle\,(\mathbf{c}_e,\, \mathbf{v}_p)] \tag{6.66}$$

where $\mathbf{c}_e$ designates the unit vector along the cutting edge, C_e.

The unit vectors $\mathbf{c}_e$ and $\mathbf{v}_p$ yield the following representation in the local coordinate system $X_T Y_T Z_T$:

$$\mathbf{c}_e = \begin{bmatrix} 0 \\ -\sin \kappa_r \\ -\cos \kappa_r \end{bmatrix} \tag{6.67}$$

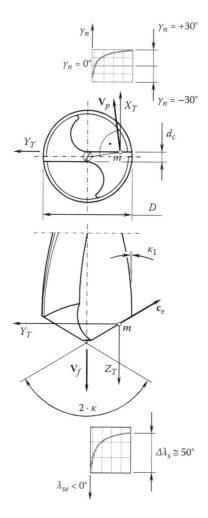

FIGURE 6.13
An example of calculation of the geometry of the cutting edge of a twist drill.

$$\mathbf{v}_p = \begin{bmatrix} \cos \mu \\ \sin \mu \\ 0 \end{bmatrix} \tag{6.68}$$

This immediately yields an equation for the calculation of the inclination angle, λ_{se}:

$$\lambda_{se} = -\arcsin \left(\frac{d}{d_i} \sin \kappa_r \right) \tag{6.69}$$

Equation 6.69 for the calculation of the angle λ_{se} of inclination reveals that for a twist drill, the cutting edge inclination angle, λ_{se}, is always negative. The actual value of the angle λ_{se} depends upon a distance at which current point of interest, m, within the cutting edge, C_e, is remote from the axis of rotation of the twist drill. It is easy to verify that the angle λ_{se} is of the smallest magnitude at the outside diameter of the twist drill where it is equal $\lambda_{se} = -7.464°$. The magnitude of the angle λ_{se} increases toward the axis of rotation of the twist drill. The angle λ_{se} of the largest magnitude occurs for the point within the cutting edge that is closest to the axis of rotation where it is as $\lambda_{se} \approx -55°$. The difference of the angle of inclination λ_{se} for various points within the cutting edge is about $\Delta\lambda_{se} \approx 50°$. That large difference ($\Delta\lambda_{se} \approx 50°$) causes problems in optimization of the angle of inclination along the cutting edge of the twist drill. There is no proper solution to this challenging problem so far. The afore-mentioned problem becomes more complex if variation of cutting speed, variation of volume of the stock to be removed, and so forth is put into account.

Example 6.3

Consider that same twist drill as that considered in Example 6.2. The helix angle at the major diameter of the drill is $\omega = 30°$. It is necessary to determine the tool normal rake angle γ_n at a point of the cutting edge.

By definition of the normal rake angle γ_n, the rake angle γ_n of the twist drill can be specified in terms of two unit normal vectors $\mathbf{n}_{ps}$ and $\mathbf{n}_{rs}$:

$$\tan \gamma_n = \frac{\mathbf{n}_{ps} \cdot \mathbf{n}_{rs}}{|\mathbf{n}_{ps} \times \mathbf{n}_{rs}|} \tag{6.70}$$

where $\mathbf{n}_{ps}$ is the unit normal vector to the *tool cutting edge reference plane* P_s and $\mathbf{n}_{rs}$ is the unit normal vector to the twist drill rake face R_s.

The unit normal vector $\mathbf{n}_{ps}$ can be calculated as cross product $\mathbf{n}_{ps} = \mathbf{c}_e \times \mathbf{v}_p$:

$$\mathbf{n}_{ps} = \begin{bmatrix} \mathbf{i} & \mathbf{j} & \mathbf{k} \\ 0 & -\sin \kappa_r & -\cos \kappa_r \\ \cos \mu & \sin \mu & 0 \end{bmatrix} \tag{6.71}$$

This makes the equation for the unit normal vector $\mathbf{n}_{ps}$ possible:

$$\mathbf{n}_{ps} = \begin{bmatrix} \cos \kappa_r \sin \mu \\ -\cos \kappa_r \cos \mu \\ \sin \kappa_r \cos \mu \end{bmatrix} \tag{6.72}$$

The unit normal vector $\mathbf{n}_{rs}$ can be calculated as the cross product $\mathbf{n}_{rs} = \mathbf{c}_e \times \mathbf{t}_{rs}$. Here, $\mathbf{t}_{rs}$ designates a unit vector that is a tangent to the rake surface R_s of the twist drill. It is convenient to employ the vector $\mathbf{t}_{rs}$ that is tangent to a helix line of the rake surface. In this case, the following equation is valid:

$$\mathbf{t}_{rs} = \begin{bmatrix} -\sin \omega_y \sin \mu \\ \sin \omega_y \cos \mu \\ \cos \omega_y \end{bmatrix} \tag{6.73}$$

where ω_i designates the helix angle of the screw rake surface, R_s, at a current point of interest, m, of the cutting edge C_e of the twist drill.

The above expressions for the unit vectors $\mathbf{c}_e$ and $\mathbf{t}_{rs}$ make the equation for the unit normal vector $\mathbf{n}_{rs}$ in matrix form possible :

$$\mathbf{n}_{rs} = \begin{bmatrix} \mathbf{i} & \mathbf{j} & \mathbf{k} \\ 0 & \sin \kappa_r & \cos \kappa_r \\ -\sin \omega_y \sin \mu & \sin \omega_y \cos \mu & \cos \omega_y \end{bmatrix} \tag{6.74}$$

This result immediately returns the equation for the unit normal vector $\mathbf{n}_{rs}$:

$$\mathbf{n}_{rs} = \begin{bmatrix} \sin \kappa_r \cos \omega_y - \cos \kappa_r \cos \mu \sin \omega_y \\ -\cos \kappa_r \sin \mu \sin \omega_y \\ \sin \kappa_r \sin \mu \sin \omega_y \end{bmatrix} \tag{6.75}$$

Substituting the derived expressions for the unit normal vectors $\mathbf{n}_{ps}$ and $\mathbf{n}_{rs}$ into Equation 6.70 for $\tan \gamma_n$, one can come up with the expression for $\tan \gamma_n$:

$$\tan \gamma_n = \frac{1 - \sin^2 \kappa_r \sin^2 \mu}{\sin \kappa_r \cos \mu} \cdot \tan \omega_y - \cos \kappa_r \tan \mu \tag{6.76}$$

Example 6.4

Consider that same twist drill as considered in the Example 6.1 and Example 6.2. It is necessary to determine the tool normal clearance angle α_n at a point of the cutting edge.

By definition of the normal clearance angle α_n, the clearance angle α_n of the twist drill can be specified in terms of two unit normal vectors $\mathbf{n}_{ps}$ and $\mathbf{n}_{rs}$:

$$\tan \alpha_n = \frac{\mathbf{n}_{ps} \cdot \mathbf{n}_{cs}}{|\mathbf{n}_{ps} \times \mathbf{n}_{cs}|} \tag{6.77}$$

where $\mathbf{n}_{cs}$ designates normal unit vector to the clearance surface C_s.

The unit normal vector $\mathbf{n}_{ps}$ was defined in Equation 6.72. The unit normal vector $\mathbf{n}_{cs}$ can be calculated as the cross product $\mathbf{n}_{cs} = \mathbf{n}_{ps} \times \mathbf{t}_{cs}$. Here $\mathbf{t}_{cs}$ designates a vector that is tangent to the clearance surface C_s of the twist drill at a point of interest, m, within the cutting edge C_e.

In the coordinate system $X_T Y_T Z_T$, the unit tangent vector $\mathbf{t}_{cs}$ yields representation in the following form:

$$\mathbf{t}_{cs} = \begin{bmatrix} -\sin \alpha_i \sin \mu \\ \sin \alpha_i \cos \mu \\ \cos \alpha_i \end{bmatrix} \tag{6.78}$$

In Equation 6.78, α_i designates clearance angle that is measured in a section of the twist drill by a cylinder coaxial with the twist drill. The diameter of the cylinder is chosen so the point of interest, m, is located on the surface of the cylinder.

Following a method similar to that implemented in Example 6.3, the equation for the calculation of the normal clearance angle, α_n, can be derived as well.

The cutting edge radius of curvature is measured within the rake surface R_s. To determine the curvature of the cutting edge, C_e, an additional reference plane is constructed. This reference plane, P_{ce}, is perpendicular to the *cutting edge plane* P_s, and is a tangent to the cutting edge, C_e, at a point of interest, m. In this reference plane, the radius of normal curvature of the generating surface T of a form-cutting tool corresponds to radius of curvature R_T^* of the cutting edge, C_e. Mensnier's equation (see Equation 5.24) together with Euler's equation (see Equation 1.79) yield the equation for the radius R_T in terms of the radius R_T^*:

$$R_T = \frac{R_T^*}{\cos \gamma_n} = \frac{R_{1.T} \sin^2 \lambda_s + R_{2.T} \cos^2 \lambda_s}{R_{1.T} R_{2.T} \cos \gamma_n} \tag{6.79}$$

The cutting edge of a cutting tool is not absolutely sharp. Actually, there exists a transition surface that connects the rake surface R_s and the flank C_s. This transition surface is supposed to have a circular profile of a certain radius. The radius is considered as radius ρ_n of the *cutting edge roundness* (Figure 6.14).

The roundness of the cutting wedge can be determined from experiments (for example, from etching tests). The roundness of the cutting wedge of a cutting tool made of HSS (high-speed steel) is usually in the range of $\rho_n = 20$–$50\ \mu m$, while that of a cutting tool made of sintered carbide is as low as $\rho_n = 10$–$30\ \mu m$. For diamond inserts, roundness drops down to $\rho_n = 5$–$8\ \mu m$ and, in a particular case, can even be reduced to $\rho_n \cong 2\ \mu m$.

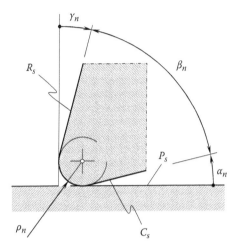

FIGURE 6.14
Roundness, ρ_n, of the cutting edge in the normal plane section, P_n.

The cutting edge roundness ρ_n affects a material removal process in metal cutting. The effect of the cutting edge roundness is more substantial when thin chip is removed, especially when the thickness of the stock to be removed is of the same range as the cutting edge roundness ρ_n. When the stock thickness is about 0.003 mm or less, the rake angle γ_n does not affect the material removal process; thus, it can be neglected in the calculations.

In the reference plane P_o through the velocity vector $\mathbf{V}_p$ that is perpendicular to the cutting edge plane P_s, roundness ρ of the cutting edge, C_e, can be expressed in terms of the normal roundness ρ_n by the Mensnier equation:

$$\rho = \rho_n \cos \lambda_{se} \tag{6.80}$$

The torsion of the cutting edge is one more geometric parameter of the active part of a cutting tool to be considered.

The shapes of the rake surface R_s and the clearance surface C_s affect the material removal process in metal cutting. This is because in many cases the geometry of the active part of a cutting tool is measured in reference plane configuration, which depends upon configuration of the tangent planes to the surfaces R_s and C_s. The effect of the cutting edge torsion onto the material removal process in metal cutting has not yet been profoundly investigated.

A cutting edge, C_e, can be interpreted as a line of intersection of three surfaces: the generating surface T of the cutting tool, the rake surface R_s, and the clearance surface C_s. The equation of the cutting edge can be derived as a result of mutual consideration of the equation of one of three pairs of surfaces:

a. $\mathbf{r}_{rs}$ (U_{rs}, V_{rs}) and $\mathbf{r}_{cs}$ (U_{cs}, V_{cs}),

b. $\mathbf{r}_T$ (U_T, V_T) and $\mathbf{r}_{rs}$ (U_{rs}, V_{rs}),

c. $\mathbf{r}_T$ (U_T, V_T) and $\mathbf{r}_{cs}$ (U_{cs}, V_{cs}).

The solution to any of three pairs of equations can be reduced to the equation of the cutting edge, C_e, which yields representation in matrix form:

$$\mathbf{r}_{ce} = \mathbf{r}_{ce}(t_{ce}) = \begin{bmatrix} X_T(t_{ce}) \\ Y_T(t_{ce}) \\ Z_T(t_{ce}) \end{bmatrix} \tag{6.81}$$

where t_{ce} denotes the parameter of the cutting edge, C_e.

In a particular case, length S_{ce} of the cutting edge can be chosen as the parameter of the cutting edge (that is, $t_{ce} \equiv S_{ce}$). Under such a scenario, torsion τ_{ce} of the cutting edge, C_e, can be calculated from the expression

$$\tau_{ce} = \left[\rho_{ce}^2 \left(\frac{d\,\mathbf{r}_{ce}}{d\,S_{ce}} \cdot \frac{d^2\mathbf{r}_{ce}}{d\,S_{ce}^2} \cdot \frac{d^3\mathbf{r}_{ce}}{d\,S_{ce}^3} \right) \right]^{-1} \tag{6.82}$$

where the sign of the torsion τ_{ce} is not in compliance with the direction of the angle of inclination λ_s.

6.2.6 Diagrams of Variation of the Geometry of the Active Part of a Cutting Tool

The use of analytical methods for the calculation of actual values of the geometry of the active part of a cutting tool returns accurate results of the calculations. These methods are capable of computing the distribution of a geometrical parameter of a cutting tool

- At a given point of the cutting edge in different reference cross sections
- Within the active part of the cutting edge in similar cross sections

Results of such calculations are accurate and are of critical importance to the cutting tool designer.

Graphical methods for the determination of the geometrical parameters at a point of the cutting edge are used along with the analytical methods. The implementation of diagrams of variations of the geometrical parameters has proven to be useful for the preliminary analysis of the geometry of the active part of a cutting tool.

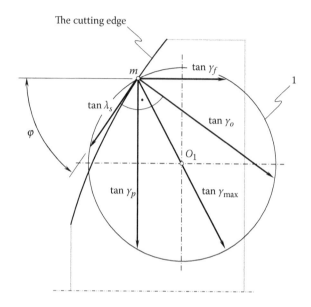

FIGURE 6.15
Example of the circular diagram of the distribution of the function tan γ of a form-cutting tool rake angle, γ.

Distribution of the function tan γ of the rake angle in different reference planes through the point of interest, m, within the cutting edge, C_e, of a form-cutting tool is illustrated in Figure 6.15. Figure 6.15 shows a type of Mohr* diagram. Once the values of the rake angle in two different reference planes are determined, the distribution of the function tan γ follows the circle (that is, it follows the Mohr diagram). The circle constructed on any two known vectors through the point of interest m enables easy determination of the function tan γ_i in any direction through the point m. The magnitude of the vector is equal to the actual value of the function tan γ in the corresponding direction.

Similarly, the distribution of the function cot α of the clearance angle α in different reference planes through the point of interest m within the cutting edge, C_e, of a form-cutting tool can be constructed. The corresponding circular diagram of the distribution of the function cot α is depicted in Figure 6.16. Again, a circle constructed on two known values of the function cot α in two different directions reflects the distribution of the function cot α in all other reference planes.

One more example of implementation of the circular diagrams for the preliminary analysis of the cutting tool geometry at a point m of the cutting edge, C_e, of a form milling cutter is illustrated in Figure 6.17.

---

* Christian Otto Mohr (October 8, 1835–October 2, 1918), a German civil engineer.

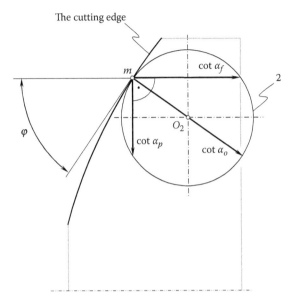

FIGURE 6.16
An example of the circular diagram of the distribution of the function $\cot\alpha$ of a form-cutting tool clearance angle, α.

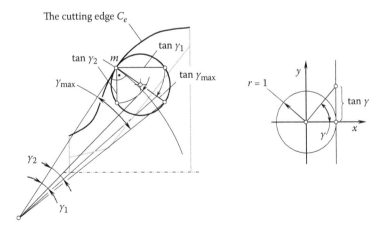

FIGURE 6.17
Example of implementation of the circular diagram at a point m of the cutting edge, C_e, from the milling cutter.

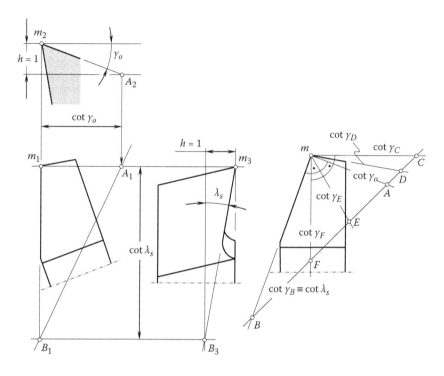

FIGURE 6.18
Example of the diagram of distribution of the function, cot γ, of a cutting tool rake angle, γ.

The diagram of another type [4], shown in Figure 6.18, is helpful for the analysis of the distribution of the function $\cot \gamma$. Two known values of the function $\cot \gamma$ in prespecified corresponding directions make possible construction of the straight line AB. Ultimately, the actual value of the function $\cot \gamma_i$ in the reference plane through a current direction is specified by the corresponding point within the straight line AB.

All the diagrams are in perfect correlation with the results of the analytical calculations.

6.3 Geometry of the Active Part of Cutting Tools in the *Tool-in-Use* System

When machining a part surface, the actual direction of the primary motion as well as of the feed rate motion can differ from the assumed directions of these motions, for example, in the *tool-in-hand* reference system. Moreover, the actual kinematics of a machining operation can be made up not only of the primary and the feed rate motions but also of motions of other nature, namely, the orientational motions of the cutting tool (see Chapter 2) and so forth. The actual geometry at a point of the cutting edge of a cutting tool also can be affected by vibrations. To precisely specify the geometric parameters of the active part of a form-cutting tool, all the elementary motions that compose the resultant motion of the cutting tool relative to the work must be taken into consideration.

There are two possible ways to represent the machined part surface P.

First, the machined part surface P can be considered as an enveloping surface to consecutive positions of the generating surface T of the form-cutting tool when the cutting tool is moving relative to the blank. The enveloping surface is supposed to be the nominal part surface.

Second, the machined part surface P can be considered as a set of discrete surfaces of cut P_{se}. Each of the surfaces of cut is generated by a corresponding cutting edge of the cutting tool when the cutting edge, C_e, is traveling in relation to the work.

At a certain instance of time when the part surface P is generated, both the cutting tool surface T and the surface P_{se} are tangents to P either at the contact point K or along the characteristic line E. Because of this, the *tool-in-hand* reference system can be associated either with the assumed surface of the cut or with the cutting tool [122]. The two options are identical in the sense of the *tool-in-hand* reference system.

When machining a part surface, it is necessary to consider the kinematic geometric parameters of the active part of the cutting tool in a reference system associated with the surface of the cut. For this purpose, the *tool-in-use* reference system is used.

Commonly, rake surface R_s as well as flank surface C_s of a form-cutting tool are shaped in the form of three-dimensional surfaces having complex geometry. Due to this, consideration of the surfaces R_s and C_s at a distinct point of the cutting edge is necessary. Contact of the cutting wedge with the work is considered at the distinct point of the cutting edge—at the point of interest, m. Because the size of the area of contact of the cutting wedge and the work is small, the rake surface, R_s as well as the flank surface, C_s, are locally approximated by corresponding planes—by the planes that are tangents to the surfaces R_s and C_s at the point of interest, m, of the cutting edge.

The geometry of the active part of a form-cutting tool must be determined for an elementary cutting edge of the length $d\,l$ (that is, in differential vicinity of the point of interest, m, within the cutting edge, C_e, of the cutting tool). It is also necessary to consider the geometry of the active part of a form-cutting tool at a given instant of time, for example, for the velocity vector $\mathbf{V}_\Sigma$ of known magnitude and direction. Such an approach would enable one to determine the distribution curves of geometric parameters along the cutting edge and the distribution curves of geometric parameters in time. To perform such an analysis, a generalized method of calculation of geometry of the active part of a form-cutting tool must be implemented.

In particular cases, the actual values of geometric parameters of the active part of a form-cutting tool can impose certain constraints onto parameters of kinematics of the machining operation. For example, a variation in the actual value of geometric parameters either along the cutting edge C_e or in time may impose certain restrictions onto the parameters of feed rate motion, orientational motion of the cutting tool, and so forth. If the parameters of kinematics of the machining operation exceed the limits, the machining operation becomes infeasible.

The capability to determine critical feasible values of parameters of geometry of the active part of a form-cutting tool is critically important for the cutting tool designer.

6.3.1 Resultant Speed of the Relative Motion in the Cutting of Materials

As follows from the above analysis, the direction of the resultant velocity vector V_Σ of relative motion in the cutting of materials is a critical issue for establishing the *tool-in-use* reference system.

Usually, the relative motion of the cutting tool is of a complex nature. In the general case of part surface machining, this motion is composed of the actual primary motion V_p, the surface generation motion V_{gen}, one or more feed rate motions $V_{f.i}$, the orientational motions of the first V_{or}^I and second V_{or}^{II} types and of other motions. This makes possible the following equation for the velocity vector V_Σ:

$$V_\Sigma = V_p + V_{gen} + \sum_{i=1}^{n} V_{f.i} + V_{or}^I + V_{or}^{II} + \ldots = \sum_{j=1}^{m} V_j \qquad (6.83)$$

where V_p is the vector of the primary motion, V_{gen} is the vector of the motion of part surface generation, $V_{f.i}$ is the vector of the feed rate motion, n is the total number of feed rate motions, V_{or}^I is the vector of the orientation motion of the first type of the form-cutting tool, V_{or}^{II} is the vector of the orientation motion of the second type of the form-cutting tool, V_j is the j-th elementary relative motion of the form-cutting tool, and m is the total number of elementary relative motions of the form-cutting tool.

When determining the velocity vector V_Σ of the resultant relative motion, the vectors of all particular relative motions that significantly affect vector V_Σ must be taken into account. Relative motions that cause sliding of the part surface P and/or the generating surface T of the form-cutting tool over itself must be incorporated as well.

Velocity vectors V_{or}^I and V_{or}^{II} of the orientational motions of the form-cutting tool as well as the velocity vectors of the feed rate motions $V_{f.i}$ are usually significantly smaller compared with the velocity vector V_p of the primary motion. However, all must be incorporated for determination of the velocity vector $\mathbf{V}_\Sigma$. In particular cases, some of these motions are comparable with the velocity vector $\mathbf{V}_p$. Moreover, in special cases, they can even exceed the velocity vector $\mathbf{V}_p$ of primary motion.

When cutting a material, vibration of the cutting tool is often observed. The vibration may result in positive or negative clearance angles (Figure 6.19a). For certain frequencies and magnitudes of the vibration, neglecting of the velocity vector $\mathbf{V}_{vib}$ of vibration is not allowed [27,31,114]. Due to vibrations, the rake and the clearance angles vary within a certain interval $\pm\sigma_o$. The current value of the angle σ_o is

$$\sigma_o = \arctan \frac{|\mathbf{V}_{vib}|}{|\mathbf{V}_p|}. \tag{6.84}$$

If the velocity vector $\mathbf{V}_{vib}$ is pointed toward the part surface P, the corresponding rake angle γ_o raises to the range of $\gamma_o' = \gamma_o + \delta_o$. At this instant,

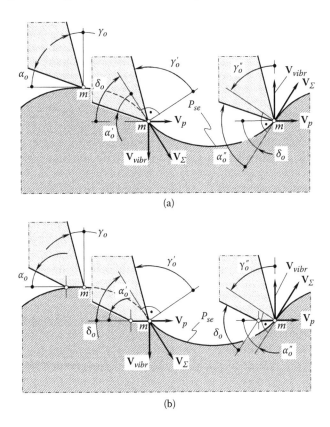

(a)

(b)

FIGURE 6.19
Effect of vibration on the actual geometry of the active part of a form-cutting tool.

the clearance angle α_o reduces to $\alpha'_o = \alpha_o - \delta_o$. If the velocity vector $\mathbf{V}_{vib}$ is directed oppositely, then the corresponding rake angle γ_o and the clearance angle α_o can be calculated from the equations $\gamma''_o = \gamma_o - \delta_o$ and $\alpha''_o = \alpha_o + \delta_o$ (see Figure 6.19b).

When a partly worn cutting tool is used, then the clearance angle within a narrow land on the clearance surface, C_s, next to the cutting edge, C_e, reduces to $0°$. Cutting tool wear causes the effect of vibrations onto the cutting tool geometry to be more severe (see Figure 6.19b).

The performed brief analysis of the impact of vibration onto the resultant instant motion of the cutting tool reveals the importance of careful analysis of all elementary motions: the cutting tool is performed when machining a part surface.

6.3.2 *Tool-in-Use* Reference System

The actual configuration of reference planes is a critical issue for the establishment of the *tool-in-use* reference system [112]. Because the efficiency of the chip removal process closely depends upon the actual orientation of the cutting wedge in relation to the surface of the cut, then the surface of the cut must be introduced as a base element for the determination of instant geometry of the active part of a cutting tool.

The surface of cut P_{se} can be represented as a locus of consecutive positions of the cutting edge, C_e, which is traveling with the resultant speed $\mathbf{V}_\Sigma$ relative to the work. The plane of cut is a tangent to the surface of cut P_{se} at the point of interest, m, within the cutting edge, C_e. In a particular case, the surface of the cut and the plane of the cut are congruent. The last case is the degenerate one.

The main reference plane P_{re} is perpendicular to the velocity vector $\mathbf{V}_\Sigma$. The working plane P_{fe} is the plane through the directions of the primary motion and of the feed rate motion. Because of this, the working plane P_{fe} is perpendicular to the main reference plane P_{re}. The tool back plane P_{pe} is perpendicular to the reference planes P_{re} and P_{fe}.

Other reference planes are important for the establishment of the *tool-in-use* system. They are plane of cut P_{se}, rake surface plane R_s, and clearance plane C_s.

To determine the geometry of the active part of the cutting tool, it is convenient to employ three reference planes P_{se}, R_s, and C_s in conjunction with the velocity vector $\mathbf{V}_\Sigma$ of the resultant motion of the cutting tool. The current orientation of the reference planes P_{se}, R_s, and C_s is specified by the unit normal vectors $\mathbf{n}_{rs}$, $\mathbf{n}_{cs}$, and $\mathbf{c}_e$.

Prior to running the analysis, it is necessary to represent the equation $\mathbf{r}_P = \mathbf{r}_P (U_P, V_P)$ of the part surface P as well as the equation $\mathbf{r}_T = \mathbf{r}_T (U_T, V_T)$ of the generating surface T of the form-cutting tool in a common coordinate system $X_T Y_T Z_T$. For this purpose, the implementation of the operator $\mathbf{Rs}\,(P \rightarrow T)$ of the resultant coordinate system transformation is helpful. Equations for the position vectors of a point $\mathbf{r}_{P.tp}$ and $\mathbf{r}_{T.tp}$ of the planes tangent to the surfaces P and T at the point of interest m can be represented in a vector form:

$$\left[\mathbf{r}_{P.tp} - \mathbf{r}_P^{(m)}\right] \times \mathbf{n}_P = 0 \tag{6.85}$$

and

$$\left[\mathbf{r}_{T.tp} - \mathbf{r}_T^{(m)}\right] \times \mathbf{n}_T = 0 \tag{6.86}$$

The kinematic method can be employed for the derivation of the equation of the surface of the cut P_{se}. For this purpose, it is necessary to know the equation of the cutting edge, C_e, and the velocity vector, $\mathbf{V}_\Sigma$, of the resultant relative motion of the cutting tool relative to the work.

The equation of the surface of the cut P_{se} can be obtained in the following way.

Consider a form-cutting tool. Cutting edge, C_e, of the form-cutting tool is determined as the line of intersection of the rake surface R_s by the clearance surface C_s. Therefore, in the coordinate system $X_T Y_T Z_T$, the cutting edge, C_e, of the form-cutting tool can be described analytically by a set of two vector equations:

$$C_e \implies \begin{cases} \mathbf{r}_{rs} = \mathbf{r}_{rs}(U_{rs}, V_{rs}) \\ \mathbf{r}_{cs} = \mathbf{r}_{cs}(U_{cs}, V_{cs}) \end{cases} \tag{6.87}$$

An auxiliary *Cartesian* coordinate system $X_{ce} Y_{ce} Z_{ce}$ is associated with the cutting edge, C_e. Initially, axes of the coordinate system $X_{ce} Y_{ce} Z_{ce}$ align with corresponding axes of the coordinate system $X_T Y_T Z_T$. Then, consider the motion that the cutting edge, C_e, is performing together with the coordinate system $X_{ce} Y_{ce} Z_{ce}$ in relation to the coordinate system $X_T Y_T Z_T$. The parameters of this relative motion of the cutting edge, C_e, are identical to the corresponding parameters of motion of the form-cutting tool relative to the work. The equation of the cutting edge in a current configuration of the coordinate system $X_{ce} Y_{ce} Z_{ce}$ with respect to the coordinate system $X_T Y_T Z_T$ can be represented in the form

$$\begin{cases} \mathbf{r}_{rs} = \mathbf{r}_{rs}(U_{rs}, V_{rs}, \Xi_\Sigma) \\ \mathbf{r}_{cs} = \mathbf{r}_{cs}(U_{cs}, V_{cs}, \Xi_\Sigma) \end{cases} \tag{6.88}$$

where Ξ_Σ designates the parameter of the resultant relative motion of the form-cutting tool.

On the premise of Equation 6.88, one of the two curvilinear parameters, either the U_{rs} or V_{cs} parameter can be expressed in terms of another parameter. For example, the U_{rs} parameter is expressed in terms of the V_{cs} parameter. This relationship makes possible an analytical representation in the form $U_{rs} = U_{rs} (V_{cs})$. Ultimately, this results in a vector equation of the surface of cut P_{se} in the form

$$\mathbf{r}_{se} = \mathbf{r}_{se} [U_{cs} (V_{cs}), V_{cs}, \varXi_\Sigma] = \mathbf{r}_{se}[V_{cs}, \varXi_\Sigma] \tag{6.89}$$

Similarly, the equation of the surface of the cut P_{se} can be expressed in terms of the V_{cs} and $\varXi_\Sigma$ parameters.

For many purposes, the generating surface T of the form-cutting tool can be considered as a good approximation of the surface of cut P_{se} in the local vicinity of the point of interest, m, within the cutting edge, C_e.

To establish the *tool-in-use* reference system for machining a part surface on a conventional machine tool, two vectors are of principal importance, namely, the velocity vector $\mathbf{V}_\Sigma$ of the resultant relative motion of the cutting tool with respect to the work and the unit normal vector $\mathbf{n}_{se}$ to the surface of cut, P_{se}.

The velocity vector $\mathbf{V}_\Sigma$ is calculated from Equation 6.83. The unit normal vector $\mathbf{n}_{se}$ can be calculated as the cross-product $\mathbf{n}_{se} = \mathbf{u}_{se} \times \mathbf{v}_{se}$. For the derivation of the unit tangent vectors $\mathbf{u}_{se}$ and $\mathbf{v}_{se}$, Equation 6.89 of the surface of the cut P_{se} can be used. That same unit normal vector $\mathbf{n}_{se}$ can also be calculated as the cross-product (Figure 6.20):

$$\mathbf{n}_{se} = \mathbf{v}_\Sigma \times \mathbf{c}_e \tag{6.90}$$

where the unit velocity vector $\mathbf{v}_\Sigma = \mathbf{V}_\Sigma / |\mathbf{V}_\Sigma|$.

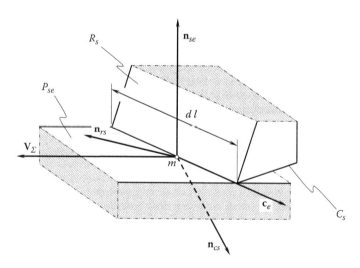

FIGURE 6.20
Elementary cutting wedge of infinitesimal length $d\,l$.

Equation 6.90 for the unit normal vector $\mathbf{n}_{se}$ is often preferred for performing the calculations.

Other equations for the calculation of the unit normal vector $\mathbf{n}_{se}$ can be used as well:

$$\mathbf{n}_{se} = \frac{\mathbf{V}_{\Sigma} \times [\mathbf{n}_{rs} \times \mathbf{n}_{cs}]}{|\mathbf{V}_{\Sigma} \times [\mathbf{n}_{rs} \times \mathbf{n}_{cs}]|} \tag{6.91}$$

It should be noted here that the approximation $\mathbf{n}_{se} \cong \mathbf{n}_T$ is valid in most practical cases when performing the calculations.

Unit vectors $\mathbf{v}_{\Sigma}$ and $\mathbf{n}_{se}$ are convenient for the analytical representation of the *tool-in-use* system.

6.3.3 Reference Planes

Investigation of the impact of kinematics of a machining operation on actual (kinematical) values of geometry of the active part of a cutting tool can be traced back to the research undertaken by A. V. Pankin [50] or even to the earlier works.

A proper *tool-in-use* reference system is necessary but not sufficient for determining geometric parameters of the active part of a cutting tool. The specification of the configuration of reference planes is also of critical importance in this concern.

For free orthogonal cutting of materials, the reference plane for the rake angle γ, the clearance angle α, the cutting wedge angle β, and the angle of cutting δ is the plane through the velocity vector $\mathbf{V}_{\Sigma}$. This reference plane is orthogonal to the plane of the cut P_{se}.

For free oblique cutting of materials, there are several reference planes for the specification of the angles γ, α, β, and δ.

The configuration of the reference planes for non-free cutting of materials cannot be specified in general terms. Mechanics of non-free cutting of materials has not yet been thoroughly investigated.

6.3.3.1 *Plane of Cut*

The plane of cut, P_{se}, is a tangent to the surface of cut at a point of interest m. For the specification of the configuration of the plane of the cut P_{se}, the velocity vector $\mathbf{V}_{\Sigma}$ of the resultant motion of the cutting tool relative to the work and the unit vector $\mathbf{c}_e$ that is a tangent to the cutting edge, C_e, at the point m can be employed (Figure 6.21). These two vectors immediately yield the vector equation for the tangent plane:

$$\left[\mathbf{r}_{se.tp} - \mathbf{r}_{se}^{(m)} \right] \times [\mathbf{c}_e \times \mathbf{v}_{\Sigma}] = 0 \tag{6.92}$$

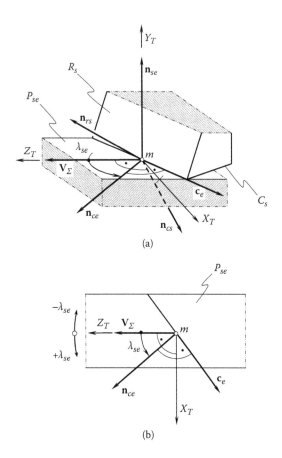

FIGURE 6.21
Definition of the angle of inclination λ_{se} of the cutting edge, C_e.

6.3.3.1.1 The Angle of Inclination of the Cutting Edge

The orientation of the cutting edge, C_e, relative to the velocity vector $\mathbf{V}_\Sigma$ of the resultant motion of the cutting edge is specified by the angle of inclination of the cutting edge, λ_{se}. The angle of inclination λ_{se} is measured within the plane of the cut P_{se}. This is the angle between the velocity vector $\mathbf{V}_\Sigma$ and the unit normal vector $\mathbf{n}_{ce}$. The vector $\mathbf{n}_{ce}$ is orthogonal to the cutting edge, C_e (Figure 6.21a), and is located within the plane of cut P_{se}. If observing from the end of unit normal vector $\mathbf{n}_{se}$ to the surface of cut P_{se}, the positive angle λ_{se} is measured in counterclockwise direction, and the negative angle λ_{se} is measured in clockwise direction as schematically depicted in Figure 6.21b. The sign of the inclination angle, λ_{se}, can be uniquely specified in terms of the unit vectors $\mathbf{v}_\Sigma$ and $\mathbf{n}_{ce}$: the inclination angle λ_{se} is positive ($\lambda_{se} > 0$) if the unit vector $\mathbf{v}_\Sigma \times \mathbf{n}_{ce}$ is pointed at that same direction as the unit normal vector $\mathbf{n}_{se}$

is pointed. Otherwise, the inclination angle λ_{se} is negative ($\lambda_{se} < 0$). Other vectors can be used for the determination of the sign of the inclination angle λ_{se}.

In the case when the equality $\lambda_{se} = 0°$ is valid, the cutting is *orthogonal cutting*. Otherwise, when $\lambda_{se} \neq 0°$, a more general case of cutting—oblique cutting—is observed.

Major frictions of the cutting tool (that is, chip deformation, direction of chip flow over the rake surface, and so forth) depend upon the actual value of the angle of inclination λ_{se}.

The algebraic value of the inclination angle λ_{se} can be calculated from the following equation (Figure 6.21b):

$$\lambda_{se} = \angle(\mathbf{c}_e, \mathbf{v}_\Sigma) - 90° = -\arctan \frac{\mathbf{c}_e \cdot \mathbf{v}_\Sigma}{|\mathbf{c}_e \times \mathbf{v}_\Sigma|} \tag{6.93}$$

For the form-cutting tools of various designs, the actual value of the inclination angle, λ_{se}, varies within the interval $\lambda_{se} = \pm 80°$.

6.3.3.2 Normal Reference Plane

Configuration of the normal reference plane P_{ne} of a cutting tool in the *tool-in-use* reference system is identical to its configuration in the *tool-in-hand* system.

The normal plane, P_{ne}, is orthogonal to several reference planes simultaneously, namely

- To the rake surface, R_s
- To the clearance surface, C_s
- To the plane of the cut P_{se}
- To the cutting edge C_e itself (Figure 6.22)

The unit normal vector $\mathbf{n}_{ce}$ to the cutting edge, C_e, is entirely located within the normal reference plane P_{ne}. Therefore, the configuration of the normal reference plane P_{ne} can be specified in terms of any two unit vectors $\mathbf{n}_{rs}$, $\mathbf{n}_{cs}$, $\mathbf{n}_{se}$, and $\mathbf{n}_{ce}$ at the point of interest m (Figure 6.22), or by the point m and the unit vector $\mathbf{c}_e$ along the cutting edge, C_e. Evidently, there are more options for the specification of configuration of the normal reference plane in the *tool-in-use* reference system rather than in the *tool-in-hand* system.

6.3.3.2.1 Normal Rake Angle

The orientation of the rake surface R_s of a cutting tool relative to the plane of the cut, P_{se}, depends upon the actual value of normal rake angle γ_{ne}. A normal rake angle is measured in the normal reference plane, P_{ne}. This is the angle that forms the rake surface R_s and the unit normal vector $\mathbf{n}_{se}$ to the plane of

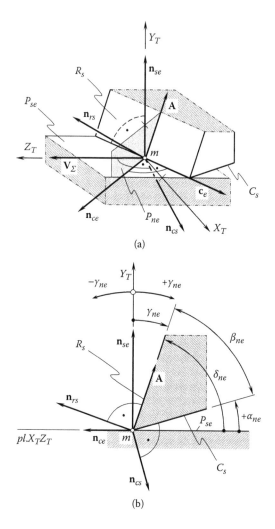

FIGURE 6.22
Geometry of the active part of a cutting tool in the *normal reference plane, P_{ne}.*

the cut P_{se}. The value of the angle γ_{ne} is measured from the vector $\mathbf{n}_{se}$ toward the rake surface R_s. The normal rake angle γ_{ne} is positive when the unit normal vector $\mathbf{n}_{se}$ does not pass through the cutting wedge of the cutting tool, and it is negative when the vector $\mathbf{n}_{se}$ passes through the cutting wedge of the cutting tool (Figure 6.22b).

It is convenient to determine the normal rake angle γ_{ne} as the angle that complements to 90° the angle between the unit normal vectors $\mathbf{n}_{se}$ and $\mathbf{n}_{rs}$ (see Figure 6.22b):

$$\gamma_{ne} = 90° - \angle(\mathbf{n}_{rs}, \mathbf{n}_{se}) = \arctan \frac{\mathbf{n}_{rs} \cdot \mathbf{n}_{se}}{|\mathbf{n}_{rs} \times \mathbf{n}_{se}|} \tag{6.94}$$

For cutting tools of various designs, the optimal value of the normal rake angle γ_{ne} is usually within the interval $\gamma_{ne} = -10°–30°$.

6.3.3.2.2 Normal Clearance Angle

The orientation of the clearance surface C_s relative to the plane of the cut P_{se} depends upon the normal clearance angle α_{ne}. This angle is measured in the normal reference plane, P_{ne}. The normal clearance angle α_{ne} is the angle that the unit normal vector $\mathbf{n}_{se}$ to the plane of cut, P_{se}, forms with the opposite direction of the unit normal vector, $\mathbf{n}_{cs}$, to the clearance surface C_s. The value of the clearance angle α_{ne} is measured from the plane of cut P_{se} toward the clearance surface C_s. The normal rake angle α_{ne} is always positive ($\alpha_{ne} > 0°$). Only within a narrow strip along the cutting edge, C_e, can the normal clearance angle α_{ne} be zero or even negative ($\alpha_{ne} \le 0°$).

It is convenient to determine the normal clearance angle α_{ne} as the angle that complements to 180° the angle between the unit normal vectors $\mathbf{n}_{se}$ and $\mathbf{n}_{cs}$ (see Figure 6.22b):

$$\alpha_{ne} = 180° - \angle(\mathbf{n}_{cs}, \mathbf{n}_{se}) = -\arctan \frac{|\mathbf{n}_{cs} \times \mathbf{n}_{se}|}{\mathbf{n}_{cs} \cdot \mathbf{n}_{se}} \tag{6.95}$$

For cutting tools of various designs, the optimal value of the normal clearance angle α_{ne} is usually within the interval $\alpha_{ne} = 10°–30°$.

The uncut chip thickness a is the predominant factor that affects the optimal value of the clearance angle. On the premise of the analysis of impact of chip thickness a, Larin proposed [34] an empirical formula for the calculation of a reasonable value of the clearance angle:

$$\alpha_{ne} = \arcsin \frac{0.13}{a^{0.3}} \tag{6.96}$$

After a short period of cutting, zero clearance angle $\alpha_{ne} = 0°$ is always observed within a narrow worn strip along the cutting wedge.

6.3.3.2.3 Mandatory Relationship

For a workable cutting tool, relationship $\mathbf{N}_{se} \cdot \mathbf{N}_{ce} < 0$ (or of the equivalent relationship $\mathbf{n}_{se} \cdot \mathbf{n}_{ce} = -1$) should be satisfied (see Figure 6.22). Violation of the relationship is allowed only within a narrow strip along the cutting wedge of the form-cutting tool.

6.3.3.2.4 Normal Cutting Wedge Angle

The normal cutting wedge angle β_{ne} is measured within the normal reference plane, P_{ne}. The normal cutting wedge angle is the angle that is formed by the rake plane R_s and by the clearance plane C_s. The value of the angle β_{ne} can be calculated from a simple equation (see Figure 6.22b):

$$\beta_{ne} = 90° - (\alpha_{ne} + \gamma_{ne}) \tag{6.97}$$

6.3.3.2.5 Normal Cutting Angle

The normal cutting angle δ_{ne} is measured in the normal reference plane, P_{ne}. The normal cutting angle is the angle that forms the plane of cut P_{se} and the clearance plane C_s. The value of this angle, δ_{ne}, is equal (see Figure 6.22b) to

$$\delta_{ne} = 90° - \gamma_{ne} \tag{6.98}$$

Definitely, both angles β_{ne} and δ_{ne} can be expressed in terms of unit normal vectors to the corresponding planes of the cutting wedge and to the appropriate reference surfaces.

6.3.3.3 Major Section Plane

The configuration of the major section plane, P_{ve}, is determined by two directions through the point of interest, m, within the cutting edge, C_e. One of the directions is specified by the unit normal vector $\mathbf{n}_{se}$ to the plane of the cut P_{se}, and another direction is specified by the velocity vector $\mathbf{V}_\Sigma$ of the resultant motion of the cutting tool with respect to the work as illustrated in Figure 6.23a. The major section plane P_{ve} is perpendicular to the plane of cut P_{se}.

An expression for the major section plane P_{ve} in terms of the vectors $\mathbf{V}_\Sigma$ and $\mathbf{n}_{se}$ yields representation in vector form:

$$\left[\mathbf{r}_{ve.tp} - \mathbf{r}_{se}^{(m)} \right] \times [\mathbf{n}_{se} \times \mathbf{v}_\Sigma] = 0 \tag{6.99}$$

where $\mathbf{r}_{ve.tp}$ designates the position vector of a point of the major section plane.

6.3.3.3.1 Rake Angle

The rake angle, γ_{ve}, is measured in the major section plane P_{ve} (Figure 6.23b). The rake angle γ_{ve} is equal to the angle between two unit vectors. The unit normal vector $\mathbf{n}_{se}$ to the plane of the cut is one of two vectors. The unit vector $\mathbf{b}$ that is a tangent to the rake surface R_s and is located within the reference plane P_{ve} is the other unit vector. The rake angle γ_{ve} can be calculated from the following expression:

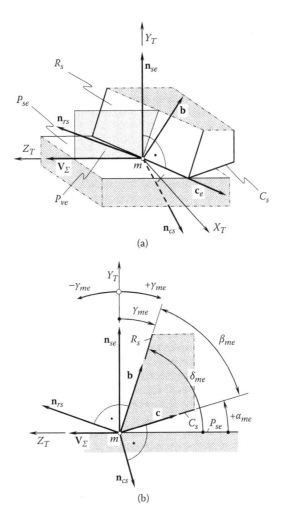

FIGURE 6.23
Geometry of the active part of a cutting tool in the *major section plane, P_{ve}.*

$$\gamma_{ve} = \angle(\mathbf{n}_{se}, \mathbf{b}) = \arctan \frac{|\mathbf{n}_{se} \times \mathbf{b}|}{\mathbf{n}_{se} \cdot \mathbf{b}} \qquad (6.100)$$

The rake angle γ_{ve} is positive when the unit vector $\mathbf{n}_{se}$ does not penetrate the cutting wedge, and it is negative when it does (Figure 6.23b).

6.3.3.3.2 Clearance Angle

The clearance angle, α_{ve}, is the angle that the unit normal vector $\mathbf{n}_{se}$ forms with the unit vector c. Here, the unit vector c is a tangent to the line of intersection of the clearance surface C_s by the major section plane P_{ve} (see Figure 6.23b):

$$\alpha_{ve} = \angle(\mathbf{n}_{se}, \mathbf{c}) = \arctan \frac{|\mathbf{n}_{se} \times \mathbf{c}|}{\mathbf{n}_{se} \cdot \mathbf{c}} \tag{6.101}$$

6.3.3.3.3 Cutting Wedge Angle

The cutting wedge angle, β_{ve}, is the angle between the unit vectors $\mathbf{b}$ and $\mathbf{c}$ (see Figure 6.23b):

$$\beta_{ve} = \angle(\mathbf{b}, \mathbf{c}) = \arctan \frac{|\mathbf{b} \times \mathbf{c}|}{\mathbf{b} \cdot \mathbf{c}} \tag{6.102}$$

In the major section plane P_{ve}, the following equality is always valid:

$$\beta_{ve} = 90° - (\alpha_{ve} + \gamma_{ve}) \tag{6.103}$$

6.3.3.3.4 Angle of Cutting

The angle of cutting, δ_{ve}, is the angle that the unit vector $\mathbf{b}$ forms with the velocity vector $\mathbf{V}_\Sigma$ of the resultant motion of the cutting tool relative the surface of cut, P_{se} (see Figure 6.23b):

$$\delta_{ve} = 180° - \angle(\mathbf{b}, \mathbf{V}_\Sigma) = -\arctan \frac{|\mathbf{b} \times \mathbf{v}_\Sigma|}{\mathbf{b} \cdot \mathbf{v}_\Sigma} \tag{6.104}$$

The following equality is always observed in the major section plane P_{ve}:

$$\delta_{ve} = (90° - \gamma_{ve}) \tag{6.105}$$

6.3.3.4 Correspondence between the Geometric Parameters Measured in Different Reference Planes

When geometric parameters of the active part of a cutting tool are known in the plane of cut P_{se} and in the normal reference plane P_{ne}, the corresponding geometric parameters can be calculated in the major section plane P_{ve}, and vice versa.

Consider the calculation of the rake angle γ_{ve} as an example of the proposed approach.

The origin of the *Cartesian* coordinate system $X_T Y_T Z_T$ is at the point of interest, m (see Figure 6.21) within the cutting edge, C_e. Construct a vector $\mathbf{A}$ that is a tangent to the line of intersection of the rake surface R_s by the normal reference plane P_{ne} (see Figure 6.22). The projection length of the vector $\mathbf{A}$ onto the coordinate plane $X_T Z_T$ is equal to one ($Pr_{zx}\, \mathbf{A} = 1$). This yields the following equation:

$$\mathbf{A} = \begin{bmatrix} \sin \lambda_{se} \\ \cot \gamma_{ne} \\ -\cos \lambda_{se} \end{bmatrix} \qquad (6.106)$$

The unit vector $\mathbf{b}$ is tangent to the line of intersection of the rake surface R_s by the major section plane P_{ve} (see Figure 6.23). For vector $\mathbf{b}$, the following expression:

$$\mathbf{b} = \begin{bmatrix} 0 \\ \cos \gamma_{ve} \\ -\sin \gamma_{ve} \end{bmatrix} \qquad (6.107)$$

is valid.

Ultimately, the unit vector $\mathbf{c}_e$ is tangent to the cutting edge, C_e, at the point of interest m (see Figure 6.21). The following expression is valid for the unit vector $\mathbf{c}_e$:

$$\mathbf{c}_e = \begin{bmatrix} 0 \\ \cos \lambda_{se} \\ -\sin \lambda_{se} \end{bmatrix} \qquad (6.108)$$

By construction, all three vectors $\mathbf{A}$, $\mathbf{b}$, $\mathbf{c}_e$ are located within a certain plane that is tangent to the rake surface R_s at the point of interest m. This means that the vectors $\mathbf{A}$, $\mathbf{b}$, $\mathbf{c}_e$ represent a set of coplanar vectors. Therefore, the triple scalar product of these vectors is identical to zero ($\mathbf{A} \times \mathbf{b} \cdot \mathbf{c}_e \equiv 0$). Consequently, the following equality is valid:

$$[\mathbf{A}, \mathbf{b}, \mathbf{c}_e] = \begin{vmatrix} -\sin \lambda_{se} & \cot \gamma_{ne} & -\cos \lambda_{se} \\ 0 & \cos \gamma_{ve} & -\sin \gamma_{ve} \\ \cos \lambda_{se} & 0 & -\sin \lambda_{se} \end{vmatrix} \equiv 0 \qquad (6.109)$$

After the required formula transformations are performed, one can come up with equations for the calculation of the rake angle γ_{ve}

$$\tan \gamma_{ve} = \frac{\tan \gamma_{ne}}{\cos \lambda_{se}} \qquad (6.110)$$

or in the form

$$\cot \gamma_{ve} = \cot \gamma_{ne} \cos \lambda_{se} \tag{6.111}$$

Following a method similar to that disclosed above, the equation for the calculation of the clearance angle α_{ve} can be derived:

$$\tan \alpha_{ve} = \tan \alpha_{ne} \cos \lambda_{se} \tag{6.112}$$

Equations 6.110 through 6.112 for the calculation of the rake angle γ_{ve} and the clearance angle α_{ve} have been known since the publication by Stabler [157].

The roundness ρ of the cutting edge, C_e, in the major section plane P_{ve} can be calculated from the equation

$$\rho_{ve} = \rho_{ne} \cdot \cos \lambda_{se} \tag{6.113}$$

where ρ_{ne} denotes the roundness of the cutting edge in the normal reference plane P_{ne}. The equation for the roundness ρ_{ve} is another example of the correlation between the geometric parameters of the active part of a cutting tool measured in different reference planes.

6.3.3.5 Main Reference Plane

The *main reference plane*, P_{re}, is orthogonal to the velocity vector $\mathbf{V}_\Sigma$ of resultant motion of the cutting tool with respect to the surface of the cut as schematically illustrated in Figure 6.24. This reference plane can also be determined as a plane through the unit normal vector $\mathbf{n}_{se}$ to the surface of cut P_{se} and through the unit vector $\mathbf{m}_e$, which is orthogonal to the velocity vector $\mathbf{V}_\Sigma$. The unit normal $\mathbf{m}_e$ belongs to the surface of cut P_{se} as illustrated in Figure 6.24a. In the coordinate system $X_T Y_T Z_T$ (see Figure 6.24), the unit normal vector $\mathbf{m}_e$ is identical to the unit vector $\mathbf{i}$ along X_T axis, namely, $\mathbf{m}_e = \mathbf{i}$.

6.3.3.5.1 The Major Cutting Edge Approach Angle

The velocity vector $\mathbf{V}_f$ of the feed rate motion and the projection of the major cutting edge, C_e, of a cutting tool form the major cutting edge approach angle φ_e as is schematically illustrated in Figure 6.25. The angle φ_e (κ_r in other designation) is an acute angle ($0° < \varphi_e \le 90°$).

The major cutting edge approach angle φ_e can be calculated from the equation

$$\varphi_e = \arctan \frac{\left| \mathbf{V}_f^{(\varphi)} \times \mathbf{C}^{(\varphi_e)} \right|}{\mathbf{V}_f^{(\varphi)} \cdot \mathbf{C}^{(\varphi_e)}} \tag{6.114}$$

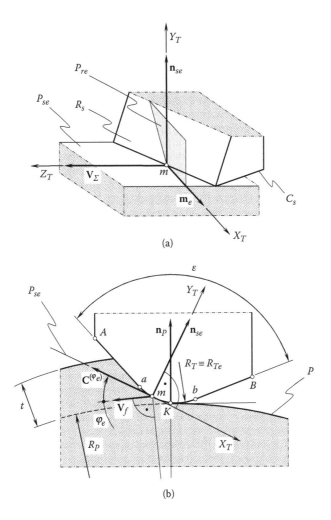

FIGURE 6.24
Geometry of the active part of a cutting tool in the *main reference plane, P_{re}.*

where $C^{(\varphi_e)}$ is a vector along the major cutting edge, C_e. In the case of curvilinear cutting edge, the vector $C^{(\varphi_e)}$ is tangent to the curved cutting edge, C_e, at a point of interest *m*.

6.3.3.5.2 Minor Cutting Edge Approach Angle

Similarly, the minor cutting edge approach angle φ_{1e} can be measured between the projection of the minor cutting edge, C_{1e}, and the velocity vector V_f of the feed rate motion, as schematically depicted in Figure 6.25. The angle φ_{1e} (κ_{r1} in another designation) is also an acute angle ($0° \leq \varphi_{1e} \leq 90°$). Moreover, usually, the value of the angle φ_{1e} does not exceed the value of the corresponding angle φ_e (the inequality $\varphi_{1e} < \varphi_e$ is commonly valid).

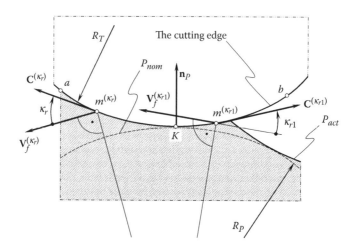

FIGURE 6.25
Cutting edge angle, κ_r (φ_e in another designation), and the minor (end) cutting edge angle, κ_{r1} (φ_{1e} in another designation), for a curved cutting edge in the *tool-in-use* reference system (the *main reference plane*, P_{re}).

For calculation of the minor cutting edge approach angle φ_{1e}, the following formula can be employed:

$$\varphi_{1e} = \arctan \frac{\left| \mathbf{V}_f^{(\varphi_{1e})} \times \mathbf{C}^{(\varphi_{1e})} \right|}{\mathbf{V}_f^{(\varphi_{1e})} \cdot \mathbf{C}^{(\varphi_{1e})}} \tag{6.115}$$

where $\mathbf{C}^{(\varphi_{1e})}$ denotes the vector that aligns with the minor cutting edge, C_{1e}. In the case of a curved cutting edge, the vector $\mathbf{C}^{(\varphi_{1e})}$ is a tangent to the minor cutting edge, C_{1e}, at the point of interest m.

The minor cutting edge approach angle φ_{1e} is calculated for a portion of the cutting edge within the height of the residual cusps. On the rest of the portion of the cutting edge C_{1e}, it does not affect the material removal process.

In the event of small rake angle γ_{ne} and a small angle of inclination λ_{se}, the analysis of actual values of the angles φ_e and φ_{1e} can be performed within the rake plane, R_s, of the cutting tool and not in the main reference plane P_{re}. In this case, instead of actual values of the angles φ_e and φ_{1e}, projections of these angles onto the rake plane, R_s, can be used.

6.3.3.5.3 Tool Tip (Nose) Angle

The tool tip (nose) angle ε_e is determined for the tip of a cutting tool. The angle ε_e can be calculated from the expression

$$\varepsilon_e = 180° - (\varphi_e + \varphi_{1e}) \tag{6.116}$$

The tip of a form-cutting tool coincides with the point of contact K of the generating surface T of the cutting tool and the part surface P being machined (see Figure 6.25). At the contact point K the tool tip angle $\varepsilon_e = 180°$.

The cutting edge approach angle φ_e affects the parameters of the uncut chip, for example, thickness a and width b of the uncut chip. If the depth of cut t and the feed rate V_f (or S) are of constant value, then the following two equations:

$$a = S \cdot \sin \varphi_e \qquad (6.117)$$

$$b = \frac{t}{\sin \varphi_e} \qquad (6.118)$$

are valid, and a lower cutting edge approach angle φ_e, a higher width b of the uncut chip, and larger tool-nose angle ε_e are observed.

Both angles φ_e and φ_{1e} affect parameters of residual cusps on the machined part surface. Larger angles φ_e and/or φ_{1e} result in higher residual cusps on the machined part surface.

6.3.3.6 Chip Flow Reference Plane

In the case of free orthogonal cutting of materials (when the inclination angle $\lambda_{se} = 0°$), the velocity vector of chip motion over the rake surface is orthogonal to the cutting edge. Kinematic geometric parameters of the cutting edge are specified in the plane that is orthogonal to the cutting edge. The correctness of that approach is comprehensively validated experimentally.

Oblique cutting (when the angle of inclination $\lambda_{se} \neq 0°$) is a much more complex phenomenon than orthogonal cutting of materials. This is, first, because deformation of material does not occur in the major reference plane P_m, but within a certain volume, and thus deformation of material occurs in a three-dimensional space. Oblique cutting of materials is much less understood than orthogonal cutting.

However, approximate results of investigation of orthogonal cutting of materials can be adjusted for implementation for the analysis of oblique cutting as well.

For oblique cutting, it is necessary to specify the rake angle taking into consideration the direction of chip flow over the rake face.

Lots of research has been carried out to determine the actual direction of chip flow over the rake surface. The research was summarized by Stabler [157]. Without going into detail, consider Stabler's chip flow law.

6.3.3.6.1 Stabler's Chip Flow Law

It is convenient to specify the direction of chip flow over the rake surface in terms of the chip flow angle η. The chip flow angle η is measured within the rake plane. This is the angle that the velocity vector $\mathbf{V}_{cf}$ of the chip flow forms with the perpendicular to the cutting edge C_e within the rake plane [157].

The chip flow angle η can be expressed in terms of width of cut b_{cf}, width b of the machined plane, and inclination angle λ_{se}:

$$\cos \eta = \frac{b_{cf}}{b} \cos \lambda_{se} \qquad (6.119)$$

Equation 6.119 is derived under the assumption that there is no deformation within the chip width. It has been proven that this assumption is valid for orthogonal cutting [153].

In compliance with the *chip flow law*, the chip flow angle η is approximately equal to the angle of inclination of the cutting edge λ_{se}. This correlation can be analytically expressed by the following approximate equality:

$$\eta \cong \lambda_{se}. \qquad (6.120)$$

This equation is based on the assumption that $b_{cf} = b$, and it is valid for all cutting tools having the inclination angle $\lambda_{se} < 45°$. When the inclination angle $\lambda_{se} \geq 45°$, the difference between the angles λ_{se} and η remains within $(\lambda_{se} - \eta) \leq 5$–$6°$.

Later, Stabler has modified the chip flow law and represented it in the form

$$\eta \cong (1.0 \dots 0.9)\lambda_{se}. \qquad (6.121)$$

Along with the chip flow angle η, the direction of chip flow over the rake surface can be specified by the projection of this angle onto the main reference plane [20].

6.3.3.6.2 Chip Flow Rake Angle

In an attempt to specify (approximately) the poorly understood oblique cutting in terms of the comprehensively investigated orthogonal cutting of materials, the term *chip flow reference plane* is introduced. The chip flow rake angle, γ_{cf}, is measured in the *chip flow reference plane*, P_{cf}. The rake angle γ_{cf} (as well as the depth of cut, t_{cf}) in the chip flow reference plane differs from the analogue parameters measured in other reference planes.

The *chip flow reference plane* P_{cf} is the plane through the velocity vectors $\mathbf{V}_{\Sigma}$ and $\mathbf{V}_{cf}$. Here $\mathbf{V}_{\Sigma}$ designates the velocity vector of the resultant motion of the cutting edge in relation to the surface of the cut and $\mathbf{V}_{cf}$ designates the velocity vector of chip flow over the rake surface. The velocity vector $\mathbf{V}_{cf}$ is located within the rake plane, R_s. It forms the chip flow angle η with the perpendicular to the cutting edge as shown in Figure 6.26.

The velocity vector $\mathbf{V}_{cf}$ is orthogonal to the unit normal vector $\mathbf{n}_{rs}$ to the rake surface R_s. Therefore, the equality $\mathbf{V}_{cf} \cdot \mathbf{n}_{rs} = 0$ is valid.

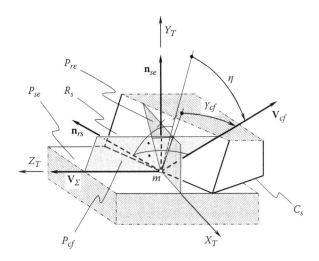

FIGURE 6.26
Geometry of the active part of a cutting tool in the *chip flow reference plane, P_{cf}*.

At the point of interest, m, the chip flow reference plane P_{cf} is a plane through the velocity vectors $\mathbf{V}_\Sigma$ and $\mathbf{V}_{cf}$ at the point m. This yields

$$[\mathbf{r}_{cf} - \mathbf{r}^{(m)}] \cdot \mathbf{V}_\Sigma \times \mathbf{V}_{cf} = 0 \tag{6.122}$$

where $\mathbf{r}_{cf}$ designates the position vector of a point of the chip flow reference plane P_{cf} and $\mathbf{r}^{(m)}$ designates the position vector of the point of interest m.

The chip flow rake angle γ_{cf} is the angle that the velocity vector $\mathbf{V}_{cf}$ of chip flow over the rake plane, R_s, forms with the main reference plane P_{re} (Figure 6.22). Therefore, the rake angle γ_{cf} can be calculated from the equation

$$\gamma_{cf} = \angle(\mathbf{V}_\Sigma, \mathbf{V}_{cf}) - 90° = \arctan\frac{|\mathbf{V}_\Sigma \times \mathbf{V}_{cf}|}{\mathbf{V}_\Sigma \cdot \mathbf{V}_{cf}} - 90° = \arctan\frac{\mathbf{V}_\Sigma \cdot \mathbf{V}_{cf}}{|\mathbf{V}_\Sigma \times \mathbf{V}_{cf}|} \tag{6.123}$$

For the unit vector $\mathbf{v}_\Sigma = \mathbf{V}_\Sigma / |\mathbf{V}_\Sigma|$, one can derive that

$$\mathbf{v}_\Sigma = \begin{bmatrix} -\sin\lambda_{se} \\ \cos\lambda_{se} \\ 0 \end{bmatrix} \tag{6.124}$$

The projection length of the velocity vector $\mathbf{v}_{cf}$ onto the coordinate plane YZ is a unit vector. Therefore, the following equality is valid for the vector $\mathbf{v}_{cf}$

$$\mathbf{v}_{cf} = \begin{bmatrix} \tan \eta \\ -\sin \gamma_{ne} \\ \cos \gamma_{ne} \end{bmatrix} \tag{6.125}$$

Substituting Equations 6.124 and 6.125 into Equation 6.123, one can come up with the expression for the calculation of the chip flow rake angle γ_{cf}:

$$\sin \gamma_{cf} = \sin \eta \sin \lambda_{se} + \cos \eta \cos \lambda_{se} \sin \gamma_{ne} \tag{6.126}$$

Usually, the angle of inclination does not exceed $\lambda_{se} \le 45°$. Under such a scenario, the following approximate equality $\eta \cong \lambda_{se}$ is valid. This immediately leads to the Stabler equation for the calculation of the rake angle γ_{cf} [157]:

$$\sin \gamma_{cf} \cong \sin^2 \lambda_{se} + \cos^2 \lambda_{se} \sin \gamma_{ne} \cong 1 - \cos^2 \lambda_{se} (1 - \sin \gamma_{ne}) \tag{6.127}$$

Taking into account that

$$\tan \gamma_{ne} = \tan \gamma_{ve} \cos \lambda_{se} \tag{6.128}$$

and

$$\sin \gamma_n = \frac{\tan \gamma_n}{\sqrt{1 + \tan^2 \gamma_n}} = \frac{\tan \gamma_{ve} \cos \lambda_{se}}{\sqrt{1 + \tan^2 \gamma_{ve} \cos^2 \lambda_{se}}} \tag{6.129}$$

Equation 6.126 casts into

$$\sin \gamma_{cf} = \sin \eta \sin \lambda_{se} - \cos \eta \cos^2 \lambda_{se} \frac{\tan \gamma_{ve} \cos \lambda_{se}}{\sqrt{1 + \tan^2 \gamma_{ve} \cos^2 \lambda_{se}}} \tag{6.130}$$

The similar equation

$$\sin \gamma_{cf} \cong 1 - \cos^2 \lambda_{se} \left(1 - \frac{\tan \gamma_{ve} \cos \lambda_{se}}{\sqrt{1 + \tan^2 \gamma_{ve} \cos^2 \lambda_{se}}} \right) \tag{6.131}$$

can be derived from Equation 6.127.

The derived equations are valid for free oblique cutting of materials. They enable calculation of the rake angle γ_{cf} in the chip flow reference plane for any actual value of the inclination angle λ_{se}.

Calculation of the chip flow angle η is a challenging problem. A reliable value of the chip flow angle η can be obtained experimentally.

The vector method that is disclosed in this book, enables one derivation of all necessary equations for the calculation of geometry of the active part of a cutting tool in any reference plane. The method works in the *tool-in-hand* as well as in the *tool-in-use* reference systems. The method can also be used for the derivation of equations of some of the parameters that describe the material removal process (that is, the angle of cutting,* the shear angle φ, the angle of action ψ, depth of cutting, and so forth). All of the parameters can be determined in any reference plane of interest.

6.3.4 Descriptive Geometry–Based Method for Determination of the Chip Flow Rake Angle

In addition to analytical methods, the descriptive geometry–based method (DGB method, for simplicity) can also be implemented for determination of the chip flow rake angle.

To come up with a DGB solution to the problem of determination of the *chip flow rake angle*, γ_{cf}, the cutting wedge of a form-cutting tool is depicted in the set of three planes of projections HVF as shown in Figure 6.27. Here and below, the subscript "1" is assigned to projections of all elements (points, lines, planes) onto the horizontal plane of projections H. The subscript "2" is assigned to projections of those same elements onto the vertical plane of projections V. Ultimately, the subscript "3" is assigned to projections of those same elements onto the frontal plane of projections F.

When cutting, the cutting wedge 1 travels along the H/V axis. The velocity vector $\mathbf{V}_\Sigma$ of this motion is projected with no distortion onto both planes of projections, namely, on the planes of projections H and V. Due to this, no subscripts are assigned to the projections of the velocity vector $\mathbf{V}_\Sigma$.

The angle of inclination of the cutting edge ab is designated conventionally as λ_{se}. The reference plane P_{se} within which the inclination angle λ_{se} can be measured is parallel to the horizontal plane of projections H. The inclination angle λ_{se} is measured between the unit normal vector $\mathbf{n}_c$ to the cutting edge, C_e, and the velocity vector $\mathbf{V}_\Sigma$ of the cutting wedge resultant motion.

The unit normal vector $\mathbf{n}_{se}$ to the surface of cut, P_{se}, is erected at the point of interest, m, that is located within the cutting edge, C_e.

The two auxiliary planes of projections, for example, W and U, are constructed for determining the velocity vector $\mathbf{V}_{cf}$ of chip flow over the rake plane, R_s. Subscripts "4" and "5" are assigned to projections of all elements onto the planes of projections W and U, respectively. The axis of projections

---

* The effective angle of cutting δ_{cf} is the angle that can be measured between the opposite direction of the velocity vector $\mathbf{V}_\Sigma$ and the velocity vector $\mathbf{V}_{cf}$ of chip flow over the rake surface. For the case of free oblique cutting of materials, the angle of cutting δ_{cf} depends upon the angle of inclination λ_{se} and upon the *chip flow angle, η,* say, $\delta_{cf} = \arccos [\cos \lambda_{se} \cos \eta \cos \delta_{ne} + \sin \lambda_{se} \sin \eta]$. The last equation reveals that the angle of cutting δ_{cf} goes down as the inclination angle λ_{se} and the chip flow angle η become smaller. This is in perfect correlation with the results of experimental investigations of the metal-cutting process.

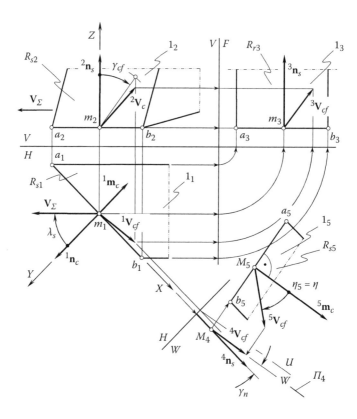

FIGURE 6.27
A descriptive geometry–based method for the determination of the actual value of the rake angle, γ_{cf}, in the *chip flow reference plane*, P_{cf}.

HW is at right angle to the projection $a_1 b_1$ of the cutting edge, C_e, onto the horizontal plane of projections *H*.

Onto the auxiliary plane of projections *W*, the cutting edge, C_e, is projected into the point $A_4 \equiv B_4 \equiv m_4$. The rake plane is projected onto the plane of projections *W* into the trace R_{s4}. Finally, the normal rake angle γ_{ne} is projected onto the plane *W* without distortions. The projection of the rake angle γ_{ne} onto the plane of projections *W* is equal to the angle γ_{ne}.

The axis of projections *W/U* is parallel to the trace R_{s4} of the rake plane. Therefore, the rake plane R_s is projected onto the auxiliary plane of projections *U* without distortions. Because of this, the projection of the *chip flow angle*, η, onto the plane of projections *U* is equal to the angle η, and the velocity vector $\mathbf{V}_{cf}$ of chip flow is at the angle η to the unit normal vector $\mathbf{m}_c$ to the cutting wedge.

Following conventional rules from descriptive geometry, one can construct projections of the velocity vector $\mathbf{V}_{cf}$ of the chip flow onto all the other planes of projections. Ultimately, the projection $\mathbf{V}_{cf}^1$ is turned about the axis that is parallel to the axis of projections V/H. In the orientation of the projection $\mathbf{V}_{cf}^1$ when it is parallel to the vertical plane of projections, the nondistorted value of the *chip flow rake angle*, γ_{cf}, is obtained in the plane of projections V (Figure 6.27).

6.4 On the Capabilities of the Analysis of Geometry of the Active Part of Cutting Tools

The implementation of the vector method for the calculation of geometry of the active part of a cutting tool is very helpful to cutting tool design engineers. To demonstrate the creative capabilities of the method, consider a few practical examples.

6.4.1 Elements of Geometry of the Active Part of a Skiving Hob

The geometry of the active part of skiving hobs for finishing or semifinishing hardened gears is significantly different compared to the geometry of hobs of conventional design. Only lateral cutting edges of the skiving hob teeth remove the stock—the top cutting edges do not cut the work material [69,93,113,114,132]. This is an important feature of skiving hobs. The negative rake angle γ_t at the top edge of the skiving hob varies within the interval $\gamma_t = (-30°--60°)$. The large negative rake angle γ_t is another important feature of the design of skiving hobs.

Because of the large negative rake angle γ_t at the top edge of the skiving hob, the inclination angle λ_{se} of the lateral cutting edges varies greatly. The estimation of the variation of the inclination angle λ_{se} can be derived from Figure 6.28a, which reveals that the inequality $\gamma_{t.a} < \gamma_{t.b} < \gamma_{t.c}$ is observed. The difference between actual values of the inclination angle λ_{se} that are measured at the outside diameter, $d_{o.T}$, of the hob and at the limit diameter, $d_{l.T}$, of the hob is in the range up to $\Delta\lambda_{se} \cong 30°$. It is known that a mere deviation of the inclination angle of a cutting tool of $\Delta\lambda_{se} \cong 5°$ from its optimal value can cause a double reduction of the cutting tool life. This means that design of a skiving hob can be significantly improved if

- The inclination angle of the hob cutting edges has a constant value within the cutting edge C_e.
- The angle of inclination, λ_{se}, would be of optimal value at every point of the cutting edge of the hob.

For this purpose, it is recommended that the rake face be shaped with the help of a convex segment of logarithmic spiral curve having the pole at the

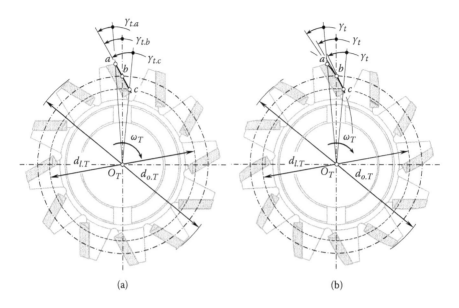

FIGURE 6.28
Elements of geometry of the active part of the skiving hob (US Patent No. 3.786.719, Azumi), having $\gamma_{t.a} < \gamma_{t.b} < \gamma_{t.c}$ (a) and the skiving hob (USSR Patent No. 1.114.543, S.P. Radzevich) having $\gamma_t = Const$ (b).

hob axis of rotation O_T as shown in Figure 6.28b. (For hobs having positive rake angle γ_t, a concave segment of logarithmic spiral curve is used.)

The equation of the generating curve of the hob rake surface can be derived on the premise of a differential equation for isogonal trajectories (see Equation 6.18):

$$\left(\frac{\partial \varphi}{\partial x}\cos \varsigma - \frac{\partial \varphi}{\partial y}\sin \varsigma\right)dx + \left(\frac{\partial \varphi}{\partial x}\sin \varsigma + \frac{\partial \varphi}{\partial y}\cos \varsigma\right)dy = 0 \qquad (6.132)$$

In the case under consideration, the solution to Equation 6.132 returns the equation of the logarithmic spiral curve (see Equation 6.12):

$$\rho = \rho_0\, e^{\varphi \tan \lambda_{opt}} \qquad (6.133)$$

Actually, Equation 6.133 is a reasonably good approximation of the requirement $\lambda_{opt} = Const$. This is because the actual value of the inclination angle λ_{se} is measured not in the transverse cross section of the hob cutting edge, but within the surface of the cut. However, it has been proven that the derived solution (see Equation 6.133) is practical.

6.4.2 Elements of Geometry of the Active Part of a Cutting Tool for Machining Modified Gear Teeth

A change in the shape of the rake surface of hob teeth can enhance the capabilities of the cutting tool. It would be possible to use the hob having a modified rake surface of the teeth for machining of gears having a modified tooth profile. Modification of the gear tooth profile is recognized to be a powerful tool for improving the performance of gear drives.

For this purpose, the rake surface of the hob teeth is composed of two portions R_{s1} and R_{s2} as shown in Figure 6.29 [64,114]. These two portions of the rake surface are at a certain angle φ to each other. To ensure the necessary value of the gear tooth profile modification, it is necessary to accurately calculate the actual value of the angle φ.

Consider a *Cartesian* coordinate system XYZ associated with the hob tooth. Origin O of the coordinate system XYZ is located within the line of intersection of the two portions of the hob rake surfaces. The axis Y aligns to the line of intersection of the two portions, R_{s1} and R_{s2}, of the rake surface. The axis Z is parallel to the velocity vector $\mathbf{V}_\Sigma$, and ultimately, the axis X complements the axes Y and Z to the right-hand–oriented *Cartesian* coordinate system XYZ.

Consider a plane through the origin O. This plane is a tangent to the clearance surface C_s of the hob teeth. Then, construct three vectors $\mathbf{A}$, $\mathbf{B}$, and $\mathbf{c}$. These three vectors are constructed so as to be within the common tangent plane to the clearance surface C_s of the hob teeth.

The magnitudes of the vectors $\mathbf{A}$ and $\mathbf{B}$ are prespecified to get projections of these vectors onto the coordinate plane XZ equal to the one. Vector $\mathbf{c}$ is a unit vector. In compliance with Figure 6.29, the following expressions for vectors $\mathbf{A}$, $\mathbf{B}$, and $\mathbf{c}$ can be composed:

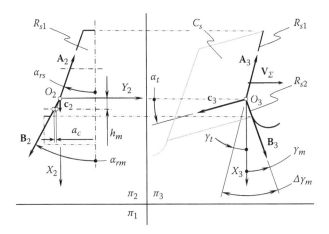

FIGURE 6.29
Elements of geometry of the active part of a gear-cutting tool having a modified rake surface (USSR Patent No. 1.017.444, S.P. Radzevich).

$$\mathbf{A} = \begin{bmatrix} -\cos\gamma_t \\ \tan\alpha_{rs} \\ \sin\gamma_t \end{bmatrix} \tag{6.134}$$

$$\mathbf{B} = \begin{bmatrix} \cos\gamma_m \\ -\tan\alpha_{rm} \\ \sin\alpha_t \end{bmatrix} \tag{6.135}$$

$$\mathbf{c} = \begin{bmatrix} \sin\alpha_t \\ 0 \\ -\cos\alpha_t \end{bmatrix} \tag{6.136}$$

where, in Equations 6.134 through 6.136, γ_t is the rake angle at the top of the hob tooth, γ_m is the rake angle at the modified portion of the hob tooth, α_t is the clearance angle of the hob tooth, α_{rs} is the nominal pressure angle of the hob tooth, and α_{rm} is the pressure angle of the modified portion of the hob tooth.

By construction, three vectors $\mathbf{A}$, $\mathbf{B}$, and $\mathbf{c}$ make up a triple of coplanar vectors. Due to this, the triple scalar product of the vectors is identical to zero: $\mathbf{A} \times \mathbf{B} \cdot \mathbf{c} \equiv 0$. The last identity makes possible the equation:

$$[\mathbf{A}, \mathbf{B}, \mathbf{c}] = \begin{bmatrix} -\cos\gamma_t & \tan\alpha_{rs} & \sin\gamma_t \\ \cos\gamma_m & -\tan\alpha_{rm} & \sin\alpha_t \\ \sin\alpha_t & 0 & -\cos\alpha_t \end{bmatrix} \equiv 0 \tag{6.137}$$

Equation 6.137 casts into the formula

$$\varphi = \gamma_t + \gamma_m = \gamma_t + \tan^{-1}\left[\frac{(1-\tan\gamma_t\tan\alpha_{rm}) - \tan\alpha_{rm}}{\tan\alpha_{rs}} \cdot \frac{\cos\gamma_t}{\sin\alpha_t} \right] \tag{6.138}$$

for calculation of the desired angle φ between the portions R_{s1} and R_{s2} of the rake surface of the hob for cutting modified gear teeth.

In a particular case, when the rake angle γ_t is equal to zero (for example, when the equality $\gamma_t = 0°$ is observed), then Equation 6.138 reduces to

$$\tan\varphi = \frac{\tan\alpha_{rs} - \tan\alpha_{rm}}{\tan\alpha_{rs}\sin\alpha_t} \tag{6.139}$$

Even a brief look through the capabilities of analysis of the cutting tool geometry reveals how much room for improvement can be found there.

6.4.3 Elements of Geometry of the Active Part of a Precision Involute Hob

Consider a precision involute hob having straight lateral cutting edges [60]. The concept of the design of the hob is based on the following considerations:

- Lateral cutting edges of one side of an involute hob tooth belong to the corresponding screw involute surface.
- Lateral cutting edges of the opposite side of the involute hob tooth belong to the opposite screw involute surface.
- The screw involute surfaces of the opposite sides of the involute hob tooth intersect each other, and a helix is the line of intersection.
- The two characteristic lines E_l and E_r through a point of the helix intersect one another at that point, and, thus, they specify a plane.
- The aforementioned plane is used as a rake face of the involute hob teeth.

The steps listed above determine the configuration of the rake surface of the precision involute hob teeth.

6.4.3.1 Auxiliary Parameter R

Lateral tooth surfaces of the auxiliary rack $\mathcal{R}$ of the hob intersect each other along a straight line through the point A as shown in Figure 6.30. This straight line is at a distance R from the hob axis of rotation. For the distance R, an analysis of Figure 6.31 yields the following formula:

$$R = 0.5 \cdot (d_h + t_c \cdot \cot \phi_n) \tag{6.140}$$

6.4.3.2 Angle ϕ_r between the Lateral Cutting Edges of the Hob Tooth

Prior to deriving the equation for the calculation of the angle ϕ_r that makes the lateral cutting edges of the gear hob tooth, it is convenient to derive an equation for the calculation of the projection ν of the angle ϕ_r onto the coordinate plane $X_h Y_h$.

The projections of the lateral cutting edges of the involute hob tooth onto the coordinate plane $X_h Y_h$ form an angle ν. For calculation of the actual value of the angle ν, the following expression can be used:

$$\tan \nu = \frac{d_{b.h}}{\sqrt{4 \cdot R^2 - d_{b.h}^2}} \tag{6.141}$$

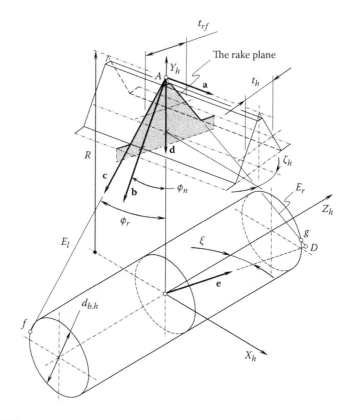

FIGURE 6.30
Elements of geometry of the active part of a precision involute hob (USSR Patent No. 990.445,
S.P. Radzevich): configuration of the rake plane, b.

Then, consider three unit vectors **a**, **b**, and **c** (unit vectors are denoted by
lowercase boldface characters). These vectors yield the following analytical
representations:

$$
\mathbf{a} = \begin{bmatrix} \cos \zeta_h \\ 0 \\ \sin \zeta_h \end{bmatrix} \tag{6.142}
$$

$$
\mathbf{b} = \begin{bmatrix} \sin \phi_n \sin \zeta_h \\ -\cos \phi_n \\ -\sin \phi_n \cos \zeta_h \end{bmatrix} \tag{6.143}
$$

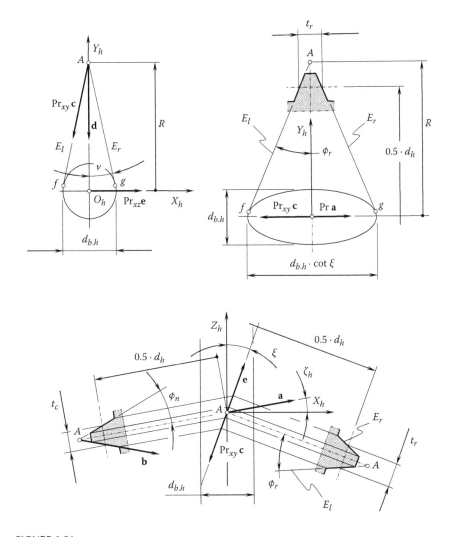

FIGURE 6.31
Elements of geometry of the active part of a precision involute hob (USSR Patent No. 990.445, S.P. Radzevich): the employed characteristic vectors.

$$
\mathbf{c} = \begin{bmatrix} -\cos \phi_r \tan v \\ -\cos \phi_r \\ -\sin \phi_r \cos \xi \end{bmatrix} \tag{6.144}
$$

where ζ_h designates the hob-setting angle of the involute hob [107,113].

The hob-setting angle ζ_h required for further calculations can be chosen by the designer of the gear hob of the design under consideration. Usually, it is recommended that the actual value of the hob-setting angle ζ_h be equal to the pitch helix angle ψ_h of the hob being designed. As proven in our earlier work [113] to satisfy the equality $\zeta_h = \psi_h$ (this condition is the most favorable), the actual value of the hob-setting angle must be calculated from the equation

$$\tan \zeta_h = \frac{m \cdot N_h}{\sqrt{(d_{o.h} - 2 \cdot 1.25 \cdot m - \Delta d_{o.h})^2 - m^2 N_h^2}} \tag{6.145}$$

where $\Delta d_{o.h}$ designates the reduction of the hob outside diameter $d_{o.h}$ due to resharpening of the worn gear hob.

Three vectors $\mathbf{a}$, $\mathbf{b}$, and $\mathbf{c}$ are located within the common lateral surface of the auxiliary rack $\mathcal{R}$ of the hob; therefore, the following identity $\mathbf{a} \times \mathbf{b} \cdot \mathbf{c} \equiv 0$ is observed. The last expression makes possible a determinant:

$$\begin{vmatrix} \cos \zeta_h & 0 & \sin \zeta_h \\ \sin \phi_n \sin \zeta_h & -\cos \phi_n & -\sin \phi_n \cos \zeta_h \\ -\cos \phi_r \tan v & -\cos \phi_r & -\sin \phi_r \cos \xi \end{vmatrix} = 0 \tag{6.146}$$

After expanding the determinant (see Equation 6.146) and after the necessary formulae transformations are performed, one can come up with the equation of two unknowns, namely, ϕ_r and ξ.

6.4.3.3 Angle ξ of Intersection of the Rake Surface and of the Hob Axis of Rotation

The rake surface of the involute hob is inclined to the hob axis at a certain angle ξ. To determine the necessary value of the angle ξ, the following three unit vectors $\mathbf{c}$, $\mathbf{d}$, and $\mathbf{e}$ can be used. For the vectors $\mathbf{d}$ and $\mathbf{e}$, Figure 6.31 yields

$$\mathbf{d} = \begin{bmatrix} 0 \\ -1 \\ 0 \end{bmatrix} \tag{6.147}$$

$$\mathbf{e} = \begin{bmatrix} \sin \xi \\ 0 \\ \cos \xi \end{bmatrix} \tag{6.148}$$

As the three vectors $\mathbf{c}$, $\mathbf{d}$, and $\mathbf{e}$ are located within the rake surface of the hob tooth, the identity $\mathbf{c} \times \mathbf{d} \cdot \mathbf{e} \equiv 0$ is observed. This makes possible a determinant:

$$\begin{vmatrix} -\cos\phi_r \tan\nu & -\cos\phi_r & -\sin\phi_r \cos\xi \\ 0 & -1 & 0 \\ \sin\xi & 0 & \cos\xi \end{vmatrix} = 0 \qquad (6.149)$$

After expanding the determinant (see Equation 6.149) and after the necessary formulae transformations are performed, one can come up with one more equation of two unknowns, namely, ϕ_r and ξ.

Further, consider the set of two equations, Equations 6.146 and 6.149, of the two unknowns ϕ_r and ξ. The solution to the set of these equations can be represented in the form

$$\tan\xi = \frac{\cos\zeta_h \cdot \tan\nu}{\tan\phi_n + \sin\zeta_h \cdot \tan\nu} \qquad (6.150)$$

$$\tan\phi_r = \frac{\tan\nu}{\sin\xi} \qquad (6.151)$$

The inclination of the gear hob axis of rotation O_h with respect to the auxiliary rack $\mathcal{R}$ is specified by the hob-setting angle ζ_h. It is necessary to point out here that the angle ζ_h is a parameter of the gear hob design and is not a parameter of the gear-hobbing operation. This angle can be positive ($+\zeta h > 0°$), negative ($-\zeta_h < 0°$), or zero ($\zeta_h = 0°$). Under special conditions, the hob-setting angle can be equal to the gear hob pitch helix angle $\psi_\mathcal{R}$ (in this particular case, the equality $\zeta_h = \psi_\mathcal{R}$ is observed).

The above examples reveal the capabilities of the vector method for the calculation of the geometry of the active part of a cutting tool.

7

Conditions of Proper Part
Surface Generation

When machining a part surface, the cutting tool is traveling relative to the work. While traveling, the cutting tool removes the operating stock from the work. The machined part surface can be considered as an enveloping surface to consecutive positions of the cutting tool relative to the work.

Due to peculiarities of the shape of the part surface to be machined, the generating surface of the form-cutting tool, and kinematics of the machining operation, the shape of the machined part surface can deviate from its desired geometry. When a portion of stock on the part surface remains uncut, an *overcut* is observed. When the cutting tool removes material beneath the part surface, the *undercut* is observed. Both the undercut and the overcut are allowed if and only if the resultant deviations of the machined part surface from desired geometry of the part surface are within the tolerance for accuracy of the part surface.

It is important to investigate the root causes for why the actual machined part surface P_{ac} deviates from the desired nominal part surface P_{nom}. This problem becomes more severe for computer-aided manufacturing (*CAD/CAM*) applications when the entire machining process must be completely formalized.

To ensure precise machining of the given part surface, it is necessary to properly orient the work on the worktable of a multiaxis numerical control (*NC*) machine, and to satisfy a set of conditions of proper part surface generation.

For the analysis and for the purposes of analytical description of conditions of proper part surface generation, it is presumed that

- The part surface P is described analytically (see Chapter 1).
- The kinematics of the machining operation is specified (see Chapter 2).
- The generating surface T of the form-cutting tool is determined (see Chapter 5).

Once the three above listed items are known, the set of conditions of proper part surface generation can be represented in an analytical form.

7.1 Optimal Workpiece Orientation on the Worktable of a Multiaxis Numerical Control (*NC*) Machine

There are many feasible configurations of a given sculptured part surface on the worktable of a multiaxis *NC* machine. Machining of the sculptured part surface can be executed for every feasible configuration. It is natural to assume that not all feasible configurations of the part surface are equivalent, and that a particular orientation of the sculptured part surface is preferred for machining purposes—that is, this configuration is the most favorable (is optimal in a certain sense).

Consider the general case of machining of a sculptured part surface on a multiaxis *NC* machine. The most favorable (that is, optimal) workpiece orientation is generally defined as orientation of the workpiece to minimize the number of setups in multiaxis *NC* machining of a given sculptured part surface P, or to allow the maximal number of surfaces to be machined in a single setup. Here, a method for calculation of such an optimal workpiece orientation is developed based on the geometry of the sculptured part surface P to be machined, the geometry of the generating surface T of the form-cutting tool, and the articulation capabilities of the multiaxis *NC* machine.

In addition, for cases in which some freedom of orientation remains after conditions for machining in a single setup are fulfilled, a second sort of optimality can also be considered: finding an orientation such that the cutting condition remains as close as possible to the conditions of machining at a surface point at which they are the most favorable. This second form of optimality is obtained by choosing an orientation (within the bounds of those allowing a single setup) in which the angle between the neutral axis of the milling tool and the area-weighted mean normal to the part surface is zero, or as small as possible.

To find this solution, use of mapping of surfaces on a unit sphere sounds promising. Mapping of a surface on a unit sphere was initially proposed by C.-F. Gauss [21]. He used this type of surface mapping for the purpose of investigating the surface topology. Since the Gauss publication [21], the mapping of a surface on a unit sphere is usually referred to as Gauss mapping of the surface. Later, Gauss' idea of surface mapping received wide implementation in both science [4] and engineering.

As early as in 1987, Radzevich* [82,103] applied Gauss' idea to the sculptured part surface orientation problem. He developed the general approach to solve this important engineering problem [82,116,136,141]. The proposed method of finding the optimal workpiece orientation maintains the minimal difference of the angle between the normal and sculptured part surface P at its central point and a milling tool axis of rotation at its optimal position

---

* Patent No. 1442371 (USSR), *A Method of Optimal Workpiece Orientation on the Worktable of a Multi-Axis NC Machine*, S. P. Radzevich. Int. Cl. B23q15/007, filed February 17, 1987.

to minimize the number of setups in rough machining. By means of Gauss mapping of surfaces, the problem of optimal workpiece orientation can be formulated as a geometric problem on a unit sphere. The method incorporates impact of the area-weighted mean normal to the surface P. This technique is applicable for form-cutting tools of any design.

7.1.1 Analysis of a Given Workpiece Orientation

Rational workpiece orientation is an important task in optimal sculptured part surface machining on a multiaxis NC machine.

When machining a sculptured part surface, conditions of cutting strongly depend upon

- The location of the point on the part surface P in the vicinity of which machining is occurring
- The relative orientation of the part surface P and the form-cutting tool while machining

If the actual workpiece orientation is far from being optimal, this leads to decreased cutting tool performance or to a situation in which machining of the part surface P is not feasible in one setup.

As an example, consider the configuration of a sculptured part surface P on the worktable of a three-axis NC machine as schematically illustrated in Figure 7.1. The *Cartesian* coordinate system $X_P Y_P Z_P$ is associated with the work. It is assumed that the Z_P-axis is parallel to the cutting tool axis of rotation, O_T. The shadowed planar region $A_{xy} B_{xy} CB^*A^*$ appears in Figure 7.1. Thus, for the given configuration, the portion $ABCB^*A^*$ of the sculptured part surface P cannot be machined in that setup.

Consider an arbitrary curve B^*IBJHS that is entirely located within the sculptured part surface P (Figure 7.1). Points I, B, I,B,J,H, I,B,J,H, and S belong to the curve B^*IBJHS. At every contact point, K, unit normal vector $\mathbf{n}_P$ to the part surface P, and unit normal vector $\mathbf{n}_T$ to the generating surface T of the form-cutting tool are pointed opposite from each other.

When the generating surface T of the cutting tool makes contact with the sculptured part surface P at the points B, I,B,J,H, I,B,J,H, and S, the designation of these points changes to K_B, K_J, K_H, and K_S respectively. Evidently, the angle that the axis of rotation, O_T, of the cutting tool forms with the unit normal vector $\mathbf{n}_P$ through the points K_B, K_J, K_H, and K_S is of different values. Without going into details of the mechanics of metal cutting, one can conclude that the conditions of the material removal at the points K_B, K_J, K_H, and K_S are different. This is mostly due to three reasons:

- First, because the angles that the tool axis of rotation, O_T, forms with the unit normal vector $\mathbf{n}_P$ at each of these contact points are of different values.

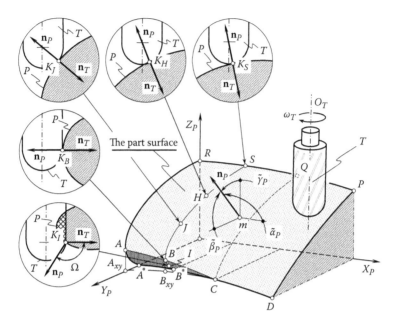

FIGURE 7.1
An arbitrary orientation of a sculptured part surface P on the worktable of a multiaxis NC machine. (From Radzevich, S.P., Goodman, E.D., *ASME J. Mech. Des.*, 124, 201–212, 2002. With permission.)

- Second, because the normal radii of curvature of the part surface P, and of the generating surface T of the form-cutting tool at the contact points K_B, K_J, K_H, and K_S are different.
- Third, local relative orientation of the part surface P, and of the generating surface T of the form-cutting tool at the contact points KB, KJ, KH, and KS is different.

Moreover, near points I and B^*, interference of the surfaces P and T occurs. Machining of the part surface portion near the curve IBB^* is not feasible in such an orientation of the part surface P.

Sculptured part surface P can be represented to be an infinite number of small (infinitesimal), flat patches. In this scenario, the criteria of *optimization* presented here are as follows:

(a) To maximize the number of flat patches that can be accessed in a single setup

(b) To minimize the mean difference in the condition of machining over all of these small flat patches

The second requirement yields interpretation in terms of the difference in angle between part surface unit normal $\mathbf{n}_P$ and the cutting tool axis of rotation in its optimal configuration.

The number of infinitesimal flat patches that approximate the part surface P is infinite. It is not possible to simultaneously minimize the angles between the cutting tool axis of rotation and the unit normal vectors to each of the infinitesimal flat patches. To summarize the situation, the infinite number of unit normal vectors to the part surface P is not considered below, but the so-called *weighted normal* to the sculptured part surface P is considered instead.

For cutting condition optimization and as a useful datum for the setup minimization evaluation, the orientation of the sculptured part surface sought is that which minimizes not the maximal angle between the cutting tool axis of rotation and the normal to the part surface P, but the angle between the *surface area–weighted* normal to the part surface P and the cutting tool axis of rotation. The results of the calculation following this approach are reasonably close to the results of the calculation performed in compliance with another procedure. In the last case, the angle between the surface area–weighted normal and the average normal to the part surface P is minimized. The use of both approaches reduces the mean difference in the condition of machining of each small portion of the given sculptured part surfaces P and thereby increases tool life.

For the evaluation of the degree of deviation of actual sculptured part surface orientation from its desired orientation, a measure of the deviation is required.

7.1.2 *Gauss* Maps of a Sculptured Part Surface *P* and of the Generating Surface *T* of the Form-Cutting Tool

C.-F. Gauss introduced the notion of mapping of surface normals onto the surface of a unit sphere by means of parallel normals, in which a point on a map is the result of the intersection of the surface normal vector, translated so as to emanate from the center of a unit sphere, with the surface of the unit sphere [21]. The point on the surface of the unit sphere so produced has come to be known as the *Gauss* map of the point on the given surface P. The map of all points on a specified sculptured part surface P is called its *Gauss* map, $Map_G(P)$. The boundary of the map $Map_G(P)$, if it exists, of a given part region or surface P is referred to as its *Gauss spherical indicatrix*, $GInd\ (P)$.

The measure proposed for the degree of deviation of the actual conditions of machining from optimal conditions of machining of a sculptured part surface patch is the mean angle between the neutral configuration of the axis of rotation of the cutting tool and the weighted normal to the sculptured part surface P.

For implementation of *Gauss* mapping of a part surface for the calculation of the parameters of optimal workpiece orientation, it is necessary to derive the corresponding equations of the *Gauss* map, $Map_G(P)$, of the sculptured part surface P and of the *Gauss* map, $Map_G(T)$, of the generating surface T of the form-cutting tool as well.

Let the position vector of a point on the $Map_G(P)$ of the sculptured part surface P be designated as $\mathbf{r}_{P0}$. A *Cartesian* coordinate system $X_{P0}Y_{P0}Z_{P0}$ is

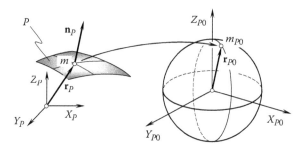

FIGURE 7.2
Spherical map, m_{P0}, of a point m on the sculptured part surface P. (From Radzevich, S.P., Goodman, E.D., *ASME J. Mech. Des.*, 124, 201–212, 2002. With permission.)

associated with a unit sphere as schematically illustrated in Figure 7.2. In the coordinate system $X_{P0}Y_{P0}Z_{P0}$, the equality $\mathbf{r}_{P0} = \mathbf{n}_P$ for the *Gauss* map is always valid (here $|\mathbf{n}_P| = 1$).

In a reference system $X_P Y_P Z_P$ associated with the sculptured part surface P, cosines of the angles α, β, γ that the unit normal vector $\mathbf{n}_P$ forms with the axes of the coordinate system $X_P Y_P Z_P$ are calculated by

$$\cos\alpha = \mathbf{i} \cdot \mathbf{n}_P, \tag{7.1}$$

$$\cos\beta = \mathbf{j} \cdot \mathbf{n}_P, \tag{7.2}$$

$$\cos\gamma = \mathbf{k} \cdot \mathbf{n}_P. \tag{7.3}$$

Equations 7.1 through 7.3 for $\cos\alpha$, $\cos\beta$, and $\cos\gamma$ make the following matrix expression possible:

$$\mathbf{r}_{P0} = \begin{bmatrix} \cos\alpha \\ \cos\beta \\ \cos\gamma \\ 1 \end{bmatrix} \tag{7.4}$$

for the position vector of a point, $\mathbf{r}_{P0}$, of the *Gauss* map of a surface point.

A similar equation holds for the generating surface T of the form-cutting tool.

The exploration of the Gauss map, *GMap* (P), of the sculptured part surface P (see Equation 7.4), such as the study of an arbitrary surface, is facilitated by expressing it in canonical form, say, in terms of the first and the second fundamental forms, $\Phi_{1.P0}$ and $\Phi_{2.P0}$, of the *Gauss* map, *GMap* (P), of the given part surface P. Any and all parameters of a surface geometry can be expressed in terms of the fundamental magnitudes of the surface the first and the second fundamental forms, $\Phi_{1.P0}$ and $\Phi_{2.P0}$.

The first fundamental form, $\Phi_{1.P0}$, of the *Gauss* map, *GMap* (*P*), is given by expression [14]

$$\Phi_{1.P0} \Rightarrow ds_{P0}^2 = e_P du_P^2 + 2 f_P du_P dv_P + g_P dv_P^2, \tag{7.5}$$

where ds_{P0} is the differential of an arc of a curve on the unit sphere, e_P, f_P, g_P are the first-order *Gauss* coefficients of the *GMap* (*P*), and u_P, v_P are the parametric coordinates of an arbitrary point of the *Gauss* map *GMap* (*P*).

Omitting bulky derivations, one can write the following equation for the first fundamental form $\Phi_{1.P0}$ of the *Gauss* map, *GMap* (*P*), of the part surface *P*:

$$\Phi_{1.P0} \Rightarrow ds_{P0}^2 = \Phi_{2.P} \mathcal{M}_P - \Phi_{1.P} \mathcal{G}_P \tag{7.6}$$

where $\mathcal{M}_P$ designates the mean curvature at a point of the sculptured part surface *P* (see Equation 1.50) and $\mathcal{G}_P$ designates *Gaussian* curvature at a point of that same sculptured part surface *P* (see Equation 1.51).

The second fundamental form $\Phi_{2.P0}$ of the *Gauss* map, *GMap* (*P*), of a given patch of the part surface *P* is calculated as [14]

$$\Phi_{2.P0} \Rightarrow -d\mathbf{r}_{P0} \cdot d\mathbf{n}_{P0} = l_P du_P^2 + 2 m_P du_P dv_P + n_P dv_P^2 \tag{7.7}$$

and is derived in a similar manner.

In Equation 7.7, the values l_P, m_P, n_P are the second-order *Gaussian* coefficients of the *Gauss* map, *GMap* (*P*), of the surface of unit sphere.

Skipping the proofs, some useful properties of the *Gauss* maps, *GMap* (*P*), and *GMap* (*T*), can be noted:

- The *Gauss* map, *GMap* (*P*), of an orthogonal net on a sculptured part surface *P*, for which mean curvature $\mathcal{M}_P$ is not equal to zero ($\mathcal{M}_P \neq 0$), is also an orthogonal net if and only if the initial net is made up of lines of curvature.

  If the mean curvature $\mathcal{M}_P$ at a point of the part surface *P* is equal to zero ($\mathcal{M}_P = 0$), then the net of coordinate lines on the *Gauss* map, *GMap* (*P*) will be orthogonal as well.

- The points on the boundaries of the part surface *P* patch and on its *Gauss* map, *GMap* (*P*), are not necessarily in one-to-one correspondence.

- *Gauss* map, *GMap* (*P*), is a many-to-one map: Each point on a smooth regular part surface *P* has a corresponding point on the *Gauss* map, *GMap* (*P*), but each point on the *Gauss* map, *GMap* (*P*), may correspond

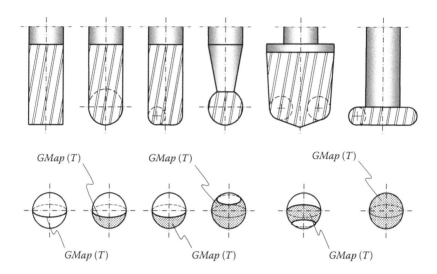

FIGURE 7.3
Examples of the *Gauss* maps, *GMap* (*T*), of the form-cutting tools of various designs. (From Radzevich, S.P., Goodman, E.D., *ASME J. Mech. Des.*, 124, 201–212, 2002. With permission.)

to more than one point on the part surface *P*. This means that in particular cases, the *Gauss* map, *GMap* (*P*), can be interpreted as having more than one layer. The *Gauss* maps, *GMap* (*P*), of this type are commonly referred to as *multilayer Gauss* maps, *GMap* (*P*). For example, the *Gauss* map, *GMap* (*P*), of a torus surface is of two layers.

The general solution to the problem of designing a form-cutting tool for machining sculptured surfaces on a multiaxis *NC* machine (see Chapter 5) reveals that the generating surface *T* of the form-cutting tool can be as complex, as a sculptured part surface *P* can be. The generating surface *T* of the cutting tool of any design has the corresponding *Gauss* map, *GMap* (*T*). Examples of the *Gauss* maps, *GMap* (*T*), of the form-cutting tools that are commonly used in industry are illustrated in Figure 7.3. The calculation of the parameters of the *Gauss* map *GMap* (*T*) of the generating surface *T* of the form-cutting tool is similar to the calculation of the parameters of the *Gauss* map, *GMap* (*P*), of a sculptured part surface *P*.

7.1.3 Area-Weighted Mean Normal to a Sculptured Part Surface *P*

The efficiency of machining a sculptured part surface on a multiaxis *NC* machine can be extremely high when the workpiece orientation is optimal. It is convenient to calculate the parameters of the (two-criterion) optimal workpiece orientation taking into consideration the orientation of the area-weighted mean normal to the part surface *P*. A point on the part surface *P* at which the surface normal is parallel to the area-weighted mean normal of the surface P is referred to as the *central point* of the part surface P.

To calculate the parameters of the area-weighted mean normal, the part surface P can be subdivided into a large number of reasonably small patches of size $S_{Pi} = \Delta U_{Pi} \times \Delta V_{Pi}$. Here i indexes the small patches on the part surface P. At the central point m_i inside of each small patch of the part surface P, the parameters of the perpendicular $\mathbf{N}_{Pi}$ to the part surface P can be calculated:

$$\mathbf{N}_{Pi} = \left.\frac{\partial \mathbf{r}_P}{\partial U_P}\right|_i \times \left.\frac{\partial \mathbf{r}_P}{\partial V_P}\right|_i \tag{7.8}$$

The perpendicular $\mathbf{N}_{Pi}$ may be considered an *area vector element* with magnitude equal to the i-th infinitesimal area of part surface P.

For the calculation of parameters of orientation of the area-weighted mean normal vector $\tilde{\mathbf{N}}_P$, the following formula is employed:

$$\tilde{\mathbf{N}}_P = \frac{\displaystyle\sum_{i=1}^{n} \mathbf{N}_{Pi} \Delta S_{Pi}}{S_P} = \frac{\displaystyle\sum_{i=1}^{n} \mathbf{N}_{Pi} \Delta U_{Pi} \Delta V_{Pi}}{S_P} \tag{7.9}$$

where n designates the total number of small patches on the part surface P and S_P designates the area of the part surface P to be machined (S_P is equal to the sum of all the areas of the separate workpiece surfaces to be machined in one setup).

Allowing the number of small patches on the part surface P to approach infinity yields

$$\tilde{\mathbf{N}}_P = \frac{\displaystyle\int_{S_P} \mathbf{N}_P(U_P; V_P) dS_P}{S_P} = \frac{\displaystyle\int_{S_P} \mathbf{N}_P(U_P; V_P) dU_P dV_P}{S_P} \tag{7.10}$$

Differential of the part surface P area is equal to

$$dS_P = \sqrt{E_P G_P - F_P^2}\, dU_P dV_P. \tag{7.11}$$

Accordingly, Equation 7.10 casts into

$$\tilde{\mathbf{N}}_P = \frac{\displaystyle\iint_P \sqrt{E_P G_P - F_P^2}\, \mathbf{N}_P(U_P; V_P) dU_P dV_P}{S_P} \tag{7.12}$$

In cases when several part surfaces P_i are to be machined on a multiaxis NC machine in one setup, Equation 7.12 yields the more general formula:

$$\tilde{\mathbf{N}}_P = \frac{\sum_{i=1}^{k} \iint_{P_i} \sqrt{E_P G_P - F_P^2}\, \mathbf{N}_P(U_P; V_P)\, dU_P\, dV_P}{\sum_{i=1}^{k} S_{Pi}} \tag{7.13}$$

where k is the total number of the part surfaces P_i to be machined in one setup.

In the last case, the area-weighted mean normal to the part surface P is not considered, but the area-weighted mean normal to the several surfaces P_i is considered instead. The last is referred to as the *area-weighted mean normal* to all part surfaces P_i. In this case, instead of a central point of the part surface, a central point of the entire part to be machined is considered. Definitely, this is a considerably more general approach.

The area-weighted mean normal to a flat portion of a part surface P is $N_{Pi}S_{Pi}$. This result can be used in Equation 7.13.

The parameters of the area-weighted mean normal at a point of the sculptured part surface P as calculated above allow the alteration of the initial orientation of the sculptured surface P to the desired orientation in the coordinate system of the multiaxis NC machine. By rotations of the workpiece, for example, through the angles of nutation ψ, precession θ, and pure rotation φ, the workpiece can be reoriented to its most favorable orientation. In its optimal orientation, the workpiece allows for machining all surfaces with a single setup.

7.1.4 Optimal Workpiece Orientation

In the initial orientation of the workpiece, the angles that the area-weighted mean normal to the surface P forms with the coordinate axes of the NC machine are denoted by α, β, γ (Figure 7.4). It is convenient to show these angles on the *Gauss* map, *GMap* (P), of the part surface P (remembering that the area-weighted mean normal to the part surface P has the same direction as the position vector of the point on the *Gauss* map, *GMap* (P), corresponding to the point on the part surface P at which the perpendicular is erected). In the case under consideration, the problem of optimal workpiece orientation reduces to a problem of coordinate system transformation.

Consider two *Cartesian* coordinate systems $X_P Y_P Z_P$ and $X_{NC} Y_{NC} Z_{NC}$. The first coordinate system is associated with the workpiece. Another is connected to the multiaxis NC machine.

In the initial orientation of the workpiece, the configuration of the coordinate system $X_P Y_P Z_P$ relative to the coordinate system $X_{NC} Y_{NC} Z_{NC}$ is defined

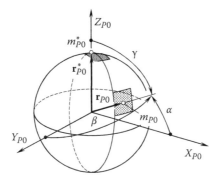

FIGURE 7.4
Spherical map m_{P0} of a point m of the part surface P in the initial orientation of the workpiece and after its optimal orientation, m_{P0}^*. (From Radzevich, S.P., Goodman, E.D., *ASME J. Mech. Des.*, 124, 201–212, 2002. With permission.)

by the angles α, β, and γ. The actual values of these angles can be calculated using one of the above-derived Equations 7.9 through 7.13. The calculated value of the area-weighted mean normal vector $\tilde{\mathbf{N}}_P$ immediately yields calculation of the angles α, β, and γ. For this purpose, the following formulae

$$\cos \alpha = \mathbf{i} \cdot \tilde{\mathbf{N}}_P, \tag{7.14}$$

$$\cos \beta = \mathbf{j} \cdot \tilde{\mathbf{N}}_P, \tag{7.15}$$

$$\cos \gamma = \mathbf{k} \cdot \tilde{\mathbf{N}}_P \tag{7.16}$$

can be used.

All the calculations must be performed in a common reference system. For this purpose, use of the coordinate system $X_{NC}Y_{NC}Z_{NC}$ is preferred.

In the optimal workpiece orientation, corresponding axes of these coordinate systems are parallel to each other and are of the same direction. To put the workpiece into the optimal orientation means to make three successive rotations, for example, by the *Euler* angles—that is, to rotate the workpiece in the coordinate $X_{NC}Y_{NC}Z_{NC}$ through the angles of nutation, ψ, precession, θ, and pure rotation, φ.

The resultant coordinate system transformation using *Euler's* angles can be analytically represented with the operator $\mathbf{Eu}(\psi,\theta,\varphi)$ of Eulerian transformation (see Equation 3.15).

In the optimal workpiece orientation, it is possible to rotate the part surface P about the area-weighted mean normal $\tilde{\mathbf{N}}_P$. Under such a rotation, the optimal orientation of the workpiece is preserved, but the orientation of the part surface P relative to the NC machine coordinate axes changes. This feasible rotation of the part surface P can be used for satisfying additional requirements to the part surface orientation on the worktable of the multiaxis NC machine.

For example, the workspace of the multiaxis *NC* machine is the bounded plane or volume within which the cutting tool and the workpiece can be positioned and through which controlled motion can be invoked. When *NC* instructions are generated by a part programmer, the geometry of the workpiece must be transformed into a coordinate system that is consistent with the workspace origin and coordinate reference frame. That is why after the workpiece is turned to a position at which its area-weighted mean normal has an optimal orientation, it is necessary to rotate it about the weighted normal to a position in which the projection of the part surface *P* (or of the part surfaces P_i) to be machined is within the largest closed contour traced by the cutting tool on the plane of the *NC* machine worktable. In addition, the vertical position of the workpiece must conform to the capabilities of the *NC* machine to move in the vertical direction.

Proper location of the workpiece on the worktable of a multiaxis *NC* machine can be specified in terms of

- The *joint space*, which is defined by a vector whose components are the relative space displacements at every joint of a multiaxis *NC* machine
- The *working envelope*, which is understood as a surface or surfaces that bound the working space
- The *working range*, which means the range of any variable for normal operation of multiaxis *NC* machine
- The *working space*, which includes the totality of points that can be reached by the reference point of the form-cutting tool

7.1.5 Expanded *Gauss* Map of the Generating Surface of the Form-Cutting Tool

An ordinary *Gauss* map can be constructed for any generating surface of a form-cutting tool. Examples of the *Gauss* map, *GMap* (*T*), of the form-cutting tools of various designs are shown in Figure 7.3. However, when machining a sculptured part surface, the cutting tool is traveling with respect to the part surface *P*. Correspondingly, the *Gauss* map, *GMap* (*T*), is moving over the surface of the unit sphere. In such a motion, the *Gauss* map, *GMap* (*T*), is covering the an area of the unit sphere surface that exceeds the area of the original *Gauss* map, *GMap* (*T*). The *Gauss* map that is constructed for the moving generating surface *T* of the form-cutting tool in all its feasible positions is referred to as the *expanded Gauss* map, *GMape* (*T*), of the generating surface of the form-cutting tool.

Actually, when machining a sculptured part surface on a multiaxis *NC* machine, the workpiece and the form-cutting tool perform certain relative motions. For the purposes of further analysis, it is convenient to implement the principle of inversion of relative motions. On the premise of the principle of inversion of relative motions, consider the resultant motion of the form-cutting tool relative to the workpiece, which is now stationary.

At every point K of contact of the sculptured part surface P and the generating surface T of the form-cutting tool, the unit normal vectors $\mathbf{n}_P$ and $\mathbf{n}_T$ to these surfaces are of opposite directions (remember that a normal to the part surface P is pointed outward the part body, and a normal to the generating surface T of the cutting tool is pointed outward the generating body of the cutting tool; therefore, the equality $\mathbf{n}_P = -\mathbf{n}_T$ must be fulfilled). Then, employ the concept of *antipodal points* [29]. Those points on the Gauss map are usually referred to as the *antipodal points*, which are the pairs of diametrically opposed points on the unit sphere. Implementation of the antipodal points makes it possible to introduce the centrosymmetrical image of the *Gauss* map, *GMap* (*T*), of the form-cutting tool surface T. The last is referred to as the *antipodal map GMapa* (*T*) of the generating surface T of the form-cutting tool.

Analysis of possible relative positions of the *Gauss* map, *GMap* (*P*), of the sculptured part surface P and of the antipodal *Gauss* map, *GMapa* (*T*), of the generating surface T of the form-cutting tool makes the following intermediate conclusions possible:

Conclusion 7.1

If Gauss map, *GMap* (*P*), of the part surface P is entirely located within the antipodal Gauss map, *GMap*$_a$ (*T*), of the generating surface T of the form-cutting tool [that is, the *GMap* (*P*) contains no points outside *GMap*$_a$ (*T*)], the machining of the part surface P is possible. ■

This is a necessary but not sufficient condition for the machinability of the sculptured part surface with the given form-cutting tool.

Conclusion 7.2

If a portion of the Gauss map, *GMap* (*P*), of the part surface P is located outside the antipodal Gauss map, *GMap*$_a$ (*T*), of the generating surface T of the form-cutting tool, the machining of the part surface P by the given cutting tool is impossible. ■

This is a sufficient condition that the part surface P cannot be machined with the given cutting tool.

When machining a sculptured part surface on a multiaxis NC machine, the cutting tool is capable of moving along three axes of the coordinate system $X_{NC}Y_{NC}Z_{NC}$ associated with the NC machine, and rotating about one or more of the axes. These additional degrees of freedom (rotations) allow the antipodal indicatrix, $GMap_a$ (*T*), of the generating surface *T* of the form-cutting tool to expand around the surface of the unit sphere, while the *Gauss* map, *GMap* (*P*), remains fixed.

Similar to the spherical indicatrix *GInd* (*T*) of the cutting tool surface T that serves as the boundary curve for the corresponding *Gauss* map, *GMap* (*T*), the antipodal indicatrix $GInd_a$ (*T*) serves as the boundary curve to the antipodal *Gauss* map, $GMap_a$ (*T*).

For example, consider machining of a sculptured part surface *P* on a three-axis *NC* machine. The antipodal map, $GMap_a$ (*T*), of the generating surface *T* of the form-cutting tool occupies the fixed area *ABCD* shown in Figure 7.5. Then, assume that one more *NC*-axis is added somehow to the articulation capabilities of the *NC* machine. The additional fourth *NC*-axis (say, rotation of the cutting tool about an axis not coinciding with the axis of its cutter-speed rotation) causes the antipodal *Gauss* map, $GMap_a$ (*T*), to extend in direction 1 from the initial location *ABCD* to encompass A_1B_1CD. If the fifth and the sixth *NC* axes are added, these additional *NC* axes cause the antipodal map, $GMap_a$ (*T*), to extend in direction 2 and to rotate about an axis through the center of the unit sphere and through a point within the antipodal map, $GMap_a$ (*T*), of the generating surface *T* of the form-cutting tool.

A surface patch on the unit sphere that is covered by the antipodal *Gauss* map, $GMap_a$ (*T*), in such its motion over the unit sphere, is referred to as the *expanded antipodal Gauss map* $GMap_{ae}$ (*T*) of generating surface *T* of the form-cutting tool.

The expanded antipodal indicatrix $GMap_{ae}$ (*T*) is a useful tool for the investigation of workpiece orientation on the worktable of a multiaxis *NC* machine.

The boundary curve of the expanded antipodal *Gauss* map, $GMap_{ae}$ (*T*), of the generating surface *T* of the form-cutting tool serves as the expanded antipodal indicatrix $GInd_{ae}$ (*T*) of the cutting tool surface *T*. Because the part

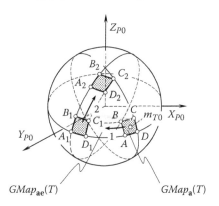

FIGURE 7.5
An example of the expanded antipodal map, $GMAP_{ae}$(*T*), of the generating surface *T* of the form-cutting tool. (From Radzevich, S.P., Goodman, E.D., *ASME J. Mech. Des.*, 124, 201–212, 2002. With permission.)

surface P is considered motionless, the expanded antipodal indicatrix $GInd_{ae}$ (T) of the generating surface T of the form-cutting tool as well as its expanded antipodal *Gauss* map, $GMap_{ae}$ (T), cannot rotate about an axis through the origin of the coordinate system $X_{P0}Y_{P0}Z_{P0}$.

When the parameters of the relative motion of the given sculptured part surface P and of the given generating surface T of the form-cutting tool are known, the parameters of the expanded initial indicatrix, $GInd_e$ (T), and of the expanded antipodal $GInd_{ae}$ (T) indicatrices of the cutting tool surface T can be calculated using the methods of spherical trigonometry developed [29].

For machining a sculptured part surface on a multiaxis (four or more axes) NC machine, the following two statements hold:

Conclusion 7.3

If Gauss map, $GMap$ (P), of the part surface P is contained entirely inside the expanded antipodal Gauss map, $GMap_{ae}$ (T), of generating surface T of the form-cutting tool, then the part surface P can be machined by the given form-cutting tool. ∎

This is a necessary but not sufficient condition for the machinability of a sculptured part surface on a give multiaxis NC machine by the given cutting tool.

Conclusion 7.4

If Gauss map, $GMap$ (P), of the part surface P contains at least one point outside the expanded antipodal Gauss map, $GMap_{ae}$ (T) of generating surface T of the form-cutting tool, then machining of the part surface P by means of the given cutting tool is not feasible. ∎

This condition is sufficient that the sculptured part surface cannot be machined on a given multiaxis NC machine by the given form-cutting tool.

7.1.6 Important Peculiarities of *Gauss* Maps of a Part Surface *P* and the Generating Surface *T* of the Form-Cutting Tool

In particular cases, a sculptured part surface P as well as the generating surface T of the form-cutting tool can have two or more points, at which unit normal vectors are parallel to each other and are pointed in that same direction. Points of this sort can be easily found out, for example, on the torus surface.

When parallel and similarly directed unit normal vectors are observed, the *Gauss* map, *GMap* (P), of the sculptured part surface P becomes "multi-layered." The number of layers of the *Gauss* map, *GMap* (P), is equal to the number of points with parallel and similarly oriented unit normal vectors. For example, parallel and similarly oriented unit normal vectors occur on the part surface P (Figure 7.6).

The part surface P is bounded by the bordering line ABCDEFG. The Gauss map, *GMap* (P), for this portion of the surface P is represented by the portion $A_0B_0G_0D_0E_0F_0$ of the unit sphere. Figure 7.6 reveals that the area $B_0C_0D_0G_0$ on the unit sphere corresponds to the *Gauss* map, *GMap* (P), of the portion BCDG of the part surface P. This means that the portion $A_0B_0G_0D_0E_0F_0$ of the *Gauss* map, *GMap* (P), is covered twice.

The number of layers of the *Gauss* map, *GMap* (P), of a sculptured part surface P can exceed two layers. Equations 7.7 through 7.10 take into account that the Gauss maps of surfaces may have two or more layers. Multiple layers affect the weight of the multilayer portion of the *Gauss* map, *GMap* (P). The weight of the multilayer portion of the *Gauss* map, *GMap* (P), is getting larger. It can result in shadowed portions on part surface P if one strictly directs the rule of optimal orientation of the sculptured part surface P in compliance with the direction of its area-weighted mean normal. (That is, it is not allowed to ignore the reachability constraint on the part surface P orientation).

As an example, consider the machining of a sculptured part surface on a three-axis NC machine. Two cases are distinguished.

In the first case, a sculptured part surface P_1 has a one-layer *Gauss* map, *GMap* (P), shown in Figure 7.7. The *Gauss* map, *GMap* (P), of the part surface P_1 occupies a portion of the unit sphere within the closed contour ABDCFE. The area-weighted mean normal $\mathbf{n}_{P0}^* \equiv \mathbf{r}_{P0}^*$ passes through the point m_{P0}^*. In

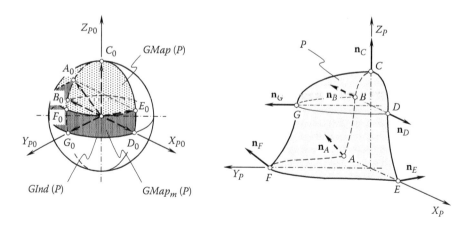

FIGURE 7.6
An example of multilayer *Gauss* map, *GMap* (P), of the sculptured part surface P. (From Radzevich, S.P., Goodman, E.D., *ASME J. Mech. Des.*, 124, 201–212, 2002. With permission.)

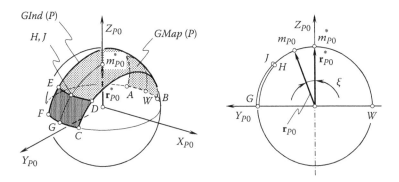

FIGURE 7.7
A multilayer (two-layer) *Gauss* map, $GMAp_m$ (P), of a sculptured part surface P that is machined on a three-axis *NC* machine. (From Radzevich, S.P., Goodman, E.D., *ASME J. Mech. Des.*, 124, 201–212, 2002. With permission.)

such an orientation, the part surface P_1 can be machined on the three-axis *NC* machine in one setup.

In the second case, a sculptured part surface P_2 has a multilayer (two-layer) *Gauss* map, $GMap_m$ (P). The subscript m here indicates that the Gauss map is multilayer. Likewise, in the first case, the *Gauss* map, $GMap_m$ (P), of the part surface P_2 occupies the portion within the closed contour *ABDCFE* on the unit sphere. In addition to the portion *ABDCFE*, the *Gauss* map, $GMap_m$ (P), is represented with the portion *CDEF* on the unit sphere (Figure 7.7). When the portion *CDEF* is added, the Gauss map is a two-layer map, so it is twice as heavily weighted. Due to the increase of weight of the *Gauss* map, $GMap_m$ (P), the area-weighted mean normal $\mathbf{n}_{P0} \equiv \mathbf{r}_{P0}$ of the part surface P_2 turns about the center of the unit sphere through a certain angle ε. In this position, the unit normal vector passes through the point m_{P0}.

In such a position of the workpiece, it is infeasible to machine the sculptured part surface P_2 on the three-axis *NC* machine in one setup. It is necessary to consider the tradeoffs between the orientation of the part surface P_2 in compliance with the position of its area-weighted mean normal and the orientation of the part surface P_2 to avoid shadowed areas. One could consider, for example, whether it is preferred to machine the sculptured part surface P_2 on a three-axis *NC* machine (that is cheaper) with nonoptimal workpiece orientation vis-à-vis cutting conditions or to machine the part surface P_2 in the optimal workpiece orientation but on a more costly four (or more) axis *NC* machine. Generally, machining of a part surface in a single setup with some loss of optimality of cutting condition is preferable to machining in two or more setups. Thus, the generally favored situation is to orient the workpiece such that the difference in angle between the area-weighted normal to the part surface to be machined and the tool axis of rotation changes as little as possible, without requiring more setups than necessary.

After the analysis of Figure 7.7 is performed, it is important to focus here again on the properties of Gauss mapping of the surfaces. Figure 7.6 provides a good example to illustrate the property (b) of the Gauss map, *GMap* (*P*), (see Section 7.1.2). The Gauss map of the bordering contour *ABCDE* of the part surface *P* is represented by the circular arc $A_0B_0C_0D_0E_0$ (Figure 7.6). In this case, all points of the bordering contour *ABCDE* of the part surface *P* and all points of the boundary $A_0B_0C_0D_0E_0$ of the Gauss map are in one-to-one correspondence. On the other hand, the Gauss map of the bordering contour *AFE* of the part surface *P* is represented by the circular arc $A_0F_0E_0$. It is evident that the Gauss map $A_0F_0E_0$ of the bordering contour *AFE* is not a border for the Gauss map, *GMap* (*P*), of the sculptured part surface *P*. Moreover, boundary arc $B_0G_0D_0$ of the Gauss map, *GMap* (*P*), of the part surface *P* is merely an image of the curve *BGD* on the part surface *P*. However, the *BGD* is not a boundary of the part surface *P*. This example illustrates that a boundary of the Gauss map, *GMap* (*P*), of a sculptured part surface *P* may or may not be a boundary of its Gauss map, *GMap* (*P*), and vice versa.

7.1.7 Spherical Indicatrix of Machinability of a Sculptured Part Surface

The above-considered Gauss maps of a sculptured part surface and a generating surface of the form-cutting tool provide engineers with a powerful analytical tool for optimization of the workpiece orientation of the worktable of a multiaxis *NC* machine. Among others, the implementation of this tool is helpful for determining whether or not a given sculptured part surface *P* can be machined by the given cutting tool on the *NC* machine with the specified articulation. For this purpose, spherical indicatrices, *GInd* (*P*) and *GInd* (*T*), of the surfaces *P* and *T* can be used.

It is inconvenient to treat simultaneously two separate characteristic curves *GInd* (*P*) and *GInd* (*T*) to determine whether the sculptured part surface *P* can or cannot be machined in one setup by the given cutting tool on the *NC* machine with the specified articulation. For this purpose, a characteristic curve of another nature was proposed by Radzevich [136]. This characteristic curve is referred to as the *spherical indicatrix of machinability Mch* (*P/T*) of a given sculptured part surface *P* by the given cutting tool *T* on the *NC* machine with the given articulation. For *CAD/CAM* application, it is necessary to represent this characteristic curve analytically.

Without loss of generality, one can consider for simplicity the machining of a sculptured part surface *P* with a ball-end milling cutter. For this case, the *Gauss* map, *GMap* (*T*), of the generating surface of the cutting tool occupies a hemisphere of the unit sphere (Figure 7.8). The *Gauss* map, *GMap* (*P*), of the sculptured part surface *P* is represented with a certain patch on the unit sphere. The great circle of the unit sphere serves as the *GInd* (*T*) of the generating surface *T* of the cutting tool. Ultimately, the indicatrix *GInd* (*P*) is represented with the boundary of the *Gauss* map, *GMap* (*P*).

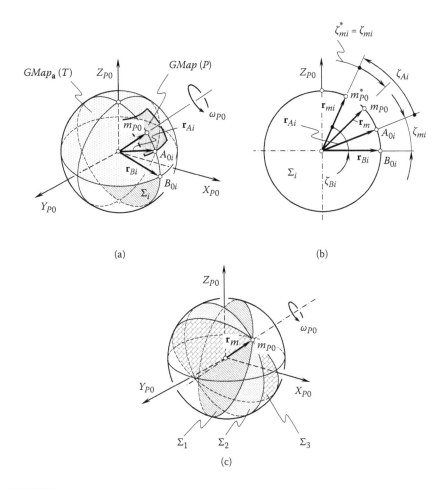

FIGURE 7.8
Derivation of an equation of the spherical indicatrix of machinability, *Mch* (*P/T*), of a sculptured part surface *P* by the given form-cutting tool. (From Radzevich, S.P., Goodman, E.D., *ASME J. Mech. Des.*, 124, 201–212, 2002. With permission.)

An arbitrary point m_{P0} is chosen within the *Gauss* map, *GMap* (*P*), of the part surface *P*. A cross section of the unit sphere by the plane Σ_i through the origin of the coordinate system $X_{P0}Y_{P0}Z_{P0}$ and the chosen point m_{P0} is considered. The plane Σ_i intersects the spherical indicatrices *GInd* (*P*) and *GInd* (*T*) at the points A_{0i} and B_{0i}, respectively. The angle between the position vector $\mathbf{r}_{Ai}$ of the point A_{0i} and the position vector $\mathbf{r}_m$ of the chosen point m_{P0} is designated as ς_{Ai}. The similar angle between the position vector $\mathbf{r}_{Bi}$ of the point B_{0i} and the position vector $\mathbf{r}_m$ is designated as ς_{Bi}. The difference of the angles ς_{Ai} and ς_{Bi} is equal to $(\varsigma_{Ai} - \varsigma_{Bi}) = \varsigma_{mi}$. Angle ς_{mi}^*, which determines the location of the point m_{0i}^* on the indicatrix of machinability *Mch* (*P/T*) of the part surface *P* by means of the cutting tool *T*, is equal to $\varsigma_{mi}^* = \varsigma_{mi}$. This angle is measured within the plane Σ_i through the vectors $\mathbf{r}_m$, $\mathbf{r}_{Ai}$, $\mathbf{r}_{Bi}$ either in the clockwise direction if $\varsigma_{Ai} > \varsigma_{Bi}$ (in this case, $\varsigma_{mi}^* < 0°$), or in the counterclockwise direction if $\varsigma_{Ai} < \varsigma_{Bi}$ (in this case, $\varsigma_{mi}^* < 0°$).

Rotating, ω_{P0}, the plane Σ_i about the position vector $\mathbf{r}_m$ to positions Σ_1, Σ_2, Σ_3, and so on, all points of the indicatrix of machinability, *Mch* (P/T) can be obtained.

Conclusion 7.5

A sculptured part surface *P* is machinable using the given generating surface *T* the form-cutting tool if and only if the indicatrix of machinability Mch (P/T) has no negative diameters, i.e., it is not a self-intersecting curve on the unit sphere.■

The equation of the indicatrix of machinability, *Mch* (P/T), immediately follows from the analysis below.

The position vector $\mathbf{r}_m$ of the point m_{0P} is

$$\mathbf{r}_m = \begin{bmatrix} \cos\alpha_m \\ \cos\beta_m \\ \cos\gamma_m \\ 1 \end{bmatrix} \tag{7.17}$$

The equation of the *Gauss* map, *GMap* (T), of the generating surface *T* of the form-cutting tool yields two equations. The first equation is for the unit vector $\mathbf{r}_{Ai}$

$$\mathbf{r}_{Ai} = \begin{bmatrix} \cos\alpha_{Ai} \\ \cos\beta_{Ai} \\ \cos\gamma_{Ai} \\ 1 \end{bmatrix} \tag{7.18}$$

and another equation is for the unit vector $\mathbf{r}_{Bi}$

$$\mathbf{r}_{Bi} = \begin{bmatrix} \cos\alpha_{Bi} \\ \cos\beta_{Bi} \\ \cos\gamma_{Bi} \\ 1 \end{bmatrix} \tag{7.19}$$

The vectors $\mathbf{r}_m$ and $\mathbf{r}_{Ai}$ form an angle ς_{Ai}. The actual value of the angle ς_{Ai} can be calculated from the following formula:

$$\varsigma_{Ai} = \angle(\mathbf{r}_m, \mathbf{r}_{Ai}) = \arctan \frac{|\mathbf{r}_m \times \mathbf{r}_{Ai}|}{\mathbf{r}_m \cdot \mathbf{r}_{Ai}} \tag{7.20}$$

The vectors $\mathbf{r}_m$ and $\mathbf{r}_{Bi}$ make an angle ς_{Bi}. The actual value of the angle ς_{Bi} is

$$\varsigma_{Bi} = \angle(\mathbf{r}_m, \mathbf{r}_{Bi}) = \arctan \frac{|\mathbf{r}_m \times \mathbf{r}_{Bi}|}{\mathbf{r}_m \cdot \mathbf{r}_{Bi}} \tag{7.21}$$

With these results, the value of the angle ς_{mi} that determines the location of an arbitrary point of the indicatrix of machinability, *Mch* (*P/T*), of the part surface *P* using the specified cutting tool surface *T* is

$$\varsigma_{mi} = \angle(\mathbf{r}_{Ai}, \mathbf{r}_{Bi}) = \varsigma_{Ai} - \varsigma_{Bi} = \arctan \frac{|\mathbf{r}_m \times \mathbf{r}_{Ai}|}{\mathbf{r}_m \cdot \mathbf{r}_{Bi}} - \arctan \frac{|\mathbf{r}_m \times \mathbf{r}_{Bi}|}{\mathbf{r}_m \cdot \mathbf{r}_{Bi}} \tag{7.22}$$

The location of a current point of the spherical indicatrix of machinability, *Mch* (*P/T*), is specified by Equation 7.22. This characteristic curve is convenient for determining whether the sculptured part surface *P* can or cannot be machined using the given generating surface *T* of the form-cutting tool.

As an illustration of implementation of the spherical indicatrix of machinability, *Mch* (*P/T*), consider machining various cylindrical local portions of a sculptured part surface *P* on a three-axis *NC* machine (Figure 7.9). In the case under consideration, the antipodal *Gauss* indicatrix *GInd*$_a$ (*T*) of the generating surface *T* of the ball-end milling cutter is represented with a hemisphere. An arc of a great circle of the unit sphere serves as the spherical *Gauss* indicatrix *GInd* (*P*) of the part surface *P*.

Analysis of the relative configuration of the spherical indicatrices *GInd* (*P*) and the *GInd* (*T*) shows that the part surface *P* can be machined in the first two cases schematically illustrated in Figure 7.9a and b, and it cannot be machined in the third case depicted in Figure 7.9c. One can come up with that same result via analysis of the shape of the indicatrix of machinability, *Mch* (*P/T*), of the part surface *P* using that same cutting tool surface *T*.

The analysis shows that in the first case (Figure 7.9a), the part surface *P* can be machined on a three-axis *NC* machine. Moreover, a certain freedom in orienting of the workpiece remains, that is, the part surface *P* can be turned about its axis in opposite directions through a certain angle of $\xi > 0°$.

In the second case (Figure 7.9b), the part surface *P* can also be machined on a three-axis *NC* machine. However, in this case, no degree of freedom remains, that is, the part surface *P* cannot be turned about it axis because the angle *GMap* (*P*) is equal to zero $\xi = 0°$.

In the third case (Figure 7.9c), the part surface *P* cannot be machined on a three-axis *NC* machine. The arc of the great circle through the spherical map *GMap* (*P*) includes points for which the angle ξ is negative ($\xi < 0°$). The

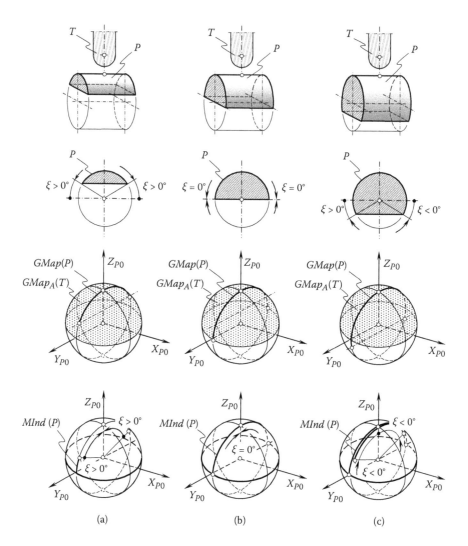

FIGURE 7.9
Generation of a various cylindrical local portions of a sculptured part surface, P, by the ball-end milling cutter having spherical generating surface, T. (From Radzevich, S.P., Goodman, E.D., _ASME J. Mech. Des._, 124, 201–212, 2002. With permission.)

last indicates that negative diameters of the indicatrix of machinability _Mch_ (P/T) are observed there.

For convenience in implementation, the indicatrix of machinability _Mch_ (P/T) can be depicted in _Cartesian_ coordinates. No formulae transformations are required in this concern. The spherical parameters must be considered as the _Cartesian_ coordinates.

Actually, using the indicatrix of machinability, _Mch_ (P/T), we came up with that same result as that previously obtained, when two spherical _Gauss_ indicatrices, *GInd* (P) and *GInd* (T), were implemented. As follows from the above analysis, it is preferred to treat one spherical indicatrix of machinability, *Mch* (P/T), rather than two characteristic curves *GInd* (P) and *GInd* (T) together.

Example 7.1

Consider machining of a portion P of a torus surface shown in Figure 7.10. The radius r_p of the generating circle of the surface P is $r_p = 100$ mm. The radius R_p of the directing circle of the surface P is $R_p = 200$ mm. *Gaussian* (curvilinear) coordinates of a point on the surface P are designated as θ_p and φ_p correspondingly. They vary in intervals: $0° \leq \theta_p \leq 90°$ and $0° \leq \varphi_p \leq 90°$.

The goal is to define an optimal orientation of the part surface P on the table of a three-axis *NC* milling machine.

Position vector R of a point of the surface P can be expressed as the summa $\mathbf{R} = \mathbf{R}^* = \mathbf{r}^*$ of two position vectors. Here $\mathbf{R}^*$ is the position vector of the center of the generating circle that rotates about the Z_p-axis, and $\mathbf{r}^*$ is the position vector of a point on the generating circle. Due to lack of space, the vectors $\mathbf{R}$, $\mathbf{R}^*$, and $\mathbf{r}^*$ are not depicted on Figure 7.10.

Representation of the position vector R in the form $\mathbf{R} = \mathbf{R}^* + \mathbf{r}^*$ makes expanded equation for the position vector of a point of the surface P possible:

$$\mathbf{R} = \begin{bmatrix} -R_p \cos\theta_p + r_p \cos\varphi_p \cos\theta_p \\ -R_p \sin\theta_p + r_p \cos\varphi_p \sin\theta_p \\ r_p \sin\varphi_p \\ 1 \end{bmatrix} \qquad (7.23)$$

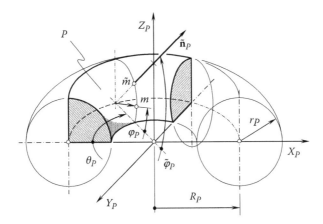

FIGURE 7.10
An example of a part surface, P, to be machined on a multiaxis *NC* machine. (From Radzevich, S.P., Goodman, E.D., *ASME J. Mech. Des.*, 124, 201–212, 2002. With permission.)

The perpendicular vector $\mathbf{N}_p$ to the surface P at an arbitrary point m can be calculated as

$$\mathbf{N}_p = \mathbf{U}_p \times \mathbf{V}_p, \qquad (7.24)$$

where tangent vectors $\mathbf{U}_p$ and $\mathbf{V}_p$ are given by

$$\mathbf{U}_p = \frac{\partial \mathbf{R}}{\partial \theta_p} = \begin{bmatrix} R_p \sin\theta_p - r_p \cos\varphi_p \sin\theta_p \\ -R_p \cos\theta_p + r_p \cos\varphi_p \cos\theta_p \\ r_p \cos\theta_p \\ 0 \end{bmatrix} \qquad (7.25)$$

$$\mathbf{V}_p = \frac{\partial \mathbf{R}}{\partial \varphi_p} = \begin{bmatrix} -r_p \sin\varphi_p \cos\theta_p \\ -r_p \sin\varphi_p \sin\theta_p \\ 0 \\ 0 \end{bmatrix} \qquad (7.26)$$

Equations 7.25 and 7.26 yield the formula

$$\mathbf{N}_p = \begin{vmatrix} \mathbf{i} & \mathbf{j} & \mathbf{k} \\ R_p \sin\theta_p - r_p \cos\varphi_p \sin\theta_p & -R_p \cos\theta_p + r_p \cos\varphi_p \cos\theta_p & r_p \cos\theta_p \\ -r_p \sin\varphi_p \cos\theta_p & -r_p \sin\varphi_p \sin\theta_p & 0 \end{vmatrix} \qquad (7.27)$$

for the calculation of the normal vector $\mathbf{N}_p$ to the surface P.

The angles α_p, β_p, and γ_p, which the vector $\mathbf{N}_p$ forms with the axes of the coordinate system $X_p Y_p Z_p$ associated to the part to be machined ($\cos\alpha_p = \mathbf{i} \cdot \mathbf{N}_p$, $\cos\beta_p = \mathbf{j} \cdot \mathbf{N}_p$, and $\cos\gamma_p = \mathbf{k} \cdot \mathbf{N}_p$), can be calculated as

$$\alpha_p = \arccos \frac{r_p \sin\theta_p \cos\theta_p}{\sqrt{r_p^2 \cos^2\theta_p + R_p^2 (1 - \cos\varphi_p)^2}} \qquad (7.28)$$

$$\beta_p = \arccos \frac{r_p \cos^2\theta_p}{\sqrt{r_p^2 \cos^2\theta_p + R_p^2 (1 - \cos\varphi_p)^2}} \qquad (7.29)$$

$$\gamma_p = \arccos \frac{R_p (\cos\varphi_p - 1)}{\sqrt{r_p^2 \cos^2\theta_p + R_p^2 (\cos\varphi_p - 1)^2}} \qquad (7.30)$$

For the calculation, it is convenient to consider discrete values of the parameters θ_p and φ_p with certain increments $\delta\theta_p$ and $\delta\varphi_p$, respectively. Under such a scenario, the surface P could be represented in $\delta\theta_p = 1°$ and $\delta\varphi_p = 1°$ increments by $90 \cdot 90 = 8100$ points, which provide sufficient accuracy for the calculation schematically illustrated in Figure 7.11.

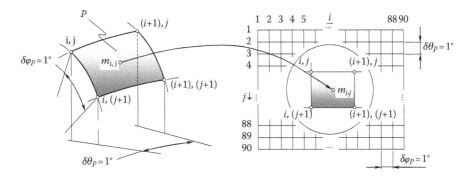

FIGURE 7.11
A grid of nearly rectangular patches on the part surface P. (From Radzevich, S.P., Goodman, E.D., *ASME J. Mech. Des.*, 124, 201–212, 2002. With permission.)

Thereby, the surface P is covered with nearly rectangular patches, the vertices of which may be enumerated in the following manner: (i, j), $(i + 1, j)$, $(i, j + 1)$ and $(i + 1, j + 1)$. Each patch of the surface P can be considered as a nearly flat patch that can be inscribed in a circle.

The optimal orientation of the part surface P is calculated in the following six steps:

Step 1. Using the enumeration of the rectangular patches from above, Equation 7.9 yields

$$\tilde{\mathbf{N}}_P = \frac{\sum_{i-1}^{n} \mathbf{N}_{P.i.j} \ S_{P.i.j}}{S_P} \tag{7.31}$$

In Equation 7.31, the perpendicular $\mathbf{N}_{P.i.j}$ is calculated by

$$\mathbf{N}_{P.i.j} =$$

$$\begin{vmatrix} \mathbf{i} & \mathbf{j} & \mathbf{k} \\ R_p \sin(1°\cdot j) - r_p \cos(1°\cdot i)\sin(1°\cdot j) & -R_p \cos(1°\cdot j) + r_p \cos(1°\cdot i)\cos(1°\cdot j) & r_p \cos(1°\cdot j) \\ -r_p \sin(1°\cdot i)\cos(1°\cdot j) & -r_p \sin(1°\cdot i)\sin(1°\cdot j) & 0 \end{vmatrix}$$

$$\tag{7.32}$$

Equation 7.32 casts into the matrix form:

$$\mathbf{N}_{P.i.j} = \begin{bmatrix} -r_p^2 \sin(1°\cdot i)\sin(1°\cdot j)\cos(1°\cdot j) \\ r_p^2 \sin(1°\cdot i)\cos^2(1°\cdot j) \\ r_p(-R_p + r_p \cos(1°\cdot i))\sin(1°\cdot i) \\ 1 \end{bmatrix} \tag{7.33}$$

Step 2. The length of each side of one of the nearly rectangular patches is

$$a_{i,j} = \left| \mathbf{r}_{(i+1),j} - \mathbf{r}_{i,j} \right| \tag{7.34}$$

$$b_{i,j} = \left| \mathbf{r}_{(i+1),(j+1)} - \mathbf{r}_{(i+1),j} \right| \tag{7.35}$$

$$c_{i,j} = \left| \mathbf{r}_{i,(j+1)} - \mathbf{r}_{(i+1),(j+1)} \right| \tag{7.36}$$

$$d_{i,j} = \left| \mathbf{r}_{i,j} - \mathbf{r}_{i,(j+1)} \right| \tag{7.37}$$

Step 3. The semiperimeter of one of the nearly rectangular patches is

$$p_{i,j} = \frac{a_{i,j} + b_{i,j} + c_{i,j} + d_{i,j}}{2} \tag{7.38}$$

Step 4. Because each of the nearly rectangular patches is nearly flat and can be inscribed in a circle, area $S_{P.i,j}$ of the nearly rectangular patches can be calculated by

$$S_{P.i,j} = \sqrt{(p_{i,j} - a_{i,j}) \cdot (p_{i,j} - b_{i,j}) \cdot (p_{i,j} - c_{i,j}) \cdot (p_{i,j} - d_{i,j})} \tag{7.39}$$

Step 5. Area of the part surface P is calculated by

$$S_P = \frac{1}{16} \cdot 4\pi^2 R_p r_p = 4.4674 R_p r_p \tag{7.40}$$

Step 6. The angles $\tilde{\alpha}_p$, $\tilde{\beta}_p$, and $\tilde{\gamma}_p$ that the area-weighted mean normal $\tilde{\mathbf{N}}_p$ makes with the axes of the coordinate system $X_p Y_p Z_p$ associated to the part surface P are calculated by the formulae

$$\cos \tilde{\alpha} = \mathbf{i} \cdot \tilde{\mathbf{n}}_P, \tag{7.41}$$

$$\cos \tilde{\beta} = \mathbf{j} \cdot \tilde{\mathbf{n}}_P, \tag{7.42}$$

$$\cos \tilde{\gamma} = \mathbf{k} \cdot \tilde{\mathbf{n}}_P. \tag{7.43}$$

To orient the part surface P optimally for machining on a three-axis mill, it must be rotated about three axes as shown, such that the arc-weighted mean normal vector $\tilde{\mathbf{N}}_p$ is in a vertical orientation (Figure 7.12) (that is, aligned with the cutting tool axis of rotation).

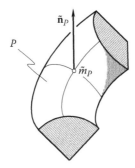

FIGURE 7.12
The optimal orientation of the part surface *P* on the worktable of three-axis *NC* milling machine. (From Radzevich, S.P., Goodman, E.D., *ASME J. Mech. Des.*, 124, 201–212, 2002. With permission.)

7.2 Necessary and Sufficient Conditions of Proper Part Surface Generation

Once the optimal (or at least a feasible) workpiece orientation is determined, it is necessary to establish the rest of the necessary and sufficient conditions of proper part surface generation (further *conditions of proper PSG*) [100,101,109, 112,116,117,136].

7.2.1 First Condition of Proper Part Surface Generation

When machining, the sculptured part surface is generated by the generating surface of the form-cutting tool. A cutting tool of a certain design is necessary for the machining of a given sculptured part surface on a multiaxis *NC* machine. A form-cutting tool of any design can be designed on the premise of the generating surface *T* of the cutting tool. This means that existence of the generating surface *T* of the cutting tool is a prerequisite for the feasibility of machining a given sculptured part surface.

Accuracy of the generated part surface by the generating surface of the cutting tool (when the cutting tool surface *T* exists) is ensured by an appropriate geometry of the cutting tool surface *T*. Deviation of the actually machined part surface from the desired part surface *P* must be within the prespecified tolerance for the accuracy of the part machining. This is the second constraint on the shape and design parameters of generating surfaces of form-cutting tools.

7.2.1.1 Existence of the Generating Surface of a Form-Cutting Tool

The principal methods for generation of the generating surface *T* of the form-cutting tool are disclosed in Chapter 5.

Regardless of whether the generation of the cutting tool surface T as an enveloping surface is convenient or not, the generating surface T of the form-cutting tool can be considered as the enveloping surface to consecutive positions of the part surface P in their relative motion on a multiaxis NC machine. Due to this, in the machining operation, the sculptured part surface P, and the generating surface T of the form-cutting tool are conjugate to one another.

In particular cases, the enveloping surface does not exist. As an example, consider a plane surface π that is passing through the axis Z of a stationary reference system XYZ as illustrated in Figure 7.13a. The plane π is rotated about this Z axis with a certain angular velocity ω_π. The vector $\mathbf{V}_m$ of linear velocity of an arbitrary point m of the surface π is aligned to the unit normal vector $\mathbf{n}_\pi$ to the plane π at the point m. Both vectors $\mathbf{V}_m$ and $\mathbf{n}_\pi$ are

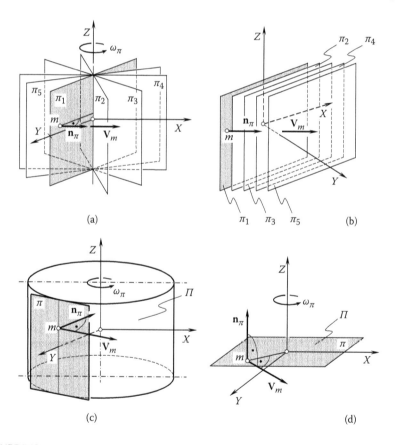

(a)

(b)

(c)

(d)

FIGURE 7.13
Examples of violation (a) and (b) and of satisfaction (c) and (d) of the *first necessary condition* of proper part surface generation on a machine tool. (Reprinted from *Computer-Aided Design, 34*, Radzevich, S.P., Conditions of proper sculptured surface machining, 727–740, Copyright 2002, with permission from Elsevier.)

perpendicular to the plane surface π, and, thus, they are parallel to each other. At no instant of time, the vectors $\mathbf{V}_m$ and $\mathbf{n}_\pi$ can be perpendicular to one another. Because of this, an enveloping surface to consecutive positions: $\pi_1, \pi_2, ..., \pi_i$ of the plane surface π does not exist.

A similar situation is observed when a plane π is performing straight motion $\mathbf{V}_m$ in the direction of the perpendicular $\mathbf{n}_\pi$ to the plane π (Figure 7.13b). Again, the vectors $\mathbf{V}_m$ and $\mathbf{n}_\pi$ can be perpendicular to one another at no instant of time. In this second case, the enveloping surface to the consecutive positions $\pi_1, \pi_2, \pi_4, \pi_5$ of the plane π does not exist either.

An enveloping surface $\varPi$ exists if and only if at a certain instant of time the velocity vector $\mathbf{V}_m$ of is perpendicular to the unit normal vector $\mathbf{n}_\pi$ to the plane surface π. For example, a plane surface π is parallel to the axis Z (Figure 7.13c). The plane is rotated about Z axis with a certain angular velocity. Thus, for certain points of the plane π, vectors $\mathbf{V}_m$ and $\mathbf{n}_\pi$ are perpendicular to each other ($\mathbf{n}_\pi \cdot \mathbf{V}_m = 0$). The points at which the condition $\mathbf{n}_\pi \cdot \mathbf{V}_m = 0$ is met are located within the characteristic line. Therefore, in the case shown in Figure 7.13c, the enveloping surface $\varPi$ exists—this is the surface of a round cylinder.

The same condition ($\mathbf{n}_\pi \cdot \mathbf{V}_m = 0$) is satisfied when the plane surface π is perpendicular to the axis of rotation (Figure 7.13d). The condition $\mathbf{n}_\pi \cdot \mathbf{V}_m = 0$ is met at all points of the plane π. The plane π is sliding over itself when traveling with the velocity vector $\mathbf{V}_m$. In the case depicted in Figure 7.13d, the enveloping surface is represented with a plane that is congruent to the originally given plane Figure 7.13d.

The plane $\mathbf{n}_\pi$ can form a certain angle with the Z axis. In the last case, a circular cone surface would be the enveloping surface to consecutive positions of the plane $\mathbf{n}_\pi$.

It is convenient to illustrate the effectiveness of the first necessary condition of proper part surface generation with an example of machining of an involute working surface of a cam (Figure 7.14). Working surface P of the cam is shaped in the form of involute surface that has the base circle of diameter d_b. When machining the part surface P, the cam swings about the axis O_P with an angular velocity $\pm\omega_P$.

The grinding wheel has a flat working surface, T_G. When machining the part surface P, the grinding wheel rotates about the axis O_T with an angular velocity of ω_T. In addition to the rotation ω_T, the grinding wheel reciprocates along the axis O_T with a certain speed $\pm V_T$.

The swinging motion $\pm\omega_P$ of the cam and the reciprocation $\pm V_T$ of the grinding wheel are timed in a certain manner. Due to the timing, an imaginary (phantom) pitch line associated with the grinding wheel rolls without sliding over an imaginary (phantom) pitch circle associated with the part surface P of the cam. The location of the pitch line and the actual value of the pitch diameter depend upon the timing of the motions $\pm\omega_P$ and $\pm V_T$. This means that the pitch diameter can vary depending on the actual ratio of the motions $\pm\omega_P$ and $\pm V_T$ when machining the part surface P.

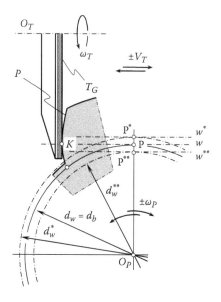

FIGURE 7.14
Examples of satisfaction and of violation of the *first necessary condition* of proper part surface generation when machining the involute working surface P of a cam. (Reprinted from *Computer-Aided Design*, 34, Radzevich, S.P., Conditions of proper sculptured surface machining, 727–740, Copyright 2002, with permission from Elsevier.)

If motions $\pm\omega_p$ and $\pm V_T$ are timed such that pitch diameter is equal to $(d_w = d_b)$ or exceeds $(d_w^* > d_b)$, the base diameter d_b of the involute part surface P, then the first condition of proper part surface generation is satisfied.

In the first case under consideration, the pitch line w is tangent to the base cylinder d_b of the cam at the pitch point P. In the second case, the pitch line w^* makes contact with the pitch circle of diameter d_w^* at the pitch point P^*. In both cases, the enveloping surface exists—this is the plane surface T_G of the grinding wheel. The surfaces P and T_G make contact at point K.

If the motions $\pm\omega_p$ and $\pm V_T$ are timed such that pitch diameter d_w^{**} is less than the base diameter d_b, (i.e. $d_w^{**} < d_b$), the pitch line w^{**} makes contact with the pitch cylinder at the pitch point P^{**}. In this last case, the enveloping surface does not exist and, therefore, the machining of the involute part surface P is impossible.

The above-considered examples, those illustrated in Figures 7.13 and 7.14, make the following statement possible:

Statement 7.1

To design a form-cutting tool for machining a given part surface P, existence of the generating surface T of the form-cutting tool is a necessity; the generating surface T must be conjugate to the given part surface P to be machined. ■

No cutting tool can be designed in cases when a generating surface T does not exist.

7.2.1.2 Accuracy of Shape of the Generating Surface of a Form-Cutting Tool

The above-performed analysis reveals that if the enveloping condition is met,* then a moving smooth, regular part surface generates the corresponding generating surface of the form-cutting tool. Generation of enveloping surfaces is illustrated in Figures 5.13, 5.16, 5.17, 5.20, and others. In these and similar cases, if all the motions in the kinematic scheme of part surface generation are reversed, the generating surface T of a form-cutting tool generates the part surface P_{act} that is identical to the initially given part surface P (that is, $P_{act} \equiv P$). Two smooth regular surfaces P and T possessing this property (that is, to accurately generate each other when the kinematics of the part surface generation is reversed) are referred to as *reversibly enveloping surfaces* (or R_e-surfaces for simplicity). The surfaces P and T, those depicted in Figures 5.13, 5.16, 5.17, and 5.20, are good examples of R_e-surfaces.

In many cases of part surface machining (gear shaping, gear hobbing, and so forth), rolling motion is incorporated into the kinematic scheme of part surface generation. Due to the rolling motion, an axode that is associated with the part surface P and an axode that is associated with the generating surface T of a form-cutting tool roll over one another with no slippage between the surfaces.

When a rolling motion occurs, then regular enveloping condition ($\mathbf{n} \cdot \mathbf{V} = 0$) can be met, the enveloping surface to a moving part surface P can exist, but the generated envelope is not capable of generating the part surface identical to the initially given part surface P. This means that in the case under consideration, the actually generated part surface P_{act} differs from the originally specified part surface P, that is, inequality $P_{act} \neq P$, is valid. Under such a scenario, the part surface P and the enveloping surface T are not a type of R_e-surface. The derived enveloping surface can serve as the generating surface T of a form-cutting tool if and only if deviation of the actually machined part surface P_{act} from the part surface P that is specified by the part blueprint does not exceed tolerance for accuracy of the part surface.

Consider an example.

A schematic of generation of the generating surface T of the hob for cutting a multispline shaft is depicted in Figure 7.15. The multispline shaft is rotated about its axis of rotation, O_P, at a uniform angular velocity, ω_P. The auxiliary generating rack of the hob is traveling with at a constant linear velocity, V_T. The velocities ω_P and V_T are timed with one another so that pitch circle, associated with the spline-shaft, and pitch line, associated with the auxiliary generating rack roll over each other with no slippage (the pitch circle and the pitch line are tangents to one another at a pitch point). The tooth profile,

---

* Remember that the enveloping condition can be expressed in the form of the so-called equation of contact $\mathbf{n} \cdot \mathbf{V} = 0$. This equation is commonly referred to as *Shishkov's equation of contact*.

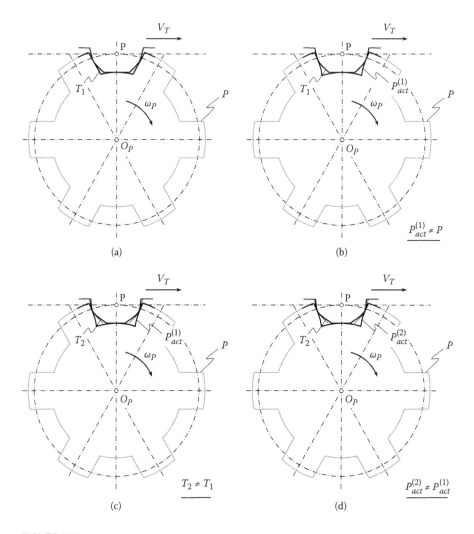

FIGURE 7.15
Generation of the generating surface T of the hob for cutting a multispline shaft.

T_1, of the auxiliary generating rack of the hob is an envelope to consecutive positions of the spline profile, P, in the relative motion of the auxiliary rack in relation to the spline-shaft.

In the example under consideration (Figure 7.15a), the part surface P and the cutting tool surface T_1 are not a type of R_e-surface. This is clearly illustrated by the following. Assume that the inverse problem of part surface generating is considered, that is the cutting tool surface T_1 is given and it is necessary to determine the actually machined part surface. The solution to the inverse problem of part surface generation in the particular case under consideration is illustrated in Figure 7.15b. The actually machined part surface $P_{act}^{(1)}$ is an envelope to consecutive positions of the cutting tool surface T_1 in the relative motion of the auxiliary rack in relation to the spline-shaft. Because the part surface P and the cutting tool surface T_1 are not a type of R_e-surface, then the actually machined part surface $P_{act}^{(1)}$ deviates from the initially given (nominal) part surface P [that is, $P_{act}^{(1)} \neq P$].

Further, consider generation of the auxiliary generating surface of the hob by means of the actually machined part surface $P_{act}^{(1)}$ as schematically illustrated in Figure 7.15c. The cutting tool surface T_2 developed in this particular case is an envelope to consecutive positions of the part surface, $P_{act}^{(1)}$, in the relative motion of the auxiliary rack in relation to the spline-shaft. Again, because the part surface P and the cutting tool surface T_2 are not a type of R_e-surface, the cutting tool surface T_2 is not identical to the cutting tool surface T_1 (that is, $T_1 \neq T_2$).

The performed analysis can be continued further (Figure 7.15d). It can be shown that generated in this way actually machined part surface $P_{act}^{(2)}$ deviates from both, from the actually machined part surface $P_{act}^{(1)}$ as well as from the initially given part surface P [that is, $P_{act}^{(2)} \neq P_{act}^{(1)} \neq P$]. The root cause for the deviation is that the part surface and the generation surface of the cutting tool are not a type of R_e-surface. Examples of R_e-surfaces in cases of simple kinematic schemes of part surface generation are the following:

- When axes of rotation of the part to be machined and of the form-cutting tool are parallel, then only involute surfaces (both, involute part surface, and involute generating surface of the cutting tool) are a type of R_e-surface.
- When axes of rotation of the part to be machined and of the form-cutting tool intersect, then only those involute surfaces developed from appropriate base cones (both, cone involute part surface, and cone involute generating surface of the cutting tool) are a type of R_e-surface.
- When axes of rotation of the part to be machined and of the form-cutting tool cross one another, then only surfaces of special geometry (that is, tooth flanks of the driving and of the driven members in R gearing*) are a type of R_e-surface.

In cases of more complex kinematics of part surface generation, corresponding investigation must be performed to answer the question of whether or not a particular pair of surfaces P and T belongs to R_e-surfaces.

In the event the part surface P and the cutting tool surface T_1 are not a type of R_e-surface, the generating surface of a form-cutting tool can be generated only of approximately. Part surface, P_{act}, that is machined by an approximate generating surface of the cutting tool, always deviates from the originally specified part surface P. Therefore, an approximate generating surface of a form-cutting tool can be used for designing the form-cutting tool if and only if deviations of

---

* Patent pending.

the actually machined part surface, P_{act}, from the nominal part surface, P, are within the tolerance for accuracy that is specified by the part blueprint.

The above-considered examples, those illustrated in Figure 7.15, make the following statement possible:

Statement 7.2

To design a form-cutting tool, the generating surface T of a form-cutting tool, and a given part surface P to be machined must be reversibly enveloping surfaces (R_e-surfaces). Otherwise, the shape and design parameters of the generating surface T of a form-cutting tool must ensure deviation of the machined part surface from the desired geometry within the tolerance for accuracy of the part surface P.   ∎

No cutting tool can be designed in cases when the generating surface T does not ensure the required accuracy of the actually machined part surface.

Summarizing the discussion in Sections 7.2.1.1 and 7.2.1.2 allows the formulation of the following:

> *The First Condition of Proper Part Surface Generation.* To machine a given part surface P, a corresponding generating surface of the form-cutting tool, T, must exist. In machining surfaces by a continuously indexing method, the cutting tool surface T is conjugate to the given part surface P to be machined (to be generated). The conjugate surfaces must be a type of R_e-surface. Otherwise, the approximate generating surface of the form-cutting tool must ensure the accuracy of the actually machined part surface within the tolerance for accuracy that is specified by the part blueprint.

In the cases when the first condition of proper part surface generation is violated, that is, the generating surface T does not exist or the accuracy of the approximate generating surface T is insufficient, a form-cutting tool cannot be designed; therefore the sculptured surface P cannot be machined to meet the blueprint requirements. When a form cutter has been selected or is given, then verification of the first condition of proper part surface generation might be omitted.

Satisfaction of the first condition of proper part surface generation is necessary but not sufficient for correct handling of the machining operation.

When the generating surface T of the form-cutting tool does not exist, then the first necessary condition of proper part surface generation is violated. In particular cases, not the entire generating surface T but merely a portion does not exist. For example, the partial violation of the first necessary condition of proper part surface generation is observed when hobbing an involute gear with a multistart hob of small diameter. If the base diameter of the hob exceeds the hob limit diameter ($d_{b.h} > d_{l.h}$), then the portion of the generating surface T of the involute hob, that should be located between the diameters $d_{b.h}$ and $d_{l.h}$, does not exist. The screw involute surface cannot be extended inside the base cylinder $d_{b.h}$ of the involute hob. This causes partial violation of the first necessary condition of proper part surface generation.

In a similar way, the first necessary condition of proper part surface generation is violated when the involute hob is shifted too much in its axial direction, and thus, there are not enough hob threads within the line of action of the hob and of the gear being machined.

The gear-shaving operation is another good example of when the first necessary condition of proper part surface generation can be violated. For all methods of gear shaving, if the face width of the shaving cutter is narrow and, thus insufficient for proper finishing of the gear, this causes violation of the first necessary condition of proper part surface generation.

7.2.2 Second Condition of Proper Part Surface Generation

When machining a part, the generating surface T of the form-cutting tool must be in contact with the part surface P to be machined. The surfaces P and T can be either in permanent contact with one another or they could make contact in a certain instant of time. In the first case, generation of the sculptured part surface P is referred to as *continuous part surface generation*. In the second case, generation of the sculptured part surface P is referred to as *instantaneous part surface generation*.

In the instant when the part surface is generated, the sculptured part surface P and the generating surface T of the form-cutting tool must be in tangency to each other. For the analytical interpretation of this requirement, the so-called *equation of contact* $\mathbf{n}_{P/T} \cdot \mathbf{V}_\Sigma = 0$ must be satisfied.* Here, in the equation of contact, $\mathbf{n}_{P/T}$ designates a common unit normal vector (the unit vector $\mathbf{n}_{P/T}$ can be interpreted either as the $\mathbf{n}_{P/T} \equiv \mathbf{n}_P$ or as the $\mathbf{n}_{P/T} \equiv -\mathbf{n}_P$), and $\mathbf{V}_\Sigma$ designates a velocity vector of the resultant relative motion of the surfaces P and T.

To satisfy the equation of contact, the unit normal vectors $\mathbf{n}_P$ and $\mathbf{n}_T$ at all points of contact of the surfaces P and T have to be aligned to each other and directed in opposite directions. Both of these two necessary requirements are satisfied in Figure 7.16a. Here, the equality $\mathbf{n}_P + \mathbf{n}_T = 0$ is observed. When the unit vectors $\mathbf{n}_P$ and $\mathbf{n}_T$ are aligned to each other but are of the same direction (Figure 7.16b), the surfaces P and T interfere with each other. Ultimately, when the unit normal vector $\mathbf{n}_P$ is not aligned to the unit normal vector $\mathbf{n}_T$ and they intersect each other at a certain angle $\varepsilon \neq 180°$ (Figure 7.16c), surfaces P and T cannot be in tangency and interference of the surfaces is unavoidable.

The above-mentioned equality $\mathbf{n}_P + \mathbf{n}_T = 0$ in common sense engineering applications is equivalent to the equality $\mathbf{n}_P \cdot \mathbf{n}_T = -1$. Any of these two equations can be interpreted as the analytical representation of the second necessary condition of proper part surface generation.

---

* Prof. V. A. Shishkov was the first to propose (in the second half of 1940s) to represent the equation of contact in the form of dot product $\mathbf{n} \cdot \mathbf{V} = 0$. Therefore this equation is commonly referred to as *Shishkov's equation of contact*.

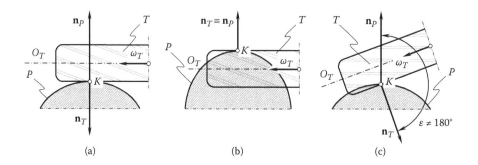

FIGURE 7.16

Examples of satisfaction (a), and of violation (b) and (c) of the *second necessary condition* of proper part surface generation. (Reprinted from *Computer-Aided Design*, 34, Radzevich, S.P., Conditions of proper sculptured surface machining, 727–740, Copyright 2002, with permission from Elsevier.)

Other forms of the equalities can be drawn up as well.

As an example, consider machining of an *Archimedean* screw part surface P by the milling cutter having a cylindrical generating surface T of diameter d_T. A schematic of the machining operation is illustrated in Figure 7.17a. When machining the part surface P, the work is rotated about the axis of rotation, O_P, at a uniform angular velocity, ω_P. The milling cutter is rotated about its axis of rotation, O_T, at a uniform angular velocity, ω_T. The axes of rotations intersect at right angles. In addition, the milling cutter travels along the work axis of rotation, O_P, with a uniform linear velocity, V_T. Magnitudes of the angular velocity, ω_T, and of the linear velocity, V_T, are set so the milling cutter travels over the distance P_x per revolution of the work. Here, the axial pitch of the Archimedean screw part surface P is denoted by P_x. When planning the machining operation, it is assumed that the straight generating line of the cutting tool surface T and the straight generating line of the *Archimedean* screw part surface P are aligned to one another. In other words, it is assumed that points a_T, b_T, and c_T coincide with the corresponding points a_P, b_P, and c_P of the *Archimedean* screw part surface P. Only in this case can the part surface P be generated geometrically accurately. Unfortunately, geometrically accurate generation of a *Archimedean* screw part surface P by means of a cylindrical milling cutter is infeasible due to violation the second condition of proper part surface generation.

The unit normal vectors $\mathbf{n}_{T.a}$, $\mathbf{n}_{T.b}$, and $\mathbf{n}_{T.c}$ to the cutting tool surface T at the points a_T, b_T, and c_T are parallel to one another ($\mathbf{n}_{T.a}||\mathbf{n}_{T.b}||\mathbf{n}_{T.c}$) as schematically shown in Figure 7.17b. The unit normal vectors $\mathbf{n}_{P.a}$, $\mathbf{n}_{P.b}$, and $\mathbf{n}_{P.c}$ to the *Archimedean* screw part surface P at the same points a_T, b_T, and c_T are not parallel to each other. This unit normal vectors are perpendicular to the helices through the corresponding points on the part surface P. Directions of the tangents to the helices are specified in Figure 7.17b by the unit tangent vectors $\mathbf{t}_{P.a}$, $\mathbf{t}_{P.b}$, and $\mathbf{t}_{P.c}$. Therefore, the following conditions $\mathbf{n}_{P.a} \perp \mathbf{t}_{P.a}$, $\mathbf{n}_{P.b} \perp \mathbf{t}_{P.b}$, and $\mathbf{n}_{P.c} \perp \mathbf{t}_{P.c}$ are observed. The condition $\mathbf{n}_{P/T} \equiv -\mathbf{n}_T$ is met with no pairs of the unit normal vectors, namely, it neither meets $\mathbf{n}_{P.a}$ and $\mathbf{n}_{T.a}$, nor $\mathbf{n}_{P.b}$ and

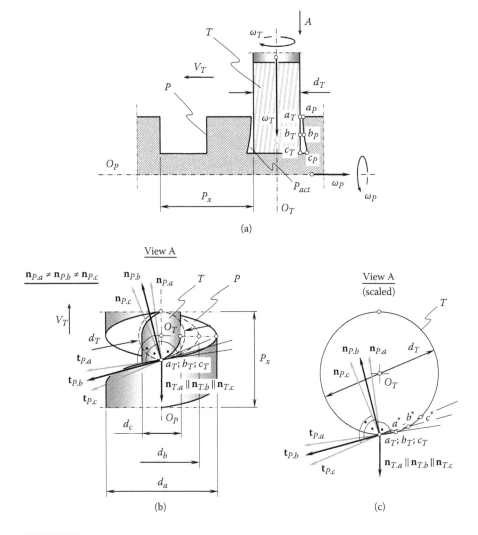

FIGURE 7.17
Analysis of satisfaction/violation of the *second necessary condition* of proper part surface generation when milling an *Archimedean* screw part surface P by a cylindrical milling cutter having the generating surface T.

$\mathbf{n}_{T.b}$, nor $\mathbf{n}_{P.c}$ and $\mathbf{n}_{T.c}$. Because of this, in the case under consideration, the generating surface T of the cylindrical milling cutter intersects the part surface P at the points a^*, b^*, c^*, and so forth (Figure 7.17c). The actually machined part surface P_{act} is generated instead of the desired nominal part surface P. Due to the interference of the cutting tool surface T into the part surface P the *Archimedean* screw part surface P cannot be geometrically accurately generated by the cylindrical milling cutter. The last statement can be generalized: neither *Archimedean* screw part surface, nor *convolute* screw part surface can be geometrically accurately generated by means of a cutting tool having cylindrical generating surface T. Screw surfaces of these types also cannot be generated by means of a plane surface. Only screw *involute* surfaces can be generated by a plane that is performing a screw motion. As a consequence, a screw involute surface can also be generated by a cylinder of revolution that is performing a corresponding screw motion.

The second necessary condition of proper part surface generation can be formulated in the following way:

> *The Second Condition of Proper Part Surface Generation.* The unit normal vectors to the sculptured part surface P to be machined and to the generating surface T of a form-cutting tool at each point of surface contact must be aligned to each other and be directed oppositely.

Usually, it is not difficult to satisfy the second condition of proper part surface generation when developing software for machining a sculptured surface on a multiaxis *NC* machine.

7.2.3 Third Condition of Proper Part Surface Generation

To ensure proper contact of the part surface P and of generating surface T of the form-cutting tool without penetration of one of the surfaces into another surface, it is also necessary to satisfy certain correspondence of their normal radii of curvature at every cross section of the surfaces by a plane through the common perpendicular. Evidently, no problem arises when two convex surfaces P and T must be put in contact. It is also evident that it is not feasible to put in contact two concave surfaces P and T or two surfaces one of which is concave and another one is saddle-like. These cases are evident, and thus, they do not need careful analysis.

The problem of critical importance for the machining of a sculptured part surface on a multiaxis *NC* machine is to establish necessary and sufficient conditions of proper contact of the surfaces P and T near an arbitrary contact point K when

- One of the contacting surfaces P and T is a convex surface and another is a concave surface.
- One of the contacting surfaces P and T is a convex surface and another is a saddle-like surface.
- Both of the contacting surfaces P and T are the saddle-like surfaces.

Analytical interpretation of the condition of contact of the surfaces is the most critical for the machining of a sculptured part surface on a multiaxis *NC* machine.

For the analysis of the actual correspondence between the radii of normal curvature of the surfaces, the normal cross sections of all possible types of contact of the part surface P and of generating surface T of the form-cutting tool have been analyzed (Table 7.1).

When the proper correspondence is observed between the radii of normal curvature R_P of the sculptured part surface P and of the radii of normal curvature R_T of the generating surface T of the form-cutting tool (that is, when the inequality $|R_P| > R_T$ is observed for concave surface P, or when the inequality $|R_T| > R_P$ is observed for concave cutting tool surface T), then the sculptured part surface P can be generated with the cutting tool surface T near every point K of their contact (Figure 7.18a). Otherwise, when the inequality $|R_P| < R_T$ is valid, an interference with surfaces P and T occurs, and thus surface P cannot be geometrically accurately generated in the differential vicinity of the contact point K (Figure 7.18b).

The inequality $|R_P| > R_T$ can be used for the analytical description of satisfaction of the third necessary condition of proper part surface generation.

To verify the satisfaction or violation of the third necessary condition of proper part surface generation, the indicatrix of conformity $Cnf_R(P/T)$ at a point of contact of the part surface P and the generating surface T of the cutting tool can be implemented (see Chapter 4 for details on this characteristic curve).

In polar coordinates, the equation of the indicatrix of conformity $Cnf_R(P/T)$ [73,78] of the surfaces P and T can be represented in the following form (see Equation 4.87):

$$Cnf_R(P/T) \Rightarrow r_{cnf} = \sqrt{\left| \frac{E_P G_P}{L_P G_P \cos^2 \varphi - M_P \sqrt{E_P G_P} \sin 2\varphi + N_P E_P \sin^2 \varphi} \right|} \, \text{sgn} \, \Phi_{2.P} +$$

$$\sqrt{\left| \frac{E_T G_T}{L_T G_T \cos^2(\varphi + \mu) - M_T \sqrt{E_T G_T} \sin 2(\varphi + \mu) + N_T E_T \sin^2(\varphi + \mu)} \right|} \, \text{sgn} \, \Phi_{2.T} \geq 0$$

$$(7.44)$$

(See Chapter 4 for details of Equation 7.44.)

When the third necessary condition of proper part surface generation is satisfied, then all the diameters* $d_{cnf} = 2r_{cnf}$ of the indicatrix of conformity $Cnf_R(P/T)$ are nonnegative. Thus, within the common tangent plane in any direction through the contact point K, the inequality $r_{cnf} \geq 0$ is satisfied. The set of two equations sgn $r_{cnf} = 0$ and sgn $r_{cnf} = +1$ is equivalent to the inequality $r_{cnf} \geq 0$. When the third necessary condition of proper part surface generation is satisfied, either the equation sgn $r_{cnf} = 0$ or the equation sgn $r_{cnf} = +1$ is valid.

---

* A diameter of a centrosymmetrical curve can be defined as a distance between two points of the curve measured along a straight line through the center of symmetry of the curve.

TABLE 7.1

A Classification of Possible Kinds of Contact of the Surfaces P and T (in Normal Cross Sections)

		Normal Plane Section of the Part Surface P (Local Representation)		
		Convex	Linear	Concave
Normal Plane Section of the Generating Surface T of a Form Cutting Tool (Local Representation)		1. $R_P > 0$	2. $R_P \to \infty$	3. R_P
Convex	1. $0 < R_T < +\infty$	1.1 $R_P > 0$ $R_T > 0$	1.2 $0 < R_T < R_P \to \infty$	$|R_P| > |R_T|$ $|R_P| = |R_T|$ $|R_P| < |R_T|$

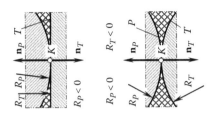

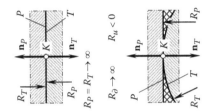

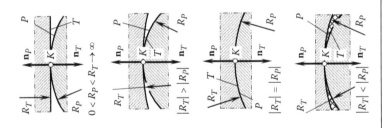

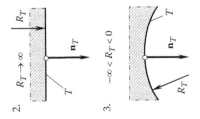

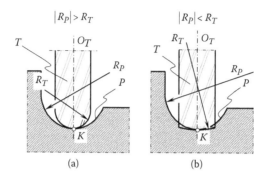

FIGURE 7.18

Examples of satisfaction (a) and of violation (b) of the *third necessary condition* of proper part surface generation. (Reprinted from *Computer-Aided Design*, 34, Radzevich, S.P., Conditions of proper sculptured surface machining, 727–740, Copyright 2002, with permission from Elsevier.)

When the minimal diameter d_{cnf}^{min} of the indicatrix of conformity $Cnf_R(P/T)$ at a point of contact of the sculptured part surface P and the generating surface T of the form-cutting tool is nonnegative, say, when the inequality $d_{cnf}^{min} \geq 0$ is observed, all other diameters d_{cnf} of this characteristic curve are nonnegative as well. Therefore, the third necessary condition of proper part surface generation is satisfied when the minimal diameter d_{cnf}^{min} is equal to or exceeds zero $\left[d_{cnf}^{min} \geq 0 \right]$.

The above performed analysis enables the following formulation:

> *The Third Condition of Proper Part Surface Generation.* The condition of proper contact of a sculptured part surface P to be machined and of the generating surface T of the form-cutting tool without their mutual penetration, (that is, without their mutual interference in the differential vicinity of the point of contact) is the third necessary condition of proper part surface generation.

As an illustration of the necessity of proper correspondence between the radii of normal curvature of the contacting surfaces, consider the contact of a concave portion of the sculptured part surfaces P and of a convex portion of the generating surface T of a form-cutting tool as shown in Figure 7.19a. In Figure 7.19, the configuration of the principal plane sections $C_{1.P}$ and $C_{2.P}$ is specified by the unit tangent vectors $\mathbf{t}_{1.P}$ and $\mathbf{t}_{2.P}$ of the principle directions on the sculptured part surfaces P. The configuration of the principal plane sections $C_{1.T}$ and $C_{2.T}$ is specified by the unit tangent vectors $\mathbf{t}_{1.T}$ and $\mathbf{t}_{2.P}$ of the principle directions on the generating surface of the form-cutting tool T. The unit tangent vectors $\mathbf{t}_{1.P}$, $\mathbf{t}_{2.P}$ and $\mathbf{t}_{1.T}$, $\mathbf{t}_{2.P}$ of the principal directions are aligned with the axis of symmetry of *Dupin's* indicatrix $Ind(P)$ and $Ind(T)$ of the surfaces P and T.

Radii of normal curvature of the part surface P and the generating surface T of the form-cutting tool at the point K of their contact are measured in a plane section through the unit normal $\mathbf{n}_P$ and through an arbitrary direction

$$\rho_{T.c} = \frac{R_{T.c}}{\sin(\beta - \gamma_t)} \tag{7.46}$$

where $R_{T.c}$ designates the distance of the point c from the axis O_T of the grinding wheel and β designates the angle that forms the axes O_P and O_T.

To satisfy the third necessary condition of proper part surface generation, satisfaction of the inequality $\rho_{T.c} \leq |\rho_{P.c}|$ is a necessity. Therefore, the inequality

$$\frac{r_{P.c}}{\sin\gamma_t} \geq \frac{R_{T.c}}{\sin(\beta - \gamma_t)} \tag{7.47}$$

must be satisfied when regrinding the rake surface of the broach.

The inequality (7.47) can be cast into the formula for calculation of the maximal feasible outside radius $R_{T.c}$ (diameter) of the grinding wheel:

$$R_{T.c} \leq \frac{r_{P.c}\sin(\beta - \gamma_t)}{\sin\gamma_t} \tag{7.48}$$

In a similar way, a formula for the calculation of the maximal feasible outside radius $R_{T.c}$ (diameter) of the grinding wheel for grinding broaches having helical grooves can be derived.

The indicatrix of conformity $Cnf_R(P/T)$ at a point of contact of the part surface P and the generating surface T of the cutting tool (see Equation 7.44) can be used for the purpose of verification whether the third necessary condition of proper part surface generation is satisfied or violated. The equation of this characteristic curve can also be interpreted as an analytical representation of the third necessary condition of proper part surface generation. Implementation of this characteristic curve is of particular importance for the development of software for the machining of a sculptured part surface on a multiaxis *NC* machine.

When the radii of normal curvature R_P and R_T of the surface P and T are of the same magnitude and of opposite sign (say, when the equality $R_P = -R_T$ is valid), the scenario is of special interest for the theory of part surface generation. Due to deviations in relative configuration when machining a sculptured part surface on a multiaxis *NC* machine (the deviations are inevitable), types of contact for which the condition $R_P = -R_T$ is valid, must be eliminated. However, the condition $R_P = -R_T$ could be satisfied in one of quasi-types of contact of the part surface P and the generating surface T of the form-cutting tool.

For example, consider contact of a curve *ABC* having continuously decreasing radii of curvature from point A toward point C. For this curve, the inequalities $R_A > R_B > R_C$ are satisfied. At point B, the curve *ABC* makes tangency with a circle of radius R. Evidently at points a, b, and c of the circular arc abc, the equality $R_a = R_b = R_c$ is valid. Figure 7.21a reveals either the arcs AB and ab are in proper contact or the arcs BC and bc interfere with one another ($R_A > R_a$, $R_B = R_b$, and $R_C < R_c$), or vice versa, the arcs BC and bc are in proper contact and the arcs AB and ab interfere in one another ($R_A < R_a$, $R_B = R_b$, and $R_C > R_c$). The interference is observed regardless of whether at point B the equality $R_B = R_b$ is valid.

Because of this, when the equality $R_P = -R_T$ is valid, then for the proper contact of the surfaces of P and T, it is necessary to properly orient them with respect to one another, and to ensure corresponding gradients of radii of normal curvature (Figure 7.21b). Otherwise, violation of the third necessary condition of proper part surface generation is inevitable (Figure 7.21c).

Violation of the third necessary condition of proper part surface generation under a scenario when the equality $R_P = -R_T$ is valid, could be observed in the practice of sculptured part surface machining on a multiaxis NC machine. Figure 7.22 illustrates a few cases when the third condition is violated.

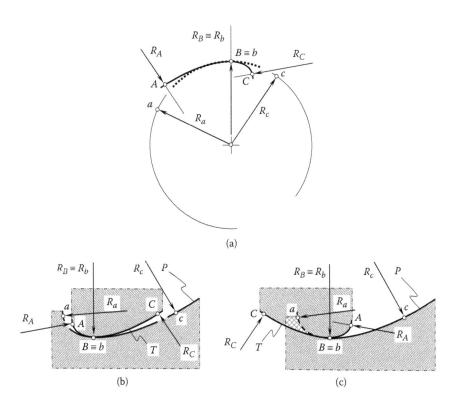

FIGURE 7.21
Examples of satisfaction and of violation of the *third necessary condition* of proper part surface generation in the cases when the condition $R_P = -R_T$ is observed.

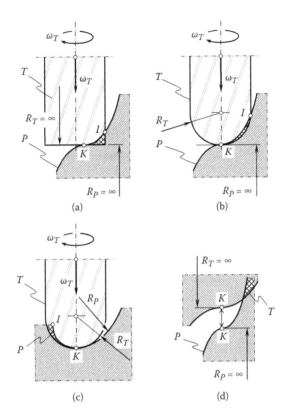

FIGURE 7.22
Examples of special cases of the violation of the *third necessary condition* of proper part surface generation.

The machining of a sculptured part surface P near a point of inflection (Figure 7.22a) cannot be performed with a flat-end milling cutter (Figure 7.22a). The curvature k_P of the part surface P and the curvature k_T of the cutting tool surface T at the contact point K are equal to each other ($k_P = k_T = 0$). However, because the equality $R_P = -R_T$ is valid, satisfaction of the condition $k_P = k_T = 0$ is not sufficient for the satisfaction of the third necessary condition of proper part surface generation in this case.

Similarly, when a gradient of increasing curvature k_P of the part surface P exceeds the limits, then the part surface P cannot be properly machined near the point of inflection (Figure 7.22b) or near the concave point (Figure 7.22c). Again, at the contact point K the ratio between the curvatures k_P and k_T yields the assumption, the machining of the part surface P is feasible. However, because of insufficient gradient of increasing curvature k_P of the part surface P, the third necessary condition of proper part surface generation in this case is not satisfied.

A similar analysis can be performed for the case, when at the point of contact K of the part surface P and the generating surface T of the form-cutting tool, both of them are contacting with the point of inflection (Figure 7.22d). Evidently, in this case, the third necessary condition of proper part surface generation can be easily violated.

7.2.4 Fourth Condition of Proper Part Surface Generation

When no local interference of the part surface P and the generating surface T of the form-cutting tool is observed, surfaces P and T can interfere with each other out of the local vicinity of the point K of their contact. Interference of this type occurs when milling parts by cylindrical milling cutters as shown in Figure 7.23a, conical milling cutters (Figure 7.23b), and so forth. This type of interference of the surfaces P and T is referred to as *global interference* of the part surface P and the generating surface T of the cutting tool.

To verify satisfaction or violation of global interference of the sculptured part surface P and the generating surface T of the form-cutting tool, equation $\mathbf{r}_P = \mathbf{r}_P(U_P, V_P)$ of the part surface P, and equation $\mathbf{r}_T = \mathbf{r}_T(U_T, V_T)$ of the generating surface T must be represented in a common reference system. For the satisfaction of the fourth necessary condition of proper part surface generation, no real solutions to the set of two equations

$$\begin{cases} \mathbf{r}_P = \mathbf{r}_P(U_P, V_P) \\ \mathbf{r}_T = \mathbf{r}_T(U_T, V_T) \end{cases} \tag{7.49}$$

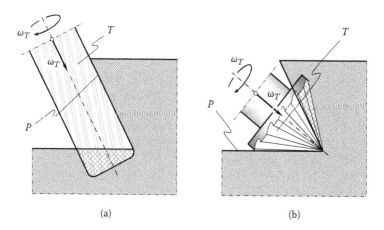

(a)   (b)

FIGURE 7.23
Examples of violation of the *fourth necessary condition* of proper part surface generation. (Reprinted from *Computer-Aided Design*, 34, Radzevich, S.P., Conditions of proper sculptured surface machining, 727–740, Copyright 2002, with permission from Elsevier.)

must be observed out of point(s) at which surfaces P and T make contact of a regular type.

> *The Fourth Condition of Proper Part Surface Generation.* The fourth necessary condition of proper part surface generation is satisfied if and only if no global interference of the sculptured part surface P to be machined and of the generating surface T of the form-cutting tool is observed.

As an example of violation of the fourth necessary condition of proper part surface generation, hobbing of involute gears having low tooth count may be mentioned.

A gear is hobbed by the hob having entering cutting edges $a_c b_c$, and exiting cutting edges $c_c d_c$ as schematically illustrated in Figure 7.24. When hobbing a gear, the work is rotated about its axis of rotation O_g at a uniform angular velocity ω_g. In the rolling motion, the cutting rack of the hob travels straight with a certain linear velocity V_{cutter}. The velocities are synchronized in a timely manner.

When the entering cutting edge $a_c b_c$ is generating the involute tooth profile $a_g b_g$ of the gear, at that same time the lateral cutting edge $c_c d_c$ (to be more exactly, the rack corner c_c) is undercutting the not yet generated gear tooth profile $c_g d_g$. Further, when the rotations progress, the exiting lateral cutting edge $c_c d_c$ starts generating the gear tooth profile $c_g d_g$. Simultaneously, the already generated involute tooth profile $a_g b_g$ is undercut by the cutting edge $a_c b_c$ (to be more exact, the rack corner a_c). Ultimately, instead of a true involute tooth profile $a_g b_g$, the undercut tooth profile $b_g e_g$ is generated. The same happens to the opposite gear tooth profile $c_g d_g$. Finally, the large undercut profile $c_g f_g e_g$ is generated in the gear.

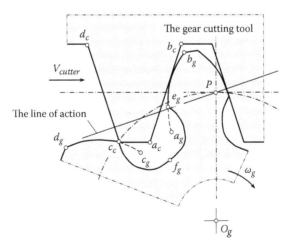

FIGURE 7.24
Example of violation of the *fourth necessary condition* of proper part surface generation when hobbing an involute gear having low tooth count.

The undercut reduces the strength of the gear teeth, and, thus, it must be eliminated. For this purpose, the hob, as well as the entire gear hobbing operation must be designed properly.

7.2.5 Fifth Condition of Proper Part Surface Generation

Mechanism and machine components are usually bounded not by a single surface P to be machined, but by several part surfaces P_i, for example, having a sculptured part surface P component, can also be bounded by walls next to the surface P. A sculptured surface P_i may have not only two but usually more neighboring surfaces $P_{i\pm1}$.

To machine the part, the form-cutting tool has to reproduce all the corresponding generating surfaces T_i. For machining of the neighboring part surface $P_{i\pm1}$, the form-cutting tool has to be capable of generating corresponding surface $T_{i\pm1}$. Various types of relative configuration of the neighboring generating surfaces T_i and $T_{i\pm1}$ are feasible.

The generating surfaces T_i and $T_{i\pm1}$ of the form-cutting tool can be apart from each other (Figure 7.25a). In this case, boundaries ab and cd of the surface portions T_i and $T_{i\pm1}$ have no common points.

The generating surfaces T_i and $T_{i\pm1}$ of the form-cutting tool may have a common boundary curve $ab \equiv cd$ (Figure 7.25b) or they can intersect each other (Figure 7.25c) at a segment ef of a curved line.

Finally, the generating surface T_i of the form-cutting tool can be located within the interior of the cutting tool body, which is bounded by the generated surface $T_{i\pm1}$ (Figure 7.25d).

Various feasible relative configurations of the generating surfaces T_i and $T_{i\pm1}$ of the cutting tool cause different conditions of generation of the sculptured part surface P_i and $P_{i\pm1}$.

When the surface portions T_i and $T_{i\pm1}$ are apart from each other (Figure 7.25a), both the portions T_i and $T_{i\pm1}$ of the generating surface of the form-cutting tool can be reproduced by the cutting tool. Under such a scenario, the given sculptured part surface can be machined without deviating from the desired geometry.

The same occurs when the surface portions T_i and $T_{i\pm1}$ share a boundary curve $ab \equiv cd$ (Figure 7.25b). In this case, the given sculptured part surface can be machined without any deviations from desired its shape.

In a particular case, the surface portions T_i and $T_{i\pm1}$ may intersect each other (Figure 7.25c). When the intersection is observed, the entire generating surface of the form-cutting tool cannot be reproduced by the form-cutting tool. The portion $abfe$ of the surface T_i and the portion $cdef$ of the surface $T_{i\pm1}$ cannot be reproduced by the form-cutting tool edges. Therefore, corresponding portions of the sculptured part surfaces P_i and $P_{i\pm1}$, which have to be generated by the portions $abfe$ and $cdef$ of the surfaces T_i and $T_{i\pm1}$, cannot be machined—this is caused by the form-cutting tool not being capable of reproducing the portions $abfe$ and $cdef$ of the surfaces T_i and $T_{i\pm1}$. Portions of

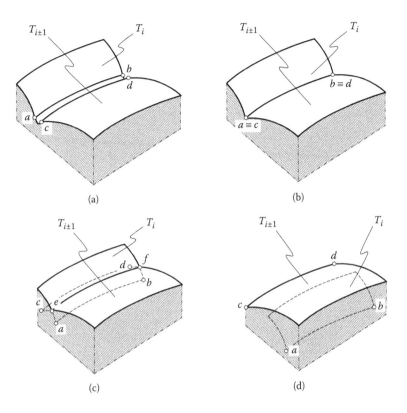

FIGURE 7.25
Feasible relative configurations of two neighboring portions of the generating surface of a form-cutting tool. (Reprinted from *Computer-Aided Design*, 34, Radzevich, S.P., Conditions of proper sculptured surface machining, 727–740, Copyright 2002, with permission from Elsevier.)

the part surfaces P_i and $P_{i\pm1}$, those that have to be generated by the portions *abfe* and *cdef* of the cutting tool surfaces T_i and $T_{i\pm1}$, are actually replaced with some type of *transient surface* instead.

As an example, consider two neighboring portions P_i and P_{i+1} of a sculptured surface. The surface portions P_i and $P_{i\pm1}$ share the mutual boundary curve *AB* (Figure 7.26a). For proper generation of the part surface P_i, the Shishkov equation of contact $\mathbf{n}_i \cdot \mathbf{V}_\Sigma = 0$ must be satisfied. Similarly, for proper machining of the surface $P_{i\pm1}$, the equation of contact $\mathbf{n}_{i\pm1} \cdot \mathbf{V}_\Sigma = 0$ must be satisfied. For machining both surfaces P_i and $P_{i\pm1}$ simultaneously, both the equations of contact $\mathbf{n}_i \cdot \mathbf{V}_\Sigma = 0$ and $\mathbf{n}_{i\pm1} \cdot \mathbf{V}_\Sigma = 0$ must be satisfied simultaneously.

When machining a convex portion of the sculptured part surface as illustrated in Figure 7.26b, the equations of contact $\mathbf{n}_i \cdot \mathbf{V}_\Sigma = 0$ and $\mathbf{n}_{(i\pm1)} \cdot \mathbf{V}_\Sigma = 0$ can easily be satisfied at every instant of machining. In this case, the cutting tool surfaces T_i and $T_{i\pm1}$ can be located apart from each other (as shown on Figure 7.25a). The lines E_i and $E_{i\pm1}$ (and so forth) of contact of the surfaces P_i and T_i, $P_{i\pm1}$ and $T_{i\pm1}$, and so forth are known as *characteristic lines*.

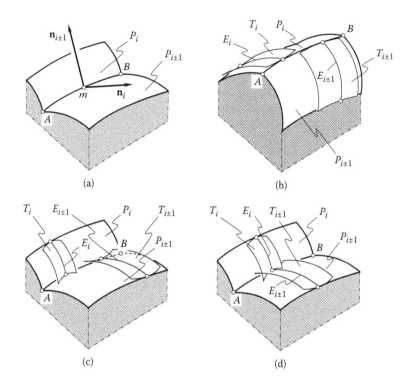

(a)   (b)

(c)   (d)

FIGURE 7.26
Various conditions of generating of the two neighboring portions, P_i and $P_{i\pm 1}$, of a sculptured part surface. (Reprinted from *Computer-Aided Design*, 34, Radzevich, S.P., Conditions of proper sculptured surface machining, 727–740, Copyright 2002, with permission from Elsevier.)

It is much more difficult to satisfy simultaneously the equations of contact $\mathbf{n}_i \cdot V_\Sigma = 0$ and $\mathbf{n}_{i\pm 1} \cdot V_\Sigma = 0$ when machining concave portions of a sculptured part surface (Figure 7.26c). The cutting tool surfaces T_i and $T_{i\pm 1}$ intersect each other along a certain curve AB as shown in Figure 7.26c. The entire surfaces P_i and $P_{i\pm 1}$ cannot be machined in this way. Some portions of the surfaces P_i and $P_{i\pm 1}$ are replaced with a certain transient curved surface instead.

For proper machining of the surface (Figure 7.26c), there is only one possible way to satisfy the equations of contact $\mathbf{n}_i \cdot V_\Sigma = 0$ and $\mathbf{n}_{i\pm 1} \cdot V_\Sigma = 0$ simultaneously. For this purpose, it is necessary to eliminate crossing of the characteristic lines E_i and $E_{i\pm 1}$ as shown in Figure 7.26c, and to ensure the characteristic lines E_i and E_i share common end-points (Figure 7.26d).

For the analytical description of the fifth necessary condition of proper part surface generation, the following set of two equations is considered:

$$\begin{cases} \mathbf{r}_T^{(i)} = \mathbf{r}_T^{(i)}\left[U_T^{(i)}; V_T^{(i)}\right] \\ \mathbf{r}_T^{(i\pm 1)} = \mathbf{r}_T^{(i\pm 1)}\left[U_T^{(i\pm 1)}; V_T^{(i\pm 1)}\right] \end{cases} \tag{7.50}$$

where $\mathbf{r}_T^{(i)}$ and $\mathbf{r}_T^{(i\pm1)}$ designate position vectors of the cutting tool surfaces T_i and $T_{i\pm1}$, and $U_T^{(i)}$ and $V_T^{(i)}$, and $U_T^{(i\pm1)}$ and $V_T^{(i\pm1)}$ are the curvilinear (*Gaussian*) coordinates of a point on the surfaces T_i and $T_{i\pm1}$, respectively.

The fifth necessary condition of proper part surface generation is satisfied if and only if the set of two Equations 7.50) has no real solution. A solution (if any) is allowed only along the common boundary curve (similar to the curve AB shown in Figure 7.26d).

> *The Fifth Condition of Proper Part Surface Generation.* The fifth necessary condition of proper part surface generation could be satisfied if and only if the neighboring portions of the generating surface of the form-cutting tool do not intersect each other, and none of them is located within the interior of the generating body of the cutting tool beneath the other neighboring surface portion.

In other words, to satisfy the fifth necessary condition of proper part surface generation, no transient surfaces are allowed to be observed on the machined part surface P.

A particular case of part surface generation occurs when the adjacent portions of the generating surface of the cutting tool overlap. Such a scenario is illustrated with an example.

Consider machining a part having two adjacent surfaces P_1 and P_2 as schematically illustrated in Figure 7.27. The conical surface P_1 and the cylindrical surface P_2 of the part are machined in one setup with the cutting tool having the conical generated surface T. Both portions T_1 and T_2 of the generating surface of the cutting tool are congruent to each other (say, $T_1 \equiv T_2$). When machining the part, the axis of rotation of the work is parallel to the generating

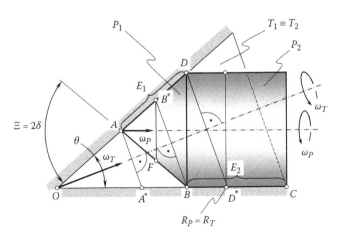

FIGURE 7.27
Generation of conical P_1 and of cylindrical P_2 part surfaces with the generating surface of the cutting tool shaped in the form of a surface $T_1 \equiv T_2$ of internal cone.

straight line of the cone surface $T_1 \equiv T_2$. The work rotates about the part axis of rotation, while the cutting tool rotates about its axis of rotation. The cone angle of the cutting tool surfaces $T_1 \equiv T_2$ is designated as θ, the rotation of the work is designated as ω_P, and rotation of the cutting tool is designated as ω_T.

For the generation of the conical surface P_1 having generating straight line AD, the characteristic line E_1 is necessary. Cylindrical surface P_2 is generated by straight line BC. To be generated, the characteristic line E_2 is required. However, due to the symmetry, certain portions BD^* and B^*D of the characteristic lines E_1 and E_2 overlap. Because the characteristics E_1 and E_2 do not intersect each other, no transient surface is generated. This is because the fifth necessary condition of proper part surface generation is not violated in the case under consideration.

Violation of the fifth necessary condition of proper part surface generation is illustrated with another example.

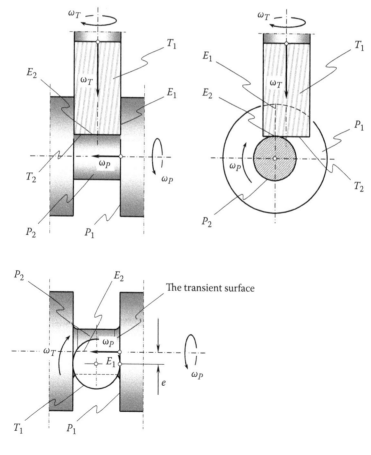

FIGURE 7.28
Machining of two surfaces P_1 and P_2 in one setup with the flat-end milling cutter.

Consider milling a part having two adjacent surfaces P_1 and P_2 as schematically shown in Figure 7.28. The flat part surface P_1 and the cylindrical part surface P_2 are machined in one setup with the flat-end milling cutter. When machined the part, the axis of rotation of the cutting tool is perpendicular to the axis of rotation of the part. The work rotates about the part axis of rotation, while the cutting tool rotates about its axis of rotation. The rotation of the work is designated as ω_P and the rotation of the cutting tool is designated as ω_T.

For the generation of the flat portion P_1, the characteristic line E_1 is necessary. Cylindrical portion P_2 is generated by the characteristic line E_2. However, the characteristic lines E_1 and E_2 intersect each other. Because of this, neither the entire flat surface P_1, nor the entire cylindrical surface P_2 can be generated to the blueprint. The form transient surface is machined instead. This is due to violation of the fifth necessary condition of proper part surface generation.

7.2.6 Sixth Condition of Proper Part Surface Generation

When machining a sculptured part surface, a point contact of the part surfaces P and of the generating surface T of the form-cutting tool is usually observed. Due to point contact of the surfaces, the so-called *discrete* generation of the sculptured part surface often occurs.* Representation of the generating surface T by distinct cutting edges of the form-cutting tool is the other reason the discrete generation of the sculptured part surface takes place.

In an instant of time, it is physically impossible to generate the sculptured part surface P by a single moving point. When the discrete surface generation occurs, the nominal smooth regular sculptured part surface P_{nom} and the actual machined part surface P_{act} are not identical. The actual part surface P_{act} can be interpreted as the nominal sculptured surface P_{nom} that is covered by cusps (Figure 7.29) or may have other deviations from the desired surface P_{nom}.

The sixth necessary condition of proper part surface generation is formulated as follows:

> *The Sixth Condition of Proper Part Surface Generation.* The actual part surface P_{act} with cusps, if any, must remain within the tolerance for part surface accuracy.

The cusps on the machined sculptured part surface P must be within the tolerance for part surface accuracy.

Then maximal height h_Σ of the cusps must not exceed the tolerance $[h]$ for the sculptured part surface accuracy.

---

* Discrete generation of a sculptured part surface in more detail is illustrated in Figures 8.1 and 8.2 (see Chapter 8).

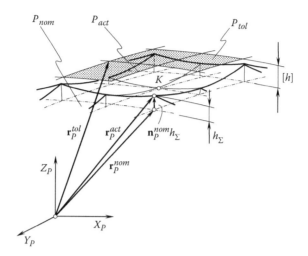

FIGURE 7.29
Cusps on the machined sculptured part surface P. (Reprinted from *Computer-Aided Design*, 34, Radzevich, S.P., Conditions of proper sculptured surface machining, 727–740, Copyright 2002, with permission from Elsevier.)

Consider a *Cartesian* coordinate system $X_P Y_P Z_P$ associated with the sculptured part surface P.

The sixth necessary condition of proper part surface generation is satisfied if and only if the following condition is satisfied at every point of the nominal sculptured part surface P:

$$\mathbf{n}_P^{nom} \cdot h_\Sigma = \mathbf{r}_P^{act} - \mathbf{r}_P^{nom} \leq \mathbf{r}_P^{tol} = \mathbf{n}_P^{nom} \cdot [h] \qquad (7.51)$$

where the position vector of a point of a nominal sculptured part surface P_{nom} is designated as $\mathbf{r}_P^{nom}$; the position vector of the corresponding point* of the actual part surface P_{act} is designated as $\mathbf{r}_P^{act}$, the position vector of a point of the surface of tolerance P_{tol} is designated as $\mathbf{r}_P^{tol}$, and, finally the unit normal vector to the surface P_{nom} is designated as $\mathbf{n}_P^{nom}$.

If the sixth necessary condition of proper part surface generation is satisfied, then the actual part surface P_{act} is entirely located within the gap between the nominal sculptured part surface P_{nom} and the surface of tolerance $P^{(t)}$. In the example shown in Figure 7.29, the surface of tolerance P_{tol} is depicted over the nominal part surface P_{nom} at a distance equal to $\mathbf{n}_P^{(n)} \cdot [h]$.

Fulfillment of the set of six conditions of proper part surface generation is necessary and sufficient to insure machining of the part surface to the part blueprint.

---

* Two points on the surfaces $\mathbf{r}_P^{nom}$ and $\mathbf{r}_P^{act}$ correspond to each other if they share a common straight line, which is aligned with the perpendicular $\mathbf{n}_P^{nom}$ to the surface $\mathbf{r}_P^{nom}$.

7.3 Global Verification of Satisfaction of the Conditions of Proper Part Surface Generation

When machining a sculptured part surface on a multiaxis *NC* machine, it is important to get to know whether the entire part surface can or cannot be machined on the given machine. It also important to detect those sculptured part surface regions not accessible by the cutting tool of a given design in a prespecified *NC* machine. In other words, it is necessary to detect regions on the sculptured part surface *P*, which the cutting tool cannot reach without being obstructed by another portion of the part. Certainly, such regions (if any) are not merely due to the geometry of the sculptured part surface *P* but also to the geometry of the generating surface *T* of the cutting tool. The particular problem under consideration now is referred to as the *cutting tool–dependent partitioning* of a sculptured part surface onto the cutting tool–accessible and onto the cutting tool–inaccessible regions.

7.3.1 Implementation of the Focal Surfaces

For solving the problem of cutting tool–dependent partitioning (further *CT*-dependent partitioning) of a sculptured part surface, the *third necessary condition* of proper part surface generation is the most critical issue. The geometry of contact of the part surface *P* and the generating surface *T* of the form-cutting tool in the infinitesimal vicinity of a cutter-contact point *K* (further *CC* point *K*) is a vital link for verification of whether the third necessary condition of proper part surface generation is globally satisfied or not.

Within the cutting tool–accessible portions of the sculptured part surface, the proper correspondence is observed between the normal curvature k_P of the part surface *P* and the normal curvature k_T of the generating surface *T* of the form-cutting tool (Table 7.1). The normal curvatures k_P and k_T are measured in a common plane section through the unit normal vector $\mathbf{n}_P$ direction of which is specified by a unit tangent vector t_P. Otherwise, when the correspondence between the normal curvatures k_P and k_T is improper (Table 7.1), the interference of the surfaces *P* and *T* occurs. Such regions of the part surface *P* cannot be machined properly.

Implementation of the indicatrix of conformity $Cnf_R(P/T)$ at a point of contact of a part surface *P* and the generating surface *T* of the form-cutting tool (see Equation 4.87) enables detection of local, not global, interference of the surfaces *P* and *T*. If negative diameters d_{cnf} of the indicatrix of conformity $Cnf_R(P/T)$ are observed, this immediately indicates that a certain portion of the part surface *P* is not machinable with the cutting tool of a given design. It is easy to conclude that within the bordering curve between the cutting tool–accessible and the cutting tool–inaccessible portions of the part surface

P, the identity $d_{cnf} \equiv 0$ is observed.* Ultimately, the problem of partitioning of a sculptured part surface reduces to the problem of finding those lines on the part surface P within which the identity $d_{cnf} \equiv 0$ is valid. For solving the problem, various approaches can be used. The implementation of focal surfaces is promising in this concern.

7.3.1.1 Focal Surfaces

The geometry of contact of the part surface P and the generating surface T of the form-cutting tool in the infinitesimal vicinity of a CC point K, turns our attention to the normal curvatures of the surfaces P and T and to the location of centers of normal curvature of these surfaces.

The direction of feasible tool approach to a surface point is defined as the direction along which a cutting tool can reach a part surface without being obstructed by another portion of the part. For a part design to be machinable, every feature of the part design should have at least one such feasible direction. For a sculptured part surface, if a point on the surface does not have at least one such feasible direction, it is not machinable.

Global analysis and the detection of the part surface P regions, those that are cutting tool–accessible as well as those that are cutting tool–inaccessible, and a visual interpretation of the global accessibility of the surface can be performed by means of focal surfaces constructed at a point of contact of the part surface P and of the cutting tool surface T.

For generating the focal surfaces, it is necessary to recall that there are two principle plane sections $C_{1.P}$ and $C_{2.P}$ through a point m of smooth regular sculptured part surface P. Principle planes $C_{1.P}$ and $C_{2.P}$ pass through the part surface P unit normal vector n_P, and through the directions specified by the principal unit tangent vectors $t_{1.P}$ and $t_{2.P}$. Principal radii of curvature $R_{1.P}$ and $R_{2.P}$ of the part surface P are measured in the principal plane sections $C_{1.P}$ and $C_{2.P}$. The centers of curvature $O_{1.P}$ and $O_{2.P}$ of the sculptured part surface at the point m (Figure 7.30) are located within the straight line that is aligned with the unit normal vector $\mathbf{n}_P$ erected at the point m. Points of this type are commonly referred to as *focal points* of a part surface P at a point m.

The locus of the focal points $O_{1.P}$, which are determined for every point of the sculptured part surface P, represents the first principal focal surface $f_{1.P}(U_P, V_P)$. Similarly, the locus of the focal points $O_{2.P}$, which are also determined for every point of the sculptured part surface P, represents the second principal focal surface $f_{2.P}(U_P, V_P)$. In particular cases, two focal surfaces $f_{1.P}(U_P, V_P)$ and $f_{2.P}(U_P, V_P)$ can be congruent to each other. Under such a scenario, both focal surfaces appear as a single surface.

A focal surface can shrink to a curved line. The surface of a cylinder of revolution is a good example in this concern. The focal surface $f_{1.P}(U_P, V_P)$ for the cylinder of revolution is degenerated to the straight line that aligns with

---

* The same is true with respect to the $\mathscr{An}_R(P/T)$–indicatrix (see Chapter 4).

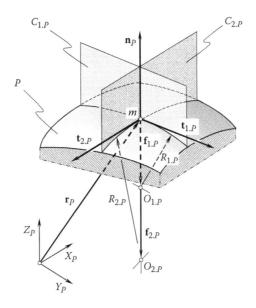

FIGURE 7.30
Determination of the focal point at the current point of the sculptured part surface *P*.

the axis of rotation of the cylinder surface. Moreover, a focal surface can even shrink to a point. The last is observed for both the focal surfaces $f_{1.P}(U_P,V_P)$ and $f_{2.P}(U_P,V_P)$ of a spherical surface *P*.

Under certain conditions, consecutive positions of a straight line that is aligned to unit normal vector $\mathbf{n}_P$ at points along lines of curvature of a part surface *P* can form enveloping curves that are space curves shown in Figure 7.31. Focal surfaces can be considered as the family of consecutive positions of the corresponding space enveloping curves [117].

The considered geometrical property of focal surfaces is used below for derivation of the equation of the focal surfaces.

The current point of the focal surface coincides with the centers of corresponding principal curvature of the part surface *P*. Thus, position vector $\mathbf{f}_{1.P}(U_P,V_P)$ of a point of the first focal surface as well as position vector and $\mathbf{f}_{2.P}(U_P,V_P)$ of a point of the second focal surface can be represented in terms of position vector $\mathbf{r}_P(U_P,V_P)$ of a point of the part surface *P*; in terms of unit normal vector $\mathbf{n}_P$ to the part surface *P* and in terms of corresponding radii of principal curvature, either $R_{1.P}$ or $R_{2.P}$, as it is illustrated in Figure 7.30:

$$\mathbf{f}_{1.P}(U_P,V_P) = \mathbf{r}_P(U_P,V_P) - R_{1.P} \cdot \mathbf{n}_P \tag{7.52}$$

$$\mathbf{f}_{2.P}(U_P,V_P) = \mathbf{r}_P(U_P,V_P) - R_{2.P} \cdot \mathbf{n}_P \tag{7.53}$$

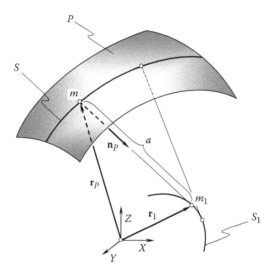

FIGURE 7.31
Representation of a focal surface as an enveloping surface to perpendiculars to the part surface P.

Elementary substitution $R_{1.P} = k_{1.P}^{-1}$ and $R_{2.P} = k_{2.P}^{-1}$ yields expression for position vectors of a point $\mathbf{f}_{1.P}$ and $\mathbf{f}_{2.P}$ of the focal surfaces (Equations 7.52 and 7.53) in terms of principal curvatures:

$$\mathbf{f}_{2.P}(U_P, V_P) = \mathbf{r}_P(U_P, V_P) - k_{2.P}^{-1} \cdot \mathbf{n}_P \tag{7.54}$$

$$\mathbf{f}_{1.P}(U_P, V_P) = \mathbf{r}_P(U_P, V_P) - k_{1.P}^{-1} \cdot \mathbf{n}_P \tag{7.55}$$

Radii of principle curvatures $R_{1.P}$ and $R_{2.P}$ in Equations 7.52 and 7.53 are calculated using one of the equations represented in Chapter 1 (see Equations 1.49 and 1.60).

Radii of principle curvatures $R_{1.P}$ and $R_{2.P}$ can be expressed in terms of the mean curvature $\mathcal{M}_P$ and of the *Gaussian* curvature $\mathcal{G}_P$ of the surface P:

$$R_{1.P} = \left(\mathcal{M}_P + \sqrt{\mathcal{M}_P^2 - \mathcal{G}_P} \right)^{-1} \tag{7.56}$$

$$R_{2.P} = \left(\mathcal{M}_P - \sqrt{\mathcal{M}_P^2 - \mathcal{G}_P} \right)^{-1} \tag{7.57}$$

Ultimately, equations for position vectors of a point $\mathbf{f}_{1.P}(U_P, V_P)$ and $\mathbf{f}_{2.P}(U_P, V_P)$ of the focal surfaces can be represented in the form:

$$\mathbf{f}_{1.P}(U_P, V_P) = \mathbf{r}_P - \frac{\mathbf{n}_P}{\mathcal{M}_P + \sqrt{\mathcal{M}_P^2 - \mathcal{G}_P}} \tag{7.58}$$

$$\mathbf{f}_{2.P}(U_P, V_P) = \mathbf{r}_P - \frac{\mathbf{n}_P}{\mathcal{M}_P - \sqrt{\mathcal{M}_P^2 - \mathcal{G}_P}} \tag{7.59}$$

The focal surfaces $\mathbf{f}_{1.P}$ and $\mathbf{f}_{2.P}$ for a saddle-like patch of a sculptured part surface P are plotted in Figure 7.29a. Such a patch of the part surface P can be machined, for example, with the convex generating surface T of a form-cutting tool. Focal surfaces $\mathbf{f}_{1.T}(U_T, V_T)$ and $\mathbf{f}_{2.T}(U_T, V_T)$ for this cutting tool surface T are depicted in Figure 7.32b. In Figure 7.32, the respective lines of curvature are designated as $\breve{C}_{1.P}$, $\breve{C}_{2.P}$, and $\breve{C}_{1.T}$, $\breve{C}_{2.T}$, respectively. Points $O_{1.T}$,

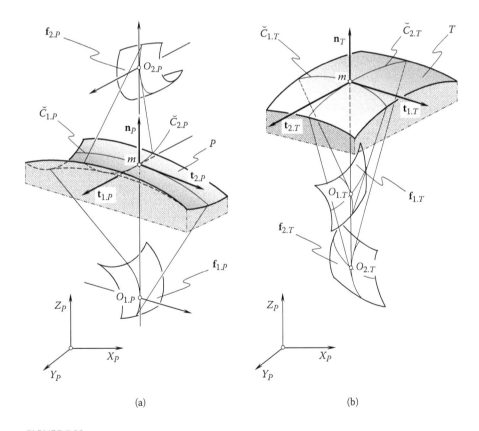

FIGURE 7.32
Examples of focal surfaces constructed for a local patch of the sculptured part surface P (a), and for a local patch of the generating surface T of the form-cutting tool (b).

and $O_{1.T}$ are the points of the corresponding focal surfaces $\mathbf{f}_{1.T}$ and $\mathbf{f}_{2.T}$ for the generating surface T of the form-cutting tool at a point m:

$$\mathbf{f}_{1.T}(U_T,V_T) = \mathbf{r}_T(U_T,V_T) - k_{1.T}^{-1} \cdot \mathbf{n}_T \qquad (7.60)$$

$$\mathbf{f}_{2.T}(U_T,V_T) = \mathbf{r}_T(U_T,V_T) - k_{2.T}^{-1} \cdot \mathbf{n}_T \qquad (7.61)$$

Focal surfaces $\mathbf{f}_{1.P}$ and $\mathbf{f}_{2.P}$ intersect the sculptured part surface P along parabolic curved lines on it—that is, along lines at which *Gaussian* curvature $\mathcal{G}_P$ at a point of the sculptured surface is equal to zero ($\mathcal{G}_P \equiv 0$).

To use focal surfaces for the purpose of verification of whether or not the third necessary condition of proper part surface generation is satisfied globally, it is necessary to plot both of the focal surfaces $\mathbf{f}_{1.P}$ and $\mathbf{f}_{2.P}$ for the sculptured part surface P, and the similar focal surfaces $\mathbf{f}_{1.T}$ and $\mathbf{f}_{2.T}$ for the generating surface T of the form-cutting tool in a common reference system.

An example of relative configuration of the focal surfaces $\mathbf{f}_{1.P}$, $\mathbf{f}_{2.P}$ and $\mathbf{f}_{1.T}$, $\mathbf{f}_{2.T}$ at the point K of contact of the given part surface P and the generating surface T of the form-cutting tool is illustrated in Figure 7.33. The saddle-like ($\mathcal{G}_P < 0$) local patch of a sculptured part surface P is machined with a convex patch ($\mathcal{G}_T < 0$, $\mathcal{M}_T < 0$) of the generating surface T of the cutting tool. In the case under consideration, angle μ of the local relative orientation of the surfaces P and T is equal to zero ($\mu = 0°$). Inspecting Figure 7.33, it easy to realize that the first principle planes $C_{1.P} \equiv C_{1.T}$ (the identity is due to $\mu = 0°$) intersect the surfaces P and T. The lines of the intersection $\breve{C}_{1.P}$ and $\breve{C}_{1.T}$ are convex lines ($k_{1.P} > 0$; $k_{1.T} > 0$). Therefore, no problem is observed to satisfy the third necessary condition of proper part surface generation in this plane section. The second principle planes $C_{2.P} \equiv C_{2.T}$ (the identity is due to $\mu = 0°$, these plane sections are also congruent to each other) intersect the surfaces P and T. The line $\breve{C}_{2.P}$ of the intersection is a concave curve. The line $\breve{C}_{2.T}$ of the intersection is the convex curve. Because the distance $KO_{2.T}$ exceeds the distance $KO_{2.P}$ (i.e., $KO_{2.T} > KO_{2.P}$), the principle curvatures $k_{2.P}$ and $k_{2.T}$ correspond to each other as $|k_{2.P}| > k_{2.T}$. Because the inequality $|k_{2.P}| > k_{2.T}$ is valid, the third necessary condition of proper part surface generation in the second principal section of the surfaces P and T is not satisfied. Summarizing, one can conclude that the third necessary condition of proper part surface generation is not satisfied in the infinitesimal vicinity of the CC -point K (Figure 7.33).

Analysis of Table 7.1 allows for an analytical expression for the criterion for verification of whether the third necessary condition of proper part surface generation is satisfied or not:

$$\operatorname{sgn} k_P \cdot \operatorname{sgn} k_T \cdot \operatorname{sgn}(k_P + k_T) = \begin{cases} 0 \\ -1 \end{cases} \qquad (7.62)$$

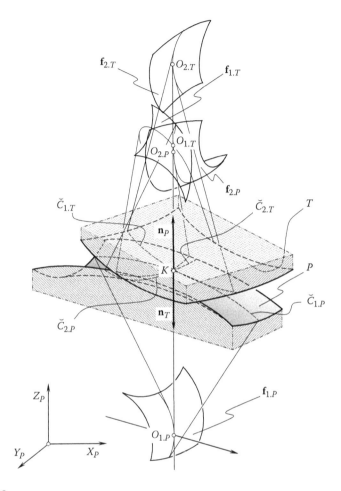

FIGURE 7.33
Configuration of the focal surfaces $\mathbf{f}_{1.P}$, $\mathbf{f}_{2.P}$ for the sculptured part surface P relative to the focal surfaces $\mathbf{f}_{1.T}$ and $\mathbf{f}_{2.T}$ for the generating surface T of the form-cutting tool.

To globally satisfy the third necessary condition of proper part surface generation, it is necessary to ensure satisfaction of Equation 7.62 at every contact point K and in every cross section of the part surface P and the generating surface T of the form-cutting tool by a plane through the unit normal vector $\mathbf{n}_P$.

The third necessary condition of proper part surface generation could be satisfied globally when each of the focal surfaces $\mathbf{f}_{1.T}$ and $\mathbf{f}_{2.T}$ is entirely located in between the convex part surface P and the corresponding focal surface $\mathbf{f}_{1.P}$ or $\mathbf{f}_{2.P}$. Focal surfaces $\mathbf{f}_{1.T}$ and $\mathbf{f}_{2.T}$ can touch one or both focal surfaces $\mathbf{f}_{1.P}$ or $\mathbf{f}_{2.P}$. In a similar way, location of the focal surfaces $\mathbf{f}_{1.T}$ and $\mathbf{f}_{2.T}$ for concave and for saddle-like local patches of a part surface P can be specified. Focal surfaces $\mathbf{f}_{1.T}$ and $\mathbf{f}_{2.T}$ must not intersect the sculptured part surface P

and the corresponding focal surfaces $\mathbf{f}_{1.P}$ and $\mathbf{f}_{2.P}$ for the generating surface of the form-cutting tool. Otherwise, the third necessary condition of proper part surface generation would be violated.

Focal surfaces $\mathbf{f}_{1.P}$ and $\mathbf{f}_{2.P}$ are the bounding surfaces of space, within which the centers of principal curvatures of the generating surface T of the form-cutting tool have been located. The portions of space bounded by the focal surfaces $\mathbf{f}_{1.P}$ and $\mathbf{f}_{2.P}$ are referred to as *cutting tool–allowed (CT-allowed) zones*. The rest of the space is referred to as the *cutting tool–prohibited (CT-prohibited) zones*.

7.3.1.2 Cutting Tool–Dependent Characteristic Surfaces

When the third necessary condition of proper part surface generation is globally satisfied, certain constraints are imposed on the actual configuration of the focal surfaces. For the purpose of verification of accessibility of the part surface P by the cutting tool, the *cutting tool–dependent (CT-dependent) characteristic surfaces* can be used. It is convenient to illustrate the concept of the CT-dependent characteristic surfaces with an example of generation of a concave patch of the sculptured part surface P.

First, the current point of the first focal surface $\mathbf{f}_{1.T}$ of the generating surface T of the cutting tool is located within the straight line along the unit normal vector $\mathbf{n}_P$ that is erected at the corresponding point of the part surface P. Second, there exists a straight-line segment within the straight line. Location of the current point of the focal surface $\mathbf{f}_{1.T}$ is allowed within the straight-line segment as well as at its end-points. Therefore, without loss of generality, instead of two focal surfaces $\mathbf{f}_{1.P}$ and $\mathbf{f}_{1.T}$, just one CT-dependent characteristic surface $\widehat{\mathbf{f}}_1(U_T, V_T)$ can be employed. This characteristic surface features the summa $(R_{1.P} + R_{1.T})$ of the first principal radii of curvature.

The locus of points, determined in the above way, forms the first *CT-dependent characteristic surface* $\widehat{\mathbf{f}}_1(U_T, V_T)$ of the sculptured part surface P and the generating surface T of the form-cutting tool. The position vector of a point of the first CT-dependent characteristic surface $\widehat{\mathbf{f}}_1$ can be expressed in terms of the parameters $\mathbf{r}_P$, $\mathbf{n}_P$, $R_{1.P}$, and $R_{1.T}$:

$$\widehat{\mathbf{f}}_1(U_T, V_T) = \mathbf{r}_P(U_T, V_T) - (R_{1.P} + R_{1.T}) \cdot \mathbf{n}_P \tag{7.63}$$

A similar analysis can be performed for the second focal surface $\mathbf{f}_{2.T}$ of the generating surface T of the form-cutting tool.

Ultimately, the position vector of a point of the second CT-dependent characteristic surfaces $\widehat{\mathbf{f}}_2$ can be expressed in terms of the parameters $\mathbf{r}_P$, $\mathbf{n}_P$, $R_{2.P}$, and $R_{2.T}$:

$$\widehat{\mathbf{f}}_2(U_T, V_T) = \mathbf{r}_P(U_T, V_T) - (R_{2.P} + R_{2.T}) \cdot \mathbf{n}_P \tag{7.64}$$

Summarizing, one can conclude that the *CT*-dependent characteristic surface is a surface, each point of which is remote from the sculptured part surface *P* perpendicular to it at a distance that is equal to the algebraic sum of the corresponding radii of principal curvature of the part surfaces *P* and of the generating surface *T* of the form-cutting tool.

When the *CT*-dependent characteristic surfaces $\hat{\mathbf{f}}_1$ and $\hat{\mathbf{f}}_2$ do not intersect the sculptured part surface *P*, then the third necessary condition of proper part surface generation is satisfied globally. Under such a scenario, the sculptured part surfaces *P* can be machined properly to the part surface blueprint. Otherwise, if the *CT*-dependent characteristic surfaces $\hat{\mathbf{f}}_1$ and $\hat{\mathbf{f}}_2$ intersect the part surface *P*, or they are entirely located within the interior of the part body, the third necessary condition of proper part surface generation cannot be satisfied. In this case, the part surfaces *P* cannot be machined properly.

Application of the *CT*-dependent characteristic surfaces for the purposes of resolving the problem of partitioning the sculptured part surface onto the cutting tool–accessible and cutting tool–inaccessible regions reduces the number of surfaces to be considered from four focal surfaces ($\mathbf{f}_{1.P}$, $\mathbf{f}_{2.P}$ and $\mathbf{f}_{1.T}$, $\mathbf{f}_{2.T}$) to two *CT*-dependent characteristic surfaces ($\hat{\mathbf{f}}_1$ and $\hat{\mathbf{f}}_2$).

The cutting tool–accessible regions are separated from the cutting tool–inaccessible regions of the sculptured part surface *P* by a corresponding boundary curve.

7.3.1.3 Boundary Curves of the CT-Dependent Characteristic Surfaces

The boundary curve for cutting tool–accessible region of the sculptured part surface *P* is the line of intersection of the part surface by the corresponding *CT*-dependent characteristic surfaces $\hat{\mathbf{f}}_1$ and $\hat{\mathbf{f}}_2$. Therefore, every point of the boundary curve $\mathbf{r}_{bc}$ satisfies the corresponding set of two equations:

$$\begin{cases} \hat{\mathbf{f}}_1 = \hat{\mathbf{f}}_1(U_P, V_P) = \mathbf{r}_P - (R_{1.P} + R_{1.T}) \cdot \mathbf{n}_P \\ \mathbf{r}_P = \mathbf{r}_P(U_P, V_P) \end{cases} \tag{7.65}$$

$$\begin{cases} \hat{\mathbf{f}}_2 = \hat{\mathbf{f}}_2(U_P, V_P) = \mathbf{r}_P - (R_{2.P} + R_{2.T}) \cdot \mathbf{n}_P \\ \mathbf{r}_P = \mathbf{r}_P(U_P, V_P). \end{cases} \tag{7.66}$$

Equations for the two-surface intersection curve can be derived from the conditions

$$\mathbf{r}_P - (R_{1.P} + R_{1.T}) \cdot \mathbf{n}_P = \mathbf{r}_P (U_P, V_P) \tag{7.67}$$

$$\mathbf{r}_P - (R_{2.P} + R_{2.T}) \cdot \mathbf{n}_P = \mathbf{r}_P (U_P, V_P) \tag{7.68}$$

The approach for determining the boundary curves, which is based on the solution to Equations 7.67 and 7.68 can be significantly simplified taking into consideration the Equation 7.62. After inserting the previously derived Equation 7.62 and rearranging Equations 7.67 and 7.68 cast into

$$\mathbf{r}_P - \text{sgn} k_{1.P} \cdot \text{sgn} k_{1.T} \cdot \text{sgn}\, (k_{1.P} + k_{1.T}) \cdot \mathbf{n}_P = \mathbf{r}_P(U_P, V_P) \tag{7.69}$$

$$\mathbf{r}_P - \text{sgn} k_{2.P} \cdot \text{sgn} k_{2.T} \cdot \text{sgn}(k_{2.P} + k_{2.T}) \cdot \mathbf{n}_P = \mathbf{r}_P(U_P, V_P) \tag{7.70}$$

Equations 7.69 and 7.70 represent an analytical description of the boundary curves that separate the cutting tool–accessible regions of the sculptured part surface P from the cutting tool–inaccessible regions on it.

Derivation of the boundary curves of the CT-dependent characteristic surfaces is illustrated below with two examples.

Consider generation of the torus surface P. A computer model of a torus surface is widely used as a convenient test case. It has been proven [116,117,136] that the torus surface provides significantly higher accuracy of approximation, and thus it is preferred for local approximation of the part surface P and the generating surface T of the form-cutting tool over quadrics. This is because the principal radii of curvature $R_{1.P}$ and $R_{2.P}$ of the part surface P (and the similar principal radii of curvature $R_{1.T}$ and $R_{2.T}$ of the cutting tool surface T) uniquely specify the corresponding torus surface.

The first principal radius of curvature $R_{1.P}$ is equal to the radius of the generating circle of the torus surface, and the second principal radius of curvature $R_{2.P}$ is equal to the radius of the outside circle of the torus surface (and therefore, the radius R of the directing circle is equal to the difference $R = R_{2.P} - R_{1.P}$). A similar condition is valid with respect the generating surface T of the form-cutting tool.

For both examples below, Equation 7.23 of the torus surface P from Example 7.1 is implemented.

Example 7.2

Consider machining of a torus surface P with the flat-end milling cutter as illustrated in Figure 7.34. The radius r of the generating circle of the surface P is equal to $r = 50$ mm, and radius R of the directing circle of the surface P is equal to $R = 90$ mm. *Gaussian* (curvilinear) coordinates θ_P and φ_P of a point on the torus surface P vary in the range of $0° \le \theta_P \le 180°$ and $0° \le \varphi_P \le 360°$. Using Equations 7.69 and 7.70 in the commercial software *MathCAD*, makes the following equation possible:

$$\mathbf{r}_{bc}(\theta_P) = \begin{bmatrix} 90 \cdot \sin \theta_P \\ 90 \cdot \cos \theta_P \\ \pm 50 \\ 1 \end{bmatrix} \tag{7.71}$$

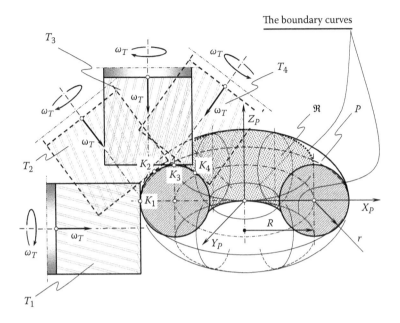

FIGURE 7.34
Partitioning of the torus surface P that is machined with a flat-end milling cutter.

for the position vector $\mathbf{r}_{bc}(\theta_P)$ of a point of the boundary curve for the
CT-dependent characteristic surface.

When machining the torus surface P, the milling cutter rotates about
its axis with a certain angular velocity ω_T. The milling cutter travels with
respect to the work occupying various positions T_1, T_2, T_3, and T_4 relative
to the surface P. The milling cutter contacts the surface P at the corre-
sponding CC points K_1, K_2, K_3, and K_4. The boundary curve $\mathbf{r}_{bc}(\theta_P)$ subdivides the surfaces P onto the cutting tool–accessible and onto the cutting
tool–inaccessible $\mathfrak{R}$ (shadowed) regions. The boundary curve $\mathbf{r}_{bc}(\theta_P)$ (see
Equation 7.71) indicates that the positions T_1 and T_2 of the milling cutter
are feasible. The cutter position T_3 is also allowed, and is limited in its
position. Due to the CC point K_4 being located within the cutting tool–
inaccessible region $\mathfrak{R}$, the position T_4 of the milling cutter is not feasible.
In that position of the milling cutter, the third necessary condition of
proper part surface generation is not satisfied; thus, the torus surface P
cannot be machined properly.

Example 7.3

Consider machining of a torus surface P with a cylindrical milling cutter as schematically shown in Figure 7.35. The same surface P as that in
Example 7.2 can be machined with a cylindrical milling cutter of the
radius of 50 mm. When machining the torus surface P, the cutter rotates

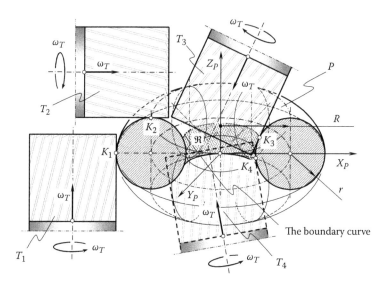

FIGURE 7.35
Partitioning of the torus surface *P* that is machined with a cylindrical milling cutter. (Reprinted from *Computer-Aided Design*, 37, Radzevich, S.P., A cutting-tool-dependent approach for partitioning of sculptured surface, 767–778, Copyright 2005, with permission from Elsevier.)

about of its axis at a constant angular velocity ω_T. The milling cutter travels with respect to the work.

Using Equations 7.69 and 7.70 in the commercial software *MathCAD* allows the equation

$$
\mathbf{r}_{bc}(\theta_P) = \begin{bmatrix} 90 \cdot \sin\theta_P \\ 90 \cdot \cos\theta_P \\ \pm 21.793 \\ 1 \end{bmatrix}
\tag{7.72}
$$

for the position vector $\mathbf{r}_{bc}(\theta_P)$ of a point of the boundary curve for the CT-dependent characteristic surface.

When machining the torus surface *P*, the milling cutter occupies various positions T_1, T_2, T_3, and T_4 relative to the torus surface *P*. It contacts the surface *P* at the CC points K_1, K_2, K_3, and K_4, respectively. The boundary curve $\mathbf{r}_{bc}(\theta_P)$ subdivides the torus surfaces *P* onto the cutting tool–accessible and cutting tool–inaccessible $\mathfrak{R}$ (shadowed) regions. The boundary curve $\mathbf{r}_{bc}(\theta_P)$ (see Equation 7.71) indicates that the configurations T_1 and T_2 of the milling cutter are allowed. The configuration T_3 of the cutting tool is also allowed, and limits the cutter location. Because the CC point K_4 is located within the cutting tool–inaccessible region $\mathfrak{R}$, the position T_4 of the milling cutter is not allowed. In that position of the milling cutter, the third necessary condition of proper part surface generation is not satisfied; thus, the torus surface *P* cannot be machined properly.

7.3.1.4 Cases of Local-Extremal Tangency of the Part Surface P and the Generating Surface T of the Form-Cutting Tool

The possible types of contact of the part surface P and the generating surface T of the form-cutting tool are investigated in Chapter 4. In the theory of part surface generation, pure local-extremal tangency of the surfaces is not of practical interest. However, this type of surface contact could be observed in the form of quasi-types of surface contact when relative displacements of the contacting surfaces are maximal.

Local-extremal types of contact of the surfaces P and T are observed when the equality $k_P = -k_T$ is valid. Under such a scenario, the focal surfaces $\mathbf{f}_{1.P}$, $\mathbf{f}_{2.P}$ and $\mathbf{f}_{1.T}$, $\mathbf{f}_{2.T}$ (or the two CT-dependent characteristic surfaces $\hat{\mathbf{f}}_1$ and $\hat{\mathbf{f}}_2$) are not helpful for solving the problem of verification of the global satisfaction of the third necessary condition of proper part surface generation. In the case under consideration, another tool must be implemented.

On the premise of the above analysis, it is recommended to use derivatives of the corresponding functions. In this way, the *derivative focal surfaces (DF surfaces)* are introduced [129]. The *DF* surfaces are analytically described by the following equation:

$$\mathbf{f}^*_{1,2.P(T)} = \mathbf{r}_{P(T)} - \frac{\partial^n R_{1,2.P(T)}}{d\tilde{C}^n_{1,2.P(T)}} \cdot \mathbf{n}_{P(T)} \tag{7.73}$$

where n designates the smallest integer number under which the any uncertainty in global satisfaction of the third necessary condition of proper part surface generation does not arise, and $\dfrac{\partial^n R_{1,2.P(T)}}{d\tilde{C}^n_{1,2.P(T)}}$ designates the derivative of $R_{1,2.P(T)}$ in the direction of $\tilde{C}^n_{1,2.P(T)}$.

In the cases under consideration, it is necessary to determine the *DF* surfaces for the sculptured part surface P:

$$\mathbf{f}^*_{1.P} = \mathbf{r}_P - \frac{\partial^n R_{1.P}}{d\tilde{C}^n_{1.P}} \cdot \mathbf{n}_P \tag{7.74}$$

$$\mathbf{f}^*_{2.P} = \mathbf{r}_P - \frac{\partial^n R_{2.P}}{d\tilde{C}^n_{2.P}} \cdot \mathbf{n}_P \tag{7.75}$$

It is necessary to determine the similar *DF* surfaces for the generating surface T of the cutting tool:

$$\mathbf{f}^*_{1.T} = \mathbf{r}_T - \frac{\partial^n R_{1.T}}{d\tilde{C}^n_{1.T}} \cdot \mathbf{n}_T \tag{7.76}$$

$$\mathbf{f}^*_{2.T} = \mathbf{r}_T - \frac{\partial^n R_{2.T}}{d\tilde{C}^n_{2.T}} \cdot \mathbf{n}_T \tag{7.77}$$

To globally satisfy the third necessary condition of proper part surface generation, the shape, the parameters, and the relative disposition of the *DF* surfaces $\mathbf{f}^*_{1.P}$, $\mathbf{f}^*_{2.P}$ and $\mathbf{f}^*_{1.T}$, $\mathbf{f}^*_{2.T}$ must be correlated with the shape, the parameters, and the relative configuration of the contacting surfaces *P* и *T*, say, in a similar way to that considered above.

Similarly, the *derivative cutting tool–dependent characteristic surfaces (DCT-dependent surfaces)* can be introduced:

$$\hat{\mathbf{f}}^*_1 = \mathbf{r}_P - \left(\frac{\partial^n R_{1.P}}{d\breve{C}^n_{1.P}} + \frac{\partial^n R_{1.T}}{d\breve{C}^n_{1.T}} \right) \cdot \mathbf{n}_P \tag{7.78}$$

$$\hat{\mathbf{f}}^*_2 = \mathbf{r}_P - \left(\frac{\partial^n R_{2.P}}{d\breve{C}^n_{2.P}} + \frac{\partial^n R_{2.T}}{d\breve{C}^n_{2.T}} \right) \cdot \mathbf{n}_P \tag{7.79}$$

The surfaces above can be used in a way that the focal surfaces $\mathbf{f}_{1.P}$, $\mathbf{f}_{2.P}$, $\mathbf{f}_{1.T}$, and $\mathbf{f}_{2.T}$ (and the *CT*-dependent characteristic surfaces $\hat{\mathbf{f}}_1$ and $\hat{\mathbf{f}}_2$) are used for the cases of regular tangency of the part surface *P* and the generating surface *T* of the cutting tool.

In cases of local-extremal tangency of the surfaces *P* and *T*, implementation of the *DF* surfaces and implementation of the *DCT*-dependent characteristic surfaces is helpful for the purpose of partitioning of the sculptured part surface *P* onto the *cutting tool–accessible*, and onto the *cutting tool–inaccessible* regions.

7.3.2 Implementation of $\mathcal{R}$ Surfaces

A proper correspondence between the normal curvatures k_P of the part surface *P* and the corresponding normal curvatures k_T of the generating surface *T* of the form-cutting tool is one of the major prerequisites for proper generation of the part surface *P* in the differential vicinity of the CC point.

7.3.2.1 Local Consideration

The geometry of contact of a part surface *P* and the generating surface *T* of the form-cutting tool can be analytically described by means of the indicatrices of conformity $Cnf_R(P/T)$ and $Cnf_k(P/T)$ [116,117,136] and by the $\mathcal{A}n_R(P/T)$–indicatrix,* or $\mathcal{A}n_k(P/T)$– indicatrix [106] (see Chapter 4). For the purpose of

---

* Equation of the characteristic curves $\mathcal{A}n_R(P/T)$ and $\mathcal{A}n_k(P/T)$ is derived by Radzevich [106] on the premise of the equation of the well-known surface—namely, of the surface of Plücker's conoid (see Chapter 4).

verification of global satisfaction of the third necessary condition of proper part surface generation, the implementation of these characteristic curves has proven to be convenient in *CAD/CAM* applications.

It is critically important to stress that ***all*** of the characteristic curves $Cnf_R(P/T)$, $Cnf_k(P/T)$, $\mathscr{A}n_R(P/T)$, and $\mathscr{A}n_k(P/T)$– specify that same directions of the extremal degree of conformity of the surfaces P and T at the current CC point. This important property of the characteristic curves is illustrated by an example of machining of a bicubic *Bezier* surface P shown in Figure 7.36.

The matrix equation for a bicubic *Bezier* patch P that is defined by a 4×4 array of points is as follows [42]:

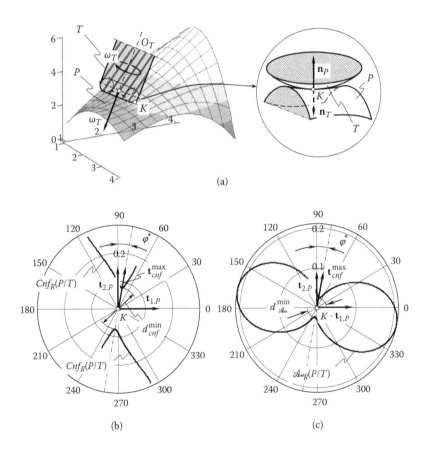

(a)

(b)

(c)

FIGURE 7.36
Planar characteristic curves $Cnf_R(P/T)$ and $\mathscr{A}n_R(P/T)$ identically determine the direction $\mathbf{t}_{cnf}^{max}$ of the maximal degree of conformity of the surfaces P and T at the current cutter–contact point K. (Reprinted from *Computer-Aided Design*, 37, Radzevich, S.P., A cutting-tool-dependent approach for partitioning of sculptured surface, 767–778, Copyright 2005, with permission from Elsevier.)

$$\mathbf{r}_P(U_P, V_P) = \begin{bmatrix} (1-U_P)^3 & 3U_P(1-U_P)^2 & 3U_P^2(1-U_P) & U_P^3 \end{bmatrix} \cdot \mathbf{P}_P \cdot \begin{bmatrix} (1-V_P)^3 \\ 3V_P(1-V_P)^2 \\ 3V_P^2(1-V_P) \\ V_P^3 \end{bmatrix} \qquad (7.80)$$

where $[\mathbf{P}_P] = [\mathbf{p}_{i,j}]_{\substack{i=1,\ldots,4 \\ j=1,\ldots,4}}$ and position vectors of the control points are denoted by $\mathbf{p}_{i,j}$.

In Equation 7.69, the bicubic patch is expressed in a form similar to the *Hermite* bicubic patch [42].

The matrix [**P**] contains the position vectors for points that define the characteristic polyhedron and, therefore, the *Bezier* surface patch. In the *Bezier* formulation, only four corner points $\mathbf{p}_{11}$, $\mathbf{p}_{41}$, $\mathbf{p}_{14}$, and $\mathbf{p}_{44}$ actually lie on the surface patch. The points $\mathbf{p}_{21}$, $\mathbf{p}_{31}$, $\mathbf{p}_{12}$, $\mathbf{p}_{13}$, $\mathbf{p}_{42}$, $\mathbf{p}_{43}$, $\mathbf{p}_{24}$, and $\mathbf{p}_{34}$ control the slope of the boundary curves. The remaining four points $\mathbf{p}_{22}$, $\mathbf{p}_{32}$, $\mathbf{p}_{23}$, and $\mathbf{p}_{33}$ control the cross slopes along the boundary curves in the same way as the twist vectors of the bicubic patch. The *Bezier* surface is completely defined by a net of design points describing two families of *Bezier* curves on the surface.

In Figure 7.36, the direction of the minimal diameter $d_{cnf}^{\min}$ of the indicatrix of conformity $Cnf_R(P/T)$ aligns with the direction $\mathbf{t}_{cnf}^{\max}$ at which the degree of conformity of the surfaces P and T is maximal.

At a current CC point, both planar characteristic curves $Cnf_R(P/T)$ (Figure 7.36b) and $\mathscr{A}n_R(P/T)$ (Figure 7.36c) identify that same direction $\mathbf{t}_{cnf}^{\max}$ within the common tangent plane, at which degree of conformity of the part surface P and generating surface T of the form-cutting tool is maximal. From this standpoint, the characteristic curves $Cnf_R(P/T)$ and $\mathscr{A}n_R(P/T)$ are equivalent. This is because the actual value of the angle φ^* that is determined from the characteristic curve $Cnf_R(P/T)$ (Figure 7.36b), is identical to the actual value of the angle φ^* that is determined from the characteristic curve $\mathscr{A}n_R(P/T)$ (Figure 7.36c). The consideration below is focused on implementation of the indicatrix of conformity $Cnf_R(P/T)$. However, that same result can be obtained using the $\mathscr{A}n_R(P/T)$–characteristic curve.

It is well known that the larger the cutting tool, the smaller the resulting scallop height, or the excessive material not being removed might result for the same tool path. The indicatrix of conformity $Cnf_R(P/T)$, and the $\mathscr{A}n_R(P/T)$–characteristic curve can be employed to directly bound the largest cutting tool radius that can be used for the machining of saddle-like and concave regions and hence aid in tool selection.

To satisfy the third necessary condition of proper part surface generation all diameters $d_{cnf} = 2r_{cnf}$ of the characteristic curve $Cnf_R(P/T)$ must be nonnegative—that is, in all directions through the contact point K the inequality $r_{cnf} \geq 0$ must be satisfied.

The indicatrix of conformity $Cnf_R(P/T)$ (see Equation 4.87) yields a conclusion on the actual type of contact of the part surface P and the generating surface T of the form-cutting tool at a current CC point (Figure 7.37a).

When surfaces P and T make a regular point contact, the minimal diameter d_{cnf}^{min} of the indicatrix of conformity $Cnf_R(P/T)$ is always positive $\left(d_{cnf}^{min} > 0\right)$ as depicted in Figure 7.37b. A CC point of that type cannot be a point of the boundary curve $\mathbf{r}_{bc}$.

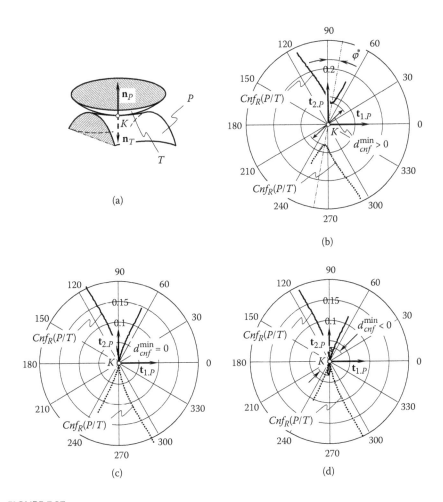

FIGURE 7.37
Examples of satisfaction and of violation of the third necessary condition of proper part surface generation when machining a sculptured part surface on a multiaxis NC machine [the current CC point K in (c) represents a point of the boundary curve $\mathbf{r}_{bc}$ that subdivides the surface P onto the *cutting tool–accessible* and onto the *cutting tool–inaccessible regions*]. (Reprinted from *Computer-Aided Design*, 37, Radzevich, S.P., A cutting-tool-dependent approach for partitioning of sculptured surface, 767–778, Copyright 2005, with permission from Elsevier.)

If the minimal diameter d_{cnf}^{min} of the indicatrix of conformity $Cnf_R(P/T)$ is equal to zero $d_{cnf}^{min} = 0$ (Figure 7.37c), then the direction of $\mathbf{t}_{cnf}^{max}$ along the diameter d_{cnf}^{min} radii of normal curvature of the surfaces P and T are of the same magnitude and of opposite sign ($R_P = -R_T$). A point at which the equality $d_{cnf}^{min} = 0$ is observed is a point of the boundary curve $\mathbf{r}_{bc}$.

Finally, the minimal diameter d_{cnf}^{min} of the indicatrix of conformity, $Cnf_R(P/T)$, can be of negative value $\left(d_{cnf}^{min} < 0\right)$ as shown in Figure 7.37d. Negative value of the diameter d_{cnf}^{min} indicates that at this CC point the cutting tool surface T interferes with the part surface P. This means that violation of the third necessary condition of proper part surface generation occurs at this CC point. A CC point, at which the inequality $d_{cnf}^{min} < 0$ is observed, is not within the boundary curve $\mathbf{r}_{bc}$.

7.3.2.2 Global Interpretation of the Results of the Local Analysis

For derivation of equations of the characteristic surfaces that enable one partitioning of a sculptured part surface P onto the cutting tool–accessible regions $\mathfrak{R}^+$, and onto the cutting tool–inaccessible regions $\mathfrak{R}^-$ (if any), geometric properties of the characteristic curve $Cnf_R(P/T)$ are employed below.

7.3.2.2.1 Characteristic Surface of the First Type

Consider unit normal vector $\mathbf{n}_P$ erected at an arbitrary point of the sculptured part surface P. A straight-line segment of the length $r_{cnf}^{min}(U_P, V_P, U_T, V_T, \mu)$ aligns with the vector $\mathbf{n}_P$.

Further, consider a vector of the length r_{cnf}^{min}, which aligns with the vectors $\mathbf{n}_P$. Positive vectors $\mathbf{n}_P \cdot r_{cnf}^{min} > 0$ are directed outward the part body and negative vectors $\mathbf{n}_P \cdot r_{cnf}^{min} < 0$ are directed inward the part body.

The loci of end-points of the vectors $\mathbf{n}_P \cdot r_{cnf}^{min}$ determine a characteristic $\mathcal{R}_1$ – surface of the first type*:

The characteristic $\mathcal{R}_1$–surface $\Rightarrow$ $\mathbf{r}_{\mathcal{R}}^{(1)}(U_P, V_P, U_T, V_T, \varphi, \mu) = \mathbf{r}_P + \mathbf{n}_P \cdot r_{cnf}^{min}$ (7.81)

Example 7.4

Consider a part surface P (Figure 7.38) that is given by the equation

$$\mathbf{r}_P(\theta_P, \varphi_P) = \begin{bmatrix} 3 \cdot \cos\varphi_P \cos\theta_P + 5 \cdot \cos\theta_P \\ 3 \cdot \cos\varphi_P \sin\theta_P + 5 \cdot \sin\theta_P \\ 3 \cdot \sin\varphi_P \\ 1 \end{bmatrix}, \ 0 \le \theta \le \frac{\pi}{2}, \ 0 \le \varphi \le \pi \quad (7.82)$$

---

* Reminder: the vectors $\mathbf{n}_P = \mathbf{n}_P(U_P, V_P)$ and $\mathbf{r}_{cnf} = r_{cnf}(U_P, V_P, U_T, V_T, \mu)$ are functions of coordinates of a point on the part surface P.

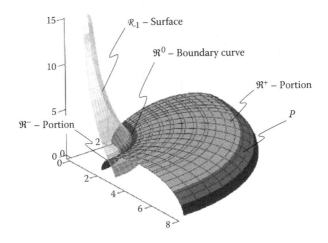

FIGURE 7.38
An example of the characteristic $\mathcal{R}_1$–surface. (Reprinted from *Computer-Aided Design*, 37, Radzevich, S.P., A cutting-tool-dependent approach for partitioning of sculptured surface, 767–778, Copyright 2005, with permission from Elsevier.)

This surface has convex and saddle-like regions.

In the example under consideration, only the second principal radius of curvature $R_{2.P}$ of the surface P is of importance. The first principal radius of curvature $R_{1.P}$ imposes no restrictions on the satisfaction or violation of the third necessary condition of proper part surface generation.

An expanded equation for the unit normal vector $\mathbf{n}_P$ at a the surface P point can be easily derived from Equation 7.82 using the equality $\mathbf{n}_P = \mathbf{u}_P \times \mathbf{v}_P$.

For the machining of surface P, the cylindrical milling cutter of diameter $d_T = 3''$ is used. The equation for the generating surface T of the cylindrical milling cutter is trivial.

Equation 7.82 of surface P along with the parameters of the second principal radius of curvature $R_{2.P}$, the unit normal vector $\mathbf{n}_P$, and the equation of the cylindrical cutting tool surface T yields derivation of the equation of the $\mathcal{R}_1$- surface. A routing formulae transformation leads to the following expression for the position vector of a point, $\mathbf{r}_\mathcal{R}$, of the $\mathcal{R}_1$- surface:

$$
\mathbf{r}_\mathcal{R}(\theta_P, \varphi_P) =
\begin{bmatrix}
\left(3 + \dfrac{5 + 3 \cdot \cos\varphi_P}{|\cos\varphi_P|}\right) \cdot \cos\varphi_P \cos\theta_P + 5 \cdot \cos\theta_P \\[4mm]
\left(3 + \dfrac{5 + 3 \cdot \cos\varphi_P}{|\cos\varphi_P|}\right) \cdot \cos\varphi_P \sin\theta_P + 5 \cdot \sin\theta_P \\[4mm]
\left(3 + \dfrac{5 + 3 \cdot \cos\varphi_P}{|\cos\varphi_P|}\right) \cdot \sin\varphi_P \\[4mm]
1
\end{bmatrix}
\qquad (7.83)
$$

The surface r_R (see Equation 7.83) is depicted in Figure 7.38. It is evident that the surfaces P and $\mathcal{R}_1$ intersect one another. The line of intersection serves as the boundary curve $\mathfrak{R}^0$. The major feature of the boundary curve $\mathfrak{R}^0$ is that it is represented by a locus of points at which minimal diameter d_{cnf}^{min} of the indicatrix of conformity $Cnf_R(P/T)$ is identical to zero $\left(d_{cnf}^{min} \equiv 0\right)$.

The boundary curve $\mathfrak{R}^0$ subdivides the surface P onto two (or more) portions $\mathfrak{R}^+$ and $\mathfrak{R}^-$. The region $\mathfrak{R}^+$ of the surfaces P is located above the boundary curve $\mathfrak{R}^0$, and thus, it is the cutting tool–accessible region. At all points within the portion $\mathfrak{R}^+$, the minimal diameter d_{cnf}^{min} of the indicatrix of conformity, $Cnf_R(P/T)$, is positive $\left(d_{cnf}^{min} > 0\right)$. The portion $\mathfrak{R}^+$ of the surface P can be machined on a multiaxis NC to the part blueprint.

The portion $\mathfrak{R}^-$ of the surfaces P is located below the boundary curve $\mathfrak{R}^0$. At all points within the portion $\mathfrak{R}^-$, the minimal diameter d_{cnf}^{min} of the indicatrix of conformity, $Cnf_R(P/T)$, is negative $\left(d_{cnf}^{min} < 0\right)$. The portion $\mathfrak{R}^-$ is the *cutting tool–inaccessible* region of the surface P. Therefore, the portion $\mathfrak{R}^-$ of the surface P cannot be machined on a multiaxis NC machine by the given cutting tool [99].

It is important to note that in the example under consideration, the $\mathcal{R}_1$–surface approaches infinity. This indicates that at some points of the surface P the principal radii of curvature approach infinity $(R_{2.P} \to \infty)$, and the corresponding points of the surface P are of parabolic type. Figure 7.39 reveals that the parabolic line at $\varphi = \pi/2$ is really observed on surface P (Equation 7.82). At point(s) of parabolic type on surface P the $\mathcal{R}_1$–surface always approaches infinity.

Use of the characteristic $\mathcal{R}_1$–surface enables verification of whether the third necessary condition of proper part surface generation is globally satisfied or not.

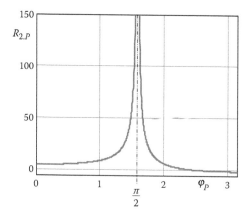

FIGURE 7.39

Parabolic lines of surface P: The second principal radius of curvature $R_{2.p}$ versus φ-curvilinear coordinate. (Reprinted from *Computer-Aided Design*, 37, Radzevich, S.P., A cutting-tool-dependent approach for partitioning of sculptured surface, 767–778, Copyright 2005, with permission from Elsevier.)

7.3.2.2.2 Elements of Local Topology of the $\mathcal{R}_1$– Surfaces

For the implementation of the characteristic surfaces of the first type, it is convenient to express all the elements of local topology of the $\mathcal{R}_1$– surface in terms of the corresponding elements of local topology of the surface P, and of the minimal radius r_{cnf}^{min} of the indicatrix of conformity $Cnf_R(P/T)$ at the current CC point.

Consider a sculptured part surface P that is given by vector equation $\mathbf{r}_P = \mathbf{r}_P(U_P, V_P)$.

Components $g_{R.ij}$ of the metric tensor of the first order of the $\mathcal{R}_1$– surface can be expressed in terms of the components of the fundamental tensors g_{ij} of the first order and b_{ij} of the second order and of the mean $\mathcal{M}_P$ and full (*Gaussian*) $\mathcal{G}_P$ curvatures at a current point of part surface P:

$$g_{R.ij} = \left[1 - \mathcal{G}_P \cdot \left(r_{cnf}^{min} \right)^2 \right] g_{ij} - 2 r_{cnf}^{min} \left[1 + \mathcal{M}_P \cdot r_{cnf}^{min} \right] b_{ij} \tag{7.84}$$

The mean $\mathcal{M}_P$ and the *Gaussian* $\mathcal{G}_P$ curvatures at a point of a smooth regular part surface P can be calculated from the expressions:

$$\mathcal{M}_P = \frac{E_P N_P - 2 F_P M_P + G_P L_P}{2 \left(E_P G_P - F_P^2 \right)} \tag{7.85}$$

$$\mathcal{G}_P = \frac{L_P N_P - M_P^2}{E_P G_P - F_P^2} \tag{7.86}$$

The determinant of the metric tensor of the $\mathcal{R}_1$–surface is equal to $\det(g_{R.ij}) = A^2 \cdot \det(g_{ij})$, where*

$$A = \mathcal{G}_P \cdot \left(r_{cnf}^{min} \right)^2 + 2 \mathcal{M}_P \cdot r_{cnf}^{min} + 1 = \left(1 + k_{1.P} \cdot r_{cnf}^{min} \right) \cdot \left(1 + k_{2.P} \cdot r_{cnf}^{min} \right) \tag{7.87}$$

Singularities of the $\mathcal{R}_1$–surface are observed at points that correspond to points of the part surface P, at which one of the principal curvatures is equal to $-\left(r_{cnf}^{min} \right)^{-1}$.

The unit normal vector $\mathbf{n}_{\mathcal{R}}$ at a point of the $\mathcal{R}_1$–surface can be analytically represented as

$$\mathbf{n}_{\mathcal{R}} = \frac{A}{|A|} \mathbf{n}_P \tag{7.88}$$

---

* This derivation immediately follows from the definition of principal curvatures at a point of a smooth regular surface [117].

The second fundamental tensor of the $\mathcal{R}_1$-surface can be expressed in terms of the components of the fundamental tensors g_{ij} and b_{ij}, the mean $\mathcal{M}_P$ and full (*Gauss'*) $\mathcal{G}_P$ curvature at a point of part surface P:

$$b_{\mathcal{R}.ij} = \frac{A}{|A|}\left[\mathcal{G}_P \cdot r_{cnf}^{min} \cdot g_{ij} + \left(1 + 2\mathcal{M}_P \cdot r_{cnf}^{min}\right) \cdot b_{ij}\right] \qquad (7.89)$$

As for the metric tensor, the determinant of the second fundamental tensor

$$\det(b_{\mathcal{R}.ij}) = A \cdot \det(b_{ij}) \qquad (7.90)$$

has singularities at points that correspond to points of the sculptured part surface P, at which one of the principal curvatures is equal to $-(r_{cnf}^{min})^{-1}$.

Gaussian curvature $\mathcal{G}_{\mathcal{R}}$ and mean curvature $\mathcal{M}_{\mathcal{R}}$ of the $\mathcal{R}_1$-surface can be calculated from the equations:

$$\mathcal{G}_{\mathcal{R}} = \frac{\mathcal{G}_P}{A} \qquad (7.91)$$

$$\mathcal{M}_{\mathcal{R}} = \frac{\mathcal{M}_P + \mathcal{G}_P \cdot r_{cnf}^{min}}{|A|} \qquad (7.92)$$

Principal curvatures $k_{1.\mathcal{R}}$ and $k_{2.\mathcal{R}}$ of the $\mathcal{R}_1$-surface at a point, at which the principal curvatures $k_{1.P}$ and $k_{2.P}$ of the sculptured part surface P are known, can be calculated from the expression:

$$k_{1,2.\mathcal{R}} = \frac{A}{|A|} \cdot \frac{k_{1,2.P}}{\left|1 + k_{1,2.P} \cdot r_{cnf}^{min}\right|} \qquad (7.93)$$

The principal curvatures* $k_{1.P}$ and $k_{2.P}$ can be expressed in terms of *Gauss'* $\mathcal{G}_P$ and mean $\mathcal{M}_P$ curvature of the part surface P

$$k_{1,2.P} = \mathcal{M}_P \pm \sqrt{\mathcal{M}_P^2 - \mathcal{G}_P} \qquad (7.94)$$

or they can be calculated as the roots of the quadratic equation:

$$\begin{vmatrix} L_P - E_P \cdot k_P & M_P - F_P \cdot k_P \\ M_P - F_P \cdot k_P & N_P - G_P \cdot k_P \end{vmatrix} = 0 \qquad (7.95)$$

---

* Reminder: the following inequality $k_{1.P} > k_{2.P}$ is always observed.

Other approaches for the calculation of the principal curvatures $k_{1.P}$, and $k_{2.P}$ can be employed as well.

7.3.2.2.3 Normalized $\bar{\mathcal{R}}_1$– Surface

For global satisfaction of the third necessary condition of proper part surface generation, the actual magnitude of the minimal radius r_{cnf}^{min} of the indicatrix of conformity $Cnf_R(P/T)$ is out important. Therefore, the minimal radius r_{cnf}^{min} can be normalized. This leads to an equation of the characteristic surface of another type

The normalized $\bar{\mathcal{R}}_1$–surface $\Rightarrow \bar{\mathbf{r}}_\mathcal{R}^{(1)}(U_P, V_P) = \mathbf{r}_P + \mathbf{n}_P \cdot \dfrac{r_{cnf}^{min}}{\left|r_{cnf}^{min}\right|} \equiv \mathbf{r}_P + \mathbf{n}_P \cdot \operatorname{sgn} r_{cnf}^{min}$

$$(7.96)$$

The characteristic surface (see Equation 7.96) is referred to as the *normalized characteristic surface of the first type* or simply as the $\bar{\mathcal{R}}_1$-surface.

Example 7.5

Consider that same part surface P that is given by Equation 7.82. Equation 7.82 of surface P along with the above-determined parameters $R_{2.P}$, $\mathbf{n}_P$, and the equation of the cylindrical cutting tool surface T, yield the equation of the normalized $\bar{\mathcal{R}}_1$-surface in matrix form:

$$\bar{\mathbf{r}}_\mathcal{R}(\theta_P, \varphi_P) = \begin{bmatrix} \left(3 + \operatorname{sgn} \dfrac{5 + 3 \cdot \cos\varphi}{|\cos\varphi|}\right) \cdot \cos\varphi_P \cos\theta_P + 5 \cdot \cos\theta_P \\[2mm] \left(3 + \operatorname{sgn} \dfrac{5 + 3 \cdot \cos\varphi}{|\cos\varphi|}\right) \cdot \cos\varphi_P \sin\theta_P + 5 \cdot \sin\theta_P \\[2mm] \left(3 + \operatorname{sgn} \dfrac{5 + 3 \cdot \cos\varphi}{|\cos\varphi|}\right) \cdot \sin\varphi_P \\[2mm] 1 \end{bmatrix} \qquad (7.97)$$

The normalized $\bar{\mathcal{R}}_1$-surface (see Equation 7.97) is depicted in Figure 7.40. The $\bar{\mathcal{R}}_1$-surface is shown together with the part surface P. The performed analysis of Figure 7.40 reveals that the characteristic $\bar{\mathcal{R}}_1$-surface intersects part surfaces P. The line of intersection is the boundary curve $\mathfrak{R}^0$. The boundary curve $\mathfrak{R}^0$ is identical to the boundary curve that is shown in Figure 7.38. This indicates that application of Equation 7.96 of the normalized $\bar{\mathcal{R}}_1$-surface returns results that are identical to the result obtained from the Equation 7.81 of the ordinary $\bar{\mathcal{R}}_1$-surface.

Use of the normalized characteristic $\bar{\mathcal{R}}_1$-surface enables verification of whether the third necessary condition of proper part surface generation is globally satisfied or not.

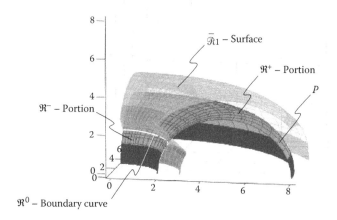

FIGURE 7.40
An example of the normalized characteristic $\bar{\mathcal{R}}_1$–surface of the first type. (Reprinted from *Computer-Aided Design*, 37, Radzevich, S.P., A cutting-tool-dependent approach for partitioning of sculptured surface, 767–778, Copyright 2005, with permission from Elsevier.)

7.3.2.2.4 *Equation of the Boundary Curve*

Boundary curve(s) $\mathfrak{R}^0$ subdivides the part surface P onto the *cutting tool–accessible* regions $\mathfrak{R}^+$ and onto the *cutting tool–inaccessible* regions $\mathfrak{R}^-$. The equation of the boundary curve $\mathfrak{R}^0$ is employed for marking these curves on the part surface P to identify the surface regions that require additional care.

Two approaches for the derivation of equation of the boundary $\mathfrak{R}^0$-curve can be employed.

First, the equation of the boundary curve $\mathfrak{R}^0$ can be derived from consideration of the line of intersection of the characteristic $\bar{\mathcal{R}}_1$-surface and of the part surface P itself. Coordinates of points of the boundary curve $\mathfrak{R}^0$ satisfy equations of both the surfaces, say, of the part surface P and of the $\bar{\mathcal{R}}_1$-surface simultaneously.

Second, the equation of the boundary curve can be derived in quite a different way. At points of the boundary curve $\mathfrak{R}^0$, the minimal diameter of the indicatrix of conformity is identical to zero $\left(d_{cnf}^{min} \equiv 0\right)$. The diameter d_{cnf}^{min} is positive for small changes of φ (i.e., for value of the parameter φ changed on $\pm d\ \varphi$). Then, at all points of the boundary curve, the direction along which the diameter d_{cnf}^{min} is measured, is a tangent to the $\mathfrak{R}^0$-curve. Hence, the equation of the boundary curve $\mathfrak{R}^0$ can be derived from the equation

$$r_{cnf} = r_{cnf}(U_P, V_P, U_T, V_T, \mu, \varphi) \tag{7.98}$$

of the indicatrix of conformity $Cnf_R(P/T)$ at a point of contact of the sculptured part surface P and the generating surface T of the form-cutting tool. For this purpose, a set of four equations:

$$\frac{\partial r_{cnf}}{\partial U_T} = 0 \tag{7.99}$$

$$\frac{\partial r_{cnf}}{\partial V_T} = 0 \tag{7.100}$$

$$\frac{\partial r_{cnf}}{\partial \mu} = 0 \tag{7.101}$$

$$\frac{\partial r_{cnf}}{\partial \varphi} = 0 \tag{7.102}$$

that represents the necessary conditions of the function extremum must be considered together with the well-known sufficient conditions for extremum of the function $r_{cnf} = r_{cnf}(U_P, V_P, U_T, V_T, \mu, \varphi)$ of four variables (actually, for a given CC point, the values of the parameters U_P and V_P are specified). The prerequisite to the obtained solution is that it must satisfy the condition $r_{cnf}^{min} \equiv 0$.

Example 7.6

Consider the part surface P that is given by that same Equation 7.82 as in the previous examples. Equation 7.82 of the part surface P along with the above-determined parameters $R_{2.P}$, $\mathbf{n}_P$, and the equation of the cylindrical surface T of the cutting tool, yield derivation of the computer code* for the computation of parameters of the boundary curve $\mathfrak{R}^0$. Results of computer modeling are shown in Figure 7.41. The boundary curve $\mathfrak{R}^0$ (Figure 7.41) is identical to the boundary curve shown in Figure 7.38, and to that shown in Figure 7.40.

In particular cases, derivation of the equation of the $\mathfrak{R}^0$–boundary curve can be significantly simplified.

The use of the boundary curve $\mathfrak{R}^0$ enables verification of whether the third necessary condition of proper part surface generation is globally satisfied or not.

7.3.2.2.5 Algorithm for the Computation of Parameters of the Boundary Curve $\mathfrak{R}^0$

Computation of parameters of the $\mathfrak{R}^0$–boundary curve for machining of a given sculptured part surface on a multiaxis NC machine can be performed following the algorithm shown in Figure 7.42.

The input information for the computations is available from the blueprint (1) of the sculptured part surface P. The equation $\mathbf{r}_P = \mathbf{r}_P(U_P, V_P)$ for the

---

* The equation of the boundary curve $\mathfrak{R}^0$ is not represented here due to space constraints.

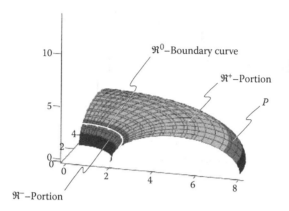

FIGURE 7.41
The boundary $\mathfrak{R}^0$-curve on the part surface P constructed with implementation of the normalized characteristic $\bar{\mathcal{R}}_1$-surface of the first type.

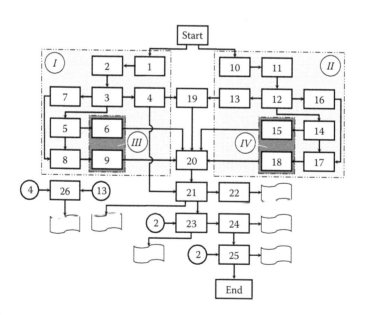

FIGURE 7.42
Flowchart of the algorithm for the computation of parameters of the boundary $\mathfrak{R}^0$-curve on the sculptured part surface P.

position vector of a point of the part surface P is derived (2) based on the information available from the part blueprint (1). The derived equation of the part surface P allows computation (3) of the first derivatives $\mathbf{U}_P = \partial \mathbf{r}_P / \partial U_P$ and $\mathbf{V}_P = \partial \mathbf{r}_P / \partial V_P$ of the part surface P as well as the computation (4) of the unit normal vector $\mathbf{n}_P$ at a point of the part surface P. Then, fundamental magnitudes $E_P = \mathbf{U}_P \cdot \mathbf{U}_P$, $F_P = \mathbf{U}_P \cdot \mathbf{V}_P$, and $G_P = \mathbf{V}_P \cdot \mathbf{V}_P$ of the first order are computed (5), and the first fundamental form $\Phi_{1.P}$ of the part surface P is composed (6). Further, the computed first derivatives of the part surface P (3) yield computation (7) of the second derivatives $\partial^2 \mathbf{r}_P / \partial U_P^2$, $\partial^2 \mathbf{r}_P / \partial U_P \partial V_P$, and $\partial^2 \mathbf{r}_P / \partial V_P^2$ of the surface P. Fundamental magnitudes L_P, M_P, and N_P of the second order are computed (8) based on the results obtained on the previous steps (3) and (7). The second fundamental form $\Phi_{2.P}$ of the part surface P is composed (9) using the derived equations for the fundamental magnitudes L_P, M_P, and N_P. Steps (1) through (9) form the sculptured part surface P block I that is shaded in Figure 7.42.

Steps (10) through (18) (block II, that is also shaded in Figure 7.42) of the generating surface T of the form-cutting tool, are similar to the corresponding steps (1) through (9) of the sculptured part surface P in block I.

The fundamental forms $\Phi_{1.P}$ (6) and $\Phi_{2.P}$ (9) represent the natural form of the sculptured part surface P parameterization (shaded in Figure 7.42 Block III). The similar block IV (also shaded in Figure 7.42) is represented by steps (15) and (18) on which the fundamental forms $\Phi_{1.T}$ (15) and $\Phi_{2.T}$ (18) are computed.

The fundamental magnitudes of the first (5) and (14) and of the second (8) and (17) order of the part surface P and the generating surface T of the form-cutting tool yield the composition (20) of the equation

$$r_{cnf} = r_{cnf}(U_P, V_P, U_T, V_T, \varphi, \mu) \tag{7.103}$$

of the indicatrix of conformity $Cnf_R(P/T)$ at a current point of contact of the part surface P and the generating surface T of the form-cutting tool. Further, the minimal radius r_{cnf}^{min} of the indicatrix of conformity $Cnf_R(P/T)$ is computed at a current CC point. The computed value of the unit normal vector $\mathbf{n}_P$ (4), together with the minimal radius r_{cnf}^{min} of the indicatrix of conformity $Cnf_R(P/T)$, yields the equation of the characteristic $\mathcal{R}_1$-surface (21).

7.3.2.3 Characteristic Surfaces of the Second Type

It is important to note the perfect correlation between the characteristic $\mathcal{R}_1$-surface, the normalized $\mathcal{R}_1$-surface, and the boundary $\mathfrak{R}^0$-curve from one side, and, from another side, between the spherical indicatrix of machinability $Mch\,(P/T)$ of the sculptured part surface P with the given cutting tool (see Section 7.1.7) [116,117,136]. The perfect correlation becomes more evident when in addition to the characteristic surfaces of the first type, the characteristic surfaces of the second type are considered.

7.3.2.3.1 Determination of the Characteristic Surface of the Second Type

To globally satisfy the third necessary condition of proper part surface generation, the characteristic $\mathcal{R}_1$–surface of the first type has to be located outside the part body, and not intersect the sculptured part surface P. Only tangency of the surfaces P and T is allowed.

Further, we go to a characteristic surface, which is determined by equation

The characteristic $\mathcal{R}_2$-surface $\;\Rightarrow\; \mathbf{r}_{\mathcal{R}}^{(2)}(U_P,V_P,U_T,V_T,\mu) = \mathbf{r}_{\mathcal{R}}^{(1)} - \mathbf{r}_P = \mathbf{n}_P \cdot r_{cnf}^{min}$

$$(7.104)$$

The characteristic surface that is given by Equation 7.104 is referred to as the *characteristic surface of the second type* or simply as the $\mathcal{R}_2$-surface.

Elementary analysis of Equation 7.104 reveals that location of the $\mathcal{R}_2$-surface yields an answer to the question of whether the third necessary condition of proper part surface generation is globally satisfied or not. Consideration of the $\mathcal{R}_2$-surface alone is sufficient for making an appropriate conclusion. If the $\mathcal{R}_2$-surface intersects the $X_P Y_P$–coordinate plane, then one can make a conclusion that certain portion of the part surface P is not accessible by the cutting tool of the given design. The line of intersection of the $\mathcal{R}_2$-surface with the $X_P Y_P$–coordinate plane is analogous of the boundary curve $\mathfrak{R}^0$.

The following three scenarios could be observed depending on the sign of Z_P–coordinate of the $\mathcal{R}_2$-surface:

When $Z_P \geq 0$, then the corresponding portion of the part surface P represents the cutting tool–accessible region $\mathfrak{R}^+$ of the surface P,

When $Z_P \equiv 0$, then the corresponding portion of the sculptured part surface P represents the boundary $\mathfrak{R}^0$–curve on the surface P, and finally

When $Z_P < 0$, then the corresponding portion of the sculptured part surface P represents the cutting tool–inaccessible region $\mathfrak{R}^-$ of the surface P.

Example 7.7

Consider the surface P that is given by Equation 7.82 as in the previous examples. Substitute the required parameters from Equation 7.82, the above-determined parameters $R_{2.P}$, $\mathbf{n}_P$, and the equation of the cylindrical surface T of the cutting tool into Equation 7.104. This yields an equation of the characteristic $\mathcal{R}_2$-surface in matrix representation.* A visualized image of the $\mathcal{R}_2$-surface is depicted in Figure 7.43. The line of intersection of the $\mathcal{R}_2$-surface exactly corresponds to the boundary $\mathfrak{R}^0$–curve on the part surface P shown in Figure 7.41. So, application of only one $\mathcal{R}_2$-

---

* The equation of the characteristic $\mathcal{R}_2$ - surface is not represented here due to space constraints.

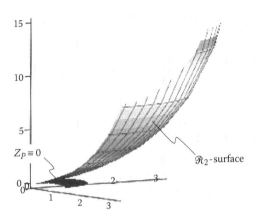

FIGURE 7.43
An example of the characteristic $\mathcal{R}_2$-surface of the second type.

surface (and not two surfaces P and $\mathcal{R}_1$) yields an answer to the question of whether the whole sculptured part surface P can be machined with the cutting tool of a given design or it cannot be machined under such a scenario.

Use of the characteristic $\mathcal{R}_2$-surface solely enables verification of whether the third necessary condition of proper part surface generation is globally satisfied or not.

Elements of local topology of the characteristic $\mathcal{R}_2$-surface can also be expressed in terms of the corresponding elements of local topology of the sculptured part surface P and of the minimal radius $r_{cnf}^{min}(U_P, V_P, U_T, V_T, \mu)$ of the indicatrix of conformity $Cnf_R(P/T)$ at a current CC point K as done with respect to the characteristic $\mathcal{R}_1$-surfaces of the first type (see Section 7.3.2.2.2).

7.3.2.3.2 Normalized Characteristic Surface of the Second Type

Following the method similar to that used to derive the equation of the normalized $\overline{\mathcal{R}}_1$-surface of the first type, the equation of the normalized characteristic $\overline{\mathcal{R}}_2$-surface of the second type can be derived. The equation of this characteristic surface is represented as follows:

$$\text{Normalized } \overline{\mathcal{R}}_2\text{-surface} \Rightarrow \overline{\mathbf{r}}_{\mathcal{R}}^{(2)}(U_P, V_P) = \mathbf{r}_{\mathcal{R}}^{(1)} - \mathbf{r}_P = \mathbf{n}_P \cdot \frac{r_{cnf}^{min}}{\left| r_{cnf}^{min} \right|} \equiv \mathbf{n}_P \cdot \text{sgn } r_{cnf}^{min}$$

(7.105)

The characteristic surface that is described by Equation 7.105 is referred to as the *normalized characteristic surface of the second type* or simply as to the $\overline{\mathcal{R}}_2$-surface.

Analysis of Equation 7.105 reveals that the $\bar{\mathcal{R}}_2$-surface is equivalent to the spherical indicatrix of machinability Mch (P/T) of the sculptured part surface P with the given cutting tool proposed [116,117,136] by the author (see Section 7.1.7). The line of intersection of the $\bar{\mathcal{R}}_2$-surface with $X_P Y_P$-coordinate plane perfectly corresponds to the boundary curve of the indicatrix of machinability Mch (P/T).

The characteristic surfaces of the first and second types can be employed for any feasible value of the angle μ of the local relative orientation of the sculptured part surface P and the generating surface T of the form-cutting tool.

Much room remains for further developments in the field of characteristic surfaces of the discovered type.

7.3.3 Selection of the Form-Cutting Tool of Optimal Design

For the verification of satisfaction of the third necessary condition of proper part surface generation globally, use of $\mathbb{K}$-mapping of a sculptured part surface P and of the generating surface T of the form-cutting tool has proven to be useful [117,137]. Two types of the surfaces P and T $\mathbb{K}$-mapping are of importance in this concern. First, the local $\mathbb{K}_{LR}$-mapping of the surfaces P and T, and, second, the global $\mathbb{K}_{GR}$-mapping of the surfaces P and T. The local $\mathbb{K}_{LR}$-mapping of the surfaces gives an insight into the global $\mathbb{K}_{GR}$-mapping of the surfaces.

7.3.3.1 Local $\mathbb{K}_{LR}$-Mapping of the Part Surface P and the Generating Surface T of the Form-Cutting Tool

Analysis of the local interference of a sculptured part surface P and the generating surface T of the form-cutting tool can be performed with application of the local $\mathbb{K}_{LR}$-mapping of the surfaces [124].

Consider the coordinate plane $k_{1.P}$ $k_{2.P}$ (Figure 7.44). For a given smooth, regular sculptured part surface P, the first $R_{1.P}$ and the second $R_{2.P}$ principal radii of curvature and the corresponding principal curvatures $k_{1.P}$ and $k_{2.P}$ can be determined from Equation 1.49. Principal curvatures are measured in the respective principal plane-sections $C_{1.P}$ and $C_{2.P}$ through the unit tangent vectors $\mathbf{t}_{1.P}$ and $\mathbf{t}_{2.P}$ of the principal directions at a point of the surface P.

Every point of the sculptured part surface P has a corresponding point $m(k_{1.P}, k_{2.P})$ in the coordinate plane $k_{1.P}$ $k_{2.P}$ (and not vice versa: each point of the coordinate plane $k_{1.P}$ $k_{2.P}$ may have one or more corresponding points on the sculptured part surface P). A point with coordinates $(k_{1.P}, k_{2.P})$ in the coordinate plane $k_{1.P}$ $k_{2.P}$ represents the local $\mathbb{K}_{LR}$-map of the corresponding point of the sculptured part surface P—that is, it represents the local $\mathbb{K}_{LR}$-map of the sculptured part surface point.

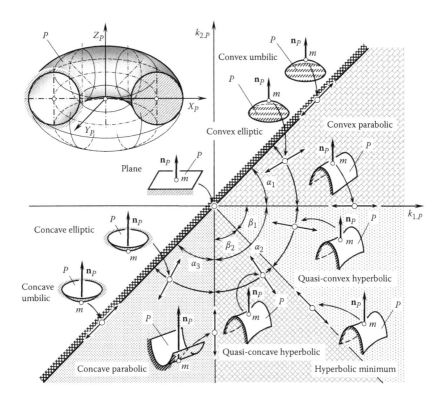

FIGURE 7.44
The local $\mathbb{K}_{LR}$-mapping of a sculptured part surface P.

The location of the local $\mathbb{K}_{LR}$-map of the sculptured part surface point within the coordinate plane $k_{1.P} k_{2.P}$ depends on parameters of a part surface P local topology in differential vicinity of the surface point m.

All the local $\mathbb{K}_{LR}$-maps are located within the first, the third, and the fourth quadrants in the so-called allowed region of the coordinate plane $k_{1.P} k_{2.P}$—that is, along a boundary straight line specified by the equation $k_{2.P} = k_{1.P}$, and below this boundary line.

The origin of the coordinate system $k_{1.P} k_{2.P}$ represents the local $\mathbb{K}_{LR}$-map of the planar local region of a sculptured part surface P. Within the planar local region of the surface, its normal curvature identically equal to zero, $k_P \equiv 0$. The mean curvature $\mathcal{M}_P$ and the full curvature (*Gaussian* curvature) $\mathcal{G}_P$ are also equal to zero: $\mathcal{M}_P = 0$, and $\mathcal{G}_P = 0$.

The local $\mathbb{K}_{LR}$-maps of umbilical local patches of a sculptured part surface P are located within the boundary straight line $k_{2.P} = k_{1.P}$. The local $\mathbb{K}_{LR}$-maps of convex ($\mathcal{G}_P > 0$, $\mathcal{M}_P > 0$) patches are located in the first quadrant, and the similar local $\mathbb{K}_{LR}$-maps of concave ($\mathcal{G}_P > 0$, $\mathcal{M}_P < 0$) patches of the part surface P are located in the fourth quadrant (see Figure 7.44).

The allowed region is subdivided onto three sectors α_1, α_2, and α_3 as illustrated in Figure 7.44.

The local $\mathbb{K}_{LR}$-maps of convex elliptical local patches $(\mathcal{G}_P > 0, \mathcal{M}_P > 0)$ of a sculptured part surface P are located within the sector α_1. Saddle-like hyperbolic local patches $(\mathcal{G}_P < 0)$ are locally $\mathbb{K}_{LR}$-mapped within the α_2 sector. Finally, the local $\mathbb{K}_{LR}$-maps of concave elliptical local patches $(\mathcal{G}_P > 0, \mathcal{M}_P < 0)$ of a sculptured part surface P are located within the α_3 sector.

Within the boundary line $k_{2.P} = 0$, that separates sector α_1 from sector α_2—that is, within the axis of abscissas—the local $\mathbb{K}_{LR}$-maps of convex parabolic local patches $(\mathcal{G}_P = 0, \mathcal{M}_P > 0)$ of a part surface P are located. The similar local $\mathbb{K}_{LR}$-maps of concave parabolic patches $(\mathcal{G}_P = 0, \mathcal{M}_P < 0)$ are located within the boundary line $k_{1.P} = 0$ that separates the sector P from the sector α_3, (that is, within the axis of ordinates).

Mean curvature $\mathcal{M}_P$ of a saddle-like local patch of a sculptured part surface P can be positive or negative or it can be equal to zero. Due to this, the sector α_2 is subdivided onto two symmetrical subsectors β_1 and β_2. The local $\mathbb{K}_{LR}$-maps of quasi-convex hyperbolical local patches $(\mathcal{G}_P < 0, \mathcal{M}_P > 0)$ of a part surface P are located within the subsector β_1, and the local $\mathbb{K}_{LR}$-maps of quasi-concave hyperbolical local patches $(\mathcal{G}_P < 0, \mathcal{M}_P < 0)$ of a part surface P are located with the subsector β_2 as shown in Figure 7.44. The boundary line $k_{2.P} = -k_{1.P}$ that separates the subsectors β_1 and β_2 is the locus of the local $\mathbb{K}_{LR}$-maps of minimum hyperbolical local patches $(\mathcal{G}_P < 0, \mathcal{M}_P = 0)$ of a sculptured part surface P.

Arrows that are coming out from each the local $\mathbb{K}_{LR}$-map, schematically depicted in Figure 7.44, indicate the directions in which parameters of a sculptured part surface P local topology could be changed. While parameters of the surface topology are changing, the type of a sculptured surface local patch remains the same.

Evidently, the local $\mathbb{K}_{LR}$-mapping is also available for the generating surface T of a form-cutting tool.

The difference in topology of local patches of a sculptured part surface P within sectors α_1, α_2, α_3, and within sectors β_1, β_2 (Figure 7.44) can be demonstrated with implementation of the following analysis.

7.3.3.2 Global $\mathbb{K}_{GR}$-Mapping of the Part Surface P and the Generating Surface T of the Form-Cutting Tool

Application of the local $\mathbb{K}_{LR}$-mapping of the surfaces enables development of a global approach to verify whether or not the third necessary condition of proper part surface generation is satisfied within the entire sculptured part surface P. The analysis below is based on application of the global $\mathbb{K}_{GR}$-mapping of surfaces.

A global $\mathbb{K}_{GR}$-map of the entire sculptured part surface can be obtained if every local region of the surface P is locally $\mathbb{K}_{LR}$-mapped onto the coordinate plane $k_{1.P} k_{2.P}$ (Figure 7.45). It is evident that the following are true:

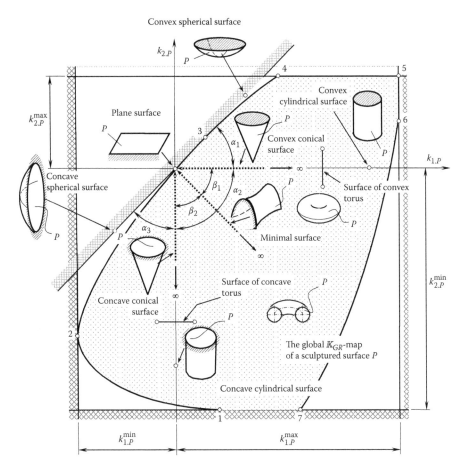

FIGURE 7.45

The global $\mathbb{K}_{GR}$-mapping of a sculptured part surface P.

- The global $\mathbb{K}_{GR}$-map is located within the allowed region—that is, on the boundary straight line $k_{2.P} = k_{1.P}$, and below this straight line as shown in Figure 7.45. Similar to that above, the allowed region is subdivided onto three sectors α_1, α_2, and α_3. Sector α_2 is subdivided onto two subsectors β_1, and β_2 that are also similar to that depicted in Figure 7.44.

- The global $\mathbb{K}_{GR}$-maps of concave patches ($\mathcal{G}_P > 0$, $\mathcal{M}_P > 0$) of a sculptured part surface P are located within sector α_1; saddle-type local patches ($\mathcal{G}_P < 0$) of a sculptured surface P could be globally $\mathbb{K}_{GR}$-mapped within sector α_2; finally, the global $\mathbb{K}_{GR}$-maps of concave patches ($\mathcal{G}_P > 0$, $\mathcal{M}_P < 0$) of a sculptured part surface P are located within sector α_3.

- Boundaries of sectors α_1, α_2, and α_3 represent the locus of global $\mathbb{K}_{GR}$-maps of local patches of a sculptured part surface P, *Gaussian* curvature of which is equal to zero ($\mathcal{G}_P = 0$): global $\mathbb{K}_{GR}$-maps of convex patches ($\mathcal{M}_P > 0$) are located along the axis of abscissas, and global $\mathbb{K}_{GR}$-maps of concave patches ($\mathcal{M}_P < 0$) are located along the axis of ordinates.

- Equations 1.55 and 1.56 for the computation of the values of the principal curvatures $k_{1.P}$ and $k_{2.P}$ at a point of a sculptured part surface P can be generalized as

$$k_{1.P} = k_{1.P}(U_P, V_P) \tag{7.106}$$

$$k_{2.P} = k_{2.P}(U_P, V_P) \tag{7.107}$$

- Application of a known method (see Chapter 1) allows computation of extremal values of the curvatures—that is, maximum $k_{1.P}^{max}$ and $k_{2.P}^{max}$ and minimum $k_{1.P}^{min}$ and $k_{2.P}^{min}$ values of the principal curvatures $k_{1.P}$ and $k_{2.P}$ within the intervals of variation of the parameters U_P and V_P, (that is, within the intervals $U_{1.P} \le U_P \le U_{2.P}$ and $V_{1.P} \le V_P \le V_{2.P}$).

- In the coordinate plane $k_{1.P} k_{2.P}$, two pairs of the straight lines $k_{1.P} = k_{1.P}^{max}$, $k_{1.P} = k_{1.P}^{min}$ and $k_{2.P} = k_{2.P}^{max}$, $k_{2.P} = k_{2.P}^{min}$ can be plotted as shown in Figure 7.45. The first pair of straight lines is parallel to the axis of abscises, and the second pair of the straight lines is parallel to the axis of ordinates of the coordinate system $k_{1.P} k_{2.P}$. These two pairs of straight lines form the *allowed rectangle*. In particular cases, the allowed rectangle can be open from one, from two, from three, and (in a completely degenerate case) even from four sides.

- The global $\mathbb{K}_{GR}$-maps of any and all sculptured part surfaces P are located within the allowed rectangle. A boundary curve of the global $\mathbb{K}_{GR}$-map of a sculptured part surface P shares at least one point with each side of the allowed rectangle above (or they are approaching infinity in the case of a corresponding side of the allowed rectangle being open). Boundaries of the global $\mathbb{K}_{GR}$-map of a sculptured part surface P can coincide in full or in part with corresponding side of the allowed rectangle.

- Generally, there are just a few types of local patches of a sculptured part surface P. They are convex, concave, and saddle-like local patches (see Chapter 1). For this reason, the global $\mathbb{K}_{GR}$-map of a sculptured part surface P can be located either within one, two or even three sectors α_1, α_2, and α_3 simultaneously. The motion of a point over a sculptured surface can be represented with a respective motion within a global $\mathbb{K}_{GR}$-map. It is evident that transition from one of sectors α_1, α_2, and α_3 to another can be performed only with crossing of a corresponding axis of the coordinate system $k_{1.P}$ $k_{2.P}$, or with passing through the origin of this coordinate system.*

- The global $\mathbb{K}_{GR}$-map of a sculptured part surface P can have multi-layer portions (say, portions with two or more layers) that are observed in a case where a sculptured surface P consists of two or more patches with the same values of principal curvatures.

Points of the boundary curve of the global $\mathbb{K}_{GR}$-map of a sculptured part surface may or may not correspond to points of boundary of the sculptured surface. This is because the extremal values of a sculptured part surface P principal curvatures might be observed not on the boundary of a surface P, but located within the part surface. For this reason, there is no one-to-one correspondence between points of the global $\mathbb{K}_{GR}$-map of a sculptured part surface and points of surface P.

In general, the global $\mathbb{K}_{GR}$-map of a sculptured part surface is represented with a portion of the coordinate plane $k_{1.P}$ $k_{2.P}$ that is bounded by a closed or semiclosed line. An example of the global $\mathbb{K}_{GR}$-map of a sculptured part surface P is shown in Figure 7.45. The global $\mathbb{K}_{GR}$-map depicted in Figure 7.45, satisfies all of the above-listed necessary requirements:

- This global $\mathbb{K}_{GR}$-map is located in the allowed region.
- It is located within an allowed rectangle.
- It is located within sectors α_1, α_2, and α_3.
- It shares common points with every side of the allowed rectangle (at points 1, 2, 3).
- It shares straight-line segments 4–5, 5–6, and 1–7 with the sides of the allowed rectangle.

In particular cases, the global $\mathbb{K}_{GR}$-map of a sculptured part surface degenerates to a straight line or even to a point (see Figure 7.45).

To derive equations of the boundary curves of the global $\mathbb{K}_{GR}$-map of a sculptured part surface P, the following approach can be employed.

Equations 7.106 and 7.107 can be rewritten in the form

---

* Remember that points located on the axis of the coordinates of the coordinate system $k_{1.P}$ $k_{2.P}$, represent the global $\mathbb{K}_{GR}$-maps of parabolic local patches of a sculptured part surface P. The origin of the coordinate system $k_{1.P}$ $k_{2.P}$ reflects the global $\mathbb{K}_{GR}$-map of planar local patch of a sculptured part surface to be machined. This statement is in good agreement with the well-known statement proven in differential geometry of surfaces [14,17]: if a sculptured surface P consists of convex and concave local patches, there is a corresponding number of parabolic curves on it.

$$k_{1.P}(U_P,V_P) = \mathcal{M}_P(U_P,V_P) + \sqrt{\mathcal{M}_P^2(U_P,V_P) - \mathcal{G}_P(U_P,V_P)} \qquad (7.108)$$

$$k_{2.P}(U_P,V_P) = \mathcal{M}_P(U_P,V_P) - \sqrt{\mathcal{M}_P^2(U_P,V_P) - \mathcal{G}_P(U_P,V_P)} \qquad (7.109)$$

The first principal curvature $k_{1.P}$ is not constant within the part surface P patch. As follows from Equation 7.108), it is a function of *Gaussian* coordinates—that is, $k_{1.P} = k_{1.P}(U_P, V_P)$. The extremal values (the maximal and the minimal values) of the first principal curvature of the sculptured part surface are equal to the roots of the set of two equations:

$$k_{1.P} = k_{1.P}(U_P, V_P) \qquad (7.110)$$

$$\frac{\partial}{\partial U_P} k_{1.P}(U_P, V_P) = 0 \qquad (7.111)$$

Equation 7.111 can be solved with respect to the variable U_P. The solution of the equation can be represented in the form $U_P = U_P(k_{1.P}, V_P)$. Substitution of the solution $U_P = U_P(k_{1.P}, V_P)$ into the Equation 7.110 after performing the necessary formula transformation can result in the equation

$$k_{1.P}^{\min/\max} = k_{1.P}^{\min/\max}(V_P) \qquad (7.112)$$

that is derived from Equation 7.110.

Similar manipulations can be performed with Equation 7.109. Consequently, one can obtain

$$k_{2.P}^{\min/\max} = k_{2.P}^{\min/\max}(V_P) \qquad (7.113)$$

Equation 7.110 can be solved with respect to the variable V_P. The obtained solution has to be substituted into the Equation 7.111. After performing necessary formulae transformations the following equations can be obtained:

$$k_{1.P}^{\min/\max} = k_{1.P}^{\min/\max}(k_{2.P}) \qquad (7.114)$$

or

$$k_{2.P}^{\min/\max} = k_{2.P}^{\min/\max}(k_{1.P}) \qquad (7.115)$$

Equations 7.114 and 7.115 describe the boundary curves of the global $\mathbb{K}_{GR}$-map of the sculptured part surface P. It is evident that the same result can be obtained in the case of differentiation of Equation 7.110 not with respect to U_P–parameter, but at first, with respect to V_P–parameter instead.

Application of the same approach makes possible plotting global $\mathbb{K}_{GR}$-maps of the generating surface T of the form-cutting tool. An example is shown in Figure 7.46. As evident, the represented global $\mathbb{K}_{GR}$-maps of the cutting tool surface T are in perfect correspondence to the global $\mathbb{K}_{GR}$-maps of similar surfaces to those depicted in Figure 7.45.

7.3.3.3 Implementation of the $\mathbb{K}_{GR}$-Mapping of Surfaces

Global $\mathbb{K}_{GR}$-mapping of surfaces was developed for the purpose of the verification of whether the third necessary condition of proper part surface generation is globally satisfied or not (Figure 7.46). For this purpose, the allowed rectangle has to be plotted onto the coordinate plane $k_{1.P}\,k_{2.P}$. The size and location of the allowed rectangle are completely determined by the maximal

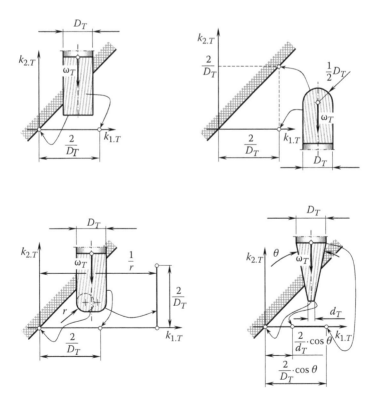

FIGURE 7.46
Examples of the global $\mathbb{K}_{GR}$-map of the generating surface T of the milling cutters.

$k_{1.P}^{max}$, $k_{2.P}^{max}$ and by the minimal $k_{1.P}^{min}$, $k_{2.P}^{min}$ values of the principal curvatures of the sculptured part surface P. The global $\mathbb{K}_{GR}$-map of the surface part P is located within the interior of the allowed rectangle and shares at least one point with each of its sides. The constructed $\mathbb{K}_{GR}$-map of the part surface P helps to visualize major restrictions that are imposed on the parameters of the generating surface of the form-cutting tool by the topology of the sculptured part surface P.

The limited case for the satisfaction of the third necessary condition of proper part surface generation corresponds to the case of the maximal degree of conformity of the generating surface of the form-cutting tool to the sculptured part surface—that is, when normal radii of curvature of the contacting surfaces in a current section by normal plane are of the same value and of the opposite sign.

7.3.3.4 Selection of an Optimal Cutting Tool for Sculptured Part Surface Machining

Further, the maximal principal curvatures can be employed to directly bound the largest cutting tool radius that can be used for the machining of concave and saddle-like regions and hence to aid in the cutting tool selection. Productivity of sculptured part surface machining depends on the type of machining tool employed. To investigate the machining tool settings, the machining tool should first be studied. The larger the machining tool, the smaller the resulting scallop height (or the excessive material not removed) that might result for the same tool-path. Moreover, larger cutting tools reduce machining times. A smooth finish usually results by using flat-end tools. Machining using ball-end tools slow because of a vanishing cutting speed at the tip of the tool—an impediment that shows up neither in the five-axis flat-end milling mode nor in five-axis side milling approach proposed herein. However, a systematic approach for the selection of cutting tools is yet to be attempted. Limited research has been conducted into choosing an optimal form-cutting tool for four-axis and five-axis freeform surface machining.

Application of the global $\mathbb{K}_{GR}$-mapping of a sculptured part surface P to be machined and of the generating surface T_i of cutting tools to be applied (here $i = 1...N$ is an integer number and N is the number of cutting tools available for machining a given sculptured part surface P) allows one to select the proper form-cutting tool for machining a given sculptured part surface (Figure 7.47). For this purpose, taken with opposite sign, the global $\mathbb{K}_{GR}$-map of all cutting tools available have to be plotted onto the same coordinate system on which the global $\mathbb{K}_{GR}$-map of the given sculptured part surface P is already plotted. The global $\mathbb{K}_{GR}$-map of the best cutting tool (among available cutting tools) contains a point that is closest to the point with coordinates $\left(k_{1.P}^{max}, k_{2.P}^{max}\right)$—that is, the one closest to the corner point $\mathcal{G}$ of the allowed rectangle of the global $\mathbb{K}_{GR}$-map of the given sculptured part surface P. A cutting

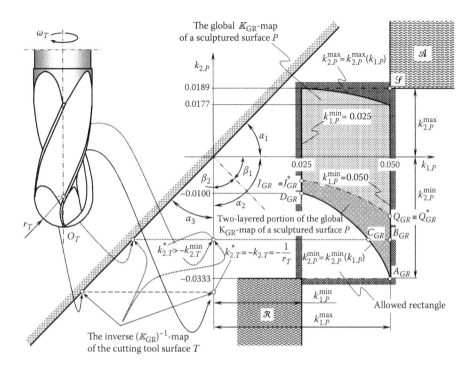

FIGURE 7.47
Example of implementation of the $\mathbb{K}_{GR}$-mapping of a sculptured part surface P (Figure 7.43) for determining the allowed parameters of the generating surface T of a cutting tool.

tool of such design would have highest degree of conformity of the cutting tool surface T to the sculptured part surface P at every CC point. Evidently, this would allow for the smallest cusp height and highest productivity of the machining operation.

The implementation of the obtained results enables one to increase efficiency of machining of a sculptured part surface in die/mold production, in high speed machining in conventional and rapid prototyping, and so forth.

It can be observed that application of the global $\mathbb{K}_{GR}$-mapping of a sculptured part surface P and the generating surface T_i of form-cutting tools allows one to select the design of cutting tool most fitted for machining a given sculptured part surface. The capabilities of the global $\mathbb{K}_{GR}$-mapping for selecting a proper cutting tool for machining a given sculptured surface were demonstrated.

8

Accuracy of Surface Generation

The accuracy of machined part surfaces is a critical issue for many reasons. Deviations of the actual part surface from the desired part surface are investigated in this chapter from the perspective of capabilities of the theory of part surface generation. This means that only geometrical and kinematical parameters of the machining process are taken into account.

Two major reasons often cause surface deviation:

- When machining a part surface, not all of generating surface of the cutting tool actually exists. In all cases of implementation of wedge-cutting tools, the generating surface of the cutting tool is not represented entirely but by a limited number of cutting edges. In other words, the generating surface of the form-cutting tool is represented discretely. The discrete representation of the surface T of the cutting tool causes deviations of the actually machined part surface P_{ac} from the desired (say, from the nominal) part surface P_{nom}.

- The point contact of the part surface and the generating surface of the cutting tool are usually observed when machining a sculptured part surface on a multiaxis NC machine. When the surfaces make point contact, then articulation capabilities of the multiaxis NC machine can be utilized in full. From this perspective, the point contact of the surfaces can be considered as the most general type of surface contact. However, point contact of the part surface P and generating surface T of the form-cutting tool also causes deviations of the actually machined part surface P_{ac} from the desired part surface P_{nom}.

Ultimately, when the generating surface T of a cutting tool is represented discretely and the part surface P and the generating surface T of the form-cutting tool make point contact, then the deviations of the actually machined part surface P_{ac} from the desired part surface P_{nom} increase.

Sources for the deviations of the machined part surface from the desired part surface are limited to two major reasons only in a simplified case of part surface machining. In the simplified cases of part surface machining, no deviations in configuration the part surface P and generating surface T of the form-cutting tool are observed. In reality, deviations in configuration of surfaces P and T are inevitable in nature. Therefore, the impact of deviations of the configuration of surfaces P and T onto the resultant deviation of the surface P_{ac} from the surface P_{nom} must be investigated as well.

8.1 Two Principal Types of Deviations of the Machined Part Surface from the Nominal Part Surface

In the discrete representation of the generating surface of the cutting tool and the point contact of the part surface P and generating surface T of the form-cutting tool, during certain limited periods, it is impossible to generate the part surface precisely, without deviations of the actually machined part surface from the desired part surface. Deviations of two principal types are distinguished when machining a part surface by the cutting tool having point contact with the surface to be machined.

8.1.1 Principal Deviations of the First Type

For the proper generation of a part surface, the entire generating surface of the cutting tool is required to be represented by the cutting tool. Actually, the surface T of a cutting tool is represented as a certain number of cutting edges. The number of cutting edges of the cutting tools of conventional design is limited, and the total number of them could be easily counted. The generating surface T of a cutting tool of this type is discontinuous.

The number of the cutting edges of grinding wheels and other abrasive tools actually is also limited. However, it is not that easy to count all the cutting edges of a grinding wheel as can be done with respect to wedge-cutting tools. Therefore, in most cases of part surface machining, the generating surface of abrasive cutting tools can be considered as a continuous surface T.

When machining a part surface, for example, with a milling cutter (Figure 8.1), the cutting tool rotates about its axis O_T with a certain angular velocity ω_T. In addition, the milling cutter travels across the part surface P with a certain

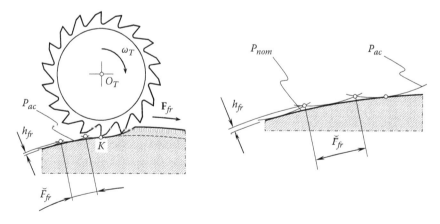

FIGURE 8.1
Deviation, h_{fr}, of the machined part surface, P_{ac}, from the desired part surface, P_{nom}, that is caused by the limited number of cutting edges of the form-cutting tool.

feed rate $\mathbf{F}_{fr}$. The generating surface T of the milling cutter contacts the nominal part surface P at a contact point K. The actual machined part surface P_{ac} is formed as consecutive positions of trajectories of the cutting edge. Usually, the trajectories can be represented by prolate cycloids. In particular cases, the trajectories are represented by pure cycloids and even by curtate cycloids. In any case, the actual part surface P_{ac} becomes wavy. The length of the waves is equal to the feed rate per tooth $\tilde{F}_{fr}$ of the milling cutter, and the wave height (cusp) is specified by h_{fr}. The elementary surface deviation h_{fr} (the surface waviness) is measured along the unit normal vector $\mathbf{n}_p$ to the nominal part surface P_{nom} and is equal to the distance between the surfaces P_{ac} and P_{nom}.

If the part surface to be machined and the generating surface of the form-cutting tool are in line contact, then the cusp height h_{fr} is the only source of the resultant deviation h_Σ of the part surface P_{ac} from the part surface P_{nom}.

Figure 8.1 reveals that the cusp height h_{fr} strongly depends upon the feed rate per tooth $\tilde{F}_{fr}$ of the milling cutter. For milling cutters of most conventional designs, those that work at high rotation ω_T, the cusp height h_{fr} is negligibly small. However, this does not mean that the elementary deviation h_{fr} may always be eliminated from the analysis of part surface P accuracy. The elementary deviation h_{fr} contributes more or less to the resultant deviation h_Σ of the actual part surface P_{ac} from the nominal part surface P_{nom}.

8.1.2 Principal Deviations of the Second Type

The point contact of the desired part surface P and the generating surface T of the form-cutting tool is observed when machining a sculptured part surface on a multiaxis NC machine. Generating of the part surface is performed by a series of consequent tool paths. After the generation of a certain tool path is completed, the cutting tool is then moved in side step to generate the next tool path.

Consider milling of a sculptured part surface on multiaxis NC machine as schematically illustrated in Figure 8.2. The cutting tool rotates about its axis O_T at a certain angular velocity ω_T. In addition, the milling cutter shifts across the tool path at a certain side step $\mathbf{F}_{ss}$. The generating surface T of the milling cutter contacts the nominal part surface P at a point K. The actual machined part surface P_{ac} is formed as consecutive positions of axial profiles of the surface T of the milling cutter. The length of the facets is equal to the side step $\tilde{F}_{ss}$ of the milling cutter, and the cusp height is specified by h_{ss}. The surface deviation h_{ss} is measured along the unit normal vector $\mathbf{n}_p$ to the nominal part surface P_{nom} and is equal to the distance between the surfaces P_{ac} and P_{nom}. Figure 8.2 reveals that the cusps height h_{ss} strongly depends upon the side step $\tilde{F}_{ss}$ of the milling cutter.

The cusp height h_{ss} is another source of deviation that contributes to the resultant deviation h_Σ of the machined sculptured part surface P_{ac} from the desired part surface P_{nom}.

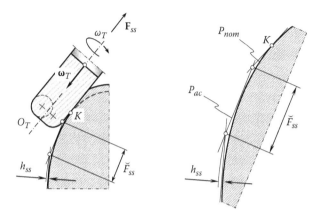

FIGURE 8.2
Deviation, h_{ss}, of the machined part surface, P_{ac}, from the desired part surface, P_{nom}, that is caused by the point type of contact of the part surface P and the generating surface T of the form-cutting tool.

8.1.3 Resultant Deviation of the Machined Part Surface

The resultant deviation h_Σ of the actual part surface P_{ac} from the nominal surface P_{nom} is measured along the unit normal vector $\mathbf{n}_P$ to the nominal part surface P_{nom}, and is equal to the distance between the surfaces P_{ac} and P_{nom}. The value of the resultant deviation h_Σ depends upon the elementary deviations h_{fr} and h_{ss}.

Consider a portion of the actual part surface P_{ac} schematically depicted in Figure 8.3. This portion of the part surface is bounded by two neighboring arc segments m and $(m + 1)$ and by two arc segments n and $(n + 1)$. The distance between the arc segments m and $(m + 1)$ is equal to the feed rate per tooth $\breve{F}_{fr}$ of the cutting tool, while the distance between the arc segments n and $(n + 1)$ is equal to the side step $\breve{F}_{ss}$. The surface P portion that is bounded by the arc segments m, $(m + 1)$ and n, $(n + 1)$ is referred to as the *elementary surface cell* of the sculptured part surface P.

The major parameters h_{fr}, $\breve{F}_{fr}$, h_{ss}, and $\breve{F}_{ss}$ of the elementary surface cell are not constant within part surface P. They vary in certain intervals within the sculptured surface. The current values of the major parameters of the elementary surface cell depend on

- The principal radii or curvature $P_{1.P}$, $P_{2.P}$ of the part surface P
- The principal radii or curvature $P_{1.T}$, $P_{2.T}$ of the generating surface T of the form-cutting tool
- The angle μ of surfaces P and T local relative orientation
- The instant parameters of kinematics of the part surface machining

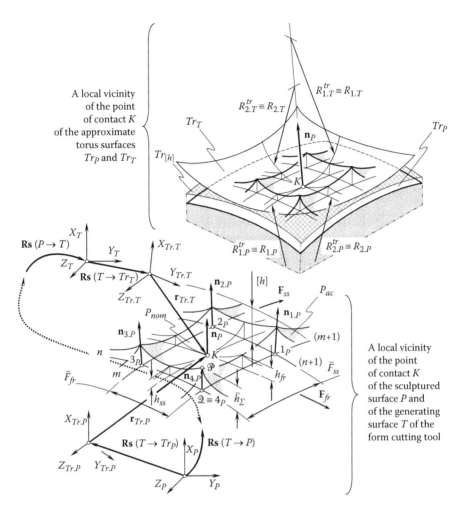

FIGURE 8.3
Generation of an elementary surface cell on the machined part surface P_{ac}.

Therefore, the resultant deviation $h_{\Sigma,i}$ at a current i-point of the part surface P is

$$h_{\Sigma,i} = h_{\Sigma,i}(h_{fr,i}, h_{ss,i}) \qquad (8.1)$$

where the elementary deviations h_{fr} and h_{ss} have to be considered as functions of coordinates of the point on the part surface P, of the corresponding point of the generating surface of the cutting tool, of the angle φ that specifies the direction of the feed rate motion $\mathbf{F}_{fr}$, and the angle μ of surfaces

P and T local relative orientation. This relationship is expressed by two functions:

$$h_{fr} = h_{fr}(U_P, V_P, U_T, V_T, \varphi, \mu) \tag{8.2}$$

$$h_{ss} = h_{ss}(U_P, V_P, U_T, V_T, \varphi, \mu) \tag{8.3}$$

The feed rate $\tilde{F}_{fr}$ and the side step $\tilde{F}_{ss}$ are not incorporated into Equations 8.2 and 8.3.

The maximal resultant deviation h_Σ^{max} of the part surface P_{ac} from the desired surface P_{nom} is often used for the quantitative evaluation of accuracy of the machined part surface.

It is widely recognized that in sculptured part surface machining on a multiaxis *NC* machine, the principle of superposition of the elementary deviations h_{fr} and h_{ss} is valid. Implementation of the principle of superposition to the elementary deviations h_{fr} and h_{ss} is questionable. This is mostly because the elementary deviations h_{fr} and h_{ss} are significantly nonlinear depending on parameters of the process of machining of the sculptured part surface. This issue requires further investigation.

If the principle of superposition of the elementary deviations h_{fr} and h_{ss} is assumed to be valid, then for the calculation of the resultant deviation h_Σ, the following equation can be used:

$$h_\Sigma = a_h \cdot h_{fr} + b_h \cdot h_{ss} \tag{8.4}$$

where a_h, and b_h designate certain constants for a given contact point K. The constants a_h and b_h are within the intervals $0 \le a_h \le 1$ and $0 \le b_h \le 1$.

The resultant deviation of the part surface h_Σ becomes of maximal value h_Σ^{max} when the equality $a_h = b_h = 1$ is observed. In this particular case, the deviation h_Σ^{max} can be calculated from the equation:

$$h_\Sigma^{max} = h_{fr}^{max} + h_{ss}^{max} \tag{8.5}$$

Generally, the function $h_\Sigma = h_\Sigma(h_{fr}, h_{ss})$ is complex.

In compliance with the sixth necessary condition of proper part surface generation [112], the deviation h_Σ must be within the tolerance for accuracy of the part surface machining (see Section 7.2.6). The maximal value of the resultant deviation h_Σ is limited by the tolerance $[h]$ for accuracy of the part surface machining. It is recommended that sculptured part surface machining operation be designated in such a way that the maximal deviation h_Σ^{max} of the machined part surface P_{ac} from the desired part surface P_{nom} is equal to the tolerance $[h]$. A significant reduction in machining time can be achieved if the equality $h_\Sigma^{max} = [h]$ is satisfied within the entire part surface P being machined.

Both the elementary deviations h_{fr} and h_{ss}, and the resultant deviation h_Σ can be calculated on the premise of methods developed in the theory of part surface generation.

8.2 Local Approximation of the Contacting Part Surface *P* and the Generating Surface *T* of the Form-Cutting Tool

The major surface deviations h_{fr}, h_{ss}, and h_Σ can be interpreted in terms of geometry of the nominal part surface *P* and of the surface P_{ac} of the elementary surface cell. To solve the problem, an analytical local representation of the surfaces is helpful.

It is assumed that the nominal part surface is given. Locally, at a current contact point *K*, the part surface *P* is specified by the principal radii of curvature $R_{1.P}$ and $R_{2.P}$. In a direction $\mathbf{t}_P$ at the surface point *K*, which differs from the principal directions $\mathbf{t}_{1.P}$ and $\mathbf{t}_{2.P}$, the torsion of a surface curve τ_P through the point *K* can be calculated as well.

The actual part surface P_{ac} within the elementary surface cell is congruent to the surface of cut (see Section 8.6). When machining a part, the cutting edge of the cutting tool moves relative to the work. Consecutive positions of the moving cutting edge form the surface of cut S_c. The portion of the surface of cut that is located within the elementary surface cell is congruent to the actual part surface ($S_c \equiv P_{ac}$). At a current contact point *K*, the surface of cut S_c can also be locally specified in terms of the principal radii of curvature $R_{1.c}$ and $R_{2.c}$. For calculation of the parameters $R_{1.c}$ and $R_{2.c}$, the equation of the surface of cut S_c is necessary. The equation can be derived on the premise of the geometry of the cutting edge of the cutting tool, the kinematics of the relative motion of the cutting edge with respect to the work, and the operators of coordinate systems transformations (see Chapter 3 for details). Fortunately, for most types of part surface machining, the surface of cut S_c is very close to the generating surface *T* of the form-cutting tool (as long as the elementary surface cell is considered). Therefore, beside the major parameters $R_{1.c}$ and $R_{2.c}$ of the surface of cut S_c cannot be calculated, the equivalent major parameters $R_{1.T}$ and $R_{2.T}$ of the generating surface *T* of the form-cutting tool can be calculated instead.

8.2.1 Local Approximation of Surfaces *P* and *T* by Portions of Torus Surfaces

Commonly, a part surface *P* and the generating surface *T* of the cutting tool are given in a complex analytical form that is not convenient for calculation of the major parameters of the surfaces. The solutions to many geometrical problems in the theory of part surface generation can be more easily derived from local consideration of the surfaces rather than from consideration of the entire surfaces.

For the local analysis, the surfaces are often represented by quadrics. As it was shown in our previous works [116,117,136], from the perspective of local approximation of surface patches, helical canal surfaces feature important advantages over other candidates.

As shown in Figure 8.4, a helical canal surface is a particular case of swept surface. Monge was the first to investigate the class of surfaces formed by sweeping a sphere, in 1850 [41]. He named them *canal surfaces*. In the particular case when the path on which the sphere is swept along is a helix and the sphere has constant radius, the surface swept out is referred to as a *helical canal surface*. A surface of this type is of particular interest for engineers.

A canal surface is the envelope of a one-parametric family of spheres. The envelope is defined as the union of all circles of intersection of infinitesimally neighboring pairs of spheres. These circles are referred to as the *composing circles*.

As shown in Figure 8.4, helical canal surfaces P_{hc} can fit the principal curvatures of the local patch of sculptured part surface P as well as of the generating surface T of the cutting tool. The principal radii of curvature $R_{1.hc}$ and $R_{2.hc}$ of the helical canal surfaces P_{hc} are measured in the plane sections through the unit vectors $\mathbf{n}_{hc}$ and $\mathbf{t}_{1.hc}$ (for the radius $R_{1.hc}$) and through the unit vectors $\mathbf{n}_{hc}$ and $\mathbf{t}_{2.hc}$ (for the radius $R_{2.hc}$), respectively. At a point of tangency of the approximating helical canal surfaces P_{hc} and the part surface P, the unit vectors $\mathbf{n}_{hc}$, $\mathbf{t}_{1.hc}$, and $\mathbf{t}_{2.hc}$ align to the corresponding unit vectors $\mathbf{n}_P$, $\mathbf{t}_{1.P}$, and $\mathbf{t}_{2.P}$ of the sculptured part surface P. The equality of the principal radii of curvatures $R_{1.hc} = R_{1.P}$ and $R_{2.hc} = R_{2.P}$ is observed. The same is valid with respect to the unit vectors $\mathbf{n}_{hc}$, $\mathbf{t}_{1.hc}$, and $\mathbf{t}_{1.hc}$ of the approximating helical canal

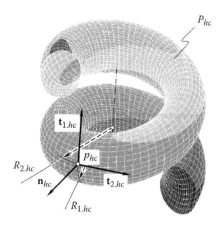

FIGURE 8.4
Local portion of a helical canal surface.

surfaces P_{hc} and the corresponding unit vectors $\mathbf{n}_T$, $\mathbf{t}_{1.T}$, and $\mathbf{t}_{2.T}$ of the generating surface T of the cutting tool.

As illustrated in Figure 8.5, in a direction specified by unit tangent vector $\mathbf{t}_{hc}$ through the surface point p_{hc} that differs from the principal directions $\mathbf{t}_{1.hc}$ and $\mathbf{t}_{1.hc}$, torsion τ_{hc} of a surface curve $\mathbf{c}(t)$ through the point p_{hc} can be also calculated. Configuration of the osculating plane C at the point p_{hc} can be specified in terms of unit vectors $\mathbf{n}_{hc}$ and $\mathbf{t}_{hc}$.

A further simplification of local approximation of the part surface P and the generating surface T of the form-cutting tool becomes possible when torus surfaces are used for the purposes of local surface approximation of the surfaces.

The torus surface can be considered as the above-discussed helical canal surface, the screw parameter of which is put equal to zero. Implementation of torus surfaces makes possible perfect approximation of bigger surface areas, not just in differential vicinity of a surface point like quadrics do. A possibility of implementation of a torus surface for the local approximation of a sculptured part surface P as well as of the generating surface T of the form-cutting tool is based on the schematic depicted in Figure 8.6. Here, in Figure 8.6, the major parameters associated with a torus surface are designated. Unit normal vector $\mathbf{n}_{tr}$, unit tangent vectors $\mathbf{t}_{1.tr}$ and $\mathbf{t}_{2.tr}$, principal plane sections $C_{1.tr}$ and $C_{2.tr}$, and principal curvatures $k_{1.tr}$ and $k_{2.tr}$ at the approximating torus point p_{tr} are among them.

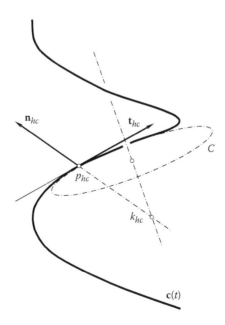

FIGURE 8.5
Configuration of a surface curve $\mathbf{c}(t)$ with the oscillating circle c at point p_{hc}.

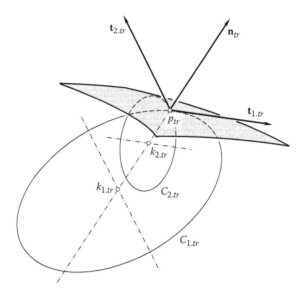

FIGURE 8.6
Principal frame, and principal circles at a surface point.

The local approximation of the part surface P by the torus surface Tr_P is illustrated in Figure 8.7. The part surface P is given in a *Cartesian* coordinate system $X_P Y_P Z_P$. The position vector of an arbitrary point p_1 of the surface P is designated as $\mathbf{r}_{p1}$. Here, in Figure 8.7, the surfaces P and Tr_P share common unit tangent vectors $\mathbf{t}_{1.tr} \equiv \mathbf{t}_{1.P}$ and $\mathbf{t}_{2.tr} \equiv \mathbf{t}_{2.P}$ of surfaces P and Tr_P principal

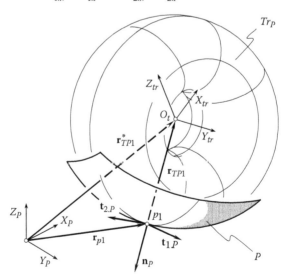

FIGURE 8.7
Construction of the torus surface T_{P1} at the point p_1 of the part surface P.

directions and they also share common unit normal vectors $\mathbf{n}_{tr} \equiv \mathbf{n}_p$. The vectors $\mathbf{t}_{1.P}$, $\mathbf{t}_{2.P}$ and $\mathbf{n}_p$ together with the calculated values of the principal radii of curvature $R_{1.P}$ and $R_{2.P}$ make possible calculation of the position vector $\mathbf{r}_{TP1}$. The last vector together with the position vector $\mathbf{r}_{p1}$ of point p_1 allow for calculation of the position vector $\mathbf{r}_{TP1}^*$ in the coordinate system $X_p Y_p Z_p$ associated with the part surface P. Ultimately, a *Cartesian* coordinate system $X_{tr} Y_{tr} Z_{tr}$ is associated with the torus surface Tr_P.

For further consideration, it is important to stress that not just any point of the approximating torus surface is used for the local approximation of the part surface P and the generating surface T of the form-cutting tool. For this purpose, only points that are within the circle either of the biggest meridian of the approximating torus or of the smallest meridian are used.

The geometry of a torus surface can be expressed in terms of radius r_{tr} of its generating circle and in terms of radius R_{tr} of its directing circle.

Depends on the actual ratio between the radii r_{tr} and R_{tr}, the torus radius r_{tr} can be equal to the first principal radius of curvature $R_{1.P}$ of the part surface P ($r_{tr} = R_{1.P}$), while the torus radius R_{tr} in this case is equal to the difference $R_{tr} = R_{2.P} - R_{1.P}$. Thus, at a current part surface point, the approximating torus surface can be constructed on the premise of the following equalities:

$$r_{tr} = R_{1.P} \tag{8.6}$$

$$R_{tr} = R_{2.P} - R_{1.P} \tag{8.7}$$

For another ratio between the radii r_{tr} and R_{tr}, other equalities are valid:

$$r_{tr} = R_{2.P} - R_{1.P} \tag{8.8}$$

$$R_{tr} = R_{1.P} \tag{8.9}$$

At a current surface point, principal radii of curvature $R_{1.P}$ and $R_{2.P}$ can be calculated as discussed in Chapter 1.

A schematic for derivation of an equation of a torus surface Tr_P is depicted in Figure 8.8.

In the coordinate system $X_{tr} Y_{tr} Z_{tr}$ associated with the torus surface (Figure 8.8), the position vector $\mathbf{r}_{tr}(\theta_{tr}, \varphi_{tr})$ of a point of the approximating torus surface can be represented in the following way:

$$\mathbf{r}_{tr}(\theta_{tr}, \varphi_{tr}) = \mathbf{R}(\theta_{tr}) + \mathbf{r}_{tr}(\theta_{tr}, \varphi_{tr}) \tag{8.10}$$

where $\mathbf{r}(\theta_{tr}, \varphi_{tr})$ designates the position vector of a point on the generating circle of radius r_{tr} in its current location (Figure 8.8) and $\mathbf{R}(\theta_{tr})$ designates the position vector of the center of the generating circle, which rotates about the Z_{tr} axis.

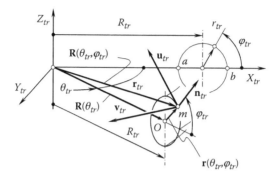

FIGURE 8.8
Generation of a torus surface T_{rP} as a locus of consecutive positions of the generating circle of radius r_{tr} that is rotated about Z_{tr} axis. (Reprinted from *Computer-Aided Design*, 37, Radzevich, S.P., A cutting-tool-dependent approach for partitioning of sculptured surface, 767–778, Copyright 2005, with permission from Elsevier.)

A routine transformation yields the following expression for the position vector $r_{tr}(\theta_{tr}, \varphi_{tr})$:

$$r_{tr}(\theta_{tr}, \varphi_{tr}) = \begin{bmatrix} -(R_{2.P} - R_{1.P})\cos\theta_{tr} + R_{1.P}\cos\varphi_{tr}\cos\theta_{tr} \\ -(R_{2.P} - R_{1.P})\sin\theta_{tr} + R_{1.P}\cos\varphi_{tr}\sin\theta_{tr} \\ R_{1.P}\sin\varphi_{tr} \\ 1 \end{bmatrix} \tag{8.11}$$

The unit normal vector n_{tr} to the torus surface r_{tr} can be calculated by the formula $n_{tr} = u_{tr} \times v_{tr}$, where $u_{tr} = U_{tr}/|U_{tr}|$, and $v_{tr} = V_{tr}/|V_{tr}|$, and the tangent vectors U_{tr} and V_{tr} are given by the equations $U_{tr} = \partial r_{tr}/\partial U_{tr}$ and $V_{tr} = \partial r_{tr}/\partial V_{tr}$.

Only points within the plane section of the torus surface by the coordinate plane $X_{tr}Y_{tr}$ are used for the local approximation of the part surface P and the generating surface T of the form-cutting tool. These points are within the circles through the points a and b (Figure 8.8), which are centering at the origin of the reference system $X_{tr}Y_{tr}Z_{tr}$.

To specify the configuration of the torus surface r_{tr}, the unit tangent vectors $t_{1.tr}$ and $t_{2.tr}$ can be employed. At contact point K, the unit tangent vectors $t_{1.tr}$ and $t_{2.tr}$ are identical to the unit tangent vectors $t_{1.P}$ and $t_{2.P}$ of the sculptured part surface P.

It is important to stress here that the patches of torus surfaces that locally approximate the part surface P and the generating surface T of the form-cutting tool on the one hand and the torus portion of the torus portion of the generating surface of the cutting tool on the other hand are completely different entities. The last is clearly illustrated in Figure 8.9 where a portion of a filleted-end milling cutter is shown. The torus portion of the generating surface of the cutting tool is specified by the radius r_{tr} of the generating circle

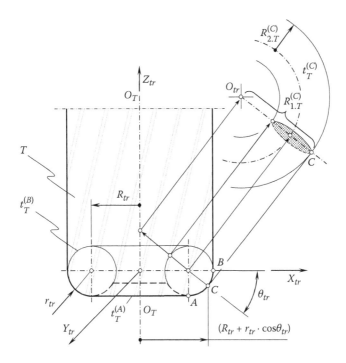

FIGURE 8.9
Analysis of the local geometry of the generating surface T of a filleted-end milling cutter.

of the torus, and by the radius R_{tr} of the directing circle of the torus surface. Three arbitrary points A, B, and C are chosen within the generating circle of the torus portion of the surface T. Approximating torus surfaces can be constructed at all of the points A, B, and C.

The approximating torus surface Tr_A through point A can be specified by the radius r_{tr} of the generating circle of the torus and by the radius $R_{tr} \to \infty$ of the directing circle of the torus surface (the radius $R_{tr} \to \infty$ is not indicated in Figure 8.9).

The approximating torus surface Tr_B through the point B is congruent to the toroidal portion of the generating surface of the milling cutter, and thus, it can be specified by the radius $r_{tr.B} \equiv r_{tr}$ of the generating circle of the torus and by the radius $R_{tr.B} \equiv R_{tr}$ of the directing circle of the torus surface.

Ultimately, the approximating torus Tr_C through point C is specified by the radius r_{tr} of the generating circle of the torus and by the radius $(R_{tr} + r_{tr} \cdot \cos \theta_{tr})$ of the directing circle of the torus surface (here the angle θ_{tr} specifies the location of the point C on the arc of the generating circle of radius r_{tr}).

Note that all 10 types of local patches of smooth regular surfaces (see Chapter 1, Figure 1.16) can be found on the torus surface Tr. Figure 8.10 illustrates this important property of the torus surface.

Consider the points on surface Tr that occupy various positions m_1, m_2, m_3, m_i, and so forth. The part body can be located either inside the torus surface Tr

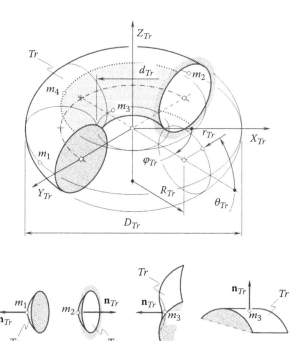

FIGURE 8.10
Major elements of the local topology of a torus surface.

or outside the surface Tr. Depending upon the chosen location of the point m_i either within the convex surface Tr or within the concave surface Tr, all 10 types of local patches of smooth regular surface can be found on the torus surface Tr.

The major advantage of implementation of the torus surface for the local approximation of a sculptured part surface is that a patch of the torus surface can provide a perfect approximation for bigger surfaces area compared with the approximation using quadrics, which is valid only within a differential vicinity of the surface point.

8.2.2 Local Configuration of the Approximating Torus Surfaces

When a sculptured part surface P and the generating surface T of a cutting tool contact each other at a certain point K, the approximating torus surfaces also contact each other at that same point K. Moreover, when parameterization of the approximating torus surfaces is chosen in the form of Equation 8.11, then the unit tangent vectors $t_{1.P}$ and $t_{2.P}$ at the point K of the surface P and the unit tangent vectors $t_{1.T}$ and $t_{2.T}$ at the point K of the surface T are identical to the corresponding unit tangent vectors of the approximating torus surfaces Tr_P and Tr_T. The last is convenient for the development of the analytical description of local configuration of the approximating torus surfaces.

It is assumed that the sculptured part surface P and the generating surface T of the cutting tool are in proper tangency at a certain contact point K as illustrated in Figure 8.11. The unit tangent vectors $\mathbf{u}_P$ and $\mathbf{v}_P$ as well as the unit normal vector $\mathbf{n}_P$ for the part surface P at the contact point K are calculated from the equation $\mathbf{r}_P = \mathbf{r}_P(U_P, V_P)$ of the part surface P (see Chapter 1.1). The equation of the part surface P also yields calculation of the unit tangent vectors $\mathbf{t}_{1.P}$ and $\mathbf{t}_{2.P}$ of the principal directions at a point of the surface P. The vectors $\mathbf{t}_{1.P}$, $\mathbf{t}_{2.P}$, and $\mathbf{n}_P$ make up the Darboux frame. The Darboux frame is implemented here for the construction of the local left-hand–oriented *Cartesian* coordinate system $x_P y_P z_P$ having origin at the contact point K.

The configuration of the sculptured part surface P as well as the configuration of the generating surface T in the coordinate system $X_{NC}Y_{NC}Z_{NC}$ associated with the machine tool is known. Therefore, corresponding operators of the coordinate system transformations, the operator $\mathbf{Rs}(NC \mapsto P)$ of the resultant transformation from the coordinates system $X_{NC}Y_{NC}Z_{NC}$ to the coordinate system $X_P Y_P Z_P$, and further, the operator $\mathbf{Rs}(NC \mapsto P)$ of the resultant

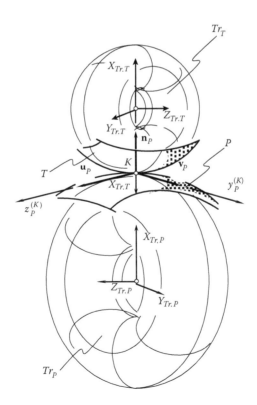

FIGURE 8.11
Example of relative disposition of the approximating torus surfaces Tr_P and Tr_T.

transformation from the coordinates system $X_pY_pZ_p$ to the local coordinate system $x_py_pz_p$ can be composed. Ultimately, the operator $\mathbf{Rs}(NC \mapsto K_p)$ of the resultant coordinate systems transformation can be composed.

Similar operators $\mathbf{Rs}(NC \mapsto T)$, $\mathbf{Rs}(T \mapsto K_T)$, and $\mathbf{Rs}(NC \mapsto K_T)$ of the consequent coordinate systems transformations are composed for the generating surface T of the cutting tool. Ultimately, the operators of the direct $\mathbf{Rs}(K_T \mapsto K_p)$ and the inverse $\mathbf{Rs}(K_p \mapsto K_T)$ coordinate system transformations can be composed as well. The operators $\mathbf{Rs}(K_T \mapsto K_p)$ and $\mathbf{Rs}(K_p \mapsto K_T)$ complement the earlier composed operators of the coordinate systems transformation to a closed loop of the coordinate system transformation (see Chapter 3).

The derived operators of the coordinate system transformations make possible representation of the surfaces $\mathbf{r}_p$, $\mathbf{r}_T$, and of all major elements of their geometry in a common reference system. Implementation of the local coordinate system $x_py_pz_p$ for this purpose is convenient.

8.3 Calculation of the Elementary Surface Deviations

The earlier performed analysis shows that the resultant deviation h_Σ of the machined part surface P_{ac} from its desired shape can be evaluated using the formula (see Equation 8.4).

$$h_\Sigma = a_h \cdot h_{fr} + b_h \cdot h_{ss} \tag{8.12}$$

For the calculation of the resultant deviation h_Σ of the machined part surface P_{ac} from the nominal part surface P, the actual values of the elementary deviations h_{fr} and h_{ss} are necessary. As will be shown, for the calculation of both elementary deviations h_{fr} and h_{ss}, similar equations can be used. Therefore, it is not necessary to investigate both elementary deviations separately. It is sufficient to investigate only one of them and write similar equations for the calculation of another one.

8.3.1 Waviness of the Machined Part Surface

Consider, for example, calculation of the elementary deviation h_{fr}. Figure 8.12 illustrates a cross section of a sculptured part surface P by a plane through the unit normal vector $\mathbf{n}_p$ and through the feed rate vector $\mathbf{F}_{fr}$. Depending on the chosen point of interest on the part surface P, the cross section of the surface P could have either a straight profile KK_1 or a convex profile K_1K_2 or a concave profile K_2K_3.

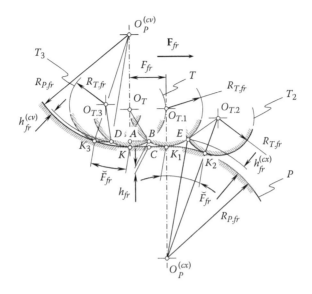

FIGURE 8.12
Calculation of the elementary deviation h_{fr} (the waviness height) on the sculptured part surface P.

It is convenient to mention here that degree of conformity of the generating surface T of the cutting tool is the lowest at the convex point K_2; it is larger at the point of inflection K_1 (or at the similar point K); and it is highest at the concave point K_3. This yields to the conclusion that when a higher degree of conformity of the cutting tool surface T to the part surface P is observed, there is a higher accuracy of the machined part surface and vice versa.

For calculation of the elementary deviation h_{fr}, the following equation is derived by Radzevich [118,119]:

$$
h_{fr} \cong \frac{R_{P.fr} \cdot (R_{P.fr} + R_{T.fr}) \cdot \left(1 - \cos\dfrac{\breve{F}_{fr}}{2 \cdot R_{P.fr}}\right)}{R_{P.fr} - (R_{P.fr} + R_{T.fr}) \cdot \cos\left(\dfrac{\breve{F}_{fr}}{2 \cdot R_{P.fr}}\right)} \tag{8.13}
$$

where the radii of normal curvature at a point of contact of surfaces P and T are designated as $R_{P.fr}$ and $R_{T.fr}$, respectively, and the arc segment $\breve{F}_{fr}$ designates the feed rate per tooth of the cutting tool.

The radii $R_{P.fr}$ and $R_{T.fr}$ are measured in the direction of the feed rate vector $\mathbf{F}_{fr}$.

For the calculation of the radius of normal curvature $R_{P.fr}$, the equation is derived in [118,119]

$$R_{P.fr} = \frac{E_P G_P}{G_P L_P \sin^2 \xi + M_P \sqrt{E_P G_P} \sin 2\xi + E_P N_P \cos^2 \xi} \tag{8.14}$$

where angle ξ specifies the direction of the feed rate vector $\mathbf{F}_{fr}$ relative to the principal directions $\mathbf{t}_{1.P}$ and $\mathbf{t}_{2.P}$ at a point of the sculptured part surface P.

An equation similar to Equation 8.14 is derived [118,119] for computation of the radius of normal curvature $R_{T.fr}$ of the cutting tool surface T:

$$R_{T.fr} \cong \frac{E_T G_T}{G_T L_T \sin^2(\xi + \mu) + M_T \sqrt{E_T G_T} \sin 2(\xi + \mu) + E_T N_T \cos^2(\xi + \mu)} \tag{8.15}$$

where μ is the angle of the local relative orientation surfaces P and T.

It is assumed in Equation 8.15 that the radius of normal curvature of the surface of cut is approximately equal to the corresponding radius of normal curvature of the generating surface T of the cutting tool.

In particular cases, Equation 8.13 can be significantly simplified. For example, when a flat portion of a part surface P is machined with the milling cutter of diameter d_T, then the cusp height is equal to

$$h_{fr} = R_{T.fr} - \sqrt{R_{T.f}^2 - 0,25 \breve{F}_{T.f}^2} \tag{8.16}$$

Equation 8.16 is well known from practice.

8.3.2 Elementary Deviation h_{ss} of the Machined Part Surface

For calculation of the elementary surface deviation h_{ss}, a plane through the unit normal vector $\mathbf{n}_P$ and through the vector $\mathbf{F}_{ss}$ of side step of the cutting tool is employed. The plane through vectors $\mathbf{n}_P$ and $\mathbf{F}_{ss}$ is orthogonal to the plane through the vectors $\mathbf{n}_P$ and F_{fr} [116–119,136].

The derivation of the equation for the calculation of the elementary deviation h_{ss} is similar to the derivation of Equation 8.13. Therefore, without going into details of derivation, the final equation for the calculation of the elementary deviation h_{ss} is represented as

$$h_{ss} \cong \frac{R_{P.ss} \cdot (R_{P.ss} + R_{T.ss}) \cdot \left(1 - \cos \dfrac{\breve{F}_{ss}}{2 \cdot R_{P.ss}}\right)}{R_{P.ss} - (R_{P.ss} + R_{T.ss}) \cdot \cos \left(\dfrac{\breve{F}_{ss}}{2 \cdot R_{P.ss}}\right)} \tag{8.17}$$

where

$$R_{P.ss} = \frac{E_P G_P}{G_P L_P \cos^2 \xi + M_P \sqrt{E_P G_P} \sin 2\xi + E_P N_P \sin^2 \xi} \qquad (8.18)$$

$$R_{T.ss} \cong \frac{E_T G_T}{G_T L_T \cos^2(\xi + \mu) + M_T \sqrt{E_T G_T} \sin 2(\xi + \mu) + E_T N_T \sin^2(\xi + \mu)} \qquad (8.19)$$

In Equation 8.17, the radii of normal curvature of the part surface P and the generating surface T of the form-cutting tool are designated as $R_{P.ss}$ and $R_{T.ss}$, respectively, and the arc segment $\tilde{F}_{ss}$ designates the side step of the cutting tool.

The radii $R_{P.ss}$ and $R_{T.ss}$ of normal curvature are measured in the direction of the side step vector $\mathbf{F}_{ss}$.

8.3.3 Alternative Approach for Calculation of the Elementary Part Surface Deviations

Reasonable assumptions make possible simplification of equations for the calculation of elementary part surface deviations. As an example, an alternative approach for calculation of the elementary surface deviation h_{fr} is illustrated in Figure 8.13.

An elementary analysis of Figure 8.13 makes possible calculation of coordinates of centers $O_T^{(1)}$ and $O_T^{(2)}$ in two consecutive positions of the cutting tool relative to the work. The equations of the circular arcs of the radius $R_{T.fr}$ together with the equation of the circular arc of the radius $R_{P.fr}$ yield

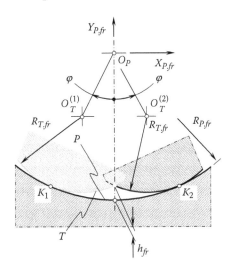

FIGURE 8.13
Another approach for calculation of the part surface waviness h_{fr}.

computation of cusp height h_{fr}. For convenience, the equation for the calculation of cusp height h_{fr} of the part surface P waviness is expanded in the Taylor series. Ultimately, this yields the approximate equation

$$h_{fr} \cong \frac{\varphi^2}{2R_{T.fr}} R_{P.fr}(R_{P.fr} - R_{T.fr}) \tag{8.20}$$

for calculation of the elementary surface deviation h_{fr}.

The interested reader may wish to go to Reference [117] for details on the derivation of Equation 8.20.

The considered approach can be enhanced to the situation when radii of normal curvature of the part surface P and the generating surface T of the form-cutting tool are significantly different between two consequent tool paths, namely, when radii of the circular arcs centering at $O_T^{(1)}$ and $O_T^{(2)}$ in Figure 8.13 significantly differ from one another.

8.4 Total Displacement of the Cutting Tool with Respect to the Part Surface

No absolute accuracy is observed in machining of sculptured part surfaces on a multiaxis *NC* machine. Both the *NC* machine and the cutting tool are the major sources of unavoidable deviations of the machined part surface from the desired sculptured surface. Actual relative motion of the cutting tool is performed with certain deviations of its parameters with respect to the desired relative motion of the cutting tool. The last is also a source of significant part surface deviations.

The displacements of the generating surface T of the cutting tool with respect to the desired part surface P are inevitable. Two types of problems arise in this concern.

First, it is important to calculate how much the displacement of a cutting tool contributes to the resultant deviation of the actually machined part surface from the desired part surface.

Second, to avoid the cutter penetration into the part surface P, it is of critical importance to determine the maximal allowed dimensions of the cutting tool to avoid violation of the necessary conditions of proper surface generation (see Chapter 7).

To solve both types of problems, the calculation of the *closest distance of approach* (*CDA*) of the part surface P and the generating surface T of the form-cutting tool is necessary. The minimal separation between objects is a fundamental problem that has application in a variety of arenas. The problem of calculation of the closest distance of approach of two surfaces is sophisticated. However, it can be solved using methods developed in the theory of part surface generation.

8.4.1 Actual Configuration of the Cutting Tool with Respect to the Part Surface

It is convenient to begin the analysis from the ideal case, when the part surface P and the generating surface T of the form-cutting tool are in proper tangency at a certain point K as illustrated in Figure 8.14. For the ideal case of part surface generation, the closed chain of consequent coordinate system transformations (Figure 3.13) can be constructed.

The unit tangent vectors $\mathbf{u}_P$ and $\mathbf{v}_P$, the unit normal vector $\mathbf{n}_P$, as well as the unit tangent vectors $\mathbf{t}_{1.P}$ and $\mathbf{t}_{2.P}$ of the principal directions at a point the sculptured part surface P can be calculated. An analytical representation [vector equation $\mathbf{r}_P = \mathbf{r}_P(U_P, V_P)$] of the part surface P is used for this purpose (see Chapter 1).

The Darboux frame can be constructed on the premise of the unit vectors $\mathbf{t}_{1.P}$, $\mathbf{t}_{2.P}$, and $\mathbf{n}_P$. The Darboux frame is used for the construction of the left-hand–oriented local *Cartesian* coordinate system $x_P y_P z_P$ having origin at the contact point K as illustrated in Figure 8.14.

The similar left-hand–oriented local *Cartesian* coordinate system $x_T y_T z_T$ is associated with the generating surface T of the form-cutting tool. The local coordinate system $x_T y_T z_T$ is turned about z_T axis with respect to the local coor-

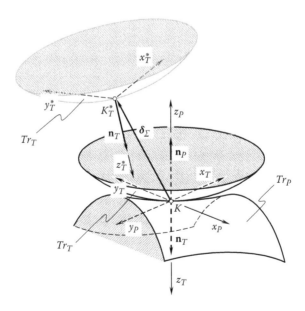

FIGURE 8.14
Actual configuration of local patches of the approximating torus surfaces Tr_P and Tr_T.

dinate system $x_P y_P z_P$ through the angle μ of the local relative orientation of surfaces P and T. The axes z_P and z_T are pointed oppositely to each other.

A left-hand–oriented *Cartesian* coordinate system $X_{NC} Y_{NC} Z_{NC}$ is associated with the multiaxis *NC* machine.

The configuration of the sculptured part surface P as well as configuration of the generating surface T of the cutting tool in the reference system $X_{NC} Y_{NC} Z_{NC}$ is specified. Therefore, corresponding operators of the coordinate system transformations—the operator $\mathbf{Rs}(NC \mapsto P)$ of the resultant transformation from the coordinate system $X_{NC} Y_{NC} Z_{NC}$ to the coordinate system $X_P Y_P Z_P$, and the operator $\mathbf{Rs}(P \mapsto K_P)$ of the resultant transformation from the coordinate system $X_P Y_P Z_P$ to the local coordinate system $x_P y_P z_P$—can be composed. The operators $\mathbf{Rs}(NC \mapsto P)$ and $\mathbf{Rs}(P \mapsto K_P)$ of the resultant coordinate system transformations are expressed in terms of the operators of elementary coordinate system transformations (see Chapter 3). Ultimately, the operator $\mathbf{Rs}(NC \mapsto K_P)$ of the resultant coordinates systems transformation can be composed as well.

Similar operators $\mathbf{Rs}(NC \mapsto T)$, $\mathbf{Rs}(T \mapsto K_T)$, and $\mathbf{Rs}(NC \mapsto K_T)$ of the consequent coordinate systems transformations are composed for the generating surface T of the form-cutting tool. Ultimately, the operators of the direct $\mathbf{Rs}(K_T \mapsto K_P)$ and the inverse $\mathbf{Rs}(K_P \mapsto K_T)$ coordinate system transformations can be composed as well. It is important to note that the equality $\mathbf{Rs}(K_T \mapsto K_P) = \mathrm{Rs}^{-1}(K_P \mapsto K_T)$ is always observed. The operators $\mathbf{Rs}(K_T \mapsto K_P)$ and $\mathbf{Rs}(K_P \mapsto K_T)$ of the coordinate system transformations complement the earlier composed operators of the coordinate systems transformation to the closed chain of the consequent coordinate system transformations (see Figure 3.13).

Equation $\mathbf{r}_P = \mathbf{r}_P(U_P, V_P)$ for the position vector of a point of the part surface P, and equation $\mathbf{r}_T = \mathbf{r}_T(U_T, V_T)$ for the position vector of a point of the generating surface T of the form-cutting tool together with the above-mentioned operators of the coordinate systems transformations make possible representations of the surfaces $\mathbf{r}_P$ and $\mathbf{r}_T$ in a common reference system. Below, the local *Cartesian* coordinate system $x_P y_P z_P$ is used for this purpose.

When the generating surface T of the cutting tool is in proper tangency with the sculptured part surface P (Figure 8.14), then the origin of both local coordinate systems $x_P y_P z_P$ and $x_T y_T z_T$ coincide with the point of contact K of surfaces P and T. In reality, surfaces P and T do not make proper contact. Actually, the surfaces are either slightly apart or the cutting tool surface T penetrates into the part surface P. This is due to the inevitable deviations in configuration of the cutting tool with respect to the part surface P.

The deviations cause a displacement of the local coordinate system $x_T y_T z_T$ from its desired position to the actual position $x_T^* y_T^* z_T^*$. Deviations of this type are unavoidable.

The resultant linear displacement $\boldsymbol{\delta}_\Sigma$ of the cutting tool in relation to the part surface P can be expressed in terms of the elementary linear displacements δ_x, δ_y, and δ_z of the cutting tool along the axes x_P, y_P, z_P:

$$\boldsymbol{\delta}_{\Sigma} = \begin{bmatrix} \delta_x \\ \delta_y \\ \delta_z \\ 1 \end{bmatrix} \tag{8.21}$$

In addition to the linear displacements δ_x, δ_y, and δ_z, the elementary angular displacements θ_x, θ_y, and θ_z of the local coordinate system $x_T y_T z_T$ with respect to the local coordinate system $x_P y_P z_P$ are always observed.

The resultant angular displacement $\boldsymbol{\theta}_{\Sigma}$ of the cutting tool in relation to the part surface P can be expressed in terms of the elementary angular displacements of the cutting tool through the angles* θ_x, θ_y, and θ_z about the axes x_P, y_P, z_P

$$\boldsymbol{\theta}_{\Sigma} = \begin{bmatrix} \theta_x \\ \theta_y \\ \theta_z \\ 1 \end{bmatrix} \tag{8.22}$$

Ultimately, the local coordinate system $x_T y_T z_T$ associated with the cutting tool moves to a position $x_T^* y_T^* z_T^*$.

Because the displacements $\boldsymbol{\delta}_{\Sigma}$ and $\boldsymbol{\theta}_{\Sigma}$ are always observed, then either a gap is observed between local patches of surfaces P and T or the cutting tool surface T interferes with the part surface P.

The resultant linear $\boldsymbol{\delta}_{\Sigma}$ and the resultant angular $\boldsymbol{\theta}_{\Sigma}$ displacements can be expressed in terms of the corresponding elementary displacements of all the local coordinate systems between point $K_P \equiv K$ and point K_T. Here K_P and K_T designate origins of the local coordinate systems $x_P y_P z_P$ and $x_T^* y_T^* z_T^*$. No closed chain of the consequent coordinate systems transformations can be constructed at this point. The chain of the consequent coordinate systems transformations is not closed yet. To make the chain close, it is necessary to compose the operator $\mathbf{Rs}(K_T^* \mapsto K_P)$ of the resultant coordinate systems transformation, and the operator $\mathbf{Rs}(K_P \mapsto K_T^*) = \mathbf{Rs}^{-1}(K_T^* \mapsto K_P)$ of the inverse coordinate systems transformation. For composing the operators $\mathbf{Rs}(K_T^* \mapsto K_P)$ and $\mathbf{Rs}(K_P \mapsto K_T^*)$, the earlier developed operators $\mathbf{Rs}(P \mapsto K_P)$ and $\mathbf{Rs}(P \mapsto K_T)$ are helpful:

$$\mathbf{Rs}(K_T^* \mapsto K_P) = \mathbf{Rs}^{-1}(P \mapsto K_T) \cdot \mathbf{Rs}(P \mapsto K_P) \tag{8.23}$$

To get the problem solved, the closest distance of approach between the part surface P and the generating surface T of the form-cutting tool must be calculated.

---

* As rotations are not vectors by nature, special care is necessary when treating the angular displacements θ_x, θ_y, and θ_z as vectors and matrices.

In the ideal case of part surface generation, when no displacement of the cutting tool surface T with respect to the part surface P occurs, surfaces P and T make contact at a point K. Actually, it is allowed to interpret the ideal surface contact in the way that the point K_P of the part surface P and the point K_T of the cutting tool surface T are snapped into a common point K. Therefore, the identity $K_P \equiv K_T \equiv K$ is valid for the ideal case of part surface generation. For convenience, the designation $K_P \equiv K_T \equiv K$ is not used for the point of contact of the surfaces in the ideal case of part surface generation, but the designation K is used instead.

Because the identity $K_P \equiv K_T \equiv K$ is valid, the closest distance of approach between surfaces P and T is identical to the closest distance of approach between the approximating torus surfaces Tr_P and Tr_T, and it is identical to zero. The closest distance of approach between surfaces P and T can be interpreted as the distance between points K_P and K_T. Therefore, for the ideal case of part surface generation, the equality $K_P K_T = 0$ is valid.

In reality, the generating surface T of the cutting tool is displaced in relation to the part surface P. The total linear displacement of the surface T with respect to the surface P is equal to the magnitude of the vector $\boldsymbol{\delta}_\Sigma$ (see Equation 8.21). The total angular displacement of the surface T with respect to the surface P is equal to the magnitude of the vector $\boldsymbol{\theta}_\Sigma$ (see Equation 8.22). The closest distance of approach of surfaces P and T is not equal to zero. It can be positive or negative. In the first case, the cutting tool surface T is located apart from the part surface P. In the second case, the cutting tool surface T interferes with part surface P.

It is critical to realize that the following theorem is correct:

Theorem

The closest distance of approach of two smooth regular surfaces is perpendicular to both the surfaces simultaneously. ∎

The theorem is proven analytically. We are not going into the details of the proof of the theorem here. The interested reader may wish to exercise this concern on his or her own.

Due to occurrence of the displacements $\boldsymbol{\delta}_\Sigma$ and $\boldsymbol{0}_\Sigma$, the closest distance of approach of the part surface P and the generating surface T of the cutting tool is not equal to the distance between the points K_T^* and $K \equiv (KP)$. In-perpendicularity of the straight-line segment KK_T^* to surfaces P and T is the major reason for the equality not being valid.

Once the inequality of the length of the straight-line segment KK_T^* to the closest distance of approach between two surfaces P and T is understood, one can proceed with further analysis.

The analysis below is based as much on the presumption that the configuration of the local coordinate system $x_T^* y_T^* z_T^*$ with respect to the local coordinate system $x_p y_p z_p$ is known. The configuration is specified by the operator $\mathbf{Rs}(K_T^* \mapsto K_P)$ of the resultant coordinate system transformation (see Equation 8.23). Use of the operator $\mathbf{Rs}(K_T^* \mapsto K_P)$ together with the operator $\mathbf{Rs}(K_P \mapsto K_T)$ discussed earlier in this section yield introduction of the matrix $\mathbf{Ds}(T/P)$ of the displacement of the generating surface T of the cutting tool with respect to the part surface P. The displacement matrix $\mathbf{Ds}(T/P)$ specifies the actual configuration of the local coordinate system $x_T^* y_T^* z_T^*$ associated with the cutting tool in relation to the local coordinate system $x_T y_T z_T$ associated with the cutting tool's ideal configuration with respect to the part surface P.

Actually, the matrix $\mathbf{Ds}(T/P)$ of the resultant displacements can be composed in the following way. Consider all the n elements and joints between the elements, those that are involved in the closed chain of the consequent coordinate system transformations. The elementary displacement of every element and at every joint contributes to the resultant displacement of the cutting tool with respect to the part surface P. The elementary i-th displacement can be interpreted as the displacement of the actual elementary coordinate system $X_i^{ac} Y_i^{ac} Z_i^{ac}$ with respect to the nominal location of the corresponding elementary coordinate system $X_i Y_i Z_i$. Implementation of the generalized formula for the resultant coordinate system transformations (see Equation 3.19) makes it possible to derive the matrix $\mathbf{ds}_i(ac_i \mapsto nom_i)$ of a particular elementary displacement:

$$\mathbf{ds}_i(ac_i \mapsto nom_i) = \begin{bmatrix} \cos\theta_{xx}^{(i)} & \cos\theta_{xy}^{(i)} & \cos\theta_{xz}^{(i)} & \delta_x^{(i)} \\ \cos\theta_{xy}^{(i)} & \cos_{yy}^{(i)} & \cos\theta_{yz}^{(i)} & \delta_y^{(i)} \\ \cos\theta_{xz}^{(i)} & \cos\theta_{yz}^{(i)} & \cos_{zz}^{(i)} & \delta_z^{(i)} \\ 0 & 0 & 0 & 1 \end{bmatrix} \tag{8.24}$$

In particular cases, the i-th displacement can be either linear or angular. Encompassing all the elementary displacements between the local coordinate systems $x_T^* y_T^* z_T^*$ and $x_p y_p z_p$, the following equation for the calculation of the matrix $\mathbf{Ds}(T/P)$ of the resultant displacements can be obtained:

$$\mathbf{Ds}(T/P) = \prod_{i=1}^{n} \mathbf{ds}_i(ac_i \mapsto nom_i) \tag{8.25}$$

where ac_i and nom_i designate the actual and the nominal location of the i-th coordinate system.

When the operators $\mathbf{Rs}(K_T^* \mapsto K_P)$ and $\mathbf{Rs}(K_P \mapsto K_T)$ are calculated, the displacement matrix $\mathbf{Ds}(T/P)$ can be expressed in terms of the operators $\mathbf{Rs}(K_T^* \mapsto K_P)$ and $\mathbf{Rs}(K_P \mapsto K_T)$ of the resultant coordinate systems transformations:

$$\mathbf{Ds}(T/P) = \mathbf{Rs}(K_P \mapsto K_T) \cdot \mathbf{Rs}(K_T^* \mapsto K_P) \tag{8.26}$$

Ultimately, the displacement matrix $\mathbf{Ds}(P/T)$ can be expressed in terms of the elementary linear and angular displacements of the local coordinate system $x_T^* y_T^* z_T^*$ in relation to the desired location of the local coordinate system $x_T y_T z_T$:

$$\mathbf{Ds}(P/T) = \begin{bmatrix} \cos\theta_{xx} & \cos\theta_{xy} & \cos\theta_{xz} & \delta_x \\ \cos\theta_{xy} & \cos\theta_{yy} & \cos\theta_{yz} & \delta_y \\ \cos\theta_{xz} & \cos\theta_{yz} & \cos\theta_{zz} & \delta_z \\ 0 & 0 & 0 & 1 \end{bmatrix} \tag{8.27}$$

For the inverse transformation, the displacement matrix $\mathbf{Ds}(P/T)$ can be used. The matrix $\mathbf{Ds}(P/T)$ can be either composed similar to the way the displacement matrix $\mathbf{Ds}(T/P)$ is composed (see Equation 8.26) or it can be calculated from

$$\mathbf{Ds}(P/T) = \mathbf{Ds}^{-1}(T/P) \tag{8.28}$$

All the elements of the displacements matrix $\mathbf{Ds}(T/P)$ can be expressed in terms of the actual elementary displacements of all the elements that make up the closed chain of the consequent coordinate systems transformations. The elementary displacements include

- The linear and the angular displacements in all mechanical joints
- All the deflections caused by elasticity of material of the components involved in the closed chain of the consequent coordinate systems transformations
- All the thermal extensions of the components involved in the closed chain of the consequent coordinate systems transformations, and so forth

The displacements of the generating surface T of the form-cutting tool with respect to the part surface P are not known. Theoretically, components of all of the above-listed matrices of the actual elementary displacements can be determined through the direct measurements of the system composed of "Work/NC Machine Tool/Cutting Tool". Actually, it is not practical to perform such complex measurements.

Under such a scenario, the tolerances matrix $\mathbf{Tl}(T/P)$ can be used for calculation of the actual configuration of the generating surface T of the cutting tool in relation to the part surface P instead of the displacement matrix $\mathbf{Ds}(T/P)$.

The tolerances matrix $\mathbf{Tl}(T/P)$ is composed similar to the displacements matrix $\mathbf{Ds}(T/P)$. The only difference is that the elementary displacements

$ds_i(ac_i \mapsto nom_i)$ of the surface T with respect to the surface P are not employed for the computations, but the corresponding tolerances $tl_i(ac_i \mapsto nom_i)$ are used instead:

$$\mathbf{Tl}(T/P) = \prod_{i=1}^{n} \mathbf{tl}_i(ac_i \mapsto nom_i) \tag{8.29}$$

Based on the last statement, the following approximate equality occurs:

$$\mathbf{Ds}(T/P) \cong \mathbf{Tl}(T/P) \tag{8.30}$$

The required elementary tolerances for composing the tolerance matrix $\mathbf{Tl}(T/P)$ can be determined much more easily. Therefore, if the displacements matrix $\mathbf{Ds}(T/P)$ is not known, the tolerance matrix $\mathbf{Tl}(T/P)$ can be used instead.

The approximating torus surface Tr_T is associated with the local coordinate system $x_T^* y_T^* z_T^*$.

Once the displacements matrix $\mathbf{Ds}(T/P)$ is composed, then the equation $\mathbf{r}_{tr.T}(\theta_{tr.T}, \varphi_{tr.T})$ (see Equation 8.11) of the approximating torus surface Tr_T can be represented in the local coordinate system $x_P y_P z_P$:

$$\mathbf{r}_{tr.T}^{(P)}(\theta_{tr.T}, \varphi_{tr.T}) = \mathbf{Ds}(T/P) \cdot \mathbf{r}_{tr.T}(\theta_{tr.T}, \varphi_{tr.T}) \tag{8.31}$$

Equation $\mathbf{r}_{tr.P}(\theta_{tr.P}, \varphi_{tr.P})$ of the approximating torus surface Tr_P is initially determined in the local coordinate system $x_P y_P z_P$. Equation 8.31 describes analytically the approximating torus surface Tr_T in that same local coordinate system $x_P y_P z_P$. This yields the conclusion that the actual configuration of the torus surfaces Tr_P and Tr_T is determined.

8.4.2 Closest Distance of Approach between the Part Surface P and the Generating Surface T of the Form-Cutting Tool

Generally, the problem of calculation of the closest distance of approach between two smooth regular surfaces is sophisticated and challenging. To the author's knowledge, no general solution to the problem of calculation of the closest distance of approach between two smooth regular surfaces has been published. For the purpose of calculation of the deviation δ_P of the actual part surface P_{ac} with respect to the desired part surface P_{nom}, the problem under consideration can be reduced to the problem of computation of the closest distance of approach between two torus surfaces Tr_P and Tr_T.

Consider the part surface P and the generating surface T of the form-cutting tool that initially are given in a common coordinate system $X_{NC}Y_{NC}Z_{NC}$ (Figure 8.15) associated with the NC machine. Surfaces P and T are locally

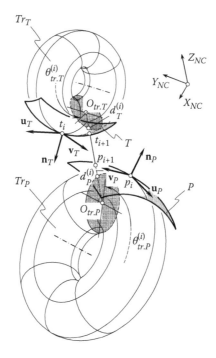

FIGURE 8.15
Computation of the closest distance of approach of the part surface P and the generating surface T of the form-cutting tool.

approximated by portions of torus surfaces Tr_P and Tr_T, respectively. Again, not all points of the torus surfaces Tr_P and Tr_T can be used for the local approximation of surfaces P and T. Only points that are located either within the biggest meridian or within the smallest meridian of the torus surface are employed for this purpose.

The points K_P and K_T^* are chosen as the first guess points on the torus surfaces Tr_P and Tr_T. For the analysis below, it is convenient to relabel the points K_P and K_T^* to p_i and t_i, respectively.

For a given configuration of the torus surfaces Tr_P and Tr_T, the closest distance of approach between these surfaces can be used as a first approximation to the closest distance of approach between the original surfaces P and T.

The closest distance of approach between the torus surfaces Tr_P and Tr_T is measured along the common perpendicular to these surfaces. The following equations can be composed on the premise of this property of the closest distance of approach.

Unit normal vector $\mathbf{n}_{Tr.P}$ to the torus surface Tr_P is within a plane through the axis of rotation of the surface Tr_P. In the coordinate system $X_{tr.P}Y_{tr.P}Z_{tr.P}$, which is associated with the torus surface Tr_P, the equation of a plane through the axis of rotation of the torus surface Tr_P can be expressed in the form [14,116,117,136]

$$\left[\mathbf{r}_{\tau P} - \mathbf{r}_{tr.P}^{(0)}\right] \times \mathbf{k}_{tr.P} \times \mathbf{R}_{tr.P} = 0 \qquad (8.32)$$

where $\mathbf{r}\tau_P$ is the position vector of a point of the plane through the axis of rotation of the torus Tr_P, $\mathbf{r}_{tr.P}^{(0)}$ is the position vector of a point within the plane $\mathbf{r}_{\tau P}$ (it is assumed below that this point coincides with the origin of the coordinate system $X_{tr.P}Y_{tr.P}Z_{tr.P}$), and $\mathbf{k}_{tr.P}$ is the unit vector of the $z_{tr.P}$ axis.

Equation 8.32 is expressed in terms of the radius $\mathbf{R}_{tr.P}$. This indicates that the set of all planes through the fixed $z_{tr.P}$ axis forms a pencil of planes. The equation of the pencil of planes $\mathbf{r}_{\tau P}$ in the common coordinate system $X_{NC}Y_{NC}Z_{NC}$ can be represented in the form

$$\mathbf{r}_{\tau P}(Z_{tr.P}, V_{\tau r.P}, \theta_{tr.P}) = \mathbf{Rs}(Tr_P \mapsto NC) \cdot \begin{bmatrix} V_{tr.P} \cdot \cos\theta_{tr.P} \\ V_{tr.P} \cdot \sin\theta_{tr.P} \\ Z_{tr.P} \\ 1 \end{bmatrix} \qquad (8.33)$$

The unit normal vector $\mathbf{n}_{Tr.T}$ to the torus surface Tr_P is located within a plane through the axis of rotation of the surface Tr_P. In the coordinate system $X_{tr.T}Y_{tr.T}Z_{tr.T}$, which is associated with the surface Tr_P, the equation of a plane through the axis of rotation of the torus surface Tr_P can be represented in the form

$$\left[\mathbf{r}_{\tau T} - \mathbf{r}_{tr.T}^{(0)}\right] \times \mathbf{k}_{tr.T} \times \mathbf{R}_{tr.T} = 0 \qquad (8.34)$$

where $\mathbf{r}_{\tau T}$ is the position vector of a point of the plane through the torus Tr_P axis of rotation, $\mathbf{r}_{tr.T}^{(0)}$ is the position vector of a point within the plane $\mathbf{r}_{\tau T}$ (it is assumed below that this point coincides with the origin of the coordinate system $X_{tr.T}Y_{tr.T}Z_{tr.T}$), and $\mathbf{k}_{tr.T}$ is the unit vector of the $z_{tr.T}$ axis.

Equation 8.34 is expressed in terms of the radius $\mathbf{R}_{tr.T}$. This indicates that the set of all planes through the fixed $z_{tr.T}$ axis forms a pencil of planes. The equation of this pencil of planes $\mathbf{r}_{\tau T}$ in the common coordinate system $X_{NC}Y_{NC}Z_{NC}$ can be represented in the form

$$\mathbf{r}_{\tau T}(Z_{tr.T}, V_{tr.T}, \theta_{tr.T}) = \mathbf{Rs}(Tr_T \mapsto NC) \cdot \begin{bmatrix} V_{tr.T} \cdot \cos\theta_{tr.T} \\ V_{tr.T} \cdot \sin\theta_{tr.T} \\ Z_{tr.T} \\ 1 \end{bmatrix} \qquad (8.35)$$

A straight line through points $d_P^{(i)}$ and $d_T^{(i)}$ along which the shortest distance of approach $d_{PT}^{\min}$ of the torus surfaces Tr_P and Tr_P is measured is the line of intersection of the planes $\mathbf{r}_{tP}$ and $\mathbf{r}_{tT}$. Therefore, this line $d_{PT}^{\min}$ must be aligned with both unit normal vectors $\mathbf{n}_{tr.P}$ and $\mathbf{n}_{tr.T}$.

In the coordinate system $X_{NC}Y_{NC}Z_{NC}$, the equation for the unit normal vector $\mathbf{n}_{tr.P}$ to the torus surface Tr_P yields representation in matrix form

$$
\mathbf{n}_{tr.P} = \mathbf{Rs}(Tr_P \mapsto NC) \cdot
\begin{bmatrix}
(C_{tr.P} + \cos\varphi_{tr.P}) \cdot \cos\varphi_{tr.P} \cdot \cos\theta_{tr.P} \\
(C_{tr.P} + \cos\varphi_{tr.P}) \cdot \cos\varphi_{tr.P} \cdot \sin\theta_{tr.P} \\
(C_{tr.P} + \cos\varphi_{tr.P}) \cdot \sin\varphi_{tr.P} \\
1
\end{bmatrix}
\tag{8.36}
$$

where $C_{tr.P}$ designates the parameter $C_{tr.P} = 1 - \dfrac{R_{2.P}}{R_{1.P}}$.

Similarly, in the coordinate system $X_{NC}Y_{NC}Z_{NC}$, the equation for the unit normal vector $\mathbf{n}_{tr.T}$ to the torus surface Tr_T yields matrix representation in the form

$$
\mathbf{n}_{tr.T} = \mathbf{Rs}(Tr_T \mapsto NC) \cdot
\begin{bmatrix}
(C_{tr.T} + \cos\varphi_{tr.T}) \cdot \cos\varphi_{tr.T} \cdot \cos\theta_{tr.T} \\
(C_{tr.T} + \cos\varphi_{tr.T}) \cdot \cos\varphi_{tr.T} \cdot \sin\theta_{tr.T} \\
(C_{tr.T} + \cos\varphi_{tr.T}) \cdot \sin\varphi_{tr.T} \\
1
\end{bmatrix}
\tag{8.37}
$$

where $C_{tr.T}$ designates the parameter $C_{tr.T} = 1 - \dfrac{R_{2.T}}{R_{1.T}}$.

Evidently, the points $O_{tr.P}$, $O_{tr.T}$, $d_P^{(i)}$, and $d_T^{(i)}$ (Figure 8.15) are located within the straight line through the centers $O_{tr.P}$, $O_{tr.T}$. The position vector $\mathbf{r}_{cd}$ of this straight line can be calculated from the equation

$$
(\mathbf{r}_{cd} - \mathbf{r}_{cp}) \times (\mathbf{r}_{ct} - \mathbf{r}_{cp}) = 0
\tag{8.38}
$$

where $\mathbf{r}_{cp}$ is the position vector of a point on the circle of radius $R_{tr.P}$ and $\mathbf{r}_{ct}$ is the position vector of a point on the circle of radius $R_{tr.T}$.

It is necessary that the straight line $\mathbf{r}_{cd}$ be along the unit normal vectors $\mathbf{n}_{tP}$ and $\mathbf{n}_{tT}$ to the torus surfaces Tr_P and Tr_T.

Considered together, Equations 8.33, 8.35, and 8.38 make it possible to calculate the closest distance of approach between the torus surfaces Tr_P and Tr_T. Then, the line $d_{PT}^{\min}$ intersects the part surface P and the generating surface T of the form-cutting tool at the points p_{i+1} and t_{i+1}, respectively. The points p_{i+1} and t_{i+1} serve as the second guess to the closest distance of approach between surfaces P and T.

The cycle of recursive calculations is repeated as many times as necessary for making the deviation of the calculation of the closest distance of approach between surfaces P and T smaller than the maximal permissible value.

There is an alternative approach for calculation of the closest distance of approach between two torus surfaces. The direction of the unit normal vector to an offset surface to Tr_P is identical to the direction of the unit normal vector $\mathbf{n}_{tr.P}$ to the torus surface Tr_P. This statement is also valid for the unit normal vector $\mathbf{n}_{tr.T}$ to the torus surface Tr_T. This property of the unit normal vectors $\mathbf{n}_{tr.P}$ and $\mathbf{n}_{tr.T}$ can be used for the modification of the method of calculation of the closest distance of approach between two torus surfaces.

The equation of the circle of radius $R_{tr.P}$ yields matrix representation

$$\mathbf{r}_{cp}(\theta_{tr.P}) = \mathbf{Rs}(P \mapsto NC) \cdot \begin{bmatrix} R_{tr.P} \cdot \cos\theta_{tr.P} \\ R_{tr.P} \cdot \sin\theta_{tr.P} \\ 0 \\ 1 \end{bmatrix} \tag{8.39}$$

The equation of the circle of radius $R_{tr.T}$ can be analytically described in a similar manner:

$$\mathbf{r}_{ct}(\theta_{tr.T}) = \mathbf{Rs}(T \mapsto NC) \cdot \begin{bmatrix} R_{tr.T} \cdot \cos\theta_{tr.T} \\ R_{tr.T} \cdot \sin\theta_{tr.T} \\ 0 \\ 1 \end{bmatrix} \tag{8.40}$$

The distance d_{pt} between two arbitrary points on the circles $\mathbf{r}_{cp}(\theta_{tr.P})$ and $\mathbf{r}_{ct}(\theta_{tr.T})$ is equal:

$$d_{pt}(\theta_{tr.P}, \theta_{tr.T}) = |\mathbf{r}_{cp}(\theta_{tr.P}) - \mathbf{r}_{ct}(\theta_{tr.T})| \tag{8.41}$$

The distance d_{pt} is minimal for a specific (optimal) combination of the parameters $\theta_{tr.P}$ and $\theta_{tr.T}$. The favorable values of the parameters $\theta_{tr.P}$ and $\theta_{tr.T}$ can be calculated on solution of the set of two equations:

$$\frac{\partial}{\partial\theta_{tr.P}} \mathbf{r}_{cp}(\theta_{tr.P}) = 0 \tag{8.42}$$

$$\frac{\partial}{\partial\theta_{tr.T}} \mathbf{r}_{ct}(\theta_{tr.T}) = 0 \tag{8.43}$$

From the solution of Equations 8.42 and 8.43, the optimal values $\theta_{tr.P}^{(opt)}$ and $\theta_{tr.T}^{(opt)}$ can be determined. These angles specify the direction of the closest distance of approach of the torus surfaces Tr_P and Tr_T.

Following this method, the three-dimensional problem of calculation of the closest distance of approach of two torus surfaces is reduced to the problem of calculation of the closest distance of approach between two circles. Under a certain scenario, the last approach could possess an advantage over the previous approach.

The convergence of the disclosed algorithms for the computation of the closest distance of approach between two smooth regular surfaces is illustrated in Figure 8.16. The computation procedure is convergent regardless of the actual location of the first guess points on surfaces P and T.

It is instructive to draw attention here to the similarities between the disclosed iterative method for the computation of the closest distance of approach between two smooth regular surfaces and between the Newton–Raphson method, the iterative method of chords, and so forth. Many similarities can be found out in this comparison.

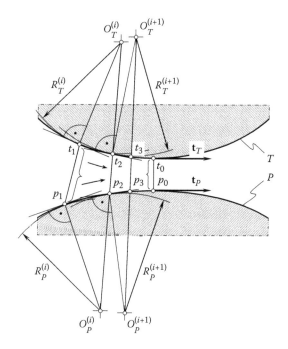

FIGURE 8.16
Convergence of the methods of computation of the closest distance of approach of the part surface P and the generating surface T of the cutting tool.

8.5 Effective Reduction of the Elementary Surface Deviations

As follows from the above consideration, the resultant deviation of the machined part surface P_{ac} from the desired surface P_{nom} depends mostly on two components: height h_Σ of the residual cusps on the machined part surface is the first component, and the deviation δ_P of the surfaces is the second component. Therefore, the resultant deviation δ_Σ of the machined part surface from the desired surface can be expressed by a simple formula:

$$\delta_\Sigma = h_\Sigma + \delta_P \tag{8.44}$$

Recall that both components h_Σ and δ_P are signed values.

Both components h_Σ and δ_P must be reduced to increase the resultant accuracy of the machined part surface. However, only the component h_Σ is under control using methods developed in the theory of part surface generation. The component δ_P is out of control when only the theory of part surface generation is used. To keep component δ_P under control, other methods can be implemented.

However, even under such restrictions, the methods developed in the theory of part surface generation are helpful when aiming for reduction of the machining error. Consider only two opportunities in this concern.

8.5.1 Method of Gradient

For the most intensive reduction of the resultant surface deviation h_Σ gradient of the function $h_\Sigma = h_\Sigma(R_{P.fr}, R_{P.ss}, R_{T.fr}, R_{T.ss}, \breve{F}_{fr}, \breve{F}_{ss}, \ldots)$ can be implemented. As an example, the implementation of the gradient function $\mathbf{grad}(h_{fr})$ for a flattened portion of a sculptured part surface (Figure 8.12) is considered below.

In the case under consideration, the height of the waviness h_{fr} is

$$h_{fr} = R_{T.fr} - \sqrt{R_{T.fr}^2 - 0,25 \cdot \breve{F}_{fr}^2} \tag{8.45}$$

By definition,

$$\mathbf{grad}(h_{fr}) = \frac{\partial h_{fr}}{\partial R_{T.fr}} \mathbf{i} + \frac{\partial h_{fr}}{\partial \breve{F}_{fr}} \mathbf{j} \tag{8.46}$$

In the case under consideration, solution to Equation 8.46 is obtained in parametric form

$$R_{T.fr}^{-3} = C \cdot t^2 \cdot (t^2 + 4)^{-3} \cdot (t^2 + 12)^2 \tag{8.47}$$

$$(t^2 + 4) \cdot \breve{F}_{fr} = -8 \cdot R_{T.fr} \cdot t \tag{8.48}$$

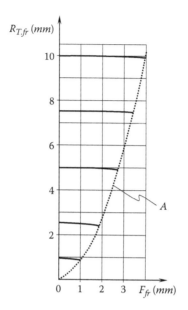

FIGURE 8.17
Interpretation of the solution obtained by implementation of the gradient method.

where the constant parameter C specifies a curve of the family of curves and t designates the parameter.

Figure 8.17 illustrates the obtained solution to the differential Equation 8.46 where the gradient curves for the function $h_{fr} = h_{fr}(R_{T.fr}, \breve{F}_{fr})$ are depicted. The boundary curve A represents the projection of the line $h_{fr} = 0.020$ mm onto the coordinate plane $R_{T.fr} \breve{F}_{fr}$.

For the most efficient reduction of waviness height h_{fr}, the parameters of the machining operation must be alternated in compliance with Equations 8.46 through 8.48.

If some parameter affects the cusp height the most, then it is recommended to alternate this parameter the first.

8.5.2 Optimal Feed Rate and Side Step Ratio

The resultant height of cusps h_Σ is a function of two elementary surface deviations—it is a function of the height of waviness h_{fr} and height of the elementary surface deviation h_{ss}. Various ratios between the feed rate $|\mathbf{F}_{fr}|$ and the side step $|\mathbf{F}_{ss}|$ result in different total cusp height h_Σ. It is natural to assume that the resultant cusp height h_Σ reaches its minimal value under certain ratio between the feed rate and the side step. This ratio is referred to as the *optimal ratio* between the feed rate and the side step.

The maximal elementary surface deviation h_{ss} is smaller than the resultant tolerance [h] for accuracy of the part surface. The deviation h_{ss} is equal to a portion of the tolerance [h], say $h_{ss} = c \cdot [h]$. Here, c designates the local parameter of distribution of the tolerance [h]. The actual value of the parameter c is within the interval $0 \le c \le 1$. At a current point of the surface P, there exists the optimal value of the parameter c. Therefore, the current value of the parameter c can be expressed in terms of *Gaussian* coordinates of the sculptured part surface P, that is, $c = c(U_P, V_P)$.

If it is assumed here that if

$$h_\Sigma = h_{fr} + h_{ss},\qquad(8.49)$$

then the equality

$$h_{fr} = (c - 1) + [h]\qquad(8.50)$$

is valid for the elementary deviation h_{fr}. Ultimately, the height of cusps h_Σ can be expressed as a function of the parameter c, say $h_\Sigma = h_\Sigma(c)$. When the parameter c is of optimal value, the following equality is observed:

$$\frac{\partial h_\Sigma(c)}{\partial c} = 0\qquad(8.51)$$

This condition (see Equation 8.51) is necessary for the minimum of the function $h_\Sigma = h_\Sigma(c)$. In addition, the inequality

$$\frac{\partial^2 h_\Sigma(c)}{\partial c^2} > 0\qquad(8.52)$$

must be observed.

For the computation, a computer code has been developed [140]. An example of the results of the computations is depicted in Figure 8.18. The major conclusion to be made here is that for every point of the sculptured part surface P, there exists the optimal distribution of the tolerance [h] between the elementary surface deviations h_{fr} and h_{ss}. The optimal distribution of the tolerance [h] is specified by the optimal value of the parameter $c = c(U_P, V_P)$. When the parameter c is equal to its optimal value ($c = c_{opt}$), then the accuracy of the machined sculptured part surface is within the prespecified tolerance [h], and the machining time in this case is the shortest possible.

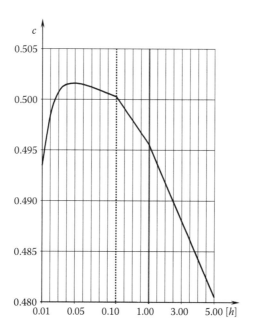

FIGURE 8.18
Parameter c against tolerance $[h]$ on accuracy of the part surface P ($\tilde{R}_{P.fr} = 100$ mm, $\tilde{R}_{P.ss} = 20$ mm, $\tilde{R}_{T.fr} = 50$ mm, and $\tilde{R}_{P.ss} = 50$ mm).

8.6 Principle of Superposition of the Elementary Surface Deviations

The resultant cusp height h_Σ depends upon the two components: waviness height h_{fr} and the elementary surface deviation h_{ss}. When the waviness height is negligibly small, the approximate equality $h_\Sigma \approx h_{ss}$ is valid. Otherwise, for the calculation of the resultant cusp height h_Σ, a corresponding formula is necessary.

For simplicity, the equation

$$h_\Sigma = h_{fr} + h_{ss} \tag{8.53}$$

is often used for this purpose. For more accurate calculation, Equation 8.4 is used instead. In particular cases when the assumptions $h_{fr} = h_{fr}^{max}$, $h_{ss} = h_{ss}^{max}$ and $h_\Sigma = h_\Sigma^{max}$ are valid, Equation 8.5 can be implemented for the calculation of the maximal value h_Σ^{max} of the resultant cusp height.

All the approximate equations, Equations 8.53, 8.4, and 8.5, are derived on the premise of the principle of superposition of the elementary surface deviations. Actually, the use of the principle of superposition is valid only for linear functions. The analysis of the earlier derived Equation 8.22 for the calculation of the elementary deviation h_{fr} and Equation 8.17 for the calculation of the elementary deviation h_{ss} reveals that the functions

$$h_{fr} = h_{fr}(R_{P.fr}, R_{T.fr}, \breve{F}_{fr}, \ldots),$$ (8.54)

$$h_{ss} = h_{ss}(R_{P.ss}, R_{T.ss}, \breve{F}_{ss}, \ldots)$$ (8.55)

are the substantially nonlinear functions. In this concern, the question "under which conditions is implementation of the principle of superposition of the elementary surface deviations valid?" naturally arises.

To answer this practical question, a comparison of results of calculation of the resultant cusp height that are performed using Equation 8.53 (or using more general Equation 8.4) with the results of precise calculations is vital. For this purpose, a local approximation of the part surface P and the generating surface T of the form-cutting tool by patches of torus surfaces is very helpful.

At the point of contact K, surfaces P and T are locally approximated by the patches of the torus surface Tr_P and of the torus surface Tr_T, respectively (Figure 8.3). The use of this scheme for the precise calculation of the resultant cusp height $\tilde{h}_\Sigma$ does not require the implementation of the principle of superposition of the elementary surface deviations h_{fr} and h_{ss}.

A torus surface $Tr_{[h]}$ is the offset surface to the torus surface Tr_P. The distance $[h]$ between the torus surfaces Tr_P and $Tr_{[h]}$ exceeds or is equal to the maximal allowed cusp height (Figure 8.3). The torus surface Tr_T is plunged into the space between the surfaces Tr_P and $Tr_{[h]}$. When the torus surface Tr_T contacts the torus surface Tr_P, then the elementary surface cell on the torus surface Tr_P can be constructed. Boundaries of the elementary surface cell on the surface Tr_P can be projected on the common tangent plane. The projection of the boundaries is located within the projection on the common tangent plane of the line of intersection of the torus surfaces $Tr_{[h]}$ and Tr_T.

The maximal deviation $\tilde{h}_\Sigma$ is measured along the unit normal vector $\mathbf{n}_P$ that passes through a vertex of the elementary surface cell on the machined part surface. For example, at surface point 2 (Figure 8.3), the maximal deviation $\tilde{h}_\Sigma$ is equal to $\tilde{h}_\Sigma = \mathcal{P}2$. The point $\mathcal{P}$ is within the torus surface Tr_T. Within the arc segments $\breve{F}_{fr}$ and $\breve{F}_{ss}$, the radii of normal curvature $R_{1.P}^{(tr)}$, $R_{2.P}^{(tr)}$ and $R_{1.T}^{(tr)}$, $R_{2.T}^{(tr)}$ are of constant value.

In Figure 8.3, the points on surface P that correspond to vertices of the elementary cell are designated as 1_P, 2_P, 3_P, and 4_P. Unit normal vectors through vertices 1_P, 2_P, 3_P, and 4_P are designated as $\mathbf{n}_{1.P}$, $\mathbf{n}_{2.P}$, $\mathbf{n}_{3.P}$, and $\mathbf{n}_{4.P}$, respectively. The unit normal vectors $\mathbf{n}_{1.P}$ and $\mathbf{n}_{2.P}$ are the coplanar vectors. The same is true with respect to the pairs of the unit normal vectors $\mathbf{n}_{1.P}$ and $\mathbf{n}_{3.P}$, $\mathbf{n}_{1.P}$ and $\mathbf{n}_{4.P}$, $\mathbf{n}_{2.P}$ and $\mathbf{n}_{3.P}$, $\mathbf{n}_{2.P}$ and $\mathbf{n}_{4.P}$, and $\mathbf{n}_{3.P}$ and $\mathbf{n}_{4.P}$—each pair of these vectors represent a pair of coplanar vectors.

When the parameters of machining of a sculptured part surface are properly calculated, the resultant surface deviation $\tilde{h}_\Sigma$ is of the same value at all four vertices of the elementary cell, and the point K is the distance $0.5\breve{F}_{fr}$ and $0.5\breve{F}_{ss}$ from the corresponding boundaries of the elementary cell.

For calculation of the resultant surface deviation $\tilde{h}_\Sigma$, it is necessary to calculate the coordinates of the point 2 to derive the equation of the unit normal vector to the torus surface Tr_P at the point 2 and to calculate the coordinates of the point $\mathcal{P}$ of intersection of the torus surface Tr_T by the unit normal vector to the torus surface Tr_P at the point 2. The resultant surface deviation $\tilde{h}_\Sigma$ is equal to the distance between points $\mathcal{P}$ and 2.

For the analysis below, the local *Cartesian* coordinate system $x_P^{(K)} y_P^{(K)} z_P^{(K)}$ (see Figure 8.11) is used.

In the local reference system $x_P^{(K)} y_P^{(K)} z_P^{(K)}$, the unit tangent vectors $\mathbf{t}_{1.P}$ and $\mathbf{t}_{2.P}$ of the principal directions at a point on the sculptured part surface P align with the coordinate axes $y_P^{(K)}$ and $z_P^{(K)}$. The unit tangent vectors $\mathbf{t}_{1.T}$ and $\mathbf{t}_{2.T}$ of the principal directions on the generating surface T of the cutting tool are at the angle μ of local relative orientation of surfaces P and T.

The vector $\mathbf{F}_{fr}$ of feed rate motion of the cutting tool is at a certain angle ε with respect to the first principal direction $\mathbf{t}_{1.P}$ of the part surface. The vector $\mathbf{F}_{ss}$ of side step motion is orthogonal to the vector $\mathbf{F}_{fr}$ of the feed rate motion. The point 2^* is within the common tangent plane through the point K. Actual location of the point 2^* is specified by the position vector

$$\mathbf{r}_{2^*} = 0.5 \cdot (\mathbf{F}_{fr} + \mathbf{F}_{ss}). \tag{8.56}$$

In the local coordinate system $x_P^{(K)} y_P^{(K)} z_P^{(K)}$, the vectors $\mathbf{F}_{fr}$ and $\mathbf{F}_{ss}$ yield representation in matrix form

$$\mathbf{F}_{fr} = \begin{bmatrix} 0.5 \cdot \breve{F}_{fr} \cdot \cos(\varepsilon + \mu) \\ 0.5 \cdot \breve{F}_{fr} \cdot \sin(\varepsilon + \mu) \\ 0 \\ 1 \end{bmatrix} \tag{8.57}$$

$$\mathbf{F}_{ss} = \begin{bmatrix} 0.5 \cdot \breve{F}_{ss} \cdot \sin(\varepsilon + \mu) \\ -0.5 \cdot \breve{F}_{ss} \cdot \cos(\varepsilon + \mu) \\ 0 \\ 1 \end{bmatrix} \tag{8.58}$$

Equations 8.57 and 8.58 yield an expression for the position vector $\mathbf{r}_{2^*}$:

$$\mathbf{r}_{Q^*} = \begin{bmatrix} 0.5 \cdot [\breve{F}_{fr} \cos(\varepsilon + \mu) + \breve{F}_{ss} \sin(\varepsilon + \mu)] \\ 0.5 \cdot [\breve{F}_{fr} \sin(\varepsilon + \mu) - \breve{F}_{ss} \cos(\varepsilon + \mu)] \\ 0 \\ 1 \end{bmatrix} \qquad (8.59)$$

Further, consider a portion of the torus surface Tr_T within the elementary surface cell. Assume that "in small," the portion of the torus surface Tr_T is developable. Under such an assumption, location of the point 2 on the torus surface Tr_T is predetermined by the location of the corresponding point 2^* within the common tangent plane. Therefore, when the tangent plane is locally bending onto the torus surface Tr_T, the point 2^* goes to the location of the point 2. Ultimately, in the local coordinate system $x_P^{(K)} y_P^{(K)} z_P^{(K)}$, this yields calculation of the parameters θ_2 and φ_2 of the point 2, which are equal

$$\theta_Q = \frac{|\mathbf{r}_{Q^*} \sin \sigma|}{r_{Tr.P}} \qquad (8.60)$$

$$\varphi_Q = \frac{|\mathbf{r}_{Q^*} \cos \sigma|}{r_{Tr.P}}. \qquad (8.61)$$

where the radius of the generating circle of the torus surface Tr_P is denoted by $r_{Tr.P}$, and the angle σ is calculated from the equation

$$\sigma = [90° - (\varepsilon + \mu)] - \arctan\left(\frac{\breve{F}_{fr}}{\breve{F}_{ss}}\right). \qquad (8.62)$$

The calculated parameters θ_2 and φ_2 yield an analytical expression for the position vector $\mathbf{r}_2$ of the point 2 in the local coordinate system $x_P^{(K)} y_P^{(K)} z_P^{(K)}$.

For the calculation of the unit normal vector $\mathbf{n}_{Tr.P}$, the matrix equation of the approximating torus surface Tr_P can be used:

$$\mathbf{r}_{Tr.P} = \begin{bmatrix} (r_{Tr.P} \cdot \cos\theta_{Tr.P} + R_{Tr.P}) \cdot \cos\varphi_{Tr.P} \\ (r_{Tr.P} \cdot \cos\theta_{Tr.P} + R_{Tr.P}) \cdot \sin\varphi_{Tr.P} \\ r_{Tr.P} \cdot \sin\theta_{Tr.P} \\ 1 \end{bmatrix} \qquad (8.63)$$

This equation yields

$$
\mathbf{n}_{Tr.P} = \begin{bmatrix} -\cos\theta_{Tr.P}\cos\varphi_{Tr.P} \\ -\cos\theta_{Tr.P}\sin\varphi_{Tr.P} \\ -\sin\theta_{Tr.P} \\ 1 \end{bmatrix} \tag{8.64}
$$

Height $\tilde{h}_\Sigma$ of the resultant cusps is measured along the straight line $\mathbf{r}_h$. The straight line $\mathbf{r}_h$ passes through the point $\mathbf{r}_2$ and is pointed along the unit normal vector $\mathbf{n}_{Tr.P}$. Therefore, equation of this straight line is

$$
\mathbf{r}_h(\theta_{Tr.P}, \varphi_{Tr.P}) = \begin{bmatrix} X_Q - t_h\cos\theta_{Tr.P}\cos\varphi_{Tr.P} \\ Y_Q - t_h\cos\theta_{Tr.P}\sin\varphi_{Tr.P} \\ Z_Q - t_h\sin_{Tr.P} \\ 1 \end{bmatrix} \tag{8.65}
$$

Equations 8.65 and 8.63 of the torus surface Tr_P uniquely specify the coordinates of the point $\mathcal{P}$ (see Figure 8.3). For this purpose, the torus surface Tr_P must be represented in that same reference system $x_P^{(K)} y_P^{(K)} z_P^{(K)}$ as the straight line $\mathbf{r}_h$ is represented.

The required coordinate systems transformation can be performed by the operator $\mathbf{Rs}(Tr_P \mapsto K)$ of the resultant coordinate systems transformation. Ultimately, the equation of the approximating torus surface Tr_P in the local reference system $x_P^{(K)} y_P^{(K)} z_P^{(K)}$ casts into the matrix form

$$
[\mathbf{r}_{Tr.P}]_{(K)} = \mathbf{Rs}(Tr_P \mapsto K) \cdot [\mathbf{r}_{Tr.P}]
$$

$$
= \begin{bmatrix} (r_{Tr.P} \cdot \cos\theta_{Tr.P} + R_{Tr.P}) \cdot \cos\varphi_{Tr.P} - R_{Tr.P} \\ (r_{Tr.P} \cdot \cos\theta_{Tr.P} + R_{Tr.P}) \cdot \sin\varphi_{Tr.P} \cdot \cos\mu + r_{Tr.P} \cdot \sin\theta_{Tr.P} \cdot \sin\mu \\ -(r_{Tr.P} \cdot \cos\theta_{Tr.P} + R_{Tr.P}) \cdot \sin\varphi_{Tr.P} \cdot \sin\mu + r_{Tr.P} \cdot \sin\theta_{Tr.P} \cdot \cos\mu \\ 1 \end{bmatrix} \tag{8.66}
$$

At the point of intersection $\mathcal{P}$ of the straight line $\mathbf{r}_h$ (see Equation 8.65) and to the torus surface Tr_P (see Equation 8.66), the equality

$$
\mathbf{r}_h(\theta_{Tr.P}, \varphi_{Tr.P}) = [\mathbf{r}_{Tr.P}]_{(K)} \tag{8.67}
$$

is observed. In expanded form, this equality casts into

$$
\begin{bmatrix}
X_Q - t_h \cos\theta_{Tr.P}\cos\varphi_{Tr.P} \\
Y_Q - t_h \cos\theta_{Tr.P}\sin\varphi_{Tr.P} \\
Z_Q - t_h \sin_{Tr.P} \\
1
\end{bmatrix}
$$

$$
=
\begin{bmatrix}
(r_{Tr.P}\cdot\cos\theta_{Tr.P}+R_{Tr.P})\cdot\cos\varphi_{Tr.P}-R_{Tr.P} \\
(r_{Tr.P}\cdot\cos\theta_{Tr.P}+R_{Tr.P})\cdot\sin\varphi_{Tr.P}\cdot\cos\mu+r_{Tr.P}\cdot\sin\theta_{Tr.P}\cdot\sin\mu \\
-(r_{Tr.P}\cdot\cos\theta_{Tr.P}+R_{Tr.P})\cdot\sin\varphi_{Tr.P}\cdot\sin\mu+r_{Tr.P}\cdot\sin\theta_{Tr.P}\cdot\cos\mu \\
1
\end{bmatrix}
\quad (8.68)
$$

which yields the computation of the three parameters specify the coordinates of point P: $\theta_{Tr.P}$, $\varphi_{Tr.P}$, and t_h. Finally, an analytical expression for the vector $\mathbf{r}_{\mathscr{P}}$ can be derived.

The calculated vectors $\mathbf{r}_{\mathscr{Q}}$ and $\mathbf{r}_{\mathscr{P}}$ yield the the resultant surface deviation $\tilde{h}_\Sigma$:

$$
\tilde{h}_\Sigma = \left|\mathbf{r}_Q - \mathbf{r}_P\right| \quad (8.69)
$$

With the above analysis, the final conclusion can be made with respect to implementation of the principle of superposition of the elementary surface deviations:

The principle of superposition of the elementary surface deviations h_{fr} and h_{ss} is valid if and only if the inequality $\tilde{h}_\Sigma - h_\Sigma \leq [\Delta h_\Sigma]$ is observed.

Here $[\Delta h_\Sigma]$ designates the tolerance for accuracy of determination of the height of the resultant cusps.

In most practical cases of sculptured part surface machining on a *NC* multiaxis machine, the principle of superposition of the elementary surface deviations is valid.

Section III

Application

Application of the disclosed *DG/K*-based method of part surface generation is final and the most important consideration in this book. The selection of the criterion of optimization and the local, regional, and global synthesis of the most favorable operations of part surface machining are covered in this section of the book and are illustrated with examples of machining of sculptured part surfaces on a multiaxis *NC* machine as well as of part surfaces on conventional machine tools.

9

Selection of the Criterion of Optimization

For machining a part surface, a machine tool, a cutting tool, a fixturing, and so forth are necessary. All these elements are together referred to as the *technological system*. For lubrication and for cooling purposes, liquids and gas substances are often used. The coolants and the lubricants create the *technological environment*. When a part surface machining process is designed properly, the capabilities of both the technological system and the technological environment are fully utilized. When the capabilities of the technological system and technological environment are fully utilized, the manufacturing processes of this kind are usually called the *extremal manufacturing processes*. Ultimately, the use of the extremal manufacturing processes results in the most economical machining of part surfaces.

The *differential geometry/kinematics* (*DG/K*) method of part surface generation (disclosed in the previous chapters of the monograph) is capable of synthesizing extremal methods of machining of sculptured part surfaces on a multiaxis numerical control (*NC*) machine as well as synthesizing extremal methods of machining part surfaces having relatively simple geometry on conventional machine tools [116,117,136].

Machining the part surface in the most economical way is the main goal when designing a manufacturing process—this should be the rule for the future developments in the field.

To synthesize the most efficient machining operation, appropriate input information is required. The capabilities of a theoretical approach that is used for the synthesis can be estimated by the amount of input information the approach requires for its implementation, and by the amount of output information the method is capable of creating. A more powerful theoretical approach requires less input information to solve a problem and its use enables more output information in comparison with the less powerful theoretical approach. Ultimately, the target can be formulated by using as little input information as possible and producing as much output information as possible.

The *DG/K* method requires a minimum of input information: only the geometrical information on the part surface to be machined. The geometrical information on the part surface to be machined is the least possible input information for solving of a problem of synthesis of optimal machining operation. Based only on the geometrical information on part surface *P*, use of the *DG/K* method yields calculation of the optimal parameters of the machining process. No selection of parameters of the machining operation is required when the DG/K method is used. This makes it possible to conclude that the DG/K method of part surface generation is the most powerful theoretical method capable of solving problems of synthesis of optimal machining operations on the premise of the least possible input information. No other theoretical method is capable of solving problems of this sort only on the premise of geometrical information on the part surface to be machined.

An accurate analytical description of the part surface to be machined (see Chapter 1) is followed by the selection of an appropriate criterion of optimization. The selection of an appropriate criterion of optimization* is critical for the implementation of the DG/K method of part surface generation.

9.1 Criteria of the Efficiency of Part Surface Machining

The design of a sculptured part surface machining process is an example of a problem having multivariant solutions. To solve a problem of this sort, a criterion of optimization is necessary.

Various criteria of optimization are used in industry for the optimization of parameters of part surface machining. The productivity of part surface machining, tool life, accuracy, and quality of the machined part surface are among them. Other criteria of optimization of parameters of machining operations are used as well.

The economical criteria of optimization are the most general and the most preferred criteria of optimization of machining processes. However, analytical description of economical criteria of optimization is complex and makes them very inconvenient for practical calculations. For particular cases of part surface machining, the equivalent criteria of optimization of significantly simpler structures can be proposed.

The productivity of part surface machining and the productivity of part surface generation are the important criteria of optimization. Both are often used for creating the more general criteria of optimization of part surface machining processes. Therefore, it is reasonable to use the productivity of part surface machining as the criterion of optimization to demonstrate the potential capabilities of the DG/K method of part surface generation. The results of the

---

* Use of the word *optimization* is a bit confusing here. The *optimization* is capable of finding the optimal combination of known parameters of the machining process. In other words, all the parameters of a process to be optimized must be known before the optimization starts. Capabilities of the synthesis are much wider. The *synthesis* is capable of
  - Finding all of the important parameters the part surface machining process depending upon
  - Determination of the most favorable values of all the parameters.

  This difference between the *optimization* and the *synthesis* must be kept in mind when reading this book.

synthesis of favorable part surface machining operations can be generalized for the case of implementation of another criterion of optimization.

There are many ways to increase the productivity of part surface machining on machine tools. Here, mostly the geometrical and kinematical aspects of the optimization of part surface machining are considered.

In the theory of part surface generation, three aspects of the part surface generation process are distinguished:

- The local part surface generation
- The regional part surface generation
- The global part surface generation [116,117,136]

The local analysis of the part surface generation process encompasses generation of part surface P in the differential vicinity of the point K of contact of part surface P and generating surface T of the form-cutting tool. The generation of the part surface within a single tool-path is investigated from the perspective of the regional part surface generation. Ultimately, partial interference of the neighboring tool-paths, coordinates of the start-point for the part surface machining, and impact of shape of the contour of part surface P patch are investigated from the perspective of the global part surface generation. Consequently, three types of productivity of the part surface machining process are distinguished

- Local productivity of part surface generation
- Regional productivity of part surface generation
- Global productivity of part surface generation

All three types of productivity of the part surface machining process are considered below.

9.2 Productivity of Part Surface Machining

The intensity of generation of the nominal part surface in time is reflected by the productivity of part surface generation. The productivity of part surface generation can be used for the purpose of synthesis of favorable part surface machining operations (for example, of machining a sculptured part surface on a multiaxis *NC* machine).

9.2.1 Major Parameters of the Part Surface Machining Operation

It is natural to begin the investigation of the major parameters of the part surface machining operation by considering part surface generation locally.

When machining a sculptured surface on a multiaxis *NC* machine, all major parameters of the machining operation and the instantaneous productivity of part surface generation vary in time. This makes reasonable the consideration of instantaneous (current) values of the part surface generation process.

The instantaneous productivity of part surface generation $P_{sg}(t)$ is determined by current values of the feed rate $\breve{F}_{fr}$ and the side step $\breve{F}_{ss}$ (t designates time). Usually, vector $\mathbf{F}_{fr}$ of the feed rate motion and vector $\mathbf{F}_{ss}$ of the side step motion are orthogonal to each other ($\mathbf{F}_{fr} \perp \mathbf{F}_{ss}$). In a particular case, vectors $\mathbf{F}_{fr}$ and $\mathbf{F}_{ss}$ are at a certain angle θ to one another.

Instantaneous productivity of part surface generation can be calculated by the following formula [118,119]:

$$P(t) = \left| \mathbf{F}_{fr} \times \mathbf{F}_{ss} \right| \tag{9.1}$$

Equation 9.1 casts into [118,119]

$$P(t) = \breve{F}_{fr} \cdot \breve{F}_{ss} \cdot \sin \theta \tag{9.2}$$

where $\breve{F}_{fr}$ is equal to $|\mathbf{F}_{fr}|$ and $\breve{F}_{ss}$ is equal to $|\mathbf{F}_{ss}|$.

Equations 9.1 and 9.2 reveal that an increase of the feed rate $\breve{F}_{fr}$ and an increase of the side step $\breve{F}_{ss}$ lead to an increase of the instantaneous productivity of part surface generation $P(t)$. Deviation of the angle θ from $\theta = 90°$ results in a corresponding reduction of the instantaneous productivity of part surface generation $P(t)$.

At a current point K of contact of part surface P and generation surface T of the cutting tool, the optimal values of parameters $\breve{F}_{fr}$, $\breve{F}_{ss}$, and θ depend on the local geometrical (differential) characteristics of surfaces P and T, on surfaces P and T local relative orientation (which is specified by the angle μ of the surfaces P and T local relative orientation), and on the tolerance for accuracy $[h]$ of the machined part surface.

The value of the tolerance for accuracy $[h]$ of part surface machining is usually constant within the patch of part surface P. However, in a more general case of part surface machining, the current value of the tolerance $[h]$ can vary within the surface patch:

$$[h] = [h](U_P, V_P) \tag{9.3}$$

Within certain portions of a part surface patch, the tolerance can be larger, and within other portions, it can be smaller depending on functional requirements to the actual part surface.

Because the resultant cusp height h_Σ is made up of two components h_{fr} and h_{ss}, it is necessary to split the tolerance $[h]$ onto two corresponding portions:

on the portion $[h_{fr}]$ for the elementary deviation h_{fr} and on the portion $[h_{ss}]$ for the elementary deviation h_{ss}. The equality

$$[h] = [h_{fr}] + [h_{ss}] \tag{9.4}$$

is always observed (see Equation 8.53). However, the equality

$$h_{\Sigma} \cong h_{fr} + h_{ss} \tag{9.5}$$

is always approximate.

For calculations of the surface deviation h_{Σ}, it is recommended that Equation 8.4 be used:

$$h_{\Sigma} \cong a_h \cdot h_{fr} + b_h \cdot h_{ss} \tag{9.6}$$

where coefficients a_h and b_h are within the intervals $0 \leq a_h \leq 1$ and $0 \leq b_h \leq 1$. The coefficients a_h and b_h can be determined at a current point K of the sculptured part surface P.

At current point K on part surface P having coordinates U_P, and V_P, the current values of coefficients a_h and b_h also depend on coordinates of point K on surface P (that is, they depend on the U_T and V_T parameters and on angle μ of the local relative orientation of surfaces P and T at contact point K). This relationship is expressed by two formulae:

$$a_h = a_h(U_P, V_P, U_T, V_T, \mu) \tag{9.7}$$

$$b_h = b_h(U_P, V_P, U_T, V_T, \mu) \tag{9.8}$$

The values of the feed rate $\breve{F}_{fr}$ per tooth of the cutting tool and of the side step $\breve{F}_{ss}$ at a current point K depend on the partial tolerances $[h_{fr}]$ and $[h_{ss}]$. One can immediately conclude from the above that both the feed rate $\breve{F}_{fr}$ and the side step $\breve{F}_{ss}$ are functions of

- Coordinates of the contact point K on part surface P
- Coordinates of the contact point K on the cutting tool surface T
- The angle μ of the local relative orientation of the surfaces P and T
- The direction of motion of the cutting tool surface T relative to part surface P

The following expressions reveal this relationship:

$$\breve{F}_{fr} = \breve{F}_{fr}([h_{fr}]) = \breve{F}_{fr}(U_P, V_P, U_T, V_T, \mu, \varphi) \tag{9.9}$$

$$\breve{F}_{ss} = \breve{F}_{ss}([h_{ss}]) = \breve{F}_{ss}(U_P, V_P, U_T, V_T, \mu, \varphi) \tag{9.10}$$

where the angle that specifies the direction of the feed-rate vector $\mathbf{F}_{fr}$ is designated as φ.

By substituting Equations 9.9 and 9.10 in Equation 9.2, it is easy to conclude that the productivity of part surface generation P_{sg} also depends on coordinates of the current point of contact K on both surfaces P and T, on angle μ of the local relative orientation of the surfaces P and T, and on the direction of the relative motion of surfaces P and T at contact point K:

$$P_{sg} = P_{sg}(U_P, V_P, U_T, V_T, \mu, \varphi) \tag{9.11}$$

Certainly, not only tolerance $[h]$ but partial tolerances $[h_{fr}]$ and $[h_{ss}]$ can be either constant within part surface patch or they can vary within the sculptured part surface P. In the first case, the actual values of the tolerances $[h]$, $[h_{fr}]$, and $[h_{ss}]$ must be given. In the second case, the functions must be known:

$$[h] = [h](U_P, V_P, U_T, V_T, \mu) \tag{9.12}$$

$$[h_{fr}] = [h_{fr}](U_P, V_P, U_T, V_T, \mu) \tag{9.13}$$

$$[h_{ss}] = [h_{ss}](U_P, V_P, U_T, V_T, \mu) \tag{9.14}$$

The principal radii of curvature $R_{1.P}$ and $R_{2.P}$ at a point of the part surface are the functions of parameters U_P and V_P of the sculptured part surface P, whereas the principal radii of curvature $R_{1.T}$ and $R_{2.T}$ of generating surface T of the form-cutting tool are the functions of the parameters U_T and V_T. It should be recalled that different points of the cutting tool surface T can be in contact with a specified point K on the sculptured part surface P.

In special cases of sculptured part surface machining, for example, when elastic deformation is applied to the work for the technological purposes as shown in Figure 2.3 or in the case of special-purpose cutting tools with a changeable generating surface used for machining [108,127,152], then in addition to parameters U_P, V_P, U_T, V_T, μ, and φ, other parameters have to be incorporated into Equation 9.11 for the calculation of the productivity of part surface generation P_{sg} (see Chapter 8 in Reference [117] for details).

9.2.2 Productivity of Material Removal

When machining a part surface, the intensity of stock removal is evaluated by the productivity of material removal. The productivity of material removal is equal to the amount of stock removed from the work in a unit of time.

9.2.2.1 Equation of the Work-Piece Surface

For the analytical description of productivity of material removal in terms of parameters of the machining operation, an equation on the work-piece surface W_{ps} must be derived. The equation for the position vector of a point $\mathbf{r}_{wp}$ of the surface W_{ps} can be composed on the premise of numerical data obtained from measurements of the actual work-piece.

The thickness of the stock to be removed b can be of constant value or its value can vary within part surface *Patch*. In the first case, the thickness of the stock b must be known. In the second case, it is necessary to know the function of the stock distribution $b(U_P, V_P)$.

In the event the equation of the work-piece surface W_{ps} is obtained on the basis of the surface measurements, the equation for the position vector of a point $\mathbf{r}_P$ of part surface P together with the equation for the position vector of a point $\mathbf{r}_{wp}$ of the work-piece surface W_{ps} yields a calculation of the stock-distribution function $b(U_P, V_P)$:

$$b(U_P, V_P) = |\mathbf{r}_{wp} - \mathbf{r}_P| \tag{9.15}$$

When the stock-distribution function $b(U_P, V_P)$ is given, the equation of the work-piece surface W_{ps} can be derived analytically. For this purpose, an equation for the position vector of a point $\mathbf{r}_P = \mathbf{r}_P(U_P, V_P)$ of the nominal part surface is employed (Figure 9.1):

$$\mathbf{r}_{wp} = \mathbf{r}_P + \mathbf{n}_P \cdot b(U_P, V_P) \tag{9.16}$$

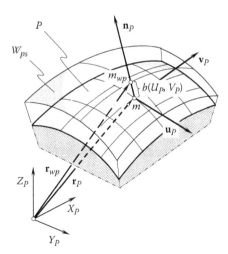

FIGURE 9.1
Derivation of equation for the position vector of a point $\mathbf{r}_{wp}$ of the work-piece surface W_{ps}.

In Figure 9.1, point m_{wp} on the surface of the work-piece W_{ps} is shown at a distance $b(U_P, V_P)$ from the point M on the nominal part surface m.

Elements of local topology of the work-piece surface W_{ps} (say the first $\Phi_{1.ps}$ and the second $\Phi_{2.ps}$ fundamental forms of the work-piece surface W_{ps}) can be expressed in terms of the elements of local topology of the nominal part surface P.

The components $g_{w.ij}$ of the metric tensor of the first order of the W_{ps} surface can be expressed in terms of the components of the fundamental tensors g_{ij} of the first order and b_{ij} of the second order and the mean M_P and full (*Gaussian*) G_P curvatures at a current point of part surface P:

$$g_{R.ij} = \left[1 - G_P \cdot \left(r_{cnf}^{min}\right)^2\right] g_{ij} - 2r_{cnf}^{min}\left[1 + M_P \cdot r_{cnf}^{min}\right] b_{ij} \tag{9.17}$$

The determinant $\det(g_{w.ij})$ of the metric tensor of the W_{ps} surface is equal to

$$\det(g_{w.ij}) = A_w^2 \cdot \det(g_{ij})$$

where

$$A_w = G_P \cdot b^2 + 2M_P \cdot b + 1 = (1 + k_{1.P} \cdot b) \cdot (1 + k_{2.P} \cdot b) \tag{9.18}$$

Singularities of the surface of work-piece W_{ps} are observed at points that correspond to points of the part surface P, at which one of the principal curvatures is equal to $-b^{-1}$.

The unit normal vector $\mathbf{n}_w$ at a point of the surface W_{ps} can be analytically represented as

$$\mathbf{n}_w = \frac{A}{|A|}\mathbf{n}_P \tag{9.19}$$

The second fundamental tensor, $b_{w.ij}$, of the surface W_{ps} can be expressed in terms of the components of the fundamental tensors g_{ij} and b_{ij}, the mean M_P, and full (*Gauss*) G_P curvature at a point of part surface P:

$$b_{w.ij} = \frac{A}{|A|}[G_P \cdot b \cdot g_{ij} + (1 + 2M_P \cdot b) \cdot b_{ij}] \tag{9.20}$$

As for the metric tensor, the determinant of the second fundamental tensor $\det(b_{w.ij})$

$$\det(b_{w.ij}) = A \cdot \det(b_{ij}) \tag{9.21}$$

has singularities at points that correspond to points of the sculptured part surface P, at which one of the principal curvatures is equal to $-b^{-1}$.

These derivations are similar to the derivations of major elements of local topology of the characteristic $\mathcal{R}_1$ surfaces (see Section 7.3.2.2.2 for details).

For the calculation of the productivity of material removal, equation $\mathbf{r}_P = \mathbf{r}_P(U_P, V_P)$ for the position vector of a point of part surface P and Equation 9.16 of the work-piece surface W_{ps} must be represented in a common reference system. When necessary, an appropriate operator $\mathbf{Rs}(W_{ps} \mapsto P)$ of the resultant coordinate system transformations can be composed for this purpose (see Chapter 3 for details).

Modified Equation 9.16 can also be helpful for the calculation of parameters of uncut chip.

Similar to Equation 9.16, the equation of the surface of tolerance $S_{[h]}$ can be written as

$$\mathbf{r}_{[h]} = \mathbf{r}_P + \mathbf{n}_P \cdot [h](U_P, V_P) \tag{9.22}$$

Equation 9.22 is used for the calculation of parameters of the critical values of feed rate, $\breve{F}_{fr}$, and side step, $\breve{F}_{ss}$.

Elements of analysis of machine tool performance can be found in Reference [149].

9.2.2.2 Mean Chip-Removal Output

For the calculation of the chip-removal output, vector equations of part surface P to be machined and work-piece surface W_{ps} are necessary. Mean chip-removal output is used for the analysis of efficiency of a global machining operation, say for the whole part surface P. The mean chip-removal output $\tilde{P}_{mr}$ can be used as an index. By definition [116,117,136,149],

$$\tilde{P}_{mr} = \frac{V_{mr}}{t_\Sigma} \tag{9.23}$$

where V_{mr} is the total volume of the stock to be removed and t_Σ is the total time required for the stock removal.

9.2.2.3 Instantaneous Chip-Removal Output

For the local analysis of efficiency of a machining operation, instantaneous chip-removal output is used. The instantaneous chip-removal output P_{mr} can also be used as an index. By definition [116,117,136,149],

$$P_{mr}(t) = \frac{dv_{mr}}{dt} \tag{9.24}$$

The volume of chip dv_{mr} to be removed in an instant of time is

$$dv_{mr} = \mathbf{b} \cdot d\mathbf{F}_P \tag{9.25}$$

where $d\mathbf{F}_P$ is the vector area element and $\mathbf{b}$ is the vector of the stock thickness. Here, $\mathbf{b} = b(U_P, V_P) \cdot \mathbf{n}_P$ (see Equation 9.15 for details) and $|\mathbf{b}| = mm_{wp}$ in Figure 9.1.

From another viewpoint, the following equation can be used for the calculation of vector $\mathbf{b}$:

$$\mathbf{b} = \mathbf{r}_{wp} - \mathbf{r}_P \tag{9.26}$$

The vector area element $d\mathbf{F}_P$ is as follows [17,116,117,136]:

$$d\mathbf{F}_P = \left(\frac{\partial \mathbf{r}_P}{\partial U_P} \times \frac{\partial \mathbf{r}_P}{\partial V_P} \right) dU_P dV_P \tag{9.27}$$

After integration with respect to the surface $F_P = |\mathbf{F}_P|$, an expression for the volume

$$V_{mr} = \iint_{F_P} \mathbf{b} \cdot d\mathbf{F}_P = \iint_{F_P} \left(\mathbf{b} \cdot \frac{\partial \mathbf{r}_P}{\partial U_P} \times \frac{\partial \mathbf{r}_P}{\partial V_P} \right) \cdot dU_P \cdot dV_P \tag{9.28}$$

can be derived.

Generally, the curvilinear coordinates U_P and V_P depend upon time t according to the relations

$$U_P = U_P(w, t) \tag{9.29}$$

$$V_P = V_P(w, t) \tag{9.30}$$

where w is a new variable.

The *Jacobian* matrix of transformation J_P for the implementation of Equations 9.24 and 9.25 is as follows [14,116,117,136]:

$$J_P = \begin{vmatrix} \dfrac{\partial U_P}{\partial w} & \dfrac{\partial U_P}{\partial t} \\ \dfrac{\partial V_P}{\partial w} & \dfrac{\partial V_P}{\partial t} \end{vmatrix} \tag{9.31}$$

The region of the part surface within which the sign of the Jacobi transformation matrix J_P is maintained constant is the one under consideration.

Further, an expression for the calculation of the instantaneous chip-removal output can be obtained [116,117,136,149]:

$$P_{mr}(t) = \int_{w_1(t)}^{w_2(t)} \left(\mathbf{b} \cdot \frac{\partial \mathbf{r}_P}{\partial U_P} \times \frac{\partial \mathbf{r}_P}{\partial V_P} \right) \cdot |J_P| \cdot dU_P \cdot dV_P \qquad (9.32)$$

Here, $w_1(t)$ and $w_2(t)$ are the boundary values of the variable w on the coordinate curve $t = Const$, which corresponds to the boundaries of part surface P (Figure 9.2).

An alternative approach for the calculation of the instantaneous chip-removal output can be used.

For this purpose, consider the surface of the cut. The surface of cut S_c is generated by the cutting edge of the cutting tool in its motion in relation to the work. The surface of cut S_c can be considered as a set of consecutive positions of the cutting edge C_e of the cutting tool that travels relative to the work. Such a consideration makes an equation for the position vector of a point $\mathbf{r}_{sc}$ of the surface of cut possible. The implementation of a corresponding operator of the resultant coordinate system transformations could be helpful when performing derivations of this type.

Two different types of analytical representation of the instantaneous chip-removal output $P_{mr}(t)$ can be derived in this case.

The first type of analytical representation of the instantaneous chip-removal output $P_{mr}(t)$ relates to implementation of the cutting tools having the whole generating surface T. In other words, it relates to implementation of grinding wheels, shaving cutters, and so forth. In this case, equation for the position vector of a point $\mathbf{r}_{sc}$ of the surface of cut S_c can be represented in the form

$$\mathbf{r}_{sc} = \mathbf{r}_{sc}(U_T, V_T, t_i) \qquad (9.33)$$

where U_T and V_T denote curvilinear coordinates on generating surface T of the form-cutting tool and t_i is a fixed moment of time.

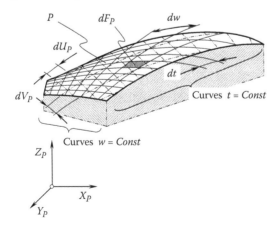

FIGURE 9.2
Definition of the chip-removal output $P_{mr}(t)$.

The second type of analytical representation of the instantaneous chip-removal output $P_{mr}(t)$ relates to the implementation of the cutting tools having a discrete generating surface T (for example, it relates to the implementation of milling cutters and so forth). In this case, the equation of the surface of cut S_c can be represented in the form

$$\mathbf{r}_{sc} = \mathbf{r}_{sc}(U_{ce}, t) \tag{9.34}$$

where U_{ce} designates a coordinate along the cutting edge C_e of the form-cutting tool.

Vector area element $d\mathbf{F}_{sc}$ of the surface of cut S_c can be calculated either from the formula

$$d\mathbf{F}_{sc} = \left(\frac{\partial \mathbf{r}_{sc}}{\partial U_P} \times \frac{\partial \mathbf{r}_{sc}}{\partial V_P} \right) \cdot dU_P \cdot dV_P \tag{9.35}$$

or from the formula

$$d\mathbf{F}_{sc} = \left(\frac{\partial \mathbf{r}_{sc}}{\partial U_{ce}} \times \frac{\partial \mathbf{r}_{sc}}{\partial t} \right) \cdot dU_{ce} \cdot dt \tag{9.36}$$

Equation 9.35 is valid for the first type and Equation 9.36 is applicable for the second type of analytical representation of the instantaneous chip-removal output.

When machining a part surface, the vector area element $d\mathbf{F}_{sc}$ travels with a certain velocity $\boldsymbol{\omega}$ through the stock to be removed. This way, a volume of the stock $dv_{mr} = \boldsymbol{\omega} \cdot d\mathbf{F}_{sc}$ is removed in a unit of time.

Hence, for the calculation of the instant chip-removal output, the following formula can be used:

$$P_{mr}(t) = \iint\limits_{(S_c)} \boldsymbol{\omega} \cdot d\mathbf{F}_{sc} \tag{9.37}$$

When surface machining is performed with a cutting tool having multiple cutting edges, Equation 9.37 acquires the form

$$P_{mr}(t) = \sum_{i=1}^{N^*} \iint\limits_{(S_{c.i})} \boldsymbol{\omega}_i \cdot d\mathbf{F}_{sc.i} \tag{9.38}$$

where N^* is the total number of cutting edges of the cutting simultaneously taking part in the chip-removal process and $d\mathbf{F}_{sc.i}$ is the vector area element of the surface of cut $S_{c.i}$ that is created by the ith cutting edge.

9.2.3 Part Surface Generation Output

When machining a part surface, the rate of increase of the machined surface area reflects the surface generation output.

The mean part surface generation output $\tilde{P}_{sg}$ can be analytically expressed by the formula:

$$\tilde{P}_{sg} = \frac{S_{sg}}{t_{\Sigma}} \tag{9.39}$$

where S_{sg} designate the machined part surface area.

Instantaneous part surface generation output P_{sg} is another characteristic of part surface machining performance. By definition, the instantaneous part surface generation output P_{sg} is:

$$P_{sg} = \frac{dS_{sg}}{dt} \tag{9.40}$$

It can be assumed that $S_{sg} = c \cdot t$, where c is a certain constant value. This immediately results in:

$$P_{sg} = \tilde{P}_{sg} = c \tag{9.41}$$

This means that the instantaneous part surface generation output P_{sg} can be of constant value. In this case, it is equal to the mean surface generation output $\tilde{P}_{sg}$. Generally speaking, the instantaneous part surface generation output is time dependent. An expression for the calculation of $P_{sg}(t)$ in this case can be derived in the following way.

For the calculation of area S_{sg} of part surface P patch, a formula

$$S_{sg} = \iint\limits_{(S_{sg})} |d\mathbf{F}_P| \tag{9.42}$$

is used.

The magnitude $|d\mathbf{F}_P|$ of vector $d\mathbf{F}_P$ that determines the element of the part surface area is

$$|d\mathbf{F}_P| = \left| \frac{\partial \mathbf{r}_P}{\partial U_P} \times \frac{\partial \mathbf{r}_P}{\partial V_P} \right| \cdot dU_P \cdot dV_P = \sqrt{E_P G_P - F_P^2} \cdot dU_P \cdot dV_P \tag{9.43}$$

This equation yields

$$S_{sg} = \iint\limits_{(S_{sg})} \sqrt{E_P G_P - F_P^2} \cdot |J_P| \cdot dt \cdot dw \tag{9.44}$$

Substituting Equation 9.44 into Equation 9.40, a formula [116,117,136,149]:

$$P_{sg}(t) = \int\limits_{w_1(t)}^{w_2(t)} \sqrt{E_P G_P - F_P^2} \cdot \left| J_P \right| \cdot dt \cdot dw \qquad (9.45)$$

for the calculation of the instantaneous part surface generation output $\tilde{P}_{sg}$ can be obtained (see Figure 9.2).

9.2.4 Limit Parameters of the Cutting Tool Motion

When part surface P and generating surface T of the cutting tool are in point contact with each other, then the feed rate motion and the side step motion of the cutting tool must both be performed when machining a sculptured part surface on a multiaxis NC machine.

The maximal allowed displacements of the cutting tool are constrained by the corresponding limit values $[\breve{F}_{fr}]$ and $[\breve{F}_{ss}]$ of the feed rate and side step, respectively. The parameters of the elementary surface cell on the machined part surface are mostly specified by these limits.

The limit values $[\breve{F}_{fr}]$ and $[\breve{F}_{ss}]$ of the cutting tool displacements $\breve{F}_{fr}$ and $\breve{F}_{ss}$ can be calculated. For this purpose, the tolerance $[h]$ for accuracy of part surface machining has been taken into consideration. In compliance with [112], it is assumed below that the maximal resultant height of cusps h_Σ is within the tolerance $[h]$. This means that inequality $h_\Sigma \leq [h]$ is valid; however, the elementary surface deviation δ_P (see Equation 8.44) is not investigated here.

9.2.4.1 Calculation of the Limit Feed-Rate Shift

Milling cutters are widely used for machining sculptured part surfaces on multiaxis NC machines. The use of milling cutters causes waviness of the machined part surface P. It is necessary to keep the waviness height h_{fr} under the corresponding portion $[h_{fr}]$ of the total tolerance $[h]$. The limit feed rate displacement $[\breve{F}_{fr}]$ strongly depends on the allowed value of the partial tolerance $[h_{fr}]$.

To calculate the limit feed rate displacement $[h_{fr}]$, it is necessary to investigate the topography of the machined part surface.

In the direction of vector $\mathbf{F}_{fr}$ of the feed rate motion, the cusps profile is shaped in the form of prolate cycloids.* The elementary machined surface cells on the part surface represent portions of the surface of cut S_c in its consecutive positions relative to the work. However, the linear speed of cutting edges of the rotating cutting tool is incomparably bigger rather than their linear speed in the feed rate motion $\mathbf{F}_{fr}$. Therefore, in most practical cases of part surface machining, the surface of cut S_c is not considered. Instead, when

---

* In special cases of part surface machining, the profile of the machined part surface in the direction of the feed rate motion of the cutting tool can be shaped in the form of a pure cycloid and even in the form of a curtate cycloid.

the limit feed rate displacement $[\breve{F}_{fr}]$ is calculated, an approximation of the surface of cut by generating surface T of the form-cutting tool is considered.

Near contact point K, the nominal sculptured part surface P and generating surface T of the cutting tool are locally approximated by torus surfaces. The approximation allows significant reduction of calculations without considerable loss of accuracy of the calculations.

The approximation is based on the assumption, in compliance with which,

> *The principal radii of curvature $R_{1.P}$ and $R_{2.P}$ of part surface P, and the principal radii of curvature $R_{1.T}$ and $R_{2.T}$ of generating surface T of the form-cutting tool do not change their values within the elementary surface cell on the machined part surface.*

The above assumption is reasonable because in the direction of vector $\mathbf{F}_{fr}$ (as well as in the direction of vector $\mathbf{F}_{ss}$), the radii of normal curvature of the surfaces P and T are much bigger compare to the corresponding parameters of the elementary surface cell on the machined part surface. The inequalities $\breve{F}_{fr} \ll R_{P.fr}$ and $\breve{F}_{fr} \ll R_{T.fr}$ are observed in the direction of the feed rate motion $\mathbf{F}_{fr}$. (The similar inequalities $\breve{F}_{ss} \ll R_{P.ss}$ and $\breve{F}_{ss} \ll R_{T.ss}$ are valid for the direction of the side step shift $\mathbf{F}_{ss}$.)

Figure 9.3 reveals that for the calculation of the instantaneous value of the feed rate $[\breve{F}_{fr}]$ per tooth of the cutting tool, the approximate formulae can be used [116,117,136,142]:

$$[\breve{F}_{fr}] \cong 2R_{P.fr} \arccos \frac{R_{P.fr}^2 + R_{T.fr} \cdot (R_{P.fr} + [h_{fr}] \cdot \mathrm{sgn}\, R_{P.fr})}{(R_{P.fr} + R_{T.fr}) \cdot (R_{P.fr} + [h_{fr}] \cdot \mathrm{sgn}\, R_{P.fr})} \qquad (9.46)$$

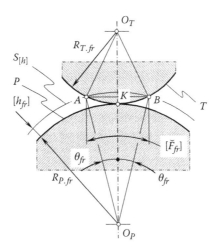

FIGURE 9.3
Schematic for the calculation of the limit feed rate $[\breve{F}_{fr}]$.

It is assumed in this equation that the inequality $\breve{F}_{fr} \ll R_{P.fr}$ is valid; therefore, $\breve{F}_{fr} \cong AB = F_{fr}$. It is assumed also that $[h_{fr}]^2$ is a reasonably small value, and therefore, it can be omitted from further analysis.

Equation 9.46 is valid for the cases when no limitations on actual value of the feed rate motion are imposed by power capacity of the multiaxis *NC* machine. Under such a scenario, the limit value of the feed rate $[\breve{F}_{fr}]$ is a function only of the corresponding radii of normal curvature of the surfaces *P* and *T*.

In Equation 9.46, the radii of normal curvature $R_{P.fr}$ and $R_{T.fr}$ of part surface *P* and generating surface *T* of the form-cutting tool in the direction of the feed rate motion $\mathbf{F}_{fr}$ are equal

$$R_{P.fr} = \frac{R_{1.P} R_{2.P}}{R_{1.P} \sin^2 \varphi + R_{2.P} \cos^2 \varphi} \tag{9.47}$$

$$R_{T.fr} = \frac{R_{1.T} R_{2.T}}{R_{1.T} \sin^2 (\varphi + \mu) + R_{2.T} \cos^2 (\varphi + \mu)} \tag{9.48}$$

Here, in Equations 9.47 and 9.48, φ designates the angle that specifies the direction of the feed rate vector $\mathbf{F}_{fr}$ and μ denotes the angle of surfaces *P* and *T* local relative orientation at contact point *K*.

For the calculation of the radii of normal curvature $R_{P.fr}$ and $R_{T.fr}$ of part surface *P* and generating surface *T* of the form-cutting tool in the direction of $\mathbf{F}_{fr}$, the following formulae can be helpful:

$$R_{P.fr} = \frac{E_P G_P}{G_P L_P \sin^2 \xi + M_P \sqrt{E_P G_P} \sin 2\xi + E_P N_P \cos^2 \xi} \tag{9.49}$$

$$R_{T.fr} = \frac{E_T G_T}{G_T L_T \sin^2 (\xi + \mu) + M_T \sqrt{E_T G_T} \sin[2 \cdot (\xi + \mu)] + E_T N_T \cos^2 (\xi + \mu)} \tag{9.50}$$

where angle ξ is equal to $\xi = \varphi + 90°$.

For further analysis, it is important to point out here that Equations 9.49 and 9.50 yield the representation of the relationships "$[\breve{F}_{fr}]$ vs. φ" and "$[\breve{F}_{fr}]$ vs. μ" in the form $[\breve{F}_{fr}] = [\breve{F}_{fr}](\varphi, \mu)$. The limit feed rate displacement $[\breve{F}_{fr}]$ depends on the direction of the feed rate motion $\mathbf{F}_{fr}$ as well as on the relative local orientation of the surfaces *P* and *T* at contact point *K*.

9.2.4.2 Calculation of the Limit Side-Step Shift

In most practical cases, the feed step displacement $\breve{F}_{ss}$ affects the part surface generation output P_{sg} the most. The calculation of the limit side step displacement $[\breve{F}_{ss}]$ is similar to the calculation of the limit feed rate displacement $[\breve{F}_{fr}]$. Without going into details of derivation of equations, in this case, we will just rewrite the equation

$$[\breve{F}_{ss}] \cong 2R_{P.ss} \arccos \frac{R_{P.ss}^2 + R_{T.ss} \cdot (R_{P.ss} + [h_{ss}] \cdot \operatorname{sgn} R_{P.ss})}{(R_{P.ss} + R_{T.ss}) \cdot (R_{P.ss} + [h_{ss}] \cdot \operatorname{sgn} R_{P.ss})} \qquad (9.51)$$

for the calculation of the limit side step displacement $[\breve{F}_{ss}]$. This equation is derived in a similar way to how Equation 9.46 has been derived. This means that the approximate Equation 9.51 is derived under similar assumptions as those made for Equation 9.46. It is assumed in Equation 9.51 that the inequality $\breve{F}_{ss} \ll R_{P.ss}$ is valid. It is assumed also that $[h_{P.ss}]^2$ is a reasonably small value, and it can be omitted from further analysis.

Equation 9.51 is valid for the cases when no limitations on actual value of the side step motion are imposed by power capacity of the multiaxis NC machine. Under such a scenario, the limit value of the side step $[\breve{F}_{ss}]$ is a function only of the corresponding radii of normal curvature of the surfaces P and T.

In Equation 9.51, the radii of the normal curvature $R_{P.ss}$ and $R_{T.ss}$ of surfaces P and T in the direction of the side step shift $\mathbf{F}_{ss}$ are equal:

$$R_{P.ss} = \frac{R_{1.P} R_{2.P}}{R_{1.P} \sin^2 \varphi + R_{2.P} \cos^2 \varphi} \qquad (9.52)$$

$$R_{T.ss} = \frac{R_{1.T} R_{2.T}}{R_{1.T} \sin^2 (\varphi + \mu) + R_{2.T} \cos^2 (\varphi + \mu)} \qquad (9.53)$$

For calculation of radii of normal curvature $R_{P.ss}$ and $R_{T.ss}$ of part surface P and generating surface T of the form-cutting tool in the direction $\mathbf{F}_{ss}$, the following formulae can be helpful:

$$R_{P.ss} = \frac{E_P G_P}{G_P L_P \cos^2 \xi + M_P \sqrt{E_P G_P} \sin 2\xi + E_P N_P \sin^2 \xi} \qquad (9.54)$$

$$R_{T.ss} = \frac{E_T G_T}{G_T L_T \cos^2 (\xi + \mu) + M_T \sqrt{E_T G_T} \sin[2 \cdot (\xi + \mu)] + E_T N_T \sin^2 (\xi + \mu)} \qquad (9.55)$$

It is necessary to stress here that Equations 9.54 and 9.55 yield the representation of the relationships "$[\breve{F}_{ss}]$ vs. φ" and "$[\breve{F}_{ss}]$ vs. μ" in the form $[\breve{F}_{ss}] = [\breve{F}_{ss}](\varphi, \mu)$. The limit side step displacement $[\breve{F}_{ss}]$ depends on the direction of the side step shift $\mathbf{F}_{ss}$ as well as on relative local orientation of the surfaces P, and T at contact point K.

9.2.5 Maximal Instantaneous Productivity of Part Surface Generation

The greatest part surface area coverage is an important output when machining a sculptured part surface on a multiaxis NC machine. The most favorable conditions of part surface generation, those determined on the premise of implementation of economical criteria of optimization and those determined on the premise of implementation of the productivity of part surface generation as the criterion, are close to each other. This is due to the impact of high cost of a multiaxis NC machine prevailing over the impact of others factors on the conditions of optimal sculptured part surface machining.

Determining the conditions under which the productivity of part surface generation is maximal is a critical issue in sculptured part surface machining. The use of the DG/K method of part surface generation enables an analytical solution to this challenging engineering problem to be found.

The productivity of part surface generation P_{sg}^{max} reaches its maximal value if and only if the instantaneous productivity of part surface generation $P_{sg}^{max}(t)$ is maximal at every point of contact K of part surface P and generating surface T of the form-cutting tool. For the calculation of conditions of the maximal instantaneous productivity of part surface generation, the modified Equation 8.1 can be used:

$$P_{sg}^{max}(t) = [\breve{F}_{fr}] \cdot [\breve{F}_{ss}] \cdot \sin\theta \tag{9.56}$$

Evidently, for the maximal instantaneous productivity $P_{sg}^{max}(t)$, the vectors of feed rate motion $\mathbf{F}_{fr}$ and side step shift $\mathbf{F}_{ss}$ must be orthogonal to each other; thus, the equality $\theta = 90°$ must be valid at every instant of the part surface machining.

Substitute Equations 9.46 and 9.51 to Equation 9.56. Under the condition $\theta = 90°$, Equation 9.56 casts into

$$P_{sg}^{max}(t) \cong 4 \cdot R_{P.fr} \cdot R_{P.ss} \cdot \arccos\left[\frac{R_{P.fr}^2 + R_{T.fr} \cdot (R_{P.fr} + [h_{fr}] \cdot \operatorname{sgn} R_{P.fr})}{(R_{P.fr} + R_{P.fr}) \cdot (R_{T.fr} + [h_{fr}] \cdot \operatorname{sgn} R_{P.fr})}\right]$$
$$\cdot \arccos\left[\frac{R_{P.ss}^2 + R_{T.ss} \cdot (R_{P.ss} + [h_{ss}] \cdot \operatorname{sgn} R_{P.ss})}{(R_{P.ss} + R_{T.ss}) \cdot (R_{P.ss} + [h_{ss}] \cdot \operatorname{sgn} R_{P.ss})}\right] \tag{9.57}$$

Despite Equation 9.57 being bulky, the analytical representation of the instantaneous maximal productivity $P_{sg}^{max}(t)$ of part surface generation is incomparably more compact in comparison to an analytical description of any economical criteria of optimization. The use of $P_{sg}^{max}(t)$ for the synthesis of the most favorable processes of sculptured part surface machining on a multiaxis NC machine returns the result that is practically equivalent to the result when an economical criterion of optimization is used. Due to significantly simpler analytical representation, the uneconomical criteria of optimization do not have to be used for the synthesis of optimal machining operations, but the instantaneous productivity of part surface generation $P_{sg}^{max}(t)$ has to be used instead.

9.3 Interpretation of the Part Surface Generation Output as a Function of Conformity

The transition from the economical criteria of optimization to the instantaneous maximal part surface generation output is a good step toward the simplification of the analytical description of a part surface machining process on a multiaxis *NC* machine. More opportunities in this concern can be found when functions of conformity of two smooth regular surfaces (see Chapter 4) are involved in the analysis.

As follows from Equation 9.56, to increase the instantaneous part surface generation output $P_{sg}^{max}(t)$, the feed rate displacement $\breve{F}_{fr}$ as well as the side step displacement $\breve{F}_{ss}$ must be of maximal value at every point K of contact of part surface P and generating surface T of the form-cutting tool. The current configuration of the cutting tool with respect to the part surface being machined and the instantaneous kinematics of the machining operation are the good tools used to control current values of the parameters $\breve{F}_{fr}$ and $\breve{F}_{ss}$ at every point of contact of the surfaces P and T.

Consider the cross-section of a sculptured part surface P and generating surface T of the form-cutting tool by a normal plane through a contact point K. The cross section is along the side step vector $\mathbf{F}_{ss}$ as shown in Figure 9.4. The radius of normal curvature of part surface P at the point K is equal to R_P. The partial tolerance for accuracy of the machined part surface P is denoted by $[h_{ss}]$.

Part surface P can be machined with the cutting tools of different designs.

Assume that at point K, part surface P is machined by the form-cutting tool having a radius of normal curvature of a certain positive value $R_{T.s}^{(1)}$. The limit side step shift in this case is equal to $\left[\breve{F}_{ss}^{(1)} \right]$ (Figure 9.4a).

That same local portion of the sculptured part surface P can be machined by the form-cutting tool having at the point K radius of normal curvature of a certain value $R_{T.s}^{(2)}$. The radius of normal curvature $R_{T.s}^{(2)}$ is also positive, but its magnitude exceeds the magnitude of the radius of normal curvature $R_{T.s}^{(1)}$ (Figure 9.4b). The limit side step shift in this case is equal to $\left[\breve{F}_{ss}^{(2)} \right]$. Because the inequality $R_{T.s}^{(2)} > R_{T.s}^{(1)}$ is observed, the limit side step shift $\left[\breve{F}_{ss}^{(2)} \right]$ is bigger

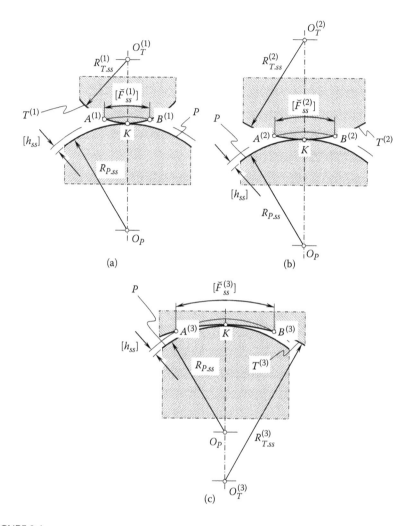

FIGURE 9.4
Various degrees of conformity of generating surface T of the cutting tool to a sculptured part surface P at a current CC point K.

than the limit side step shift $\left[\breve{F}_{ss}^{(1)}\right]$ in the first case. In the second case (see Figure 9.4b), the inequality $\left[\breve{F}_{ss}^{(2)}\right] > \left[\breve{F}_{ss}^{(1)}\right]$ is observed.

That same local portion of the sculptured part surface P can be machined with the form-cutting tool having at point K the radius of normal curvature of a certain negative value $R_{T.s}^{(3)}$ (Figure 9.4c). The limit side step shift in this case is equal to $\left[\breve{F}_{ss}^{(3)}\right]$. Evidently, the limit side step shift $\left[\breve{F}_{ss}^{(3)}\right]$ is bigger than the limit side step $\left[\breve{F}_{ss}^{(2)}\right]$ in the second case. Therefore, the inequality $\left[\breve{F}_{ss}^{(3)}\right] > \left[\breve{F}_{ss}^{(2)}\right] > \left[\breve{F}_{ss}^{(1)}\right]$ is observed.

The degree of conformity of the generating surface T of the cutting tool to the sculptured part surface P at contact point K is the smallest in the first case (Figure 9.4a), bigger in the second case (Figure 9.4b), and biggest in the third case (Figure 9.4c).

This analysis reveals that the width of the limit side step shift $[\breve{F}_{ss}]$ increases when the degree of conformity of the cutting tool surface T to part surface P increases. A similar conclusion can be derived with respect to the limit feed rate shift $[\breve{F}_{fr}]$.

Because the instantaneous productivity of part surface generation $P_{sg}^{max}(t)$ is a function of the limit feed rate shift $[\breve{F}_{fr}]$ and the limit side step shift $[\breve{F}_{ss}]$, this statement immediately makes possible a conclusion:

> *The bigger the degree of conformity of generating surface T of the cutting tool to the sculptured part surface P, the bigger instantaneous productivity of part surface generation $P_{sg}^{max}(t)$.*

It can be proven analytically that the function $P_{sg}^{max}(t)$ (see Equation 9.57) is a type of function of the conformity of two smooth regular surfaces. All the functions of conformity have extremes under the same values of the input arguments. Therefore, not only can function $P_{sg}^{max}(t)$ be used for solving the problem of synthesis of the most favorable machining operations, but any corresponding function of conformity of the surfaces P and T can be used as well. For example, the function $P_{sg}^{max}(t)$ of the instantaneous productivity of part surface generation can be substituted with the indicatrix of conformity $Cnf_R(P/T)$ at a point of contact K of part surface P and generating surface T of the form-cutting tool. Such a substitution is reasonable because the analytical expression for the indicatrix of conformity $Cnf_R(P/T)$ is simpler compared with the analytical expression for the function $P_{sg}^{max}(t)$. The indicatrix of conformity $Cnf_R(P/T)$ at a point of contact of surfaces P and T can be considered as a type of geometrical analogue of the instantaneous productivity $P_{sg}^{max}(t)$ of part surface generation.

For the practical implementation of the discussed approach for the calculation of sculptured part surface machining output, a wide application of computers is necessary.

10

Synthesis of the Part Surface
Machining Operations

Synthesis of the most favorable part surface machining operations here means the development of a procedure of computation of the optimal parameters of the part surface machining process. Minimum input information is required for solving the challenging problem of synthesis. The derived solution to the problem enables us to find the desired extremum of the criterion of optimization.

The problem of synthesizing of an optimal part surface machining operation can be solved in three steps. The local part surface generation is synthesized in the first step. Then, in the second step, the regional synthesis is performed on the premise of the results of the calculations obtained in the first step. Ultimately, the solution to the problem of global synthesis of the most favorable part surface machining operation is derived in the final third step.

10.1 Synthesis of the Most Favorable Part Surface Generation Process: The Local Analysis

Local part surface generation is considered within an elementary surface cell on the machined part surface. From the perspective of local consideration, the synthesis of the most favorable part surface generation aims to determine the following:

- Coordinates of a point within the generation surface T of the form-cutting tool, the local geometry of surface T which enables the desired geometry of contact of part surfaces P and generating surface T of the cutting tool (see Chapter 4). For convenience, this point is designate below as K_T. In the event the cutting tool is given, the coordinates of the point on the cutting tool surface T can be calculated on the premise of proper correspondence between the local geometry of surfaces P and T at the cutter–contact point K (CC point K). Otherwise, the calculated desired parameters of local geometry of the cutting tool surface T are used further for design of the optimal cutting tool as the $\mathbb{R}$-map of part surface P to be machined (see Chapter 5).

- Local configuration of the cutting tool in relation to the part surface being machined.
- Optimal direction of the instant motion of the cutting tool relative to the work.
- Optimal parameters of the instant motions of orientation of the form-cutting tool (see Chapter 2).

For illustrative purposes, the minimal diameter of the indicatrix of conformity at the point of contact of the part surface and of the generating surface of the cutting tool is used below as the criterion of optimization. Chip removal output, productivity of part surface generation, or an economical criterion of optimization can be used for this purpose as well.

10.1.1 Local Synthesis

For the analytical representation of the indicatrix of conformity Cnf_R (P/T) at a point of contact of two smooth regular surfaces P and T, the following expression is derived (see Equation 4.87):

$$Cnf_R(P/T) \quad \Rightarrow \quad r_{cnf} = \sqrt{\left| \frac{E_P G_P}{L_P G_P \cos^2 \varphi - M_P \sqrt{E_P G_P} \sin 2\varphi + N_P E_P \sin^2 \varphi} \right|} \, \text{sgn} \, \Phi_{2.P}^{-1} + $$

$$\sqrt{\left| \frac{E_T G_T}{L_T G_T \cos^2(\varphi + \mu) - M_T \sqrt{E_T G_T} \sin 2(\varphi + \mu) + N_T E_T \sin^2(\varphi + \mu)} \right|} \, \text{sgn} \, \Phi_{2.T}^{-1}$$

$$(10.1)$$

The current diameter d_{cnf} of the characteristic curve $Cnf_R(P/T)$ is equal to $d_{cnf} = 2 \cdot r_{cnf}$.

Further, it is necessary to recall that

- The fundamental magnitudes of the first and the second order of part surface P can be expressed in terms of the curvilinear (Gaussian) coordinates U_P and V_P. This means that the analytical functions $E_P = E_P(U_P, V_P)$, $F_P = F_P(U_P, V_P)$, $G_P = G_P(U_P, V_P)$, and $L_P = L_P(U_P, V_P)$, $M_P = M_P(U_P, V_P)$, $N_P = N_P(U_P, V_P)$ are known (see Chapter 1).
- The fundamental magnitudes of the first and the second order of the generating surface T of the form-cutting tool can be expressed in terms of the curvilinear (Gaussian) coordinates U_T and V_T. This

means that analytical expressions for the functions $E_T = E_T(U_T, V_T)$, $F_T = F_T(U_T, V_T)$, $G_T = G_T(U_T, V_T)$, and $L_T = L_T(U_T, V_T)$, $M_T = M_T(U_T, V_T)$, $N_T = N_T(U_T, V_T)$ are also known (see Chapter 1).

Equation 10.1, together with two items above make possible a generalized form for the diameter d_{cnf} of the indicatrix of conformity $Cnf_R(P/T)$:

$$d_{cnf} = d_{cnf}(U_P, V_P, U_T, V_T, \mu, \varphi) \tag{10.2}$$

Once a point of interest K_P on the sculptured part surface P is chosen, the Gaussian coordinates U_P and V_P of this point are known.

For a point within the generating surface T of the cutting tool being capable of making optimal contact with part surface P at the selected point of interest, the following necessary conditions have to be satisfied:

$$\frac{\partial d_{cnf}}{\partial U_T} = 0 \tag{10.3}$$

$$\frac{\partial d_{cnf}}{\partial V_T} = 0 \tag{10.4}$$

In addition to the necessary conditions for the minimum of the diameter d_{cnf} (see Equations 10.3 and 10.4), the sufficient conditions for the minimum diameter d_{cnf}^{min} must be satisfied as well:

$$\begin{vmatrix} \dfrac{\partial^2 d_{cnf}}{\partial U_T^2} & \dfrac{\partial^2 d_{cnf}}{\partial U_T \partial V_T} \\[2ex] \dfrac{\partial^2 d_{cnf}}{\partial U_T \partial V_T} & \dfrac{\partial^2 d_{cnf}}{\partial V_T^2} \end{vmatrix} > 0 \tag{10.5}$$

$$\frac{\partial^2 d_{cnf}}{\partial U_T^2} > 0 \tag{10.6}$$

The solution to the set of Equations 10.3 and 10.4 under the conditions (see Equations 10.5 and 10.6) returns Gaussian coordinates U_T^{opt} and V_T^{opt} of the optimal point K_T within the generating surface T of the form-cutting tool.

For the highest possible productivity of part surface generation, it is necessary that at the point of interest K_P, the generation surface T of the cutting tool contacts the part surfaces P with its optimal point K_T. Once points K_P and K_T are snapped together, their designation is further substituted with

K (i.e., $K_P \equiv K_T \equiv K$). It is assumed here and below that all the necessary and sufficient conditions of proper part surface generation [9,10,11,16] are satisfied (see Chapter 7).

The calculated optimal parameters U_T^{opt} and V_T^{opt} specify location of the contact point K_T within the generating surface T of the cutting tool.

The point of contact of surfaces imposes strong restrictions on feasible motions of the cutting tool surface T relative to part surface P. The only motion allowed is for surfaces P and T while they are in contact at the fixed point K. Depending on the parameters of the actual surfaces P and T at contact point K, the cutting tool surface T is allowed either to rotate about the common normal vector $\mathbf{n}_P$ or to turn through a certain angle about this unit normal vector. No other relative motions are feasible for surfaces P and T in the case under consideration.

Definitely, the local productivity of part surface generation under various angular positions of the cutting tool surface T with respect to part surface P is different. Thus, the optimal configuration of the generating surface T in relation to part surface P exists; moreover, its parameters can be calculated. For this purpose, the most favorable local relative orientation of part surface P and generating surface T of the form-cutting tool must be calculated.

Current configuration of surfaces P and T at the CC point is specified by the angle μ of their local relative orientation. To get the local relative orientation optimal, it is necessary to calculate the optimal value μ^{opt} of the angle of surfaces P and T local relative orientation. Equation 10.2 is helpful for solving this particular problem.

The most favorable value μ^{opt} of the angle of surfaces P and T local relative orientation can be calculated as a solution to the equation

$$\frac{\partial d_{cnf}}{\partial \mu} = 0 \tag{10.7}$$

The derived solution to Equation 10.7 must satisfy the sufficient condition

$$\frac{\partial^2 d_{cnf}}{\partial \mu^2} > 0 \tag{10.8}$$

for the minimum of the diameter $d_{cnf}^{\min}$.

The solution to Equation 10.7 under the condition (see Equation 10.8) specifies the most favorable local configuration of the cutting tool in relation to the part surface being machined. In compliance with the derived solution, the cutting tool must be turned about the unit normal vector $\mathbf{n}_P$ through a certain angle to its optimal configuration relative to part surface P that is specified by the calculated angle μ^{opt}.

Once the optimal configuration of the generating surface of the cutting tool in relation to the sculptured part surface is determined, the optimal parameters of the instant kinematics of local part surface generation can also be determined. To achieve the highest possible part surface machining output, the vector of instant motion of the cutting tool relative to the part surface must be orthogonal to the direction along which the minimal diameter d_{cnf}^{min} of the indicatrix of conformity is measured.

The direction along which the current diameter d_{cnf} of the indicatrix of conformity is measured forms a certain angle φ with the first principal direction $\mathbf{t}_{1.P}$ of part surface P. For the minimal diameter d_{cnf}^{min}, the equality

$$\frac{\partial d_{cnf}}{\partial \varphi} = 0 \tag{10.9}$$

is satisfied. Here, for Equation 10.9, the sufficient condition for the maximum of the diameter d_{cnf}^{min} is also observed:

$$\frac{\partial^2 d_{cnf}}{\partial \varphi^2} > 0 \tag{10.10}$$

The solution to Equation 10.9 under condition (see Equation 10.10) returns the optimal value φ^{opt} of the angle φ.

Vector $\mathbf{F}_{fr}$ of the cutting tool feed-rate motion is at the angle $\xi^{opt} = \varphi^{opt} + 90°$. Ultimately, the calculated value ξ^{opt} of the angle ξ specifies the most favorable direction of the instant motion of the cutting tool relative to the work. Vector $\mathbf{F}_{fr}$ also defines the direction at which point K travels over the generating surface of the cutting tool.

After vector $\mathbf{F}_{fr}$ of the cutting tool feed-rate motion is determined, the problem of synthesis of local part surface generation is over.

As follows from the above consideration, when solving the problem of synthesis of local part surface generation, the calculation of the first and the second derivatives of the indicatrix of conformity $Cnf_R(P/T)$ (see Equation 10.2) is necessary. This trivial mathematical operation does not cause any inconvenience when spline functions are used for the local approximation of the sculptured part surface P. The calculation of the first and second derivatives of spline functions can be performed easily.

Two examples in Figure 10.1 illustrate the results of solutions to the problem of synthesis of the local part surface generation process.

In the first example, when the saddle-like local patch of the sculptured part surface P is machined by the convex portion of the generating surface T of the cutting tool (Figure 10.1a), a solution to the problem of synthesis of the local part surface generation process is shown in Figure 10.1b. Here, the minimal diameter d_{cnf}^{min} of the indicatrix of conformity $Cnf_R(P/T)$ is at the

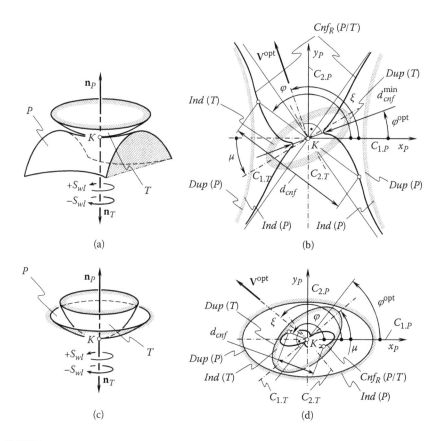

FIGURE 10.1
Two examples of solutions to the problem of local synthesis of sculptured part surface generation.

optimal angle φ^{opt} to the first principal cross section $C_{1.P}$ of the part surface P at the contact point K. The optimal direction of vector $\mathbf{V}^{\mathrm{opt}}$ of surfaces P and T instant relative motion forms the angle $\xi^{\mathrm{opt}} = \varphi^{\mathrm{opt}} + 90°$ with the first principal cross section $C_{1.P}$. When the part surface being machined is a sculptured surface, vector $\mathbf{V}^{\mathrm{opt}}$ is identical to vector $\mathbf{F}_{fr}^{\mathrm{opt}}$ of the optimal feed-rate motion of the cutting tool $\left(\mathbf{V}^{\mathrm{opt}} \equiv \mathbf{F}_{fr}^{\mathrm{opt}}\right)$. In other cases, vector $\mathbf{v}^{\mathrm{opt}}$ yields different interpretations.

Similarly, in the second example, when the convex local patch of the sculptured part surface P is machined by the convex portion of the generating surface T of the cutting tool (Figure 10.1c), the solution to the problem of synthesis of local surface generation process is illustrated in Figure 10.1d. Again, the minimal diameter $d_{cnf}^{\min}$ of the indicatrix of conformity $Cnf_R(P/T)$ is at the optimal angle φ^{opt} to the first principal cross section $C_{1.P}$ of the part surface P

at the contact point K. The optimal direction of vector $\mathbf{v}^{\mathrm{opt}}$ of surfaces P and T instant relative motion forms the angle $\xi^{\mathrm{opt}} = \varphi^{\mathrm{opt}} + 90°$ with the first principal cross section $C_{1.P}$.

When solving a problem of synthesis of the local part surface generation process, some peculiarities could be observed.

10.1.2 Indefiniteness

"Indefiniteness," whether $\dfrac{0}{0}$, $\dfrac{\infty}{\infty}$, or $0 \cdot \infty$, could occur when the optimal values of the local part surface generation process are calculated. To overcome this particular problem, the substitution of the equation of the indicatrix of conformity $Cnf_R(P/T)$ with the equation of the corresponding indicatrix of conformity $Cnf_k(P/T)$ is usually helpful.

10.1.3 Possibility of Alternative Optimal Configurations of the Cutting Tool

Particular cases of contact of a part surface P and generating surface T of the form-cutting tool could be observed when solving the problem of synthesis of the local part surface generation process. One such problem is due to the possibility of two alternative optimal configurations of the cutting tool that are actually equivalent to each other. An example of two equivalent optimal configurations of the cutting tool is schematically depicted in Figure 10.2.

The example in Figure 10.2 relates to the generation of the saddle-like local portion of the sculptured part surface P by the convex portion of the generating surface T of the form-cutting tool. The minimal radius of the indicatrix of conformity $Cnf_R(P/T^*)$ at a point of contact of part surface P and cutting tool surface T^* in its first optimal configuration is equal to zero $\left(r_{cnf}^{*\,min} = 0 \right)$. The first principal plane section $C_{1.T}^*$ of the cutting tool surface T^* is at the optimal angle μ_{opt}^* of the local relative orientation of the surfaces P and T^*. Vector $\mathbf{V}_{opt}^*$ of the optimal direction of motion of the generating surface T^* of the cutting tool relative to part surface P is orthogonal to the direction along which the minimal radius $r_{cnf}^{*\,min}$ of the indicatrix of conformity $Cnf_R(P/T^*)$ is measured. This direction is the direction of maximal degree of conformity of the generating surface of the form-cutting tool to the sculptured part surface. This direction in the common tangent plane is labeled as $\mathbf{t}_{cnf}^{*\,max}$.

That same saddle-like local portion of the sculptured part surface P can be generated with that same convex portion of the generating surface T of the form-cutting tool under the different configuration of the cutting tool in relation to the work. The minimal radius of the indicatrix of conformity $Cnf_R(P/T^{**})$ at a point of contact of part surface P and cutting tool surface T^{**} in its second optimal configuration is also equal to zero $\left(r_{cnf}^{**\,min} = 0 \right)$. The first principal plane-section $C_{1.T}^{**}$ of the cutting tool surface T^{**} is at the optimal

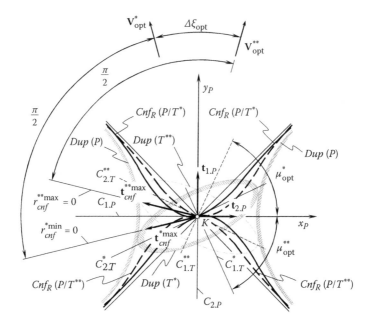

FIGURE 10.2
An example of two alternative configurations of the cutting tool, T, in relation to the part surface, P, being machined.

angle μ_{opt}^{**} of the local relative orientation of the surfaces P and T^{**}. Vector $\mathbf{V}_{opt}^{**}$ of the optimal direction of motion of the generating surface T^{**} of the cutting tool relative to part surface P is orthogonal to the direction along which the minimal radius $r_{cnf}^{**\,min}$ of the indicatrix of conformity $Cnf_R(P/T^{**})$ is measured. This direction is the direction of maximal degree of conformity of the generating surface of the cutting tool to the sculptured part surface. This direction in the common tangent plane is labeled as $\mathbf{t}_{cnf}^{**\,max}$.

Ultimately, the analysis of Figure 10.2 reveals that two optimal directions of the form-cutting tool relative motion are feasible. The first optimal direction $\mathbf{V}_{opt}^{*}$ is specified by the angle $\xi_{opt}^{*} = \varphi_{opt}^{*} + 90°$. The second optimal direction $\mathbf{V}_{opt}^{**}$ is specified by the angle $\xi_{opt}^{**} = \varphi_{opt}^{**} + 90°$. Due to lack of space, the angles $\xi_{opt}^{*}, \varphi_{opt}^{*}$ and $\xi_{opt}^{**}, \varphi_{opt}^{**}$ are not depicted in Figure 10.2. Directions $\mathbf{V}_{opt}^{*}$ and $\mathbf{V}_{opt}^{**}$ are at a certain angle $\Delta\xi_{opt}$ to each other.

However, no problem arises in this concern. One of two the most favorable directions $\mathbf{V}_{opt}^{*}$ and $\mathbf{V}_{opt}^{**}$ is selected automatically depending upon the cutting tool configuration at the previous points of contact with part surface P. The correlation between directions of the cutting tool motion along the tool path always is always observed.

A situation that is shown in Figure 10.2 requires in special attention when the angle $\Delta\xi_{opt}$ between the two directions $\mathbf{V}_{opt}^{*}$ and $\mathbf{V}_{opt}^{**}$ is reasonably small and its magnitude is comparable with the deviation $\delta\varphi_{opt}$ of the computation of the angle φ_{opt}. If the deviation $\delta\varphi_{opt}$ exceeds the angle $\Delta\xi_{opt}$, say when the inequality $\delta\varphi_{opt} \geq \Delta\xi_{opt}$ is observed, then the numerical control (*NC*) system of the machine tool generates two contradictory commands to move the cutting tool in two different directions.

Two ways to follow are possible for avoidance of such a scenario.

- To replace the actual cutting tool with the cutting tool having appropriate parameters of the generating surface *T*. In this way, either the angle $\Delta\xi_{opt}$ can be increased to the value $\Delta\xi_{opt} > \delta\varphi_{opt}$, or just one optimal direction $\mathbf{V}_{opt}$ of the cutting tool relative motion could actually exist.
- A multiaxis *NC* machine having higher resolution and capable of higher accuracy when performing the computations is recommended to be implemented.

The problem under consideration can be solved following either of the ways listed above.

10.1.4 Specific Cases of Part Surface *P* Generation

A few specific cases of synthesis of the local part surface generation process are briefly outlined below.

10.1.4.1 Cases of Multiple Points of Contact of Part Surface *P* and Generating Surface *T* of the Form-Cutting Tool

The scenario when the generating surface *T* of the form-cutting tool is contacting part surface *P* at two or more points is actually not practical, but it is possible.

Consider surfaces *P* and *T* that make contact at two distinct points K_1 and K_2. To make the machining operation feasible, the equation of contact $\mathbf{n}_P \cdot \mathbf{V}_\Sigma = 0$ (see Equation 2.22) must be fulfilled at both points K_1 and K_2. The equation of contact must be satisfied at both point K_1 (i.e., $\mathbf{n}_{P.1} \cdot \mathbf{V}_{\Sigma.1} = 0$) and point K_2 (i.e., $\mathbf{n}_{P.2} \cdot \mathbf{V}_{\Sigma.2} = 0$). Here, $\mathbf{n}_{P.1}$ and $\mathbf{n}_{P.2}$ denote the unit normal vectors to part surface *P* at points K_1 and K_2, respectively (actually, points K_1 and K_2 could be within the different portions P_1 and P_2 of the part surface). This requirement imposes a strong restriction onto the allowed direction of the relative motion $\mathbf{V}_\Sigma^{(2)}$ of the cutting tool.

When two points of contact K_1 and K_2 of surfaces *P* and *T* are observed, then

$$\mathbf{V}_\Sigma^{(2)} = \left| \mathbf{V}_\Sigma^{(2)} \right| \cdot \mathbf{n}_{P.1} \times \mathbf{n}_{P.2} \tag{10.11}$$

is the only feasible direction for $\mathbf{V}_\Sigma^{(2)}$.

If the number of points of contact exceed two, the generation of the sculptured part surface P under such a scenario is feasible if and only if the unit
normal vector at the third and at all other ith points of contact are coplanar
to the unit normal vectors $\mathbf{n}_{P.1}$ and $\mathbf{n}_{P.2}$.

10.1.4.2 Possibility of an Alternative Direction of the Feed-Rate Motion

The most favorable direction of the feed-rate motion in most cases of part surface
machining is perpendicular to the direction along which the minimum diameter d_{cnf}^{min} of the indicatrix of conformity $Cnf_R(P/T)$ is measured. This approach
is valid in all cases when linear velocity V_{cut} of the rotating cutting edges of the
cutting tool is significantly greater compared to the feed-rate motion F_{fr}.

In a particular case, this condition $(V_{cut} \gg F_{fr})$ can be violated, that is, the
magnitude of the feed-rate motion F_{fr} can be greater than the linear velocity
V_{cut} of the rotating cutting edges of the cutting tool $(F_{fr} \gg V_{cut})$. Cases like this
are far from to be practical in current practice; however, they can exist. In the
event the inequality $F_{fr} \gg V_{cut}$ is observed, the most favorable direction of the
feed-rate motion is in the direction along which the maximal diameter d_{cnf}^{max}
of the indicatrix of conformity $Cnf_R(P/T)$ is measured.

Finally, velocities V_{cut} and F_{fr} can be of comparable values. Under such a
scenario, instant part surface generation output should be used for determining the most favorable direction of the feed-rate motion.

10.1.4.3 Important Conclusions from the Synthesis of the Most
Favorable Local Part Surface Generation Process

Two important conclusions can be drawn from the solution to the problem of
synthesis of the most favorable local part surface generation:

- For computation of the most favorable directions of the tool paths on
  the sculptured part surface, the geometry of the both the sculptured
  part surface and the generating surface of the cutting tool must be
  known. It is impossible to derive the optimal tool paths only on the
  premise of geometry of part surface P. The last is possible only in particular cases of sculptured part surface machining (e.g., when machining, either the ball-nose or the flat-end milling cutters are used). In
  particular cases such as those mentioned, the optimal tool paths align
  with the lines of curvature on the sculptured part surface.

- The most favorable tool paths do not intersect each other: generally, they are not a type of transversal curves on the sculptured part
  surface.

The importance of these conclusions is twofold. First, they are important
for the practice of sculptured part surface machining on a multiaxis *NC*

machine. Second, the conclusions are important from the philosophical and gnoseological perspectives.

10.2 Synthesis of the Most Favorable Part Surface Generation Process: The Regional Analysis

The derived solution to the problem of synthesis of the most favorable local part surface generation process (see Section 10.1) returns the optimal parameters of the part surface generation process that is valid only within a surface cell on the machined part surface. This solution is the key input for the synthesis of the most favorable regional part surface generation process. The synthesis of the optimal regional part surface generation process is a targeted solution to the problem of the most efficient generation along a tool path on a sculptured part surface. Once the problem of the synthesis of the optimal local part surface generation process is solved, the problem of synthesis of the most favorable regional part surface generation process could be under condition when the local solution is obtained at every point of the tool path.

Consider the generation of a sculptured part surface P by the generating surface T of the form-cutting tool (Figure 10.3). At a given CC point K within the part surfaces P, Gaussian coordinates U_T and V_T of the point of contact within the generating surface T of the cutting tool are of optimal values U_T^{opt} and V_T^{opt}. The optimal angle μ^{opt} of the local relative orientation of surfaces P and T is measured between the principal directions $\mathbf{t}_{1.P}$ and $\mathbf{t}_{1.T}$. The direction of minimal value of the diameter d_{cnf}^{min} of the characteristic curve $Cnf_R(P/T)$ is at the angle φ^{opt} with respect to the principal direction $\mathbf{t}_{1.P}$. The optimal direction of traveling of the cutting tool over the sculptured part surface P is specified by the angle $\xi^{opt} = \varphi^{opt} + 90°$.

The local *Cartesian* coordinate system $x_P y_P z_P$ is associated with part surface P. The origin of the local coordinate system is a the CC point K. The Darboux frame can be used for the purpose of construction of the local coordinate system $x_P y_P z_P$.

In the local coordinate system $x_P y_P z_P$, vector $\mathbf{V}^{opt}$ of the most favorable direction of the cutter travel relative part surface P can be expressed in matrix form:

$$\mathbf{V}^{opt} = \begin{bmatrix} \sin \xi^{opt} \\ \cos \xi^{opt} \\ 0 \\ 1 \end{bmatrix} \tag{10.12}$$

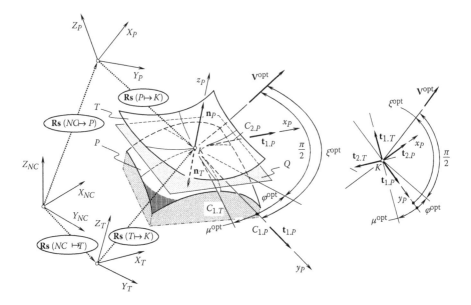

FIGURE 10.3
Configuration and relative motion of part surface *P* and generating surface *T* of the form-cutting tool in the *Cartesian* coordinate system $X_{NC}Y_{NC}Z_{NC}$ associated with the multiaxis numerical control machine. (Reprinted from *Mathematical and Computer Modeling*, 43, Radzevich, S.P., A closed-form solution to the problem of optimal tool-path generation for sculptured surface machining on multi-axis NC machine, 222–243, Copyright 2006, with permission from Elsevier.)

A *Cartesian* coordinate system $X_{NC}Y_{NC}Z_{NC}$ is associated with the multi-axis *NC* machine. The operator $\mathbf{Rs}(K \mapsto NC)$ of the resultant coordinate system transformation, say for the transition from the local coordinate system $x_P y_P z_P$ to the coordinate system $X_{NC}Y_{NC}Z_{NC}$, can be composed as disclosed in Chapter 3. The use of the operator $\mathbf{Rs}(K \mapsto NC)$ makes it possible representation of vector $\mathbf{V}^{\text{opt}}$ in the coordinate system $X_{NC}Y_{NC}Z_{NC}$:

$$\mathbf{V}_{NC}^{\text{opt}} = \mathbf{Rs}(K \mapsto NC) \cdot \mathbf{V}^{\text{opt}} = \begin{bmatrix} \left(\dfrac{\partial x_P}{\partial U_P}(t)\cos\varphi^{\text{opt}}(t) - \dfrac{\partial x_P}{\partial V_P}(t)\sin\varphi^{\text{opt}}(t) \right) \\[2ex] \left(\dfrac{\partial y_P}{\partial U_P}(t)\cos\varphi^{\text{opt}}(t) - \dfrac{\partial y_P}{\partial V_P}(t)\sin\varphi^{\text{opt}}(t) \right) \\[2ex] \left(\dfrac{\partial z_P}{\partial U_P}(t)\cos\varphi^{\text{opt}}(t) - \dfrac{\partial z_P}{\partial V_P}(t)\sin\varphi^{\text{opt}}(t) \right) \\[2ex] 1 \end{bmatrix} \tag{10.13}$$

Equation 10.13 can be interpreted as the differential form of solution to the problem of synthesis of the most favorable regional part surface generation process. A closed-form solution to the problem of the optimal tool paths can be derived on the premise of Equation 10.13. A surface tool path is a time-parameterized surface trajectory:

$$\mathbf{r}\,[u(t),\,v(t)] = \mathbf{r}(t), \tag{10.14}$$

which corresponds to a certain curve $u(t)$, $v(t)$. Thus, the closed form solution to the problem of the most favorable tool paths* can be represented in the form [4]:

$$\left[\mathbf{r}_{tp}^{\text{opt}}\right](t) = \begin{bmatrix} \displaystyle\int_{t_1}^{t_2}\left(\frac{\partial X_P}{\partial U_P}(t)\cos\varphi^{\text{opt}}(t) - \frac{\partial X_P}{\partial V_P}(t)\sin\varphi^{\text{opt}}(t)\right)\cdot dt \\[2mm] \displaystyle\int_{t_1}^{t_2}\left(\frac{\partial Y_P}{\partial U_P}(t)\cos\varphi^{\text{opt}}(t) - \frac{\partial Y_P}{\partial V_P}(t)\sin\varphi^{\text{opt}}(t)\right)\cdot dt \\[2mm] \displaystyle\int_{t_1}^{t_2}\left(\frac{\partial Z_P}{\partial U_P}(t)\cos\varphi^{\text{opt}}(t) - \frac{\partial Z_P}{\partial V_P}(t)\sin\varphi^{\text{opt}}(t)\right)\cdot dt \\[2mm] 1 \end{bmatrix} \tag{10.15}$$

which directly follows from Equation 10.13 [97].

Another form of representation of Equation 10.15 is known from References [10,11,16].

At a current *CC* point, the maximal allowed speed of travel of the cutting tool along the optimal tool path (see Equation 10.15) is restricted by the limit value of the feed-rate per tooth $[\tilde{F}_{fr}]$ of the cutting tool (see Equation 9.46). From the perspective of geometric and kinematic consideration, the maximal speed of the cutting tool travel is bigger when the concave portion of the part surface *P* is machined, and it is smaller when machining a convex portion of the part surface *P*. However, when the chip removal output is included in consideration, the maximal speed of the cutting tool travel is smaller when the concave portion of part surface *P* is machined, and it is bigger when machining a convex portion of the part surface *P*.

For the orthogonally parameterized surface *P*, Equation 10.15 for the position vector of a point $\left[\mathbf{r}_{tp}^{\text{opt}}\right]$ reduces to

$$\left[\mathbf{r}_{tp}^{\text{opt}}\right] \;\Rightarrow\; U_P = -\int_{V_{P.1}}^{V_{P.2}} \cot[\varphi^{\text{opt}}(V_P)]\,dV_P \tag{10.16}$$

---

* The author would like to mention here that many proficient researchers loosely came up with the decision that no closed-form solution to the problem of optimal tool paths is feasible.

In many particular cases of sculptured part surface machining, both Equations 10.15 and 10.16 can be integrated analytically.

In some particular cases of sculptured part surface generation, the equation for the optimal tool paths simplifies to the differential equation:

$$\left[\mathbf{r}_{tp}^{opt}\right] \Rightarrow \begin{vmatrix} E_p dU_p + F_p dV_p & F_p dU_p + G_p dV_p \\ L_p dU_p + M_p dV_p & M_p dU_p + N_p dV_p \end{vmatrix} = 0 \qquad (10.17)$$

Equation 10.17 for the optimal tool paths is applicable, for instance, when machining a sculptured part surface P either with a ball-end milling cutter or with a flat-end milling cutter and so forth. Under such a scenario, the angle μ of the local relative orientation of surfaces P and T vanishes. It becomes indefinite: no principal directions can be identified on a sphere or on the plane surface. Therefore, the optimal tool paths align with lines of curvature on part surface P (see Equation 10.17).

When machining a part surface, the coordinate system $X_T Y_T Z_T$ that is associated with the cutting tool rotates like a rigid body. This rotation is performed about a certain instant axis of rotation. The angular velocity of the rotation of the coordinate system $X_T Y_T Z_T$ is equal to

$$\Omega = \sqrt{k_{tp}^2 + \tau_{tp}^2} \; . \qquad (10.18)$$

The axis of instant rotation aligns with the Darboux vector:

$$\Omega = k_{tp} \mathbf{t}_{tp} + \tau_{tp} \mathbf{b}_{tp} \qquad (10.19)$$

(here k_{tp} and τ_{tp} denote curvature and torsion of the trajectory of CC point and $\mathbf{t}_{tp}$ and $\mathbf{b}_{tp}$ are the unit tangent vector and the bi-normal vector to the trajectory of CC point at a current contact point K). The Darboux vector is located in the rectifying plane to the trajectory of the CC point. It can be expressed in terms of the normal vector $\mathbf{n}_{tp}$ and of the tangent vector $\mathbf{t}_{tp}$ to the trajectory of CC point:

$$\Omega = \sqrt{k_{tp}^2 + \tau_{tp}^2} \, (\mathbf{t}_{tp} \cos\theta + \mathbf{n}_{tp} \sin\theta) \qquad (10.20)$$

where θ is the angle that makes the Darboux vector Ω and the tangent vector $\mathbf{t}_{tp}$ to the trajectory at CC point.

It is instructive to note that velocity $|\Omega|$ is a function of full curvature of the trajectory of the CC point.

In a particular case, the tool paths on the part surface to be machined can be prespecified. For example, tool paths can be expressed in terms of geometry of the part surface contour and so forth. Even under such a scenario, the most favorable contact geometry must be maintained at every contact point, or in other words, at every instant of time. In this case, the direction of the instant feed-rate motion $\mathbf{F}_{fr}$ is not perpendicular to the minimal diameter d_{cnf}^{min} of the indicatrix of conformity $Cnf_R(P/T)$ at a current point of contact of part surface P and generating surface T of the form-cutting tool. To solve the problem of the most favorable regional part surface generation process in this particular case, it is necessary to determine the orientation of the cutting tool in relation to the part at which minimal diameter $d_{cnf}^{*\,min}$ of the indicatrix of conformity $Cnf_R(P/T)$ at a current point of contact of surfaces P and T is measured in the direction perpendicular to the given tool path. Diameters d_{cnf}^{min} and $d_{cnf}^{*\,min}$ are commonly not equal to one another. Often, the inequality $d_{cnf}^{min} < d_{cnf}^{*\,min}$ is observed.

Further steps for solving the problem of regional synthesis of part surface generation remain the same.

10.3 Synthesis of the Most Favorable Part Surface Generation Process: The Global Analysis

The synthesis of the most favorable global part surface generation process is the final subproblem of the general problem of synthesis of the most favorable part surface generation process. The solution to the problem of optimal global part surface generation process is based much on the derived solutions to the problems of the optimal local and optimal regional part surface generation processes.

Minimal machining time is the major goal of the problem of synthesis of the most favorable global part surface generation process. To solve the problem under consideration, it is necessary to do the following:

- Minimize interference of the neighboring tool paths of the cutting tool over the part surface being machined.
- Determine the optimal parameters of placing the cutting tool into contact with the part surface, and of its departing from the contact. This subproblem is referred to as the *boundary problem* of the part surface generation process.
- Determine the location of the optimal start point of the part surface machined.

These subproblems are discussed in detail below.

10.3.1 Minimization of Partial Interference of the Neighboring Tool Paths

The actually machined part surface is represented as a set of tool paths that cover the nominal part surface P. At a current surface point, the width of the tool path is equal to the side step $\breve{F}_{ss}$ calculated at that same CC point (see Equation 9.51). The tool path width varies along the trajectory of the CC point over the sculptured part surface as well as across the trajectory. Because of this, neighboring tool paths partially interfere with each other. Ultimately, some portions of part surface P are double-covered by the tool paths. Partial interference of the neighboring tool paths causes a reduction of the part surface generation output. For the synthesis of the most favorable part surface generation operation, the interference of the neighboring tool paths must be minimized.

The trajectory of the CC point over the sculptured part surface is a three-dimensional curve. For the analysis below, it is convenient to operate with the natural parameterization of the trajectory:

$$l_{tr} = l_{tr}(r_{tr}, \tau_{tr}). \tag{10.21}$$

Here, the length l_{tr} of arc of the trajectory measured from a certain point within the trajectory. The length l_{tr} is expressed in terms of radius of curvature r_{tr} at a current trajectory point and of torsion τ_{tr} of the trajectory at that same point.

At the current CC point, the tool path width can be expressed in terms of length of the ith trajectory:

$$\breve{F}_{ss}^{(i)} = \breve{F}_{ss}^{(i)}\left[l_{tr}^{(i)}\right] \tag{10.22}$$

During the infinitesimal time dt, the cutting tool travels along the ith trajectory at a distance dl_{th}. A portion of the sculptured part surface is generated in this motion of the cutting tool:

$$dS_{tr}^{(i)} = \breve{F}_{ss}^{(i)}\left[l_{tr}^{(i)}\right] \cdot dt \tag{10.23}$$

Area of a single ith tool path is

$$S_{tr}^{(i)} = \int\limits_{\left[l_{tr}^{(i)}\right]} \breve{F}_{ss}^{(i)}\left[l_{tr}^{(i)}\right] \cdot dt \tag{10.24}$$

The total area of all the tool paths can be computed from the formula

$$S_{tr} = \sum_{i=1}^{n} S_{tr}^{(i)} = \sum_{i=1}^{n} \int_{\left[l_{tr}^{(i)}\right]} \breve{F}_{ss}^{(i)} \left[l_{tr}^{(i)}\right] \cdot dt \tag{10.25}$$

where n denotes the total number of tool paths that are necessary to cover the entire part surface.

Due to partial interference of the neighboring tool paths, the total area S_{tr} (Equation 10.25) exceeds the area S_{sg} of the actually generated part surface P (i.e., the inequality $S_{tr} > S_{sg}$ is always observed). The degree of interference of the neighboring tool paths is evaluated by the interference factor K_{int}:

$$K_{int} = \frac{S_{tr} - S_{sg}}{S_{sg}} = \frac{\sum_{i=1}^{n} \int_{\left[l_{tr}^{(i)}\right]} \breve{F}_{ss}^{(i)} \left[l_{tr}^{(i)}\right] \cdot dt - S_{sg}}{S_{sg}} \tag{10.26}$$

The interference factor K_{int} is a function of design parameters of the part surface being machined, of design parameters of the generating surface of the cutting tool, and of parameters of kinematics of the part surface machining operation. Thus, it can be minimized ($K_{int} \rightarrow$ min) using for this purpose conventional methods of minimization of analytical functions.

10.3.2 Solution to the Boundary Problem

The generation of a part surface within the area next to the surface border differs from that when machining of boundless surfaces is investigated. The shape and parameters of the part surface contour affects the efficiency of the part surface generation process. Before searching for a solution to the boundary problem, it is necessary to determine the part surface region within which the boundary effect is significant.

Consider a sculpture part surface P having tolerance $[h]$ for the accuracy of the surface machining. The surface of tolerance $S_{[h]}$ is at distance $[h]$ from part surface P. The actual machined part surface is located within the interior between surfaces P and $S_{[h]}$.

The generating surface T of the form-cutting tool is contacting the nominal part surface P at a certain point K_1 (Figure 10.4). The cutting tool surface T intersects the surface of tolerance $S_{[h]}$. Line 1 of intersection of the surfaces $S_{[h]}$ and T is a type of a closed ellipse-like curve. The curve has no common points with the part surface boundary. Therefore, no boundary effect is observed in this location of the cutting tool.

At point K_2 within the trajectory of CC point, the cutting tool surface T also intersects the surface of tolerance $S_{[h]}$. Line 2 of intersection is also a type of closed ellipse-like curve. However, in this location of the cutting tool, curve 2

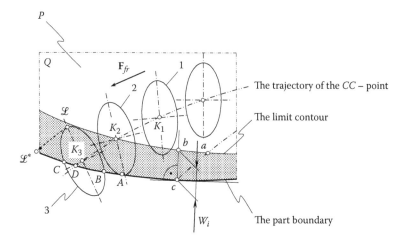

FIGURE 10.4
Boundary effect when machining a sculptured part surface P on a multiaxis numerical control machine.

becomes a tangent with the part surface boundary at point A. This indicates that starting at point K_2, the boundary affects the efficiency of part surface generation. The impact of the boundary is getting stronger toward point D on the part surface boundary curve.

At a certain point K_3 of the trajectory of CC point, the line of intersection 3 of surfaces $S_{[h]}$ and T is not a closed line. It intersects the part surface boundary at points B and C.

The departure of the cutting tool from the interaction with part surface P is over when the limit point L on the biggest diameter of curve 3 reaches the part surface boundary curve at point $\mathcal{L}^*$.

Point K_2 is constructed for point A of the part surface boundary curve. For every point A_i of the part surface boundary curve, point K_i, which is similar to point K_2, can be constructed. All points K_i specify the *limit contour* on part surface P. It is necessary to take into account the impact of the boundary effect for those arcs of CC point trajectories that are located in between the part surface boundary curve and the limit contour.

Width bc of the part surface boundary-affected region is not constant. Width W_i at a current point c is measured along the perpendicular to the part surface boundary curve. Point c is the end-point of the arc ac of the trajectory of CC point. The feed rate per tooth $\tilde{F}_{fr}$ of the cutting tool could be either constant within the arc ac of the trajectory of CC point or it can vary in compliance with current width of the tool path.

Particular features of the impact of the boundary effect could be observed:

- When the stock thickness is bigger, this causes longer trajectories of the cutting tool to enter in contact with the part surface.

- A bigger tolerance [*h*] for accuracy of the machined part surface results in longer trajectories of the cutting tool to exit from contact with the part surface.
- The smaller the area of the nominal part surface *P*, the more significant is the impact of the boundary effect on the efficiency of the machining operation.
- The impact of the boundary effect could be more significant when machining *long* surfaces.

The above-listed features must be encountered when solving the problem of global synthesis of the most favorable process of sculptured part surface machining on a multiaxis *NC* machine.

10.3.3 Optimal Location of the Starting Point

The location of the point from which machining of the sculptured part surface begins also affects the resultant part surface generation output. One can conclude from this that the most favorable location of the starting point exists and it can be determined.

Consider the machining of a sculptured part surface on a multiaxis *NC* machine as schematically illustrated in Figure 10.5. The boundary of the

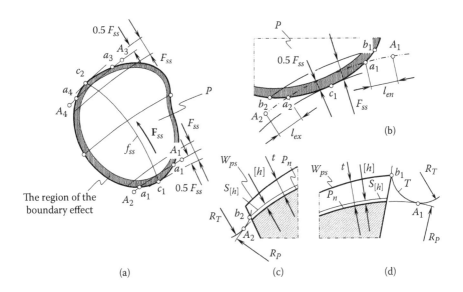

FIGURE 10.5
Location of the optimal start point for sculptured part surface machining on a multiaxis numerical control machine.

sculptured part surface P has an arbitrary shape. The region of the boundary effect is shown as the shaded strip along the boundary curve.

Part surface P can be covered by the infinite number of the optimal trajectories of the CC point. The equation of the most favorable trajectories of the CC point is the output of the subproblem of synthesis of the optimal regional part surface generation process. Two of the infinite number of trajectories are tangents to the sculptured part surface boundary curve at points c_1 and c_2 (see Figure 10.5).

Another two optimal trajectories of the CC point are at a distance $0.5F_{ss}$ from points c_1 and c_2 inward the bounded portion of the sculptured part surface P. These two last trajectories of the CC point can be used as the trajectories for the actual tool paths when machining the sculptured part surface P. They intersect the sculptured surface boundary curve at the points a_1, a_2 and a_3, a_4 respectively. The rest of the trajectories of the CC point are at the limit side-step $[\tilde{F}_{ss}]$ from each other (see Equation 9.51). It is important to point out here that length f_{ss} of the arc through points c_1 and c_2 usually is not divisible on the limit side step $[\tilde{F}_{ss}]$. However, no big problem arises in this concern, and it can be neglected at this point.

Then, outside the bounded portion of the sculptured part surface P, two points A_1 and A_2 are selected within the trajectory through points a_1 and a_2. Point A_1 is at distance l_{en} from the boundary curve of the part surface P. The l_{en} distance is sufficient for entering of the cutting tool in contact with the part surface. Another point A_2 is at a distance l_{ex} from the boundary curve of the part surface P. The l_{ex} distance is sufficient for the exiting of the cutting tool from contact with the part surface. Similarly, two more points, A_3 and A_4, are selected within the trajectory through points a_3 and a_4. The machining of part surface P begins at point A_1.

In most cases of sculptured part surface machining, the inequality $l_{en} > l_{ex}$ is observed. Therefore, if one wishes to begin the part surface machining not from point A_1 but from the opposite end of the trajectory a_1a_2, another four points A_1^*, A_2^*, A_3^*, and A_4^* (points A_1^*, A_2^*, A_3^*, and A_4^* are not shown in Figure 10.5) can be constructed instead. Points A_1^*, A_2^*, A_3^*, and A_4^* are constructed in the same manner as points A_1, A_2, A_3, and A_4 are constructed. The only difference here is that for all points A_1^*, A_2^*, A_3^*, and A_4^*, the arc segments of the length l_{en} are substituted with the arc segments of the length l_{ex} and vice versa.

When located at point A_1 of the trajectory a_1a_2, the generating surface T of the cutting tool contacts the work-piece surface W_{ps} at point b_1. The work-piece surface W_{ps} is an offset surface at distance t to part surface P. Here t designates the thickness of the stock to be removed. In the general case, a function $t = t(U_P, V_P)$ is observed (see Chapter 9).

When located at point A_2 of the trajectory a_1a_2, the generating surface T of the cutting tool contacts the surface of tolerance $S_{[h]}$ at a point b_2. The surface of tolerance $S_{[h]}$ is an offset surface at distance $[h]$ to part surface P. Here $[h]$

denotes the tolerance for accuracy of the machined part surface P. In the general case, a function $[h] = [h](U_P, V_P)$ is observed (see Chapter 9).

The distance l_{en} that is necessary for entering of the cutting tool in contact with the sculptured part surface can be expressed in terms of thickness of the stock t, radius of normal curvature R_T of the generating surface T of the cutting tool (here R_T is measured in the direction tangent to the trajectory a_1a_2 at point A_1), and radius of curvature R_{tr} of the trajectory at point A_1 through points a_1 and a_2.

The distance l_{ex} that is necessary for the exiting of the cutting tool from contact with the sculptured part surface can be expressed in terms of tolerance $[h]$ for accuracy of the part surface, radius of normal curvature R_T of the generating surface T of the cutting tool (here R_T is measured in the direction tangent to the trajectory a_1a_2 at point A_2), and radius of curvature R_{tr} of the trajectory at point A_2 through points a_1 and a_2.

Ultimately, any one of four points A_1, A_2, A_3, A_4 or one of four points A_1^*, A_2^*, A_3^*, A_4^* is selected as the starting point of the sculptured part surface machining. Practically, both sets of points are equivalent. The computation of the coordinates of the chosen point is a trivial mathematical procedure, and the interested reader may wish to exercise himself or herself in this.

Prior to machining of the given part surface, the coordinates of a contact point within the generating surface T of the cutting tool are computed. This is the point local geometry of the cutting tool surface T that corresponds to the local geometry of part surface P at point a_1. Then, the cutting tool contact point is snapped with the computed starting point, say, with point A_1. The conditions of proper part surface generation (see Chapter 7) must be satisfied.

Much room for investigation is left in the synthesis of the most favorable global part surface generation.

10.4 Rational Reparameterization of the Part Surface

The solution to the problem of optimal regional synthesis of part surface generation returns a set of optimal trajectories of the CC point on part surface P. For the purposes of the development of a computer program for sculptured part surface machining on a multiaxis NC machine, it is convenient to use the computed optimal trajectories as a set of curvilinear coordinates on the sculptured part surface P. For this purpose, it is necessary to change the initial parameterization of the surface P with a new parameterization—with the parameterization by means of the most favorable trajectories of CC point on part surface P. For the reparameterization of surface P, known methods [1,7,11,13] and others can be used.

10.4.1 Transformation of Surface Parameters

Consider a part surface P that is given by vector equation $\mathbf{r}_P = \mathbf{r}_P(U_P, V_P)$. It is assumed that surface P is a smooth regular surface. The required additional restrictions that must be imposed will be introduced later.

The initial (U_P, V_P) parameterization of the part surface can be transformed. The new parameterization of part surface P is denoted as (U_P^*, V_P^*). In the new parameters, the initial equation of part surface P is substituted with the equivalent equation $\mathbf{r}_P = \mathbf{r}_P(U_P^*, V_P^*)$. The new parameters U_P^* and V_P^* can be expressed in terms of the original parameters U_P and V_P:

$$U_P^* = U_P^*(U_P, V_P) \tag{10.27}$$

$$V_P^* = V_P^*(U_P, V_P) \tag{10.28}$$

One of the curvilinear parameters in Equations 10.27 and 10.28 (for example, U_P^* coordinate curve) can be congruent to the most favorable trajectories of CC point (see Equation 10.15), whereas another curvilinear parameter V_P^* can be directed orthogonally to the first one.

The equations for the derivatives in the new parameters are as follows:

$$\frac{\partial \mathbf{r}_P}{\partial U_P^*} = \frac{\partial \mathbf{r}_P}{\partial U_P} \cdot \frac{\partial U_P}{\partial U_P^*} + \frac{\partial \mathbf{r}_P}{\partial V_P} \cdot \frac{\partial V_P}{\partial U_P^*} \tag{10.29}$$

$$\frac{\partial \mathbf{r}_P}{\partial V_P^*} = \frac{\partial \mathbf{r}_P}{\partial U_P} \cdot \frac{\partial U_P}{\partial V_P^*} + \frac{\partial \mathbf{r}_P}{\partial V_P} \cdot \frac{\partial V_P}{\partial V_P^*} \tag{10.30}$$

The cross product of tangents is equal:

$$\frac{\partial \mathbf{r}_P}{\partial U_P^*} \times \frac{\partial \mathbf{r}_P}{\partial V_P^*} = \begin{bmatrix} U_P & V_P \\ U_P^* & V_P^* \end{bmatrix} \cdot \left(\frac{\partial \mathbf{r}_P}{\partial U_P} \times \frac{\partial \mathbf{r}_P}{\partial V_P} \right) \tag{10.31}$$

To satisfy the restriction

$$\frac{\partial \mathbf{r}_P}{\partial U_P^*} \times \frac{\partial \mathbf{r}_P}{\partial V_P^*} \neq 0 \tag{10.32}$$

for part surface P expressed in the new parameters, the Jacobian matrix of transformation J must not be equal to zero:

$$J = \begin{bmatrix} U_P & V_P \\ U_P^* & V_P^* \end{bmatrix} = \begin{vmatrix} \dfrac{\partial U_P}{\partial U_P^*} & \dfrac{\partial U_P}{\partial V_P^*} \\ \dfrac{\partial V_P}{\partial U_P^*} & \dfrac{\partial V_P}{\partial V_P^*} \end{vmatrix} \neq 0 \qquad (10.33)$$

The matrix $[D_P]$ of the first derivatives of part surface P in its original parameterization is

$$[D_P] = \begin{bmatrix} \dfrac{\partial \mathbf{r}_P}{\partial U_P} \; ; & \dfrac{\partial \mathbf{r}_P}{\partial V_P} \end{bmatrix} \qquad (10.34)$$

The similar matrix $\left[D_P^* \right]$ can be composed for the new parameterization of part surface P:

$$\left[D_P^* \right] = \begin{bmatrix} \dfrac{\partial \mathbf{r}_P}{\partial U_P^*} \; ; & \dfrac{\partial \mathbf{r}_P}{\partial V_P^*} \end{bmatrix} \qquad (10.35)$$

The following equality is true:

$$\left[D_P^* \right] = [D_P] \cdot J \qquad (10.36)$$

Matrices $[D_P]$ and $\left[D_P^* \right]$ enable the calculation of the first fundamental matrix $\left[\Phi_{1.P}^* \right]$ in the new parameters of part surface P:

$$\left[\Phi_{1.P}^* \right] = \left[D_P^* \right]^T \cdot [D_P^*] = J^T \cdot [D_P]^T \cdot [D_P] \cdot J = J^T \cdot [\Phi_{1.P}] \cdot J \qquad (10.37)$$

Similarly, the equation for the second fundamental matrix $\left[\Phi_{2.P}^* \right]$ in the new parameters of the part surface P can be derived:

$$\left[\Phi_{2.P}^* \right] = J^T \cdot [\Phi_{2.P}] \cdot J \qquad (10.38)$$

The discriminant of the first order H_P^* is computed from

$$H_P^* = J \cdot H_P \qquad (10.39)$$

The same is true with respect to the discriminant of the second order T_P^*:

$$T_P^* = J \cdot T_P \tag{10.40}$$

The rest of the major parameters of geometry in the new parameterization of part surface P can be calculated on the premise of the above discussed equations, particularly Equations 10.37 and 10.38.

10.4.2 Transformation of Surface Parameters in Connection with the Surface Boundary Contour

As an example, boundary contour C of the sculptured part surface P is made up of four smooth arcs c_{11}, c_{12}, c_{21}, and c_{22} (Figure 10.6). The plane P_0 serves as a coordinate plane. Consider a certain region Ω within part surface P. The region Ω_0 is the projection of Ω onto the coordinate plane P_0. The distance H of P to P_0 is measured in the direction of the unit normal vector $\mathbf{n}_0$ to P_0.

In a certain reference system, the region Ω_0 within the plane P_0 is represented by

$$\mathbf{r}^f = \mathbf{r}^f (\alpha^i) \tag{10.41}$$

The projection Ω_0 is not a canonical region: the contour lines C_{ij}^0 of the Ω_0 do not align with the coordinate lines $\alpha^i = Const.$

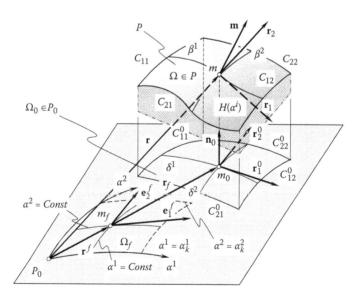

FIGURE 10.6
Transformation of parameters in connection with the boundary contour of the sculptured part surface.

The problem of the reparameterization of part surface P can be solved in two steps. Following this method, the solution to the problem would be represented as the superposition of the two consequent mappings.

On the first step, a canonical region Ω_f is constructed within the reference plane P_0. The region is bounded by the coordinate curves $\alpha^i = \alpha_k^i$ (Figure 10.6). For this purpose, the following equation is used for mapping of the region onto the region Ω_0 [2]:

$$\mathbf{r}(\alpha^i) = \mathbf{r}^f(\alpha^i) + F_k(\alpha^i)\mathbf{r}_f^k = \mathbf{r}^f(\alpha^i) + F^k(\alpha^i)\mathbf{r}_k^f \qquad (10.42)$$

where $\mathbf{r}_k^f$ and $\mathbf{r}_f^k$ are the covariant and contravariant basis vectors on part surface P and F_k and F^k are the covariant and contravariant components of the vector of the fictive displacements.

Components F_k and F^k in Equation 10.42 must be constructed based on the requirements of one-to-one correspondence between the contours.

After the necessary formulae transformations are accomplished, the following equation can be obtained:

$$\mathbf{r}^0(\alpha^1, \alpha^2) = \mathbf{r}^f(\alpha^1, \alpha^2) + F_i(\alpha^1, \alpha^2)\mathbf{r}_f^i \qquad (10.43)$$

where $\mathbf{r}^f(\alpha^i)$ is the position vector of a point m_f that is mapped into point m_0 having position vector $\mathbf{r}^0$, $\mathbf{r}_f^i$ are the reciprocal basis vectors at point m_f, F_1 and F_2 are the components of the vector of the fictive displacements m_f that can be constructed based on the shape of the region Ω_f.

At every point of the region Ω_0, the constructed functions F_i together with Equation 10.41 yield the computation of the following:

(a) Position vector $\mathbf{r}_i$:

$$\mathbf{r}_i = \left(\delta_i^k + e_i^{fk}\right)\mathbf{r}_k^f = \left(a_{ik} + e_{ik}^f\right)\mathbf{r}_f^k \qquad (10.44)$$

(b) Major basis vectors $\mathbf{r}_i^0$ at point m_0:

$$\mathbf{r}_i^0 = \left(\delta_i^k + e_i^{fk}\right)\mathbf{r}_k^f = \left(a_{ik}^f + e_{ik}^f\right)\mathbf{r}_f^k \qquad (10.45)$$

These vectors are tangents to the coordinate curves specified by the mapping (see Equation 10.43). In Figure 10.6, they are designated by lines $\delta^i = Const$.

(c) Covariant components of the first metric tensor:

$$\Phi_{1.P} \quad \Rightarrow \quad a_{ik}^0 = a_{ik}^f + 2\varepsilon_{ik}^f \qquad (10.46)$$

(d) Christoffel symbols of the second type at point m_0:

$$\Gamma_{ij}^{0k} = \Gamma_{ij}^{fk} + A_{ij}^{fk} \tag{10.47}$$

In Equation 10.37, parameters e_i^{fk} and e_{ik}^{f} can be calculated by formulae

$$e_i^{fk} = \nabla_i^{f} F^k \tag{10.48}$$

$$e_{if}^{f} = \nabla_i^{f} F_k \tag{10.49}$$

For the calculation of the parameter $2\varepsilon_{ik}^{f}$, the following formula is used:

$$2\varepsilon_{ik}^{f} = \mathbf{r}_i \cdot \mathbf{r}_k - \mathbf{r}_i^{f} \cdot \mathbf{r}_k^{f} = e_{ik}^{f} + e_{ki}^{f} + a_{f}^{js} e_{ij}^{f} e_{ks}^{f} \tag{10.50}$$

Ultimately, the parameter A_{ik}^{0j} is calculated from

$$A_{ik}^{0j} = a_0^{jn} P_{n,ik}^{f} \tag{10.51}$$

Here, for the case under consideration, the following equalities are valid:

$$a_0^{11} = \frac{a_{22}^{0}}{a_0} \tag{10.52}$$

$$a_0^{22} = \frac{a_{11}^{0}}{a_0} \tag{10.53}$$

$$a_0^{12} = -\frac{a_{12}^{0}}{a_0} \tag{10.54}$$

$$a_0 = a_{11}^{0} a_{22}^{0} - \left(a_{12}^{0}\right)^2 \tag{10.55}$$

$$P_{j,ik}^{f} = \nabla_i^{f} \varepsilon_{jk}^{f} + \nabla_k^{f} \varepsilon_{ij}^{f} - \nabla_j^{f} \varepsilon_{ik}^{f} \tag{10.56}$$

On the second step, the region Ω_0 (see Equation 10.43) is mapped onto the sculptured part surface P. For the mapping, the following vector equality is used:

$$\mathbf{r}(\alpha^i) = \mathbf{r}^0(\alpha^i) + H(\alpha^i)\,\mathbf{n}_0 \tag{10.57}$$

In this case, with the help of Equation 10.43, Equation 10.57 casts into

$$\mathbf{r}(\alpha^i) = \mathbf{r}^\Phi(\alpha^i) + F_\kappa(\alpha^i)\mathbf{r}_\Phi^\kappa + H(\alpha^i)\mathbf{n}_0 \tag{10.58}$$

Point m_0 is the projection of point m onto the reference plane P_0. Therefore, Gaussian coordinates of a certain point m_f serve as the Gaussian coordinates of point m (see Equation 10.43).

In region m_0, for the basic vectors given by Equation 10.45 as well as for $\mathbf{r}_0^i = a_0^{ik}\mathbf{r}_k^0$, the following two equalities are observed:

$$\mathbf{r}_{ik}^0 = \Gamma_{ik}^{0j}\mathbf{r}_j^0 \tag{10.59}$$

$$\mathbf{r}_k^{0i} = -\Gamma_{kj}^{0i}\mathbf{r}_0^j \tag{10.60}$$

The Christoffel symbols Γ_{ik}^{0j} in Equations 10.59 and 10.60 were determined earlier (see Equation 10.47).

If parameters α^1 and α^2 in Equation 10.41 determine two families of orthogonal coordinate curves within the plane P_0, then not Equation 10.43 but the following equality is used instead:

$$\mathbf{r}^0(\alpha^i) = \mathbf{r}^f(V) + F_1(\alpha^i)\mathbf{e}_1^f + F_2(\alpha^i)\mathbf{e}_2^f \tag{10.61}$$

Here, $\mathbf{r}_1^f$ and $\mathbf{r}_2^f$ are the unit vectors of the coordinate curves $\alpha^1 = Const$ and $\alpha^2 = Const$ at point m_f.

Under such a scenario, Equation 10.45, through Equation 10.56, is not used; instead, the following relationships are used:

$$\mathbf{r}_i^0 = A_i^f \left(\delta_{is} + e_{is}^f\right)\mathbf{e}_s^f \tag{10.62}$$

$$a_{ik}^0 = A_i^f A_k^f \left(\delta_{ki} + 2\varepsilon_{ki}^f\right) \tag{10.63}$$

$$2\varepsilon_{ik}^f = e_{ik}^f + e_{ki}^f + e_{is}^f e_{ks}^f \tag{10.64}$$

$$\mathbf{r}_0^1 = \frac{\left[\left(1+2\varepsilon_{22}^f\right)\left(1+e_{11}^f\right) - 2\varepsilon_{12}^f e_{21}^f\right]\mathbf{e}_1^f + \left[\left(1+2\varepsilon_{12}^f\right)e_{12}^f - 2\varepsilon_{12}^f\left(1+e_{22}^f\right)\right]\mathbf{e}_2^f}{A_1^f\left[\left(1+e_{11}^f\right) - e_{12}^f e_{21}^f\right]^2} \tag{10.65}$$

The following relationships used in Equations 10.45 through 10.56 are still valid:

$$\mathbf{r}_i^0 = A_i^f \left(\delta_{is} + e_{is}^f \right) \mathbf{e}_s^f \tag{10.66}$$

$$a_{ik}^0 = A_i^f A_k^f \left(\delta_{ki} + 2\varepsilon_{ki}^f \right) \tag{10.67}$$

$$2\varepsilon_{ik}^f = e_{ik}^f + e_{ki}^f + e_{is}^f e_{ks}^f \tag{10.68}$$

$$\mathbf{r}_0^1 = \frac{\left[\left(1 + 2\varepsilon_{22}^f \right)\left(1 + e_{11}^f \right) - 2\varepsilon_{12}^f e_{21}^f \right] \mathbf{e}_1^f + \left[\left(1 + 2\varepsilon_{12}^f \right) e_{12}^f - 2\varepsilon_{12}^f \left(1 + e_{22}^f \right) \right] \mathbf{e}_2^f}{A_1^f \left[\left(1 + e_{11}^f \right) - e_{12}^f e_{21}^f \right]^2} \tag{10.69}$$

Again, in these equations, the following equalities are still valid:

$$e_{11}^f = \frac{F_{1,1}}{A_1^f} + \frac{F_2}{A_1^f A_2^f} A_{1,2}^f \tag{10.70}$$

$$e_{12}^f = \frac{F_{2,1}}{A_1^f} - \frac{F_1}{A_1^f A_2^f} A_{1,2}^f \cdot \overrightarrow{1,2} \tag{10.71}$$

$$a_0 = \left[\left(1 + e_{11}^f \right) - e_{12}^f e_{21}^f \right]^2 \left(A_1^f A_2^f \right)^2 \tag{10.72}$$

In Equations 10.59 and 10.60, Christoffel symbols are calculated using the expanded formulae:

$$a\Gamma_{11}^1 = \frac{a_{22} a_{11,1}}{2} - a_{12} \left(a_{12,1} - \frac{a_{11,2}}{2} \right) \tag{10.73}$$

$$a\Gamma_{12}^1 - \frac{a_{22} a_{11,2} - a_{12} a_{22,1}}{2} \tag{10.74}$$

$$a\Gamma_{22}^1 = a_{22} \left(a_{12,2} - \frac{a_{22,1}}{2} \right) - \frac{a_{12} a_{22,2}}{2} \cdot \overrightarrow{1,2} \tag{10.75}$$

For the calculations, the expressions (see Equation 10.61) are substituted into the last equations.

10.5 Possibility of the *DG/K*-Based *CAD/CAM* System for Optimal Sculptured Part Surface Machining

The machining of a sculptured part surface on a multiaxis *NC* machine is a challenging engineering problem. The implementation of the *DG/K* method of part surface generation makes it possible to develop a *CAD/CAM* system for the most favorable sculptured part surface machining on a multiaxis *NC* machine.

10.6 Major Blocks of the *DG/K*-Based *CAD/CAM* System

The proposed concept of the *DG/K*-based *CAD/CAM* system for the most favorable sculptured part surface machining on a multiaxis *NC* machine is composed of seven major parts. In Figure 10.7, the major parts are depicted as blocks of the *DG/K*-based *CAD/CAM* system.

Only data on part surface geometry are used as the input information for the functioning of the *DG/K*-based *CAD/CAM* system.

The synthesis of the most favorable machining operation begins from analytical description of the sculptured part surface being machined (I). The initially given representation of the sculptured part surface *P* is converting to the natural parameterization of the surface *P*, when part surface *P* is expressed in terms of the first $\Phi_{1.P}$ and the second $\Phi_{2.P}$ fundamental forms (see Chapter 1). The converted analytical representation of the sculptured part surface is used for solution (II) to the problem of optimal orientation of part surface *P* on the worktable of the *NC* machine (see Chapter 7). It also yields a solution to the problem of designing the most favorable cutting tool (see Chapter 5) for machining the sculptured part surface *P* (III). To solve the above problems, the analytical description of the contact geometry between the sculptured part surface *P* and the generating surface *T* of the form-cutting tool is employed (see Chapter 4).

Further, the implementation of the analytical description of the contact geometry of part surface *P* and generating surface *T* of the form-cutting tool enables the computation of the most favorable parameters of kinematics of sculptured part surface machining. Ultimately, this makes a closed-form solution (IV) to the problem of optimal tool path generation, computation of coordinates of the optimal starting point for part surface machining, and verification of satisfaction or violation of the necessary conditions of proper part surface generation (V) possible. The cutting tool for machining of the sculptured part surface can be either designed or chosen (VI) within the available cutting tools.

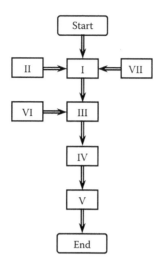

FIGURE 10.7
Principal blocks of the *DG/K*-based *CAD/CAM* system for optimal sculptured part surface machining on a multiaxis numerical control machine.

In particular cases of sculptured part surface machining, maintaining the desired type of contact of surfaces P and T is allowed (VII) and their optimal contact is not mandatory.

Finally, when the optimization is accomplished, the shortest possible machining time of sculptured part surface machining as well as the lowest possible cost of the machining operation can be achieved.

More detail about the major blocks of the *DG/K*-based *CAD/CAM* system (see Figure 10.7) are disclosed in the following sections.

10.6.1 Representation of the Input Data

A sculptured part surface P to be machined is initially represented either analytically or discretely. For the application of the *DG/K* method of part surface generation, the initial representation of part surface P converts to the natural representation in terms of the fundamental magnitudes E_P, F_P, G_P of the first order $\Phi_{1.P}$ and of the fundamental magnitudes L_P, M_P, N_P of the second order $\Phi_{2.P}$. The conversion of the initial part surface P representation to its representation in the natural form is performed by the first block (I) of the *CAD/CAM* system (Figure 10.7).

To represent part surface P analytically, the computation of the fundamental magnitudes of the first-order, $\Phi_{1.P}$, and second-order, $\Phi_{2.P}$ turns to a routing mathematical procedure (see Chapter 1). The sequence of the required steps of computation is as follows (Figure 10.8):

$$(1)\rightarrow(2)\rightarrow(3)\rightarrow(4)\rightarrow(5)\rightarrow(6)\rightarrow\ldots \tag{10.76}$$

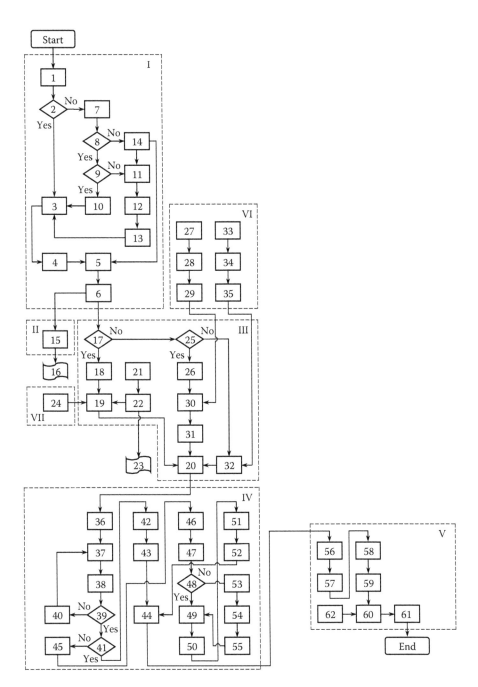

FIGURE 10.8
The generalized flowchart of the *DG/K*-based *CAD/CAM* system for optimal sculptured part surface machining on a multiaxis numerical control machine.

In case of discrete representation (7), part surface P is approximated either by a common analytical function

$$\ldots \rightarrow (7) \rightarrow (8) \rightarrow (10) \rightarrow (3) \rightarrow (4) \rightarrow (5) \rightarrow (6) \rightarrow \ldots \qquad (10.77)$$

or piecewise approximation

$$\ldots \rightarrow (7) \rightarrow (9) \rightarrow (10) \rightarrow (11) \rightarrow (12) \rightarrow (13) \rightarrow (3) \rightarrow (4) \rightarrow (5) \rightarrow (6) \rightarrow \ldots \qquad (10.78)$$

is implemented.

Fundamental magnitudes of the first order and the second order of part surface P can be determined directly from the specification of the surface P in discrete form (8). Following this approach, the determined partial derivatives (14) are used as input to

$$\ldots \rightarrow (8) \rightarrow (14) \rightarrow (5) \rightarrow (6) \rightarrow \ldots \qquad (10.79)$$

(The interested reader may wish to consult Reference [11] for more information on this concern.)

The derived natural representation of part surface P is the output of the first block (1) of the *DG/K*-based *CAD/CAM* system (Figure 10.7). It is used below as the input to the cutting tool block (III).

10.6.2 Optimal Work-Piece Configuration

The computed (6) values of the fundamental magnitudes of the first order $\Phi_{1.P}$ and the second order $\Phi_{2.P}$ are used (II) for the computation (15) of parameters of the optimal orientation of part surface P on the worktable of a multiaxis *NC* machine. For the optimization, the approach earlier developed by the author [3,5,11,16,17] is used (see Chapter 7). The computed parameters of optimal orientation of the work-piece on the worktable of the multiaxis *NC* machine are the output (16) of this subsystem of *CAD/CAM* system (Figure 10.8).

10.6.3 Optimal Design of the Form-Cutting Tool

The design parameters of the form-cutting tool for the optimal machining of part surface P on a multiaxis *NC* machine are computed in the block (III). Again, the criterion of the optimization is the lowest possible cost of the machining operation. The key problem at that point is to determine the geometry of the generating surface T of the cutting tool [6,8,14,15].

The equation of the cutting tool surface T is derived in natural parameterization—that is, in terms of the fundamental magnitudes E_T, F_T, G_T of the first order $\Phi_{1.T}$ and the fundamental magnitudes L_T, M_T, N_T of the second order $\Phi_{1.T}$ of surface T. A method for computation of geometry of the cutting tool surface T is disclosed in Chapter 5. Shown in Figure 10.7, the cutting tool block (III) utilizes the technique briefly disclosed there. The computation of the parameters of the design of the most favorable form-cutting tool for machining a given sculptured part surface P encounters principal steps (17) through (24).

The cutting tool block (III) yields the selection of a design of a form-cutting tool that best fits the requirements of machining of the given part surface P [the principal steps (27) through (29) in Figure 10.8]. Ultimately, the machining of a given sculptured part surface P can be performed by a given form-cutting tool T:

$$\ldots \to (25) \to (26) \to (30) \to (31) \to (20) \to \ldots \qquad (10.80)$$

Under such a scenario, the parameters of design of the form-cutting tool that are required for further computations can be computed following the routing

$$\ldots \to (33) \to (34) \to (35) \to (32) \to (20) \to \ldots \qquad (10.81)$$

The computed parameters of design of the form-cutting tool for machining a given sculptured part surface P are the output (20) of the cutting tool block (III).

10.6.4 Optimal Tool Paths for Sculptured Part Surface Machining

The major purpose of the block (IV) is to compute the parameters of the optimal tool paths for shortest time of machining of a sculptured part surface on a multiaxis NC machine.

For this purpose, unit normal vectors $\mathbf{n}_P$ and $\mathbf{n}_T$ to part surface P and generating surface T of the form-cutting tool are computed in (36). Then, the contact of surfaces P and T is numerically simulated in (37). At the CC point K, unit normal vectors $\mathbf{n}_P$ and $\mathbf{n}_T$ are directed opposite to each other. Further, the verification of satisfaction or violation of the necessary conditions of proper part surface generation [9,11] is performed (38). If the necessary conditions of proper part surface generation are violated (39), an appropriate correction of the cutting tool configuration relative to the work is performed in (40).

After the necessary conditions of proper part surface generation (39) are satisfied and verified, the machining of the sculptured surface is performed (41) along the tool paths of the widest possible width. This is because the parameters of the optimal tool paths are computed (42) as the curves on part surface P that are equidistant (parallel) to the longest geodesic curve on the surface P. Aiming for the best results of the computations, the boundary conditions could be incorporated (43) into the procedure of the computation. Ultimately, the most favorable parameters of tool paths transfer to (44), which is the output of the block (IV).

Usually, the sculptured part surface P and the generating surface T of the cutting tool make point contact (41). This type of sculptured part surface machining is the most commonly used in various industries. Under such a scenario, for the computation of the optimal tool paths, the indicatrix of conformity $Cnf_R(P/T)$ [11,12,16] at a point of contact of surfaces P and T at the CC point is implemented.

After the parameters of the indicatrix of conformity $Cnf_R(P/T)$ are calculated (45), the cutting tool configuration could be optimized. For this purpose, the orientation motions of the first and second types are performed by the cutting tool.

The orientation motion of the second type of cutting tool results in the point, at which the local topology of the cutting tool surface T is the optimal, is put into contact (46) with the current CC point of part surface P. The orientation motion of the first type of cutting tool yields an optimization (47) of the local orientation of surfaces P and T, at which the minimal diameter d_{cnf}^{min} of the indicatrix of conformity $Cnf_R(P/T)$ becomes the smallest possible.

For grinding sculptured part surfaces when the cutting tool represents (48) the entire generating surface T, the characteristic curve $Cnf_R(P/T)$ yields the computation (49) of the optimal direction of tool paths at every point of part surface P. Further, the computation (50) of the most favorable tool paths is made possible.

A closed-form solution to the problem of optimal tool path generation for sculptured part surface machining on a multiaxis NC machine is developed [4]. Following the derived solution to the problem, the optimal tool paths are directed orthogonally to the minimal diameter d_{cnf}^{min} of the characteristic curve $Cnf_R(P/T)$ at every CC point. Thus, the indicatrix of conformity yields directions tangential to the optimal tool paths. Further, elementary integration returns the equation of the optimal (the most favorable) tool paths for sculptured part surface machining on a multiaxis NC machine.

Without going into detail of the derived solution to the problem, the final equation of the optimal tool paths for sculptured part surface machining is represented in matrix form (see Equation 10.15).

Targeting the computation of the most accurate solution to the problem of synthesis of the most favorable machining operation, the following factors can be incorporated into the computations: the boundary effect (51), partial interference of the neighboring tool paths (52), and so forth. Finally, the computed parameters of the optimal tool paths represent (44) the output of the block (IV).

When the sculptured part surface P is machined with an edge-cutting tool (for example, by milling cutter), then (48) the generating surface T of the cutting tool is represented discretely by a certain number of distinct cutting edges. In this case, the cutting tool surface T cannot be represented as a continuous surface. The DG/K-based CAD/CAM system is capable of treating (53) all the restrictions that are imposed by the discretely represented generating surface T of the form-cutting tool. The encountered restrictions include (and are not limited to) the optimal distribution (54) of the resultant tolerance on the accuracy of surface P onto two portions, the effect of the critical values of the feed rate $\mathbf{F}_{fr}$ and of the side step $\mathbf{F}_{ss}$ on the optimization (55) of the kinematics of the machining operation and so forth.

Further computations in the *CAD/CAM* system are performed following the above considered route:

$$\ldots \rightarrow (55) \rightarrow (49) \rightarrow (50) \rightarrow (51) \rightarrow (52) \rightarrow (44) \rightarrow \ldots \qquad (10.82)$$

The most favorable tool paths of sculptured part surface machining on a multiaxis *NC* machine are the output of the block (IV) (Figure 10.8).

10.6.5 Optimal Location of the Starting Point

The coordinates of the optimal starting point for machining of the sculptured part surface are computed in the block (V). Prior to starting the machining operation, the cutter–location point (*CL* point) must be coincident with the starting point.

For the computation of coordinates of the optimal staring point, a family of nonintersecting curves on part surface P is determined (56). The family of infinite number of curves on the surface P is that same as that within which the most favorable tool paths have to be selected.

Two curves from the family are selected (57). Depending on part surface P geometry and the shape of its boundary curve, these two curves are tangents to the contour of the portion of the surface P to be machined, align with part surface P border, or share with the surface P border at least one common point each.

Then, the two selected curves are shifted (58) inside part surface P area. The distance at which the curves are shifted is half the side step $|\mathbf{F}_{ss}|$. One of four optimal starting points is located on one of two curves defined above.

Further, the four distances necessary for the cutting tool entering into the machining operation and exiting from the machining operation are computed (59). At this point, one can come up with coordinates of four points that are remote from the boundary curve of part surface P at the computed distances. Selecting (60) one gives the coordinates of the most favorable starting point for machining the sculptured part surface.

The computed coordinates of the optimal start-point transfer (61) to the output of the block (V). Computing the optimal start point can incorporate (62) the actual features of the cutting tool entering into and exiting from machining and so forth.

The *DG/K*-based *CAD/CAM* system (Figure 10.8) can be used not only for machining of sculptured part surfaces but it can also be adjusted for the machining of surfaces with simpler geometry.

11

Examples of Implementation of the DG/K-Based Method of Part Surface Generation

The *DG/K*-based approach for part surface generation is discussed in the previous chapters. This approach can be used for the development of novel advanced methods of part surface machining. Below, numerous examples that illustrate the capabilities of the *DG/K*-based approach are briefly considered. In the examples below, it is assumed that all necessary conditions of proper part surface generation [112,116,117,136] are satisfied (see Chapter 7).

11.1 Machining of Sculptured Part Surfaces on a Multiaxis Numerical Control (*NC*) Machine

Numerous methods of sculptured part surface machining on a multiaxis *NC* machine have been developed by now. A review of known methods of sculptured part surface machining is available from literature [108,127].

Below, a method of sculptured part surface machining that is developed on the premise of the *DG/K*-based approach of part surface generation is considered.

Consider machining a sculptured part surface on a multiaxis *NC* machine schematically illustrated in Figure 11.1.

Prior to machining the part surface, the work-piece has to be properly oriented on the worktable of the *NC* machine [82]. The work-piece is oriented in compliance with the method of optimal work-piece orientation on the worktable of the multiaxis *NC* machine (SU Patent No. 1442371). Optimal part surface orientation is discussed in detail in Chapter 7.

For machining the sculptured part surface P, a form-milling cutter is used. The parameters of geometry of the generating surface T of the cutting tool are computed based on the method of design of a form-cutting tool for sculptured part surface machining on a multiaxis *NC* machine (SU Patent No. 4242296/08). This method of form-cutting tool design [5] widely employs the $\mathbb{R}$-mapping of the sculptured part surface P onto the generating surface T of the form-cutting tool [135].

When machining the sculptured part surface P (Figure 11.1), the cutting tool rotates about its axis O_T with a certain angular velocity ω_T. The cutting

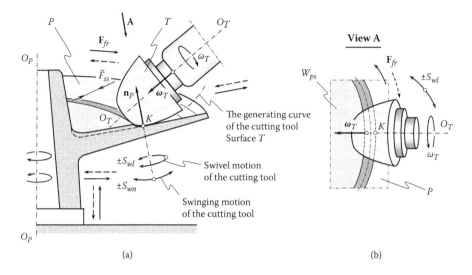

FIGURE 11.1
Optimization of machining of a sculptured part surface on a multiaxis numerical control machine. (From SU Patent Number 1185749; SU Patent Number 1249787.)

tool travels with the optimal feed rate $\mathbf{F}_{fr}$ along the most favorable tool paths. The optimal tool paths are given by Equation 10.15. After the machining of a tool path is accomplished, the cutting tool moves across the trajectory of the cutter-contact point (*CC* point) in the direction of side step $\mathbf{F}_{ss}$ at a distance $\breve{F}_{ss}$. In the new position of the cutting tool, the machining of the next tool path begins.

For the most efficient part surface machining, the width of the tool path must be the maximal possible at every instance of the part surface machining. To maintain the maximal width of the tool path, two more motions are performed by the cutting tool. These two motions are the motions of orientation of the cutting tool. One of the orientation motions is the swiveling* of the cutting tool [73]. Another orientation motion† is the swinging $\pm S_{wn}$ of the cutting tool [78].

It is convenient to consider the swinging motion of the cutting tool before considering its swivel motion.

The swinging motion $\pm S_{wn}$ is the orientation motion of the second type of the cutting tool (see Chapter 2).

Consider the cross section of the part surfaces and generating surface of the cutting tool by a plane surface through the common perpendicular $\mathbf{n}_P \equiv -\mathbf{n}_T$ at a current *CC* point K (Figure 11.2). The *CC* point K that belongs to

---

* SU Patent No. 1185749, *A Method of Sculptured Part Surface Machining on a Multi-Axis NC Machine*, S.P. Radzevich, Int. Cl. B23C 3/16, filed October 24, 1983.

† SU Patent No. 1249787, *A Method of Sculptured Part Surface Machining on a Multi-Axis NC Machine*, S.P. Radzevich, Int. Cl. B23C 3/16, filed December 27, 1984.

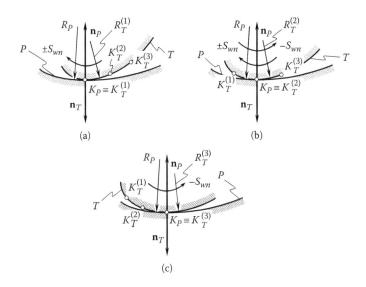

FIGURE 11.2
Swinging motion $\pm S_{wn}$ of the cutting tool: the orientation motion of the second type of the cutting tool T in relation to the part surface P. (SU Patent Number 1249787.)

part surface P is designated here as K_P. Similarly, the *CC* point K that belongs to the cutting tool surface T is designated as K_T.

When machining the sculptured part surface, various points $K_T^{(1)}$, $K_T^{(2)}$, $K_T^{(3)}$, and so forth of the generating surface T of the form-cutting tool can make contact with surface P at the given point K_P. The machining of the sculptured part surface P is performed by the cutting tool of different radii of normal curvature $R_T^{(1)}$, $R_T^{(2)}$, $R_T^{(3)}$, and so forth, of its surface T. When the radii of normal curvature are different, say the inequality $R_T^{(1)} < R_T^{(2)} < R_T^{(3)}$ is observed, this gives a possibility for controlling of degree of conformity of the cutting tool surface T to the sculptured part surface P at the current *CC* point.

To maintain the maximal possible degree of conformity of the cutting tool surface T to part surface P at every *CC* point K, it is necessary to perform a swinging motion of the cutting tool either in the direction $+S_{wn}$ or in the opposite direction $-S_{wn}$. The actual direction of the swinging motion $\pm S_{wn}$ depends upon the current configuration of the cutting tool relative to the part surface as well as upon parameters of the geometry of surfaces P and T at the current *CC* point K.

The swinging motion $\pm S_{wn}$ of the cutting tool can be decomposed onto the rotation ω_c about certain axis and onto translation $\mathbf{V}_c$ in the direction orthogonal to the axis of rotation ω_c. The direction of the translation $\mathbf{V}_c$ is orthogonal to the unit normal vector $\mathbf{n}_P$ to part surface P. The axes of rotation ω_c are orthogonal to the direction of the translation motion. Moreover, the magnitudes of rotation ω_c and translation $\mathbf{V}_c$ are timed with each other in compliance with $|\mathbf{V}_P|/|\omega_P| = R_P^{(c)}$, where $R_P^{(c)}$ is the radius of curvature of the part surface P in the normal plane surface through the vector $\mathbf{V}_c$. Because the cutting tool is performing the swinging motion, the location of the CC point on the sculptured part surface P remains the same. At that same time, the location of the CC point on the generating surface T of the cutting tool changes. The orientation motion of the second type enables control of degree of conformity of the cutting tool surface T to part surface P at the current CC point and in such a way that reduces the machining time.

For precise positioning of the cutting tool, the swinging motion is performed at a certain angle to the direction specified by the vector $\mathbf{t}_{cnf}^{max}$ (SU Patent No. 1336366). The oblique trajectory of the swinging motion reduces the impact of the deviations inherent in the multiaxis NC machine of this particular design* [79].

The swivel motion $\pm S_{wl}$ is the orientation motion of the first type of the cutting tool (see Chapter 2). This orientational motion results in rotation ω_n (or turning through a creation angle) of the cutting tool a about common perpendicular, $\mathbf{n}_P \equiv -\mathbf{n}_T$. Evidently, the speed of the orientational motion $\pm S_{wl}$ is equal to $\omega_n = \partial\mu/\partial t$. The swivel motion is performed by the cutting tool simultaneously with its swinging motion.

The radii of principal curvature $R_{1.T}$ and $R_{2.T}$ of the generating surface T of the form-cutting tool are not equal to each other, and the inequality $R_{1.T} < R_{2.T}$ is always observed. This provides an additional opportunity for controlling the degree of conformity of the cutting tool surface T to the sculptured part surface P at every CC point. To reach the maximal possible degree of conformity of surfaces P and T at the current point of their contact, it is necessary to perform a swivel motion $\pm S_{wl}$ of the form-cutting tool on the multiaxis NC machine.

Figure 11.3 reveals that the swivel motion $\pm S_{wl}$ of the cutting tool is the orientational motion of the first type. It also makes clear that the swivel motion $\pm S_{wl}$ of the cutting tool provides an additional opportunity for controlling the degree of conformity of the cutting tool surface T to part surface P at every CC point and in such a way that reaches the highest possible degree of conformity of the surfaces.

The direction and speed of both orientational motions $+S_{wl}$ and $-S_{wn}$ are timed with each other and are synchronized with other geometrical and kinematical parameters of the machining operation. This makes possible permanent maintenance of the minimal possible value of the diameter $d_{cnf}^{min} \Rightarrow min$ as well as the minimal possible value of the diameter $d_{cnf}^{max} \Rightarrow min$ of the indicatrix of conformity $Cnf_R(P/T)$ of surfaces P and T at the current CC point (both of these diameters are functions of the angle μ of local orientation of surfaces P and T as is $d_{cnf}^{min} = d_{cnf}^{min}(\mu)$ and $d_{cnf}^{max} = d_{cnf}^{max}(\mu)$). The orientational motion $\pm S_{wl}$ (Figure 11.3a) reduces the minimal diameter of

---

* SU Patent No. 1336366, *A Method of Sculptured Part Surface Machining on a Multi-Axis NC Machine*, S.P. Radzevich, Int. Cl. B23C 3/16, filed October 21, 1985.

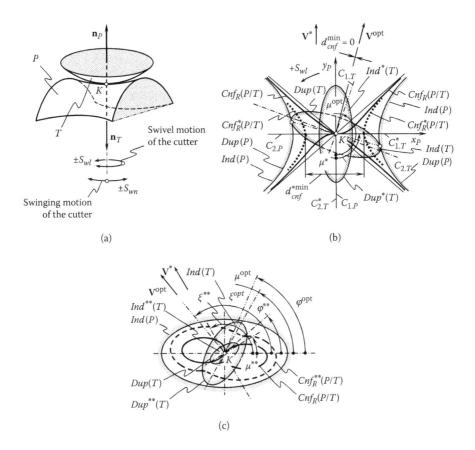

FIGURE 11.3
Swivel motion $\pm S_{wl}$ of the cutting tool: the orientational motion of the first type of the form-cutting tool. (SU Patent Number 1185749.)

the characteristic curve $Cnf_R(P/T)$ and changes its direction from $d_{cnf}^{*\min}$ to $d_{cnf}^{\min}$ (Figure 11.3b). The orientational motion $\pm S_{wn}$ allows additional reduction of the minimal diameter of the indicatrix of conformity $Cnf_R(P/T)$ of surfaces P and T at the current CC point and makes additional changes in its direction (Figure 11.3c).

Both orientational motions $\pm S_{wn}$ and $\pm S_{wl}$ allow for the optimal cutting tool posture as well as the optimal direction of the resultant motion of the cutting tool relative to the sculptured part surface being machined. The orientational motions of the cutting tool enable changing directions of the relative motion along the tool path from v^* to $v^{(opt)}$. At every CC point K, the most favorable tool paths are directed orthogonally to the direction of normal plane surface through K, at which degree of conformity of surfaces P and T is the highest possible (Figure 11.3)—that is, it is directed orthogonally to the minimal diameter $d_{cnf}^{\min}$ of the characteristic curve $Cnf_R(P/T)$.

The resultant relative motion of the part and the cutting tool could be decomposed onto several elementary rotations and translations. The rotations and translations (they are not labeled in Figure 11.1) are executed by corresponding servo-drives of a multiaxis *NC* machine. A tool path of any desired geometry can be generated in this way.

The direction along which degree of conformity of the generating surface of the cutting tool to the part surface is maximal is aligned with the direction along which diameter of the indicatrix of conformity $Cnf_R(P/T)$ is minimal. The direction of the maximal degree of conformity at a point of contact of the cutting tool surface T to part surface P is specified by the unit tangent vector $\mathbf{t}_{cnf}^{max}$.

Similarly, the direction of the minimal degree of conformity at a point of contact of the surface T to surface P can be specified by the unit tangent vector $\mathbf{t}_{cnf}^{min}$.

In the general case of sculptured part surface machining, the directions $\mathbf{t}_{cnf}^{max}$ and $\mathbf{t}_{cnf}^{min}$ of the extremal degree of conformity at a point of contact of surfaces P and T are not orthogonal to each other. This should be encountered when selecting a criterion of optimization of a sculptured part surface machining process.

Usually, the feed-rate motion $\mathbf{F}_{fr}$ is orthogonal to $\mathbf{t}_{cnf}^{max}$. Thus, this motion $\mathbf{F}_{fr}$ does not align with the direction of $\mathbf{t}_{cnf}^{min}$. Accordingly, if the direction $\mathbf{F}_{fr}$ aligns with the unit tangent vector $\mathbf{t}_{cnf}^{min}$, it is not orthogonal to the direction specified by the unit tangent vector $\mathbf{t}_{cnf}^{max}$. This is because unit tangent vectors $\mathbf{t}_{cnf}^{max}$ and $\mathbf{t}_{cnf}^{min}$ are not orthogonal to each other. Therefore, in the general case of sculptured part surface machining, it is impossible to generate the sculptured part surface under the widest tool path and with the highest possible feed rate per tooth of the cutting tool. This issue is getting critical when the magnitude of the limit feed rate $|[\mathbf{F}_{fr}]|$ and the magnitude of the limit side step $|[\mathbf{F}_{ss}]|$ are of comparable value. Under such a scenario, the surface generation output $\mathcal{P}_{sg}$ must be used for determining of the most favorable parameters of the sculptured part surface machining* process.

A method of sculptured part surface machining on a multiaxis *NC* machine [81] is illustrated in Figure 11.4.

In the common tangent plane through the *CC* point K, two principal cross sections $C_{1.P}$ and $C_{2.P}$ of the sculptured part surface P are along the unit tangent vectors $\mathbf{t}_{1.P}$ and $\mathbf{t}_{1.P}$. The unit tangent vector $\mathbf{t}_{cnf}^{max}$ is at the optimal angle φ^{opt} with respect to the unit tangent vector $\mathbf{t}_{1.P}$. The unit tangent vector $\mathbf{t}_{cnf}^{max}$ specifies the direction of the feed rate motion $\mathbf{F}_{fr}$. The last is orthogonal to the unit tangent vector $\mathbf{t}_{cnf}^{max}$.

---

* SU Patent No. 1367300, *A Method of Sculptured Part Surface Machining on a Multi-Axis NC Machine*, S.P. Radzevich, Int. Cl. B23C 3/16, filed January 30, 1986.

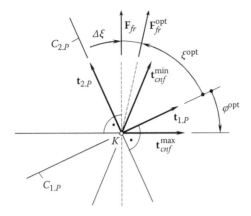

FIGURE 11.4
A method of sculptured part surface machining on a multiaxis *NC* machine. (SU Patent Number 1367300.)

The direction of the minimal degree of conformity of the part surface P and generating surface T of the form-cutting tool at point K is specified by the unit tangent vector $\mathbf{t}_{cnf}^{min}$. In the case under consideration, the unit tangent vectors $\mathbf{t}_{cnf}^{max}$ and $\mathbf{t}_{cnf}^{min}$ are not orthogonal to each other.

The instant part surface generation output $\mathcal{P}_{sg}$ (see Equation 9.1) is as follows:

$$\mathcal{P}_{sg} = |\mathbf{F}_{fr} \times \mathbf{F}_{ss}| \tag{11.1}$$

Without going into the details of the analysis, a generalized expression (Equation 9.11):

$$\mathcal{P}_{sg} = \mathcal{P}_{sg}(U_P, V_P, U_T, V_T, \mu, \varphi) \tag{11.2}$$

for the instant part surface generation output $\mathcal{P}_{sg}$ can be used in the analysis below.

When the feed rate motion $\mathbf{F}_{fr}$ is directed orthogonally to the minimal diameter d_{cnf}^{min} of the indicatrix of conformity $Cnf_R(P/T)$, the width of the tool path is maximal; however, the limit value of the feed rate per tooth $[\breve{F}_{fr}]$ of the cutting tool is getting smaller. A smaller limit value of the feed rate per tooth $[\breve{F}_{fr}]$ of the cutting tool causes a smaller instant part surface generation output $\mathcal{P}_{sg}$. Put another way, an increase in the limit feed rate $[\breve{F}_{fr}]$ requires a corresponding decrease of the limit side step $[\breve{F}_{ss}]$.

The above consideration reveals that there must be an optimal correspondence between the feed rate $\breve{F}_{fr}$ and the side step $\breve{F}_{ss}$. To determine the most favorable correspondence between parameters $\breve{F}_{fr}$ and $\breve{F}_{ss}$, the direction of the feed rate motion $\mathbf{F}_{fr}$ must be properly determined.

For the maximal part surface generation output, the optimal direction of the feed rate motion $\mathbf{F}_{fr}^{opt}$ satisfies the condition

$$\frac{\partial \mathcal{P}_{sg}}{\partial \xi} = 0 \tag{11.3}$$

where the auxiliary angle ξ is linearly dependent of the angle φ.

When the solution to Equation 11.3 returns multiple solutions, only those that satisfy the sufficient condition for the maximum of $\mathcal{P}_{sg}$ are acceptable:

$$\frac{\partial^2 \mathcal{P}_{sg}}{\partial \xi^2} < 0 \tag{11.4}$$

The computed optimal value ξ^{opt} of the auxiliary angle ξ specifies the optimal direction of the feed rate motion $\mathbf{F}_{fr}^{opt}$ in the common tangent plane through the point K. This direction is at a certain angle $\Delta \xi$ with respect to the direction $\mathbf{t}_{cnf}^{max}$ at which the minimal diameter d_{cnf}^{min} of the indicatrix of conformity $Cnf_R(P/T)$ is measured.

The efficiency of sculptured part surface machining can be increased in that way the feed rate and the side step at the current CC point K properly timed with each other. The use of the method of sculptured part surface machining [92] makes it possible to achieve this goal.

Figure 11.5 illustrates residual cusps on the machined part surface.*

When the parameters of the sculptured part surface machining are optimal, the cusps of the width $\breve{F}_{ss}^{opt}$ and of the length $\breve{F}_{fr}^{opt}$ appear on the machined part surface (Figure 11.5a).

If the value of the feed rate is increased from $\breve{F}_{fr}^{opt}$ to $\breve{F}_{fr}^{*}$, it is necessary to decrease the side step from $\breve{F}_{ss}^{opt}$ to $\breve{F}_{ss}^{*}$ (Figure 11.5b). Ultimately, the part surface generation output in this case is smaller compared with that in the first case (Figure 11.5a).

If the value of the feed rate is decreased from $\breve{F}_{fr}^{opt}$ to $\breve{F}_{fr}^{**}$, it is necessary to increase the side step from $\breve{F}_{ss}^{opt}$ to $\breve{F}_{ss}^{**}$ (Figure 11.5b). Ultimately, the part surface generation output in this case is also smaller compared with that in the first case (Figure 11.5a).

As previously discussed (see Section 8.5.2), the limit value $[h_{ss}]$ of the deviation h_{ss} is equal to a portion of the tolerance $[h]$, say $[h_{ss}] = c \cdot [h]$. Here

---

* RU Patent No. 2050228, *A Method of Sculptured Surface Machining on Multi-Axis NC Machine,* S.P. Radzevich, Int. Cl. B23C 3/16, filed December 25, 1990.

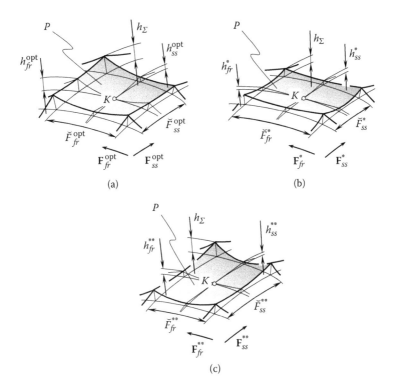

FIGURE 11.5
Cusps on the machined sculptured part surface.

c designates the local parameter of distribution of the tolerance $[h]$. Due to the equality

$$[h] = [h_{fr}] + [h_{ss}] \tag{11.5}$$

is valid, the following expression

$$[h_{fr}] = (c - 1) \cdot [h] \tag{11.6}$$

is valid for the limit value $[h_{fr}]$ of the waviness $[h_{fr}]$.

Substituting the derived formula (see Equation 11.6) for $[h_{fr}]$ to Equation 9.46, the following equation for the calculation of the limit value $[\breve{F}_{fr}]$ of the feed rate can be obtained:

$$[\breve{F}_{fr}](c) \cong 2R_{P.fr} \arccos \frac{R_{P.fr}^2 + R_{T.fr} \cdot (R_{P.fr} + (c-1) \cdot [h] \cdot \operatorname{sgn} R_{P.fr})}{(R_{P.fr} + R_{T.fr}) \cdot (R_{P.fr} + (c-1) \cdot [h] \cdot \operatorname{sgn} R_{P.fr})} \tag{11.7}$$

Similarly, substituting the derived formula for $[h_{ss}]$ to Equation 9.51, the equation for the calculation of the limit value $[\breve{F}_{ss}]$ of the side step can be obtained:

$$[\breve{F}_{ss}](c) \cong 2R_{P.ss} \arccos \frac{R_{P.ss}^2 + R_{T.ss} \cdot (R_{P.ss} + c \cdot [h] \cdot \text{sgn } R_{P.ss})}{(R_{P.ss} + R_{T.ss}) \cdot (R_{P.ss} + c \cdot [h] \cdot \text{sgn } R_{P.ss})} \qquad (11.8)$$

For the maximal possible part surface generating output $[\mathcal{P}_{sg}]$, Equation 11.1 yields

$$[\mathcal{P}_{sg}](c) = [\breve{F}_{fr}](c) \cdot [\breve{F}_{ss}](c). \qquad (11.9)$$

When the parameter c is of optimal value, the following equality is observed:

$$\frac{\partial}{\partial c}[\mathcal{P}_{sg}](c) = 0 \qquad (11.10)$$

The condition (see Equation 11.10) is necessary for the minimum of the function $h_\Sigma = h_\Sigma(c)$. In addition, the following inequality must be observed:

$$\frac{\partial^2}{\partial c^2}[\mathcal{P}_{sg}](c) > 0 \qquad (11.11)$$

The details of the solution to Equation 11.10 are available from the article [140] by the author.

The timing of the feed rate per tooth F_{fr} of the cutting tool and side step $\breve{F}_{ss}$ in compliance with Equation 11.10 ensures an increase in the part surface generation output $\mathcal{P}_{sg}$.

It is possible to achieve further improvement of the part surface generation output $\mathcal{P}_{sg}$ by means of optimization of configuration of the neighboring elementary surface cells on the machined sculptured part surface. Usually, the elementary surface cells are shaped in the form that is close to a rectangle. Under proper synchronization of the feed rate per tooth F_{fr} and side step $\breve{F}_{ss}$, the rectangular elementary surface cells (Figure 11.5) could be substituted with the approximately hexagonal elementary surface cells (Figure 11.6). When other conditions of part surface generation process remain the same, the area of hexagonal elementary surface cells is bigger in comparison with area of rectangular elementary surface cells. Therefore, the part surface generation output when machining a sculptured part surface with the hexagonal elementary surface cells could be bigger. The proper configuration of the neighboring elementary surface cells is an efficient way to increase

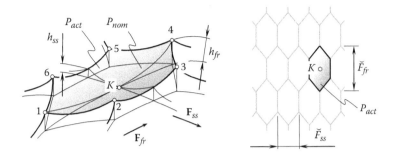

FIGURE 11.6
Hexagonal elementary surface cells on the machined part surface.

the surface generation output $\mathcal{P}_{sg}$. An appropriate method of sculptured part surface machining on a multiaxis NC machine had been proposed by the author as early as in 1991.

When the feed rate motion $\mathbf{F}_{fr}$ and the side step motion $\mathbf{F}_{ss}$ are properly timed with each other, the cusp height at all vertices 1,2,...,6 of the elementary surface cell are equal to each other and they equal to the resultant cusp height h_Σ, (that is, the equality $h_1 = h_2 = h_3 = h_4 = h_5 = h_6 = h_\Sigma$ is observed). All elementary surface cells are shaped in the form of nonregular hexagons.

For an arbitrary correspondence between motions $\mathbf{F}_{fr}$ and $\mathbf{F}_{ss}$ at a current CC point (e.g., when $\breve{F}_{fr} = f \cdot \breve{F}_{ss}$), the actual cusp heights h_i at all six vertices of the elementary surface cell are different. Evidently, the cusp height h_i can be expressed in terms of the parameter f, say as $h_i = h_i(f)$. Further, the differences $\Delta h_i(f) = h_\Sigma - h_i(f)$ are computed. The desired value of the parameter f is that under which all the differences $\Delta h_i(f)$ are of zero value. The solution to the problem of the computation of the optimal values of the parameter f is not provided here due to space limits. For details on the solution to this problem, the interested reader may wish to go to [3,35] where an example of solution is presented.

The conditions of the material removal process in conventional and in climb milling are different. When the difference is significant, then it is easier to remove the stock, for instance, by climb milling and not by conventional milling. Under such a scenario, the cutting tool travels not along each consequent tool path, but it travels over one path. For example (Figure 11.7), the machining of the sculptured part surface starts at point 1. After the machining of the first tool path is over, the tool path through point 2 is machined. The second tool path is not the neighboring one to the first tool path. Such a scheme of tool paths allows for a decrease of the load on the cutting tool when conditions of machining are inconvenient (conventional milling) and for an increase of the load on the cutting tool when conditions of machining are inconvenient (climb milling). Such a method of sculptured part surface machining has proven to be practical.

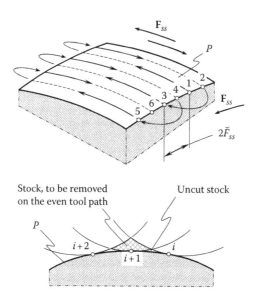

FIGURE 11.7
An overpath method of sculptured part surface machining on a multiaxis numerical control machine.

The considered examples of methods of sculptured part surface machining clearly show that the increase of the feed rate per tooth $\breve{F}_{fr}$ of the cutting tool is the reliable way to increase the efficiency of sculptured part surface machining on a multiaxis *NC* machine. For advanced methods of part surface machining, the feed rate per tooth of the cutting tool and the side step are values of the same range.

It is necessary to mention here that the *DG/K*-based method of part surface generation is useful for solving problems of synthesis of the most favorable sculptured part surface machining on a multiaxis *NC* machine not only using cutting tools or a grinding wheel but using technology of surface burnishing (by plastic subsurface deformation of the work) as well. A method* of sculptured part surface burnishing is a good example in this regard [86].

11.2 Machining of Surfaces of Revolution

The use of the *DG/K*-based method of part surface generation is fruitful for the development of novel advanced methods of machining, not only

---

* SU Patent No. 1533174, *A Method of Reinforcement of Sculptured Surface on Multi-Axis NC Machine*, S.P. Radzevich, Int. Cl. B24B 39/00, filed December 2, 1987.

of sculptured part surfaces that have already been shown above but also novel advanced methods of machining part surfaces of simpler geometry. Examples can be found in the field of machining of surfaces of revolution, cylindrical surfaces, gears, and in many other fields.

11.2.1 Turning Operations

In compliance with the conventional method of turning of a form surface of revolution, the work rotates about its axis of rotation. The cutter travels with a certain feed rate along the axis of rotation of the work. In addition to this motion, the cutter performs a motion toward the work axis of rotation or in a backward direction depending upon the actual shape of the axial contour of the part surface being machined. This is an example of a trivial turning operation that is often used for machining form surfaces of revolution.

Targeting an increase of the part surface generation output, a method of turning form surfaces of revolution was developed [94]. In this method,* the work rotates about its axis of rotation. The cutter travels along the axial profile of the form part surface with a constant peripheral feed rate. The use of this method of part surface machining ensures perfect results when the radius of curvature of the axial profile of the form part surface is of constant value or when the variation of the radius of curvature is reasonably small and thus it could be neglected.

For the machining of form part surfaces of revolution having significant variation of curvature of the axial profile, a method of turning of part surfaces of revolution was proposed [89]. In this method of part surface machining,[†] the current value of the peripheral feed rate is synchronized with the radius of curvature of the axial profile of the part surface being machined.

When machining surface P, the work rotates about its axis of rotation O_p with a certain angular velocity ω_p (Figure 11.8). The radii of curvature R_p of the axial profile 1 of the part surface P vary from one point of the profile to another. The cutting edge of the cutter is represented as the arc segment of a curve 2. The radii of curvature R_T of the cutting edge can be either of the same value at all points of the cutting edge or can vary from one point of the cutting edge to another.

The cutter travels along the axial profile 1 with the peripheral feed rate F_{prl}. In a different instance of part surface machining, the value of the feed rate F_{prl} is different. This means that the current value of the feed rate F_{prl} at the points K_1, K_2, and K_3 of the axial profile 1 of the part surface P is not of the same value.

---

* US Patent No. 4.415.977, *Method of Constant Peripheral Speed Control*, F. Hiroomi and I. Shinichi, Int. Cl. B23 15/10, 05B 19/18, National Cl. 364/474, filed June 30, 1980, No. 243928; Priority: June 30, 1979, No. 54-82779, Japan.

[†] SU Patent No. 1708522, *A Method of Turning Form Surfaces of Revolution*, S.P. Radzevich and L.V. Bondarenko, Int. Cl. B23B 1/00, filed December 13, 1988.

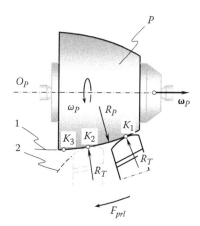

FIGURE 11.8
A method of turning of form surfaces of revolution with cusps of smallest constant height. (SU
Patent Number 1708522.)

When the cutter having constant radius $R_T = Const$ of the cutting edge is
used for machining of the part surface, the value of the peripheral feed rate
F_{prl} at the current point of contact of the axial profile 1 and of the cutting edge
can be calculated from the formula:

$$F_{prl} = 2 \cdot R_p \cdot \cos^{-1}\left\{\frac{(R_p + R_T)^2 + (R_p + [h])^2 - R_T^2}{2(R_p + R_T)(R_p + [h])}\right\} \qquad (11.12)$$

In a more general case of machining of form part surfaces of revolution—
say, when the cutter having variable radius of curvature $R_T = Var$ is used for
machining of the part surface—then the value of the peripheral feed rate F_{prl}
at the current point of contact of the axial profile 1 and of the cutting edge
can be calculated from the formula

$$F_{prl} = R_p \left(\cos^{-1}\left\{\frac{(R_p + R_{T.i})^2 + (R_p + [h])^2 - R_{T.i}^2}{2 \cdot (R_p + R_{T.i}) \cdot (R_p + [h])}\right\} \right.$$
$$\left. + \cos^{-1}\left\{\frac{[R_p + R_{T.(i+1)}]^2 + \{[R_p + [h]\}^2 - R_{T.(i+1)}^2}{2 \cdot [R_p + R_{T.(i+1)}] \cdot \{R_p + [h]\}}\right\} \right) \qquad (11.13)$$

where $R_{T.i}$ and $R_{T.(i+1)}$ designate the radii of curvature of the cutting edge of
the cutter at two neighboring contact points K_i and K_{i+1}.

Due to the optimization of the current value of the peripheral feed rate
of the cutter at every *CC* point *K*, the part surface generation output in this
method is bigger.

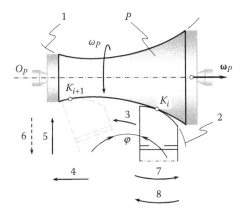

FIGURE 11.9
Utilization of the orientational motion of the second type of cutter in the method of turning of form surfaces of revolution. (SU Patent Number 1171210.)

In the method of part surface machining (Figure 11.8), the radius of curvature of the cutting edge R_T of the cutter is out of the control of the user. Analysis of Equation 11.13 reveals that the impact of variation of radius of curvature R_T on the part surface generation output can be significant. Therefore, the efficiency of turning of form part surfaces of revolution can be increased if proper control of the radius R_T is introduced.

Kinematics of part surface machining in the method of turning of form surfaces of revolution [6] is capable of properly controlling the radius of curvature R_T of the cutting edge of the cutter at the current CC point.* In this method of part surface machining, the work having axial profile 1 of variable curvature rotates about its axis of rotation O_P with a certain angular velocity ω_P (Figure 11.9). The cutting edge of the cutter, an arc segment of curve 2 with permanently variable radius of curvature R_T, is used for performing this machining operation [70].

The cutter travels along the axial profile of the part surface in the peripheral direction 3. On the lathe, this motion is obtained as the superposition of the axial motion 4 of the cutter and its reciprocal motion toward the part axis of rotation 5 and in backward direction 6.

In addition, the cutter performs the orientational motion of the second type (see Chapter 2). This motion of the cutter is performing either in direction 7 or in direction 8. The actual direction of the orientational motion of the cutter depends upon the actual geometry of the axial profile 1 of the part surface at two neighboring CC points K_i and K_{i+1}. Ultimately, due to the cutter motion being either in direction 7 or in direction 8, the cutting edge rolls with a slip over the axial profile 1 of the part surface being machined.

---

* SU Patent No. 1171210, *A Method of Turning of Form Part Surfaces of Revolution*, S.P. Radzevich, Int. Cl. B23B 1/00, filed November 3, 1984.

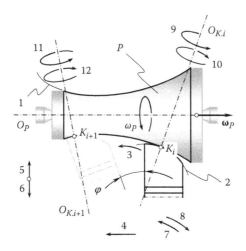

FIGURE 11.10
Utilization of the orientational motion of the first type of cutter in the method of turning of form surfaces of revolution. (SU Patent Number 1232375.)

Implementation of the orientational motion of the cutter makes possible better fit of the radius of curvature R_T of the cutting edge to the part surface radius of curvature R_P at every CC point K. In this way (see Equation 11.13), the part surface generation output increases.

It is important to note the possibility of machining of form part surfaces of revolution in compliance with the method (Figure 11.9), when not the cutter, but a milling cutter or grinding wheel having a corresponding axial profile of the generating surface of the tool is used instead.

Similarly to the utilization of the orientational motion of the second type (Figure 11.9), the orientational motion of the first type of the cutter can be utilized in turning of part surfaces of revolution as well.

The kinematics of a method of turning of the form part surfaces of revolution (Figure 11.10) uses the first type of orientational motion [76]. This method of part surface machining is similar to the earlier discussed method of part surface machining shown in Figure 11.9. For convenience, the designations of the major elements in Figure 11.10 are identical to the designations of the corresponding major elements in Figure 11.9. Therefore, there are no reasons to repeat all the details of the method under consideration.

In the method of part surface machining* (Figure 11.10), the orientational motion of the first type is utilized. The orientational motion of this type allows for the turning of the cutter about the axis $O_{K.i}$ along the unit normal vector $\mathbf{n}_P$ either in direction 9 or in direction 10. The actual direction of the orientational motion depends upon parameters of geometry of the part

---

* SU Patent No. 1232375, *A Method of Turning of Form Surfaces of Revolution*, S.P. Radzevich, Int. Cl. B23B 1/00, filed September 13, 1984.

surface P at the two neighboring CC points K_i and K_{i+1}. In the neighboring CC point K_{i+1}, the orientational motion is designated as 11/12.

Evidently, the parameters of the orientational motion of the first type are strongly constrained by the limit values of the geometrical parameters of the cutting edge of the cutter to be used, first of all, by the clearance angle of the cutting edge.

The implementation of the orientational motion of the cutter allows for better fit of the radius of curvature R_T of the cutting edge of the cutter to the part surface radius of curvature R_P at every CC point K. This way (see Equation 11.13), the part surface-generation output increases.

It is the right point to stress that it may also be possible to machine form part surfaces of revolution in compliance with the method (Figure 11.10) when not the cutter but a milling cutter or grinding wheel having a corresponding axial profile of the generating surface of the tool is used instead. Under such a scenario, no constraints are imposed by the limit values of the geometrical parameters of the cutting edge of the cutting tool to be used.

11.2.2 Milling Operations

The earlier discussed methods of turning of form part surfaces of revolution (see Figure 11.9 and Figure 11.10) allow substitution of the cutter with a milling cutter or with a grinding wheel having a corresponding profile of axial cross section of the generating surface T. These methods of part surface machining indicate that efficient methods of milling of form part surfaces of revolution can be developed.

A method of milling of form part surfaces of revolution on an NC machine tools was developed by the author [139]. In compliance with the method (Figure 11.11), a form part surface of revolution P having axial

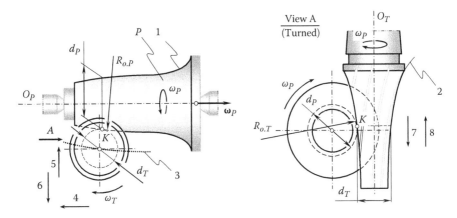

FIGURE 11.11
A method of milling of form surfaces of revolution [26]. (From *Improvement of Gear Cutting Tools*, Collected papers, Moscow, NIIMASh, 1989.)

profile 1 is machined with the milling cutter having a curved axial profile of the generating surface T. The work rotates about its axis of rotation O_p with a certain angular velocity ω_p. The axis of rotation of the work O_p and the axis of rotation of the milling cutter O_T cross at a right angle.

The milling cutter is traveling in direction 3 along the axial profile 1 of the part with a certain peripheral feed rate. This motion is a superposition of the milling cutter motion in the axial direction 4 of the work and of its motion 5 toward the work axis of rotation and in backward direction 6. In addition to the mentioned motions, the milling cutter also performs the motion of orientation of the second type. While traveling in the axial direction 4 of the work, the milling cutter simultaneously performs linear motion along its axis of rotation O_T. This motion performs either in direction 7 or in the opposite direction 8, depending upon geometry of part surface P and generating surface T of the form-cutting tool at the current CC point K.

The orientational motion of the milling cutter provides a possibility for increasing the degree of conformity of the generating surface T of the milling cutter to the form part surface of revolution P at every CC point K. In this way, the part surface generation output is increased.

The grinding of form surfaces of revolution can be performed in the same way as shown in Figure 11.11.

11.2.3 Machining of Cylinder Surfaces

Orientational motions of the cutting tool are also used for the improvement of machining of general cylinder surfaces. Such a possibility is illustrated below by the method of machining of a cam-shaft.*

The method of machining of a cam-shaft [88] targets the maximal possible material removal rate.

In compliance with the method, a grinding wheel with conical generating surface T is used for the machining of the part surface P of a cam (Figure 11.12). The grinding wheel rotates about its axis of rotation O_T with a certain angular velocity ω_T.

The part surface generation motions are performed by the work. The set of these motions includes the rotation ω_p of the work about the axis of rotation O_p and the reciprocal motion 1 in the direction of the common perpendicular to axes O_p and O_T. The rotation ω_p of the grinding wheel can be either uniform or nonuniform.

The grinding wheel performs an auxiliary straight motion 2. The direction of motion 2 is parallel to the axis of rotation of the work O_p. The

---

* SU Patent No. 1703291, *A Method of Machining of Form Surfaces*, S.I. Chukhno and S.P. Radzevich, Int. Cl. B23C 3/16, filed August 2, 1989.

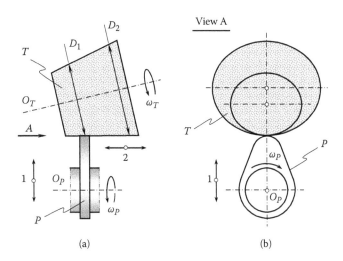

FIGURE 11.12
A method of grinding of a cam. (SU Patent Number 1703291.)

straight motion 2 is timed with work rotation ω_P in the way in which the material removal rate is constant and equal to its greatest feasible value:

$$Q_{cr}|_{max} = 0,5 \cdot [L(\varphi)]^2 \cdot v_T(\varphi) \cdot b = Const \qquad (11.14)$$

where $L(\varphi)$ is the length of the line of contact of the grinding wheel with the stock to be removed, $v_T(\varphi)$ is the rate of change of the leading coordinate, b is the width of the part surface being machined, and φ is the angular coordinate of the currently machining portion of the part surface P.

To stabilize the maximal feasible material removal output, the corresponding change of the length $L(\varphi)$ is provided by the auxiliary straight motion 2.

In addition, the grinding wheel rotation ω_T is timed with auxiliary straight motion 2. The timing of the motions is performed in inverse order to the change of working diameter of the grinding wheel. When the working diameter of the grinding wheel is D_1, the corresponding angular velocity is $\omega_{T.1}$. When the working diameter of the grinding wheel is D_2, the corresponding angular velocity is $\omega_{T.2}$. The following equality

$$\omega_{T.1} \cdot D_1 = \omega_{T.2} \cdot D_2 \qquad (11.15)$$

is observed.

The method of machining of a cam-shaft (Figure 11.12) allows for the highest possible material removal output and high quality of the machined parts.

11.2.4 Burnishing of Part Surfaces of Revolution

The burnishing of part surfaces by plastic deformation is often used for finishing of parts. The optimal parameters of such a machining operation as well as the optimal design parameters of the tool that is used for these purposes can be determined on the premise of the *DG/K*-based method of part surface generation. Examples below illustrate the capabilities of the *DG/K*-based method of part surface generation for the improvement of the finishing of part surface of revolution.

The burnishing of form part surfaces of revolution can be performed with the tool having a conical generating surface *T* [116,117]. In this method (Figure 11.13), the work rotates about its axis O_P with a certain angular velocity ω_P. The axis O_T of the conical indenter 1 (conical tool) crosses the work axis of rotation O_P at a right angle. The tool travels along the axial profile of the part surface *P* with a certain peripheral feed rate. Indenter 1 is pressed into part surface *P* by normal force P_{rnf}. In the relative motion, the *CC* point *K* traces trajectory 2 on the machined part surface.

Two configurations of indenter 1 are possible. The first configuration is shown in Figure 11.13. In such a tool configuration, its bigger diameter is below the smaller diameter. The inverse configuration of the tool, when the smaller diameter is below the bigger diameter is feasible as well.

When machining a form part surface of revolution, the portion of the part surface *P* having bigger diameter is machined with the portion of the tool having smaller diameter, and vice versa. In this way, it is possible to maintain that same pressure when machining portions of the part with different geometry of surface *P*. For this purpose, the indenter performs an auxiliary straight motion either downward 3 or upward 4, depending on the geometry of the part surface *P* being machined. The auxiliary motion requires a corresponding compensation of center distance between axes O_P and O_T of the part and of the tool, respectively. A component of the auxiliary straight motion creates the orientational motion of the second type of the tool.

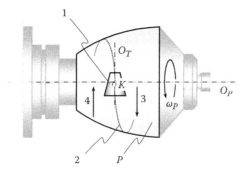

FIGURE 11.13
A method of burnishing of a form surface of revolution by a conical indenter.

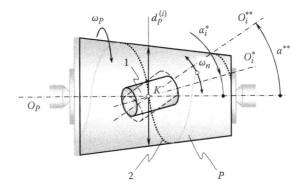

FIGURE 11.14
A method of burnishing of a form surface of revolution by a cylindrical tool. (SU Patent Number 1463454.)

The burnishing of the part surfaces under the optimal pressure that is of the same value at every CC point K enables an increase of the quality of the surface finish.

For the burnishing of form part surfaces of revolution, not only a conical tool but a cylindrical tool* can be used as well. In the method of burnishing of a form surface of revolution [84], the finishing of the part surface is performed with the cylindrical indenter. When machining part surface P, the work rotates about its axis O_p with a certain angular velocity ω_p (Figure 11.14). The cylindrical indenter 1 is pressed into part surface P by normal force P_{rnf}. Tool 1 is traveling along the axial profile of the part surface P with a certain peripheral feed rate. Simultaneously, tool 1 is performing the orientational motion of the first type ω_n about unit normal vector $\mathbf{n}_p$ to part surface P. The orientational motion of the tool is timed with part diameter $d_p^{(i)}$ at the current CC point K. Due to the orientational motion of the tool, the angle that the axis O_p of the part forms with the axis O_i^* at the current ith point is under the control of the user. At every CC point K, the angle of crossing α_i^* is of its optimal value. When the diameter d_p^i is bigger, the cross-axis angle α_i^* angle is also bigger, and vice versa. In this way, the same value of the optimal pressure is maintained at every CC point K.

In particular cases, two paths of indenter 1 are required to be performed. On the second tool path, the angle that the axis O_p of the part forms with the axis O_i^{**} at current ith point is reduced to a value α_i^{**}. On the second tool path, the angle α_i^{**} at the current CC point K is always smaller than that angle α_i^* on the first tool path ($\alpha_i^{**} < \alpha_i^*$).

---

* SU Patent No. 1463454, *A Method of Burnishing of Part Surfaces*, S.P. Radzevich, Int. Cl. B24B 39/00, 39/04, filed May 5, 1987.

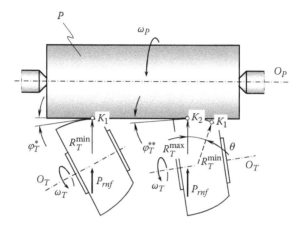

FIGURE 11.15
A method of burnishing of a surface of revolution by a form tool. (SU Patent Number 1636196.)

The burnishing of the part surfaces under the optimal pressure that is of the same value at every CC point K enables an increase in the quality of the part surface finish.

Similarly, burnishing of part surfaces of revolution can be performed with a form roller. For example, a method of burnishing of a surface of revolution features the implementation of a form tool [87].

The method of burnishing of form part surfaces of revolution is illustrated with an example of finishing of a cylindrical part surface P (Figure 11.15). However, the method of part surface finishing can be implemented for the burnishing form part surfaces of revolution as well.

In compliance with the method of surface machining,* the work is rotating about its axis of rotation with certain angular velocity ω_P (Figure 11.15). For the machining, a roller having a form axial profile is used. The radius of curvature R_T of the axial profile of the roller varies ($R_T = Var$). The roller is pressed into part surface P by normal force P_{rnf}. When machining the part surface, the roller is traveling along the axial profile of the part surface P. In this motion, the roller rotates about its axis of rotation with certain angular velocity ω_T. The roller is driven due to friction between the part surface P and the working surface of the roller.

On a rough tool path, the roller contacts the part surface with the point K_1 of the generating profile of its working surface. The radius of curvature of the axial profile of the tool at the point K_1 is minimal R_T^{min}. On the finishing tool path, the roller contacts the part surface with the point K_2 of the generating profile of its working surface. The radius of curvature of the axial profile of the tool at the point K_2 is maximal R_T^{max}.

---

* SU Patent No. 1636196, *A Method of Burnishing of Part Surfaces*, S.P. Radzevich and V.V. Novodon, Int. Cl. B24B 39/00, filed January 30, 1991.

On the rough tool path, the tool approach angle φ_T^* is bigger than that φ_T^{**} on the finishing tool path.

The turn of the roller through a certain angle θ can be interpreted as the degenerate type of orientational motion of the second type of tool.

The utilization of the degenerate orientational motion of the second type in the method of part surface finishing (Figure 11.15) makes it possible to maintain optimal conditions of part surface burnishing at every CC point K on both the rough tool path of the roller as well as on its finishing tool path. Ultimately, this improves the quality of the finished part surface.

11.3 Finishing of Involute Gears

Various methods of shaving are widely used for finishing spur and helical involute gears [144]. Most gear-shaving operations are not optimized. The calculation of the optimal parameters of diagonal shaving operation provides a perfect example of implementation of the DG/K-based method of part surface generation. In compliance with the method, it is possible to calculate the desired design parameters of the shaving cutter best suited for finishing the given involute gear. It is also possible to calculate the optimal parameters of the relative motions of the shaving cutter with respect to the gear to be finished. For this purpose, the indicatrix of conformity $Cnf_R(P_g/T_{sh})$ at a current point of contact of the generating surface T_{sh} of the shaving cutter to the screw involute tooth surface P_g of the gear is commonly employed.

In diagonal shaving (Figure 11.16), the work gear rotates about its axis O_g with a certain angular velocity ω_g. The shaving cutter rotates about its axis O_{sh} with an angular velocity ω_{sh} that is timed with the ω_g, that is, $\omega_{sh} = u \cdot \omega_g$, where u is the tooth ratio ($u = N_g/N_{sh}$), where N_g is the number of the gear teeth and N_{sh} is the number of the shaving cutter teeth). The axes of rotation O_g of the gear and O_{sh} of the shaving cutter are at a center-distance C and cross each other at an angle Σ. The angle Σ is as follows: $\Sigma = \psi_g + \psi_{sh}$. Here ψ_g is the gear helix angle. It is positive (+) to the right-hand gear and negative (–) to the left-hand gear to be machined. The same is observed with respect to the shaving cutter helix angle ψ_{sh}. In addition, the shaving machine table reciprocates relative to the shaving cutter with feed $\mathbf{F}_{diag}$. The axis of rotation O_g of the gear and direction of the feed $\mathbf{F}_{diag}$ make a certain angle θ.

The traverse path of the feed $\mathbf{F}_{diag}$ is at a certain angle θ to the gear axis of rotation O_g (Figure 11.16). The relationship between the face width of the gear B_g and the shaving cutter B_{sh} is an important consideration. It defines the value of the diagonal traverse angle.

The surface of tolerance $P_{[h]}$ is at a distance of the tolerance $[h]$ to the gear tooth surface P_g. After tooth surface P_g of a gear and tooth surface T_{sh} of the shaving cutter are put into contact at point K, then the surface T_{sh} intersects

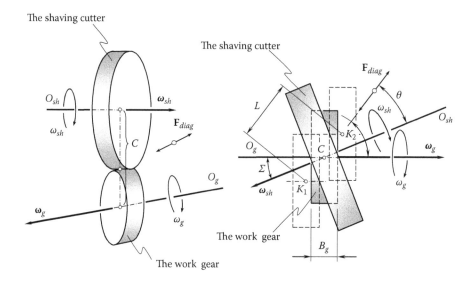

FIGURE 11.16
Schematic of diagonal shaving method. (From Radzevich, S.P. "Diagonal Shaving of an Involute Pinion: Optimization of the Geometric and Kinematic Parameters for the Pinion Finishing Operation" International Journal of Advanced Manufacturing Technology, Vol. 32, Numbers 11–12, pp. 1170–1187, May 2007.)

the surface $P_{[h]}$. The line of intersection is a certain closed three-dimensional curve $\mathscr{C}_{pt}$ shown in Figure 11.17. It bounds the spot of contact of the gear and the shaving cutter tooth. It is recommended that the area of the spot of contact $\mathscr{C}_{pt}$ be kept as small as possible (Figure 11.17).

Due to the tooth surfaces P_g and T_{sh} are making contact at a distinct point K, only discrete generation of the gear tooth flank is feasible. To increase productivity of the gear finishing operation, it is required to maintain the tool paths on the gear-tooth flank P_g as wide as possible. For this purpose, the major axis of the spot of contact $\mathscr{C}_{pt}$ has to be as long as possible and the relative motion $\mathbf{V}_\Sigma$ of the surfaces P_g and T_{sh} has to be directed orthogonally to the major axis of the spot of contact $\mathscr{C}_{pt}$.

Fortunately, it is possible to control the shape, size, and orientation of the spot of contact $\mathscr{C}_{pt}$. For this purpose, an optimal combination of the design parameters of the shaving cutter, of the direction and speed of feed $\mathbf{F}_{diag}$, of the rotation of the gear ω_g, and of the rotation of the shaving cutter ω_{sh} must be calculated. This also makes possible the control of the direction of relative motion of the tooth flanks P_g and T_{sh} and in such a way that increases the gear accuracy and to cut the shaving time.

For the analysis below, the equations of the tooth flank surfaces P_g and T_{sh} of the gear and of the shaving cutter are necessary.

The equation for the position vector of a point $\mathbf{r}_g$ of the gear tooth flank P_g can be represented in the form of the column matrix (see Equation 1.62):

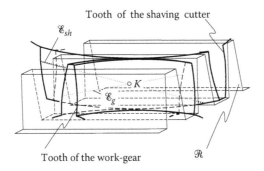

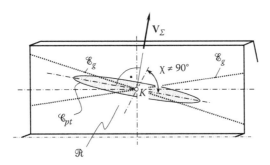

FIGURE 11.17
The problem at hand. (With kind permission from Springer Science+Business Media: *International Journal of Advanced Manufacturing Technology*, Diagonal shaving of an involute pinion: optimization of the geometric and kinematic parameters for the pinion finishing operation, 32, 2007, 1170–1187, Radzevich, S.P.)

$$
\mathbf{r}_g(U_g, V_g) = \begin{bmatrix}
r_{b.g} \cos V_g + U_g \cos \psi_{b.g} \sin V_g \\
r_{b.g} \sin V_g - U_g \sin \psi_{b.g} \sin V_g \\
r_{b.g} \tan \psi_{b.g} - U_g \sin \psi_{b.g} \\
1
\end{bmatrix}
\tag{11.16}
$$

where the gear base cylinder diameter $d_{b.g} = 2r_{b.g}$ can be calculated from the formula

$$
d_{b.g} = \frac{m \cdot N_g \cdot \cos \phi_n}{\sqrt{1 - \cos^2 \phi_n \sin^2 \lambda_{b.g}}} = \frac{25.4 \cdot N_g \cdot \cos \phi_n}{P_g \cdot \sqrt{1 - \cos^2 \phi_n \sin^2 \lambda_{b.g}}}
\tag{11.17}
$$

where m is the gear modulus, N_g is the number of gear teeth, ϕ_n is the normal pressure angle, $\lambda_{b.g}$ is the gear base lead angle ($\lambda_{b.g} = 90° - \psi_{b.g}$), $\psi_{b.g}$ is the gear base helix angle, and P_g is diametral pitch of the gear tooth flank.

The U_g parameter in Equation 11.16 can be expressed in terms of parameters of the gear design [117,143]:

$$U_g = \frac{\sqrt{d_{y.g}^2 - d_{b.g}^2}}{2 \sin \psi_{b.g}} = \frac{\sqrt{d_{y.g}^2 - d_{b.g}^2}}{2 \sin \psi_g \, \sin \phi_n} \tag{11.18}$$

where the diameter of a cylinder that is coaxial to the gear is designated as $d_{y.g}$ and ψ_g is the gear pitch helix angle.

Equations 1.14 through 1.16 yield calculation of the fundamental magnitudes of the first order:

$$E_g = 1 \tag{11.19}$$

$$F_g = -\frac{r_{b.g}}{\cos \psi_{b.g}} \tag{11.20}$$

$$G_g = \frac{U_g^2 \cos^4 \psi_{b.g} + r_{b.g}^2}{\cos^2 \psi_{b.g}} \tag{11.21}$$

for the screw involute tooth surface P_g of the gear.

For the fundamental magnitudes of the second order, the use of Equations 1.20 through 1.22 returns the expressions

$$L_g = 0 \tag{11.22}$$

$$M_g = 0 \tag{11.23}$$

$$N_g = -U_g \cdot \sin \tau_{b.g} \cdot \cos \tau_{b.g} \tag{11.24}$$

for the coefficients L_g, M_g, and N_g for the screw involute tooth surface P_g of the gear.

Equation 1.9 makes the following formula for the unit normal vector $\mathbf{n}_g$ to the gear tooth surface P_g possible:

$$\mathbf{n}_g = \begin{bmatrix} \sin \psi_{b.g} \sin V_g \\ \sin \psi_{b.g} \cos V_g \\ \cos \psi_{b.g} \\ 1 \end{bmatrix} \tag{11.25}$$

Calculations similar to those above must be performed for the generating surface T_{sh} of the shaving cutter.

Vector $\mathbf{V}_\Sigma$ of the resultant relative motion of the gear tooth flank P_g and of the shaving cutter tooth flank T_{sh} passes through the point K, and it is located in a common tangent plane to the surfaces P_g and T_{sh}.

Consider a plane through the unit normal vector $\mathbf{n}_g$ that is orthogonal to the direction of the vector $\mathbf{V}_\Sigma$. The radii of the curvature of the surfaces P_g and T_{sh} in this cross section differ from each other. The width of the tool path over the lateral tooth surface P_g depends upon the direction of the vector $\mathbf{V}_\Sigma$. By varying the direction of feed $\mathbf{F}_{diag}$, say, timing in various angular velocities ω_g and ω_{sh} with feed $\mathbf{F}_{diag}$, the tool paths of various width $\breve{F}_i$ could be obtained. The shortest shaving time and the highest accuracy of the involute gear tooth surface could be obtained if and only if the feed rate per tooth $\breve{F}_i$ of the shaving cutter remains equal to it maximal value—that is, if the equality $\breve{F}_i = \breve{F}_{cnf}^{max}$ is valid. To make the equality $\breve{F}_i = \breve{F}_{cnf}^{max}$ valid, it is necessary to remain the highest possible degree of conformity of the shaving cutter tooth surface T_{sh} to the gear tooth surface P_g at every point contact of the surfaces.

In general, the degree of conformity of an involute tooth surface T_{sh} of the shaving cutter to the involute tooth surface P_g of the gear to be finished at the point K of their contact varies, as the normal plane section rotates around the common unit vector $\mathbf{n}_g$. The direction of the major axis of the spot of contact aligns with the direction at which the highest degree of conformity of the involute tooth surfaces P_g and T_{sh} is observed.

The tangent plane to the gear tooth surface P_g at the point K is the plane through two unit tangent vectors $\mathbf{u}_g$ and $\mathbf{v}_g$. This yields the expression for the position vector of a point for the tangent plane through point K (i.e., through the point specified by the position vector $\mathbf{r}_K$) on the gear tooth surface P_g:

$$(\mathbf{r}_{g.tang} - \mathbf{r}_K) \times \mathbf{u}_g.\mathbf{v}_g = 0, \tag{11.26}$$

where $\mathbf{r}_{g.tang}$ is the position vector of a point of the tangent plane.

The angle μ of local relative orientation at a point of contact of the gear and the shaving cutter tooth surfaces (see Equations 4.1 through 4.3) is equal:

$$\sin \mu = \frac{\sin \phi_n \cdot \sin \Sigma}{\sqrt{(1 - \cos^2 \phi_n \cdot \sin^2 \psi_g) \cdot (1 - \cos^2 \phi_n \cdot \sin^2 \psi_{sh})}} \tag{11.27}$$

where ϕ_n is the normal pressure angle, ψ_g is the gear helix angle, ψ_{sh} is the shaving cutter helix angle, and Σ is the gear and the shaving cutter crossed-axes angle.

For the case under consideration, the equation of the indicatrix of conformity $Cnf_R(P_g/T_{sh})$ at a point of contact of the gear tooth flank P_g and of the shaving cutter tooth flank T_{sh} can be derived from the general form of equation of this characteristic curve (see Equation 4.87).

The first $\Phi_{1.g}$, $\Phi_{1.sh}$ and the second $\Phi_{2.g}$, $\Phi_{2.sh}$ fundamental forms are initially calculated in the coordinate systems $X_g Y_g Z_g$ and $X_{sh} Y_{sh} Z_{sh}$, respectively (see Figure 11.18). It is necessary to convert these expressions to the common local coordinate system $x_g y_g z_g$ having the origin at the contact point K. Such a

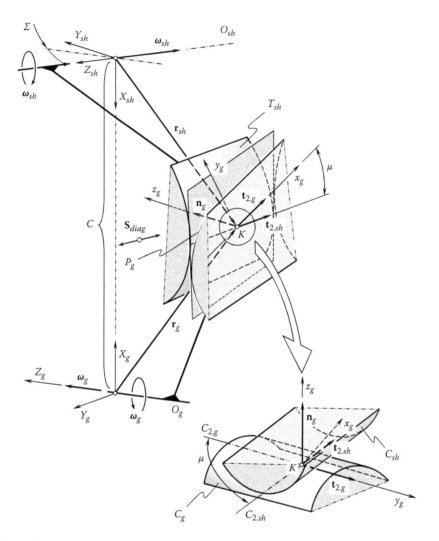

FIGURE 11.18
The major coordinate systems. (With kind permission from Springer Science+Business Media: *International Journal of Advanced Manufacturing Technology*, Diagonal shaving of an involute pinion: optimization of the geometric and kinematic parameters for the pinion finishing operation, 32, 2007, 1170–1187, Radzevich, S.P.)

transformation can be performed by means of the formula of quadratic form transformation (see Equations 3.68 and Equation 3.69):

$$[\Phi_{1.g(sh)}]_k = \mathbf{Rs}^T(1 \to 2) \cdot [\Phi_{1.g(sh)}]_{g(sh)} \cdot \mathbf{Rs}(1 \to 2) \qquad (11.28)$$

$$[\Phi_{2.g(sh)}]_k = \mathbf{Rs}^T(1 \to 2) \cdot [\Phi_{2.g(sh)}]_{g(sh)} \cdot \mathbf{Rs}(1 \to 2) \qquad (11.29)$$

where $[\Phi_{1,2.g}]_g$, $[\Phi_{1,2.g}]_k$, and $[\Phi_{1,2.sh}]_{sh}$, $[\Phi_{1,2.sh}]_k$ are the first and second fundamental forms of the contacting tooth flanks P_g and T_{sh} of the gear and the shaving cutter. These fundamental forms are initially represented in the coordinate systems $X_g Y_g Z_g$ and $X_{sh} Y_{sh} Z_{sh}$ associated with the gear and with the shaving cutter, respectively. Finally, these fundamental forms are represented in the common coordinate system $x_g y_g z_g$ associated with the contact point K.

In the local coordinate system $x_g y_g z_g$, the equation for the position vector of a point r_{cnf} for the indicatrix of conformity $Cnf_R(P_g/T_{sh})$ at a point of contact of the gear tooth flank P_g and of the shaving cutter tooth flank T_{sh} casts into

$$\text{Indicatrix of conformity } Cnf_R(P_g/T_{sh}) \Rightarrow r_{cnf}(R_{1.g}, R_{1.sh}, \mu, \varphi) = \frac{\sqrt{R_{1.g}}}{\sin\varphi} + \frac{\sqrt{R_{1.sh}}}{\sin(\mu - \varphi)}$$

(11.30)

where $r_{cnf}(R_{1.g}, R_{1.sh}, \mu, \varphi)$ is the position vector of a point of the characteristic curve $Cnf_R(P_g/T_{sh})$, wherein to finish a given gear, the function $r_{cnf}(R_{1.g}, R_{1.sh}, \mu, \varphi)$ reduces to $r_{cnf}(R_{1.sh}, \mu, \varphi)$, and φ is the polar angle (further, the argument φ is employed for determining the optimal direction of resultant relative motion $\mathbf{v}_\Sigma$ of the tooth flanks P_g and T_{sh}).

The characteristic curve $Cnf_R(P_g/T_{sh})$ is depicted in Figure 11.19. The degree of conformity of the gear tooth flank P_g and the shaving cutter tooth flank T_{sh}

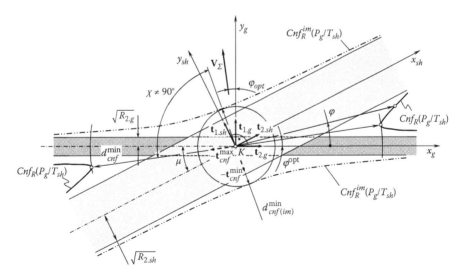

FIGURE 11.19
The indicatrix of conformity $Cnf_R(P_g/T_{sh})$ at the point of contact of the tooth flanks of the gear P_g and shaving cutter T_{sh}. (With kind permission from Springer Science+Business Media: *International Journal of Advanced Manufacturing Technology*, Diagonal shaving of an involute pinion: optimization of the geometric and kinematic parameters for the pinion finishing operation, 32, 2007, 1170–1187, Radzevich, S.P.)

in the normal cross section through the minimal diameter d_{cnf}^{min} (or, the same, through the direction $\mathbf{t}_{cnf}^{max}$ of the maximal degree of conformity of the surfaces P_g and T_{sh}) is the highest (Figure 11.20). This plane section of the contacting surfaces P_g and T_{sh} is referred to as the *optimal normal cross section*.

The equation of the indicatrix of conformity $Cnf_R(P_g/T_{sh})$ can be expressed in terms of the design parameters of the involute gear and of the shaving cutter:

$$r_{cnf}(\varphi, \mu) = \sqrt{\frac{2 \sin \psi_g \sin \phi_n}{\sqrt{d_{y.g}^2 - d_{b.g}^2} \cos \lambda_{b.g} \sin^2 \varphi}} + \sqrt{\frac{2 \sin \psi_{sh} \sin \phi_n}{\sqrt{d_{y.sh}^2 - d_{b.sh}^2} \cos \lambda_{b.sh} \sin^2 (\mu - \varphi)}}$$

(11.31)

which contains in condensed form all information that is necessary for the calculation of the most favorable design parameters of the shaving cutter and of the most favorable parameters of diagonal shaving operation.

The elements of local topology of the gear tooth flank P_g and the shaving cutter tooth flank T_{sh} relate to the lateral surface of the auxiliary phantom rack $\mathcal{R}$ of the shaving cutter. The location and relative orientation of the characteristic curve $Cnf_R(P_g/T_{sh})$ are illustrated in Figure 11.21.

The major axis of the spot of contact of the involute tooth flanks P_g and T_{sh} of the gear and of the shaving cutter aligns with the minimal diameter d_{cnf}^{min} of the characteristic curve $Cnf_R(P_g/T_{sh})$. This axis is within the angle that forms the characteristic $\mathcal{C}_g$ of the gear tooth flank P_g and the characteristic $\mathcal{C}_{sh}$ of the shaving cutter tooth flank T_{sh}. The characteristics $\mathcal{C}_g$ and $\mathcal{C}_{sh}$ (Figure 11.17) are the straight lines along which the involute tooth surfaces P_g and T_{sh} make contact with corresponding lateral plane surface of the auxiliary phantom rack $\mathcal{R}$. It is important to stress here that the major axis and the minor axis of the spot of contact are not orthogonal to each other. Generally, they are at an angle $\chi \neq 90°$. The involute gear and the shaving cutter relative motion $\mathbf{V}_\Sigma$ at contact

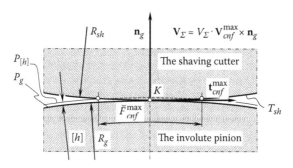

FIGURE 11.20
The cross section of the tooth surfaces P_g and T_{sh} by optimally oriented normal plane. (With kind permission from Springer Science+Business Media: *International Journal of Advanced Manufacturing Technology*, Diagonal shaving of an involute pinion: optimization of the geometric and kinematic parameters for the pinion finishing operation, 32, 2007, 1170–1187, Radzevich, S.P.)

point K is directed orthogonally to the minimal diameter $d_{cnf}^{\min}$ (to the unit tangent vector $\mathbf{t}_{cnf}^{\max}$). The cutting edge of the shaving cutter makes an angle of inclination i with the direction of the unit tangent vector $\mathbf{t}_{cnf}^{\max}$. It is important to maintain this angle equal to its optimal value i_{opt}. The same angle i_{opt} forms the vector $\mathbf{V}_{\Sigma}$ with the perpendicular to the cutting edge (Figure 11.21).

At every point of the tooth flank P_g of the gear, the first principal curvature $k_{1.g}$ is uniquely determined by the topology of the surface P_g. In the second principal curvature $k_{2.g}$ of the screw involute tooth surface, P_g is always equal to zero ($k_{2.g} \equiv 0$). Similarly, at every point of the tooth flank T_{sh} of the shaving cutter, the first principal curvature $k_{1.sh}$ is uniquely determined by the topology of the surface T_{sh}. The second principal curvature $k_{2.sh}$ of the screw involute tooth surface T_{sh} is also always equal to zero ($k_{1.sh} \equiv 0$). At this point, the rest of the parameters of the indicatrix of conformity $Cnf_R(P_g/T_{sh})$ at a point of contact of the surfaces P_g and T_{sh} (that is, the parameters $R_{1.sh}$, φ, and μ) can be considered as the variable parameters. It is necessary to determine the optimal combination of values of the parameters

$$R_{1.sh}^{opt} = R_{1.sh}^{opt}(U_g, V_g) \tag{11.32}$$

$$\varphi^{opt} = \varphi^{opt}(U_g, V_g) \tag{11.33}$$

$$\mu^{opt} = \mu^{opt}(U_g, V_g). \tag{11.34}$$

Once the proper combination of the parameters $R_{1.sh}^{opt}$, φ^{opt}, and μ^{opt} is determined, the calculation of the optimal design parameters of the shaving cutter

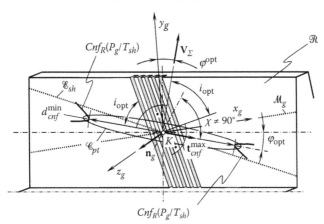

FIGURE 11.21
Elements of local topology of the tooth flanks P_g and T_{sh} referred to the lateral plane of the auxiliary phantom rack $\mathcal{R}$. (With kind permission from Springer Science+Business Media: *International Journal of Advanced Manufacturing Technology*, Diagonal shaving of an involute pinion: optimization of the geometric and kinematic parameters for the pinion finishing operation, 32, 2007, 1170–1187, Radzevich, S.P.)

and of the optimal parameters of kinematics of a diagonal shaving operation turns to the routing engineering calculations.

The indicatrix of conformity $Cnf_R(P_g/T_{sh})$ reveals how much the tooth surface T_{sh} of the shaving cutter is conformal to the gear tooth surface P_g in every cross section of the surfaces P_g and T_{sh} by normal plane through the contact point K. It enables the specification of an orientation of the normal plane section, at which surfaces P_g and T_{sh} are extremely conformal to each other—that is, the normal plane section through the unit tangent vector t_{cnf}^{max} in the direction of the maximal degree of conformity of the gear tooth flank P_g and the shaving cutter tooth flank T_{sh}. This normal plane section meets the following conditions:

$$\frac{\partial r_{cnf}}{\partial R_{1.sh}} = 0 \tag{11.35}$$

$$\frac{\partial r_{cnf}}{\partial \mu} = 0 \tag{11.36}$$

$$\frac{\partial r_{cnf}}{\partial \varphi} = 0. \tag{11.37}$$

Equations 11.30 of the indicatrix of conformity $Cnf_R(P_g/T_{sh})$ yields the following set of expressions for necessary conditions of the maximal degree of conformity of the shaving cutter tooth surface T_{sh} to the involute gear tooth surface P_g at a point of their contact:

The necessary conditions for the minimal shaving time and the maximal accuracy of the shaved involute gear

$$\Rightarrow \begin{cases} \dfrac{\partial r_{cnf}}{\partial R_{1.sh}} = \dfrac{1}{\sqrt{R_{1.sh}}\,\sin(\mu - \varphi)} = 0 \\[3mm] \dfrac{\partial r_{cnf}}{\partial \mu} = -\dfrac{\cos(\mu - \varphi)}{\sin^2(\mu - \varphi)}\sqrt{R_{1.sh}} = 0 \\[3mm] \dfrac{\partial r_{cnf}}{\partial \varphi} = -\dfrac{\cos \varphi}{\sin^2 \varphi}\sqrt{R_{1.g}} - \dfrac{\cos(\mu - \varphi)}{\sin^2(\mu - \varphi)}\sqrt{R_{1.sh}} = 0 \end{cases}$$

$$\tag{11.38}$$

The sufficient conditions for the maximum of the function $r_{cnf}(R_{1.sh}, \mu, \varphi)$ of three variables are also satisfied.

The first equality in the set of Equation 11.38 consists of the condensed form of all the necessary information on the optimal design parameters of the shaving cutter. Analysis of this equality reveals that it could be satisfied when $R_{1.sh} \to \infty$. Thus, for a conventional diagonal shaving operation when the gear and the shaving cutter are in external mesh, finishing the gear by the shaving cutter of the maximal possible pitch diameter is recommended. In the ideal case, the gear can be shaved with a rack-type shaving cutter. The application of the shaving cutter of larger pitch diameter increases the difference between pitch diameters of the gear and the shaving cutter. This ensures a larger degree of conformity of the gear tooth flank P_g and the shaving cutter tooth flank T_{sh} at the point of contact of these surfaces. Actually, the pitch diameter of the shaving cutter to be applied for a rotary shaving operation is restricted by the design of a shaving machine.

The analysis of the function $R_{1.sh} = R_{1.sh}(\phi_n, \psi_{sh})$ reveals that the degree of conformity of the tooth flanks P_g and T_{sh} increases when both normal pressure angle ϕ_n and helix angle ψ_{sh} are smaller—that is, $\phi_n \to 0°$ and $\psi_{sh} \to 0°$. The interested reader may wish to refer to Radzevich [115] for the details of the analysis.

The second and third equalities in the set of Equations 11.38 together enable one to give an answer to the question on the optimal relative orientation of the tooth flanks P_g and T_{sh} ($\mu \to 0°$; however, the inequality $\Sigma \neq 0°$ is required) and on the optimal parameters of instant kinematics of diagonal shaving operation ($\varphi = \varphi^{opt}$).

The resultant relative motion $\mathbf{V}_\Sigma$ of the tooth flanks P_g and T_{sh} of the gear and the shaving cutter is decomposed on its projections onto directions of the motions to be performed on the gear-shaving machine.

Vector $\mathbf{V}_{sl}$ of the velocity of relative slip between the tooth flanks P_g and T_{sh} of the gear and the shaving cutter is located in the common tangent plane. It is convenient to decompose the velocity vector $\mathbf{V}_{sl}$ at the contact point K into two components:

$$\mathbf{V}_{sl} = \mathbf{V}_\phi + \mathbf{V}_\psi. \tag{11.39}$$

where the first component $\mathbf{V}_\phi$ represents the slip along the tooth profile and the second component $\mathbf{V}_\psi$ represents slip in the longitudinal tooth direction.

The feed vector $\mathbf{F}_{diag}$ is directed parallel to the plane surface that is tangent to the pitch cylinders of the gear and the shaving cutter. It also affects the resultant speed $\mathbf{V}_\Sigma$ of cutting:

$$\mathbf{V}_\Sigma = \mathbf{V}_{sl} + \mathbf{F}_{diag}. \tag{11.40}$$

The varying parameters of the diagonal shaving operation and of design parameters of shaving cutter enable one to control the resultant speed $\mathbf{V}_\Sigma = \mathbf{V}_{sl} + \mathbf{F}_{diag}$ of cutting. For this purpose, the speed and direction of the shaving machine reciprocation and shaving cutter rotation have to be timed with each other.

In the local coordinate system $x_g y_g z_g$ (Figure 11.22), vector $\mathbf{V}_\Sigma$ of the resultant motion makes a certain angle φ_Σ with y_g axis. Thus,

$$\mathbf{V}_\Sigma = \begin{bmatrix} |\mathbf{V}_\Sigma| \cdot \sin \varphi_\Sigma \\ |\mathbf{V}_\Sigma| \cdot \cos \varphi_\Sigma \\ 0 \\ 1 \end{bmatrix} \tag{11.41}$$

To represent vector $\mathbf{V}_\Sigma$ in the global coordinate system $X_k Y_k Z_k$ (Figure 11.22), the operator $\mathbf{Rs}(g \rightarrow k)$ of the resultant coordinate system transformation is used:

$$\mathbf{V}_\Sigma^* = \mathbf{Rs}(g \rightarrow k) \cdot \mathbf{V}_\Sigma = \begin{bmatrix} |\mathbf{V}_\Sigma^*| \\ |\mathbf{V}_\Sigma^*| \\ |\mathbf{V}_\Sigma^*| \\ 1 \end{bmatrix} \tag{11.42}$$

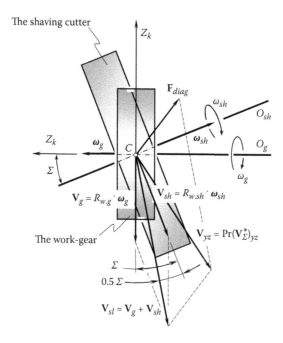

FIGURE 11.22
Timing of the feed F_{diag} with rotations of the involute gear and shaving cutter. (With kind permission from Springer Science+Business Media: *International Journal of Advanced Manufacturing Technology*, Diagonal shaving of an involute pinion: optimization of the geometric and kinematic parameters for the pinion finishing operation, 32, 2007, 1170–1187, Radzevich, S.P.)

Equation 11.42 yields the projection $\text{Pr}_{yz}(\mathbf{V}_\Sigma^*)$ of the vector $\mathbf{V}_\Sigma^*$ onto the coordinate plane $Y_k Z_k$:

$$\text{Pr}_{yz}(\mathbf{V}_\Sigma^*) = \begin{bmatrix} 0 \\ |\mathbf{V}_\Sigma^*| \\ |\mathbf{V}_\Sigma^*| \\ 1 \end{bmatrix} \tag{11.43}$$

The relative slip vector $\mathbf{V}_{sl}$ of the tooth flanks of the gear and the shaving cutter can be calculated by

$$\mathbf{V}_{sl} = \mathbf{V}_g + \mathbf{V}_{sh} = R_{w.g} \cdot \omega_g + R_{w.sh} \cdot \omega_{sh} \tag{11.44}$$

where $\mathbf{V}_g$, $\mathbf{V}_{sh}$ are the linear velocities of the rotations ω_g and ω_{sh}, respectively, $R_{w.g}$, $R_{w.sh}$ are the radii of pitch cylinders of the gear and the shaving cutter, and

$$|\mathbf{V}_{sl} = 2| \, \omega_g \, | \, R_{w.g} \cdot \cos(0.5 \cdot \Sigma) = 2 \cdot |\omega_{sh}| \cdot R_{w.sh} \cdot \cos(0.5 \cdot \Sigma) \tag{11.45}$$

In the coordinate plane $Y_k Z_k$, the vector $\mathbf{V}_{yz}$ of the resultant motion of the gear and the shaving cutter can be represented as follows:

$$\mathbf{V}_{yz} = \mathbf{V}_{sl} + \mathbf{F}_{diag} \tag{11.46}$$

Thus, reciprocation is equal to

$$\mathbf{F}_{diag} = \mathbf{V}_{sl} - \mathbf{V}_{yz} \tag{11.47}$$

This is the way the values of the shaving cutter rotation and its reciprocation are timed with each other.

The synthesized method of diagonal shaving of involute gears is disclosed in detail in References [49,115,145,146].

Conclusion

A novel method of surface generation for the purposes of part surface machining on a multiaxis numerical control (*NC*) machine as well as on a machine tool of conventional design is disclosed in this monograph. The method is developed on the premise of wide use of differential geometry of surfaces and elements of kinematics of multiparametric motion of rigid body in *Euclidian* space (E_3). Because of this, the proposed method is referred to as the *DG/K-based method of part surface generation*.

The *DG/K*-based method targets the *synthesis* of the most favorable (optimal, in a certain sense) methods of part surface machining and of the most favorable (or optimal) form-cutting tools for machining a given part surface.

A minimal amount of input information is required for the implementation of the *DG/K*-based method. Potentially, the method is capable of synthesizing an optimal part surface machining process on the premise only of the geometry of the part surface to be machined. However, any additional information on the part surface machining process can be incorporated as well. Ultimately, the use of the *DG/K*-based method of part surface generation enables one to get a maximum amount of output information on the part surface machining process while using for this purpose a minimum amount of input information, which illustrates the significant capacity of the disclosed novel method of part surface generation.

The generation of surfaces when machining parts on conventional machine tools can be interpreted as a degenerate case of sculptured part surface machining on a multiaxis numerical control machine. In this particular case, the *DG/K*-based method reduces to simpler kinematics of surface generation and simpler methods of generating surfaces of form-cutting tools. The kinematics of part surface generation in this case features the constant parameters of the relative motion of the cutting tool that do not commonly change when machining. Because of that, the kinematics of surface generation of this sort is called *rigid* *kinematics of part surface generation*. For the rigid kinematics of part surface generation, the elements of the enveloping surfaces were further developed. In particular, a novel class of surfaces is discovered. The surfaces of this class are referred to as *reversibly enveloping surfaces* (R_e surfaces, for simplicity). The involute tooth profile in parallel-axis gearing is a particular representative of R_e surfaces.

The disclosed *DG/K*-based method of part surface generation can be considered as a foundation of *theoretical manufacturing technology* that will be developed in the future. In the last few decades, the possibility of the development of *theoretical manufacturing technology* has been under discussion by leading experts in the field all around the world.

A developed *DG/K*-based method of part surface generation is the cornerstone of the subject *theoretical machining/production technology* for university students.

Appendix A: Elements of Vector Calculus

The *vector*, the key to all the theory of part surface generation, is a triple real number (in most computer languages these are usually called *floating point numbers*) and are noted in a **bold** typeface, e.g., *A* or *a*.

Care must be taken to differentiate between two types of vectors:

- *Position vector*: A position vector runs from the origin of coordinate (0, 0, 0) to a point (X, Y, Z), and its length gives the distance of the point from the origin. Its components are given by (X, Y, Z). The essential concept to understand about a position vector is that it is anchored to specific coordinates (points in space). The set of points that are used to describe the shape of all part surfaces can be thought of as position vectors.

- *Direction vector*: A direction vector differs from a position vector in that it is *not* anchored to specific coordinates. Frequently, direction vectors are used in a form where they have unit length; in this case, they are said to be *normalized*. The most common application of a direction vector in the theory of part surface generation is to specify the orientation of a surface or ray direction. For this, we use a direction vector at right angles (*normal*) and pointing away from the part surface. Such *normal* vectors are also the key in many calculations in the theory of part surface generation.

Vector calculus is a powerful tool for solving many of geometrical and kinematical problems that pertain to the design and generation of part surfaces. In this book, *vectors* are understood as quantities that have magnitude and direction and obey the law of addition.

A.1 Fundamental Properties of Vectors

The distance-and-direction interpretation suggests a powerful way to visualize a vector, and that is as a directed line segment or arrow. The length of the arrow (at some predetermined scale) represents the magnitude of the vector, and the orientation of the segment and placement of the arrowhead (at one end of the segment or the another) represent its direction.

Vectors possess certain properties, the set of which is commonly interpreted as the set of fundamental properties of vectors.

Addition. Given two vectors a and b, their sum $(\mathbf{a} + \mathbf{b})$ is graphically defined by joining the tail of b to the head of a. Then, the line from the tail of a to the head of b is the sum $\mathbf{c} = (\mathbf{a} + \mathbf{b})$.

Equality. Two vectors are equal when they have the same magnitude and direction. The position of the vectors is unimportant for equality.

Negation. The vector -a has the same magnitude as a but opposite direction.

Subtraction. From the properties of *addition* and *negation*, the following $\mathbf{a} - \mathbf{b} = \mathbf{a} + (-\mathbf{b})$ can be defined.

Scalar multiplication. The vector $k\mathbf{a}$ has the same direction as a, with a magnitude k times that of $a \cdot k$ is called a scalar as it changes the scale of the vector a.

A.2 Mathematical Operations over Vectors

The following rules and mathematical operations can be determined from the above-listed fundamental properties of vectors.

Let us assume that a set of three vectors a, b, c, and two scalars k and t are given. Then, vector addition and scalar multiplication have the following properties:

$$\mathbf{a} + \mathbf{b} = \mathbf{b} + \mathbf{a} \tag{A.1}$$

$$\mathbf{a} + (\mathbf{b} + \mathbf{c}) = (\mathbf{a} + \mathbf{b}) + \mathbf{c} \tag{A.2}$$

$$k(t\mathbf{a}) = kt\mathbf{a} \tag{A.3}$$

$$(k + t)\mathbf{a} = k\mathbf{a} + t\mathbf{a} \tag{A.4}$$

$$k(\mathbf{a} + \mathbf{b}) = k\mathbf{a} + k\mathbf{b} \tag{A.5}$$

The magnitude a of a vector a is

$$a = |\mathbf{a}| = \sqrt{a_x^2 + a_y^2 + a_z^2} \tag{A.6}$$

where a_x, a_y, and a_z are the scalar components of a.

A unit vector $\bar{\mathbf{a}}$ in the direction of a vector a is

$$\bar{\mathbf{a}} = \frac{\mathbf{a}}{|\mathbf{a}|} = \frac{\mathbf{a}}{a} \tag{A.7}$$

The components $\bar{a}_x$, $\bar{a}_y$, and $\bar{a}_z$ of a unit vector $\bar{\mathbf{a}}$ are also the direction cosines of the vector $\bar{\mathbf{a}}$:

$$\cos\alpha = \bar{a}_x \tag{A.8}$$

$$\cos\beta = \bar{a}_y \tag{A.9}$$

$$\cos\gamma = \bar{a}_z \tag{A.10}$$

It is common practice to denote the components $\bar{a}_x$, $\bar{a}_y$, and $\bar{a}_z$ by l, m, and n, respectively.

Scalar product (or dot product) of vectors: The formula

$$\mathbf{a} \cdot \mathbf{b} = a_x b_x + a_y b_y + a_z b_z = |\mathbf{a}||\mathbf{b}|\cos \angle (\mathbf{a}, \mathbf{b}) \tag{A.11}$$

is commonly used for calculation of scalar product of two vectors a and b.

Equation A.11 can also be represented in the form

$$\mathbf{a} \cdot \mathbf{b} = [\mathbf{a}]^T \cdot [\mathbf{b}] = [a_x \ a_y \ a_z] \cdot \begin{bmatrix} b_x \\ b_y \\ b_z \end{bmatrix} \tag{A.12}$$

The angle $\angle (\mathbf{a}, \mathbf{b})$ between two vectors a and b is calculated from

$$\angle (\mathbf{a}, \mathbf{b}) = \cos^{-1}\left(\frac{\mathbf{a} \cdot \mathbf{b}}{|\mathbf{a}||\mathbf{b}|}\right) \tag{A.13}$$

The scalar product of two vectors a and b features the following properties:

$$\mathbf{a} \cdot \mathbf{a} = |\mathbf{a}|^2 \tag{A.14}$$

$$\mathbf{a} \cdot \mathbf{b} = \mathbf{b} \cdot \mathbf{a} \tag{A.15}$$

$$\mathbf{a} \cdot (\mathbf{b} + \mathbf{c}) = \mathbf{b} \cdot \mathbf{a} + \mathbf{b} \cdot \mathbf{c} \tag{A.16}$$

$$(k\mathbf{a}) \cdot \mathbf{b} = \mathbf{a} \cdot (k\mathbf{b}) = k(\mathbf{a} \cdot \mathbf{b}) \tag{A.17}$$

If a is perpendicular to b, then

$$\mathbf{a} \cdot \mathbf{b} = 0 \tag{A.18}$$

Vector product (or *cross product*) **of two vectors:** The vector product of two vectors can be calculated from the formula

$$\mathbf{a} \times \mathbf{b} = (a_y b_z - a_z b_y)\mathbf{i} + (a_z b_x - a_x b_z)\mathbf{j} + (a_x b_y - a_y b_x)\mathbf{k} \tag{A.19}$$

Here, in Equation A.19, i, j, and k are unit vectors in the X, Y, and Z directions of the reference system XYZ, in which the vectors a and b are specified.

The vector product possesses the following property: in case $\mathbf{a} \times \mathbf{b} = \mathbf{c}$, then the vector c is perpendicular to a plane through the vectors a and b.

The vector product of two vectors a and b features the following properties:

$$\mathbf{a} \times \mathbf{b} = \begin{vmatrix} \mathbf{i} & \mathbf{j} & \mathbf{k} \\ a_x & a_y & a_z \\ b_x & b_y & b_z \end{vmatrix} \tag{A.20}$$

$$\mathbf{a} \times \mathbf{b} = |\mathbf{a}| |\mathbf{b}| \ \mathbf{n} \sin \angle (\mathbf{a}, \mathbf{b}) \tag{A.21}$$

where unit normal vector to the plane through the vectors a and b is denoted by n.

$$|\mathbf{a} \times \mathbf{b}| = |\mathbf{a}| |\mathbf{b}| \sin \angle (\mathbf{a}, \mathbf{b}) \tag{A.22}$$

The coordinates of the vector product $\mathbf{a} \times \mathbf{b}$ can also be expressed in the form

$$|\mathbf{a} \times \mathbf{b}| = \begin{bmatrix} 0 & -a_z & a_y \\ a_z & 0 & -a_x \\ -a_y & a_x & 0 \end{bmatrix} \cdot \begin{bmatrix} b_x \\ b_y \\ b_z \end{bmatrix} = \begin{bmatrix} -a_z b_y + a_y b_z \\ -a_x b_z + a_z b_x \\ -a_y b_x + a_x b_y \end{bmatrix} \tag{A.23}$$

$$\mathbf{a} \times \mathbf{b} = -\mathbf{b} \times \mathbf{a} \tag{A.24}$$

$$\mathbf{a} \times (\mathbf{b} + \mathbf{c}) = \mathbf{a} \times \mathbf{b} + \mathbf{a} \times \mathbf{c} \tag{A.25}$$

$$(k\mathbf{a}) \times \mathbf{b} = \mathbf{a} \times (k\mathbf{b}) = k(\mathbf{a} \times \mathbf{b}) \tag{A.26}$$

$$\mathbf{i} \times \mathbf{j} = \mathbf{k}, \quad \mathbf{j} \times \mathbf{k} = \mathbf{i}, \quad \mathbf{k} \times \mathbf{i} = \mathbf{j} \tag{A.27}$$

If a is parallel to b, then

$$\mathbf{a} \times \mathbf{b} = 0 \tag{A.28}$$

Triple scalar product **of three vectors:** The product $(\mathbf{a} \times \mathbf{b}) \cdot \mathbf{c}$ is commonly referred to as *triple scalar product* of three vectors a, b, and c.

The triple scalar product of three vectors a, b, and c features the following properties:

$$(\mathbf{a} \times \mathbf{b}) \cdot \mathbf{c} = (\mathbf{b} \times \mathbf{c}) \cdot \mathbf{a} = (\mathbf{c} \times \mathbf{a}) \cdot \mathbf{b} \qquad (A.29)$$

$$(\mathbf{b} \times \mathbf{c}) \cdot \mathbf{a} = \mathbf{a} \cdot (\mathbf{b} \times \mathbf{c}) \qquad (A.30)$$

$$(\mathbf{a} \times \mathbf{b}) \cdot \mathbf{c} = \mathbf{a} \cdot (\mathbf{b} \times \mathbf{c}) \qquad (A.31)$$

$$\mathbf{a} \cdot (\mathbf{b} \times \mathbf{c}) = \begin{vmatrix} a_x & a_y & a_z \\ b_x & b_y & b_z \\ c_x & c_y & c_z \end{vmatrix} \qquad (A.32)$$

Triple vector product **of three vectors:** The product $(\mathbf{a} \times \mathbf{b}) \times \mathbf{c}$ is commonly referred to as *triple vector product* of three vectors a, b, and c.

The product $(\mathbf{a} \times \mathbf{b}) \times \mathbf{c}$ can be evaluated by two vector products. However, it also can be evaluated in a more simple way by use of the identity

$$(\mathbf{a} \times \mathbf{b}) \times \mathbf{c} = (\mathbf{a} \cdot \mathbf{c}) \, \mathbf{b} - (\mathbf{b} \cdot \mathbf{c}) \, \mathbf{a} \qquad (A.33)$$

It should be mentioned here that, in general, the triple vector products $(\mathbf{a} \times \mathbf{b}) \times \mathbf{c}$ and $\mathbf{a} \times (\mathbf{b} \times \mathbf{c})$ are not equal

$$(\mathbf{a} \times \mathbf{b}) \times \mathbf{c} \neq \mathbf{a} \times (\mathbf{b} \times \mathbf{c}) \qquad (A.34)$$

Analytical interpretation of many problems and results in the field of geometry of surfaces becomes much simpler when vector calculus is used.

Lagrange **equation for vectors:** For the purposes of calculation of mixed product of vectors a and b, the following equation can be used.

$$(\mathbf{a} \times \mathbf{b}) \cdot (\mathbf{a} \times \mathbf{b}) = (\mathbf{a} \cdot \mathbf{a})(\mathbf{b} \cdot \mathbf{b}) - (\mathbf{a} \cdot \mathbf{b})^2 \qquad (A.35)$$

Equation A.35 is due to *Lagrange*.[*]

---

[*] *Joseph-Louis Lagrange* (January 25, 1736—April 10, 1813), a famous French mathematician, astronomer, and mechanician.

Appendix B: Change of Surface Parameters

When designing a form-cutting tool, it is often necessary to treat two or more surfaces simultaneously. For example, the cutting edge of the cutting tool can be considered as the line of intersection of the generating surface T of the form-cutting tool by the rake surface R_s. The equation of the cutting edge cannot be derived on the premise of equations of the surfaces T and R_s as long as the initial parameterization of the surfaces is improper.

If two surfaces $\mathbf{r}_i$ and $\mathbf{r}_j$ are treated simultaneously, then it is required that they are not only represented in a common reference system, but the U_i and V_i parameters of one of the surfaces $\mathbf{r}_i = \mathbf{r}_i(U_i, V_i)$ have to be synchronized with the corresponding U_j and V_j parameters of another surface $\mathbf{r}_j = \mathbf{r}_j(U_j, V_j)$. The procedure of changing of surface parameters is used for this purpose. The use of the procedure allows the representation of one of the surfaces, for example, of the surface $\mathbf{r}_j = \mathbf{r}_j(U_j, V_j)$ in the terms of the U_i and V_i parameters, say, as $\mathbf{r}_j = \mathbf{r}_j(U_i, V_i)$.

If the parameterization of a surface is transformed by the equations $U^* = U^*(U, V)$ and $V^* = V^*(U, V)$, we obtain the new derivatives

$$\frac{\partial \mathbf{r}}{\partial U^*} = \frac{\partial \mathbf{r}}{\partial U} \cdot \frac{\partial U}{\partial U^*} + \frac{\partial \mathbf{r}}{\partial V} \cdot \frac{\partial V}{\partial U^*} \tag{B.1}$$

$$\frac{\partial \mathbf{r}}{\partial V^*} = \frac{\partial \mathbf{r}}{\partial U} \cdot \frac{\partial U}{\partial V^*} + \frac{\partial \mathbf{r}}{\partial V} \cdot \frac{\partial V}{\partial V^*} \tag{B.2}$$

Thus,

$$\mathbf{A}^* = \left[\frac{\partial \mathbf{r}}{\partial U^*} \middle| \frac{\partial \mathbf{r}}{\partial V^*} \right] = \mathbf{A} \cdot \mathbf{J} \tag{B.3}$$

where

$$\mathbf{J} = \begin{bmatrix} \dfrac{\partial U}{\partial U^*} & \dfrac{\partial U}{\partial V^*} \\ \dfrac{\partial V}{\partial U^*} & \dfrac{\partial V}{\partial V^*} \end{bmatrix} \tag{B.4}$$

atrix of the transformation.

It can be shown that the new fundamental matrix $\mathbf{G}^*$ is given by

$$\mathbf{G}^* = \mathbf{A}^{*T}\,\mathbf{A}^* = \mathbf{J}^T\mathbf{A}^T\mathbf{A}\mathbf{J} = \mathbf{J}^T\mathbf{G}\mathbf{J} \tag{B.5}$$

From this equation, we see by the properties of determinants that $|\mathbf{G}^*| = |\mathbf{J}|^2|\mathbf{G}|$. Using this result and Equation B.2, we can show that the unit surface normal n is invariant under the transformation, as we could expect.

The transformation of the second fundamental matrix can similarly be shown by

$$\mathbf{D}^* = \mathbf{J}^T\mathbf{D}\mathbf{J} \tag{B.6}$$

by differentiating Equation B.2 and using the invariance of n. From Equations B.5 and B.6, it can be shown that the principal curvatures and directions are invariant under the transformation.

We conclude that the unit normal vector n and the principal directions and curvatures are independent of the parameters used and are therefore geometric properties of the surface itself. They should be continuous if the surface is to be tangent and curvature continuous.

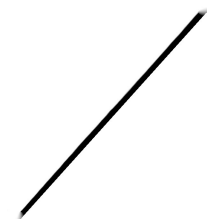

Appendix C: Non-Cartesian Reference Systems

In the theory of part surface generation, in addition to *Cartesian* coordinate systems, several types of non-*Cartesian* reference systems are used as well. Spherical polar system and cylindrical coordinate system are examples of non-*Cartesian* reference systems used in the theory of part surface generation.

C.1 Spherical Polar System

The conventional spherical polar coordinate system in relation to the *Cartesian* axes is illustrated in Figure C.1. r is a measure of the distance from the origin to a point in space. Angles θ and φ are taken relative to the Z and X axes, respectively. Unlike the *Cartesian* X, Y, and Z values, which all take the same units, spherical polar coordinates use both distance and angle measures. Importantly, there are some points in space that do not have a unique one-to-one relationship with an (r, θ, φ) coordinate value.

It is quite straightforward to change from one coordinate system to the other. When the point m in Figure C.1 is expressed as (r, θ, φ), the *Cartesian* coordinates (XYZ) are given by the trigonometric expressions:

$$X = r \sin\theta \cos\varphi \qquad\qquad (C.1)$$

$$Y = r \sin\theta \sin\varphi \qquad\qquad (C.2)$$

$$Z = r \cos\theta \qquad\qquad (C.3)$$

Conversion from *Cartesian* to spherical is a little more tricky:

$$r = \sqrt{X^2 + Y^2 + Z^2} \qquad\qquad (C.4)$$

$$\theta = \cos^{-1}\left(\frac{Z}{\sqrt{X^2 + Y^2 + Z^2}}\right) \qquad\qquad (C.5)$$

$$\varphi = \tan^{-1}\left(\frac{Y}{X}\right) \qquad\qquad (C.6)$$

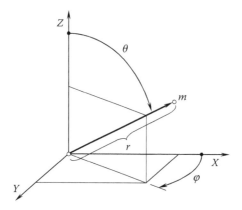

FIGURE C.1
A spherical reference system.

C.2 Cylindrical System

The conventional cylindrical coordinate system (radius ρ, azimuth φ, and elevation Z) in relation to the *Cartesian* axes is illustrated in Figure C.2. It is quite straightforward to change from one coordinate system to the other.

When the point m in Figure C.2 is expressed as (ρ, φ, Z), the *Cartesian* coordinates (XYZ) are given by the trigonometric expressions:

$$X = \rho \cos\varphi \tag{C.7}$$

$$Y = \rho \sin\varphi \tag{C.8}$$

$$Z = Z \tag{C.9}$$

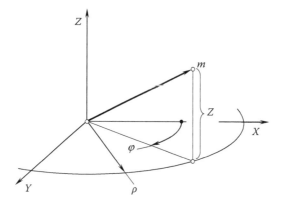

FIGURE C.2
A cylindrical reference system.

Conversion from *Cartesian* to cylindrical coordinates can be performed using the following equations:

$$\rho = \sqrt{X^2 + Y^2} \tag{C.10}$$

$$\varphi = \tan^{-1}\left(\frac{Y}{X}\right) \tag{C.11}$$

$$Z = Z \tag{C.12}$$

The cylindrical coordinates of a point m can be converted into spherical coordinates (radius r, inclination θ, and azimuth φ) of that same point by the formulae

$$r = \sqrt{\rho^2 + Z^2} \tag{C.13}$$

$$\theta = \tan^{-1}\left(\frac{\rho}{Z}\right) = \cos^{-1}\left(\frac{Z}{r}\right) \tag{C.14}$$

$$\varphi = \varphi \tag{C.15}$$

Conversely, the spherical coordinates may be converted into cylindrical coordinates by the formulae

$$\rho = r \sin\theta \tag{C.16}$$

$$\varphi = \varphi \tag{C.17}$$

$$Z = r \cos\theta \tag{C.18}$$

These formulae assume that the two reference systems have the same origin and same reference plane, measure the azimuth angle φ in the same sense from the same axis, and that the spherical angle θ is inclination from the cylindrical Z axis.

References

1. Amirouche, F.M.L., *Computer-Aided Design and Manufacturing*, Prentice Hall, Englewood Cliffs, NJ, 1993, 538 pp.
2. Ball, R.S., *A Treatise on the Theory of Screws*, Cambridge: University Press, 1990.
3. Ball, R.S., *The Theory of Screws: A Study in the Dynamics of Rigid Body*, Hodges & Foster, Dublin, 1876.
4. Banchoff, T., Gaffney, T., McCrory, C., *Cusps of Gauss Mapping*, Pitman Advanced Publishing Program, Boston, 1982, 88 pp.
5. Boehm, W., Differential geometry II, in: Farin, G., *Curves and Surfaces for Computer Aided Geometric Design. A Practical Guide*, 2nd ed., Academic Press, Boston, 1990, pp. 367–383.
6. Bonnet, P.O., *J. Ec. Polytech*, xiii, 31, 1867.
7. Cauchy, A.L., Leçons sur les Applications du Calcul Infinitésimal á la Geometrie, Imprimerie royale, Paris, 1826.
8. Chang, C.-H., Melkanoff, M.M., *NC Machine Programming and Software Design*, Prentice Hall, Englewood Cliffs, NJ, 1989, 589 pp.
9. Choi, B.K., Jerard, R.B., *Sculptured Surface Machining. Theory and Application*, Kluwer Academic Publishers, Dordrecht, 1998, 368 pp.
10. Cormac, P., *A Treaties on Screws and Worm Gear, Their Mills and Hobs*, Chapman & Hall, London, 1936, 138 pp.
11. Darboux, G., Leçons la Théorie Générale des Surfaces et ses Applications Géométiques du Calcul Infinitésimal, Vol. 1, Gauthier-Villars, Paris, 1887.
12. Darboux, G., Sur le Déplacement d'une Figure Invariable, *C. R. Acad. Sci.*, XCII, 1981, pp. 118–121 [quoted after Bottema, O., Roth, B., *Theoretical Kinematics*, Dover Publications, New York, 1990, 558 pp.].
13. Denavit, J., Hartenberg, R.S., A kinematics notation for lower-pair mechanisms based on matrices, *ASME J. Appl. Mech.*, Vol. 77, 1955, pp. 215–221.
14. doCarmo, M.P., *Differential Geometry of Curves and Surfaces*, Prentice-Hall, Englewood Cliffs, NJ, 1976, 503 pp.
15. Eisenhart, L.P., *A Treatise on the Differential Geometry of Curves and Surfaces*, Dover Publications, London, 1909; New York, reprint 1960, 474 pp.
16. Faux, L.D., Pratt, M.J., *Computational Geometry for Design and Manufacture*, Ellis Horwood, Chichester, John Wiley & Sons, New York, 1987, 331 pp.
17. Favard, J., *Course de Gèomètrie Diffèrentialle Locale*, Gauthier-Villars, Paris, 1957, viii+553 pp.
18. Ferguson, R.S., *Practical Algorithms for 3D Computer Graphics*, AK Peters, Natick, MA, 2001, 539 pp.
19. Fisher, G. (Ed.), *Mathematical Models*, Friedrich Vieweg & Sohn, Braunachweig/Wiesbaden, 1986.
20. Galimov, K.Z., Paimushin, V.N., *Theory of Shells of Complex Geometry*, Kazan', Kazan' State University Press, 1985, 164 pp.

21. Gauss, K.-F., *Disquisitions Generales Circa Superficies Curvas*, Goettingen (1828). (English translation: *General Investigation of Curved Surfaces*, by Moreheat, J.C., Hiltebeitel, A.M., Princeton, 1902, reprinted with introduction by Courant, Raven Press, Hewlett, New York, 1965, 119 pp.)

22. Gray, A., Plücker's conoid, in: *Modern Differential Geometry of Curves and Surfaces with Mathematics*, 2nd ed., CRC Press, Boca Raton, FL, 1997, pp. 435–437.

23. Hertz, H., The contact of solid elastic bodies [Über die Berührung Fester Elastischer Körper], *J. Reine Angew. Math.*, 1981, pp. 156–171; *The Contact of Solid Elastic Bodies and Their Harnesses [Über die Berührung Fester Elastischer Körper und Über die Härte]*, Berlin, 1882; Reprinted in: Hertz, H., *Collected Works [Gesammelte Werke]*, Vol. 1, pp. 155–173 and pp. 174–196, Leipzig, 1895; English translation: *Miscellaneous Papers*, translated by Jones, D.E., Schott, G.A., pp. 146–162, 163–183, McMillan, London, 1896.

24. Conoïde de Plücker, http://www.mathcurve.com/surfaces/plucker/plucker. shtml.

25. Plücker's conoid, http://www.math.hmc.edu/faculty/gu/curves_and_surfaces/ surfaces/plucker.html.

26. *Improvement of Gear Cutting Tools*, Collected Papers, NIIMASh, Moscow, 1989, 571 pp.

27. Ismail, F., Elbestawi, M.A., Du, R., Urbasik, K., Generation of milled surfaces including tool dynamics and wear, *ASME J. Eng. Ind.*, Vol. 115, August 1993, pp. 245–252.

28. Jeffreys, H., *Cartesian Tensors*, Cambridge, University Press, 1961, 93 pp.

29. Kells, L.M., Kern, W.F., Bland, J.R., *Plane and Spherical Trigonometry*, 3rd ed., McGraw-Hill, NY, 1951, 318 pp.

30. Koenderink, J.J., *Solid Shape*, The MIT Press, Cambridge, MA, 1990, 699 pp.

31. Kovacic, I., The chatter vibrations in metal cutting—theoretical approach, *Facta Univ. Mech. Eng.*, Vol. 1, No. 5, 1988, pp. 581–593.

32. Kunstetter, S., *Narzędzia skrawające do metali: Konstrukcja*, 2nd ed., Wydawnictwa Naukowo-Techniczne, Warszawa, 1970, 949 pp. [1st ed., 1961, 715 pp].

33. Lagrange, J.L., *Theorié des Fonctions Analytiques*, Impr. de la République, Paris, prairial an V, 1797, 277 pp.

34. Larin, M.N., *Optimal Geometrical Parameters of Metal Cutting Tools*, Oborongiz, Moscow, 1953.

35. Ligun, A.A., Shumeiko, A.A., Radzevich, S.P., Goodman, E.D., Asymptotically optimal disposition of tangent points for approximation of smooth convex surfaces by polygonal functions, *Comput. Aided Geom. Des.*, Vol. 14, 1997, pp. 533–546.

36. Lowe, P.G., A note on surface geometry with special reference to twist, *Math. Proc. Cambridge Philos. Soc.*, Vol. 87, 1980, pp. 481–487.

37. Lowe, P.G., *Basic Principles of Plate Theory*, Surrey University Press, Glasgow, 1982.

38. L'ukshin, V.S., *Theory of Screw Surfaces in Cutting Tool Design*, Mashinostroyeniye, Moscow, 1968, 372 pp.

39. Marciniak, K., *Geometric Modeling for Numerically Controlled Machining*, Oxford University Press, New York, 1991, 245 pp.

40. Miron, R., Observatii a Supra Unor Formule din Geometria Varietatilor Neonolonome E_3^2, *Bull. Inst. Politech. Iasi*, IPI Press, Iasi, 1958.

41. Monge, G., *Application de l'analyse à la géométrie*, Bachelier, 1850.

42. Mortenson, M.E., *Geometric Modeling*, John Wiley & Sons, NY, 1985, 763 pp.
43. Mortenson, M., *Mathematics for Computer Graphics Application*, 2nd ed., Industrial Press Inc., New York, 1999, 354 pp.
44. Mozhayev, S.S., *Analytical Theory of Twist Drills*, Mashgiz, Moscow, 1948, 136 pp.
45. Murray, R.M., Zexiang, L., Sastry, S.S., *A Mathematical Introduction to Robotic Manipulation*, CRC Press, Boca Raton, 1994, 456 pp.
46. Nutbourn, A.W., A circle diagram for local differential geometry, in Gregory, J. (ed.) *Mathematics of Surfaces*, Conference Proceedings, Institute of Mathematics and its Application, 1984, Oxford University Press, Oxford, 1986.
47. Nutbourn, A.W., Martin, R.R., *Differential Geometry Applied to Curve and Surface Design*, Volume 1: Foundations, Ellis Horwood, Chichester, 1988, 282 pp.
48. Olivier, T., *Theorie Geometrique des Engrenages*, Paris, 1842, 118 pp.
49. Palaguta, V.A., *The Development and Investigation of Methods for Increasing Productivity of Shaving of Cylindrical Gears*, Ph.D. thesis, Kiev, Kiev Polytechnic Institute, 1995, 220 pp.
50. Pankin, A.V., Kinematic acuting of the cutting wedge, *Stanki i Instrumetnt*, No. 1, 1936.
51. Patent No. 831.546, USSR, *A Method of Grinding of Relieved Surface of a Gear Hob*, Rodin, P.R., Radzevich, S.P., Int. Cl. B24b 3/00, filed February 7, 1979.
52. Patent No. 880.589, USSR, *A Gear Finishing Tool*, Radzevich, S.P., Int. Cl. B21h 5/02, filed November 6, 1979.
53. Patent No. 921.727, USSR, *A Gear Finishing Device*, Radzevich, S.P., Int. Cl. B23f 19/00, filed September 28, 1980.
54. Patent No. 933.316, USSR, *A Gear Cutting Tool*, Radzevich, S.P., Int. Cl. B23f 21/00, filed October 18, 1979.
55. Patent No. 946.833, USSR, *A Gear Hob*, Radzevich, S.P., Int. Cl. B23f 21/16, filed June 13, 1980.
56. Patent No. 965.582, USSR, *A Gear Finishing Tool*, Radzevich, S.P., Int. Cl. B21h 5/00, filed October 2, 1980.
57. Patent No. 965.728, USSR, *A Method of Grinding of Relieved Surface of a Tapered Gear Hob*, Radzevich, S.P., Int. Cl. B24b 3/12, filed 1, 1980.
58. Patent No. 969.395, USSR, *A Device for Finishing Gears*, Radzevich, S.P., Int. Cl. B21h 5/02, filed April 6, 1981.
59. Patent No. 984.744, USSR, *A Device for Finishing Gears*, Radzevich, S.P., Int. Cl. B23f 19/00, filed April 16, 1981.
60. Patent No. 990.445 USSR, *A Precision Involute Hob*, Radzevich, S.P., filed October 8, 1981, Int. Cl. B23F 21/16.
61. Patent No. 996.016, USSR, *A Device for Finishing Gears*, Radzevich, S.P., Int. Cl. B21h 5/02, filed April 16, 1981.
62. Patent No. 1.000.186, USSR, *A Device for Finishing Gears*, Radzevich, S.P., Int. Cl. B23f 19/00, filed November 6, 1981.
63. Patent No. 1.004.029, USSR, *A Gear Hob*, Radzevich, S.P., Int. Cl. B23f 21/16, filed November 6, 1979.
64. Patent No. 1.017.444, USSR, *An Involute Hob*, Radzevich, S.P., Int. Cl. B 23f21/16, filed April 26, 1982.
65. Patent No. 1.028.450, USSR, *A Device for Finishing Gears*, Radzevich, S.P., Int. Cl. B23f 19/00, filed April 26, 1982.
66. Patent No. 1.055.578, USSR, *A Device for Finishing Gears*, Radzevich, S.P., Int. Cl. B21h 5/02, filed November 6, 1981.

67. Patent No. 1.087.309, USSR, *A Method of Grinding of a Gear Hob*, Radzevich, S.P., Int. Cl. B24b 3/120, filed October 17, 1980.
68. Patent No. 1.110.566, USSR, *A Device for Finishing Gear*, Radzevich, S.P., Int. Cl. B23f 19/00, filed June 7, 1981.
69. Patent No. 1.114.543, USSR, *A Gear Hob*, Radzevich, S.P., Int. Cl. B 23f21/16, filed September 7, 1982.
70. Patent No. 1.171.210, USSR, *A Method of Turning of Form Surfaces of Revolution*, Radzevich, S.P., B23B 1/00, filed November 24, 1984.
71. Patent No. 1.174.139, USSR, *A Gear Finishing Tool*, Radzevich, S.P., Int. Cl. B21h 5/00, filed October 10, 1983.
72. Patent No. 1.174.187, USSR, *A Gear Shaper Cutter*, Radzevich, S.P., Int. Cl. B23f 21/10, filed April 23, 1981.
73. Patent No. 1.185.749, USSR, *A Method of Sculptured Surface Machining on a Multi-Axis NC Machine*, Radzevich, S.P., B23C 3/16, filed October 24, 1983.
74. Patent No. 1.194.612, USSR, *A Method of Grinding of Relieved Surface of a Gear Hob*, Radzevich, S.P., Int. Cl. B24b 3/12, filed July 20, 1984.
75. Patent No. 1.196.232, USSR, *A Method of Grinding of Relieved Surface of a Gear Hob*, Radzevich, S.P., Int. Cl. B24b 3/00, filed August 3, 1984.
76. Patent No. 1.232.375, USSR, *A Method of Turning of Form Surfaces of Revolution*, Radzevich, S.P., B23B 1/00, filed September 13, 1984.
77. Patent No. 1.240.548, USSR, *A Method of Grinding of a Gear Hob*, Radzevich, S.P., Int. Cl. B24b 3/12, filed September 22, 1984.
78. Patent No. 1.249.787, USSR, *A Method of Sculptured Surface Machining on a Multi-Axis NC Machine*, Radzevich, S.P., B23C 3/16, filed December 27, 1984.
79. Patent No. 1.336.366, USSR, *A Method of Sculptured Surface Machining on a Multi-Axis NC Machine*, Radzevich, S.P., B23C 3/16, filed October 21, 1983.
80. Patent No. 1.365.550, USSR, *A Method of Grinding of Relieved Surface of a Tapered Gear Hob*, Radzevich, S.P., Int. Cl. B24b 3/12, filed 4, 1986.
81. Patent No. 1.367.300, USSR, *A Method of Sculptured Surface Machining on Multi-Axis NC Machine*, Radzevich, S.P., B23C 3/16, filed January 30, 1986.
82. Patent No. 1.442.371, USSR, *A Method of Optimal Workpiece Orientation on the Worktable of a Multi-Axis NC Machine*, Radzevich, S.P., Int. Cl. B23q15/007, filed February 17, 1987.
83. Patent No. 1.449.246, USSR, *A Method of Experimental Simulation of Machining of a Sculptured Surface on Multi-Axis NC Machine*, S.P. Radzevich, Int. Cl. B 23 C, 3/16, filed February 17, 1987.
84. Patent No. 1.463.454 USSR, *A Method of Burnishing of Part Surfaces*, Radzevich, S.P., Int. Cl. B24B 39/00, 39/04, filed May 5, 1987.
85. Patent No. 1.504.903, USSR, *A Method of Machining of a Gear with Shaper Cutter*, Radzevich, S.P., Int. Cl. B23f 5/12, filed December 2, 1987.
86. Patent No. 1.533.174 USSR, *A Method of Burnishing of Sculptured Part Surface on a Multi-Axis NC Machine*, Radzevich, S.P., Int. Cl. B24B 39/00, filed December 2, 1987.
87. Patent No. 1.636.196, USSR, *A Method of Burnishing of Surfaces*, Radzevich, S.P., Novodon, V.V., Int. Cl. B24B 39/00, filed January 30, 1991.
88. Patent No. 1.703.291, USSR, *A Method of Machining of Form Surfaces*, Chukhno, S.I., Radzevich, S.P., Int. Cl. B23C 3/16, filed August 2, 1989.
89. Patent No. 1.708.522, USSR, *A Method of Turning of Form Surfaces of Revolution*, Radzevich, S.P., B23B 1/00, filed December 13, 1988.

90. Patent No. 1.743.810, USSR, *A Method of Grinding of Relieved Surface of a Gear Hob*, Radzevich, S.P., Int. Cl. B24b 3/12, filed September 13, 1989.

91. Patent No. 2.040.376l, Russia, *A Gear Hob*, Radzevich, S.P., Int. Cl. B23f 21/16, filed January 3, 1992.

92. Patent No. 2.050.228, Russia, *A Method of Sculptured Part Surface Machining on a Multi-Axis NC Machine*, Radzevich, S.P., B23C 3/16, filed December 25, 1990.

93. Patent No. 3.786.719, USA, *A Gear Hob*, published on January 22, 1974, AZUMI.

94. Patent No. 4.415.977, USA, *Method of Constant Peripheral Speed Control*, Hiroomi, F., Shinichi, I., Int. Cl. B23 15/10, 05B 19/18, National Cl. 700/188, 318/571, filed March 2, 1981, No. 243928; priority June 30, 1979, No. 54-82779, Japan.

95. Paul, P.R., *Robot Manipulators: Mathematics, Programming, and Control. The Computer Control of Robot Manipulators*, MIT Press, Cambridge, MA, 1981 (second printing, 1982), 279 pp.

96. Plücker, J., On a new geometry of space, *Philos. Trans. R. Soc. London*, Vol. 155, 1865, pp. 725–791.

97. Radzevich, S.P., A Closed-Form Solution to the Problem of Optimal Tool-Path Generation for Sculptured Surface Machining on Multi-Axis NC Machine, *Math. Comput. Model.*, Vol. 43, Issue 3–4, February 2006, pp. 222–243.

98. Radzevich, S.P., A crowning achievement for automotive applications, *Gear Solutions*, December 2004, pp. 16–25.

99. Radzevich, S.P., A cutting-tool-dependent approach for partitioning of sculptured surface, *Comput. Aided Des.*, Vol. 37, No. 7, July 2005, pp. 767–778.

100. Radzevich, S.P., A generalized analytical form of the conditions of proper part surface generation. Part 1, in: *Improvement of Efficiency of Metal Cutting*, Volgograd, VolgPI, 1987, pp. 70–79.

101. Radzevich, S.P., A generalized analytical form of the conditions of proper part surface generation. Part 2, in: *Improvement of Efficiency of Metal Cutting*, Volgograd, VolgPI, 1988, pp. 56–73.

102. Radzevich, S.P., *A Method for Designing of the Optimal Form-Cutting-Tool for Machining of a Given Sculptured Surface on Multi-Axis NC Machine*, Patent No. 4242296/08 (USSR), filed March 31, 1987.

103. Radzevich, S.P., A method of optimal orientation of a sculptured surface on the worktable of multi-axis NC machine, *Izv. VUZov. Mashinostroyeniye*, No. 2, 1990, pp. 140–145.

104. Radzevich, S.P., *A Method of Profiling of a Form Cutting Tool*, Patent Application No. 4242296/08 (USSR), filed March 31, 1987.

105. Radzevich, S.P., A novel method for mathematical modeling of a form-cutting-tool of the optimum design, *Appl. Math. Model.*, Vol. 31, Issue 12, December 2007, pp. 2639–2654.

106. Radzevich, S.P., A possibility of application of plücker's conoid for mathematical modeling of contact of two smooth regular surfaces in the first order of tangency, *Math. Comput. Model.*, Vol. 42, 2004, pp. 999–1022.

107. Radzevich, S.P., About hob idle distance in gear hobbing operation, *ASME J. Mech. Des.*, Vol. 124, Issue 4, December 2002, pp. 772–786.

108. Radzevich, S.P., *Advanced Methods of Sculptured Surface Machining*, VNIITEMR, Moscow, 1988, 56 pp.

109. Radzevich, S.P., Basic conditions of proper part surface generating while machining on conventional machine tool, *Facta Univ. Mech. Eng.*, Vol. 1, No. 6, 1998, pp. 637–651.

110. Radzevich, S.P., *Classification of Surfaces*, Monograph, Kyiv, UkrNIINTI, No. 1440-Uk88, 1988, 185 pp.

111. Radzevich, S.P., Concisely on the kinematical method of part surface generation and on the history of the equation of contact in the form n · V = 0, *Theory Mach. Mech.*, Vol. 8, No. 1, Issue 15, 2010, pp. 42–51. http://tmm.spbstu.ru.

112. Radzevich, S.P., Conditions of proper sculptured surface machining, *Comput.-Aided Des.*, Vol. 34, No. 10, 1 September 2002, pp. 727–740.

113. Radzevich, S.P., *Cutting Tools for Machining Hardened Gears*, VNIITEMR, Moscow, 1992, 60 pp.

114. Radzevich, S.P., *Design and Investigation of Skiving Hobs for Finishing Hardened Gears*, Ph.D. thesis, Kiev Polytechnic Institute, 1982, 286 pp.

115. Radzevich, S.P., Diagonal shaving of an involute pinion: Optimization of the geometric and kinematic parameters for the pinion finishing operation, *Int. J. Adv. Manuf. Technol.*, Vol. 32, Nos. 11–12, May 2007, pp. 1170–1187.

116. Radzevich, S.P., *Differential–Geometrical Method of Surface Generation*, Doctoral thesis, Tula, Tula Polytechnic Institute, 1991, 300 pp.

117. Radzevich, S.P., *Fundamentals of Surface Generation*, Monograph, Kiev, Rastan, 2001, 592 pp.

118. Radzevich, S.P., Generation of actual sculptured part surface on multi-axis *NC* machine. Part 1. *Izv. VUZov. Mashinostroyeniye*, No. 5, 1985, pp. 138–142.

119. Radzevich, S.P., Generation of actual sculptured part surface on multi-axis *NC* machine. Part 2. *Izv. VUZov. Mashinostroyeniye*, No. 9, 1985, pp. 141–146.

120. Radzevich, S.P., *Geometry of Surfaces: A Practical Guide for Mechanical Engineers*, Wiley, Chichester, 2013, 264 pp.

121. Radzevich, S.P., Geometry of the active part of cutting tools (in the tool-in-hand system), in: *SME Summit: Where Manufacturers, Technologies and Innovations Connect*, August 3–4, 2005, Olympia Resort and Conference Center, Oconomowoc (Milwaukee, WI), SME Paper TP06PUB37, May 15 2006, 12 pp.

122. Radzevich, S.P., Geometry of the active part of cutting tools (in the tool-in-use system), in: *SME Midwest Summit: MIDWEST 2005—Manufacturing Excellence Through Collaboration Conference*, September 13–14, 2005, Rock Financial Showplace (Novi, MI), SME Paper TP06PUB36, May 15 2006, 12 pp.

123. Radzevich, S.P., Influence of the coordinate system transformations onto fundamental magnitudes of surfaces, in: *Proceedings of Dneprodzerzhinsk State Technical University, Series: Mechanical Engineering*, Vol. 1, Dneprodzerzhinsk, Dneprodzerzhinsk State Technical University, 1995, pp. 49–54.

124. Radzevich, S.P., К-mapping of sculptured part surfaces and of the machining surface of a cutting tool, *Proc. Natl. Tech. Univ. Ukr. "Kiev Polytech. Inst." Machine-Build.*, Vol. 33, 1998, pp. 232–240.

125. Radzevich, S.P., Mathematical modeling of contact of two surfaces in the first order of tangency, *Math. Comput. Model.*, Vol. 39, Issue 9–10, May 2004, pp. 1083–1112.

126. Radzevich, S.P., *Methods for Investigation of the Conditions of Contact of Surfaces*, Monograph, Kiev, UkrNIINTI, No. 759-Uk88, 1987, 103 pp.

127. Radzevich, S.P., *Methods of Milling of Sculptured Part Surfaces*, VNIITEMR, Moscow, 1989, 72 pp.

128. Radzevich, S.P., *New Achievements in the Field of Sculptured Surface Machining on Multi-Axis NC Machine*, VNIITEMR, Moscow, 1987, 48 pp.

129. Radzevich, S.P., On a possibility of application of the R-surfaces for partitioning of a sculptured surface, *Math. Comput. Model.*, Vol. 46, Issue 7–8, October 2007.

130. Radzevich, S.P., On analytical description of contact geometry of part surfaces in highest kinematic pairs, *Theory Mech. Machines St. Petersburg Polytech. Inst.*, Vol. 3, No. 5, 2005, pp. 3–14.

131. Radzevich, S.P., Part surfaces those allowing sliding over them, in: *Research in the Field of Surface Generation*, S.P. Radzevich (ed.), Kiev, UkrNIINTI, No. 65-Uk89, 1989, pp. 29–53.

132. Radzevich, S.P., *Production of Hardened Gears*, VNIITEMR, Moscow, 1985, 56 pp.

133. Radzevich, S.P., Profiling of the form cutting tools for machining sculptured surfaces on a multi-axis *NC* machine, *Stanki I Instrum*, No. 7, 1989, pp. 10–12.

134. Radzevich, S.P., Profiling of the form cutting tools for a sculptured surface machining on a multi-axis *NC* machine, in: *Proceedings of the Conference: Advanced Designs of Cutting Tools for Agile Production and Robotic Complexes*, MDNTP, Moscow, 1987, pp. 53–57.

135. Radzevich, S.P., ℝ-mapping based method for designing of form cutting tool for sculptured surface machining, *Math. Comput. Model.*, Vol. 36, No. 7–8, 2002, pp. 921–938.

136. Radzevich, S.P., *Sculptured Surface Machining on a Multi-Axis NC Machine*, Monograph, Kiev, Vishcha Schola, 1991, 192 pp.

137. Radzevich, S.P., Selection of the cutting tool of optimum design for sculptured surface machining on multi-axis *CNC* machine, in: *Automation & Assembly Summit 2005*, April 18–20, 2005, St. Louis, MO, SME Technical Paper TP05PUB71.

138. Radzevich, S.P., *Theory of Gearing: Kinematics, Geometry, and Synthesis*, CRC Press, Boca Raton, FL, 2012, 743 pp.

139. Radzevich, S.P., Dmitrenko, G.V., Machining of form surfaces of revolution on *NC* machine tool, *Mashinostroitel'*, No. 5, 1987, pp. 17–19.

140. Radzevich, S.P. et al., On the Optimization of Parameters of Sculptured Surface Machining on Multi-Axis *NC* Machine, in: *Investigation into the Surface Generation*, Kiev, UkrNIINTI, No. 65-Uk89, pp. 57–72.

141. Radzevich, S.P., Goodman, E.D., Computation of optimal workpiece orientation for multi-axis *NC* Machining of sculptured part surfaces, *ASME J. Mech. Des.*, Vol. 124, No. 2, 2002, pp. 201–212.

142. Radzevich, S.P., Goodman, E.D., Efficiency of multi-axis *NC* machining of sculptured part surfaces, in: *Proceedings of the Sculptured Surfaces Machining Conference: Machining Impossible Shapes SSM'98*, November 9–11, 1998, Chrysler Technology Center, Auburn Hills, MI, pp. 42–58.

143. Radzevich, S.P., Goodman, E.D., Palaguta, V.A., Tooth surface fundamental forms in gear technology, *Facta Univ. Mech. Eng.*, Vol. 1, No. 5, 1998, pp. 515–525.

144. Radzevich, S.P., Palaguta, V.A., *Achievements in Finishing Cylindrical Gears*, VNIITEMR, Moscow, 1988, 52 pp.

145. Radzevich, S.P., Palaguta, V.A., *CAD/CAM* system for finishing of cylindrical gears, *Mekh. Avtom. Proizvod.*, 1988, No. 10, pp. 13–15.

146. Radzevich, S.P., Palaguta, V.A., Synthesis of optimal gear shaving operations, *Vestink Mashinostroyeniya*, No. 8, 1997, pp. 36–41.

147. Radzevich, S.P., Petrenko, T.Yu., Part surfaces and cutting tool surfaces that allow for sliding over themselves, *Mech. Eng.*, No. 1, 1999, pp. 231–240 [in Russian].

148. Radzevich, S.P., Vinokurov, I.V., *Advanced Methods of Dressing of Form Grinding Wheels*, VNIITEMR, Moscow, 1991, 88 pp.

149. Reshetov, D.N., Portman, V.T., *Accuracy of Machine Tools*, ASME Press, New York, 1988, 304 pp. [Russian ed., 1986].

150. Rodin, P.R., *Issues of the Theory of Cutting Tool Design*, Doctoral dissertation, Odessa Polytechnic Institute, Odessa, 1961, 373 pp.

151. Rodin, P.R., *Fundamentals of Theory of Cutting Tool Design*, Kiev, Mashgiz, 1960, 160 pp.

152. Rodin, P.R., Linkin, G.A., Tatarenko, G.A., *Machining of Sculptured Part Surfaces on NC Machines*, Kiev, Technica, 1976, 200 pp.

153. Shaw, M.C., *Metal Cutting Principles*, Clarendon Press, Oxford, 1984, 594 pp.

154. Shevel'ova, G.I., *Theory of Surface Generation and of Contact of Moving Bodies*, MosSTANKIN, Moscow, 1999, 494 pp.

155. Shishkov, V.A., Elements of kinematics of generating and conjugating in gearing, in: *Theory and Computation of Gears*, Vol. 6, LONITOMASH, Leningrad, 1948.

156. Shishkov, V.A., *Continuously Indexing Method of Part Surface Generation*, Mashgiz, Moscow, 1951, 152 pp.

157. Stabler, G.V., The fundamental geometry of cutting tools, *Proc. Inst. Mech. Eng.*, Vol. 165, 1951, pp. 14–21.

158. Struik, D.J., *Lectures on Classical Differential Geometry*, 2nd ed., Addison-Wesley, Boston, 1961, 232 pp.

159. Time, I.A., *Resistance of Metals and Wood to Cutting*, Dermacow Press House, St. Petersburg, Russia, 1870.

160. Vaisman, I., Unele Observatii Privind Suprafetele si Varietatile Neonolonome din S_3 Euclidian, *Mathematica*, Vol. 10, Issue 1, 195–?.

161. von Seggern, D., *CRC Standard Curves and Surfaces*, CRC Press, Boca Raton, FL, 1993, 288 pp.

162. Willis, R., *Principles of Mechanism, Designed for the Use of Students in the Universities and for Engineering Students Generally*, John W. Parker, London / J. & J.J. Deighton, West Stand, Cambridge, 1841, 446 pp.

Index

Page numbers followed by f, t, and n indicate figures, tables, and notes, respectively.

.

For Product Safety Concerns and Information please contact our EU
representative GPSR@taylorandfrancis.com
Taylor & Francis Verlag GmbH, Kaufingerstraße 24, 80331 München, Germany

www.ingramcontent.com/pod-product-compliance
Ingram Content Group UK Ltd.
Pitfield, Milton Keynes, MK11 3LW, UK
UKHW021025180425
457613UK00020B/1055